WH 200 ARCH

NHSBT
LIBRARY

HLA and MHC: genes, molecules and function

The HUMAN MOLECULAR GENETICS series

Series Advisors

D.N. Cooper, *Institute of Medical Genetics, University of Wales College of Medicine, Cardiff, UK*

S.E. Humphries, *Division of Cardiovascular Genetics, University College London Medical School, London, UK*

T. Strachan, *Department of Human Genetics, University of Newcastle-upon-Tyne, Newcastle-upon-Tyne, UK*

Human Gene Mutation
From Genotype to Phenotype
Functional Analysis of the Human Genome
Molecular Genetics of Cancer
Environmental Mutagenesis
HLA and MHC: genes, molecules and function

Forthcoming titles

Human Genome Evolution
Gene Therapy

HLA and MHC: genes, molecules and function

Michael Browning
Department of Microbiology and Immunology,
University of Leicester, Leicester, UK

Andrew McMichael
Molecular Immunology Group,
Institute of Molecular Medicine,
John Radcliffe Hospital, Oxford, UK

© **BIOS Scientific Publishers Limited, 1996**

First published 1996

All rights reserved. No part of this book may be reproduced or transmitted, in any form or by any means, without permission.

A CIP catalogue record for this book is available from the British Library.

ISBN 1 859961 15 0

BIOS Scientific Publishers Ltd
9 Newtec Place, Magdalen Road, Oxford OX4 1RE, UK.
Tel. +44 (0) 1865 726286. Fax +44 (0) 1865 246823
World Wide Web home page: http://www.Bookshop.co.uk/BIOS

DISTRIBUTORS

Australia and New Zealand
DA Information Services
648 Whitehorse Road, Mitcham
Victoria 3132

India
Viva Books Private Limited
4325/3 Ansari Road
New Delhi 110002

Singapore and South East Asia
Toppan Company (S) PTE Ltd
38 Liu Fang Road, Jurong
Singapore 2262

USA and Canada
BIOS Scientific Publishers
PO Box 605, Herndon
VA 20172-0605

Typeset by P&R Typesetters Ltd, Salisbury, UK
Printed by Butler & Tanner Ltd, Frome, UK

Contents

Contributors

Aguado, B. MRC Immunochemistry Unit, Department of Biochemistry, University of Oxford, South Parks Road, Oxford OX1 3QU, UK

Apple, R.J. Department of Human Genetics, Roche Molecular Systems Inc., 1145 Atlantic Avenue, Alameda, CA 94501, USA

Braud, V. Molecular Immunology Group, Institute of Molecular Medicine, Nuffield Department of Clinical Medicine, University of Oxford, John Radcliffe Hospital, Oxford OX3 9DU, UK

Bowness, P. Molecular Immunology Group, Institute of Molecular Medicine, Nuffield Department of Clinical Medicine, University of Oxford, John Radcliffe Hospital, Oxford OX3 9DU, UK

Browning, M. Department of Immunology and Microbiology, Medical Sciences Building, University of Leicester, Leicester LE1 9HN, UK

Campbell, R.D. MRC Immunochemistry Unit, Department of Biochemistry, University of Oxford, South Parks Road, Oxford OX1 3QU, UK

Cerundolo, V. Molecular Immunology Group, Institute of Molecular Medicine, Nuffield Department of Clinical Medicine, University of Oxford, John Radcliffe Hospital, Oxford OX3 9DU, UK

Cucca, F. Wellcome Trust Centre for Human Genetics, Nuffield Department of Surgery, Windmill Road, Headington, Oxford OX3 7BN

Davenport, M.P. Molecular Immunology Group, Institute of Molecular Medicine, Nuffield Department of Clinical Medicine, University of Oxford, John Radcliffe Hospital, Oxford OX3 9DU, UK

Erlich, H.A. Department of Human Genetics, Roche Molecular Systems Inc., 1145 Atlantic Avenue, Alameda, CA 94501, USA

Fernandez-Borja, M. Dutch Cancer Institute, Department of Cellular Biochemistry, Plesmanlaan 121, CX Amsterdam, The Netherlands

Hall, F.C. Molecular Immunology Group, Institute of Molecular Medicine, Nuffield Department of Clinical Medicine, University of Oxford, John Radcliffe Hospital, Oxford OX3 9DU, UK

Hill, A.V.S. Molecular Immunology Group, Institute of Molecular Medicine, Nuffield Department of Clinical Medicine, University of Oxford, John Radcliffe Hospital, Oxford OX3 9DU, UK

Hilton, R. Department of Immunology, Royal Postgraduate Medical School, Hammersmith Hospital, Du Cane Road, London W12 0NN, UK

Israël, A. Unité de Biologie Moléculaire de l'Expression Génique, URA 1149 CNRS, Institut Pasteur, 28 rue du Dr Roux, 75724 Paris, Cedex 15, France

Jardetzky, T. Department of Biochemistry, Molecular Biology and Cell Biology, Northwestern University, 2153 Sheridan Road, Evanston, IL 60208, USA

Kaufman, J. Institute for Animal Health, Compton, Berkshire RG20 7NN, UK

Kourilsky, P. Unité de Biologie Moléculaire du Géne, U277 INSERM-UA 535 CNRS, Institut Pasteur, 28 rue du Dr Roux, 75724 Paris, Cedex 15, France

Krausa, P. Molecular Immunology Group, Institute of Molecular Medicine, Nuffield Department of Clinical Medicine, University of Oxford, John Radcliffe Hospital, Oxford OX3 9DU, UK

Lechler, R. Department of Immunology, Royal Postgraduate Medical School, Hammersmith Hospital, Du Cane Road, London W12 0NN, UK

Lombardi, G. Department of Immunology, Royal Postgraduate Medical School, Hammersmith Hospital, Du Cane Road, London W12 0NN, UK

Lundin, K.E.A. Institute of Transplantation Immunology, The National Hospital and University of Oslo, 0027 Oslo, Norway

McMichael, A. Molecular Immunology Group, Institute of Molecular Medicine, Nuffield Department of Clinical Medicine, University of Oxford, John Radcliffe Hospital, Oxford OX3 9DU, UK

Milner, C.M. MRC Immunochemistry Unit, Department of Biochemistry, University of Oxford, South Parks Road, Oxford OX1 3QU, UK

Neefjes, J.J. Dutch Cancer Institute, Department of Cellular Biochemistry, Plesmanlaan 121, CX Amsterdam, The Netherlands

Parham, P. Department of Structural Biology and Microbiology and Immunology, Stanford University School of Medicine, Stanford, CA 94305, USA

Sollid, L.M. Institute of Transplantation Immunology, The National Hospital and University of Oslo, 0027 Oslo, Norway

Spencer Wells, R. Department of Structural Biology and Microbiology and Immunology, Stanford University School of Medicine, Stanford, CA 94305, USA

Todd, J.A. Nuffield Department of Surgery, Wellcome Trust Centre for Human Genetics, Headington, Oxford OX3 7BN, UK

Thorsby, E. Institute of Transplantation Immunology, The National Hospital and University of Oslo, 0027 Oslo, Norway

Trowsdale, J. Human Immunogenetics Laboratory, Imperial Cancer Research Fund, 44 Lincolns Inn Fields, Holborn, London WC2A 3PX, UK

Wassmuth, R., Institute of Clinical Immunology, Department of Medicine III, University Erlangen-Nurnberg, Krankenhausstr 12, D-91054 Erlangen, Germany

Wubbolts, R.W. Dutch Cancer Institute, Department of Cellular Biochemistry, Plesmanlaan 121, CX Amsterdam, The Netherlands

Abbreviations

AGE	advanced glycosylation end-products (of proteins)
APC	antigen-presenting cell
ARMS	amplification refractory mutation system
AS	ankylosing spondylitis
ASA	allele-specific amplification
β2-m	β2-microglobulin
BFA	brefeldin A
CLIP	class II-associated invariant chain peptide
CMV	cytomegalovirus
CRE	cAMP response element
CREB	cyclic AMP response element binding (protein)
CREG	cross-reactive group
CTL	cytotoxic T lymphocytes/cells
EAO	experimental allergic orchitis
EBV	Epstein–Barr virus
EC	embryonal carcinoma
EE	early endosomes
EGF	epidermal growth factor
ER	endoplasmic reticulum
FACS	fluorescence-activated cell sorting
GVH	graft-versus-host (disease)
HIV	human immunodeficiency virus
HLA	human leukocyte antigen
HPLC	high pressure liquid chromatography
HPV	human papilloma virus
HSP70	heat-shock protein of 70 kDa
HSV	herpes simplex virus
HTC	homozygous typing cell
HVR	hypervariable region
IDDM	insulin-dependent diabetes mellitus
IEF	isoelectric focusing
IFN	interferon
Ii	invariant chain
IL	interleukin (IL-1, etc.)
IRS	interferon response sequence
LCMV	lymphocytic choriomeningitis virus
LCMV-GP	lymphocytic choriomeningitis virus glycoprotein
LMP	low molecular weight proteins
LPS	lipopolysaccharide
LT	lymphotoxin
LTβR	lymphotoxin beta receptor
MBP	myelin basic protein
MHC	major histocompatibility complex

MLC	mixed lymphocyte culture
MLR	mixed lymphocyte reaction
MMTV	mouse mammary tumour virus
MOG	myelin/oligodendrocyte glycoprotein (gene)
MS	mass spectrometry
MZ	monozygotic
NK	natural killer
NMR	nuclear magnetic resonance
NOD	non-obese diabetic (mice)
NP	nucleoprotein
OKA	okadaic acid
PBD	peptide-binding domains
PBMC	peripheral blood mononuclear cells
PCR	polymerase chain reaction
PFGE	pulsed-field gel electrophoresis
PLT	primed lymphocyte test
RA	retinoic acid (Chapter 7)
RA	rheumatoid arthritis
RAR	retinoic acid receptor
ReA	reactive arthritis
RFLP	restriction fragment length polymorphism
RR	relative risk
RXR	retinoid X receptor
SBT	sequence-based typing
SDS	sodium dodecyl sulphate
SDS-PAGE	sodium dodecyl sulphate-polyacrylamide gel electrophoresis
SEB	*Staphylococcus* enterotoxin B
SLE	systemic lupus erythematosus
SSCP	single-strand conformation polymorphism
SSO	sequence-specific oligonucleotide
TAP	transporter associated with antigen processing
TCR	T lymphocyte receptor, T-cell receptor
TGN	trans-Golgi network
Th	helper T lymphocytes/cells
TM	transmembrane
TMS	tandem mass spectrometry
TNF	tumour necrosis factor
TNFR	tumour necrosis factor receptor
TNFR-RP	tumour necrosis factor receptor related protein
TRE	TPA response element
TSST-1	toxic shock syndrome toxin
URR	upstream regulatory region
UTR	untranslated region
YAC	yeast artificial chromosome

Introduction

Study of the major histocompatibility complex (MHC) and its products has been an area of considerable scientific interest for many years, due to the central role of the MHC in the fields of cellular immunology, transplantation and autoimmune diseases, as well as its relevance in relation to anthropological studies, epidemiology and forensic medicine. Data on the role of MHC in the immune response have grown exponentially in the last ten years, and much of our current understanding at the molecular level of how the MHC regulates both protective and potentially damaging immune responses has been achieved within this period. This book describes these advances, by reviewing the MHC, its genes and their origins, the molecules they encode, and their function within the immune system. The emphasis throughout the book is on the human MHC, but studies based on animal models are also included.

The clinical importance of the MHC was first identified from the vigorous immune reactions that were stimulated by transplantation of tissues between members of outbred species or between inbred strains of animals. The principal 'histocompatibility antigens' responsible for rejection of the tissues were mapped to a gene cluster, which was consequently named the Major Histocompatibility Complex. It was found subsequently that the ability of an animal's T lymphocytes to respond to certain antigens was determined by genes (called Ir, for immune response), which also mapped to the MHC, and that the function of the MHC-linked Ir gene products in antigen-induced T lymphocyte proliferation could be inhibited by the addition of specific alloantisera (Shevach *et al.*, 1972). A direct role for MHC molecules in antigen presentation to T cells was demonstrated by the classical experiments of Zinkernagel and Doherty (1975), who showed that cytotoxic T lymphocytes specific for viral antigens were restricted in their recognition of antigen by the expression of self MHC class I molecules on the virus-infected antigen presenting cell. The demonstration of MHC restriction led to speculation as to whether T cells expressed two receptors, one for antigen and one for self-MHC (the dual receptor theory), or a single receptor encompassing both functions. The controversy was resolved in the early 1980s by the cloning of a single T cell receptor (TCR) molecule (Chien *et al.*, 1984; Hedrick *et al.*, 1984) which acted both as MHC restriction element and antigen-specific receptor.

A second major question in T cell biology concerned the form in which antigen was recognized by T cells. That MHC-encoded molecules would directly interact with antigen to form an immunogenic complex capable of stimulating T cells was predicted by Benacerraf (1978). Experiments on the antigenicity of animal cytochrome-c in inbred mice showed that a single, cyanogen bromide-cleavage fragment (residues 81–104) contained the major antigenic determinant (Solinger *et al.*, 1979), and indicated that MHC class II-restricted T cells recognize peptide fragments of the intact protein antigen. The requirement for processing of the antigen by the antigen-presenting cell for T cell recognition was later demonstrated by Ziegler and Unanue (1981). These findings were extended by Townsend and colleagues (Townsend *et al.*, 1986), who showed definitively that the epitopes presented by MHC class I molecules and recognized by antigen-specific T cells could be substituted by exogenously added short synthetic peptides corresponding to amino acid sequences contained within the antigen. The broad division of T lymphocytes into two major subpopulations, both phenotypically (CD4/CD8) and functionally (helper/cytotoxic), was shown to correlate with their restriction of antigen recognition by MHC class II and class I molecules respectively. MHC class I molecules presented peptides derived primarily from endogenously synthesized, cytosolic proteins, whilst class II molecules presented peptides derived principally from exogenous proteins or from endogenously synthesized cell-surface glycoproteins.

The structure of the MHC molecule was determined by X-ray crystallography, which revealed the presence of a groove within which a short peptide could be accommodated (Bjorkman *et al.*, 1987). High-resolution crystal structures and computer analysis have allowed the interaction between peptide and MHC to be studied in detail. In the early 1990s, Rammensee and his co-workers demonstrated that the bound peptides could be eluted from MHC molecules, enabling the analysis of naturally processed MHC-binding peptides (Falk *et al.*, 1991). Each MHC specificity bound peptides that showed different, but characteristic, amino acid sequence motifs. The identification of MHC-specific ligand motifs permitted the prediction of regions within antigens which might represent T cell epitopes in individuals of appropriate MHC type, and allowed these to be tested both for their ability to bind MHC molecules and to raise T cell responses (Hill *et al.*, 1992).

Exactly how the diverse genes that constitute the MHC, and which are broadly conserved amongst vertebrate species, came together within a single complex remains an open question. The relatively strong sequence homologies between the classical (and non-classical) class I genes, and the structural homologies between the classical class I and class II genes, suggest a common ancestral 'MHC' gene. In Chapter 1, Kaufman reviews the evolution of the MHC and MHC molecules.

The human major histocompatibility complex (termed HLA, for human leukocyte antigen) contains over 150 loci in approximately 4 Mb of DNA on the short arm of chromosome 6. The complex has been divided into three regions, each containing genes that encode different functions within the immune response. In Chapter 2, Trowsdale reviews the molecular genetics of the HLA class I and class II regions. These regions contain the genes that encode molecules that present antigen, in the form of short peptides, to the antigen-specific receptor on T lymphocytes. In addition to these highly polymorphic loci, a number of other genes have been identified more recently. Amongst the most interesting of these findings was the identification of a number of genes (*TAP 1* and *2*, *LMP 2* and *7*, *DMA* and *B*) in the class II region, which are involved in the processing of antigens for presentation by the classical class I and class II molecules.

The genes of the class III region, which lies between the class I and class II regions, do not contribute directly to antigen processing or presentation for T cell recognition, but include a number of genes that encode molecules that play a role in the immune response, most notably the complement component genes, *C2*, *C4* and factor B, and the cytokine genes encoding tumour necrosis factor and lymphotoxin α and β. The molecular genetics of this region, and the function of the proteins that it encodes are reviewed by Campbell and his colleagues in Chapter 3.

One of the features of the MHC region is the high degree of polymorphism shown by many of the genes. This is particularly the case for the genes of the classical class I and class II regions, which encode the antigen presenting molecules, HLA-A, -B, -C and -DR, -DP, -DQ. The structure and diversity of these genes, and the significance of their genetic polymorphism are described by Wells and Parham, and by Apple and Erlich respectively. The high degree of polymorphism of these genes is significant, not only in terms of the cell-mediated immune response of the host, but is of particular clinical importance in organ transplantation, where it forms the basis of appropriate matching of transplant donor and recipient. In Chapter 6, Krausa and Browning review the range of methods that are currently available for detecting polymorphism in HLA genes and their molecules.

The expression of MHC class I and class II genes is regulated by a variey of mechanisms, including *cis*-acting elements within the promoter and enhancer regions of the genes, *trans*-acting nuclear binding factors, and extracellular messenger molecules including certain cytokines. The mechanisms by which MHC class I and class II molecular expression is regulated are described by Israel and Kourilsky, and by Wassmuth, respectively.

The post-transcriptional processes by which the MHC class I and class II polypeptides assemble and are transported to the cell surface are distinct, although both involve association of the MHC-encoded gene products with short peptides, derived from proteins from the cell's cytosol or extracellular environment for MHC class I and class II respectively. The processing of antigen into short peptide fragments, and their association with the MHC molecules within the cell, prior to the transport of the peptide–MHC complex to the cell surface, are described in Chapter 9 by Cerundolo and Braud, and in Chapter 10 by Neefjes and colleagues.

One of the principle breakthroughs in our understanding of MHC molecules came with the definition of the structures of MHC class I and class II molecules through the use of X-ray crystallography. These studies, reviewed by Jardetzky, gave a structural model within which to study the function of MHC gene products in the immune response, and identified the molecular basis of the interaction between peptide and MHC within the molecular complex. The relationship between peptide and MHC, and, in particular, how polymorphism within the MHC molecule influences peptide selection for presentation to the host immune system, are the subject of Chapter 12, by Davenport and Hill.

The remaining chapters of the book are concerned with the function of MHC molecules within the immune response. MHC class I molecules present antigen to T lymphocytes which express the CD8 co-receptor, and which function mainly as cytotoxic effectors, by killing the antigen-presenting cell. MHC class II molecules present antigen to $CD4^+$ T lymphocytes, which act mainly as regulatory cells through the secretion of cytokines. The roles of these T cell populations in the immune response are described by McMichael, and by Thorsby and colleagues, respectively. In addition to their role in protective immunity, the inheritance of certain MHC specificities is associated, either positively or negatively, with a variety of diseases, principally involving autoimmune organ damage. The basis of HLA-disease association, and the possible underlying molecular principles, are described in Chapter 15 by Hall and Bowness. One of the best characterized of these diseases is the complex, polygenic disease, insulin-dependent diabetes mellitus (IDDM), in which the genetic influence of HLA plays a major role in disease susceptibility, but which also involves contribution from a number of other genes. In Chapter 16, Cucca and Todd use the example of IDDM to describe in detail the complex nature of the association between HLA and disease in man.

The development of autoimmune diseases may result from a breakdown in the normal process of self-tolerance. MHC molecules play a role in the induction and maintenance of self-tolerance, both centrally through negative selection of potentially autoreactive T cell clones in the thymus, and in the periphery. The molecular and cellular processes of positive and negative selection of thymocytes also provides an explanation for the vigorous nature of T cell responses to allogeneic MHC molecules, for example following MHC-mismatched organ transplant. In the final chapter of the book, Hilton, Lombardi and Lechler review the role of MHC expression in transplantation and in the development of immunological tolerance.

In conclusion, the past decade has seen major advances in our understanding of how the T-cell-mediated immune response, and the regulatory role of the MHC within it, contribute both to protective immunity and to various autoimmune diseases. The knowledge that has been gained in these fields has led to exciting new prospects for peptide/MHC-based vaccines and therapies in a number of areas, including infection with viruses and other intracellular organisms, cancer, transplantation and autoimmune diseases. The next decade should reveal whether this potential can be realized.

M. Browning (*Leicester, UK*)
A. McMichael (*Oxford, UK*)

References

Benacerraf B. (1978) A hypothesis to relate the specificity of T lymphocytes and the activity of I region-specific Ir genes in macrophages and B lymphocytes. *J. Immunol.* **120:** 1809–1812.

Bjorkman P, Saper M, Samraoui B, Bennett W, Strominger J, Wiley D. (1987) The foreign antigen binding site and T cell recognition regions of class I histocompatibility antigens. *Nature* **329:** 512–519.

Chien Y, Becker DM, Lindsten T, Okamura M, Cohen DI, Davis MM. (1984) A third type of murine T-cell receptor gene. *Nature* **312:** 31–35.

Falk K, Rotzschke O, Stevanovic S, Jung G, Rammensee HG. (1991) Allele specific motifs revealed by sequencing of self-peptides eluted from MHC molecules. *Nature* **351:** 290–296.

Hedrick SM, Cohen DI, Nielsen EA, Davis MM. (1984) Isolation of cDNA clones encoding T cell-specific membrane-associated proteins. *Nature* **308:** 149–153.

Hill AVS, Elvin J, Willis AC, *et al.* (1992) Molecular analysis of the association of HLA-B53 and resistance to severe malaria. *Nature* **360:** 434–439.

Shevach EM, Paul WE, Green I. (1972) Histocompatibility-linked immune response gene function in guinea pigs: specific inhibition of antigen-induced lymphocyte proliferation by alloantisera. *J. Exp. Med.* **136:** 1207–1221.

Solinger AM, Ultee ME, Margoliash E, Schwartz RH. (1979) T-lymphocyte response to cytochrome c: I: Demonstration of a T-cell heteroclitic proliferative response and identification of a topographic antigenic determinant on pigeon cytochrome c whose immune recognition requires two complementing major histocompatibility complex-linked immune response genes. *J. Exp. Med.* **150:** 830–848.

Townsend ARM, Rothbard J, Gotch F, Bahadur B, Wraith D, McMichael AJ. (1986) The epitopes of influenza nucleoprotein recognized by cytotoxic T lymphocytes can be defined with short synthetic peptides. *Cell* **44:** 959–968.

Ziegler K, Unanue ER. (1981) Identification of a macrophage antigen-processing event required for I-region-restricted antigen presentation to T lymphocytes. *J. Immunol.* **127:** 1869–1875.

Zinkernagel RM, Doherty PC. (1975) H-2 compatibility requirement for T-cell mediated lysis of target cells infected with lymphocytic choriomeningitis virus. *J. Exp. Med.* **141:** 1427–1436.

1

Evolution of the major histocompatibility complex and MHC-like molecules

Jim Kaufman

1.1 Introduction

A major function determined by the *major histocompatibility complex*, and in a backhanded way the reason for which it was discovered, is the presentation of antigen to T lymphocytes of the immune system, in order to regulate and effect a response to various diseases. Like other components of the immune system, the *Mhc* is under evolutionary pressure to respond to most pathogens with all of their variants. Given that microorganisms and neoplastic cells are under intense selection to escape the immune system by any means possible, it is not surprising that the MHC is forced to front a mix of many strategies which change over widely ranging timescales, varying within an individual, between individuals of a species, between species and between vertebrate groups. In addition, other functions may become associated with the *Mhc*, taking advantage of the genetic diversity. This complexity makes the evolution of the *Mhc* interesting, but also an enormous subject to cover. Moreover, all features of the *Mhc* have some kind of an evolutionary explanation, making a comprehensive review impossible. This chapter therefore gives an overview of the *Mhc* and MHC-like molecules.

There is no consensus on what constitutes the *Mhc* and MHC molecules, especially in terms of its evolution. The *Mhc* was initially discovered as the single locus most responsible for rapid allograft rejection in mice, and it is now clear that the highly polymorphic members of the class I and class II multigene families are responsible. However, it is also generally agreed that the *Mhc* of both mice and humans (and most, if not all, mammalian species) constitutes a vast genetic region encoding many different molecules with diverse functions. In addition, there are unlinked genetic loci encoding molecules with similar structures and even with related functions. For this review, the *Mhc* is defined as the major graft rejection locus, stretching to encompass the largest contiguous genetic region containing class I and class II genes. *Mhc* molecules are all those molecules encoded in the *Mhc*, regardless of structure or function, whereas MHC molecules are the class I and class II molecules encoded in the *Mhc*. MHC-like genes and molecules are all

HLA and MHC: genes, molecules and function, edited by M.J. Browning and A.J. McMichael.
© 1996 BIOS Scientific Publishers Ltd, Oxford.

those structurally like class I and class II molecules regardless of genetic location of the genes.

This review will summarize the current information and concepts of the *Mhc* and MHC-like molecules in vertebrates, starting with the existence and structure of the molecules and their genes, followed by the existence and genetic organization of the *Mhc* and related loci, then functional strategies and polymorphism, and finally coming to the origin of MHC-like molecules. As for the immune system in general, most detailed knowledge of the *Mhc* and MHC-like molecules comes from primates and rodents, primarily human and mouse, and much of the review will reflect this fact. However, again as for the immune system in general, the *Mhc* and MHC-like molecules of other vertebrate groups and even other mammals can differ significantly, both structurally and functionally, and these differences will also be described.

1.2 Classical MHC molecules are found throughout the vertebrates

The mammalian *Mhc* was first discovered, and is still defined, as the most important genetic locus responsible for rejection of allografts (reviewed by Klein, 1975, 1986). The rapid allograft rejection is directly due to the presence of polymorphic members of the class I and class II multigene families, first called major transplantation molecules but now generally referred to as 'classical' MHC molecules. In addition, there are so-called 'non-classical' molecules that are structurally related, but do not result in very rapid allograft rejection. However, the difference between classical and non-classical molecules is, in certain cases, a matter of usage rather than of sharp definition.

Classical MHC molecules have four notable characteristics by which they might be defined: they have high polymorphism and sequence diversity; they are expressed at relatively high levels in particular cell types; they present antigenic peptides to T lymphocytes; and they have distinct structural features that are important for their function (reviewed by Kaufman *et al.*, 1994). All of these characteristics are selected so that the immune system can best tell the difference between self and non-self, and then act on this knowledge. The initial education between self and non-self occurs in the thymus, where developing T cells are selected on the basis of affinity of their T-cell receptors (TCR) to self MHC molecules bearing self peptides: those with too low or too high an affinity fail to become peripheral T cells (often called positive and negative selection). The survival of the organism then depends on the ability of those educated T cells to detect non-self peptides bound to self MHC molecules (reviewed by Lawlor *et al.*, 1990).

Mammalian classical class I and class II MHC molecules have similar but different major functions: stimulating the appropriate effector mechanisms against intracellular and extracellular pathogens, respectively. Classical class I molecules bind short peptides derived primarily from cytoplasmic proteins synthesized within the cell, utilizing a pathway involving the low molecular weight proteins (LMPs) of the proteasome and the transporters associated with antigen processing (TAPs), and then present the antigen to cytotoxic T cells bearing the co-receptor CD8 (see

Chapters 9 and 13). They are widely expressed (said to be ubiquitous on nucleated cells) to alert cytotoxic T cells to the presence of intracellular invaders, which might proliferate in any cell (reviewed by Germain and Margulies, 1993). In addition, natural killer (NK) cells detect the relative paucity of class I molecules, apparently a host adaptation to the obvious viral strategy of down-regulating the class I molecules by which they would otherwise be detected (Moretta *et al.*, 1992). In contrast, classical class II molecules bind longer peptides derived primarily from proteins acquired from outside the cell, utilizing a pathway involving non-classical class II molecules (DMA and B) and invariant chain (Ii), and then present the antigen to helper T cells bearing the co-receptor CD4 (see Chapters 10 and 14). They have a restricted tissue distribution, primarily expressed on immune system cells that present antigen to helper T cells (so-called professional antigen-presenting cells: dendritic cells, macrophages and B lymphocytes) in order to respond to extracellular invaders, in part by production of antibodies by B cells (reviewed by Germain and Margulies, 1993).

The structures of mammalian classical class I and II molecules are nearly superimposable, but both the domain organizations and the sequences are different; in particular the characteristic sequence features differ consonant with their major functions (reviewed by Madden, 1995: see also Chapter 11). Antigenic peptides are held in extended positions and defined in length by interaction of main-chain atoms with invariant residues of the MHC molecule, at the ends of the peptide-binding site in class I molecules and in the middle of the peptide-binding site in class II molecules (see Chapter 12). Polymorphic residues define the actual binding specificity for antigenic peptides; the positions of these residues are similar among class I molecules, but somewhat different from class II molecules. Co-receptors on the T cell bind MHC molecules in order to increase the affinity of interaction and signalling: CD8 binds to at least one loop in the $\alpha 3$ domain of class I molecules, and CD4 binds to the equivalent position in class II molecules (in the $\beta 2$ domain), although the actual residues involved are different, as are the structures of CD8 and CD4. The cytoplasmic tails of class II molecules are relatively short, but the cytoplasmic tails of classical class I molecules are composed of two domains (equivalent to exons 6 and 7 of the class I heavy chain gene), each with a residue that can be phosphorylated, that may be involved in intracellular movement. The transmembrane region of class I molecules is simply hydrophobic, whilst the transmembrane regions of the two chains of class II molecules interact, presumably by invariant lysines and cysteines as well as highly conserved glycines with a regular spacing. In addition, the invariant N-linked glycan at position 86 of class I molecules is apparently involved in initial binding to a chaperonin (Jackson *et al.*, 1994). Presumably there are other conserved features of classical MHC molecules, such as interaction sites for TAPs (Suh *et al.*, 1994) and invariant chain (Ii), that remain to be described.

Based on the analysis of representative species, all vertebrate groups have MHC class I and class II genes that encode molecules with classical sequence features (reviewed by Kasahara *et al.*, 1995; Kaufman *et al.*, 1994; Stet *et al.*, 1996), except that the evidence for classical class I sequences in salamanders and cartilaginous fish is not definitive (a protein band in salamanders, Kaufman *et al.*, 1995; $\alpha 3$-like and LMP sequences in sharks, Hashimoto *et al.*, 1992; M. Kasahara, personal communication). Data on polymorphism and sequence diversity, expression as proteins and tissue distribution, function in antigen presentation to T cells and in responses of NK cells are still very limited in most non-mammalian species. Indeed, it is possible that the classical MHC molecules are not polymorphic or diverse in every vertebrate group (for instance the salamanders). Thus far, there are a limited number of classical MHC genes expressed per haplotype (one to three class I molecules; one to five class II molecules), presumably because too many highly expressed MHC molecules would delete most of the T cells by negative selection in the thymus (Nowak *et al.*, 1992; Vidovic and Matzinger, 1988). Neither MHC-like nor LMP-like sequences have been reported for the next branch of the phylogenetic tree, the jawless fish (lampreys and hagfish, which are chordates but not vertebrates), so it may be that MHC-like molecules first appeared in the armoured fish that gave rise to both the bony fish and the cartilaginous fish.

Except for the characteristic sequence features mentioned above, the classical MHC molecules within each vertebrate group are more similar to each other than to classical MHC molecules of other vertebrate groups. This may be due to rapid expansion and contraction of MHC-like multigene families, as well as to segmental exchange which can homogenize sequences of related genes. This is particularly true for class I molecules, even within a vertebrate group. For example, human HLA-A, B and C sequences are most similar to each other, as are mouse H-2K, D and L sequences, as are ruminant classical class I sequences.

In contrast to class I molecules, mammalian class II molecules are divided into recognizable isotypes, like the human molecules DR, DQ, DP and DO/DN. Such isotypes are found in all mammals, but may be inactivated (for instance, DR-like Eα gene in some mice, and DO-like Aβ2 and DP-like Aβ3 pseudogenes in all mice) or expanded (DP-like genes in the mole rat). These classical class II isotypes can differ in expression level and tissue distribution as well as in sequence characteristics; it is not yet clear whether there are subtle functional differences as well. The clearest difference is that, unlike DQ and DP, DR-like α-chains are either monomorphic, or oligomorphic with little sequence diversity. Based on preferential pairing of I-A α- and β-chains, it has been suggested that, when DR-like α genes are recombinationally separated from their partner β genes, they are non-polymorphic in order to bind a variety of β-chains (Germain and Margulies 1993). (The same argument might be made for β2-microglobulin (β2-m) and class I α-chains.) The single chicken class II α gene is virtually monomorphic, has sequence features like DR and unlike DQ or DP, and is located at least 5 cM away from the partner class II β-chains (Kaufman *et al.*, 1995). The major class IIα chain in axolotis has only two apparent alleles and definite N-terminal sequence similarity with DR α chains; in the ten small families tested thus far, the polymorphism of class IIα and β proteins co-segregate (Kaufman *et al.*, 1995). This suggests that the different class II isotypes are functionally different and selected over evolutionary time.

1.3 Some non-classical MHC molecules may be found in many vertebrates; others are new in each vertebrate group

Non-classical MHC-like molecules share the domain (intron/exon) structure of at least the extracellular regions of classical MHC molecules, but have low, if any, polymorphism and sequence diversity. Such molecules span a large range of sequence relationships with classical molecules, and can present antigen or have quite different functions. They can also have tissue distributions that are like those of classical molecules, or much more restricted. There are few non-classical class II molecules, but many non-classical class I molecules, suggesting the class I multigene family is more plastic in evolution.

Of the non-classical class II molecules, the human HLA-DM and mouse H-2Mb molecules are only distantly related to classical class II MHC molecules and lack virtually all of their characteristic sequence features. They have an essential role in acquisition of peptides by classical class II molecules, apparently by binding the class II-associated invariant chain peptide (CLIP) fragment of the invariant (Ii) chain in endosomes (reviewed by Roche, 1995). Similar genes are present in chickens, and are expected in all vertebrates that have antigen presentation by class II molecules. In contrast, the human DNα/DOβ are very similar to classical molecules, but are oligomorphic and expressed only at low levels if at all. They are found in other mammals, including rodents and ruminants (Karlsson *et al.*, 1992). The human DQ2 (formerly DX) genes are closely related to DQ genes and have no obvious defects, but they are not expressed. Mouse I-A genes may be more closely related to the DQ2 pseudogenes than to DQ genes (R. Bontrop, personal communication). In chickens, there are at least three class II β genes (so-called B-LβIII, IV, V and VI genes) with low polymorphism, sequence diversity and expression, and may be in some sense non-classical molecules (Zoorob *et al.*, 1993).

The non-classical class I molecules that are closely related in sequence to classical molecules appear to be due to repeated expansions and contractions from classical molecules within the species: overall, mouse Qa, TL and Hmt molecules are most like mouse classical class I molecules, and human HLA-E, F and G are most like human classical class I molecules. These so-called non-classical genes may encode classical molecules; for example, the polymorphic and highly expressed class I molecule in some New World monkeys (e.g. the cotton top tamarin) is most closely related to HLA-G (Watkins *et al.*, 1990). Some of these non-classical class I molecules have wide tissue distributions and can present restricted peptide antigens to T lymphocytes (e.g. Qa-2 and perhaps Qa-1), including N-formylated peptides of intracellular bacteria (e.g., Hmt from the M region; reviewed by Fischer-Lindahl *et al.*, 1991; Shawar *et al.*, 1994; Soloski *et al.*, 1995; Stroynowski, 1995). Others have structural features that suggest that they should present peptides, but may not. For example, mouse TL molecules have all eight invariant residues involved in binding the ends of peptides, but have no bound peptides under normal circumstances, can be expressed on cells without functional TAP proteins (or β2-m), and when expressed on intestinal epithelium stimulate T cells best after heat shock (Cheroutre *et al.*, 1995; Obata *et al.*, 1994). Some non-classical class I molecules are secreted (mouse Q4, Q6/8 and Q10 gene products) and have unknown functions, while others are even less well understood (mouse Q1-3 and 5, human HLA-E, F and G, MR-1). A few chicken class I genes that are not expressed at high levels have been described (B-FV and B-FVI) (Guillemot *et al.*, 1988). Non-classical class I genes have been found in great abundance in the frog *Xenopus* (*XNC*) and in many bony fish. As in mammals, such non-classical molecules are most closely related to the classical molecules of that vertebrate group.

Other non-classical class I molecules described in mammals are only distantly related to classical molecules, and serve a range of functions. CD1 molecules present antigen, perhaps peptides and certainly mycobacterial lipids, to T cells. They can be expressed on cells without functional TAP proteins (or β2-m) and apparently acquire their antigen in endosomes, like classical class II molecules. They are described in humans, mice and rabbits, and are primarily expressed in the intestine (reviewed by Blumberg *et al.*, 1995; Shawar *et al.*, 1994). *MIC* genes are expressed in epithelial cells and are responsive to heat shock, but nothing further is known of their function. They have been described in humans, and cross-hybridizing DNA fragments have been found in many mammals, but not mice and rats (Bahram *et al.*, 1994). Zn-α_2-glycoprotein is a secreted protein in humans, possibly involved in transport of small molecules in the blood (Araki *et al.*, 1988). Finally within this group, the neonatal Fc receptor (FcRN) binds maternal antibody for transport across the intestinal epithelium of neonatal rodents; it has been described in humans as well (Burnmeister *et al.*, 1994; Simister and Mostov, 1989).

It seems evident that some non-classical MHC molecules fulfil specialized functions for which high levels of polymorphism offer no advantage, either for antigen presentation of highly conserved antigens (N-formylated peptides of intracellular bacteria, mycobacterial lipids, stress proteins) or for unrelated functions (binding of CLIP, transport of antibody, transport of small molecules). At least some of them may exist in other vertebrate groups, since maternal antibody is found in most vertebrate eggs and mycobacteria are not specific to mammals.

1.4 Domains in MHC-like molecules diverge at different rates depending on structural and functional constraints

Although classical MHC-like genes have identical intron/exon structures in all vertebrates (with the exception of class II α genes in a perch-like fish, which have an extra intron inserted into the α2 exon; Figueroa *et al.*, 1995), the exons diverge at different rates, both within and between vertebrate groups (reviewed by Kaufman *et al.*, 1994). As an example of divergence within a vertebrate group, the α3 domain of the mouse TL molecule is more similar to classical mouse class I molecules than classical human molecules, whereas the peptide-binding domains (PBD, that is the α1 and α2 domains) of classical mouse and human class I molecules are more similar to each other than either is to TL (Obata *et al.*, 1994). As an example of divergence between vertebrate groups, the classical class I molecules of mammals and chickens have roughly 55% amino acid identity in the α2 domain but only 40% in the α1 domain, even though the two presumably have very similar functions (Kaufman *et al.*, 1992). In fact, the PBD with the disulphide bridge in classical MHC molecules (α2 domain for class I and β1 domain for class II) are often but not always the most conserved, for reasons that remain unclear. The immunoglobulin-like domains also appear to diverge at different rates and to different extents; for example, β2-m is generally more conserved than the α3 domain of class I molecules.

These differences can be understood in terms of requirements for both domain structure and functional features (Kaufman *et al.*, 1992, 1994). For example, roughly 30% amino acid identity is found for most comparisons of the $\alpha3$ domain between vertebrate groups, which is similar to the amino acid identity of any $\alpha3$ domain with any $\beta2$-m, class II $\alpha2$ or $\beta2$ immunoglobulin-like domain, or even most antibody C domains. This is because most of the highly conserved residues are those with large hydrophobic side-chains that are buried in the centre of the domain or those in particular positions in the loops; that is, the residues that play important roles in the folding of the immunoglobulin-like domain, regardless of the particular function of that domain. Apart from a very few residues that contact the other domains of the class I heterodimer, most of the surface of $\alpha3$ is diverged. Interestingly, the CD8-binding site is not very conserved between vertebrate groups, and the CD8 molecule apparently co-evolves with this site, at least between mammals and chickens (Tregaskes *et al.*, 1995). On the other hand, $\beta2$-m has extensive contact sites with the $\alpha1$ and $\alpha2$ domains (in fact, subtle differences in mouse $\beta2$-m alleles can affect antigen presentation; Perarnau *et al.*, 1990), and therefore there are many more residues on the surface of the domain (although mostly buried in the heterodimer) that are conserved.

The divergence to a structural limit in $\alpha3$ domains of class I molecules is true for both classical and non-classical class I molecules. In contrast, there are few residues of $\alpha1$ and $\alpha2$ that are similar between classical class I molecules and non-classical CD1, MIC or FcRN molecules. Also in the three-dimensional structures, classical class I molecules and FcRN are more similar in $\alpha3$ than $\alpha1$ and $\alpha2$ domains (Burmeister *et al.*, 1994). This suggests that, compared to the $\alpha3$ domain, the PBD (sometimes called finger-clasp or open-faced sandwich) domains may be under less stringent structural pressures and more stringent functional pressures, in the case of classical molecules, to fulfil conserved functions related to antigen presentation, and for certain non-classical molecules, to fulfil other unrelated functions.

1.5 There is a standard *Mhc* organization for tetrapods, but with many variations

The *Mhc* in all mammals and apparently also in amphibians is divided into three distinct regions, with great variation in numbers of genes, in size and in some details of organization. In birds, the *Mhc* is smaller, simpler and organized differently, but the remains of three regions are evident. Nothing is yet known about the reptiles, or about the closest relatives to the ancestors of the tetrapods, lungfish and lobe-finned fish. In fact, nothing is known of the *Mhc* organization in any fish.

Based on detailed analysis, the human *Mhc* is a vast genetic region (spanning approximately 4 Mb), divided by frequent recombination into (at least) three regions (reviewed by Trowsdale, 1993, 1995). One region contains many class I α-chain genes, some of which encode highly polymorphic classical molecules and others of which encode oligo- or non-polymorphic non-classical molecules, with genes for myelin-oligodendrocyte glycoprotein (MOG) and butyrophilin near the end. Some non-classical class I genes (CD1, Zn-α_2-glycoprotein, FcRN, MR-1), as well as the gene encoding $\beta2$-m, are located on other chromosomes. A second region contains class II genes which are organized in pairs (or groups) encoding the α- and β-chains that associate in particular heterodimers. In addition, some genes involved in antigen processing, including the non-classical class II, TAP and LMP genes, as well as the RING3 gene encoding a putative transcription factor, are embedded in the middle of the class II region (see Chapter 2). The invariant chain (Ii) involved in class II biosynthesis is located on another chromosome. Finally, a region in between the class I and class II regions encodes many structurally unrelated genes that collectively have been called class III region genes (see Chapter 3). Some of these genes, such as the complement components factor B, C2 and C4, the cytokines tumour necrosis factor (TNF) α and β, and perhaps the inducible chaperonin HSP70 (heat shock protein of 70 kDa), have important functions in the immune system; but most class III region genes encode enzymes, adhesion molecules, transcription factors and so forth, a sampling of vertebrate genes with no obvious reason to be in the *Mhc*. The fact that the β2-m gene maps close to the gene for the complement factor C5 in mice has led to the interesting suggestion (Klein and Figueroa, 1986) that part of an ancestral *Mhc* that included the β2-m gene was swapped for a large region of an ancestral complement complex that included factor B, C2 and C4.

The *Mhc* of most mammals are related to the human model, but with great variation in detail, particularly due to contractions and expansions, and the appearance of pseudogenes, even in different haplotypes in the same species (reviewed by Trowsdale, 1995). In general, the class II and class III regions are the most similar, with the class I region being hardly recognizable between different mammalian groups. The existing data suggest that most primates are overall similar to humans (reviewed by Bontrop *et al.*, 1995).

Among rodents, the organization of class II and class III regions are similar to humans, but the class I regions so far appear to be fairly different. In humans, there are few non-classical class I genes (HLA-E, -F and -G, and MIC) flanked by the classical class I genes HLA-A and the closely linked HLA-B and -C. In contrast, there are many more class I genes in mice, and they are arranged in regions (K and D, Q, TL and M) containing related genes, with the classical class I genes are split into H-2K on the one side of the class II region and the H-2D region on the other side of the class III region. (However, it has been suggested that HLA-C corresponds to H-2Q region, HLA-E to T24 in the T region, HLA-A to genes in the M region next to the MOG gene; see legend to Figure 4 in Trowsdale, 1995.) Even between strains of laboratory mice, there are expansions and contractions, with few of these genes known to be expressed. The non-classical genes can exhibit pseudo-allelism, in which the expressed alleles in different *Mhc* haplotypes are encoded by different genes. The rat *Mhc* has the single reported classical class I molecule encoded by a gene in the equivalent position of H-2K (although there may be up to three classical class I molecules expressed in a rat haplotype; Etienne Joly, personal communication), and a large number of non-classical genes and pseudogenes on the other side of the class III region, encoding alloantigens detected by NK cells, as well as genes important for growth. Other rodents, like the mole-rat, have enormously expanded class I and DP-like genes (Nizetic *et al.*, 1987).

Among ruminants, the class II isotype DP is replaced by another isotype called DY, which is expressed at least in sheep (Wright *et al.*, 1994). In cattle, the DY α gene is next to DO β and TAP genes and is separated by 17 cM from the remainder of the MHC, including DR-like and variable numbers of DQ-like genes, the class III region, and as many as four classical class I genes (Andersson and Davies, 1994; Ellis *et al.*, 1996; Shalhevet *et al.*, 1995). In rabbits, as a representative of lagomorphs, the class II region is organized like that of humans, but with large distances between DO-like and DQ-like and DR-like genes, which vary between haplotypes. There are only eight to 12 class I genes, and some number of these are located close to the DR-like genes, without intervening class III genes (Chouchane and Kindt, 1992).

The chicken *Mhc* is simpler and smaller than those of mouse and man, as is much of the chicken genome. The *B-F/B-L* region is responsible for rapid allograft rejection and MLR stimulation, and is closely linked with the *B-G* region that encodes a very polymorphic multigene family of serologically detected B-G molecules, that are most closely related to MOG. The 4 Mb covered by the class II, class III and class I regions in mammals are reduced to less than 50 kB within the *B-F/B-L* region that contains a pair of classical class II β gene, a C4 gene and a pair of classical class I α genes (reviewed by Kaufman *et al.*, 1995). Thus, the multigene families of class I and class II molecules have been reduced, and most genes located in the mammalian class III region are not present between the classical class I and class II genes, and may have been deleted from the chicken genome altogether (Guillemot *et al.*, 1988; Koch 1986). The region of genes involved in antigen presentation, that in mammals is embedded in the class II region, has also been reduced and split up. The non-classical class II genes responsible for presentation by classical class II genes as well as the putative transcription factor (RING3) are found next to the class II β genes, as might be expected. No LMP genes that encode components of the proteasome have been identified at all. The TAP genes are located in between the two class I genes. The single classical class II α gene is located 5 cM away from the assigned class II β gene partners, rather than tightly linked as in all other vertebrates (Kaufman *et al.*, 1995). In addition, there is at least one other locus (*Rfp-Y*) that contains two class I genes and two or three class II β genes; the class II β genes have low polymorphism, sequence diversity and expression levels, and are therefore possibly non-classical (Miller *et al.*, 1994; Zoorob *et al.*, 1993). This region is not responsible for MLR stimulation, but does determine a moderate speed of skin graft rejection (H. Hunt, personal communication). Pheasants, which are closely related to chickens, apparently have a similar organization, but nothing is yet reported for other birds. As in mammals, the chicken β2-m gene is located outside of the *Mhc* (Kaufman *et al.*, 1995).

At the present level of resolution, the *Mhc* of amphibians appears much more like the typical mammal than the chicken. For the frog *Xenopus laevis*, a single classical class I α-chain, a number of classical class II α- and β-chains, and the complement component C4 are determined by the major locus responsible for skin graft rejection (reviewed by Du Pasquier *et al.*, 1989). In addition, genes for factor B, inducible HSP70 and LMP7 are linked to this *Mhc*, and a recombinant frog separates the *Mhc* into two regions, one that contains class II β and LMP7 genes and the other that contains the genes for HSP70 and the classical class I molecule (Kato *et al.*, 1994; Namikawa *et al.*, 1995; Salter-Cid *et al.*, 1994). A gene for LMP2 is located outside of the *Mhc*. The locations of the β2-m and invariant chain genes are unknown. However, a large number of related non-classical class I genes are unlinked to the *Mhc*, probably together on a different chromosome. These so-called XNC genes are somewhat polymorphic but not very diverse, as though they have a specialized function. Nothing is yet reported about their expression level, tissue distribution or ontogeny (Flajnik *et al.*, 1993). A final point is that most *Xenopus* species are polyploid but have the *Mhc* expressed in diploid, as though there is a strong pressure to limit the number of *Mhc*s. An exception is *X. ruwensoriensis*, an octaploid frog that has a number of MLR stimulation loci, consistent with Southern blots showing multiple loci with class I genes and/or class II genes. However, it is not yet clear whether there are multiple loci responsible for rapid graft rejection in *X. ruwensoriensis* (L. Du Pasquier, personal communication). The only other amphibian to be examined is the neotenic salamander from Mexico City, the axolotl (Kaufman *et al.*, 1995). Thus far, it is known that the axolotl has two alleles of class II molecules that include DR-like isotypes and two alleles of a class I-like α-chain (one allele of which is null). The class I-like α-chain and both the class II α- and β-chains co-segregate with the fastest locus for graft rejection in F2 families of axolotls.

Nothing is yet known about the structure of an *Mhc* of any fish. In some species, the most rapid allorejection of scales is determined by a single locus (reviewed by McCumber *et al.*, 1982); in other polyploid species a number of MLR stimulation loci have been found (Kaastrup *et al.*, 1988), as for the frog *X. ruwensoriensis*. The many MHC-like genes, as well as LMP genes, have not yet been shown to co-segregate, but cosmid clones with both class II α and β genes have been isolated in a bony fish, the zebrafish (Sültmann *et al.*, 1994).

It is probable that most fish, like most tetrapods, will have a single *Mhc* that is responsible for the most rapid allograft rejection: a single genetic region with polymorphic classical class I and class II genes. However, it is not clear why polymorphic class I and class II genes should travel together in evolution, since many families of genes with similar functions are located in several genetic loci (globins, antibodies, T cell receptors, to name a few). One possibility (Kaufman *et al.*, 1995) is that the *Mhc* is a location in the genome where high levels of polymorphism are tolerated or even encouraged, and that elsewhere in the genome polymorphic sequences tend to be purged. From this viewpoint, many different kinds of genes for which polymorphism confers some advantage would collect in the *Mhc*, while genes for which polymorphism confers no advantage would not be actively excluded, but might escape (e.g., β2-m). Based on linkage disequilibrium of *Mhc* loci in humans, it was suggested long ago that whole allelic haplotypes are selected in evolution, including particular alleles of class I, class II and class III genes that function optimally together (Bodmer, 1978).

1.6 *Mhc* function in different vertebrates reflects different immune strategies

The central role of T cells in the immune response of mammals suggests that classical class I and class II molecules are indispensable for resistance to infectious pathogens. Experience with certain human genetic immunodeficiency diseases, as well as experiments with knock-out mice, show that MHC-like molecules are important for long-term survival. However, there has long been evidence that different vertebrate groups differ widely in the speed of allograft rejection and in other correlates of T cell function, as though the *Mhc*-directed T-cell response varies substantially (reviewed by Cohen, 1980). Even in animals with robust functions, there are differences that are a result of different lifestyles. For example, frogs like *Xenopus* do not express classical polymorphic class I molecules in the tadpole stage, with clear functional consequences (Du Pasquier *et al.*, 1989). There is also evidence to suggest that there are more subtle differences in immune strategy as well; for instance, particular chicken *Mhc* haplotypes can confer decisive resistance or susceptibility to particular infectious pathogens, whereas most human *Mhc* haplotypes appear to be more-or-less resistant to all small infectious pathogens (Kaufman *et al.*, 1995). Finally, there is evidence that MHC molecules, and the *Mhc* as a genetic region, affects more than just immune responses (reviewed by Boyse *et al.*, 1987; Edidin, 1983; Potts and Wakeland 1993; and see below).

Vertebrate groups vary considerably in *Mhc*-directed T-cell-dependent functions, based on graft rejection, which is a relatively simple and robust assay (Cohen, 1980). Mammals, birds, frogs and bony fish have rapid and definitive responses to allogeneic grafts, as well as a high level of polymorphism of the transplantation antigens. In addition, they show reasonable graft-versus-host reactions, proliferation in MLR, and for tetrapods, good evidence for isotype switch in antibody response. In contrast, reptiles, salamanders and cartilaginous fish (as well as jawless fish) are generally reported to exhibit much slower (so-called chronic) rejection of grafts. Other T-cell-dependent responses are generally, but not always, reported to be poor. It appears that there may be a different reason for the reduced responses in each of these groups.

At least some, and probably most, reptiles have variable responses that depend both on temperature and season. Unlike mammals, reptiles are ectotherms that use behavioural mechanisms to maintain adequate body temperatures. Lower temperatures certainly affect the immune response. In addition, during colder seasons, certain reptiles become much less active and, in parallel, the corticosteroid levels in the blood increase, the thymus regresses and the number of T cells in the blood drops precipitously. As might be expected, proliferation of blood and spleen cells in the MLR drops and graft rejection slows to a halt. In the late spring, the corticosteroid levels drop, the thymus regrows and T cells increase; MLR responses rise and grafts are rejected relatively rapidly, including those which had stopped in mid-rejection. This may be true of many ectotherms (Zapata *et al.*, 1992).

Salamanders have subdued immune responses compared to frogs, and yet they can have very similar lifestyles and may inhabit very similar environments. They have a thymus that produces large numbers of cells in the periphery. These presumptive T cells respond well in culture to T-cell mitogens, but not to allogeneic cells in MLR. Graft rejection follows the *Mhc* that has been defined by class I and class II antigens, but is very slow. These MHC molecules have low polymorphism and unusual tissue distributions. It would appear that most of these presumptive T cells do not recognize the defined MHC antigens, which, as a result, are not driven to diversify or be expressed with particular tissue distributions. The T cells might recognize non-classical class I molecules with low polymorphism, or perhaps $\gamma\delta$ cells or NK cells make up the bulk of the presumptive T cells. It is also possible that lack of classical class I molecules renders skin graft rejection very slow (Kaufman *et al.*, 1995).

Cartilaginous fish clearly have a thymus and T cells as well as polymorphic class II genes, but it is not clear whether or what kind of class I genes they possess (Bartl and Weissman, 1994; Hashimoto *et al.*, 1992; Kasahara *et al.*, 1993). Their immunoglobulin gene organization differs from mammals (Shamblott and Litman, 1989), and they possess, in addition, some otherwise undescribed antibody-like molecules (Greenberg *et al.*, 1995). It is therefore possible that they have a more primitive immune system, but it seems more likely that they have evolved a somewhat different, but no less advanced, immune system to those of most vertebrates.

Adult frogs like *Xenopus* have a very robust immune system with most of the attributes of mammalian immune responses, including cytotoxic and helper T cells. However, tadpoles do not express polymorphic classical class I molecules, although they express polymorphic class II molecules like adults. A clear consequence is that grafts with only minor histocompatibility differences are not rejected by tadpoles. The eggs of both amphibians and fish are immediately subject to pathogens and must develop a functioning immune system relatively rapidly. One suggestion is that tadpoles are simply too small to have enough T cells to recognize both class I and class II molecules, and so the less important class I molecules are only expressed when the frog is large enough (Du Pasquier *et al.*, 1989). It is not yet clear whether this is a general phenomenon in amphibians and fish, whose progeny are generally exposed to pathogens starting with the egg. Some odd phenomena in birds and mammals may be evolutionary relicts originating in the two amphibian lifestyles, for instance the waves of T cells that leave the thymus (Turpen and Smith 1989).

A more subtle difference may be exemplified by the chicken. In humans, strong associations with the *Mhc* are all with autoimmune diseases, while the few detectable associations with natural infectious pathogens are relatively weak. The human *Mhc* is very complex, encoding multigene families of class I and class II molecules as well as structurally unrelated molecules involved in innate immunity. This may be the reason why all *Mhc* haplotypes are more-or-less responsive for any given pathogen (and to all but the most simple antigens as well) giving protection from most infectious diseases as well as potential overreaction leading to autoimmune diseases.

In contrast (Kaufman *et al.*, 1995) there are a number of examples of chicken *Mhc* haplotypes that confer susceptibility to particular infectious pathogens. Not only is the chicken *Mhc* smaller and simpler than the *Mhc* of the typical mammal, but there is a single dominantly expressed class I molecule in many common *Mhc* haplotypes. The limited peptide-binding specificity of these dominantly expressed class I molecules can explain the resistance and susceptibility of these haplotypes to small pathogens, such as Rous sarcoma virus, in a very simple way. In certain other haplotypes, there is a lower expression of class I molecules which correlates with resistance to the lethal tumours caused by Marek's disease virus, a much larger pathogen. It is not clear what immune mechanism is involved in this correlation, but NK cells are one obvious possibility. The simplicity of the chicken *Mhc* may allow the selection by these pathogens to be observed, but it is not clear why chickens would have adopted an apparently more dangerous strategy than that used by mammals.

1.7 The polymorphism of classical MHC molecules is selected

One of the defining characteristics of classical MHC molecules is the large number of common alleles present in interbreeding populations; this property is virtually unique for known vertebrate genes. Theoretically, high polymorphism of a gene can be the result of mutation rate, selection, genetic hitch-hiking or a combination of the three. Most explanations for this high level of polymorphism are based on selection of MHC molecules due to their immune functions. A continuous high mutation rate, as in rearranged antibody genes of mammalian B cells or chicken bursal cells, has never been found, although relatively rapid events of some recombinational complexity can occur. Genetic hitch-hiking, in which the change in frequency of a variant gene in a population depends on the change in frequency of a closely linked gene (presumably under selection), might play a role in the moderate level of polymorphism of certain *Mhc* genes, such as the human C4 gene, which shows a level of heterozygosity (F) expected for an unselected gene by a neutralist model of evolution (Potts and Wakeland, 1990). However, alleles of other *Mhc* genes may be selected, for instance the two alleles of TAP in rats (Momberg *et al.*, 1994).

Almost immediately upon the discovery of antigen presentation (as so-called *Mhc*-restricted recognition), it was suggested that the high polymorphism of MHC molecules was driven by changes in pathogens, based on the fact that a particular MHC molecule could present many but not all antigens (as so-called immune response genes; Doherty and Zinkernagel, 1975; Zinkernagel and Doherty, 1979). This can happen in at least two ways: heterozygous advantage (or overdominant selection), in which an *Mhc* heterozygote responds quantitatively better than either homozygote; and rare allele advantage (or frequency-dependent selection), in which individuals with a rare *Mhc* allele respond better to new pathogen variants that have evolved to evade the common *Mhc* alleles.

There are both theoretical and experimental challenges to this simple and elegant model. Theoretically, heterozygous advantage in a freely breeding population of reasonable size cannot support the large number of alleles reported for most species, since after a point the number of heterozygotes does not rise much as the number of alleles increases. Also, in relatively large populations, the elimination of disadvantageous alleles by selection is slow (since heterozygotes are still protected) and the loss of alleles by genetic drift is also slow. Experimentally, monomorphic populations of mice on islands exist without evidence of frequent decimation by pathogens (Figueroa *et al.*, 1986). Also, there is very little evidence for the association of mammalian *Mhc* haplotypes with differential resistance to common infectious pathogens: the strongest associations of the human *Mhc* are with autoimmune diseases (reviewed by Tiwari and Terasaki, 1985; and see Chapter 15), and investigations of the best association of a human *Mhc* haplotype with a particular infectious pathogen in fact shows that all *Mhc* haplotypes are more-or-less resistant, with large numbers of patients and careful statistics needed to see any difference (Hill *et al.*, 1991).

There is, however, evidence that classical MHC molecules are under selection, at least at some time. The prevalence of certain alleles in large populations might be explained by sudden strong selection, for example, HLA-A2 selected by the black plague in northern Europe. Also, the presence of virus isolates with changes in peptides bound by particular MHC alleles may also be due to direct selection, for example, paucity of certain Epstein-Barr virus variants in populations with high proportion of HLA-A11 (Decamposlima *et al.*, 1993). Most significantly, the number of replacement nucleotide substitutions far outweighs the number of silent nucleotide substitutions in the codons for the polymorphic residues contacting the peptide, although the magnitude varies markedly between loci (Hughes and Nei, 1988, 1989). However, the intensity of natural selection on each human *Mhc* gene (based on these same data) is estimated to be rather low (Satta *et al.*, 1994). This accords well with the idea that most human *Mhc* haplotypes are more-or-less resistant to all pathogens, because of the presence of multigene families of MHC molecules along with close linkage to other disease-resistance genes. Selection might be directly observed in animals with a simple *Mhc*, such as the chicken (Kaufman *et al.*, 1995). In fact, the best example of an association of an *Mhc* haplotype with an infectious disease is the resistance of the chicken *B21* haplotype to Marek's disease, which is caused by a relatively large pathogen, a herpesvirus. In any case, the fact that there is generally a low level of selection on particular MHC molecules in mammals has led to at least three alternative explanations for their high level of polymorphism.

One explanation is the accumulation of alleles over evolutionary time (the so-called trans-species hypothesis, Klein 1987). In both rodents and primates, lineages of alleles with very similar (but non-identical) sequences can be found in a number of species with common ancestors as much as 58 million years ago. Particular alleles apparently have lifetimes around the age of a species (roughly a million years), although the appearance of new alleles is faster in some loci than others (reviewed by Bontrop *et al.*, 1995; Parham *et al.*, 1995).

Another explanation is the accumulation of alleles by fusion of many small populations each with only a few alleles. Certain Amerindian tribes have only a few alleles and, in each tribe, there are different alleles that are otherwise unknown (Belich *et al.*, 1992; Watkins *et al.*, 1992). If the population structures of both humans and mice were originally small groups that evolved somewhat independently, so that selection of relatively few alleles by heterozygous advantage and rapid change by drift operated effectively, then the large number of alleles for humans and mice are essentially artefactual, caused by urbanization in humans and by sampling many separate demes in mice.

A third explanation is selection of alleles based on other functions besides disease resistance. For both *Mhc*-congenic inbred mice in laboratory experiments and relatively outbred mice with natural behaviours in experimental mouse barns, mate selection depends somewhat on *Mhc* haplotype (Boyse *et al.*, 1987; Potts and Wakeland 1993). In general, mates with a different *Mhc* are chosen, leading to increased number of heterozygotes in the progeny, as though outbreeding was the driving force. Also, kin selection has been suggested, based on the recognition of odours, particularly in urine, by both mice and rats. In birds, the visual system might be more important, and if the effectiveness of secondary sexual characteristics depends on overall fitness, which, in turn, depends on parasite load (Hamilton and Zuk 1982), then sexual selection in birds may magnify the effects of those *Mhc* alleles that give slightly better responses to the current pathogens (Kaufman *et al.*, 1990).

It is generally accepted that classical *Mhc* genes can evolve both by point mutation and by genetic events involving recombination (primarily gene conversion or segmental exchange), and do not change in the manner of a regular molecular clock (reviewed by Bontrop *et al.*, 1995; Parham *et al.*, 1995). Based on comparison of sequences, recombination involving relatively small segments up to whole exons have been inferred. Different loci of class I and class II genes evolve in different ways. The human *HLA-B* gene exhibits frequent interallelic exchange of small sequence segments and therefore evolves much faster than *HLA-A* which does not. There is little, if any, interlocus exchange in humans. On the other hand, mouse class I genes show relatively frequent interlocus shuffling, and therefore exhibit few locus-specific characteristics. The bm mutants of H-2K^b were found by screening mice in defined crosses, and are due to small segments that mostly derive from non-classical genes in the *Q* and *TL* regions (reviewed by Nathenson *et al.*, 1986). The human *DQ* genes evolve slowly and mostly by point mutations, whereas *DP* genes evolve rapidly and mostly by apparent segmental exchange involving six small regions with a few variants in each region. Different *DRβ* genes change in various ways; some evolve slowly by point mutation (yielding long-lived allelic lineages), others evolve rapidly by point mutation, and still others evolve rapidly by frequent segmental exchanges. It has, however, been strongly argued (Klein and O'hUigin, 1995) that nearly all such apparent recombination events are due to point mutation and convergence of peptide-binding sites.

Even though particular alleles have relatively short lifespans (roughly one million years, in the order of the lifetime of a species), in certain loci allelic lineages composed of closely related sequences can be long-lived, from two to as much as 58 million years (reviewed by Bontrop *et al.*, 1995). Genes in these long-lived allelic lineages all differ in residues in the peptide-binding site, so the existence of the lineage is not determined on the basis of binding specificity; rather, they may represent loci that do not have frequent recombination. Even long-lived lineages can be lost from a particular species, or, infrequently, a new lineage might begin by some recombinational event. Even pseudogenes can have long lifespans, for instance, the human class II pseudogenes *DQA2* and *DQB2* apparently have no stop mutations in humans or any primate.

1.8 The evolutionary origins of the MHC and MHC-like molecules are only inferred

The complex system of cell biology and cellular immunology that pivots on MHC molecules must have evolved in steps, but there is as yet no phylogenetic evidence to show when or how this evolution took place. All vertebrates apparently have both class I and class II molecules (along with at least some of the important molecules involved in antigen presentation and recognition, including LMP, TAP, Ii, TCR and CD8), but no such molecules have yet been isolated from the previous step of the phylogenetic tree, the jawless fish, despite many attempts (Kasahara *et al.*, 1995). There has also been no molecular confirmation of the apparent functional similarities with the allorecognition system of invertebrates, such as colonial tunicates (Scofield *et al.*, 1982). Thus, there is as yet only inference about the origins of MHC-like molecules, and the molecules and cells that are involved in their function. A series of fairly compelling hypotheses have been advanced based on the most important molecular features of MHC and TCR molecules.

Most MHC-like molecules bind peptides. The fact that chaperonins like HSP70 bind peptide stretches of proteins in the process of folding led to the suggestion they gave rise to the PBDs of MHC-like molecules, so that the primordial MHC-like molecule arose from the fusion of a molecule like HSP70 with an immunoglobulin-like domain. Certain sequence features of the PBD of a frog non-classical class I molecule (XNC1) are similar to the region of HSP70 that binds peptides, and secondary sequence predictions based on this HSP70 sequence are similar to the structure of the PBD of human classical class I molecules (Flajnik *et al.*, 1991; Rippmann *et al.*, 1991). However, the secondary structure topology of the relevant 18 kDa domain of Hsc70 (a member of the HSP70 family) determined by nuclear magnetic resonance (NMR) is not convincingly like the $\alpha 1$ and $\alpha 2$ PBD of MHC-like molecules: the eight β-strands meandered between two separate sheets with only one α-helix present (Morshauser *et al.*, 1995). However, a detailed structure may show similarities that are not yet apparent, and other members of the HSP70 family might be more similar to MHC-like molecules.

Most MHC-like molecules are recognized by TCR molecules, which are clearly related to antibody molecules. The fact that antibody molecules generally bind antigen with three loops (called complementarity-determining regions, or CDR1, CDR2 and CDR3) in the variable (V) domain led to the suggestion that the primordial recognition event involved a monomorphic MHC-like molecule that bound peptides and a TCR-like molecule with no diversity, which only later was split into multiple segments in the genome. It was found that molecular models of the TCR based on antibody structures could be docked on to structures of human class I molecules, with the loops corresponding to CDR1 and CDR2 contacting the α-helices of the PBD, and CDR3 contacting the bound peptide (Davis and Bjorkman, 1988). The properties of antibody V gene recombination led to the suggestion that, in evolution, the portion of the TCR-like gene corresponding to CDR3 was invaded by a transposon resulting in a V gene split into V and J and eventually D, giving rise to a variable loop (Tonegawa, 1983). As a group, TCR-like molecules derived from such a split gene would be able to recognize far more peptides of different sequence bound to the monomorphic MHC-like molecule. This scheme suggests that cell-bound $\alpha\beta$-like TCR recognizing peptides in MHC-like molecules came first, followed by cell-bound $\gamma\delta$-like TCR that could recognize shapes and finally antibodies secreted by B cells (Schild *et al.*, 1994) . This would explain why antibody genes are split only in CDR3 and not at every CDR. An alternate scheme suggests that archaic ligands (e.g. molecules involved in transcytosis in the intestine) were first recognized as shapes by the ancestor of both antibodies and $\gamma\delta$-like TCR, that after this divergence a subset of ligands acquired the capacity to bind peptides, and that a special subset of $\gamma\delta$-like TCR evolved into $\alpha\beta$-like TCR to recognize these peptide-binding ligands (Hein 1994).

MHC-like molecules recognized by T cells are generally multigenic and the classical MHC molecules are polymorphic as well. The fact that not all antigens can be presented by a single MHC molecule has led to the notion that both multigenicity and polymorphism are driven by the need of host defences to cope with ever-changing pathogens (Doherty and Zinkernagel, 1975; Zinkernagel and Doherty, 1979). Long ago, as the recognition system between TCR-like and MHC-like molecules became important for primitive immune recognition, there would be a selection for different MHC-like molecules to bind different kinds of peptides, and multigene families would appear. Presumably education in the thymus would be the constraint leading to the appearance of polymorphism, since a very large multigene family of classical molecules would lead to deletion of most T cells during thymic maturation in an individual (Nowak *et al.*, 1992; Vidovic and Matzinger, 1988). Some structural features of potential antigens might be so constrained (*N*-formyl methionines in bacterial proteins, mycobacterial lipids, sequences of stress proteins) that specialized binding motifs in non-polymorphic (that is, non-classical) molecules might also be selected, perhaps quite early. It is also possible that the FcRN-like molecules appeared when it became important to transport protective maternal antibody into eggs. The reason for isotypes of class II molecules conserved at least since the time of amphibians has, as yet, no explanation.

Class I and class II molecules have very similar structures, but with the domains connected differently. The organization of the domains and their exons led to the suggestion that class I and class II molecules diverged from their common ancestor by an exon shuffling event. The original idea was that the primordial MHC-like molecule was a class II β-like homodimer encoded by a single gene, which duplicated and diverged to give the class II heterodimer (Kaufman *et al.*, 1984). The disulphide bridge present in class II $\alpha1$ domains of bony fish might be taken to indicate that the PBD disulphide in a primordial class II α-chain was lost in all other class II α-chains, as well as in class I α-chains (Stet *et al.*, 1996). Since mammalian class II α/β gene pairs are located in opposite transcriptional orientation, an inversion around a central point with asymmetric break points would, in one step, generate a class I gene and a $\beta2$-m gene with a hydrophobic tail (Kaufman *et al.*, 1990). Some class II α/β genes in fish are located in the same transcriptional orientation, so another possibility is a simple deletion to produce a class I gene (Figueroa *et al.*, 1995). Since it is now clear that MHC-like molecules are not symmetrical, a third possibility is that the primordial MHC-like molecule was more like a class I molecule, with both PBD exons from a chaperonin-like gene located near a single immunoglobulin-like exon that became $\alpha3$ (Flajnik *et al.*, 1991). In any case, such events would probably have taken place after some gene duplication. However, it is not clear whether there was a selective advantage for the different organizations of PBDs of class I and class II molecules: $\alpha1$ and $\alpha2$ on the same class I chain versus $\alpha1$ and $\beta1$ on two different class II chains.

The very sophisticated cell biology of antigen presentation and cellular immunology of immune response depends on the interactions of MHC-like molecules with many other molecules. These other molecules, TAPs, LMPs and CD8 for the class I pathway and Ii, DM and CD4 for the class II pathway, are not similar, and presumably were recruited independently from existing cellular processes (Germain and Margulies 1993). For instance, CD4 and CD8 may interact with equivalent regions in class II and class I molecules, but they are so distantly related in structure as to rule out duplication of a primordial co-receptor molecule. The adaptation of these other molecules to help MHC-like molecules resist pathogens depends on the order and timing of the appearance of class I-like and class II-like molecules.

Acknowledgements

I would like to thank Ken Baker, Ronald Bontrop, Shirley Ellis and Martin Flajnik for critical comments on the manuscript.

References

Andersson L, Davies CJ. (1994) The major histocompatibility complex. In: *Cell Mediated Immunity in Ruminants* (eds BM Goddeeris, WI Morrison). CRC Press, Boca Raton FL, pp. 37–58.

Araki T, Geyyo F, Takagaki K, *et al.* (1988) Complete amino acid sequence of human plasma Zn-α2-glycoprotein cDNA and its homology to histocompatibility antigens. *Proc. Natl Acad. Sci. USA* **85:** 679–683.

Bahram S, Bresnahan M, Geraghty DE, Spies T. (1994) A second lineage of mammalian major histocompatibility complex class I genes. *Proc. Natl Acad. Sci. USA* **91:** 6259–6263.

Bartl S, Weissman I. (1994) Isolation and characterization of major histocompatibility complex class II B genes from the nurse shark. *Proc. Natl Acad. Sci. USA* **91:** 262–266.

Belich M, Madrigal JA, Hildebrand W, Zemmour J, Williams R, Luz R, Petzl-Erler ML, Parham P. (1992) Unusual HLA-B alleles in two tribes of Brazilian Indians. *Nature* **357:** 326–329.

Blumberg RS, Gerdes D, Chott A, Porcelli SA, Balk SP. (1995) Structure and function of the CD1 family of MHC-like cell surface proteins. *Immunol. Rev.* **147:** 5–30.

Bodmer WF. (1978) HLA: a super supergene. *Harvey Lect.* **72:** 91–110.

Bontrop RE, Otting N, Slierendregt BL, Lanchbury JS. (1995) Evolution of major histocompatibility complex polymorphisms and T-cell receptor diversity in primates. *Immunol. Rev.* **143:** 33–62.

Boyse EA, Beauchamp GK, Yamazaki K.(1987) The genetics of body scent. *Trends Genetics* **3:** 97–102.

Burmeister WP, Gatinel LN, Simister NE, Blum LB, Bjorkman PJ. (1994) Crystal structure at 2.2 Å resolution of the MHC-related neonatal Fc receptor. *Nature* **373:** 336–343.

Cheroutre H, Holcombe HR, Tangri S, et al. (1995) Antigen-presenting function of the TL antigen and mouse CD1 molecules. *Immunol. Rev.* **147:** 31–52.

Chouchane L, Kindt TJ. (1992) Mapping of the rabbit MHC reveals that class I genes are adjacent to the DR subregion and defines an insertion/deletion-related polymorphism in the class II region. *J. Immunol.* **149:** 1216–1222.

Cohen N. (1980) Salamanders and the evolution of the major histocompatibility complex. In: *Contemporary Topics in Immunology*, Vol. 9 (eds J Marchalonis, N Cohen). Plenum Press, New York, pp. 109-140.

Davis MM , Bjorkman PJ. (1988) T-cell antigen receptor genes and T-cell recognition. *Nature* **334:** 395–402.

Decamposlima PO, Gavioli R, Zhang QJ, Wallace LE, Dolcetti R, Rowe M, Rickinson AB, Masucci MG. (1993) HLA-A11 epitope loss isolates of Epstein-Barr-virus from a highly A11+ population. *Science* **260:** 98–100.

Doherty PC, Zinkernagel RM. (1975) Enhanced immunological surveillance in mice heterozygous at the H-2 gene complex. *Nature* **256:** 50–52.

Du Pasquier L, Schwager J, Flajnik M. (1989) The immune system of *Xenopus. Annu. Rev. Immunol.* **7:** 251–275.

Edidin M. (1983) MHC antigens and nonimmune functions. *Immunol. Today* **4:** 269–270.

Ellis SA, Staines KA, Morrison WI. (1996) cDNA sequences of cattle MHC class I genes transcribed on serologically defined halotypes A18 and A31. *Immunogenetics* **43:** 156–159.

Figueroa F, Tichy H, Berry RJ, Klein J. (1986) MHC polymorphism in island populations of mice. *Contemp. Top. Microbiol. Immunol.* **127:** 100–105.

Figueroa F, Ono H, Tichy H, O'hUigan C, Klein J. (1995) Evidence for insertion of a new intron into an MHC gene of perch-like fish. *Proc. Roy. Soc. Lond.* **B259:** 325–330.

Fischer-Lindahl KF, Hermael E, Loveland BE, Wang C. (1991) Maternally-transmitted antigen of mice – a model transplantation antigen. *Annu. Rev. Immunol.* **9:** 351–372.

Flajnik MF, Canel C, Kramer J, Kasahara M. (1991) Which came first, MHC class I or class II? *Immunogenetics* **33:** 295–300.

Flajnik MF, Kasahara M, Shum BP, Salter-Cid L, Taylor E, Du Pasquier L. (1993) A novel type of class I gene organization in vertebrates: a large family of non-MHC-linked class I genes is expressed at the RNA level in the amphibian *Xenopus. EMBO J* **12:** 4385–4396.

Germain RN, Margulies DH. (1993) The biochemistry and cell biology of antigen processing and presentation. *Annu. Rev. Immunol.* **11:** 403–450.

Greenberg AS, Avila D, Hughes M, Hughes A, McKinney EC, Flajnik MF. (1995) A new antigen receptor gene family that undergoes rearrangement and extensive somatic diversification in sharks. *Nature* **374:** 168–175.

Guillemot F, Billault A, Pourquie O, Behar G, Chausse A-M, Zoorob R, Kreiblich G, Auffray C. (1988) A molecular map of the chicken major histocompatibility complex: the class II β genes are closely-linked to the class I genes and the nucleolar organizer. *EMBO J.* **7:** 2775–2785.

Hamilton WD, Zuk M. (1982) Heritable true fitness and bright birds: a role for parasites? *Science* **218:**

384–387.

Hashimoto K, Nakanashi KT, Kurosawa Y. (1992) Isolation of a shark sequence resembling the major histocompatibility complex antigens. *Proc. Natl Acad. Sci. USA* **89:** 2209–2212.

Hein WR. (1994) Structural and functional evolution of the extracellular regions of T cell receptors. *Sem. Immunol.* **6:** 361–372.

Hill AVS, Allsopp C, Kwiatowski D, Antsey N, Twumasi P, Rowe A, Bennet S, Brewster D, McMichael A, Greenwood B. (1991) Common West African HLA antigens are associated with protection from severe malaria. *Nature* **352:** 595–600.

Hughes AL, Nei M. (1988) Pattern of nucleotide substitution at major histocompatibility complex class I loci reveals overdominant selection. *Nature* **335:** 167–170.

Hughes AL, Nei M. (1989) Nucleotide substitution at major histocompatibility complex class II loci: evidence for overdominant selection. *Proc. Natl Acad. Sci. USA* **86:** 958–962.

Jackson MR, Cohendoyle MF, Peterson PA, Williams DB. (1994) Regulation of MHC class-I transport by the molecular chaperone, Calnexin (P88, Ip90). *Science* **263:** 384–387

Kaastrup P, Nielsen B, Hoerlyck V, Simonsen M. (1988) Mixed lymphocyte reactions (MLR) in rainbow trout (*Salmo gairdneri*) sibling. *Devel. Comp. Immunol.* **12:** 801–808.

Karlsson L, Surh CD, Sprent J, Peterson P. (1992) An unusual class II molecule. *Immunogenetics* **29:** 411–413.

Kasahara M, McKinney EC, Flajnik MF, Ishibashi T. (1993) The evolutionary origin of the major histocompatibility complex: polymorphism of class II α chain genes in the cartilaginous fish. *Eur. J. Immunol.* **23:** 2160–2165.

Kasahara M, Flajnik MF, Ishibashi T, Natori T. (1995) Evolution of the major histocompatibility complex: a current overview. *Transplant. Immunol.* **3:** 1–20.

Kato Y, Salter-Cid L, Flajnik MF, Kasahara M, Namikawa C, Sasaki M, Nonaka M. (1994) Isolation of the Xenopus complement factor B complementary DNA and linkage to the frog MHC. *J. Immunol.* **153:** 4546–4554.

Kaufman JF, Auffray C, Korman AJ, Schackelford DA, Strominger JL. (1984) The class II molecules of the human and murine major histocompatibility complex. *Cell* **36:** 1–13.

Kaufman J, Skjoedt K, Salomonsen J. (1990) The MHC molecules of nonmammalian vertebrates. *Immunol. Rev.* **113:** 83–117.

Kaufman J, Andersen R, Avila D, Engberg J, Lambris J, Salomonsen J, Welinder K, Skjødt K. (1992) Different features of the MHC class I heterodimer have evolved at different rates: chicken B-F and β_2-microglobulin sequences reveal invariant surface residues. *J. Immunol.* **148:** 1532–1546.

Kaufman J, Salomonsen J, Flajnik M. (1994) Evolutionary conservation of MHC class I and class II molecules – different yet the same. *Sem. Immunol.* **6:** 411–424.

Kaufman J, Völk H, Wallny H-J. (1995) A "minimal essential MHC" and an "unrecognized MHC": two extremes in selection for polymorphism. *Immunol. Rev.* **143:** 63–88.

Klein J. (1975) *The Biology of the Mouse Histocompatibility-2 Complex.* Springer-Verlag, New York.

Klein J. (1986) *Natural History of the Major Histocompatibility Complex.* John Wiley and Sons, New York.

Klein J. (1987) Origin of major histocompatibility complex polymorphism: the trans-species hypothesis. *Hum. Immunol.* **19:** 155–162.

Klein J, Figueroa F. (1986) Evolution of the major histocompatibility complex. *CRC Crit. Rev. Immunol.* **6:** 295–386.

Klein J, O'hUigin C. (1995) Class II B MHC motifs in an evolutionary perspective. *Immunol. Rev.* **143:** 89–112.

Koch C. (1986) A genetic polymorphism of the complement component factor B in chickens not linked to the major histocompatibility complex (MHC). *Immunogenetics* **23:** 364–367.

Lawlor D, Zemmour J, Ennis P, Parham P. (1990) Evolution of class I MHC genes and proteins: from natural selection to thymic selection. *Annu. Rev. Immunol.* **8:** 23–64.

Madden DR. (1995) The three-dimensional structure of peptide MHC complexes. *Annu. Rev. Immunol.* **13:** 587–622.

McCumber LJ, Sigel MM, Trauger RJ, Cuchens MA. (1982) RES structure and function of the fishes. In: *The Reticuloendothelial System, Vol. 3. Phylogeny and Ontogeny* (eds N. Cohen, MM Sigel). Plenum Press, New York pp. 393–450.

Miller MM, Goto R, Bernot A, Zoorob R, Auffray C, Bumstead N, Briles WE. (1994) Two MHC class I and two MHC class II genes map to the chicken Rfp-Y system outside the B complex. *Proc. Natl Acad. Sci. USA* **91:** 4397–4401.

Momberg F, Roelse J, Howard J, Butcher G, Hämmerling G, Neefjes J. (1994) Selectivity of

MHC-encoded peptide transporters from human, mouse and rat. *Nature* **367:** 648–651.

Moretta L, Ciccone E, Moretta A, Höglund P, Öhlen C., Kärre K. (1992) Allorecognition by NK cells: non-self or no self? *Immunol. Today* **13:** 300–306.

Morshauser RC, Wand H, Flynn GC, Zuiderweg ERP. (1995) The peptide-binding domain of the chaperone protein Hsc70 has an unusual secondary structure topology. *Biochemistry* **34:** 6261–6266.

Namikawa C, Salter-Cid L, Flajnik MF, Kato Y, Nonaka M, Sasaki M. (1995) Isolation of Xenopus LMP-7 homologues. Striking allelic diversity and linkage to MHC. *J. Immunol.* **155:** 1964–1971.

Nathenson SG, Geliebter J, Pfaffenbach GM, Zeff RA. (1986) Murine major histocompatibility complex class I mutants: molecular analysis and structure-function implications. *Annu. Rev. Immunol.* **4:** 471–502.

Nizetic D, Figueroa F, Dembic Z, Nevo E, Klein J. (1987) Major histocompatibility complex gene organization in the mole rat *Spalax ehrenbergi*: evidence for transfer of function between class II genes. *Proc. Natl Acad. Sci. USA* **84:** 5828–5832.

Nowak MA, Tarczy-Hornoch K, Austyn JM. (1992) The optimal number of major histocompatibility complex molecules in an individual. *Proc. Natl Acad. Sci. USA* **89:** 10896–10899.

Obata Y, Satta Y, Moriwaki K, Shiroishi T, Hasegawa H, Takahashi T, Takahata N. (1994) Structure, function and evolution of mouse TL genes, nonclassical class I genes of the major histocompatibility complex. *Proc. Natl Acad. Sci. USA* **91:** 6589–6593

Parham P, Adams EJ, Arnett KL. (1995) The origins of HLA-A,B,C polymorphisms. *Immunol. Rev.* **143:** 141–180.

Perarnau B, Siegrist CA, Gillet A, Vincent C, Kimura S, Lemonnier FA. (1990) Beta-2-microglobulin restriction of antigen presentation. *Nature* **346:** 751–754

Potts WK, Wakeland EK. (1990) Evolution of diversity at the major histocompatibility complex. *Trends Ecol. Evol.* **5:** 181–187.

Potts W, Wakeland E. (1993) Evolution of MHC genetic diversity: a tale of incest, pestilence and sexual preference. *Trends Genet.* **9:** 408–412.

Rippmann F, Taylor W, Rothbard J, Green NM. (1991) A hypothetical model for the peptide binding domain of hsp70 based on the peptide binding domain of HLA. *EMBO J.* **10:** 1053–1059.

Roche PA. (1995) HLA-DM: an in vivo facilitator of MHC class II peptide loading. *Immunity* **3:** 259–262.

Salter-Cid L, Kasahara M, Flajnik MF. (1994) HSP70 genes are linked to the major histocompatibility complex. *Immunogenetics* **39:** 1–7.

Satta Y, O'hUigin C, Takahata N, Klein J. (1994) Intensity of natural selection at the major histocompatibility complex loci. *Proc. Natl Acad. Sci. USA* **91:** 7184–7188.

Scofield VL, Schlumpberger JM, West LA, Weissman IL. (1982) Protochordate allorecognition is controlled by an MHC-like gene system. *Nature* **295:** 449–502.

Shalhevet D, Da Y, Beever JE, van Eijk MJT, Ma R, Lewin HA, Gaskins HR. (1995) Genetic mapping of the LMP2 proteosome subunit gene to the BoLA class IIb region. *Immunogenetics* **41:** 44–46.

Shamblott MJ, Litman GW. (1989) Complete nucleotide sequence of primitive vertebrate immunoglobulin light chain genes. *Proc. Natl Acad. Sci. USA* **86:** 4684–4688.

Shawar SM, Vyas JM, Rodgers JR, Rich RR. (1994) Antigen presentation by major histocompatibility complex class I-B molecules. *Annu. Rev. Immunol.* **12:** 839–880.

Schild H, Mavaddat N, Litzenberger C, Ehrich EW, Davis MM, Bluestone JA, Matis L, Draper RK, Chien Y. (1994) The nature of major histocompatibility complex recognition by $\gamma\delta$ T cells. *Cell* **76:** 29–37.

Simister N, Mostov K. (1989) An Fc receptor structurally related to MHC class I antigens. *Nature* **337:** 184–187.

Soloski MJ, DeCloux A, Aldrich CJ, Forman J. (1995) Structural and functional characteristics of the class IB molecule Qa-1. *Immunol. Rev.* **147:** 67–90.

Stet RJM, Dixon B, van Erp SHM, van Lierop M-JC, Rodrigues PNS, Egbert E. (1996) Inference of structure and function of fish major histocompatibility complex (MHC) molecules from expressed genes. *Fish Shellfish Immunol.* (In press.)

Stroynowski I. (1995) Tissue-specific, peptide-binding transplantation antigens: lessons from the Qa-2 system. *Immunol. Rev.* **147:** 91–108.

Suh WK, Cohendoyle MF, Fruh K, Wang K, Peterson PA, Williams DB. (1994) Interaction of MHC class-I molecules with the transporter associated with antigen-processing. *Science* **264:** 1322–1326

Sültmann H, Mayer WE, Figueroa F, O'hUigin C, Klein J. (1994) Organization of MHC class II B genes in the zebrafish (*Brachydanio rerio*). *Genomics* **23:** 1–14.

Tiwari J, Terasaki P. (1985) *HLA and Disease Associations.* Springer Verlag, New York.

Tonegawa S. (1983) Somatic generation of antibody diversity. *Nature* **302:** 5575–5581.

Tregaskes C, Kong F, Paramithiotis E, Chen C-L, Ratcliffe M, Davidson TF, Young J. (1995) Identification and analysis of the expression of CD8-alpha-beta and CD8-alpha-alpha isoforms in chickens reveals a major Tcr-gamma-delta CD8-alpha-beta subset of intestinal intraepithelial lymphocytes. *J. Immunol.* **154:** 4485–4494.

Trowsdale J. (1993) Genomic structure and function in the MHC. *Trends Genet.* **9:** 117–122.

Trowsdale J. (1995) "Both man & bird & beast": comparative organization of MHC genes. *Immunogenetics* **41:** 1–17.

Turpen J, Smith P. (1989) Precursor immigration and thymocyte succession during larval development and metamorphosis in *Xenopus*. *J. Immunol.* **142:** 41–47.

Vidovic D, Matzinger P. (1988) Unresponsiveness to a foreign antigen can be caused by self-tolerance. *Nature* **336:** 222–225.

Watkins DI, Chen ZW, Hughes AL, Lagos A, Lewis AM, Shadduck JA, Letvin NL. (1990) Evolution of the MHC class I genes of a New World primate from ancestral homologues of human non-classical genes. *Nature* **346:** 60–63.

Watkins D, McAdam S, Liu X, *et al.* (1992) New recombinant HLA-B alleles in a tribe of South American Amerindians indicate rapid evolution of MHC class I loci. *Nature* **357:** 329–333.

Wright H, Ballingall KT, Redmond J. (1994) The D–Y subregion of the sheep MHC contains an A/B gene pair. *Immunogenetics* **40:** 230–234.

Zapata AG, Vara A, Torroba M. (1992) Seasonal variations in the immune system of lower vertebrates. *Immunol. Today* **13:** 142–147.

Zinkernagel RM, Doherty PC. (1979) MHC-restricted cytotoxic T cells: studies on the biological role of polymorphic major transplantation antigens determining T cell restriction specificity, function and responsiveness. *Adv. Immunol.* **27:** 52–177.

Zoorob R, Bernot A, Renoir DM, Choukri F, Auffray C. (1993) Chicken major histocompatibility complex class II B genes: analysis of interallelic and interlocus sequence variation. *Eur. J. Immunol.* **23:** 1139–1145.

NHSBT
LIBRARY

2

Molecular genetics of HLA class I and class II regions

John Trowsdale

2.1 Introduction

The human major histocompatibility complex [MHC; also known as the human leukocyte antigen (HLA) region in humans] is contained within about 4 Mbp of DNA (~1% of the genome) on the short arm of chromosome 6 at 6p21.3 (Campbell and Trowsdale, 1993). The human MHC has been extensively characterized and more is known about this region of the genome than most regions of comparable size. More than 100 genes have been located in the MHC, many of which have immunological functions in antigen processing and presentation. This clustering of genes with interrelated functions may be significant, reflecting a selective advantage for such an arrangement in evolution. In addition to its fundamental importance, the HLA region is strongly associated with most, if not all, diseases of autoimmune aetiology (Davies *et al.*, 1994; Lechler, 1994; Sinha *et al.*, 1990; Tiwari and Terasaki, 1985).

The human MHC is divided into class I, class II and class III regions. The class I region, at the telomeric end of the complex, contains genes encoding the classical transplantation antigens, *HLA-A, B* and *C*. The class II region is at the opposite (centromeric) end of the complex. The products of this region were initially identified as immune-response loci, known as *HLA-D*. Upon DNA cloning, it became apparent that the products of the class II loci, *HLA-DP*, *-DQ* and *-DR*, were structurally and functionally related to the class I loci as they were all members of the immunoglobulin superfamily. The class III region, lying between the other two clusters, is densely packed with genes of a variety of functions, including some components of the complement system, *C2*, *C4* and factor B (*Bf*). The class III region is the subject of a separate chapter in this volume (see Chapter 3).

The layout of the human MHC is shown in *Figure 2.1a*. The region has been cloned in yeast artificial chromosomes (YACs) and mapped by pulse-field gel electrophoresis (Abderrahim *et al.*, 1994; Bronson *et al.*, 1991; Geraghty *et al.*, 1992a, b; Ragoussis *et al.*, 1991). There is considerable coverage by overlapping cosmid clones. The complete DNA sequence of some regions has been determined (Beck *et al.*, 1992a; Iris *et al.*, 1993). This information and many more details are maintained on database (Newell *et al.*, 1994).

HLA and MHC: genes, molecules and function, edited by M.J. Browning and A.J. McMichael.
© 1996 BIOS Scientific Publishers Ltd, Oxford.

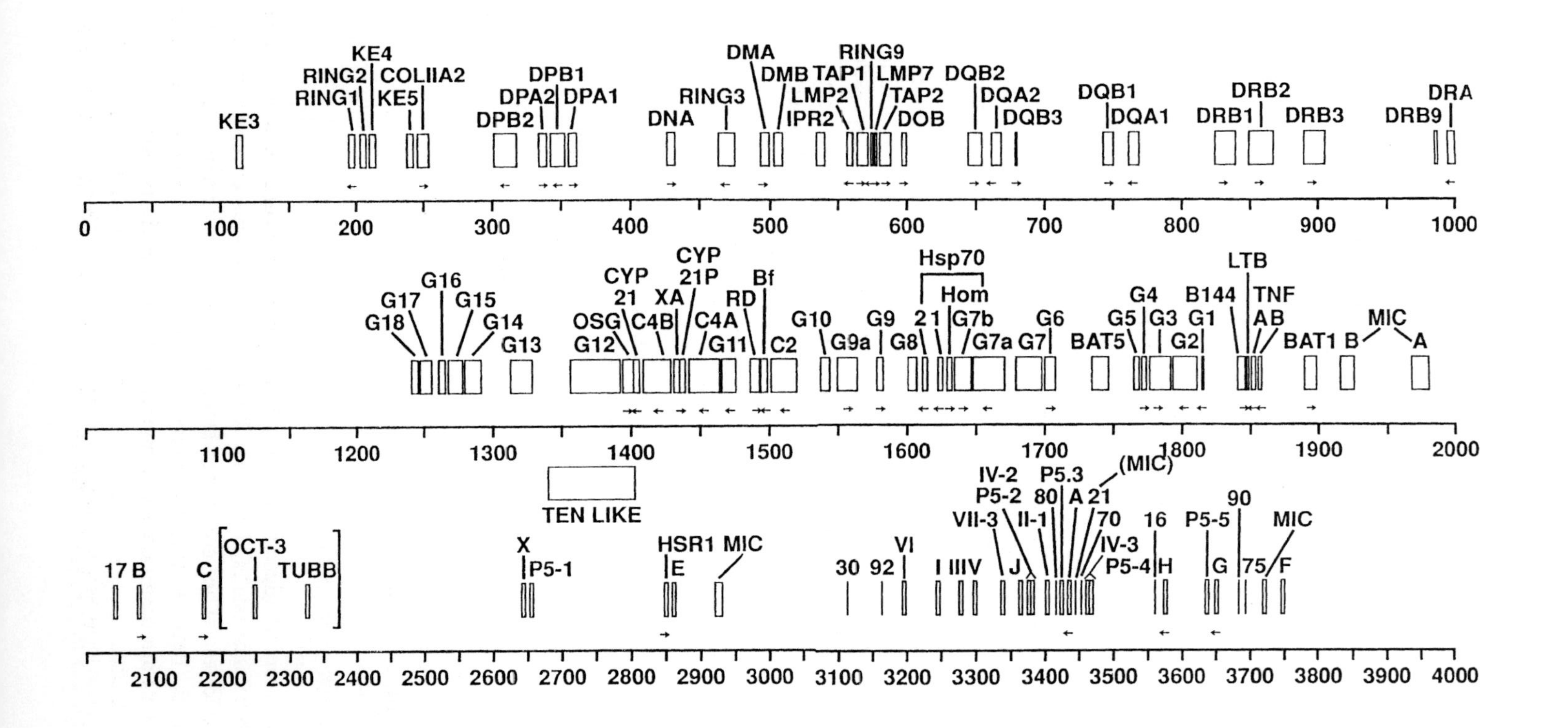

Figure 2.1. (a) Map of the human MHC. (The map is taken from Campbell and Trowsdale (1993) with modifications.)

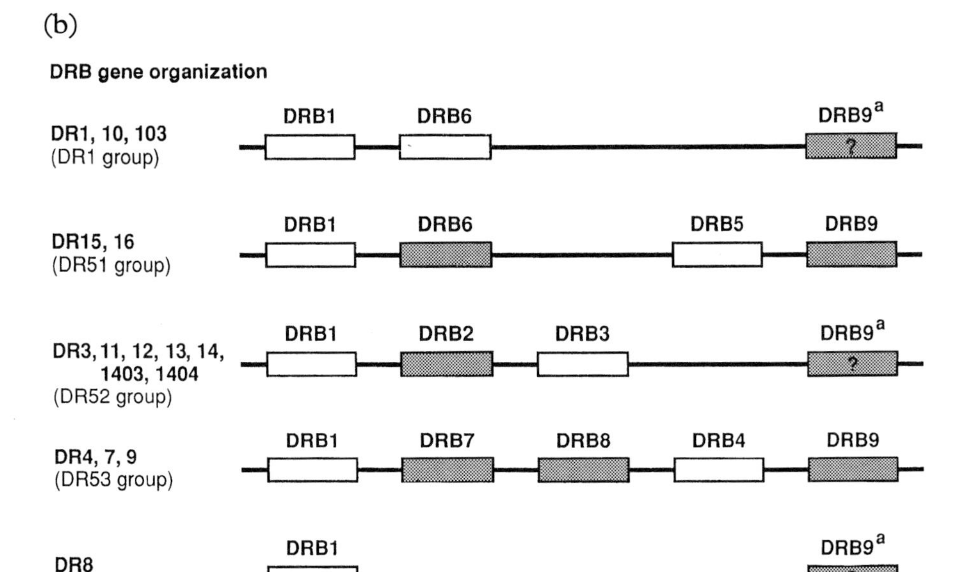

Figure 2.1. (b) Note the alternative arrangements of DR loci (shaded bars indicate loci believed to encode pseudogenes).

In immunological terms, the main products of the MHC are the classical class I and class II molecules. Both sets of products present antigens to T-cell receptors, arising from the same variable pool, but there are fundamental differences in the way they function. In most cases, class I molecules present intracellular antigens to $CD8^+$ cytotoxic T cells (see Chapters 9 and 13). This ensures that cytotoxic T cells kill cells harbouring infectious agents. It is appropriate that class I molecules are expressed on most nucleated cells. Class II molecules, on the other hand, generally present extracellular antigens to helper T cells which express the CD4 molecule (see Chapters 10 and 14). A key function of $CD4^+$ T cells, stimulated through class II molecules, is to promote production of appropriate antibodies to the offending extracellular antigen. Class II molecules tend to be expressed on professional antigen-presenting cells such as macrophages, B cells and dendritic cells, but their expression is inducible with agents such as γ-interferon on fibroblasts or epithelial cells.

2.2 Class I genes

2.2.1 Classical class I genes

The class I region of the human MHC contains three main functional loci, *HLA-A*, *HLA-B* and *HLA-C*. Each of these genes encodes an α-chain of a class I molecule. These genes are expressed on a wide range of somatic tissues at varying levels. Some tissues may express low or negligible levels of class I proteins, including sperm and oocytes, placenta and cells of the central nervous system. Interesting data have been reported recently, namely, that class I molecules are expressed selectively on electrically silent neurons which are then targeted for destruction. Thus immunosurveillance by cytotoxic T cells may be focused on to functionally impaired neurons (Neumann *et al.*, 1995). The lack of classical HLA class I expression at the materno-fetal interface (villous and extravillous cytotrophoblast and syncytiotrophoblast) is proposed to facilitate survival of the fetal tissue as an allograft in the maternal host. Levels of HLA-C molecules are generally about 10-fold lower than those of -A or -B at the cell's surface. Nevertheless, HLA-C molecules are functional. There is increasing evidence for a role of HLA-C in target recognition by natural killer (NK) cells (Colonna *et al.*, 1993). The function of MHC class I molecules is dealt with in detail elsewhere in this book (see Chapter 13).

2.2.2 Non-classical class I genes

The other functional class I loci include *HLA-E*, *HLA-F* and *HLA-G* (Geraghty, 1993). These genes, generally much less polymorphic than *HLA-A*, *-B* or *-C*, potentially lead to the production of class I-related molecules, of unknown function and restricted tissue distribution. *HLA-E* is expressed in a number of tissues at low surface levels. The molecule may be retained in the endoplasmic reticulum (Ulbrecht *et al.*, 1992). *HLA-G* is expressed as two different transcripts in placental tissues and its function may relate to the survival of the fetus as semi-allogeneic tissue already referred to (Loke and King, 1991). In addition to these loci, the class I region contains a number of other class I genes or gene fragments, most likely pseudogenes, e.g. *HLA-H*, *-J* and *-K* (Geraghty, 1993; Geraghty *et al.*, 1992b; Le Bouteiller, 1994).

Quite recently, a new family of class I-related sequences was identified at the telomeric end of the class III region. These genes, called *MIC* or *PERB*, are only very weakly related to conventional class I sequences and were not revealed by cross-hybridization (Bahram *et al.*, 1994; Leelayuwat *et al.*, 1994). They have a limited tissue distribution on epithelial cells, particularly of the gut, and their functions are unknown (see also Chapter 3).

2.2.3 Other genes in the class I region

Based on its size alone, it is likely that the class I region will contain a large number of genes in addition to the MHC class I loci already described. Several such genes have already been identified in this region, including a β-tubulin gene, a gene expressed at high levels in skin and the *OCT3* oncogene (Geraghty *et al.*, 1992b; Vernet *et al.*, 1993; Volz *et al.*, 1994; Zhou and Chaplin, 1993). The *S* (skin) gene is restricted in expression to differentiating keratinocytes and the granular layer of the epidermis (Zhou and Chaplin, 1993). The S protein product contains a high content of serine, glycine and proline and showed structural homology with other skin components such as loricrin and keratin components of the differentiating epidermis. The *HSR1* gene is a member of a unique family of GTP-binding proteins (Vernet *et al.*, 1994). The *MOG* (myelin/oligodendrocyte glycoprotein) gene maps to the M region of the mouse and its location in the human class I region provides a marker relating the distal MHC areas in the two species.

Another recent finding in the class I region is a family of olfactory receptor genes (Fan *et al.*, 1995). These are the only such members of the family located outside

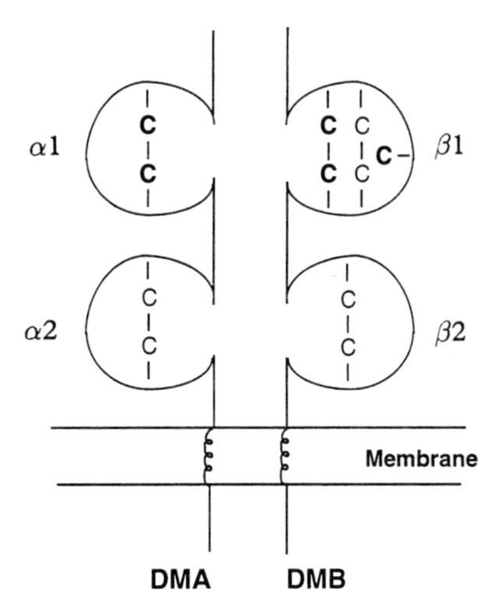

Figure 2.2. Schematic structure of the DM molecule. It is clearly similar in arrangement to that of conventional class II loci, except for the extra cysteine residues in the α1 and β1 domains, which are highlighted in bold type.

the olfactory receptor gene complex on chromosome 17. Mice appear to be able to distinguish potential mates on the basis of H-2 types, through their urine odour, and it is speculated that the olfactory receptor genes in the class I region may be responsible for this phenomenon (Potts *et al.*, 1991).

cDNA selection using YAC and cosmid clones spanning the class I region calls for a minimum of 40 novel genes to be located there.

2.3 Class II genes

The class II region extends for about 800 kbp and encompasses all of the known class II α and β genes. The main, surface-expressed molecules in humans are *HLA-DP,-DQ* and *-DR*. The α- and β-chain genes are all arranged as matched pairs (i.e. *DRA* and *DRB*, *DQA* and *DQB*; *DPA* and *DPB*) but the number of *DRB* genes and pseudogenes can differ, depending on the haplotype (*Figure 2.1b*; Andersson *et al.*, 1987; Gorski *et al.*, 1987; Kawai *et al.*, 1989; Rollini *et al.*, 1985). Both DQ and DP regions include a pseudogene pair (*DQA2* and *DQB2*, *DPA2* and *DPB2*), although the DQ pair of sequences does not contain any obvious deleterious coding region mutations (Gustafsson *et al.*, 1987; Jonsson *et al.*, 1987; Trowsdale *et al.*, 1984). As proteins, DRα, DQα and DPα preferentially pair with their respective β-chains, although there is evidence for cross-pairing, both *in vivo* and *in vitro*.

The class II region contains two other pairs of class II genes. *HLA-DNA* and *-DOB* are exceptional in that they are probably a pair, and indeed, the mouse equivalents, *Oa* and *Ob* form a heterodimeric protein, but they are not adjacent to each other, being separated by at least seven other genes (Karlsson *et al.*, 1992;

	DMA	DPA	DQA	DRA	DNA	DMB	DPB	DQB	DRB	DOB	A2	B7	Cw3	β_2M
DMA	-	34	33	32	31	24	26	23	25	23	17	18	18	27
DMB	24	33	33	38	32	-	35	33	35	35	29	33	33	28

Figure 2.3. Homology of Ig ($\alpha2$ and $\beta2$) domains in DM with Ig-like domains of other MHC encoded genes (numbers are % amino acid identity).

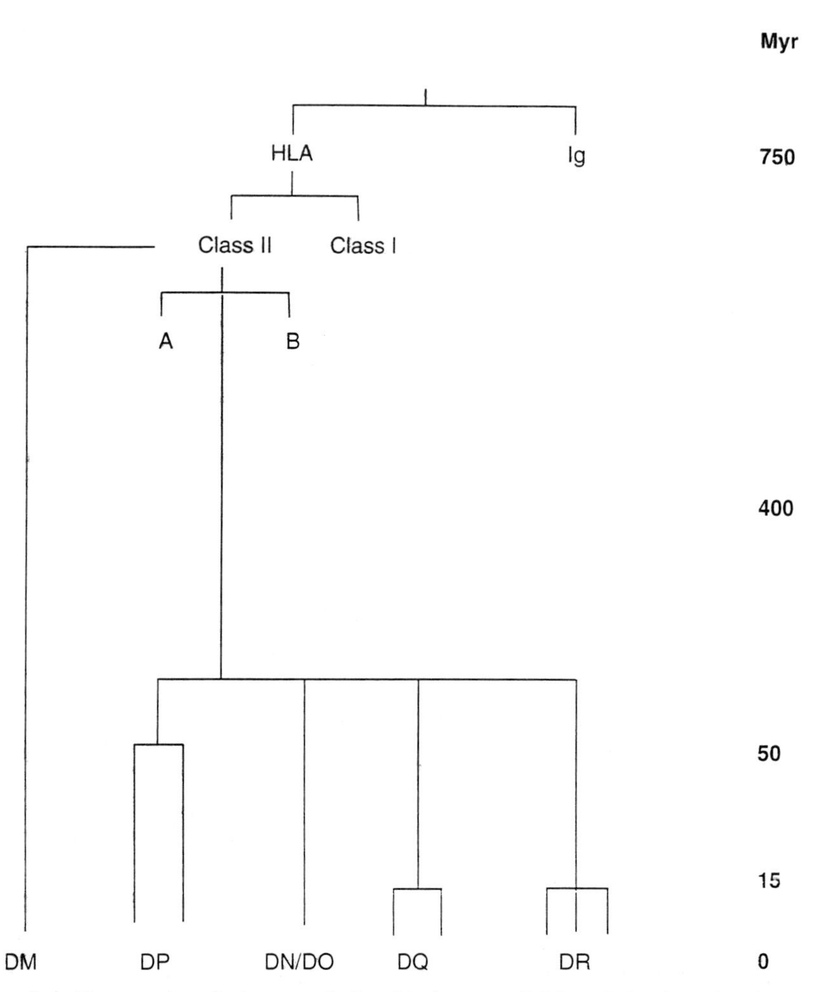

Figure 2.4. Proposed evolutionary relationship between DM and classical class I and class II sequences. The dates (in Myr) are crude estimates based on sequence similarities.

Tonnelle *et al.*, 1985; Trowsdale *et al.*, 1984; Young and Trowsdale, 1990). The *HLA-DNA* product is specified by mRNAs of 1.2 kbp and 3.5 kbp. The longer, more abundant, mRNA is due to a 2 kbp extension on the 3′ end, the significance of which is not understood. No human DNA/DOB protein has been identified.

HLA-DMA and *-DMB* are two linked genes that are only distantly related to other class II sequences (Cho *et al.*, 1991; Kelly *et al.*, 1991a). The putative

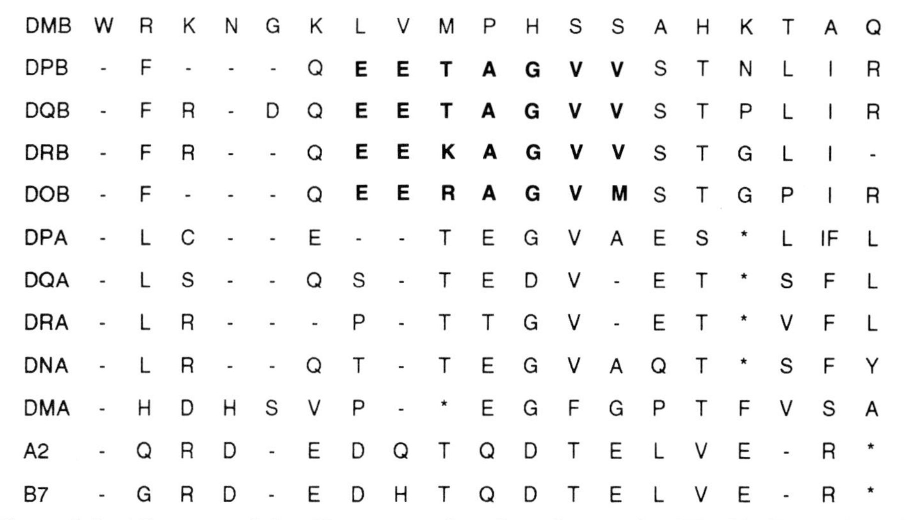

DMB	W	R	K	N	G	K	L	V	M	P	H	S	S	A	H	K	T	A	Q
DPB	-	F	-	-	-	Q	**E**	**E**	**T**	**A**	**G**	**V**	**V**	S	T	N	L	I	R
DQB	-	F	R	-	D	Q	**E**	**E**	**T**	**A**	**G**	**V**	**V**	S	T	P	L	I	R
DRB	-	F	R	-	-	Q	**E**	**E**	**K**	**A**	**G**	**V**	**V**	S	T	G	L	I	-
DOB	-	F	-	-	-	Q	**E**	**E**	**R**	**A**	**G**	**V**	**M**	S	T	G	P	I	R
DPA	-	L	C	-	-	E	-	-	T	E	G	V	A	E	S	*	L	IF	L
DQA	-	L	S	-	-	Q	S	-	T	E	D	V	-	E	T	*	S	F	L
DRA	-	L	R	-	-	-	P	-	T	T	G	V	-	E	T	*	V	F	L
DNA	-	L	R	-	-	Q	T	-	T	E	G	V	A	Q	T	*	S	F	Y
DMA	-	H	D	H	S	V	P	-	*	E	G	F	G	P	T	F	V	S	A
A2	-	Q	R	D	-	E	D	Q	T	Q	D	T	E	L	V	E	-	R	*
B7	-	G	R	D	-	E	D	H	T	Q	D	T	E	L	V	E	-	R	*

Figure 2.5. Alignment of class II sequences based on the putative CD4 binding region of the $\beta 2$ domain of the class II β chain (Cammarota *et al.* 1992; Konig *et al.* 1992). The core CD4 binding sequences (residues 137–143; Konig *et al.* 1992) are highlighted in **bold type**. Note the lack of a recognizable CD4 binding site in the DMB sequence. - indicates identity. * indicates residue missing. (Equivalent sequences from HLA class I molecules HLA-A2 and -B7 are included for comparison.)

structure of the DM molecule is similar to that of classical class II molecules, such as DR, DQ or DP, but the $\alpha 1$ and $\beta 1$ domains each contain an additional disulphide bridge (*Figure 2.2*), which is believed to make the structure more rigid. The membrane-proximal $\beta 2$ and $\alpha 2$ domains are almost as similar to class I $\alpha 3$ as they are to $\beta 2$ and $\alpha 2$ class II sequences (*Figure 2.3*). This relationship suggests that DM sequences arose at around the same time that class I and class II sequences diverged from each other, several hundred million years ago (*Figure 2.4*). The products of the DM genes form a heterodimer which is not expressed on the cell surface and does not contain a recognizable CD4-binding site (*Figure 2.5*). DM is, in fact, mostly found in intracellular vesicles, thought to be the sites of loading of peptides on to class II molecules (Sanderson *et al.*, 1994, and see Chapter 10). This finding is consistent with data from DM mutants that show an essential role for the DM molecule in the formation of the class II/peptide complex (Fling *et al.*, 1994; Morris *et al.*, 1994). Computer modelling studies suggest that the DM groove is shallow, or non-existent, being blocked by bulky hydrophobic residues, at least at one end (C. Thorpe and P. Travers, personal communication). Recent data implicate a role for DM in loading peptides on to class II molecules, possibly by inducing a conformational change (Cresswell, 1994; Karlsson *et al.*, 1994).

DM genes are expressed in similar circumstances to other class II molecules, as detected by Northern blotting, consistent with a complementary function in peptide loading. Upstream of the DM genes are sequences, X and Y boxes, similar to those found in other class II genes (*Figure 2.6*).

CONSENSUS	X BOX TYTNCCYAGNRACAGATGA	← 18/19 BASE PAIRS →	Y BOX CTGATTGGYY
DMA	-------------T-----	← 19 base pairs →	----------
DMB	-------------T-----		C-----T---
DPA	----------------GA-		-----A---G
DQA	--C-G--------T--GAT		--A-------
DQA 2	-GC-A-G-----ACA--AT		----------
DRA	--C---------------C		----------
DNA	-----------------AC		G---------
Ea	C-----------------T		----------
Aa	-GCAG--G-----G-TGAC		----------
DPB	------------GCA----	← 18 base pairs →	TCC-------
DPB 2	------------GCA----		-CC-------
DQB	-----------------T-		----------
DQB 2	-----------G-------		----------
DRB	----A-------T------		----------
Eb	----A-------T------		----------
Ab	----------------C--		----------
DOB	C---------C-T------	← 17 base pairs →	T---------

Figure 2.6. DM genes: transcriptional control regions. The figure shows a comparison of X and Y box elements from human (DPA/B; DQA/B; DRA/B; DNA, DOB) and mouse (Ea, b; Aa, b) class II genes. Sequences are aligned against consensus elements and identity is indicated by dashes. Y = T/C, R = G/A. The number of nucleotides separating the elements is as indicated for each sequence. (Data from A. Kelly, personal communication.)

2.4 The *TAP* and *LMP* genes

Remarkably, a tight cluster of genes involved in processing antigens for presentation by class I molecules is inserted in the class II region (*Figure 2.7*) between the DM and DQ gene loci (see *Figure 2.1*). The products of the two *TAP* (transporter associated with antigen processing) genes, *TAP1* and *TAP2*, are members of the ABC(ATP-binding cassette) transporter superfamily (Townsend and Trowsdale, 1993). Members of this family of transmembrane molecules are involved in transport of a wide selection of different substances across membranes, including oligopeptides, proteins and ions (Higgins *et al.*, 1990). The products of the two *TAP* genes form a complex in the endoplasmic reticulum (ER) membrane which translocates peptides from the cytoplasm into the lumen of the ER (Androlewicz and Cresswell, 1994; Momburg *et al.*, 1994, and see Chapter 9). Once they have arrived in the lumen, the transported peptides can participate in the assembly of class I molecules, and there is strong evidence for an association

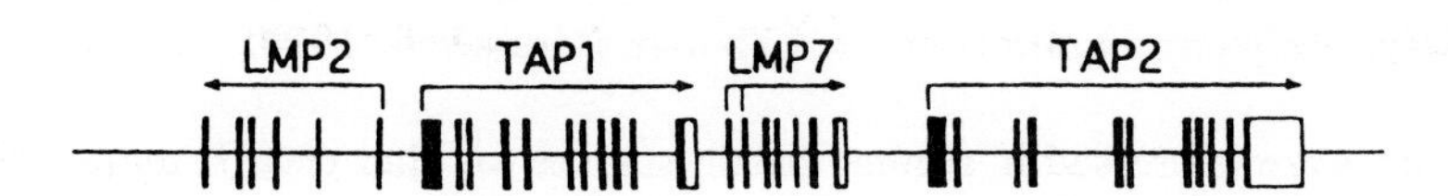

Figure 2.7. The *LMP/TAP* cluster. Note the close proximity of the *LMP2* and *TAP1* genes which share a bidirectional promoter (Wright *et al.*, 1995).

between the TAPs and class I since they may be co-immunoprecipitated (Ortmann *et al.*, 1994). Cells defective in either *TAP1* or *TAP2* have a reduction in the level of cell surface class I expression and are unable to present the usual intracellular antigens to cytotoxic T cells (Kelly *et al.*, 1992; Spies and DeMars, 1991).

The *LMP* genes, *LMP2* and *LMP7*, that map very close to the *TAP* genes (*Figure 2.7*), encode components of a large complex known as the proteasome (Glynne *et al.*, 1991; Goldberg, 1995; Kelly *et al.*, 1991b). This structure has been implicated in proteolytic digestion of a large number of short-lived cytoplasmic proteins. Proteasome subunits are part of either the α (structural) or β (thought to be catalytic) families and the LMPs belong to the latter group. The β subunits are produced with leader propeptide sequences that are cleaved off (Glynne *et al.*, 1993). They form a ring structure of seven different β subunits and they take part in cleaving peptides which enter the hole in the proteasome barrel structure. A free threonine at the N terminus of the mature subunits provides part of the active site of each subunit (Glynne *et al.*, 1993). LMP2 and LMP7 replace two other constitutive proteasome β subunits, δ and ε (MB1) (Belich *et al.*, 1994). The production of the two novel LMP subunits is thought to alter the proteolytic activity of the proteasome to favour peptides appropriate for binding to the grooves of class I molecules (Gaczynska *et al.*, 1994). Consistent with these ideas is the finding that expression of *LMP2* and *LMP7*, like *TAP*, is inducible with γ interferon. The up-regulation of the LMPs by this cytokine may reflect what happens *in vivo*, following infection with a virus, to permit efficient antigen processing, appropriate for the immune system.

2.5 RING3 and other sequences in the class II region

Except for pseudogenes, *RING3* is the only gene in the class II region of the MHC without a function in the immune system. The gene maps between *HLA-DNA* and *-DMA*. It shares homology with the *Drosophila* gene, female sterile homeotic, *fsh*. *RING3* is expressed ubiquitously (Beck *et al.*, 1992b).

There are two pseudogenes of interest in the class II region, between the *DMB* and *LMP2* loci. This region is thought to have arisen as an insertion of genetic material into the class II region from elsewhere, as there is no trace of a similar region in the mouse MHC between the *Ob* and *LMP2* genes. One pseudogene, lying about 40 kbp centromeric of LMP2, is a copy of a phosphatase inhibitor gene, *IPP-2* (Sanseau *et al.*, 1994). The other pseudogene, lying between the *IPP-2* gene and *DMB* is a short fragment of a class I sequence. Its origins are obscure but its presence in the class II region suggests that the DNA that became inserted into the class II region, carrying *IPP-2* and the class I fragment, came from the class I region (Beck and Trowsdale, 1996).

2.6 Polymorphism and the MHC as a gene cluster

The HLA class I and class II products are the most polymorphic human proteins known. There may turn out to be well over 200 alleles at the *HLA-B* locus. This variation in MHC class I and class II genes constitutes the basis of tissue rejection in transplantation. Combinations of alleles at different class I and class II loci tend to occur in a non-random fashion on what have been called ancestral haplotypes or supratypes. Many of these haplotypes contain distinctive sets of alleles, deletions, duplications and other features (Zhang *et al.*, 1990).

The class I region has obviously undergone expansion and contraction in different species and variation in size of the class I region can be extensive (Trowsdale, 1994). An interesting finding was recently made in respect to the human MHC class I region, namely several sequences centromeric to *HLA-B* were also present in a duplicate copy, near to *HLA-A*. These include sequences of the *P5, BAT1* and *MIC(PERB11)* family, as well as class I sequences. The data suggest that an inverted duplication of >200 kbp took place at some stage in the evolution of the human class I region (Leelayuwat *et al.*, 1995).

There is a case for the co-evolution of combinations of alleles at different MHC loci, on haplotypes. Indeed, it has been argued that the positioning of genes such as the *TAP*s and *LMP*s within the MHC, as well as some other immune system genes, may not have been by chance and may serve to maintain successful combinations of alleles in *cis* (Trowsdale, 1993, 1994). Studies of the rat *TAP*/class I allelic combinations within a single haplotype are particularly informative on this issue (Howard, 1993; Joly *et al.*, 1994). Of course, there are other explanations for keeping combinations of loci together on extended haplotypes. One is co-regulation of expression of genes. Another is gene conversion; keeping class I or class II genes together may be advantageous in permitting microsequence exchange (Zangenberg *et al.*, 1995).

2.7 Recombination and linkage disequilibrium

Genetic mapping based on recombination fractions places *HLA-B* 0.2 cM centromeric of *HLA-C* and *HLA-A* 0.8 cM telomeric of *HLA-C* (*Figure 2.8a,b*). This is in reasonable agreement with the physical map. There are regions of the MHC where there is not such a good agreement between physical distance and recombination. Linkage disequilibrium refers to the presence of two alleles at different loci occurring together more frequently than would be expected by chance. An example from the Caucasian population is *HLA-A1* and *HLA-B8* which occur at individual gene frequencies of about 27.5% and 15.7%, respectively. The expected frequency of the two alleles being together on the same chromosome would be around 4.3% (27.5% × 15.7%), although the haplotype is in fact found in 9.8% of the population. The phenomenon of linkage disequilibrium is common elsewhere throughout the MHC. Recombination has never been observed between *HLA-DQ* and *-DR*, for example. However, there is only weak linkage disequilibrium between *DP* and *DQ*, suggesting hotspots of recombination, and indeed several crossovers have been mapped in the class II region between *DP* and *DQ*; in the second intron of the *TAP1* gene, for example. An extreme example of this kind of distortion between the physical map and recombination data is to be found telomeric of the class I region, where studies of the iron uptake disease haemochromatosis and HLA class I markers reveal a region of over 3 Mbp in which recombination is extremely rare.

One explanation for linkage disequilibrium is that of selection for apt combinations of alleles at linked loci. Another is of population bottlenecks which

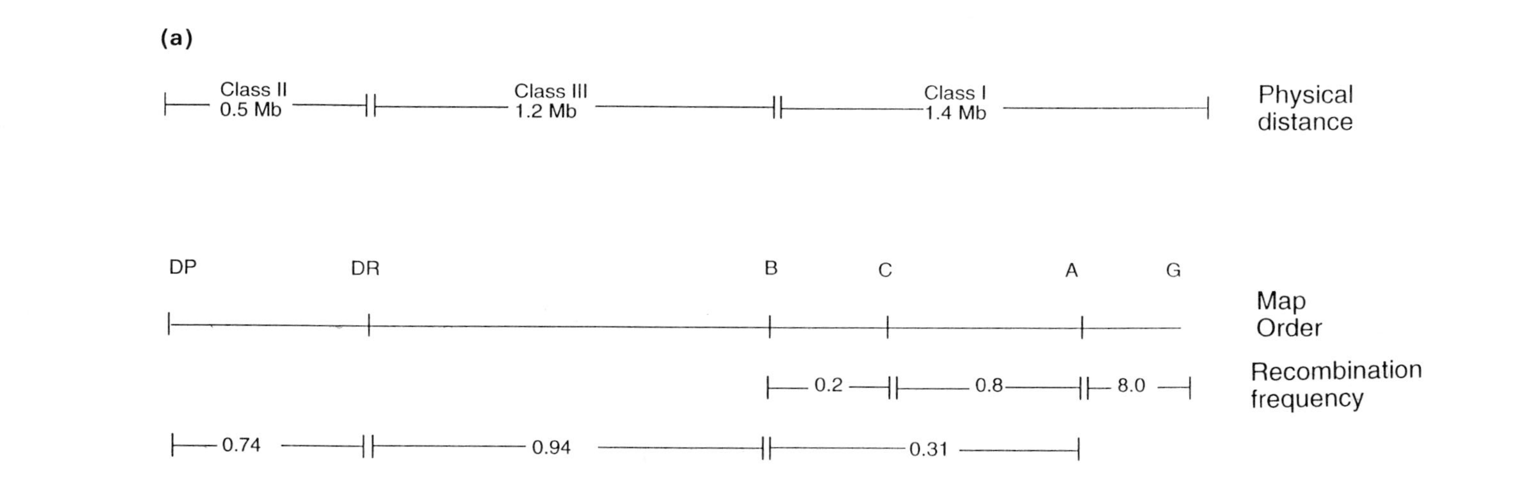

Figure 2.8. (a) Comparison of recombination rates within the human MHC and physical map. The data are taken from 59 CEPH pedigrees (Koller *et al.*, 1989; Martin *et al.*, 1995). The reason for the unusually high recombination frequency reported between HLA-A and -G is unknown, but needs to be substantiated in further studies.

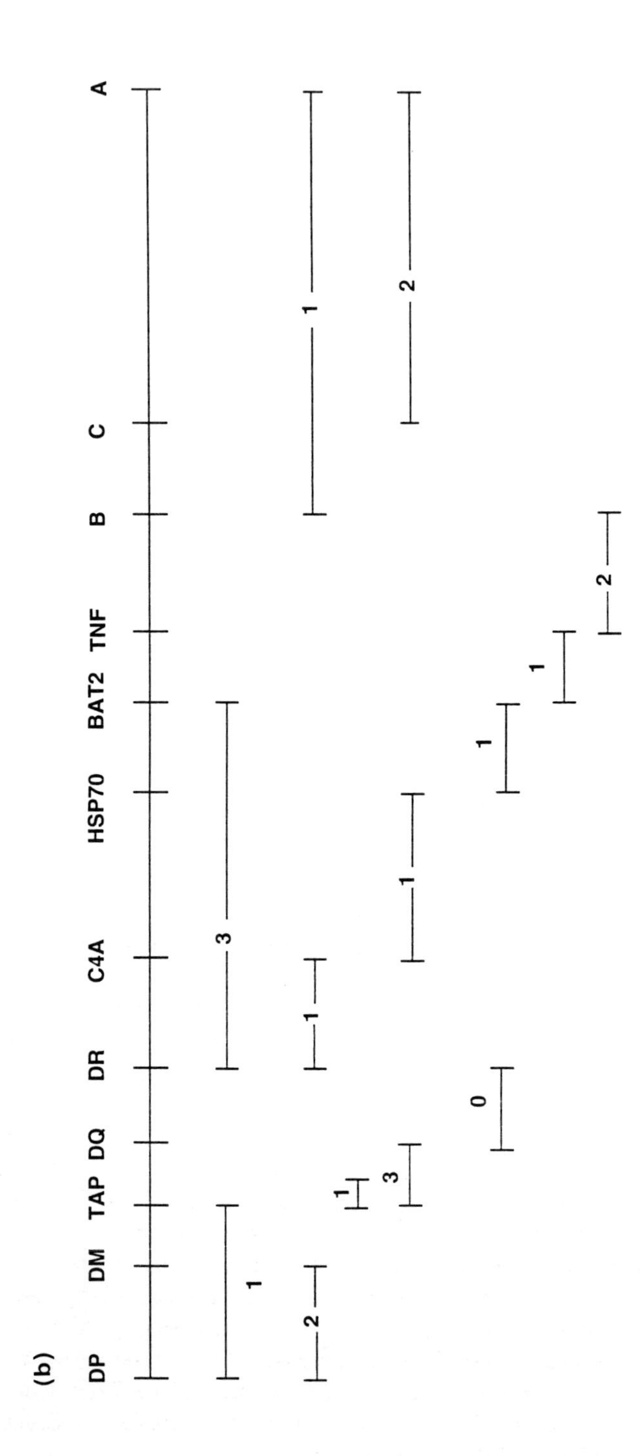

Figure 2.8. (b) Boundaries of identified recombination points in the map shown in *Figure 2.8a* overleaf. Note the lack of recombination between DQ and DR.

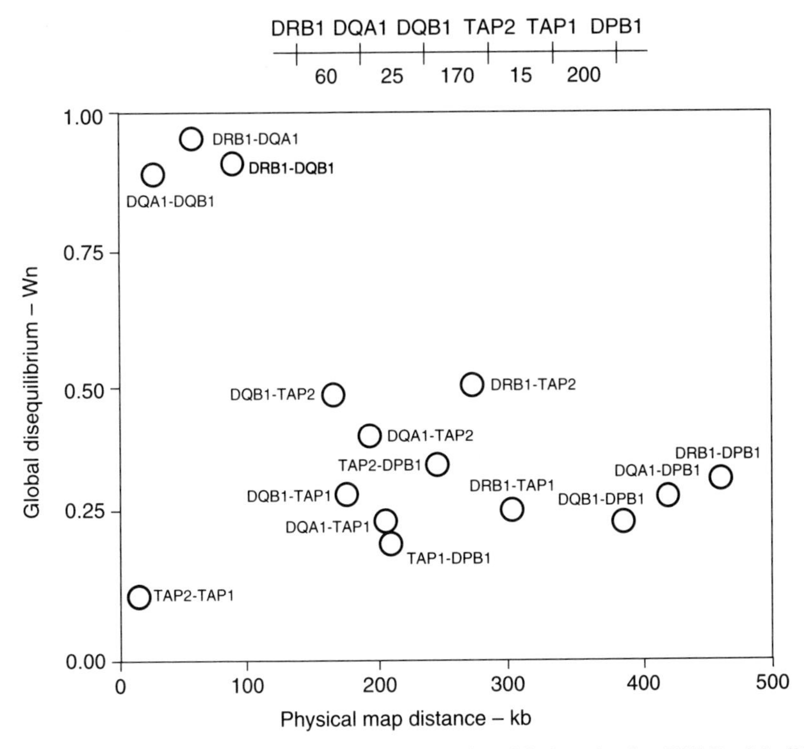

Figure 2.9. Irregularities in linkage disequilibrium/equilibrium in the MHC. Modified from Klitz *et al.* (1995) *Am. J. Hum. Genet.* **57:** 1436–1444. Reproduced with permission from the University of Chicago Press.

have not had time to have become randomized and a third is recombination hotspots.

Recent measurements of recombination rates have been made using microsatellite probes. The recombination rate within the class II region (*DRB1-DPB1*) was 0.74%. In class III (*HLA-B-DRB1*) the rate was 0.94%. Both of these rates are within the expected range, given the generally observed 1% recombination per Mbp of DNA per meiosis. The recombination rate between *HLA-A* and *HLA-B* was 0.31% though, much smaller than expected for the 1.4 Mbp.

A puzzling finding is that some regions with high levels of linkage disequilibrium in the MHC may encompass loci which are in complete equilibrium with each other. *Figure 2.9* shows an example of the equilibrium of the *TAP* loci, physically close to each other, flanked by loci in weak linkage disequilibrium, such as *DP* and *DR*. The reason for this is not clear but may relate to the fact that there are many more alleles at the *DP* and *DR* loci than at *TAP*. Linkage disequilibrium is more pronounced on specific haplotypes that are not necessarily informative for *TAP* (i.e. all have identical *TAP* alleles). Alternatively, selection may take place for combinations of certain class II alleles, irrespective of the *TAP* phenotype.

In conclusion, the MHC provides an interesting cluster of genes, of various degrees of relatedness. MHC features such as the high levels of variation, the association with disease and the various genetic phenomena described above, are of particular interest.

Acknowledgement

I would like to thank Mary Carrington for helpful information.

References

Abderrahim H, Sambucy JL, Iris F, Ougen P, Billault A, Chumakov IM, Dausset J, Cohen D, Le Paslier D. (1994) Cloning the human major histocompatibility complex in YACs. *Genomics* **23:** 520–527.

Andersson G, Larhammar D, Widmark E, Servenius B, Peterson PA, Rask L. (1987) Class II genes of the human major histocompatibility complex. Organization and evolutionary relationship of the DRbeta genes. *J. Biol. Chem.* **262:** 8748–8758.

Androlewicz MJ, Cresswell P. (1994) Human transporters associated with antigen processing possess a promiscuous peptide-binding site. *Immunity* **1:** 7–14.

Bahram S, Bresnahan M, Geraghty DE, Spies T. (1994) A second lineage of mammalian major histocompatibility complex class I genes. *Proc. Natl Acad. Sci. USA* **91:** 6259–6263.

Beck S, Trowsdale J, Abdull S, *et al.* (1996) Evolutionary dynamics of non-coding sequences of the human MHC. *J. Mol. Biol.* **255:** 1–13.

Beck S, Alderton R, Kelly A, Khurshid F, Radley E, Trowsdale J. (1992a) DNA sequence analysis of 66kb of the human MHC class II region encoding a cluster of genes for antigen processing. *J. Mol. Biol.* **228:** 433–441.

Beck S, Hanson I, Kelly A, Pappin DJC, Trowsdale J. (1992b) A homologue of the *Drosophila* female sterile homeotic (*fsh*) gene in the class II region of the human MHC. *DNA Sequence* **2:** 203–210.

Belich MP, Glynne RJ, Senger G, Sheer D, Trowsdale J. (1994) Proteasome components with reciprocal expression to that of the MHC-encoded LMP proteins. *Curr. Biol.* **4:** 769-776.

Bronson SK, Pei J, Taillon-Miller P, Chorney M, Geraghty DE, Chaplin DD. (1991) Isolation and characterization of yeast artificial chromosome clones linking the HLA-B and HLA-C loci. *Proc. Natl Acad. Sci. USA* **88:** 1676–1680.

Cammarota G, Schierle A, Takacs B, Doran DM, Knorr R, Bannwarth W, Guardiola J, Sinigaglia F. (1992) Identification of a CD4 binding site on the $\beta 2$ domain of HLA-DR molecules. *Nature* **356:** 799–801.

Campbell RD, Trowsdale J. (1993) Map of the human MHC. *Immunology Today* **14:** 349-352.

Cho S, Attaya M, Monaco JJ. (1991) New class II-like genes in the murine MHC. *Nature* **353:** 573–576.

Colonna M, Brooks EG, Falco M, Ferrara GB, Strominger JL. (1993) Generation of allospecific natural killer cells by stimulation across a polymorphism of HLA-C. *Science* **260:** 1121–1124.

Cresswell P. (1994) Assembly, transport and function of MHC class II molecules. *Annu. Rev. Immunol.* **12:** 259–293.

Davies JL, Kawaguchi Y, Bennett ST, *et al.* (1994) A genome-wide search for human type 1 diabetes susceptibility genes. *Nature* **371:** 130–136.

Fan W, Liu Y-C, Parimoo S, Weissman SM. (1995) Olfactory receptor-like genes are located in the human major histocompatibility complex. *Genomics* **27:** 119–123.

Fling SP, Arp B, Pious D. (1994) HLA-DMA and -DMB genes are both required for MHC class II/peptide complex formation in antigen-presenting cells. *Nature* **368:** 554–558.

Gaczynska M, Rock KL, Spies T, Goldberg AL. (1994) Peptidase activities of proteasomes are differentially regulated by the major histocompatibility complex-encoded genes for LMP2 and LMP7. *Proc. Natl Acad. Sci. USA* **91:** 9213–9217.

Geraghty DE. (1993) Structure of the HLA class I region and expression of its resident genes. *Curr. Opin. Immunol.* **5:** 3–7.

Geraghty DE, Pei J, Lipsky B, Handsen JA, Taillon-Miller P, Bronson SK, Chaplin DD. (1992a) Cloning and physical mapping of the HLA class I region spanning the HLA-E to HLA-F interval by using yeast artificial chromosomes. *Proc. Natl Acad. Sci. USA* **89:** 2669–2673.

Geraghty DE, Koller BH, Hansen JA, Orr HT. (1992b) The HLA class I gene family includes at least six genes and twelve pseudogenes and gene fragments. *J. Immunol.* **149:** 1934–1946.

Glynne R, Powis SH, Beck S, Kelly A, Kerr L-A, Trowsdale J. (1991) A proteasome-related gene between the two ABC transporter loci in the class II region of the human MHC. *Nature* **353:** 357–

360.

Glynne R, Kerr LA, Mockridge I, Beck S, Kelly A, Trowsdale J. (1993) The MHC-encoded proteasome component LMP7: alternative first exons and post-translational processing. *Eur. J. Immunol.* **23:** 860–866.

Goldberg AL. (1995) Functions of the proteasome: the lysis at the end of the tunnel. *Science* **268:** 522–523.

Gorski J, Rollini P, Mach B. (1987) Structural comparisons of the genes of two HLA-DR supertypic groups: the loci encoding DRw52 and DRw53 are not truly allelic. *Immunogenetics* **25:** 397–402.

Gustafsson K, Widmark E, Jonsson A-K, Servenius B, Sachs DH, Larhammar D, Rask L, Peterson PA. (1987) Class II genes of the human major histocompatibility complex. Evolution of the DP region as deduced from nucleotide sequences of the four genes. *J. Biol. Chem.* **262:** 8778–8786.

Higgins CF, Hyde SC, Mimmack MM, Gileadi U, Gill DR, Gallagher MP. (1990) Binding protein-dependent transport systems. *J. Bioerg. Biomem.* **22:** 571–592.

Howard JC. (1993) Restrictions on the use of antigenic peptides by the immune system. *Proc. Natl Acad. Sci. USA* **90:** 3777–3779.

Iris FJM, Bougueleret L, Prieur S, *et al.* (1993) Dense ALU clustering and a potential new member of the NFkB family within a 90 kb class III segment. *Nature Genetics* **3:** 137–145.

Joly E, Deverson EV, Coadwell JW, Gunther E, Howard JC, Butcher GW. (1994) The distribution of Tap2 alleles among laboratory rat RT1 haplotypes. *Immunogenetics* **40:** 45–53.

Jonsson A-K, Hyldig-Nielsen J-J, Servenius B, Larhammar D, Andersson G, Jorgensen F, Peterson PA, Rask L. (1987) Class II genes of the human major histocompatibility complex. comparisons of the DQ and Dx alpha and beta genes. *J. Biol. Chem.* **262:** 8767–8777.

Karlsson L, Surh CD, Sprent J, Peterson PA. (1992) An unusual class II molecule. *Immunology Today* **13:** 469–470.

Karlsson L, Peleraux A, Lindstedt R, Liljedahl M, Peterson PA. (1994) Reconstitution of an operational MHC class II compartment in nonantigen-presenting cells. *Science* **266:** 1569–1573.

Kawai J, Ando A, Sato T, Nakatsuji T, Tsuji K, Inoko H. (1989) Analysis of gene structure and antigen determinants of DR2 antigens using DR gene transfer into mouse L cells. *J. Immunol.* **142:** 312–317.

Kelly AP, Monaco JJ, Cho S, Trowsdale J. (1991a) A new human HLA class II-related locus, DM. *Nature* **353:** 571–573.

Kelly A, Powis SH, Glynne R, Radley E, Beck S, Trowsdale J. (1991b) Second proteasome-related gene in the human MHC class II region. *Nature* **353:** 667–668.

Kelly A, Powis SH, Kerr LA, Mockridge I, Elliott T, Bastin J, Uchanska-Ziegler B, Ziegler A, Trowsdale J, Townsend A. (1992) Assembly and function of the two ABC transporter proteins encoded in the human major histocompatibility complex. *Nature* **355:** 641–644.

Klitz W, Stephens J, Grote M, Carrington M. (1995) Discordant patterns of linkage disequilibrium of the peptide-transporter loci within the HLA class II region. *Am. J. Hum. Genet.* **57:** 1436–1444.

Koller BH, Geraghty DE, DeMars R, Duvick L, Rich SS, Orr HT. (1989) Chromosomal organization of the human major histocompatibility complex class I gene family. (Published erratum appears in *J. Exp. Med.* 1989 Apr 1;**169**(4):1517.) *J. Exp. Med.* **169:** 469–480.

Konig R, Huang L-Y, Germain R. (1992) MHC class II interaction with CD4 mediated by a region analagous to the MHC class I binding site for CD8. *Nature* **356:** 796–798.

Le Bouteiller P. (1994) HLA class I chromosomal region, genes and products: facts and questions. *Crit. Rev. Immunol.* **14:** 89–129.

Lechler R. (1994) *HLA and Disease.* Academic Press, London.

Leelayuwat C, Townend DC, Degliesposti MA, Abraham LJ, Dawkins RL. (1994) New polymorphic and multicopy MHC gene family related to non-mammalian class I. *Immunogenetics* **40: 339–351.**

Leelayuwat C, Pinelli M, Dawkins RL. (1995) Clustering of diverse replicated sequences in the MHC: evidence for en bloc duplication. *J. Immunol.* **155:** 692–698.

Loke YW, King A. (1991) Recent developments in the human maternal–fetal immune interaction. *Curr. Opin. Immunol.* **3:** 762–766.

Martin M, Mann D, Carrington M. (1995) Recombination rates across the HLA complex: use of microsatellites as a rapid screen for recombinant chromosomes. *Hum. Mol. Genet.* **4:** 423–428.

Momburg F, Roelse J, Hammerling GJ, Neefjes JJ. (1994) Peptide size selection by the major histocompatibility complex-encoded peptide transporter. *J. Exp. Med.* **179:** 1613–1623.

Morris P, Shaman J, Attaya M, Amaya M, Goodman S, Bergman C, Monaco JJ, Mellins E.

(1994) An essential role for HLA-DM in antigen presentation by class II major histocompatibility molecules. *Nature* **368:** 551–554.

Neumann H, Cavalie A, Jenne DE, Wekerle H. (1995) Induction of MHC class I genes in neurons. *Science* **269:** 549–552.

Newell WR, Trowsdale J, Beck S. (1994) MHCDB-database of the human MHC. *Immunogenetics* **40:** 109–115.

Ortmann B, Androlewicz MJ, Cresswell P. (1994) MHC class I/beta2-microglobulin complexes associate with TAP transporters before peptide binding. *Nature* **368:** 864–867.

Potts WK, Manning CJ, Wakeland EK. (1991) Mating patterns in seminatural populations of mice influenced by MHC genotype. *Nature* **352:** 619–621.

Ragoussis J, Monaco A, Mockridge I, Kendall E, Campbell RD, Trowsdale J. (1991) Cloning of the HLA class II region in yeast artificial chromosomes. *Proc. Natl Acad. Sci. USA* **88:** 3753–3757.

Rollini P, Mach B, Gorski J. (1985) Linkage map of three HLA-DR beta chain genes: evidence for a recent duplication event. *Proc. Natl Acad. Sci. USA* **82:** 7197–7201.

Sanderson F, Kleijmeer MJ, Kelly AP, Verwoerd D, Tulp A, Neefjes JJ, Geuze HJ, Trowsdale J. (1994) Accumulation of HLA-DM, a regulator of antigen presentation, in MHC class II compartments. *Science* **266:** 1566–1569.

Sanseau P, Jackson A, Alderton RP, Beck S, Senger G, Sheer D, Kelly A, Trowsdale J. (1994) Cloning and characterization of human phosphatase inhibitor-2 (IPP-2) sequences. *Mammalian Genome* **5:** 490–496.

Sinha AA, Lopez MT, McDevitt HO. (1990) Autoimmune diseases: the failure of self tolerance. *Science* **248:** 1380–1388.

Spies T, DeMars R. (1991) Restored expression of major histocompatibility class I molecules by gene transfer of a putative peptide transporter. *Nature* **351:** 323–324.

Tiwari JL, Terasaki PI. (1985) *HLA and Disease.* Springer-Verlag, New York.

Tonnelle C, DeMars R, Long EO. (1985) DO beta: a new beta chain gene in HLA-D with a distinct regulation of expression. *EMBO J* **4:** 2839–2847.

Townsend A, Trowsdale J. (1993) The transporters associated with antigen processing. *Semin.Cell Biol.* **4:** 53–61.

Trowsdale J. (1993) Genomic structure and function in the MHC. *Trends Genet.* **9:** 117–122.

Trowsdale J. (1994)'Both man & bird & beast': comparative organization of MHC genes. *Immunogenetics* **41:** 1–17.

Trowsdale J, Kelly A, Lee J, Carson S, Austin P, Travers P. (1984) Linkage map of two HLA-SBbeta-related genes: an intron in one of the SBbeta-related genes contains a processed pseudogene. *Cell* **38:** 241–249.

Ulbrecht M, Kellermann J, Johnson JP, Weiss EH. (1992) Impaired intracellular transport and cell surface expression of nonpolymorphic HLA-E: evidence for inefficient peptide binding. *J. Exp. Med.* **176:** 1083–1090.

Vernet C, Ribouchon MT, Chimini G, Mauvieux V, Sidibe I, Pontarotti P. (1993) A novel coding sequence belonging to a new multicopy gene family mapping within the human MHC class I region. *Immunogenetics* **38:** 47–53.

Vernet C, Ribouchon MT, Chimini G, Pontarotti P. (1994) Structure and evolution of a member of a new subfamily of putative GTP-binding proteins mapping to the human MHC class I region. *Mamm. Genome* **5:** 100–105.

Volz A, Weiss E, Trowsdale J, Ziegler A. (1994) Presence of an expressed beta-tubulin gene (TUBB) in the HLA class I region may provide the genetic basis for HLA-linked microtubule dysfunction. *Hum. Genet.* **93:** 42–46.

Wright KL, White LC, Kelly A, Beck S, Trowsdale J, Ting JPY. (1995) Coordinate regulation of the human TAP1 and LMP2 genes from a shared bi-directional promoter. *J. Exp. Med.* **181:** 1459–1471.

Young JA, Trowsdale J. (1990) The HLA-DNA (DZA) gene is correctly expressed as a 1.1 kb mature mRNA transcript. *Immunogenetics* **31:** 386–388.

Zangenberg G, Huang M-M, Arnheim N, Erlich H. (1995) New HLA-DPB1 alleles generated by interallelic gene conversion detected by analysis of sperm. *Nature Genetics* **10:** 407–414.

Zhang WJ, Degli-Esposti MA, Cobain TJ, Cameron PU, Christiansen FT, Dawkins RL. (1990) Differences in gene copy number carried by different MHC ancestral haplotypes. Quantitation after physical separation of haplotypes by pulsed-field gel electrophoresis. *J. Exp. Med.* **171:** 2101–2114.

Zhou Y, Chaplin DD. (1993) Identification in the HLA class I region of a gene expressed late in keratinocyte differentiation. *Proc. Natl Acad. Sci. USA* **90:** 9470–9474.

3

Genes of the MHC class III region and the functions of the proteins they encode

Begoña Aguado*, Caroline M. Milner* and R. Duncan Campbell

3.1 Introduction

The human major histocompatibility complex (MHC) constitutes a 4 Mbp region on the short arm of chromosome 6 in the 6p21.3 band. It is conveniently divided into three regions. Genes in the class I and class II regions encode highly polymorphic families of cell-surface glycoproteins that are involved in the presentation of antigenic peptides to T cells during an immune response (see Chapter 2). Interspersed between these genes are a large number of other genes that encode proteins with a variety of different functions. Some of these have been found to be immune related; for example, the *LMP* and *TAP* genes in the class II region encode proteins involved in the production and transport of peptides destined for presentation by class I molecules, while the *DM* genes (also in the class II region) encode a heterodimeric molecule involved in the loading of peptides on to class II molecules. The class I and class II regions are separated by the central class III region that spans 1100 kb of DNA. The class III region also contains genes that encode proteins with important known immune-related functions such as the complement system proteins C2, C4 and Factor B, and tumour necrosis factor (TNF), and lymphotoxin (LT) α and β. Recently it has become apparent that the class III region is one of the most gene-dense regions of the human genome with, on average, one gene every 15 kb of DNA.

Many studies have indicated that genetic predisposition to a large number of diseases of both an autoimmune and non-immune aetiology is associated with the products of genes located in the MHC (Campbell and Milner, 1993; see Chapters 15 and 16). These studies have mainly focused on the polymorphic class I and class II antigens, and also the complement gene products. However, it is clear that in many diseases the proteins encoded by these genes do not fully explain all the disease susceptibility that maps to the MHC. Thus characterization of the

* BA and CMM contributed equally to this review.

HLA and MHC: genes, molecules and function, edited by M.J. Browning and A.J. McMichael.
© 1996 BIOS Scientific Publishers Ltd, Oxford.

functions of the proteins encoded by the many other genes located within the MHC, and in particular in the class III region, has assumed increased importance as it will be through a complete understanding of the functions of these proteins that we will fully appreciate MHC and disease association. In this chapter we summarize our current knowledge of the genes located in the class III region (*Figure 3.1*, see page 255) and the functions of the proteins they encode.

3.2 Molecular mapping of the class III region

The molecular characterization of the class III region has involved the physical mapping of the region by pulsed-field gel electrophoresis (PFGE) and the construction of a contig of genomic clones, comprising both cosmids (Fukagawa *et al.*, 1995; Kendall *et al.*, 1990; Sargent *et al.*, 1989a; Spies *et al.*, 1989) and yeast artificial chromosomes (YACs) (Abderrahim *et al.*, 1994; for review see Campbell, 1993). This analysis established the location of the complement and *TNF* genes in the class III region and the physical distances between them (Carroll *et al.*, 1987; Dunham *et al.*, 1987). Most of the new genes subsequently located in the class III region were discovered by analysing the cloned genomic DNA using a variety of different techniques. The PFGE mapping studies carried out to generate the physical map of the class III region identified 33 clusters of sites, or single sites, for rare-cutting endonucleases (e.g. *Bss*HII, *Sac*II, *Eag*I, *Not*I; Kendall *et al.*, 1990; Sargent *et al.*, 1989a). The sites for these enzymes are not randomly distributed, but are clustered at CpG-islands, short stretches of DNA that (unlike the rest of the human genome) are not depleted in unmethylated CpG dinucleotides (Bickmore and Sumner, 1989; Bird, 1987). As 80% of CpG-islands are associated with the 5′ ends of genes, particularly housekeeping genes, they act as excellent markers in gene-hunting experiments. However, as it has been estimated that only about 50% of the genes in the human genome are associated with CpG-islands, other gene-hunting techniques have had to be used to increase the chances of detecting all the genes in the class III region. These include Zoo blot analysis (Sargent *et al.*, 1989a), Northern blot analysis (Sargent *et al.*, 1989a; Spies *et al.*, 1989), cDNA selection on immobilized cloned genomic DNA (Fan *et al.*, 1993), exon trapping (Burfoot and Campbell, 1994), and the direct screening of cDNA libraries using cosmid genomic DNA inserts after competing out repetitive DNA sequences (Kendall *et al.*, 1990; Spies *et al.*, 1989). This last technique has proved to be the most productive, and, in addition to isolating cDNA clones corresponding to the CpG-island associated genes, has led to the isolation of cDNAs corresponding to genes that are more restricted in their pattern of expression. Furthermore, DNA sequence analysis of some of the cosmid genomic DNA inserts has led to the identification of potential coding regions due to the detection of identities or similarities in the derived amino acid sequences of putative open-reading frames with the sequences of known proteins (Aguado and Campbell, 1995; Marshall *et al.*, 1993; Matsumoto *et al.*, 1992; Sugaya *et al.*, 1994). We have now begun to sequence the cosmid clones covering the class III region using automated fluorescent sequencing technology. This will aid in the identification of genes in the intergenic gaps through the use of exon prediction programs such as GRAIL (Uberbacher and Mural, 1991).

The combined analyses described above have resulted in the localization of at least 53 transcriptional units in the class III region (*Figure 3.1*, see p. 255), and there are a sufficient number of intergenic gaps still to be analysed to suggest that this number will increase significantly. Many of the genes are very closely spaced; for example, the intergenic gap separating the *G11* and *C4* genes is 611bp (Sargent *et al.*, 1994; Shen *et al.*, 1994), the intergenic gap separating the *RAGE* and *PBX2* genes is about 500 bp (Sugaya *et al.*, 1994), while the *C2* and *Factor B* genes are separated by only 421bp (Wu *et al.*, 1987). Thus it is possible to imagine that transcription of one gene could interfere with the transcription of a closely linked downstream gene. Recently it has been shown that the intergenic gaps separating the *C2* and *Factor B* genes, and also the *G11* and *C4* genes, both contain a sequence motif which is bound by the MAZ zinc finger protein and which elicits transcription termination of the upstream gene (Ashfield *et al.*, 1991, 1994). Terminating transcription of an upstream gene in this way helps prevent interference with transcription of the downstream gene and may be a general strategy utilized by other closely spaced genes. In addition, it has recently become apparent that the 120 kb segment of DNA encoding the duplicated *C4/P450c21/X/Y* genes is one of the most complex in the human genome, featuring overlapping genes and genes-within-genes (Bristow *et al.*, 1993a,b; Tee *et al.*, 1995).

3.3 Genes encoding proteins of known functions

3.3.1 The complement genes C2, Factor B *and* C4

The complement system is the principal effector mechanism of humoral immunity and consists of at least 24 serum proteins and 11 membrane-bound proteins that interact in a complex cascade leading to a wide variety of important biological responses (reviewed in Law and Reid, 1995; Whaley *et al.*, 1993). These include cell lysis, opsonization, clearance of immune complexes, regulation of B cell responses, and the generation of peptides that have potent anaphylactic and chemotactic activities. There are two pathways for activating the complement system. The classical pathway consists of a recognition protein complex, C1 (Loos *et al.*, 1993), and two effector proteins C2 and C4 (Campbell and Law, 1992), and is activated under normal circumstances by the binding of C1 to IgG or IgM bound to antigen. This results in the generation of proteolytic activity in the C1s subcomponent of the C1 complex which can cleave and activate both C4 and C2. Cleavage of C4 by C1s at a single peptide bond to yield C4b (190 kDa) and the anaphylatoxin C4a (10 kDa) results in the generation of a reactive acyl group released from an internal thiolester bond which allows activated C4 (C4b) to bind covalently to the activating particle. C1s also cleaves the proenzyme C2 into fragments of 73 kDa (C2a) and 34 kDa (C2b). The C2a fragment which has serine protease activity when non-covalently associated with C4b, forms the classical pathway C3 convertase (C4bC2a). C4 and C2 can also be activated by a plasma serine protease MASP in complex with mannan binding protein (MBP) once the MBP has bound to suitable carbohydrate ligands present on the surfaces of bacteria or viruses (Ji *et al.*, 1993; Takada *et al.*, 1993). This activation route, which is independent of the generation of specific antibodies, could provide rapid defence against pathogens via innate immune mechanisms.

The alternative pathway consists of Factors B (Pangburn, 1992) and D (Pascual and Schifferli, 1992), and properdin (Reid, 1992), as well as the activated portion of the C3 component (C3b) (Burger, 1992). Under normal circumstances activation of the alternative pathway is antibody independent and can be mediated by a diverse set of substances including components of yeast and bacterial cell walls. Formation of a C3bB complex results in the cleavage of Factor B by Factor D into fragments of 35 kDa (Ba) and 65 kDa (Bb) and the generation of the alternative pathway C3 convertase (C3bBb), which can be stabilized by the binding of properdin. The active site of the convertase resides in the Bb fragment. Subsequent binding of C3b to both the classical and alternative pathway C3 convertases shifts the specificity of the complex from cleavage of C3 to cleavage of C5 (Wetsel, 1992), although the enzymatic active site remains on the C2a and Bb subunits, respectively. Once C5 has been cleaved to C5b this allows assembly of the C5b-C9 terminal complex which results in the lysis of target cells or infecting organisms.

Proenzymes C2 and Factor B are encoded by single copy genes that lie 421bp apart (Wu *et al.*, 1987) in the class III region (*Figure 3.1*, see page 255). Component C4 on the other hand is usually encoded by two genetic loci labelled *C4A* and *C4B*, though haplotypes that have a single *C4* gene or three *C4* genes are relatively common (*Figure 3.2*; Campbell *et al.*, 1990; Hauptmann *et al.*, 1993; Mauff *et al.*, 1990). The duplicated *C4* loci are approximately 10 kb apart and are organized in the same transcriptional orientation as the *C2* and *Factor B* genes (*Figure 3.1*, see p. 255). Both *C4* genes are transcribed into mRNA molecules of approximately 5.5 kb that are greater than 99% identical in sequence (Belt *et al.*, 1984, 1985; Yu *et al.*, 1986). However, the small number of amino acid differences that do exist result in C4 molecules that have a profound difference in their covalent binding activities. Between amino acid residues 1101–1106 the C4A isotype has the sequence *Pro-Cys*-Pro-Val-*Leu-Asp* and has a preference for transacylation to amino groups on activating surfaces, while the C4B isotype has the sequence *Leu-Cys*-Pro-Val-*Ile-His* and reacts comparably with amino and hydroxyl groups on activating surfaces (Dodds *et al.*, 1985; Yu *et al.*, 1986). This difference in the binding properties of the two C4 isotypes has been correlated with the His/Asp substitution at position 1106 (Carroll *et al.*, 1990; Sepp *et al.*, 1993). At the protein level the two C4 isotypes exhibit extensive structural polymorphism. Four C4A allotypes, C4A2, C4A3, C4A4 and C4A6 and three C4B allotypes, C4B1, C4B2 and C4B3 occur at frequencies in the Caucasian population greater than 0.05. In addition eight C4A and 18 C4B rare structural variants have also been recognized (Mauff *et al.*, 1990).

Length variation for individual *C4* genes has been recorded. In general *C4A* genes are 'long' (22 kb in length), whereas *C4B* genes may be 22 kb in size (long) or 16 kb in size (short) (Schneider *et al.*, 1986; Yu *et al.*, 1986; *Figure 3.2*). The size difference between long and short *C4* genes has been demonstrated to be due to insertion of a 6.5 kb segment of DNA in intron 9 of long *C4* genes (Yu, 1991). Sequencing of this segment of DNA has revealed that it is a member of a novel family of human endogenous retroviruses and has a typical retrovirus structure with elements of *gag*, *pol*, and *env* domains, flanked by two long terminal repeats (Dangel *et al.*, 1994; Tassabehji *et al.*, 1994). However, the presence of multiple termination codons in all the reading frames suggests that this retroviral element is no longer transcriptionally active. One of the other major features in the genetics of

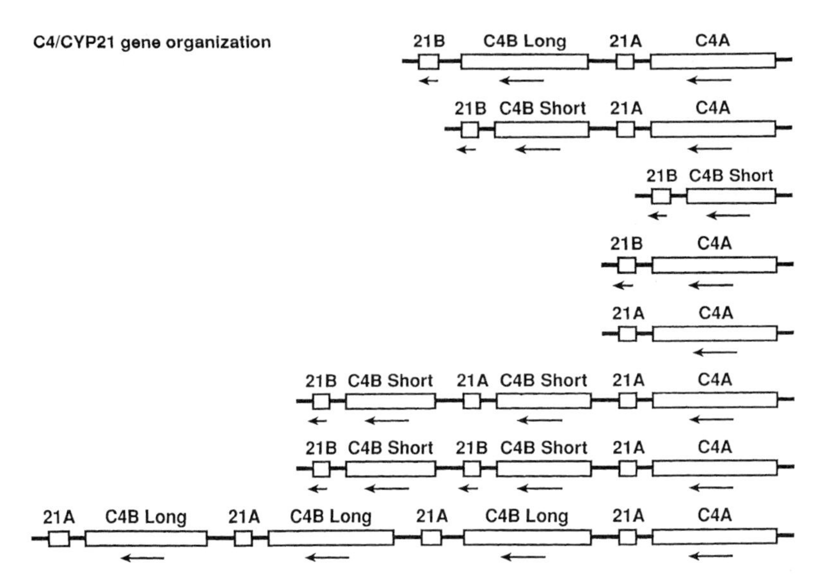

Figure 3.2. Gene organization of the *C4* and *P450c21* (*21A*/*21B*) loci in the class III region. The size difference between long and short *C4* genes has been demonstrated to be due to insertion of a 6.5 kb segment of DNA in intron 9 of long *C4* genes. Adapted from Campbell, 1993, pp. 1–33, in *Regional Physical Mapping* © 1993, with permission from Cold Spring Harbor Laboratory Press, Cold Spring Harbor, NY.

C4 is the unusually high frequency of null alleles at the two *C4* loci forming half null *C4A* or *C4B* haplotypes (*Figure 3.2*). The frequency of *C4A* null alleles is 5–15% with about 60% being due to gene deletion, while the frequency of *C4B* null alleles is 10–20% (Hauptmann *et al.*, 1993). Characterization of a number of non-expressed *C4A* genes has revealed that in the majority of cases the defect is due to a 2 bp insertion in exon 29 of the gene which alters the reading frame and results in the introduction of a premature termination codon (Barba *et al.*, 1993). Complete deficiency of C4 is a rare event and is correlated with severe immune complex disease. Partial deficiency of one of the two C4 isotypes (C4A or C4B) is much more common and is also associated with an increased susceptibility to a number of different autoimmune or immune complex diseases such as systemic lupus erythematosus (SLE), scleroderma and primary biliary cirrhosis (Hauptmann *et al.*, 1993).

Phenotypic genetics have also established that *Factor B* and *C2* are polymorphic, but to a much lesser extent than *C4* (Cross and Thomson, 1990). *C2* null alleles have also been recognized, the most common form being due to a 28 bp deletion that results in the skipping of exon 6 during splicing thus leading to generation of a truncated protein (Johnson *et al.*, 1992).

3.3.2 The P450c21 A/B, XA/XB *and* YA/YB *genes*

The *P450c21A* and *P450c21B* genes are located about 3 kb from the 3′ end of the *C4A* and *C4B* genes (*Figures 3.1*, see page 255, and *3.2*) and share 97% overall nucleotide sequence identity (Higashi *et al.*, 1986; Rodrigues *et al.*, 1987; White *et al.*, 1986). The *P450c21B* gene is transcribed uniquely in the adrenal cortex and encodes the adrenal enzyme steroid 21-hydroxylase. This enzyme is involved in the conversion of progesterone and 17-hydroxyprogesterone to 11-deoxycorticosterone and 11-deoxycortisol, the intermediate steps in mineralocorticoid and glucocorticoid biosynthesis, respectively (Miller, 1988). The *P450c21A* gene cannot encode steroid 21-hydroxylase protein as it contains deleterious mutations that render it non-functional (Higashi *et al.*, 1986; White *et al.*, 1986). However, the gene still appears to be transcribed in an adrenal-specific fashion (Bristow *et al.*, 1993a).

Deficiency of P450c21 due to deletion of the *P450c21B* gene, or to the presence of point mutations in the gene resulting in lack of *P450c21B* mRNA, or the expression of protein that has diminished or no activity, is the most frequent cause (95%) of congenital adrenal hyperplasia (CAH) (for review see Morel and Miller, 1991). The defect in *P450c21B* prevents conversion of 17-hydroxyprogesterone to 11-deoxycortisol leading to accumulation of 17-hydroxyprogesterone and to the production of excess androgen. Classical CAH can be divided into the milder simple virilizing form and the more severe, and life-threatening, salt wasting form. In addition there is a non-classical, late-onset form where individuals remain asymptomatic or develop the symptoms of an excess of androgens in childhood or at puberty.

An intriguing feature of this part of the class III region is the presence of a second pair of tandemly duplicated genes that overlap the *P450c21A* and *P450c21B* genes and encode adrenal-specific transcripts. The *YA* and *YB* transcriptional units (*Figure 3.1*, see page 255) utilize the *P450c21* promoters and although they are transcribed in the same orientation they use a different array of exons and introns (Bristow *et al.*, 1993a). However, it has not been established whether the *YA* and *YB* transcripts are translated into protein.

A third pair of tandemly duplicated genes, *XA* and *XB*, have also been discovered that overlap the 3′ ends of the *P450c21A* and *P450c21B* genes by 481 bp, respectively, and are encoded on the opposite strand of DNA from the *C4*, *P450c21* and *Y* transcriptional units (Morel *et al.*, 1989; *Figure 3.1*, page 255). The *XA* gene is about 4.5 kb in length and encodes a 2.6 kb adrenal-specific transcript of unknown function (Gitelman *et al.*, 1992). The *XB* gene (originally called *G12* and *OSG*), which is about 65 kb in length, consists of 39 exons that encode a 12 kb mRNA (Bristow *et al.*, 1993b). The predicted protein product of over 400 kDa consists of five distinct domains: a signal peptide, a hydrophobic domain containing three heptad repeats, a series of 18.5 epidermal growth factor (EGF)-like repeats, 29 fibronectin type III repeats, and a carboxy-terminal fibrinogen-like domain. Since the structure of the protein encoded by the *XB* gene closely resembles tenascin (TN-C) (also called cytotactin), the protein has been termed tenascin-X (TN-X). RNase protection experiments have shown that the *TN-X* transcript is expressed ubiquitously in human fetal tissues, with the greatest expression in the fetal testis and in fetal skeletal, cardiac, and smooth muscle.

Tenascins are a family of related proteins, encoded by related genes. Each family member has the same general structure as TN-C. TN-C, which is encoded by a gene on chromosome 9, is a large glycoprotein of the extracellular matrix (for reviews see Chiquet-Ehrismann *et al.*, 1986; Schenk and Chiquet-Ehrismann, 1994). Each of the six monomers that constitute TN-C exists in multiple size variants of 220–320 kDa and contains four domains. The amino terminus comprises the hydrophobic 'head group', which facilitates polymerization into the TN-C hexabrachion. This is followed in turn by a series of EGF-like repeats, a series of fibronectin type III repeats and a carboxy-terminal fibrinogen-like domain. However, there are variations on this theme, as illustrated by the structure of undulin, which contains a von Willebrand factor domain at the N-terminus followed by fibronectin type III repeats (Just *et al.*, 1991). The closely related protein restrictin (TN-R) (Norenberg *et al.*, 1992), also termed J1-160 (Fuss *et al.*, 1993), is a trimer composed of similar, but considerably smaller units than in TN-C, forming a tribrachion. The *XB* gene encodes a similar, but much larger, protein with the prediction that it will also polymerize into a 'brachion' comprising an unknown number of arms (Bristow *et al.*, 1993b). One interesting difference between TN-X and other members of this gene family is the sequence divergence at the extreme amino terminus of the mature protein. In TN-C, this region appears to mediate dimerization of TN-C trimers joined in the heptad region to form the final hexabrachion. Cysteine residues are notably absent from this region of TN-X, and hence it is predicted that TN-X may form pairs or trimers, but not higher polymers. It has been suggested that *TN-C*, *TN-R* and *TN-X* arose from a smaller primordial gene resembling *TN-R*, and that subsequent internal exon duplications produced the central repeats of *TN-C* and *TN-X* after their divergence from one another.

TN-C appears to mediate interactions between cells and the extracellular matrix (and possibly between cells), through an Arg-Gly-Asp-dependent (Bourdon and Ruoslahti, 1989; Leahy *et al.*, 1992) or -independent (Prieto *et al.*, 1992; Spring *et al.*, 1989) receptor. Initial studies suggested a crucial role in embryonic development (Chiquet-Ehrismann *et al.*, 1986; Tan *et al.*, 1991; Weller *et al.*, 1991), although recent *TN-C* gene knockout experiments in transgenic mice suggest that TN-C serves no mandatory, irreplaceable role in development or in the adult (Rüegg *et al.*, 1989). TN-C and/or related molecules have neurite-regulation properties. In this connection, it is of interest that as yet unidentified disease-susceptibility genes for neurological disorders such as narcolepsy may be present within or near the MHC class II or III regions. TN-C is thought to be involved in modulating epithelial–mesenchymal and neuronal–glial interactions (Chiquet-Ehrismann *et al.*, 1986; Grumet *et al.*, 1985). Rüegg *et al.* (1989) have proposed that TN-C could exert immunomodulatory activities because it alters the adhesion properties of cells of the immune system.

The mouse *TN-X* gene lies in the class III region of the murine H-2 complex, as is the case for its human counterpart. Matsumoto *et al.* (1994) have isolated a cDNA encoding mouse TN-X, and have found that the distribution of TN-X is distinct from and often reciprocal to that of TN-C. On Northern blots a *TN-X* mRNA of approximately 13 kb is detectable in most tissues analysed, whereas in various mouse cell lines two mRNAs of approximately 11 and 13 kb were detected, suggesting the possibility of alternative splicing of *TN-X* transcripts. Antibodies to TN-X cross-react with a subunit of molecular weight approximately 500 kDa, suggesting that the protein may contain up to 40 fibronectin type III repeats, making it the largest tenascin family member so far known. Although the *TN-X* mRNA is found in most tissues at a low level, the mRNA as well as the protein are predominantly expressed in heart and skeletal muscle where neither *TN-C* nor *TN-R* are expressed. Immunostaining has shown the protein to be associated with the extracellular matrix of the muscle tissues and with blood vessels in all of the tissues analysed.

The nucleotide sequence of the *XA* gene is greater than 99% identical to the corresponding region of the *XB* gene. The genetic crossover that led to the duplication of the *C4/P450c21/X* locus occurred at a CAAG sequence 248 bp upstream from the transcriptional start site of the *XA* gene (Gitelman *et al.*, 1992) in a region that corresponds to intron 26 in the *XB* gene (Bristow *et al.*, 1993b), so that the *XA* gene has lost the 5′ flanking DNA and the first 26 exons of the *XB* gene. The *XA* gene also contains a deletion of 121 bp which introduces premature stop codons making it unlikely that the adrenal-specific *XA* mRNA could encode a protein (Gitelman *et al.*, 1992). Recently Tee *et al.* (1995) have reported a homologue of *XA* that is transcribed from sequences identical to the *XA* promoter that lie within intron 26 of the *XB* gene. This transcript, termed short *XB* (*XB-S*), is expressed in an adrenal-specific fashion. It encodes a 673 amino acid protein of molecular weight 74 kDa that is identical to the 673 carboxy-terminal residues of TN-X and consists of the five carboxy-terminal fibronectin type III repeats and the fibrinogen-like domain. While the low abundance of its mRNA suggests that this protein might be present in small quantities, it was readily detected by immunoblotting. The function of this protein is unknown, but its amino terminus does not contain obvious signal sequences; hence it may be cytoplasmic. This suggests that the function of the XB-S protein is unrelated to the interactions among cells, or between cells and the extracellular matrix, that are mediated by the various tenascins. It is possible that XB-S or a related structure may represent an evolutionary precursor to the tenascin family.

3.3.3 The hsp70 proteins

The human genome contains at least ten genes encoding members of the 70 kDa heat shock protein (hsp70) family, which are expressed both constitutively and in response to stress (reviewed by Gunther and Walter, 1994). Three intronless *HSP70* genes are located in the class III region approximately 92 kb telomeric of the *C2* gene (*Figure 3.1*, see page 255; Milner and Campbell, 1990; Sargent *et al.*, 1989b). *HSP70-1* corresponds to the major heat-inducible *HSP70* (Hunt and Morimoto, 1985) which has been shown by Northern blot analysis to have a low level of basal expression (in HeLa cells, but not in peripheral blood mononuclear cells (PBMC)) and to be expressed at high levels following heat shock (Gunther and Walter, 1994; Milner and Campbell, 1990). *HSP70-2* differs from *HSP70-1* at the nucleotide sequence level, most significantly in the 3′ untranslated region (UTR), but encodes an identical protein product. There is no evidence for basal expression of *HSP70-2* at the mRNA level, but it is highly up-regulated by heat shock (Milner and Campbell, 1990). The expression of *HSP70-1* is also known to be elevated in response to a variety of other stress factors such as oxidative injury, inflammation and fever due to bacterial or viral infection. At least in the case of heat shock, elevated expression is regulated by heat shock factors due to the presence of denatured proteins (Morimoto *et al.*, 1994). Nucleotide sequence differences between the 5′ UTRs of *HSP70-1* and *-2* may result in variation in their transcriptional regulation in response to some forms of stress. The 3′ UTR of *HSP70-1* has been shown to be important in the post-transcriptional regulation of expression following heat shock by controlling mRNA stability (Moseley *et al.*,1993), and the substantial differences in the 3′ UTR of *HSP70-2* may have an effect on this.

The *HSP70-HOM* gene encodes a protein product 90% identical to hsp70-1/-2 at the amino acid level (Milner and Campbell, 1990). In man, low level basal expression of hsp-HOM has been observed in HeLa cells, but not PBMCs, by Northern blot analysis (Gunther and Walter, 1994; Milner and Campbell, 1990). Heat shock has no effect on *HSP70-HOM* expression. The orthologous genes in mouse (*Hsc70t*) and rat (*Hsc70-3*) have been found to be expressed at high levels in testicular germ cells during postmeiotic spermatogenesis, suggesting that in these species expression is tissue-specific and under developmental control (Gunther and Walter, 1994; Matsumoto *et al.*, 1993).

hsp70-1 is expressed in the cytosol under normal conditions, but following heat shock hsp70-1/-2 is found predominantly in the nucleus and nucleolus where it associates with damaged and denatured proteins (Welch and Feramisco, 1984). The amino acid sequence of hsp70-HOM indicates that this protein is also cytosolic and the preliminary results of cell staining suggest that this is the case. The importance of the heat-inducible hsp70 in protecting cells from stress is well known, but the role of hsp70-HOM has not been defined. Due to the high level of sequence similarity between the human hsp70s it is not always clear which protein is associated with a given property and there is still much to be learnt in this respect.

One of the best characterized properties of the members of the hsp70 protein family is their ability to bind unfolded proteins and peptides (Flynn *et al.*, 1989; Georgopoulos and Welch, 1993). This allows them to act as chaperones during the synthesis, folding, assembly, translocation and degradation of proteins, both in relation to normal cellular processes and following stress. Hsp70s seem to interact with extended structures, probably representing regions that are inaccessible in the native protein and there is evidence that the members of the hsp70 family differ in their binding specificities; for example, the endoplasmic reticulum protein, BiP, shows a preference for peptides with alternate large hydrophobic residues (Blond-Elguindi *et al.*, 1993). The binding and release of peptides by hsp70s is ATP-dependent, but although the hsp70s have weak ATPase activity which is upregulated in the presence of peptides or unfolded proteins it seems that ATP hydrolysis is not required for peptide release, since this has been observed in the presence of non-hydrolysable ATP analogues (Palleros *et al.*, 1993). Hsp70s can bind tightly to ATP and ADP, with the ADP-bound form having the highest affinity for unfolded protein substrates (Palleros *et al.*, 1994). Following heat shock, ATP levels fall, whilst ADP levels remain constant, resulting in an increase in hsp70–ADP complexes, which are able to bind and protect denatured proteins. During recovery, ATP levels rise allowing ADP/ATP exchange and the subsequent release of proteins which are then able to adopt their soluble and active states.

The ATP/ADP- and peptide-binding properties of the hsp70s are mediated by two distinct structural domains. The sequences of the hsp70s are highly conserved both within and between species (50–90% amino acid sequence identity between the human proteins) and this is particularly pronounced over the 44 kDa N-terminal ATPase domain (Chappell *et al.*, 1987). The latter is directly linked to an 18 kDa region, shown to be essential for peptide binding (Wang *et al.*, 1993).

Sequence variation between different hsp70s over this region, although not great, may account for differences in peptide-binding specificity. The C-terminal regions of the hsp70s are the most divergent (between family members within species) and may be responsible for specific functions, for example, subcellular localization.

The three-dimensional (3-D) structure of the ATPase domain from the constitutively expressed protein hsp70 has been determined by X-ray crystallography and shown to be highly related to the ATPase domains of hexokinase and actin (Flaherty *et al.*, 1990). On the basis of similarity in function and size, and very limited amino acid sequence similarity, it had been proposed that the peptide-binding domains of the hsp70s may adopt a similar structure to the MHC class I peptide-binding domain (Flajnik *et al.*, 1991; Rippmann *et al.*, 1991). However, recent studies by nuclear magnetic resonance (NMR) on hsc70 suggest that although the hsp70 and MHC peptide-binding domains contain similar secondary structural elements they have quite different topology, with the hsp70 domain adopting a previously undescribed topology consisting of antiparallel β-sheet and α-helix (Morshauser *et al.*, 1995). The tertiary structure of this domain still remains to be established.

Under normal conditions constitutively expressed hsp70s contribute to the protection of nascent proteins from improper folding or aggregation after synthesis, and also mediate their folding and translocation across intracellular membranes. In cells that are stressed, for example, due to infection, the inducible hsp70s prevent the aggregation of denatured proteins and facilitate their renaturation or degradation during recovery, thus preventing cell damage. This may be the basis for the protective effects of fever against oxidative injury during an immune response (Jacquier-Sarlin *et al.*, 1994). These protective effects are enhanced in cells that have previously been subject to mild stress. It has been demonstrated that heat shock and other stresses that induce hsp70 expression inhibit the production of the cytokines interleukin-1 (IL-1) and TNF, both potent mediators of inflammation, in activated monocytes and macrophages (reviewed by Hall, 1994). In addition, anti-inflammatory drugs such as dexamethasone and the gold salt auranofin, used in the therapy of rheumatoid arthritis, up-regulate hsp70 expression (Bugiel and Betts, 1993; Schmidt and Abdulla, 1988). It is probably not the hsp70s themselves that directly inhibit cytokine expression in monocytes and macrophages, since reduction, for example, in *TNF* mRNA has been seen to precede increased *HSP70* expression (Snyder *et al.*, 1992). It is more likely that a common mechanism enhances *HSP70* expression and inhibits cytokine expression, an understanding of which would be very valuable in defining targets for down-regulating cytokine production in inflammatory diseases. However, there is evidence that hsp70 can inhibit TNF-mediated cytotoxicity in tumour cells by inhibiting the activation of phospholipase A_2 and thus inhibiting arachidonic acid metabolism (Jaattela, 1993).

There is evidence that hsp70s play a variety of roles in disease. First, *HSP70* expression has been shown to be up-regulated in the context of a number of inflammatory autoimmune diseases, including scleroderma and SLE (Deguchi and Kishimoto, 1990; Twomey *et al.*, 1993). It is not always clear whether this up-regulation is a contributory factor to disease or merely a consequence of it. One interesting finding is that exposure of keratinocytes to the cytotoxic prostaglandin Δ^{12}–PGJ_2, or to ultraviolet (UV) light, results in the up-regulation and accumulation in the nucleoli of hsp70 (Furukawa *et al.*, 1993). This is concomitant with an increase in binding sites for autoantibodies to the nuclear antigens U_1RNP, SS-A/Ro and SS-B/La. These antibodies are characteristic of SLE and the related condition, cutaneous lupus, in which epidermal keratinocytes are major targets of immunological damage, suggesting that hsp70 may be involved in the development of erythematosus lesions in the skin in response to environmental stimuli. The hsps are immunodominant antigens, for example, in mycobacterial infections, and their high level of sequence conservation during evolution might be expected to result in antibodies or T cells cross-reactive with bacterial and human epitopes leading to an autoimmune response (reviewed by Winfield and Jarjour, 1991; Young, 1990). However, although autoantibodies to self hsps have been observed, there is no strong evidence that these are causative agents in autoimmune disease. In a tumour environment nutrient depletion, low pH and hypoxia are likely to induce hsp70 expression. Analysis of tumour-infiltrating lymphocytes has shown that these include $CD4^+$ cells that recognize hsp70-derived peptides in the context of HLA-DR molecules on heat-shocked B cells (Yoshino *et al.*, 1994). This suggests that recognition of stressed cells by T cells is important in anti-tumour immunity and in immune surveillance in general.

There is substantial evidence that members of the hsp70 family may act as chaperones in the generation and/or protection of peptides during antigen processing (reviewed by Pierce, 1994). Heat shock of B cells and monocytes has been shown to up-regulate the processing of antigens presented by MHC class II molecules, but it is not known whether this is due to a change in the types of peptides being presented or the way in which antigens are processed (Cristau *et al.*, 1994; Mariethoz *et al.*, 1994). It is not clear how cytosolic hsp70s could participate in MHC class II-associated antigen processing, although the constitutively expressed HSC70 does target proteins for lysosomal degradation during serum starvation (reviewed by Terlecky, 1994). It is perhaps more likely that if the MHC-linked hsp70s are involved in antigen processing it will be in the generation of cytosolic peptides for MHC class I binding. Cytosolic hsp70s isolated from tumour cells have been shown to bind peptides, and the observation that these hsp70s can be used to induce tumour-specific immunity in mice suggests that they are binding tumour-specific antigenic peptides capable of priming cytotoxic T lymphocyte (CTL) (Udono and Srivastava, 1993).

The MHC-linked hsp70s have been investigated for the presence of polymorphism, but whilst allelic variation has been found, only in one instance is this associated with altered amino acid sequence. A point mutation in the *HSP70-HOM* gene, associated with an *Nco*I restriction fragment length polymorphism (RFLP), results in a Met to Thr substitution at amino acid position 493 in HLA-DR15 and some -DR13 haplotypes (Milner and Campbell, 1992). Amino acid 493 lies within the peptide-binding domain of hsp70-HOM, so variation at this position could result in altered binding specificity. The *HSP70-HOM* gene lies within the region of the MHC shown to contain *Orch-1*, a susceptibility gene for experimental allergic orchitis (EAO; an autoimmune disease of the testes) in mice (Snoek *et al.*, 1993). Since *Hsc70t*, the mouse equivalent of *HSP70-HOM*, is expressed at a high level in testicular germ cells, this gene was a good candidate for *Orch-1*. However, analysis of *Hsc70t* from orchitis-susceptible and resistant strains has revealed no differences in the amino acid sequence or mRNA levels of *Hsc70t* (Snoek *et al.*, 1994). An A to G substitution at position 1267 in the *HSP70-2* gene results in a *Pst*I RFLP (Milner and Campbell, 1990).

The distribution of the 8.5 kb and 9.6 kb alleles associated with this RFLP has been investigated in patients with insulin-dependent diabetes (IDDM) and coeliac disease (Partanen *et al.*, 1993; Pociot *et al.*, 1993a). The results obtained suggest that the 8.5 kb allele is part of an extended haplotype associated with these diseases, but as there is no associated change in the hsp70-2 amino acid sequence this polymorphism is unlikely to be a contributory factor. A two allele pentanucleotide duplication in the 3′ UTR of *HSP70-2* has been described (Dressel and Gunther, 1994). The presence of the duplication associates most commonly with the presence of the *Pst*I site in *HSP70-2*. Three nucleotide substitutions have been identified in the 5′ UTR of *HSP70-1* resulting in three electrophoretically different alleles (Cascino *et al.*, 1993; Milner and Campbell, 1992). Analysis of HLA extended haplotypes has shown that in the majority of cases a given haplotype is associated with a particular *HSP70-1* allele. Sequence variations in the untranslated regions of the hsp70s provide useful markers for the identification of susceptibility loci and may have an effect on message stability or regulation of transcription.

3.3.4 The TNF, LTα and LTβ ligands

TNF and LTα (also called TNF-β) are related cytokines involved in many regulatory activities (for review see Guss and Dower, 1995). TNF is synthesized in response to various insults by a variety of cell types, including both haematopoietic and non-haematopoietic cells, and this is generally regarded as one of the primary initiating events in the inflammatory cascade. LTα, in contrast, is made specifically by lymphocytes, and its biological role is not understood. The *TNF* and *LTα* genes lie approximately 1 kb apart, within the class III region of the MHC (*Figure 3.1*, page 255), yet they are clearly independently regulated. In general, LTα and TNF display similar spectra of activities in *in vitro* systems, although LTα is often less potent or displays apparent partial agonist activity. Moreover, the two known TNF receptors do not appear to discriminate between the TNF and LTα ligands. These observations suggest that LTα is either a poorly redundant cytokine or that there are further facets, as yet unknown, to this cytokine.

TNF was originally defined by its antitumour activity, but it is also a major mediator of inflammation and cellular immune responses (Guss and Dower, 1995). TNF was found to be cytotoxic to a number of transformed cell lines *in vitro* and also to induce cachexia in lipopolysaccharide (LPS)-treated mice, with profound effects on general cellular metabolism and the development of weight loss, fever, acute phase reaction, infection, or neoplasia. TNF enhances the proliferation of T cells, modulates T-cell receptor expression, enhances natural killer (NK) cell activity, and regulates human B cell function. TNF also has marked effects on neutrophils, eosinophil recruitment, monocyte/macrophage activation, fibroblast growth stimulation, and endothelial cell/leukocyte interactions. TNF is produced by many cell types, including monocytes/macrophages, lymphocytes and fibroblasts. Activated macrophages have the highest TNF production.

As a result of the biological effects of TNF on different normal cell types this protein has an important role in several physiological and pathological conditions;

for example TNF is a crucial mediator in septic shock and cerebral malaria. Although host-mediated effects (i.e. actions of TNF on normal cells) are certainly also involved in the antitumour activity of TNF *in vivo*, the direct cytotoxic activity of TNF on many types of malignant cells has been studied more extensively (for review see Beyaert and Fiers, 1994). The therapeutic value of TNF in the treatment of cancer is limited by toxic side effects occurring at high TNF doses, and by a wide variation in TNF sensitivity of tumour cells. Indeed, on some tumour cells TNF at most exerts a 'cytostatic' activity, while other tumour cells do not respond to TNF at all. The TNF sensitivity of tumour cells is not correlated with the type or the histological origin of the cells. As a rule, untransformed cell lines are resistant to the cytotoxic effect of TNF, although embryonal fibroblastic cells may be an exception. A large number of tumour cells, which are normally resistant to TNF do become sensitive when TNF is used in combination with interferon-γ or some chemotherapeutic agents. However, such combined treatments may also result in an enhanced toxicity for the host. Understanding of the signalling pathways leading to TNF cytotoxicity *in vitro* will greatly contribute to the search for strategies to improve the therapeutic value of TNF (Beyaert and Fiers, 1994).

TNF is a type II membrane protein that is retained on the cell surface in both macrophages and T cells. However, surface expression of LTα does not result from the presence of a transmembrane region, but rather is due to association with a 33 kDa integral membrane glycoprotein. Cloning of the cDNA encoding this associated protein, called lymphotoxin β (LTβ), revealed it to be a type II membrane protein with significant amino acid sequence similarity to TNF, LTα, and the ligand for the CD40 receptor (Browning *et al.*, 1993). The gene for LTβ was found next to the *TNF* and *LTA* loci in the MHC class III region 2 kb centromeric of *TNF* (*Figure 3.1*, see page 255). These observations raise the possibility that a surface LTα–LTβ complex may have a specific role in immune regulation distinct from the functions ascribed to TNF.

The identification of LTβ and its ability to target LTα to the cell surface confirms the previous supposition that surface LT was a result of a heteromeric complex formed between the normally secreted LTα and the earlier defined p33 molecule. Secreted forms of TNF and LTα are homotrimers that initiate similar spectra of cellular responses by binding to two specific cell surface receptors of 60 and 80 kDa (TNFR$_{60}$ or TNFR type I, and TNFR$_{80}$ or TNFR type II). Membrane TNF is the precursor of secreted TNF; in contrast, cell surface LTα is structurally different from secreted LTα and TNF, which suggests that it may not bind to TNFR$_{60}$ or TNFR$_{80}$ (Guss and Dower, 1995). Recently a receptor specific for human LTβ has been identified (LTβR or TNFR related protein, TNFR-RP; Crowe *et al.*, 1994). LTβR contains a cysteine-rich motif characteristic of the TNF-nerve growth factor family of transmembrane proteins which are homologous to TNFR$_{60}$ and TNFR$_{80}$. The gene that encodes LTβR spans approximately 9 kb, with an exon organization similar to that of *TNFR$_{60}$* and is located on chromosome 12p13 as part of a gene cluster including *CD27* (another member of the TNFR superfamily) and *TNFR$_{60}$*. TNFR$_{60}$ does not bind the major cell surface LT complex (LT$\alpha_1\beta_2$), but rather binds a distinct complex with a predicted subunit ratio of LT$\alpha_2\beta_1$. In contrast, LTβR binds to the major cell surface LT complex and to a soluble LTβ molecule (Crowe *et al.*, 1994). Targeted knockout of the murine *LTα* gene disrupts peripheral lymphoid organ development (De-Togni *et al.*, 1994). In contrast, knockout of either the *TNFR₆₀* gene or *TNFR₈₀* gene does not result in a similar phenotype, which suggests a role for the surface LTα–LTβ complex and the LTβR in immune development (Crowe *et al.*, 1994).

LTα is synthesized primarily by T cells, although some Epstein–Barr virus (EBV)-transformed B-cell lines and tonsil B cells produce it. The restricted expression of LTα relative to TNF has tantalized workers in the field with the idea that LTα has specific and important immunoregulatory functions (Paul and Ruddle, 1988; Ruddle and Homer, 1988). Delineation of the LTα–LTβ complex poses the possibility of immunoregulatory activities unique to the complex that cannot be mimicked by the LTα homotrimer. It is possible that the relatively poor activity of the LTα homotrimer relative to TNF in many systems indicates that the secreted LTα phenomenon is only peripherally related to the true function of LTα. The tethering of soluble LTα to the cell surface via complexation with LTβ raises the speculation that cell–cell contact-specific signalling through LTα–LTβ is an important aspect of immune regulation (Browning *et al.*, 1993).

The mouse *TNF* and *LTA* genes are located in the equivalent position in the H-2S region of the murine MHC. Recently, the mouse *LTB* gene has been identified close to the *TNF/LTA* loci (Lawton *et al.*, 1995). The mouse and human genomic structures were found to be similar, but the mouse gene lacks one intron found in most members of the family. The three closely linked genes *TNF*, *LTA* and *LTB* have very similar genomic structures, each having four exons (except the mouse *LTB* gene) and spanning roughly 2 kb. Both the cDNA and genomic sequences revealed an altered splice donor for the mouse *LTB* gene in the conventional intron 2 position, rendering it non-functional. The altered gene retains an open reading frame such that an additional 66 amino acids are inserted into the stalk region connecting the transmembrane domain with the receptor binding domain encoded by exon 4 in this type II membrane protein (Lawton *et al.*, 1995).

The location of the *LTB* gene next to the *TNF/LTA* genes suggests more extensive duplication of a primordial *LT* gene than was previously realized. As TNF plays important roles in modulation of the immune response, much interest has focused on the correlation between *TNF* polymorphisms that affect inducibility of the gene, and their association with different MHC alleles and haplotypes (for review see Campbell and Milner, 1993). A G to A transition 308 nucleotides upstream of the *TNF* gene has been described, which lies in a region thought to be important in the regulation of transcription of the gene (Wilson *et al.*, 1993). This polymorphism shows a strong association with the HLA-A1, B8, DR3 haplotype. This haplotype also showed a significant correlation with increased levels of TNF secretion by lymphoblastoid cell lines transformed with EBV, whereas DR2 and DR5 haplotypes correlated with low levels of TNF secretion (Abrahám *et al.*, 1993). In a study of the LPS-induced secretion of TNF by monocytes (Pociot *et al.*, 1993b), which included typing for *TNF* RFLPs and microsatellite alleles in addition to DR subtypes, DR3 and DR4 haplotypes showed a significant correlation with high levels of TNF secretion, whereas DR2 and DR5 haplotypes again correlated with low levels of TNF secretion. It was noted that all the markers studied map to a region with strong linkage disequilibrium, leaving open the possibility that other MHC genes could regulate the TNF response. No correlation with the LTα response was observed for any of the markers, suggesting that TNF and LTα are independently regulated.

Several polymorphic microsatellite sequences within the *TNF/LTA/LTB* gene cluster have been characterized (Nedospasov *et al.*, 1991). Two of them, $(AC/GT)_n$ and $(TC/GA)_n$, designated TNFa and TNFb, are closely linked and located 3.5 kb upstream of the *LTA* gene (Jongeneel *et al.*, 1991; Nedospasov *et al.*, 1991). Another microsatellite-containing region has been mapped to a position 8–10 kb downstream of the *TNF* gene, and this was found to consist of a pair of linked TC/GA and TC/GA-like repeats (designated TNFe and TNFd). Finally, three biallelic polymorphic markers have been described, a TC/GA microsatellite (designated TNFc) in the intron of the *LTA* gene (Jongeneel *et al.*, 1991; Nedospasov *et al.*, 1991); an *Nco*I RFLP (TNFn) due to a substitution of a G for an A in intron 1 of the *LTA* gene (Fugger *et al.*, 1989; Messer *et al.*, 1991; Webb and Chaplin, 1990); and an *Eco*RI RFLP in the 3′ UTR of the *LTA* gene (Partanen and Koskimies, 1988). These microsatellites and RFLPs are now being used in disease association studies to assess the involvement of the *TNF/LTA/LTB* genes in disease susceptibility (Badenhoop *et al.*, 1992; Messer *et al.*, 1994; Mizuki *et al.*, 1995; Ruddle, 1992; Shimura *et al.*, 1994, 1995; Vendrell *et al.*, 1995).

3.3.5 A receptor for advanced glycosylation end products of proteins (RAGE)

Advanced glycosylation end products of proteins (AGEs; Brownlee *et al.*, 1988) are non-enzymatically glycosylated proteins that accumulate in vascular tissue as a result of ageing and at an accelerated rate in diabetes. AGEs result from the prolonged exposure of proteins to aldoses, such as glucose and ribose. Although the AGEs are a heterogeneous class of compounds, their ability to form cross-links to and between proteins, and their interaction with a class of binding sites on endothelial cells and monocytes, as well as other cell types, suggests two mechanisms through which they could contribute to diabetic complications:

(i) by altering the architecture of the extracellular matrix through the formation of cross-links between basement membrane components;
(ii) by modulating cellular function following interaction with cell-surface binding sites.

AGEs perturb a broad range of cellular functions, especially in endothelial cells and macrophages; for example, in cultured endothelium AGEs increase permeability and expression of procoagulant activity. AGEs also induce migration of mononuclear phagocytes, as well as the production of platelet-derived growth factor and cytokines (Kirstein *et al.*, 1990; Vlassara *et al.*, 1988). Isolation of two endothelial cell-surface-associated proteins from bovine lung which mediate, at least in part, the interaction of AGEs with endothelium was performed by Schmidt *et al.* (1992). N-terminal sequence analyses indicated that the approximately 35 kDa protein was novel, whereas the N-terminal sequence of the approximately 80 kDa protein was identical to that of lactoferrin. Human and bovine *RAGE* cDNA sequences were identified by Neeper *et al.* (1992). The bovine cDNA encodes a 394 amino acid mature protein comprised of an extracellular domain of 332 amino acids, a single hydrophobic membrane-spanning domain of 19 amino acids, and a carboxy-terminal domain of 43 amino acids corresponding to the 35 kDa protein purified by Schmidt *et al.* (1992). A partial clone encoding the human counterpart of *RAGE* was isolated and was found to encode a protein approximately 90% identical to the bovine molecule at the amino acid level. Sugaya *et al.* (1994) identified the human gene for *RAGE* located in the MHC class III region between the *G16* and *PBX2* genes (*Figure 3.1*, page 255) by DNA sequence analysis of a cosmid genomic insert. Comparison of the cosmid genomic DNA sequence with that of the reported cDNA showed that the gene was composed of 11 exons. All intron/exon junctions are in agreement with the GT/AG rule. The previously reported human cDNA clone was shown to be a truncated form, lacking the initiator methionine codon ATG, but the genomic sequence showed the ATG codon to be present in a position exactly corresponding to the initiator ATG codon of bovine *RAGE* cDNA (Neeper *et al.*, 1992; Sugaya *et al.*, 1994). The *RAGE* gene is present only once in the haploid human genome.

RAGE is a member of the immunoglobulin superfamily of cell surface molecules and shares significant amino acid sequence similarity with MUC18, N-CAM, and the cytoplasmic domain of CD20 (Neeper *et al.*, 1992). This suggests that RAGE may subserve functions beyond binding and subsequent uptake of AGEs. Two lines of evidence implicate RAGE as a cell adhesion molecule or growth factor receptor. Pilot studies have demonstrated that RAGE contributes to the enhanced adherence of diabetic red cells to endothelium. In addition AGEs can stimulate endothelial proliferation, and preliminary experiments have shown that antibodies to RAGE also directly stimulate endothelial growth. This suggests the hypothesis that AGEs are accidental and potentially pathogenic ligands for this receptor and indicate that an important challenge for future studies will be to identify the natural ligand for RAGE, potentially a growth factor or cell-surface ligand.

3.3.6 A sialidase enzyme encoded by the G9 gene

The *G9* gene (*Figure 3.1*, page 255) encodes a protein product of 415 amino acids that shows 30% amino acid sequence identity with the cytosolic sialidase from rat skeletal muscle (Miyagi *et al.*, 1993) and the hamster cell line CHO (Ferrari *et al.*, 1994), and 28% identity with the sialidase of *Salmonella typhimurium* (Hoyer *et al.*, 1992) over the entire lengths of these proteins (Milner *et al.*, 1996). In addition G9 contains three copies of the Asp-block motif (Ser-X-Asp-X-Gly-X-Thr-Trp) which is characteristic of bacterial sialidases (Roggentin *et al.*, 1989). Sialidase enzymes catalyse the removal of sialic acid residues from glycoproteins and glycolipids and thus play an important role in modulating cellular events. Several forms of mammalian sialidase are known to exist, which differ in their subcellular localization and biochemical properties (Miyagi and Tsuiki, 1984; Miyagi *et al.*, 1990). The expression of G9 in insect cells has confirmed that this protein has sialidase activity which is optimal at pH 4.6. The pH optimum and substrate specificity of G9 are consistent with it being a lysosomal sialidase, and this is further supported by the presence of a predicted N-terminal signal sequence and lysosomal targeting sequences in the G9 protein (Milner *et al.*, 1996).

The presence of a gene (*Neu-1*) encoding liver sialidase activity in the S region of the mouse H-2 complex, between *C2* and *BAT5*, has been previously reported (Figueroa *et al.*, 1982; Lafuse *et al.*, 1992). It seems most likely that *G9* corresponds to *Neu-1* and this is further supported by the results of Northern blot analysis, which shows high levels of *G9* mRNA in hepatocytes compared to other cell lines tested, and the observation that Neu-1 regulates the sialic acid content of a number of lysosomal enzymes. Neu-1 activity has been shown to be up-regulated in activated T cells and this is accompanied by a reduction in cell-surface sialic acid content (Landolfi *et al.*, 1985). This is probably due to hyposialylation of molecules expressed on the cell surface (Landolfi and Cook, 1986) and is necessary for T cell responsiveness to alloreactive B cells (Taira and Nariuchi, 1988).

The inherited human diseases sialidosis and galactosialidosis (each occurring in multiple forms) are associated with deficiencies in sialidase and/or galactosialidase activity. These lysosomal storage disorders are characterized by various developmental and neurological abnormalities (Lowden and O'Brien, 1979). There is evidence for at least two forms of lysosomal sialidase in mammals, one membrane-bound and one intralysosomal (Miyagi and Tsuiki, 1984; Miyagi *et al.*, 1990). The intralysosomal sialidase forms complexes with β-galactosidase and a carboxypeptidase (the protective protein) which are necessary for its activity (Verheijen *et al.*, 1985). Mutations in the gene encoding the protective protein have been shown to be responsible for lack of β-galactosidase activity in galactosialidosis (Shimmoto *et al.*, 1993) and there is evidence from studies with somatic cell hybrids that there may be a sialidase gene on chromosome 10 associated with one form of sialidosis (Mueller *et al.*, 1986). The role of G9 in susceptibility to the lysosomal storage disorders remains to be determined. However, there has been one report of a patient with combined sialidosis and deficiency in the steroid 21-hydroxylase (P450c21; Harada *et al.*, 1987). The close proximity of the *G9* and *P450c21B* genes suggests that the genetic defects responsible for these two conditions may be linked. There was no gross deletion of DNA in the patient concerned, but the possibility of linked-point mutations cannot be ruled out.

The presence of allelic variation in human *G9* has not been investigated. However, in the mouse, three alleles of *Neu-1*, *Neu-1*^{a}, *Neu-1*^{b} and *Neu-1*^{c}, have been identified on the basis of differential electrophoretic mobility of certain lysosomal proteins (Klein and Klein, 1982). *Neu-1*^{b} is the most common allele, occurring in the majority of mouse strains, and is associated with normal liver sialidase activity. The *Neu-1*^{a} allele, which is characteristic of the strain SM/J, is associated with sialidase deficiency (approximately 17% of normal liver sialidase activity). The third allele, *Neu-1*^{c}, is found in the strains A.SW and B.10S, but has not been well characterized. Although they are small in size, SM/J mice do not show any of the symptoms characteristic of the human lysosomal storage disorders in spite of their low sialidase activity. Understanding of the defect in these mice may be helpful in elucidating the roles of G9, for example during T-cell activation.

3.3.7 A valyl tRNA synthetase encoded by the G7a *gene*

Nucleotide sequence analysis of a cDNA corresponding to the *G7a* (*BAT6*) gene, which is located between the *HSP70* and *TNF* gene clusters (*Figure 3.1*, page 255), has shown that it encodes a protein product of 1265 amino acids (Hsieh and Campbell, 1991). The predicted amino acid sequence of G7a shows a high level of similarity with prokaryotic and eukaryotic valyl-tRNA synthetases (Trs^{Val}), the most significant similarity being with the Trs^{Val} of *Saccharomyces cerevisiae* (48.3% identity over 1043 amino acids). G7a also contains the consensus motifs: His-Ile-Gly-His (at His 152), which is typical of the class I tRNA synthetases; Lys-Met-Ser-Lys-Ser (at Lys 660), which represents a tRNA 3′ end binding motif; and Glu-Trp-Cys-Ile-Ser-Arg-Gln (at Glu 516), which corresponds to a putative valine-binding motif. These features provide strong evidence that G7a corresponds to human Trs^{Val} and this has been confirmed by functional characterization of the protein expressed in COS-7 cells (Vilata *et al.*, 1993).

The predicted molecular weight of the G7a protein (140.5 kDa) is close to that of other mammalian Trs^{Val}, but larger than the corresponding enzyme in prokaryotes (e.g. 108 kDa in *Escherichia coli*) and lower eukaryotes (e.g. 125 kDa in *S. cerevisiae*). The unique *N*-terminal region of G7a, responsible for this larger size, shows 27.8% amino acid sequence similarity over 154 amino acids with brine shrimp elongation factor 1 γ chain. This region contains two stretches of hydrophobic amino acids and may be important in the interaction of mammalian Trs^{Val} with elongation factor 1H, leading to the formation of high molecular weight aggregates. A basic region in human Trs^{Val} (amino acids 228–278), which is also conserved close to the N terminus of the yeast enzyme (but absent in prokaryotic enzymes), may be involved in interactions of Trs^{Val} with acidic surfaces such as membranes or RNAs. G7a contains five potential protein kinase C phosphorylation sites clustered between the His-Ile-Gly-His and Lys-Met-Ser-Lys-Ser motifs. The phosphorylation of mammalian Trs^{Val} by protein kinase C has been reported and this may be important in the regulation of its activity.

The *G7a* gene lies within the segment of the MHC (between *G7* and *HSP70-Hom*) shown, in mice, to contain *Orch-1*, a susceptibility locus for the autoimmune condition EAO (Snoek *et al.*, 1993; 1994). In this respect it is interesting to note that autoantibodies to Trs^{His} have been identified in the sera of 25–30% of patients with the autoimmune disease myositis and 70% of patients with myositis and interstitial lung disease (Bernstein *et al.*, 1984), so it is not impossible that *G7a* may act as an autoantigen in EAO.

3.4 Genes encoding proteins with a putative function

3.4.1 The Notch *homologue* NOTCH3

Nucleotide sequence analysis of the terminal portions of six overlapping cosmids by Sugaya *et al.* (1994), covering a region extending about 250 kb from the *P450c21B* gene towards the class II region, revealed significant sequence similarity with various portions of *Notch* homologues from species such as *Drosophila*, *Xenopus*, zebrafish, rat, mouse, and human. The highest nucleotide sequence identity, about 80%, was noted for the mouse mammary tumour gene *int-3*. Mouse mammary tumour development results from clonal outgrowth of tumour cells that frequently contain mouse mammary tumour virus (MMTV) integrated at one or more specific regions of the genome called *int* loci. Genetic linkage analysis in mice has shown that one *int* locus, *int-3*, maps between the *C4* gene and the MHC class II gene *H-2Aa* (Siracusa *et al.*, 1991). A consequence of MMTV integration at *int-3* is activation of expression of a 2.3 kb RNA species corresponding to the sequence downstream of the viral insertion. Nucleotide sequence of *int-3* RNA suggests that it corresponds to the intracellular domain of a *Notch*-like gene (Robbins *et al.*, 1992). *Notch* was first found as a *Drosophila* neurogenic gene required for correct segregation of epidermal cells from neuronal cell precursors during embryogenesis, and subsequent studies demonstrated its roles in proper formation of the mesoderm, germline, ovarian follicle cells, adult wing, and adult peripheral nervous system structures. Notch is thus widely expressed during embryonic and adult development and participates in different cell–cell interactions, mediating a variety of regulatory events (Fortini and Artavanis-Tsakonas, 1993). The *Notch* gene encodes a 2703 amino acid transmembrane protein with a large extracellular domain containing 36 EGF-like repeats, three Notch/lin 12 repeats and a 938 amino acid intracellular domain bearing six tandem ankynin (ANK) repeats and Pro-Glu-Ser-Thr (PEST) regions. Notch homologues have been isolated from a variety of vertebrates, and in human, two genes have been characterized. The *TAN-1* gene on human chromosome 9, for which cDNA sequence has been reported, is involved in the translocation t(7;9)(q34;q34.3) in acute T-cell lymphoblastic leukaemia (T-ALL; Ellisen *et al.*, 1991). The other human gene, *hN*, was isolated as a cDNA using polymerase chain reaction (PCR) primers based on the *Drosophila* and *Xenopus Notch* sequences, and its intracellular domain has been sequenced (Stifani *et al.*, 1992).

The partial sequences obtained from cosmids by Sugaya *et al.* (1994) include regions of sequence similarity with virtually all the functional domains of the Notch protein family. The size of this human *Notch*-like gene (*Figure 3.1*) is estimated to be more than 30 kb, but its full sequence has not been determined. A phylogenetic tree was constructed based on comparison of the amino acid sequences of the ANK repeats which were available from various species (Sugaya *et al.*, 1994). Human TAN-1, rat Notch 1, mouse Notch 1, zebrafish Notch, and *Xenopus* Notch are closely related and appear to constitute one group (Notch subfamily *1*). Human Notch, hN, and rat Notch 2 compose another group (Notch subfamily *2*). Although the intracellular domain sequence of Motch B (from mouse) has not been determined, Motch B probably belongs to subfamily 2 since it could be assigned to the mouse counterpart of rat Notch 2 based on comparison of the EGF and Notch/lin-12 repeat sequences. The human *Notch*-like gene in the MHC and mouse *int-3* compose another group (Notch subfamily *3*), and the human gene has thus been designated *NOTCH3* (Sugaya *et al.*, 1994). The copy number of human *NOTCH3* was determined by hybridization to YACs containing yeast and human DNAs and only one copy was found in the human genome.

In all the tumours having an MMTV-induced rearrangement of the *int-3* locus, the transcriptional orientations of the integrated viral genomes are the same and the integration site is restricted to a 500 bp region of the locus. The primary mechanism of activation of the other *int* genes by MMTV is enhancer insertion, while *int-3* expression is activated by promoter insertion. Although the activated *int-3* encoded protein does not appear to have an extracellular domain, nucleotide sequence analysis of genomic cellular DNA adjacent to the 5′ end of MMTV genomes integrated in the *int-3* locus reveals three potential exons which encode five EGF-like repeats and three Notch/lin 12 repeats (Robbins *et al.*, 1992). The potential importance of the extracellular domain of the Notch-like protein encoded by *int-3* in mediating specific adhesive interactions between cells during development has represented a major focus of attention. However, it has also been suggested that the Notch-like protein is a receptor and that its intracellular domain plays a role in signal transduction in response to ligand binding to the extracellular domain (Fehon *et al.*, 1990). If the uninterrupted *int-3* gene (*NOTCH3*) encodes a transmembrane protein (receptor), then a primary consequence of MMTV integration in this locus is activation of expression of its intracellular domain and its release from extracellular control. In this scenario, MMTV-induced activation of *int-3* has the same molecular consequences as the t(7;9)(q34;q34.3) translocations that affect the *TAN-1* gene in human T-ALL (Ellisen *et al.*, 1991), in which the transmembrane and intracellular domains are overexpressed. It has been demonstrated in HC11 cell lines that the activated *int-3* gene product can perturb the growth properties of mammary epithelial cells in a manner that could contribute to mammary tumorigenesis. In fact, expression of this genomic fragment *in vivo* in a transgenic mouse strain is associated with arrest of normal mammary gland development, intraductal hyperplasia of mammary epithelium, and a high incidence of focal mammary tumours (Jhappan *et al.*, 1992).

3.4.2 PBX2, a homeodomain-containing protein

Sequence analysis of a 10 108 nucleotide segment of genomic DNA between the *G16* and *G18* genes by Sugaya *et al.* (1994) revealed the location of a gene encoding PBX2 [named *HOX12* by Sugaya *et al.* (1994; *Figure 3.1*, page 255)]. In an independent study, DNA sequence analysis of 5.5 kb of DNA corresponding to the previously described *G17* gene (Kendall *et al.*, 1990) revealed that it encodes PBX2 (Aguado and Campbell, 1995). Comparison of this genomic DNA sequence with the published *PBX2* cDNA sequence, revealed 99.7% identity over the coding regions, and indicated that the *G17* gene is split into nine exons, with the intron/exon boundaries conforming to the normal pattern (AG/.../GT) for splice sites. *PBX2* is expressed in a variety of different cell types, including immune-related cell lines, showing two different mRNAs of 3.2 and 2.4 kb in Northern blot analysis (Aguado and Campbell, 1995). Comparison of genomic and cosmid Southern blots clearly indicates that another copy(ies) of the *PBX2* gene exist(s) in the genome. PCR amplification of exons III and IX of the *PBX2* gene, corresponding to the coding and 3′ untranslated regions, respectively, using as template genomic DNA from a panel of monochromosomal somatic human–rodent cell hybrids, gave specific products in hybrids that contain human chromosomes 6, 3 and 1. These results confirm that copies of the *PBX2* gene are located on human chromosomes 6 and 3 and indicate that a gene homologous to *PBX2* could exist on human chromosome 1. Further PCR analysis of the genes and reverse transcribed mRNA from the hybrid cell lines has revealed that the copies of the *PBX2* gene on human chromosomes 6 and 1 are expressed, while the copy on human chromosome 3 may be a processed pseudogene (Aguado and Campbell, 1995).

The novel human homeobox gene, *PBX1*, was identified because of its involvement in t(1;19) chromosomal translocations in acute pre-B cell leukaemias (Kamps *et al.*, 1990; Nourse *et al.*, 1990). This translocation results in the formation of fusion transcripts that code for E2a-PBX1 chimeric proteins, in which the C-terminal region of E2a, a transcription factor gene which contains basic and helix-loop-helix (bHLH) DNA-binding and dimerization motifs, is replaced by sequences from PBX1 that contain a highly divergent homeodomain. cDNA clones corresponding to two additional human genes *PBX2* and *PBX3*, with 79% and 83% nucleotide sequence identity, respectively, to *PBX1*, have been isolated using a 259 bp cDNA probe containing the homeobox of *PBX1* (Monica *et al.*, 1991). These cDNAs were used to localize the corresponding genes by *in situ* hybridization analysis of human metaphase chromosome spreads. *PBX2* was localized to the chromosome 3 bands 3q22–23, while *PBX3* was localized to the chromosome 9 bands 9q33–34.

Homeoboxes code for a conserved protein motif, termed the homeodomain, and the degree of similarity between different homeodomains has been used to group them into related classes (Scott *et al.*, 1989). While most of the *Drosophila* homeodomain-containing proteins have been implicated in developmental processes, the functions of the majority of non-*Drosophila* homeodomain-containing proteins remain unknown. However, because of the presence of homeodomains in known transcription factors and the structural similarity between the homeodomain and DNA-binding motifs in bacterial proteins, the current view is that most, if not all, of these proteins function as sequence-specific DNA-binding proteins that are likely to play an important role in transcriptional regulation (for reviews see Gehring and Hiromi, 1986; Scott *et al.*, 1989).

The homeodomain of the PBX proteins is extremely well conserved and is closely related to the homeodomain of the yeast a1/α2 protein (Kamps *et al.*, 1990; Nourse *et al.*, 1990). However, the a1/α2 protein and the PBX proteins exhibit no sequence similarity outside their homeodomains. The biological function of the PBX proteins is unknown, but it is possible that they bind to a common DNA sequence (Kissinger *et al.*, 1990). Flegel *et al.* (1993) have postulated a possible repressor function for PBX proteins based on the fact that polyalanine stretches have been noted previously in transcriptional repressors, and that in acute pre-B cell leukaemias the repressor domain of PBX is replaced with an activator domain derived from E2a.

The human proto-oncogene *PBX1* codes for a homologue of *Drosophila* extradenticle, a divergent homeodomain protein that modulates the developmental and DNA-binding specificity of select HOM proteins. Wild-type PBX proteins and chimeric E2a-PBX1 oncoproteins have been shown cooperatively to bind a consensus DNA probe with a class of Hox/HOM proteins. The PBX homeodomain is necessary, but not sufficient, for cooperativity which requires conserved amino acids carboxy-terminal of the homeodomain. These findings demonstrate that interactions between Hox and PBX proteins modulate their DNA-binding properties, suggesting that PBX and Hox proteins act in parallel as heterotopic complexes to regulate expression of specific subordinate genes (Chang *et al.*, 1995).

The t(1;19) chromosomal translocation in acute lymphoblastic leukaemias creates chimeric E2a-PBX1 oncoproteins that can act as DNA-binding activators of transcription. Structural analysis of the functional domains of E2a-PBX1 showed that portions of both E2a and PBX1 were essential for transformation of NIH 3T3 cells and transcriptional activation of synthetic reporter genes containing PBX1 consensus binding sites. Hyperexpression of wild-type or experimentally truncated PBX1 proteins is insufficient for transformation, consistent with their inability to activate transcription. When fused with E2a, the PBX-related proteins PBX2 and PBX3 are also transformation competent, demonstrating that all known members of this highly related subfamily of homeodomain proteins have latent oncogenic potential. The homeodomain is not essential for transformation, since a mutant E2a-PBX1 protein lacking the homeodomain efficiently transforms fibroblasts and induces malignant lymphomas in transgenic mice. Thus, transformation mediated by the chimeric oncoprotein E2a-PBX1 is absolutely dependent on motifs acquired from E2a, but the PBX1 homeodomain is optional. The latter finding suggests that E2a-PBX1 may interact with cellular proteins that assist or mediate interactions in gene expression responsible for oncogenesis, even in the absence of homeodomain–DNA interactions (Monica *et al.*, 1994).

Cytochrome P450c17 (encoded by the *CYP17* gene), which is expressed in response to peptide hormones via cAMP, is required for cortisol and sex hormone biosynthesis, thereby playing a key role in biological processes including sexual differentiation. Utilizing the cAMP-regulatory sequence CRS1 of the bovine *CYP17* gene as an affinity ligand, four proteins, from nuclear extracts of mouse adrenocortical Y1 cells shown to enhance the *in vitro* transcription of a reporter gene promoted by CRS1, were isolated. Microsequencing of these four proteins showed two of them to be the homeodomain proteins PBX1a and PBX1b. Overexpression of PBX1 in Y1 cells enhances cAMP-dependent transcription of the CRS1-dependent reporter gene. These results identify the CRS1 of bovine *CYP17* as a cellular target for PBX1 and suggest that one role of this homeodomain protein is in the regulation of steroidogenesis and, subsequently, sexual development (Kagawa *et al.*, 1994). Due to the significant sequence identity between PBX2 and PBX1 it is interesting to speculate that PBX2 could perform similar functions to PBX1.

3.4.3 A protein containing ANK repeats: G9a

The *G9a* gene, which lies between the *C2* and *HSP70* genes (*Figure 3.1*, page 255), encodes a predicted protein product of 1001 amino acids (Milner and Campbell, 1993). Northern blot analysis indicates that *G9a* is ubiquitously expressed at the mRNA level. The C-terminal region of G9a (amino acids 819–971) shows 35% sequence identity with a 149 amino acid segment from the C-terminus of the *Drosophila* trithorax protein, which is involved in the regulation of expression of homeotic genes during development of the *Drosophila* thorax (Mazo *et. al.*, 1990), and 32% identity with the C terminus of the human trithorax homologue, HRX, the disruption of which is a common feature of human multilineage leukaemias (Tkachuk *et al.*, 1992). The C terminus is the region of maximum sequence conservation (60% over 210 amino acids) between trithorax and HRX, and a highly related sequence is also found in *Drosophila* enhancer of zeste, a negative regulator of transcription during development (Jones and Gelbart, 1993). The conservation of this domain in a number of proteins involved in the regulation of transcription suggests that it may be important in binding to a common target. The *Drosophila* trithorax protein contains nine cysteine-rich zinc-finger-like domains which bind zinc. The C-terminal region (amino acids 3611–3759), which shows similarity to G9a, does not conform to a zinc finger or zinc twist consensus, but has also been shown to bind zinc *in vitro*. The C-terminal domain of G9a, expressed in E. coli has also been shown to bind zinc under reducing conditions *in vitro* (Milner, 1991).

Amino acids 469–720 of G9a show 32% sequence identity with amino acids 1895–2109 of the Notch protein of *Drosophila* (Wharton *et al.*, 1985), a region of Notch that contains six ANK repeats (Michaely and Bennett, 1992). Comparison with other available ANK repeat sequences confirmed that residues 475–673 of G9a constitute six contiguous ANK repeats preceded by a possible poorly conserved seventh repeat. ANK repeats have been identified in a large number of proteins with diverse functions, including the determination of cell fate (e.g. Notch) and the regulation of transcription (e.g. NF-κB), which occupy a wide variety of subcellular locations. However, they all have in common the ability to bind to other proteins; for example, in the interaction of human erythrocyte ankyrin with the erythroid anion exchanger (Davis *et al.*, 1991), or to DNA as in the case of the β-subunit of the transcription factor, human guanine adenine-binding protein (LaMarco *et al.*, 1991).

Preliminary results of cell-staining studies with antisera raised against a G9a polypeptide suggest that this protein is localized in the cytoplasm. However, G9a does contain a potential bipartite nuclear localization signal (residues 130–153) so it may be localized in the nucleus under appropriate conditions. The transcription factor NF-κB is known to be retained in the cytoplasm due to interaction with the inhibitor I-κB, mediated via ANK repeats present in both of these proteins (Liou and Baltimore, 1993). Following cell activation (e.g. by mitogens), NF-κB is released and migrates to the nucleus where it brings about transcriptional activation. The features of the G9a protein described above suggest that it may play a role in the regulation of transcription.

3.4.4 ikbl: a member of the IκB family of proteins?

Characterization of the *ikbl* gene, which is located approximately 12 kb telomeric of the *LTA* gene (*Figure 3.1*, page 255), has revealed that it spans approximately 13 kb of DNA and encodes a protein product of 381 amino acids with molecular weight of 43 kDa (Albertella and Campbell, 1994). It contains two copies of the 33 amino acid ANK repeat between amino acids 60 and 123 that closely resemble the second and third repeats of the IκB family (Gilmore and Morin, 1993; Nolan and Baltimore, 1992), and together are 44% identical (58% similar) to these repeats in NF-κB p110. Immediately following these two well-conserved repeats is a third repeat (between amino acids 124 and 151) that corresponds poorly to the consensus sequence of the ANK repeat, and a short region rich in acidic amino acids (nine glutamic and aspartic acids over a 13-residue stretch between residues 152 and 164). This pattern of several ANK repeats followed by a poorly fitting, incomplete repeat and an adjacent acidic domain is highly redolent of the IκB family, with the exception of Bcl3 which lacks the acidic region, suggesting that the protein encoded by the *ikbl* gene represents a divergent member of this family.

The transcription factor NF-kB belongs to the Rel family of transcription factors (Blank *et al.*, 1992) and is important for the inducible expression of a wide variety of cellular genes including cytokines, cytokine receptors and stress proteins (Baeuerle and Henkel, 1994; Thanos and Maniatis, 1995). In addition it is a key element in the replication of viruses such as human immunodeficiency virus (HIV) and cytomegalovirus (CMV). A common feature of the regulation of transcription factors belonging to the Rel family is that in unstimulated cells the majority of these proteins are in the cytoplasm, but cannot bind to DNA because they are further complexed with inhibitory molecules belonging to the IκB family of proteins (Beg and Baldwin, 1993; Gilmore and Morin, 1993). Stimulation of cells by a number of agents including PMA, IL-1β, TNF and LPS results in the phosphorylation and subsequent degradation of the IκB protein, leading to dissociation of the cytoplasmic complexes and translocation of the free Rel proteins to the nucleus (Arenzana-Seisdedos *et al.*, 1995; Beg and Baldwin, 1993; Ghosh and Baltimore, 1990; Henkel *et al.*, 1993; Mellits *et al.*, 1993; Rodriguez *et al.*, 1995; Sun *et al.*, 1993). Thus it is interesting to speculate that the ikbl protein may interact with NF-κB/rel molecules involved in the regulation of cellular genes transcribed in immune and inflammatory responses to challenges such as microbial infection or the presence of inflammatory cytokines.

3.4.5 BAT1: a putative nuclear RNA helicase

The *BAT1* gene, which spans approximately 10 kb of DNA, is located approximately 30 kb telomeric of the *LTA* gene (*Figure 3.1*, page 255) and encodes a protein product of 428 amino acids with a molecular weight of 48 kDa (Peelman *et al.*, 1995). Transfection and expression of the BAT1 cDNA after tagging with a monoclonal antibody epitope revealed a nuclear localization of the hybrid protein. The BAT1 protein contains nine blocks of amino acid sequence that are similar to blocks of highly conserved sequence found in the DEAD-box protein family of ATP-dependent RNA helicases, suggesting that BAT1 is a member of the RNA helicase superfamily. Members of this superfamily have diverse cellular functions, such as initiation of translation, RNA splicing, ribosome assembly, spermatogenesis, oogenesis, cell growth and division (reviewed in Schmid and Linder, 1992). However, the precise function of BAT1 has yet to be established.

3.4.6 The MHC class I chain-related (MIC) genes

Recently, a family of five sequences in the human MHC, which are distantly related to class I heavy chains, has been identified (Bahram *et al.*, 1994). These *MIC* genes are thought to have evolved in parallel with the human class I genes and with those in most, if not all, mammalian species. *MICA* and *MICB* were identified by isolation of cDNA clones corresponding to two locations 40 and 110 kb centromeric of *HLA-B* (*Figure 3.1*, page 255), respectively. *MICA* and *MICB* are closely related, sharing 91% nucleotide sequence identity in their coding sequences (Bahram *et al.*, 1994). Comparison of the published genomic DNA sequence for *PERB11* (a potential gene identified in an independent study) and the cDNA sequence for *MICA* shows 100% identity over a defined exon in the *PERB11* sequence (Leelayuwat *et al.*, 1994).

The *MICA* sequence was derived from fibroblast and keratinocyte cDNA clones. The single long open-reading frame of 1149 bp encodes a polypeptide of 383 amino acids with a relative molecular mass of 43 kDa (including the putative leader peptide of 23 amino acids). Both the sequence and predicted domain structure of MICA are similar to those of class I heavy chains, including three extracellular domains ($\alpha1$, $\alpha2$, $\alpha3$), a putative transmembrane segment, and a carboxy-terminal cytoplasmic tail. As in the class I genes, each of these domains and a predicted leader peptide correspond to discrete exons in *MICA*. However, the leader sequence and the $\alpha1$ exon are separated by a large intron in *MICA*, and the cytoplasmic tail and 3′ untranslated sequences are fused in a single exon. The genomic organization of *MICA* is thus distinct from all known class I genes. Moreover, in contrast to the typical class I genes, expression of *MICA* mRNA was neither detected in B and T cell lines nor regulated by γ-interferon in HeLa cells, but was equally abundant in various fibroblast and epithelial cell lines (Bahram *et al.*, 1994).

Under conditions of low stringency, the *MICA* cDNA hybridized to three fragments in total genomic DNA which were not accounted for by the *MICA* and *MICB* genes. By using a series of overlapping YAC clones, these segments were mapped to distinct locations in the vicinities of the *HLA-E*, *-A*, and *-F* genes. These results demonstrate the existence of a gene family and indicate that a primordial *MIC* gene originated from a prototypic class I gene before the evolution of the class I gene family (Bahram *et al.*, 1994).

3.5 Genes encoding proteins of unknown function

cDNA or genomic DNA fragments corresponding to the genes *G1*, *G2* (*BAT2*), *G3* (*BAT3*), *G8*, *G11*, *G13*, *RD* and *B144* have been cloned and sequenced (Banerji *et al.*, 1990; Khanna and Campbell, 1993; Levi-Strauss *et al.*, 1988; Olavesen *et al.*, 1993; Partanen and Campbell, 1993; Sargent *et al.*, 1994; Shen *et al.*, 1994; Speiser and White, 1989; Tsuge *et al.*, 1987). The predicted amino acid sequences of the proteins encoded by these genes have been screened against databases of known protein sequences to identify any similarities that might give an indication as to the functions of the novel proteins. In some cases, for example G8, there is no significant sequence similarity with other proteins (Partanen and Campbell, 1993). In others there are interesting structural features or defined regions of the novel protein which show similarity to a known protein or protein domain as described below, but the function of these proteins remains to be determined.

The *G1* gene, located approximately 36 kb centromeric of the *TNF* gene (*Figure 3.1*, page 255), encodes a 93 amino acid protein which contains two potential 'EF-hand'-type calcium-binding domains in its N-terminal half. One of these (residues 4–15) shows an exact match with the consensus sequence, whilst the second (residues 40–51) contains some mismatches. This region of G1 shows 36% amino acid sequence identity with the calcium binding protein calmodulin, which also contains 'EF-hand' domains (Olavesen *et al.*, 1993). Northern blot analysis has shown that *G1* is only expressed at significant levels in cell lines representing monocytes/macrophages and T cells. This restricted expression pattern and its similarity to calcium binding proteins suggests that G1 may have a role in an intracellular pathway during T-cell activation or macrophage stimulation, when intracellular calcium levels rise.

The *BAT2* (*G2*) and *BAT3* (*G3*) genes (*Figure 3.1*, page 255) encode protein products of 228 kDa and 110 kDa, respectively (Banerji *et al.*, 1990). Both proteins are very rich in proline residues and both contain novel repetitive elements, but they do not appear to be members of any known protein family. The BAT3 protein contains a 75 amino acid region close to its N terminus which is 35% identical to ubiquitin, but the significance of this is not clear.

Two novel genes, *RD* and *G11*, have been identified within the complement gene cluster, between the *C4* and *Factor B* genes (*Figure 3.1*, page 255; Sargent *et al.* 1989a). This region is of interest because it is thought to contain a gene associated with susceptibility to IgA deficiency (IgAD) and common variable immunodeficiency (CVID; French and Dawkins, 1991; Volanakis *et al.*, 1992). The *RD* gene encodes a 41 kDa protein (Levi-Strauss *et al.*, 1988; Speiser and White, 1989), the central region of which contains 24 consecutive pairs of alternating basic (Arg) and acidic (Asp or Glu) residues. The functional significance of this repetitive element is not known, although similar repeats have been seen in other proteins; for example, the 70 kDa subunit of small nuclear ribonucleoprotein contains a 110 amino acid region which is rich in arginines and includes dipeptide repeats similar to, but less strictly alternating than, those in RD.

The *G11* gene lies just 611 bp telomeric of the *C4A* gene and genomic sequence analysis has shown that the 3′ end of *G11* was involved in the duplication event giving rise to the two *C4/P450c21/X/Y* gene clusters (Sargent *et al.*, 1994; Shen *et al.*, 1994). Nucleotide sequence analysis of cDNAs corresponding to *G11* has shown the existence of multiple forms of *G11* at the mRNA level, perhaps due to the presence of more than one promoter. The longest *G11* cDNA that has been sequenced encodes a protein product of 368 amino acids (Shen *et al.*, 1994). A cDNA corresponding to a shorter mRNA, *G11-Y*, lacks 392 nucleotides corresponding to the 5′ end of the longer *G11* mRNA and encodes a protein product of 258 amino acids (Sargent *et al.*, 1994). The predicted protein products encoded by both G11 cDNAs contain bipartite nuclear localization signals, suggesting that they may function in the nucleus. However, the G11 amino acid sequence does not show any strong similarity with other proteins that makes it obvious what this function might be. Probably the most significant similarity is with the tyrosine kinase transforming protein from the fujinami sarcoma virus, which shows 18.7% amino acid sequence identity (32.5% similarity) with G11 over 157 amino acids. This may suggest that G11 has a novel kinase-like activity and in this regard it is interesting to note the presence of potential sites of phosphorylation at Ser/Thr and Tyr residues in G11.

The *G13* gene, lying 100 kb centromeric of the *C4* gene (*Figure 3.1*, page 255), is ubiquitously expressed and encodes a predicted protein product of 703 amino acids which is rich in proline (12.1%), serine (12.1%) and leucine (10.9%) (Khanna and Campbell 1993). The N-terminal region of the G13 protein shows 22% amino acid sequence identity over 330 amino acids with human cyclic AMP response element binding protein (CREB). The CREB family of transcription factors are sequence-specific DNA binding proteins that use dimerization to control function. Protein dimerization is mediated via a leucine zipper domain, whilst DNA-binding occurs via a region rich in basic amino acids. The G13 protein contains a characteristic leucine zipper consensus sequence between amino acids 362 and 389 and this is preceded by two basic clusters, rich in arginine and lysine (residues 335–340 and 351–356), separated by an alanine spacer. This cluster-spacer-cluster and leucine zipper motif, together known as the bZip domain, are conserved in many transcription factors including the CREB family, of which G13 may represent a novel member.

The *B144* gene lies just centromeric of the *TNF* gene cluster (*Figure 3.1*, page 255) (Sargent *et al.*, 1989a). The mouse homologue of *B144* has been partially sequenced at the DNA level, but the encoded protein does not show significant similarity to any known protein (Tsuge *et al.*, 1987). However, *B144* does show a restricted pattern of expression, being found in monocytes, macrophages and B cells, but not T cells or liver.

The 220 kb region between the *TNF* and *HLA-B* genes is thought to be important in determining susceptibility to IDDM and myasthenia gravis. Genomic sequence analysis and probing of Northern blots with genomic YAC clones has led to the identification of a number of potential coding sequences in this region, termed the *PERB* genes (*Figure 3.1*, page 255; Marshall *et al.*, 1993). *PERB1* includes two potential exons, one of which has 80% nucleotide sequence identity over 300 bases with the immunoglobulin domain 1 of the fibroblast growth factor receptor 3 gene, a member of the tyrosine kinase receptor family. Expression of *PERB1* has not been demonstrated in human tissue, but the mouse orthologue is expressed at low levels in the testis. A series of ubiquitously expressed transcripts that cross-hybridize with a *BAT1* probe has been designated *PERB2*. This may represent a second *BAT1*-like gene in the class III region. A number of transcripts, ranging in size from 3.4 to 9.5 kb, which are expressed primarily in the liver, are thought to be derived from an alternatively spliced gene of at least 72 kb in length, designated *PERB6*. Cross-hybridization of a 267 bp probe from the extracellular portion of the TNF-related activation protein (*TRAP*) gene with *PERB6* transcripts and genomic DNA has led to the suggestion that this gene may encode a transmembrane regulatory protein. *PERB10* has been defined as a duplicated locus, at least one copy of which is expressed as a 0.5 kb message in a wide variety of tissues. One potential exon of *PERB10* encodes amino acid sequence with 60% identity to CD75 over 27 residues. Several sequences with similarity to genes in the MHC class I region have been identified between *TNF* and *HLA-B*. Two sequences defined as *PERB3* and *PERB4* show similarity with the non-coding regions of *HLA-H* and *HLA-75*, respectively, whilst *PERB5* shows similarity to exonic sequence from *HLA-17*. However, there is no evidence for the expression of any of these species. Five copies of the *P5* gene have been mapped to the class I region, each lying adjacent to a class I-like gene (Vernet *et al.*, 1993). The *PERB7* sequence, in the class III region, shows similarity to the *P5* gene family and has been denoted *P5-6*. *PERB7* lies adjacent to *PERB5* (which shows similarity to *HLA-17*) and this conforms with the idea of tandem duplications involving *P5* and class I-like genes (Vernet *et al.*, 1993).

3.6 Summary

Fifty-three transcriptional units have now been located in the 1100 kb of DNA that separates the class I and class II regions, and more will undoubtedly be found in the future as the complete DNA sequence of the class III region is determined. These genes encode proteins with a variety of different functions including the complement system proteins C4, C2 and factor B, the cytokines TNFα and the heteromeric complex LTα–LTβ, a tenascin-like molecule, steroid 21-OHase, valyl-tRNA synthetase, three members of the hsp70 family, RAGE, a sialidase, a Notch homologue, and the homeodomain containing protein PBX2. As C4, C2, and factor B play important roles in the humoral immune response, and TNFα and the heteromeric complex LTα–LTβ are involved in the inflammatory response it is clear that they could directly contribute to some of the autoimmune diseases that are associated with the MHC. Others, such as the *ikbl* gene product which is a putative transcription factor inhibitor, and the *G13* and *G9a* gene products, which are putative transcription factors, although not directly involved in the immune response, could be involved in transcriptional regulation of immunologically important genes. Thus it is possible to envisage that variation in expression of genes such as *ikbl*, *G13* and *G9a*, or variation in the amino acid sequence of the protein products, could make an indirect contribution to disease susceptibility. However, as the functions of most of the class III region gene products remain to be determined it will take time to unravel which of these play a major role in disease susceptibility. In this respect genetic analysis of the class III region using microsatellite and other polymorphic markers will help to establish the location of genes that show significant association with different diseases and this will allow those genes and their protein products to be characterized in detail.

References

Abderrahim H, Sambucy J-L, Iris F, Ougen P, Billault A, Chumakov IM, Dausset J, Cohen D, Le Paslier D. (1994) Cloning the human major histocompatibility complex in YACs. *Genomics* **23:** 520–527.

Abraham LJ, French MA, Dawkins RL. (1993) Polymorphic MHC ancestral haplotypes affect the activity of tumour necrosis factor-alpha. *Clin. Exp. Immunol.* **92:** 14–18.

Aguado B, Campbell RD. (1995) The novel gene G17, located in the Human Major Histocompatibility Complex, encodes PBX2, a homeodomain-containing protein. *Genomics* **25:** 650–659.

Albertella MR, Campbell RD. (1994) Characterization of a novel gene in the human major histocompatibility complex that encodes a potential new member of the I kappa B family of proteins. *Hum. Mol. Genet.* **3:** 793–799.

Arenzana-Seisdedos F, Thompson J, Rodriguez MS, Bachelerie F, Thomas D, Hay RT. (1995) Inducible nuclear expression of newly synthesized I kappa B alpha negatively regulates DNA-binding and transcriptional activities of NF-kappa B. *Mol. Cell. Biol.* **15:** 2689–2696.

Ashfield R, Enriquez-Harris P, Proudfoot NJ. (1991) Transcriptional termination between the closely linked human complement genes *C2* and factor B: common termination factor for *C2* and c-myc? *EMBO J.* **10:** 4197–4207.

Ashfield R, Patel AJ, Bossone SA, Brown H, Campbell RD, Marcu KB, Proudfoot NJ. (1994) MAZ-dependent termination between closely spaced human complement genes. *EMBO J.* **13:** 5656–5667.

Badenhoop K, Schwarz G, Schleusener H, Weetman AP, Recks S, Peters H, Bottazzo GF, Usadel KH. (1992) Tumor necrosis factor beta gene polymorphisms in Graves' disease. *J. Clin. Endocrinol. Metab.* **74:** 287–291.

Baeuerle PA, Henkel T. (1994) Function and activation of NFκB in the immune system. *Annu. Rev. Immunol.* **12:** 141–179.

Bahram S, Bresnahan M, Geraghty DE, Spies T. (1994) A second lineage of mammalian major histocompatibility complex class I genes. *Proc. Natl Acad. Sci. USA* **91:** 6259–6263.

Banerji J, Sands J, Strominger JL, Spies T. (1990) A gene pair from the human major histocompatibility complex encodes large proline-rich proteins with multiple repeated motifs and a single ubiquitin-like domain. *Proc. Natl Acad. Sci. USA* **87:** 2374–2378.

Barba G, Rittner C, Schneider PM. (1993) Genetic basis of human complement C4A deficiency. Detection of a point mutation leading to nonexpression. *J. Clin. Invest.* **91:** 1681–1686.

Beg AA, Baldwin A Jr. (1993) The I kappa B proteins: multifunctional regulators of Rel/NF-kappa B transcription factors. *Genes Dev.* **7:** 2064–2070.

Belt KT, Carroll MC, Porter RR. (1984) The structural basis of the multiple forms of human complement component C4. *Cell* **36:** 907–914.

Belt KT, Yu CY, Carroll MC, Porter RR. (1985) Polymorphism of human complement component C4. *Immunogenetics* **21:** 173–180.

Bernstein RM, Morgan SH, Chapman J, Bunn CC, Mathews MB, Turner, Warwick M, Hughes GRV. (1984) Anti-Jo-1 antibody: A marker for myositosis with interstitial lung disease. *Br. Med. J.* **289:** 151–152.

Beyaert R, Fiers W. (1994) Molecular mechanisms of tumor necrosis factor-induced cytotoxicity. What we do understand and what we do not. *FEBS Lett.* **340:** 9–16.

Bickmore WA, Sumner AT. (1989) Mammalian chromosome banding-an expression of genome organization. *Trends Genet.* **5:** 144–148.

Bird AP. (1987) CpG islands as gene markers in the vertebrate nucleus. *Trends Genet.* **3:** 342–347.

Blank V, Kourilsky P, Israel A. (1992) NF-κB and related proteins: Rel/dorsal homologies meet ankyrin-like repeats. *Trends Biochem. Sci.* **17:** 135–140.

Blond-Elguindi S, Cwirla SE, Dower WJ, Lipshutz RJ, Sprang SR, Sambrook JF, Gething MJ. (1993) Affinity panning of a library of peptides displayed on bacteriophage reveals the binding specificity of BiP. *Cell* **72:** 717–728.

Bourdon MA, Ruoslahti E. (1989) Tenascin mediates cell attachment through an RGD-dependent receptor. *J. Cell. Biol.* **108:** 1149–1155.

Bristow J, Gitelman SE, Tee MK, Staels B, Miller WL. (1993a) Abundant adrenal-specific transcription of the human P450c21A "pseudogene". *J. Biol. Chem.* **268:** 12919–12924.

Bristow J, Tee MK, Gitelman SE, Mellon SH, Miller WL. (1993b) Tenascin-X: a novel extracellular matrix protein encoded by the human XB gene overlapping P450c21B. *J. Cell. Biol.* **122:** 265–278.

Browning JL, Ngam-ek A, Lawton P, DeMarinis J, Tizard R, Chow EP, Hession C, O'Brine-Greco B, Foley SF, Ware CF. (1993) Lymphotoxin beta, a novel member of the TNF family that forms a heteromeric complex with lymphotoxin on the cell surface. *Cell* **72:** 847–856.

Brownlee M, Cerami A, Vlassara H. (1988) Advanced glycosylation end products in tissue and the biochemical basis of diabetic complications. *N. Engl. J. Med.* **318:** 1315–1321.

Bugiel G, Betts WH. (1993) Anti-inflammatory drugs facilitate heat shock protein synthesis. *Aust. N. Z. J. Med.* **23:** 556.

Burfoot MS, Campbell RD. (1994) Improved method of gene detection using exon amplification. *Nucleic Acids Res.* **22:** 5510–5511.

Burger R. (1992) C3 complement protein. In: *Human Protein Data* (ed. A. Haeberli). VCH, Weinheim.

Campbell RD. (1993) The human Major Histocompatibility Complex: a 4000kb segment of the human genome replete with genes. In: *Regional Physical Mapping* (eds KE Davies, SM Tilghman). Cold Spring Harbor Laboratory Press, Cold Spring Harbor, NY, pp. 1–33.

Campbell RD, Law SKA. (1992) C4 complement protein. In: *Human Protein Data* (ed. A. Haeberli). VCH, Weinheim.

Campbell RD, Milner CM. (1993) MHC genes in autoimmunity. *Curr. Opin. Immunol.* **5:** 887–893.

Campbell RD, Dunham I, Kendall E, Sargent CA. (1990) Polymorphism of the human complement component C4. *Exp. Clin. Immunogenet.* **7:** 69–84.

Carroll MC, Katzman P, Alicot EM, Koller BH, Geraghty DE, Orr HT, Strominger JL, Spies T. (1987) Linkage map of the human major histocompatibility complex including the tumor necrosis factor genes. *Proc. Natl Acad. Sci. USA* **84:** 8535–8539.

Carroll MC, Fathallah DM, Bergamaschini L, Alicot EM, Isenman DE. (1990) Substitution of a single amino acid (aspartic acid for histidine) converts the functional activity of human complement C4B to C4A. *Proc. Natl Acad. Sci. USA* **87:** 6868–6872.

Cascino I, Sorrentino R, Tosi R. (1993) Strong genetic association between HLA-DR3 and a polymorphic variation in the regulatory region of the HSP70–1 gene. *Immunogenetics* **37:** 177–182.

Chang CP, Shen WF, Rozenfeld S, Lawrence HJ, Largman C, Cleary ML. (1995) Pbx proteins display hexapeptide-dependent cooperative DNA binding with a subset of Hox proteins. *Genes Dev.* **9:** 663–674.

Chappell TG, Konforti BB, Schmid SL, Rothman JE. (1987) The ATPase core of a clathrin uncoating protein. *J. Biol. Chem.* **262:** 746–751.

Chiquet-Ehrismann R, Mackie EJ, Pearson CA, Sakakura T. (1986) Tenascin: an extracellular matrix protein involved in tissue interactions during fetal development and oncogenesis. *Cell* **47:** 131–139.

Cristau B, Schafer PH, Pierce SK. (1994) Heat shock enhances antigen processing and accelerates the formation of compact class II $\alpha\beta$ dimers. *J. Immunol.* **152:** 1546–1556.

Cross SJ, Thomson W. (1990) DNA polymorphism of the C2 and Factor B genes. *Exp. Clin. Immunogenet.* **7:** 53–63.

Crowe PD, VanArsdale TL, Walter BN, Ware CF, Hession C, Ehrenfels B, Browning JL, Din WS, Goodwin RG, Smith CA. (1994) A lymphotoxin-beta-specific receptor. *Science* **264:** 707–710.

Dangel AW, Mendoza AR, Baker BJ, Daniel CM, Carroll MC, Wu LC, Yu CY. (1994) The dichotomous size variation of human complement C4 genes is mediated by a novel family of endogenous retroviruses, which also establishes species-specific genomic patterns among Old World primates. *Immunogenetics* **40:** 425–436.

Davis LH, Otto E, Bennett V. (1991) Specific 33-residue repeat(s) of erythrocyte ankyrin associate with the anion exchanger. *J. Biol. Chem.* **266:** 11163–11169.

Deguchi Y, Kishimoto S. (1990) Elevated expression of heat-shock protein gene in the fibroblasts of patients with scleroderma. *Clin. Sci.* **78:** 419–422.

De-Togni P, Goellner J, Ruddle NH, *et al.* (1994) Abnormal development of peripheral lymphoid organs in mice deficient in lymphotoxin. *Science* **264:** 703–707.

Dodds AW, Law SK, Porter RR. (1985) The origin of the very variable haemolytic activities of the common human complement component C4 allotypes including C4–A6. *EMBO J.* **4:** 2239–2244.

Dressel R, Gunther E. (1994) A pentanucleotide tandem duplication polymorphism in the 3′ untranslated region of the HLA-linked heat-shock protein 70–2 (HSP70–2) gene. *Hum. Genet.* **94:** 585–586.

Dunham I, Sargent CA, Trowsdale J, Campbell RD. (1987) Molecular mapping of the human major histocompatibility complex by pulsed-field gel electrophoresis. *Proc. Natl Acad. Sci. USA* **84:** 7237–7241.

Ellisen LW, Bird J, West DC, Soreng AL, Reynolds TC, Smith SD, Sklar J. (1991) TAN-1, the human homolog of the Drosophila notch gene, is broken by chromosomal translocations in T lymphoblastic neoplasms. *Cell* **66:** 649–661.

Fan WF, Wei X, Shukla H, Parimoo S, Xu H, Sankhavaram P, Li Z, Weissman SM. (1993) Application of cDNA selection techniques to regions of the human MHC. *Genomics* **17:** 575–581.

Fehon RG, Kooh PJ, Rebay I, Regan CL, Xu T, Muskavitch MA, Artavanis-Tsakonas S. (1990) Molecular interactions between the protein products of the neurogenic loci Notch and Delta, two EGF-homologous genes in Drosophila. *Cell* **61:** 523–534.

Ferrari J, Harris R, Warner TG. (1994) Cloning and expression of a soluble sialidase from Chinese hamster ovary cells: sequence alignment similarities to bacterial sialidases. *Glycobiology* **4:** 367–373.

Figueroa F, Klein D, Tewarson S, Klein J. (1982) Evidence for placing the Neu-I locus within the mouse H-2 complex. *J. Immunol.* **129:** 2089–2093.

Flaherty KM, DeLuca-Flaherty C, McKay DB. (1990) Three-dimensional structure of the ATPase fragment of a 70K heat-shock cognate protein. *Nature* **346:** 623–628.

Flajnik MF, Canel C, Kramer J, Kasahara M. (1991) Which came first, MHC class I or class II. *Immunogenetics* **33:** 295–300.

Flegel WA, Singson AW, Margolis JS, Bang AG, Posakony JW, Murre C. (1993) Dpbx, a new homeobox gene closely related to the human proto-oncogene pbx1 molecular structure and developmental expression. *Mech. Dev.* **41:** 155–161.

Flynn GC, Chappell TG, Rothman JE. (1989) Peptide binding and release by proteins implicated as catalysts of protein assembly. *Science* **245:** 385–390.

Fortini ME, Artavanis-Tsakonas S. (1993) Notch: neurogenesis is only part of the picture. *Cell* **75:** 1245–1247.

French MAH, Dawkins RL. (1991) Central MHC genes, IgA deficiency and autoimmune disease. *Immunol. Today* **11:** 271–274.

Fugger L, Morling N, Ryder LP, Platz P, Georgsen J, Jakobsen BK, Svejgaard A, Dalhoff K, Ranek L. (1989) NcoI restriction fragment length polymorphism (RFLP) of the tumour necrosis factor (TNF alpha) region in primary biliary cirrhosis and in healthy Danes. *Scand. J. Immunol.* **30:** 185–189.

Fukagawa T, Sugaya K, Matsumoto K, Okumura K, Ando A, Inoko H, Ikemura T. (1995) A boundary of long-range G+C% mosaic domains in the human MHC locus: pseudoautosomal boundary-like sequence exists near the boundary. *Genomics* **25:** 184–191.

Furukawa F, Ikai K, Matsuyoshi N, Shimizu K, Imamura S. (1993) Relationship between heat shock protein induction and the binding of antibodies to the extractable nuclear antigens on cultured human keratinocytes. *J. Invest. Dermatol.* **101:** 191–195.

Fuss B, Wintergerst ES, Bartsch U, Schachner M. (1993) Molecular characterization and in situ mRNA localization of the neural recognition molecule J1–160/180: a modular structure similar to tenascin. *J. Cell. Biol.* **120:** 1237–1249.

Gehring WJ, Hiromi Y. (1986) Homeotic genes and the homeobox. *Annu. Rev. Genet.* **20:** 147–173.

Georgopoulos C, Welch WJ. (1993) Roles of the major heat shock proteins as chaperones. *Annu. Rev. Cell Biol.* **9:** 601–634.

Ghosh S, Baltimore D. (1990) Activation in vitro of NF-κB by phosphorylation of its inhibitor IκB. *Nature* **344:** 678–682.

Gilmore TD, Morin PJ. (1993) The I kappa B proteins: members of a multifunctional family. *Trends Genet.* **9:** 427–433.

Gitelman SE, Bristow J, Miller WL. (1992) Mechanism and consequences of the duplication of the human C4/P450c21/gene X locus. *Mol. Cell Biol.* **12:** 2124–2134.

Grumet M, Hoffman S, Crossin KL, Edelman GM. (1985) Cytotactin, an extracellular matrix protein of neural and non-neural tissues that mediates glia-neuron interaction. *Proc. Natl Acad. Sci. USA* **82:** 8075–8079.

Gunther E, Walter L. (1994) Genetic aspects of the hsp70 multigene family in vertebrates. *Experientia* **50:** 987–1001.

Guss H-J, Dower S. (1995) Tumor necrosis factor ligand superfamily: Involvement in the pathology of malignant lymphomas. *Blood* **85:** 3378–3404.

Hall TJ. (1994) The role of hsp70 in cytokine production. *Experientia* **50:** 1048–1053.

Harada F, Nishimura Y, Suzuki K, Matsumoto H, Oohira T, Matsuda I, Sasazuki T. (1987) The patient with combined deficiency of neuraminidase and 21–hydroxylase. *Hum. Genet.* **75:** 91–92.

Hauptmann G, Tappeiner G, Schifferli JA. (1993) Inherited deficiency of the fourth component of human complement. In: *Immunodeficiencies* (eds FS Rosen, M Seligmann). Harwood Academic Publishers, Chur, pp. 253–266.

Henkel T, Machleidt T, Alkalay I, Kronke M, Ben-Neriah Y, Baeuerle PA. (1993) Rapid proteolysis of I kappa B-alpha is necessary for activation of transcription factor NF-kappa B. *Nature* **365:** 182–185.

Higashi Y, Yoshioka H, Yamane M, Gotoh O, Fujii-Kuriyama Y. (1986) Complete nucleotide sequence of two steroid 21–hydroxylase genes tandemly arranged in human chromosome: a pseudogene and a genuine gene. *Proc. Natl Acad. Sci. USA* **83:** 2841–2845.

Hoyer LL, Hamilton AC, Steenbergen SM, Vimr ER. (1992) Cloning, sequencing and distribution of the *Salmonella typhimurium* LT2 sialidase gene, nanH, provides evidence for interspecies gene transfer. *Mol. Microbiol.* **6:** 873–884.

Hsieh SL, Campbell RD. (1991) Evidence that gene G7a in human major histocompatibility complex encodes valyl-tRNA synthetase. *Biochem. J.* **278:** 809–816.

Hunt C, Morimoto RI. (1985) Conserved features of eukaryotic HSP70 genes revealed by comparison with the nucleotide sequence of human HSP70. *Proc. Natl Acad. Sci. USA* **82:** 6455–6459.

Jaattela M. (1993) Overexpression of major heat shock protein hsp70 inhibits tumor necrosis factor-induced activation of phospholipase A2. *J. Immunol.* **151:** 4286–4294.

Jacquier-Sarlin MR, Fuller K, Dinh-Xuan AT, Richard MJ, Polla BS. (1994) Protective effects of hsp70 in inflammation. *Experientia* **50:** 1031–1038.

Jhappan C, Gallahan D, Stahle C, Chu E, Smith GH, Merlino G, Callahan R. (1992) Expression of an activated Notch-related int-3 transgene interferes with cell differentiation and induces neoplastic transformation in mammary and salivary glands. *Genes Dev.* **6:** 345–355.

Ji YH, Fujita T, Hatsuse H, Takahashi A, Matsushita M, Kawakami M. (1993) Activation of the C4 and C2 components of complement by a proteinase in serum bactericidal factor, Ra reactive factor. *J. Immunol.* **150:** 571–578.

Johnson CA, Densen P, Hurford R Jr, Colten HR, Wetsel RA. (1992) Type I human complement C2 deficiency. A 28-base pair gene deletion causes skipping of exon 6 during RNA splicing. *J. Biol. Chem.* **267:** 9347–9353.

Jones RS, Gelbart WM. (1993) The Drosophila polycomb-group Enhancer of zeste contains a region of sequence similarity to trithorax. *Mol. Cell Biol.* **13:** 6357–6366.

Jongeneel CV, Briant L, Udalova IA, Sevin A, Nedospasov SA, Cambon-Thomsen A. (1991)

Extensive genetic polymorphism in the human tumour necrosis factor region and relation to extended HLA haplotypes. *Proc. Natl Acad. Sci. USA* **88:** 9717–9721.

Just M, Herbst H, Hummel M, Durkop H, Tripier D, Stein H, Schuppan D. (1991) Undulin is a novel member of the fibronectin-tenascin family of extracellular matrix glycoproteins. *J. Biol. Chem.* **266:** 17326–17332.

Kagawa N, Ogo A, Takahashi Y, Iwamatsu A, Waterman MR. (1994) A cAMP-regulatory sequence (CRS1) of CYP17 is a cellular target for the homeodomain protein Pbx1. *J. Biol. Chem.* **269:** 18716–18719.

Kamps MP, Murre C, Sun XH, Baltimore D. (1990) A new homeobox gene contributes the DNA binding domain of the t(1;19) translocation protein in pre-B ALL. *Cell* **60:** 547–555.

Kendall E, Sargent CA, Campbell RD. (1990) Human major histocompatibility complex contains a new cluster of genes between the HLA-D and complement C4 loci. *Nucleic Acids Res.* **18:** 7251–7257.

Khanna A, Campbell RD. (1993) Characterisation of a novel gene, G13, in the class III region of the human MHC. In *Proceedings of the 11th International Histocompatibility Workshop* (eds K Tsuji, M Aizawa, T Sasazuki). Oxford University Press, Oxford, vol. 2, pp. 198–201.

Kirstein M, Brett J, Radoff S, Ogawa S, Stern D, Vlassara H. (1990) Advanced protein glycosylation induces transendothelial human monocyte chemotaxis and secretion of platelet-derived growth factor: role in vascular disease of diabetes and aging. *Proc. Natl Acad. Sci. USA* **87:** 9010–9014.

Kissinger CR, Liu BS, Martin-Blanco E, Kornberg TB, Pabo CO. (1990) Crystal structure of an engrailed homeodomain-DNA complex at 2.8 A resolution: a framework for understanding homeodomain-DNA interactions. *Cell* **63:** 579–590.

Klein D, Klein J. (1982) Polymorphism of the Apl (Neu-I) locus in the mouse. *Immunogenetics* **16:** 181–184.

Lafuse WP, Lanning D, Spies T, David CS. (1992) PFGE mapping and RFLP analysis of the S/D region of the mouse H-2 complex. *Immunogenetics* **36:** 110–120.

LaMarco K, Thompson CC, Byers BP, Walto EM, McKnight SL. (1991) Identification of Ets- and Notch-related subunits in GA binding protein. *Science* **253:** 789–792.

Landolfi NF, Cook RG. (1986) Activated T lymphocytes express class I molecules which are hyposialylated compared to other lymphocyte populations. *Mol. Immunol.* **23:** 297–309.

Landolfi NF, Leone J, Womack JE, Cook R. (1985) Activation of T lymphocytes results in an increase of H-2-encoded neuraminidase. *Immunogenetics* **22:** 159–167.

Law SKA, Reid KBM. (1995) *Complement* (ed. D Male). IRL Press, Oxford, pp. 1–84.

Lawton P, Nelson J, Tizard R, Browning JL. (1995) Characterization of the mouse lymphotoxin-beta gene. *J. Immunol.* **154:** 239–246.

Leahy DJ, Hendrickson WA, Aukhil I, Erickson HP. (1992) Structure of a fibronectin type III domain from tenascin phased by MAD analysis of the selenomethionyl protein. *Science* **258:** 987–991.

Leelayuwat C, Townend DC, Degli-Esposti MA, Abraham LJ, Dawkins RL. (1994) A new polymorphic and multicopy MHC gene family related to nonmammalian class I. *Immunogenetics* **40:** 339–351.

Levi-Strauss M, Carroll MC, Steinmetz M, Meo T. (1988) A previously undetected MHC gene with an unusual periodic structure. *Science* **240:** 201–204.

Liou HC, Baltimore D. (1993) Regulation of the NK-kB/rel transcription factor and IkB inhibitor system. *Curr. Opin. Cell Biol.* **5:** 477–487.

Loos M, Reid KBM, Colomb M. (1993) Structure-function-relationship of C1q and collectins, C1-esterases and C1-inhibitor in health and disease. *Behring Institute Mitteilungen* **93:** 1–328.

Lowden JA, O'Brien JS. (1979) Sialidosis: A review of human neuraminidasé deficiency. *Am. J. Hum. Genet.* **31:** 1–18.

Mariethoz E, Tacchini-Cottier F, Jacquier-Sarlin M, Sinclair F, Polla BS. (1994) Exposure of monocytes to heat shock does not increase class II expression but modulates antigen-dependent T cell responses. *Int. Immunol.* **6:** 925–930.

Marshall B, Leelayuwat C, Degli-Esposti MA, Pinelli M, Abraham LJ, Dawkins RL. (1993) New major histocompatibility complex genes. *Hum. Immunol.* **38:** 24–29.

Matsumoto K, Ishihara N, Ando A, Inoko H, Ikemura T. (1992) Extracellular matrix protein tenascin-like gene found in human MHC class III region. *Immunogenetics* **36:** 400–403.

Matsumoto K, Saga Y, Ikemura T, Sakakura T, Chiquet-Ehrismann R. (1994) The distribution of tenascin-X is distinct and often reciprocal to that of tenascin-C. *J. Cell. Biol.* **125:** 483–493.

Matsumoto M, Kurata S, Fujimoto H, Hoshi M. (1993) Haploid-specific applications of protemine 1 and hsc70t genes in mouse spermatogenesis. *Biochem. Biophys. Acta.* **1174:** 274–278.

Mauff G, Alper CA, Dawkins R, Doxiadis G, Giles CM, Hauptmann G, Rittner C, Schneider PM. (1990) C4 nomenclature statement (1990). *Complement Inflamm.* **7:** 261–268.

Mazo AM, Huang DH, Mozer BA, David IB. (1990) The trithorax gene, a trans-acting regulator of the bithorax complex in Drosophila, encodes a protein with zinc-binding domains. *Proc. Natl Acad. Sci. USA* **87:** 2112–2116.

Mellits KH, Hay RT, Goodbourn S. (1993) Proteolytic degradation of MAD3 (I kappa B alpha) and enhanced processing of the NF-kappa B precursor p105 are obligatory steps in the activation of NF-kappa B. *Nucleic Acids Res.* **21:** 5059–5066.

Messer G, Spengler U, Jung MC, Honold G, Blomer K, Pape GR, Riethmuller G, Weiss EH. (1991) Polymorphic structure of the tumor necrosis factor (TNF) locus: an NcoI polymorphism in the first intron of the human TNF-beta gene correlates with a variant amino acid in position 26 and a reduced level of TNF-beta production. *J. Exp. Med.* **173:** 209–219.

Messer G, Kick G, Ranki A, Koskimies S, Reunala T, Meurer M. (1994) Polymorphism of the tumor necrosis factor genes in patients with dermatitis herpetiformis. *Dermatology* **189:** 135–137.

Michaely P, Bennett V. (1992) The ANK repeat: a ubiquitous motif involved in macromolecular recognition. *Trends Cell. Biol.* **2:** 127–129.

Miller WL. (1988) Molecular biology of steroid hormone synthesis. *Endocr. Rev.* **9:** 295–318.

Milner CM. (1991) Characterisation of novel genes in the human major histocompatibility complex: the HSP70 and G9a genes. D.Phil Thesis, Oxford University, UK.

Milner CM, Campbell RD. (1990) Structure and expression of the three MHC-linked human HSP70 genes. *Immunogenetics* **32:** 242–251.

Milner CM, Campbell RD. (1992) Polymorphic analysis of the three MHC-linked HSP70 genes. *Immunogenetics* **36:** 357–362.

Milner CM, Campbell RD. (1993) The G9a gene in the human Major Histocompatibility Complex encodes a novel protein containing ankyrin-like repeats. *Biochem. J.* **290:** 811–818.

Milner CM, Smith SV, Carrillo MB, Taylor GL, Campbell RD. (1996) Characterisation of a sialidase encoded in the human Major Histocompatibility Complex. *J. Biol. Chem.* (submitted).

Miyagi T, Tsuiki S. (1984) Rat liver lysosomal sialidase. Solubilization, substrate specificity and comparison with the cytosolic sialidase. *Eur. J. Biochem.* **141:** 75–81.

Miyagi T, Sagawa J, Konno K, Tsuiki S. (1990) Immunological discrimination of intralysosomal, cytosolic and two membrane sialidases present in rat tissues. *J. Biochem.* **107:** 794–798.

Miyagi T, Konno K, Emori Y, Kawasaki H, Suzuki K, Yasui A, Tsuiki S. (1993) Molecular cloning and expression of cDNA encoding rat skeletal muscle cytosolic sialidase. *J. Biol. Chem.* **268:** 26435–26440.

Mizuki N, Ohno S, Sato T, Ishihara M, Miyata S, Nakamura S, Naruse T, Mizuki H, Tsuji K, Inoko H. (1995) Microsatellite polymorphism between the tumor necrosis factor and HLA-B genes in Behcet's disease. *Hum. Immunol.* **43:** 129–135.

Monica K, Galili N, Nourse J, Saltman D, Cleary ML. (1991) PBX2 and PBX3, new homeobox genes with extensive homology to the human proto-oncogene PBX1. *Mol. Cell Biol.* **11:** 6149–6157.

Monica K, LeBrun DP, Dedera DA, Brown R, Cleary ML. (1994) Transformation properties of the E2a-Pbx1 chimeric oncoprotein: fusion with E2a is essential, but the Pbx1 homeodomain is dispensable. *Mol. Cell. Biol.* **14:** 8304–8314.

Morel Y, Miller WL. (1991) Clinical and molecular genetics of congenital adrenal hyperplasia due to 21-hydroxylase deficiency. *Adv. Hum. Genet.* **20:** 1–68.

Morel Y, Bristow J, Gitelman SE, Miller WL. (1989) Transcript encoded on the opposite strand of the human steroid 21-hydroxylase/complement component C4 gene locus. *Proc. Natl Acad. Sci. USA* **86:** 6582–6586.

Morimoto RI, Jurivich DA, Kroeger PE, Mathur SK, Murphy SP, Nakai A, Sarge K, Abravaya K, Sistonen LT. (1994) Regulation of heat shock gene transcription by a family of heat shock factors. In: *The Biology of Heat Shock Proteins and Molecular Chaperones* (eds RI Morimoto, A Tissieres, C Georgopoulos). Cold Spring Harbor Laboratory Press, Cold Spring Harbor, pp. 417–455.

Morshauser RC, Wang H, Flynn GC, Zuiderweg ERP. (1995) The peptide-binding domain of the chaperone protein hsc70 has an unusual secondary structure topology. *Biochemistry* **34:** 6261–6266.

Moseley PL, Wallen ES, McCafferty JD, Flanagan S, Kern JA. (1993) Heat stress regulates the human 70 kDa heat shock gene through the 3′ untranslated region. *Am. J. Physiol.* **264:** L533–L537.

Mueller OT, Henry WM, Haley LL, Byers MG, Eddy RL, Shows TB. (1986) Sialidosis and galactosialidosis: Chromosomal assignment of two genes associated with neuraminidase-deficiency disorders. *Proc. Natl Acad. Sci. USA* **83:** 1817–1821.

Nedospasov SA, Udalova IA, Kuprash DV, Turetskaya RL. (1991) DNA sequence polymorphism at the human tumor necrosis factor (TNF) locus. Numerous TNF/lymphotoxin alleles tagged by two closely linked microsatellites in the upstream region of the lymphotoxin (TNF-beta) gene. *J. Immunol.* **147:** 1053–1059.

Neeper M, Schmidt AM, Brett J, Yan SD, Wang F, Pan YC, Elliston K, Stern D, Shaw A. (1992) Cloning and expression of a cell surface receptor for advanced glycosylation end products of proteins. *J. Biol. Chem.* **267:** 14998–15004.

Nolan GP, Baltimore D. (1992) The inhibitory ankyrin and activator Rel proteins. *Curr. Opin. Genet. Dev.* **2:** 211–220.

Norenberg U, Wille H, Wolff JM, Frank R, Rathjen FG. (1992) The chicken neural extracellular matrix molecule restrictin: similarity with EGF-, fibronectin type III-, and fibrinogen-like motifs. *Neuron* **8:** 849–863.

Nourse J, Mellentin JD, Galili N, Wilkinson J, Stanbridge E, Smith SD, Cleary ML. (1990) Chromosomal translocation t(1;19) results in synthesis of a homeobox fusion mRNA that codes for a potential chimeric transcription factor. *Cell* **60:** 535–545.

Olavesen MG, Thomson W, Cheng J, Campbell RD. (1993) Characterisation of a novel gene (G1) in the class III region of the human MHC. In: *Proceedings of the 11th International Histocompatibility Workshop* (eds K Tsuji, M Aizawa, T. Sasazuki). Oxford University Press, Oxford, vol. 2, pp. 190–193.

Palleros DR, Reid KL, Shi L, Welch WJ, Fink AL. (1993) ATP-induced protein-hsp70 complex dissociation requires K+ but not ATP hydrolysis. *Nature* **365:** 664–666.

Palleros DR, Shi L, Reid KL, Fink AL. (1994) Hsp70–protein complexes. Complex stability and conformation of bound substrate protein. *J. Biol. Chem.* **269:** 13107–13114.

Pangburn MK. (1992) Factor B. In: *Human Protein Data* (ed. A. Haeberli). VCH, Weinheim.

Partanen J, Campbell RD. (1993) Characterisation of the G8 gene, a novel MHC class III gene. In: *Proceedings of the 11th International Histocompatibility Workshop* (eds K Tsuji, M Aizawa, T. Sasazuki). Oxford University Press, Oxford, vol. 2, pp. 196–198.

Partanen J, Koskimies S. (1988) Low degree of DNA polymorphism in the HLA-linked lymphotoxin (tumour necrosis factor beta) gene. *Scand. J. Immunol.* **28:** 313–316.

Partanen J, Milner CM, Campbell RD, Maki M, Lipsanen V, Koskimies S. (1993) HLA-linked heat shock protein 70 (HSP70-2) gene polymorphism and celiac disease. *Tissue Antigens* **41:** 15–19.

Pascual M, Schifferli JA. (1992) Factor D. In: *Human Protein Data* (ed. A Haeberli). VCH, Weinheim.

Paul NL, Ruddle NH. (1988) Lymphotoxin. *Annu. Rev. Immunol.* **6:** 407–438.

Peelman LJ, Chardon P, Nunes M, *et al.* (1995) The BAT1 gene in the MHC encodes an evolutionarily conserved putative nuclear RNA helicase of the DEAD family. *Genomics* **26:** 210–218.

Pierce SK. (1994) Molecular chaperones in the processing and presentation of antigen to helper T cells. *Experientia* **50:** 1026–1030.

Pociot F, Ronningen KS, Nerup J. (1993a) Polymorphic analysis of the human MHC-linked heat shock protein 70 (Hsp70–2) and Hsp70–Hom genes in insulin-dependent Diabetes Mellitus (IDDM). *Scand. J. Immunol.* **38:** 491–495.

Pociot F, Briant L, Jongeneel CV, Molvig J, Worsaae H, Abbal M, Thomsen M, Nerup J, Cambon-Thomsen A. (1993b) Association of tumor necrosis factor (TNF) and class II major histocompatibility complex alleles with the secretion of TNF-alpha and TNF-beta by human mononuclear cells: a possible link to insulin-dependent diabetes mellitus. *Eur. J. Immunol.* **23:** 224–231.

Prieto AL, Andersson-Fisone C, Crossin KL. (1992) Characterization of multiple adhesive and counteradhesive domains in the extracellular matrix protein cytotactin. *J. Cell. Biol.* **119:** 663–678.

Reid KBM. (1992) Properdin. In: *Human Protein Data* (ed. A Haeberli). VCH, Weinheim.

Rippmann F, Taylor WR, Rothbard JB, Green NM. (1991) A hypothetical model for the peptide-binding domain of hsp70 based on the peptide binding domain of HLA. *EMBO J.* **10:** 1053–1059.

Robbins J, Blondel BJ, Gallahan D, Callahan R. (1992) Mouse mammary tumour gene int-3: a member of the notch gene family transforms mammary epithelial cells. *J. Virol.* **66:** 2594–2599.

Rodrigues NR, Dunham I, Yu CY, Carroll MC, Porter RR, Campbell RD. (1987) Molecular characterization of the HLA-linked steroid 21–hydroxylase B gene from an individual with congenital adrenal hyperplasia. *EMBO J.* **6:** 1653–1661.

Rodriguez MS, Michalopoulos I, Arenzana-Seisdedos F, Hay RT. (1995) Inducible degradation of I kappa B alpha in vitro and in vivo requires the acidic C-terminal domain of the protein. *Mol. Cell Biol.* **15:** 2413–2419.

Roggentin P, Rothe B, Kaper KB, Galen J, Lawrisuk L, Vimr ER, Schauer R. (1989) Conserved sequences in bacterial and viral sialidases. *Glycoconj. J.* **6:** 349–353.

Ruddle NH. (1992) Tumor necrosis factor (TNF-alpha) and lymphotoxin (TNF-beta). *Curr. Opin. Immunol.* **4:** 327–332.

Ruddle NH, Homer R. (1988) The role of lymphotoxin in inflammation. *Prog. Allergy.* **40:** 162–182.

Rüegg CR, Chiquet-Ehrismann R, Alkan SS. (1989) Tenascin, an extracellular matrix protein, exerts immunomodulatory activities. *Proc. Natl Acad. Sci. USA* **86:** 7437–7441.

Sargent CA, Dunham I, Campbell RD. (1989a) Identification of multiple HTF-island associated genes in the human major histocompatibility complex class III region. *EMBO J.* **8:** 2305–2312.

Sargent CA, Dunham I, Trowsdale J, Campbell RD. (1989b) Human major histocompatibility complex contains genes for the major heat shock protein HSP70. *Proc. Natl Acad. Sci. USA* **86:** 1968–1972.

Sargent CA, Anderson MJ, Hsieh SL, Kendall E, Gomez-Escobar N, Campbell RD. (1994) Characterisation of the novel gene G11 lying adjacent to the complement C4A gene in the human major histocompatibility complex. *Hum. Mol. Genet.* **3:** 481–488.

Schenk S, Chiquet-Ehrismann R. (1994) Tenascins. *Methds. Enzymol.* **245:** 52–61.

Schmid SR, Linder P. (1992) D-E-A-D protein family of putative RNA helicases. *Mol. Microbiol.* **6:** 283–292.

Schmidt, JA., Abdulla E. (1988). Down-regulation of IL-1β biosynthesis by inducers of the heat-shock response. *J. Immunol.* **141:** 2027–2034.

Schmidt AM, Vianna M, Gerlach M, *et al.* (1992) Isolation and characterization of two binding proteins for advanced glycosylation end products from bovine lung which are present on the endothelial cell surface. *J. Biol. Chem.* **267:** 14987–14997.

Schneider PM, Carroll MC, Alper CA, Rittner C, Whitehead AS, Yunis EJ, Colten HR. (1986) Polymorphism of the human complement C4 and steroid 21–hydroxylase genes. Restriction fragment length polymorphisms revealing structural deletions, homoduplications, and size variants. *J. Clin. Invest.* **78:** 650–657.

Scott MP, Tamkun JW, Hartzell III GW. (1989) The structure and function of the homeodomain. *Biochim. Biophys. Acta* **989:** 25–48.

Sepp A, Dodds AW, Anderson MJ, Campbell RD, Willis AC, Law SK. (1993) Covalent binding properties of the human complement protein C4 and hydrolysis rate of the internal thioester upon activation. *Protein Sci.* **2:** 706–716.

Shen L, Wu LC, Sanlioglu S, Chen R, Mendoza AR, Dangel AW, Carroll MC, Zipf WB, Yu CY. (1994) Structure and genetics of the partially duplicated gene RP located immediately upstream of the complement C4A and the C4B genes in the HLA class III region. Molecular cloning, exon–intron structure, composite retroposon, and breakpoint of gene duplication. *J. Biol. Chem.* **269:** 8466–8476.

Shimmoto M, Fukuhara Y, Itoh K, Oshima A, Sakuraba H, Suzuki Y. (1993) Protective protein gene mutations in galactosialidosis. *J. Clin. Invest.* **91:** 2393–2398.

Shimura T, Hagihara M, Takebe K, Munkhbat B, Odaka T, Kato H, Nagamachi Y, Tsuji K. (1994) The study of tumor necrosis factor beta gene polymorphism in lung cancer patients. *Cancer* **73:** 1184–1188.

Shimura T, Hagihara M, Takebe K, Munkhbat B, Ogoshi K, Mitomi T, Nagamachi Y, Tsuji K. (1995) 10.5-kb Homozygote of tumor necrosis factor-beta gene is associated with a better prognosis in gastric cancer patients. *Cancer* **75:**1450–1453.

Siracusa LD, Rosner MH, Vigano MA, Gilbert DJ, Staudt LM, Copeland NG, Jenkins NA. (1991) Chromosomal location of the octamer transcription factors, Otf-1, Otf-2, and Otf-3, defines multiple Otf-3-related sequences dispersed in the mouse genome. *Genomics* **10:** 313–326.

Snoek M, Jansen M, Olavesen MG, Campbell RD, Tuescher C, van Vugt H. (1993) Three hsp70 genes are located in the C4–H-2D region: possible candidates for the Orch-1 locus. *Genomics* **15:** 350–356.

Snoek M, Olavesen MG, Van Vugt H, Milner CM, Tuescher C, Campbell RD. (1994) Coding sequences and levels of expression of hsc70t are identical in mice with different Orch-1 alleles. *Immunogenetics* **40:** 159–162.

Snyder YM, Guthrie L, Evans GF, Zuckerman SH. (1992) Transcriptional inhibition of endotoxin-induced monokine synthesis following heat shock in murine peritoneal macrophages. *J. Leuk. Biol.* **51:** 181–187.

Speiser PW, White PC. (1989) Structure of the human RD gene: a highly conserved gene in the class III region of the major histocompatibility complex. *DNA* **8:** 745–751.

Spies T, Bresnahan M, Strominger JL. (1989) Human major histocompatibility complex contains a minimum of 19 genes between the complement cluster and HLA-B. *Proc. Natl Acad. Sci. USA* **86:** 8955–8958.

Spring J, Beck K, Chiquet-Ehrismann R. (1989) Two contrary functions of tenascin: dissection of the active sites by recombinant tenascin fragments. *Cell* **59:** 325–334.

Stifani S, Blaumueller CM, Redhead NJ, Hill RE, Artavanis-Tsakonas S. (1992) Human homologs of a Drosophila Enhancer of split gene product define a novel family of nuclear proteins. *Nature Genetics* **2:** 119–127.

Sugaya K, Fukagawa T, Matsumoto K, Mita K, Takahashi T-I, Ando A, Inoko H, Ikemura T. (1994) Three genes in the human MHC class III region near the junction with the class II: gene for receptor of advanced glycosylation end products, PBX2 homeobox gene and a notch homolog, human counterpart of mouse mammary tumor gene int-3. *Genomics* **23:** 408–419.

Sun SC, Ganchi PA, Ballard DW, Greene WC. (1993) NF-kappa B controls expression of inhibitor I kappa B alpha: evidence for an inducible autoregulatory pathway. *Science* **259:** 1912–1915.

Taira S, Nariuchi H. (1988) Possible roles of neuraminidase in activated T cells in the recognition of allogeneic Ia. *J. Immunol.* **141:** 440–446.

Takada F, Takayama Y, Hatsuse H, Kawakami M. (1993) A new member of the C1s family of complement proteins found in a bactericidal factor, Ra-reactive factor, in human serum. *Biochem. Biophys. Res. Commun.* **196:** 1003–1009.

Tan SS, Prieto AL, Newgreen DF, Crossin KL, Edelman GM. (1991) Cytotactin expression in somites after dorsal neural tube and neural crest ablation in chicken embryos. *Proc. Natl Acad. Sci. USA* **88:** 6398–6402.

Tassabehji M, Strachan T, Anderson M, Campbell RD, Collier S, Lako M. (1994) Identification of a novel family of human endogenous retroviruses and characterization of one family member, HERV-K(C4), located in the complement C4 gene cluster. *Nucleic Acids Res.* **22:** 5211–5217.

Tee MK, Thomson AA, Bristow J, Miller W. (1995) Sequences promoting the transcription of the human XA gene overlapping P450c21A correctly predict the presence of a novel, adrenal-specific, truncated form of Tenascin-X. *Genomics* **28:** 171–178.

Terlecky SR. (1994) Hsp70s and lysosomal proteolysis. *Experientia* **50:** 1021–1025.

Thanos D, Maniatis T. (1995) NF-kappa B: a lesson in family values. *Cell* **80:** 529–532.

Tkachuk DC, Kohler S, Cleary ML. (1992) Involvement of a homolog of Drosophila trithorax by 11q23 chromosomal translocations in acute leukemias. *Cell* **71:** 691–700.

Tsuge I, Shen FW, Steinmetz M, Boyse EA. (1987) A gene in the H-2S:H-2D interval of the major histocompatibility complex which is transcribed in B cells and macrophages. *Immunogenetics* **26:** 378–380.

Twomey BM, Amni V, Isenberg DA, Latchman DS. (1993) Elevated levels of the 70 kD heat shock protein in patients with systemic lupus erythematosus are not dependent on enhanced transcription of the hsp70 gene. *Lupus* **2:** 297–301.

Uberbacher EC, Mural RJ. (1991) Locating protein-coding regions in human DNA sequences by a multiple sensor-neural network approach. *Proc. Natl Acad. Sci. USA* **88:** 11261–11265.

Udono H, Srivastava PK. (1993) Heat shock protein 70–associated peptides elicit specific cancer immunity. *J. Exp. Med.* **178:** 1391–1396.

Vendrell J, Gutierrez C, Pastor R, Richart C. (1995) A tumor necrosis factor-beta polymorphism associated with hypertriglyceridemia in non-insulin-dependent diabetes mellitus. *Metab. Clin. Exp.* **44:** 691–694.

Verheijen FW, Palmeri S, Hoogeveen AT, Galjaard H. (1985) Human placental neuraminidase. Activation, stabilisation and association with β-galactosidase and its protective protein. *Eur. J. Biochem.* **149:** 315–321.

Vernet C, Ribouchon M-Tm, Chimini G, Jouanolle AM, Sidibe I, Pontarotti P. (1993) A novel coding sequence belonging to a new multicopy gene family within the human MHC class I region. *Immunogenetics* **38:** 47–53.

Vilata A, Donovan D, Wood L. Vogeli G, Yang DCH. (1993) Cloning, sequencing and expression of a cDNA encoding mammalian valyl-tRNA synthetase. *Gene* **123:** 181–186.

Vlassara H, Brownlee M, Manogue KR, Dinarello CA, Pasagian A. (1988) Cachectin/TNF and IL-1 induced by glucose-modified proteins: role in normal tissue remodeling. *Science* **240:** 1546–1548.

Volanakis JE, Zhu Z, Schaffer FM, Macon KJ, Paleros J, Barger BO, Go R, Campbell RD,Schroeder HW, Cooper MD. (1992) Major histocompatibility complex class III genes and susceptibility to immunoglobulin A deficiency and common variable immunodeficiency. *J. Clin.*

Invest. **89:** 1914–1922.

Wang TF, Chang J, Wang C. (1993) Identification of the peptide-binding domain for hsc70. *J. Biol. Chem.* **35:** 26049–26051.

Webb GC, Chaplin DD. (1990) Genetic variability at the human tumor necrosis factor loci. *J. Immunol.* **145:** 1278–1285.

Welch WJ, Feramisco JR. (1984) Nuclear and nucleolar localization of the 72,000-dalton heat shock protein in heat-shocked mammalian cells. *J. Biol. Chem.* **259:** 4501–4513.

Weller A, Beck S, Ekblom P. (1991) Amino acid sequence of mouse tenascin and differential expression of two tenascin isoforms during embryogenesis. *J. Cell. Biol.* **112:** 355–362.

Wetsel RA. (1992) C5 complement protein. In: *Human Protein Data* (ed. A Haeberli). VCH, Weinheim.

Whaley K, Loos M, Weiler JM. (eds) (1993) *Complement in Health and Disease.* Kluwer Academic Publishers, Dordrecht, pp. 1–375.

Wharton KA, Jahansen KM, Xu T, Artavanis-Tsakonas S. (1985) Nucleotide sequence from the neurogenic locus Notch implies a gene product that shares homology with proteins containing EGF-like repeats. *Cell* **43:** 567–581.

White PC, New MI, Dupont B. (1986) Structure of human steroid 21–hydroxylase genes. *Proc. Natl Acad. Sci. USA* **83:** 5111–5115.

Wilson AG, de-Vries N, Pociot F, di-Giovine FS, van-der-Putte LB, Duff GW. (1993) An allelic polymorphism within the human tumor necrosis factor alpha promoter region is strongly associated with HLA A1, B8, and DR3 alleles. *J. Exp. Med.* **177:** 557–560.

Winfield J, Jarjour W. (1991) Do stress proteins play a role in arthritis and autoimmunity? *Immun. Rev.* **121:** 193–220.

Wu LC, Morley BJ, Campbell RD. (1987) Cell-specific expression of the human complement protein factor B gene: evidence for the role of two distinct 5′-flanking elements. *Cell* **48:** 331–342.

Yoshino I, Goedegebuure PS, Peoples GE, Lee KY, Eberlein TJ. (1994) Human tumor-infiltrating $CD4^+$ T cells react to B cell lines expressing heat shock protein 70. *J. Immunol.* **153:** 4149–4158.

Young RA. (1990) Stress proteins and immunology. *Annu. Rev. Immunol.* **8:** 401–420.

Yu CY. (1991) The complete exon-intron structure of a human complement component C4A gene. DNA sequences, polymorphism, and linkage to the 21-hydroxylase gene. *J. Immunol.* **146:** 1057–1066.

Yu CY, Belt KT, Giles CM, Campbell RD, Porter RR. (1986) Structural basis of the polymorphism of human complement components C4A and C4B: gene size, reactivity and antigenicity. *EMBO J.* **5:** 2873–2881.

4

HLA class I genes: structure and diversity

R. Spencer Wells and Peter Parham

4.1 Introduction

Molecular polymorphism is a dominant characteristic of the vertebrate immune system. Through specific interactions with pathogenic organisms, the immunological machinery recognizes these invaders and successfully fights off infection. The mechanisms by which polymorphism is generated and maintained differ depending on the system involved. While immunoglobulin and T-cell receptor variation is generated within an individual through somatic mechanisms of recombination and mutation, the major histocompatibility complex (MHC) locus exhibits allelic variation between individuals at the population level. Each individual therefore expresses a maximum of two alleles at any given MHC locus, and the specificity of interactions between antigen-presenting cells (APC) and T-lymphocytes or natural killer (NK) cells is determined by their products (allotypes).

The human MHC is found in a 4-megabase region on the short arm of chromosome 6. Three groups of MHC loci have been described: class I, class II and class III. The class I loci were first detected in the late 1950s as a result of blood transfusion studies (Dausset, 1959). Antisera isolated from multiply transfused patients were capable of leukoagglutinating blood from unrelated individuals, demonstrating that these antigens were segregating at the population level. Later studies revealed that these so-called human leukocyte antigen (HLA) molecules were encoded by multiple loci, with the first three to be identified assigned the letters *A*, *B* and *C*. These three loci are referred to as the 'classical' class I (or class Ia) loci, to differentiate them from the 'non-classical' (class Ib) loci (*HLA-E*, *-F* and *-G*) found in the MHC class I region of chromosome 6. A map of the class I region is shown in *Figure 4.1*. The *MICB* locus was arbitrarily used as the centromeric end of the class I region, while *HLA-F* represents its telomeric terminus.

4.2 Classical class I diversity

The three HLA 'classical' class I molecules, HLA-A, -B and -C, are expressed on nearly every cell type in the body. These molecules present antigens of intracellular origin, in contrast to class II loci, which present extracellular antigens (Townsend and Bodmer, 1989; Unanue and Allen, 1987). *Figure 4.2* shows the top of an HLA class I molecule, with the α_1 and α_2 domains forming the peptide binding groove.

HLA and MHC: genes, molecules and function, edited by M.J. Browning and A.J. McMichael.
© 1996 BIOS Scientific Publishers Ltd, Oxford.

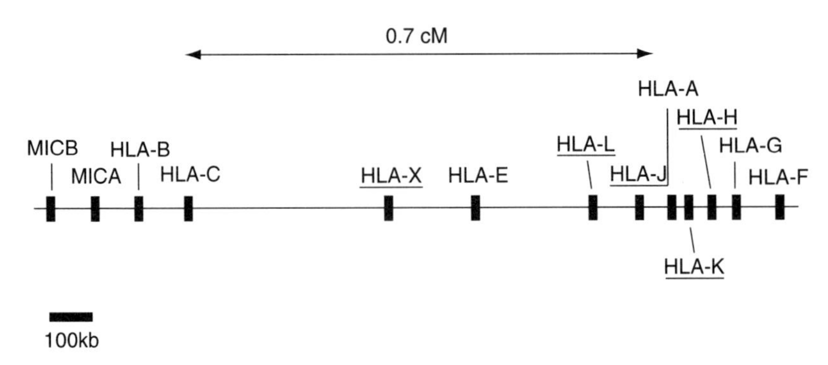

Figure 4.1. A map of the class I region of the human major histocompatibility complex (MHC). The major loci in the region are shown, including the *MICA* and *MICB* genes, which constitute a family of extremely divergent HLA class I-like loci (Bahram *et al.*, 1994; Leelayuwat *et al.*, 1994). *HLA-H*, *-J*, *-K*, *-L* and *-X* are class I pseudogenes; these are underlined. [This map is based on a more extensive one compiled by Campbell and Trowsdale (1993). The rate of recombination between *HLA-C* and *HLA-A* was taken from Thomsen *et al.* (1989).]

Most of the variation in classical class I loci is found in this region, where it can directly affect the types of peptides bound by the molecule, as well as the interactions with T-cell receptors. HLA diversity has been studied primarily by using allo-antisera specific to particular allelic gene products, making use of the fact that the exposed regions of the class I molecules are the most variable. With these methods, 24 HLA-A, 49 HLA-B and 8 HLA-C antigens have been identified (Juji *et al.*, 1992), with further resolution possible within some antigenic specificities. Recent work, however, has been aimed at sequence-based methods of HLA typing, with the knowledge that DNA sequences provide the ultimate in allelic discrimination (see Chapter 6). Without having to rely on the availability of specific antisera, we are now able to determine an individual's HLA type quickly and reliably at the sequence level. To date (Bodmer *et al.*, 1995), nucleotide sequences of 59 *HLA-A*, 118 *HLA-B* and 36 *HLA-C* alleles have been determined. As new alleles are routinely found in population screenings, these numbers will certainly increase in the future.

4.2.1 Alleles at the HLA-A, -B *and* -C *loci*

Two methods can be used to show relationships among alleles, a difference matrix and a graphical tree of the alleles. While matrices are useful for studying the details of allelic similarity, it is often difficult to see the broader pattern of relationships using this method. For this reason, phylogenetic methods are used to draw trees showing the inferred evolutionary history of the alleles. There is a complicating factor, however: because of interlineage exchange (recombination and gene conversion), phylogenetic trees of allelic sequences do not necessarily represent the actual evolutionary relationships among the alleles. In spite of this, they are a useful heuristic tool for discussing patterns of variation and degrees of similarity. Using phylogenetic methods, we are able to define allelic lineages representing

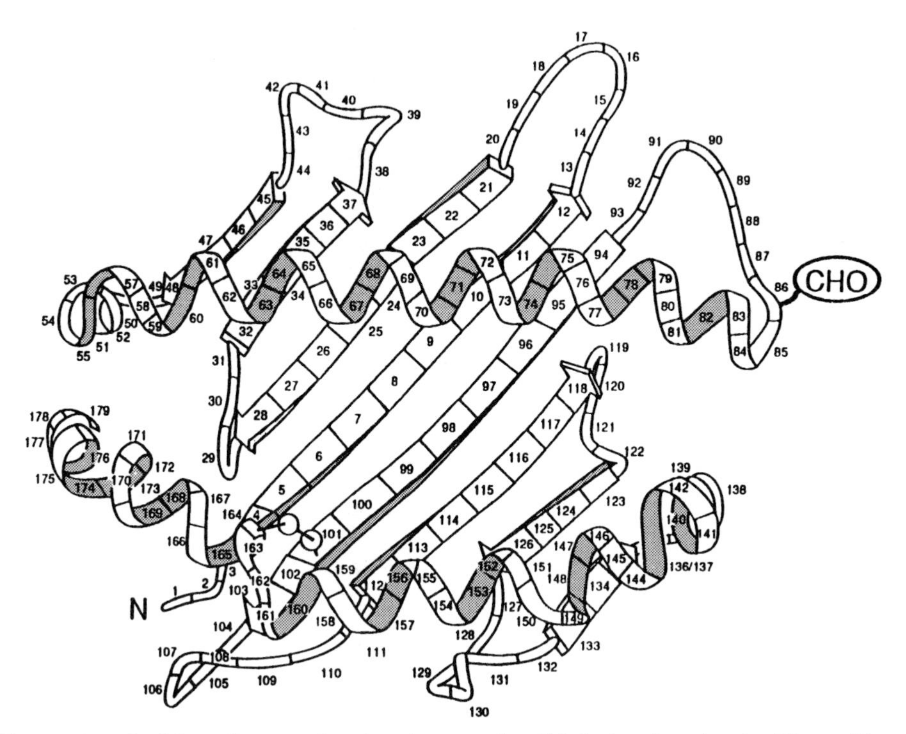

Figure 4.2. A ribbon diagram showing the top of an HLA class I molecule. The residues are numbered from the amino terminus of the mature protein, which is the first codon of exon 2.

common ancestry, and to map mutational events on to the tree, allowing some insight into the mechanism of diversification.

Figure 4.3 shows the result of a significance test of the serologically defined allelic lineages, using one representative allele from each serotype for clarity. If all alleles within a serotype (e.g. *B*2701*, *B*2702*, *B*2703*, etc.) are used, they cluster together in the tree, showing a minimal amount of variation around the *leitmotif* of the serologically defined lineage. The method of Rzhetsky and Nei (1992) was used to infer significance values for the nodes on the tree. Each locus will be discussed separately. References to specific sequences can be found in Bodmer *et al.* (1995).

HLA-A. The phylogenetic tree in *Figure 4.3a* shows clearly that there are six phylogenetically distinct families of alleles at the *HLA-A* locus, referred to here by their patterns of serological cross-reactivity (Rodey and Fuller, 1987). The A2/A28 family includes *A*02*, *A*68* and *A*69* alleles. The A1/A3/A11 family includes *A*01*, *A*03* and *A*11* alleles. The A9 family includes *A*23* and *A*24* alleles. A10 includes *A*25*, *A*26*, *A*34*, *A*43* and *A*66* alleles. A19 includes *A*29*, *A*31*, *A*32*, *A*33* and *A*74* alleles. The A1/A3/A11 group is supported at the 2% significance level, while the other groups are supported at the 1% level; alternative groupings are therefore extremely unlikely. Higher-order structure is reflected in the grouping of the A10, A19 and A2/A28 families on the tree (1% significance level).

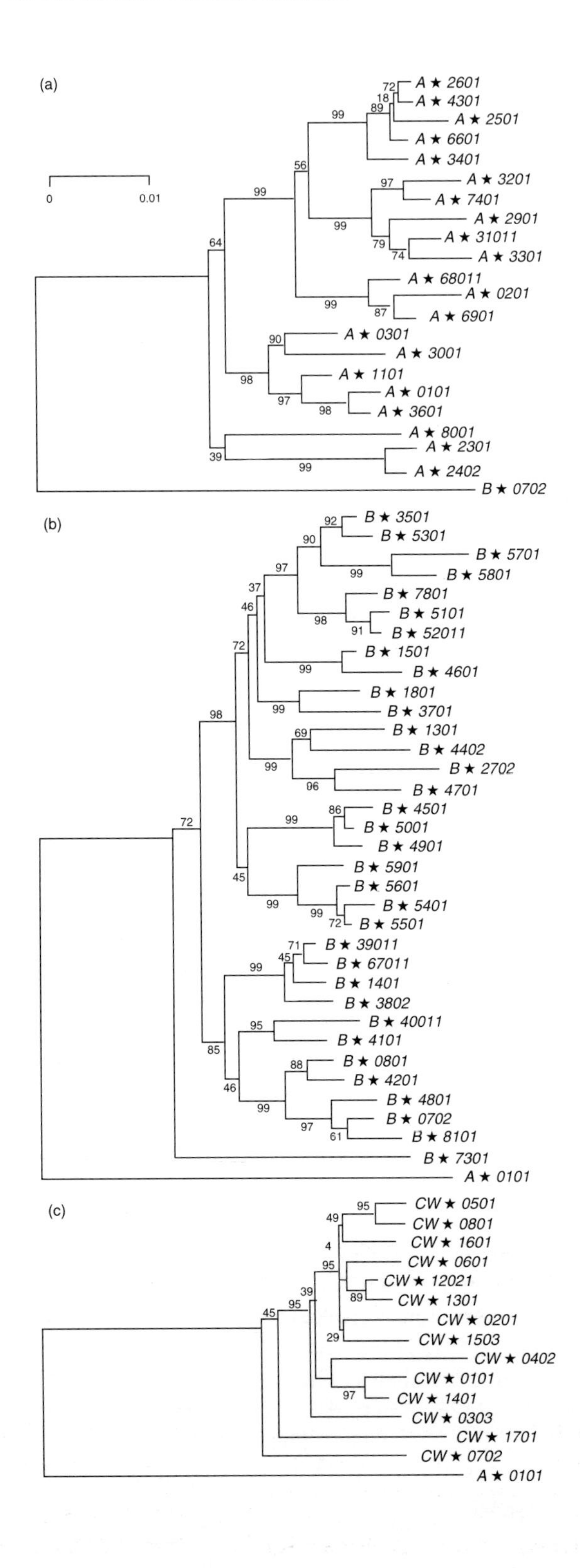
(a)
0
0.01
A★2601
A★4301
A★2501
A★6601
A★3401
A★3201
A★7401
A★2901
A★31011
A★3301
A★68011
A★0201
A★6901
A★0301
A★3001
A★1101
A★0101
A★3601
A★8001
A★2301
A★2402
B★0702
(b)
B★3501
B★5301
B★5701
B★5801
B★7801
B★5101
B★52011
B★1501
B★4601
B★1801
B★3701
B★1301
B★4402
B★2702
B★4701
B★4501
B★5001
B★4901
B★5901
B★5601
B★5401
B★5501
B★39011
B★67011
B★1401
B★3802
B★40011
B★4101
B★0801
B★4201
B★4801
B★0702
B★8101
B★7301
A★0101
(c)
CW★0501
CW★0801
CW★1601
CW★0601
CW★12021
CW★1301
CW★0201
CW★1503
CW★0402
CW★0101
CW★1401
CW★0303
CW★1701
CW★0702
A★0101

There are two unusual alleles on the tree: *A*30* and *A*80*. *A*30* is actually part of the A19 cross-reactive group (CREG) by serology, but it clearly groups with *A*03* at the sequence level (Kato *et al.*, 1989). *A*80*, on the other hand, constitutes a unique lineage of A alleles, combining a 5′ sequence motif similar to the A1/A3/A11 family with a distinctive sequence motif at the 3′ end of the molecule (Domena *et al.*, 1993; Starling *et al.*, 1994; Wagner *et al.*, 1993).

HLA-B. The analysis of *HLA-B* alleles (*Figure 4.3b*) shows that there is some superstructure among *B* alleles, although it is almost exclusively determined by variation at the 5′ end of the molecule (see below). The patterns of serological cross-reactivity at the *B* locus are far more complex than those at *A*, with six major CREGs defined: Bw4, Bw6, B5, B7, B8 and B12 (Rodey and Fuller, 1987). The allotypes included in each CREG are shown in *Figure 4.4*. There is a reasonable level of concordance overall between serological CREGs and relationships among the allelic sequences on the tree. There are glaring exceptions, however; for instance, the most widespread CREGs, Bw4 and Bw6 (Ayres and Cresswell, 1976; Muller *et al.*, 1985), are scattered around the tree at random. Within the group at the top of the tree in *Figure 4.3b*, *B*5301*, *B*5701*, *B*5801*, *B*5101* and *B*52011* are Bw4, while *B*3501* and *B*7801* are Bw6. This epitope has probably been shuffled back and forth among B locus alleles through gene conversion, resulting in the patchwork pattern seen in the tree.

Other CREGs fare somewhat better than Bw4 and Bw6 in the phylogenetic analysis. B5 is evident clearly on the tree, although the closest relative of *B*1801* is *B*3701*, a serologically unrelated member of the B12 CREG. Most of the alleles of the B12 CREG group together, although *B*4101* and *B*4801* are closer to the members of the B40 CREG, to which they also belong. Three of the four alleles of the B7 CREG clearly group together on the tree (*B*5401*, *B*5501* and *B*5601*), but *B*0702* is closer to the B40 CREG alleles than to these alleles. The B27 CREG is particularly weak, with none of its members (*B*2702*, *B*4201* and *B*0702*) grouping as closest relatives on the tree. The B8 CREG is split on the tree as well: *B*39011*, *B*67011*, *B*1401* (B64) and *B*3802* group together, but *B*5901* and *B*0801* group with members of other CREGs.

The HLA-B CREGs, while somewhat representative of the relationships between the sequences at the nucleotide level, are clearly not as good at predicting these relationships as the HLA-A CREGs. This is demonstrated most clearly by the Bw4 and Bw6 CREGs, but this same theme seems to be common to all of the defined *B* locus CREGs. This lack of concordance between serology, which infers relationships from short regions of sequence similarity, and the relationships between complete allelic sequences, supports the hypothesis that there is a higher rate of gene conversion and recombination at the *B* locus (see below). If the short regions responsible for the serological affinities have been

---

Figure 4.3. A neighbour-joining tree (Saitou and Nei, 1987) of the alleles at the HLA-A (a), HLA-B (b) and HLA-C (c) loci. One allele from each serological classification was chosen. The complete nucleotide sequence of the coding region was analysed in MEGA (Kumar *et al.*, 1993) using Kimura's 2-parameter distance measure (Kimura, 1980). Significance values were calculated using the method of Rzhetsky and Nei (1992), and are shown next to the branches to which they refer.

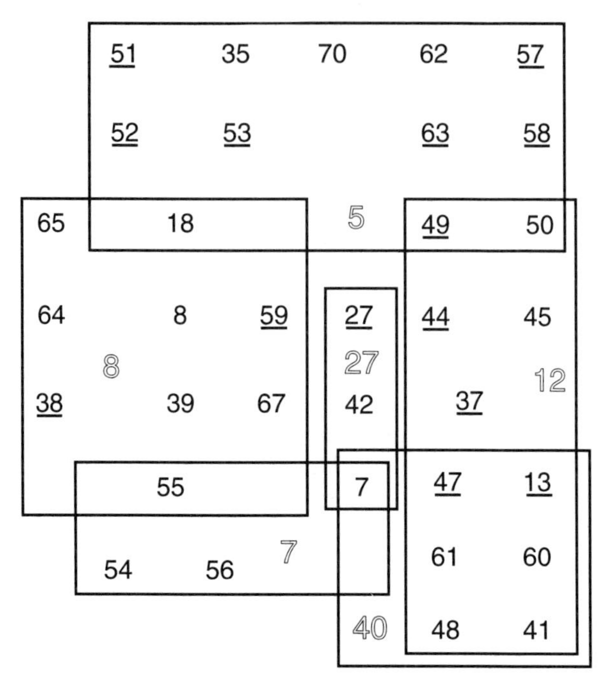

Figure 4.4. A diagram showing the pattern of cross-reactivity among HLA-B alleles. The CREG designations are shown in outline, except for Bw4 and Bw6; Bw4 alleles are underlined, Bw6 are not. [This figure is based on one in Rodey and Fuller (1987).]

shuffled among the sequences, we would expect to find a lack of concordance between serology and overall sequence similarity.

HLA-C. From an examination of *Figure 4.3c*, there appear to be three main groups of *HLA-C* alleles: *Cw*07*, *Cw*17* and all other HLA-C alleles. In addition, there is a significant node differentiating the *Cw*02*, *Cw*05*, *Cw*06*, *Cw*08*, *Cw*12*, *Cw*13*, *Cw*15* and *Cw*16* lineages from other alleles. The divergence between *Cw*07* and the other alleles is well-documented (Zemmour and Parham, 1992), and encompasses multiple unique substitutions along the length of the sequence. The divergent *Cw*07* family is analogous to the *B*7301* allele at the *HLA-B* locus in that there are a large number of lineage-specific substitutions clustered in the 3′ end of the sequence. The reasons for this are unknown, although the extensive divergence along the full length of the sequence would tend to argue for heterotic selection acting to maintain diversity outside of the peptide-binding region; frequency-dependent selection tends to favour divergence in the peptide-binding region of the molecule (see below). Perhaps *Cw*07* presents a unique set of peptides which has been present in the selective environment for a long period of time, and the changes at the 3′ end are important for this function. The *Cw*17* lineage is represented by a novel *HLA-C* allele which is distantly related to *Cw*07* (RSW, unpublished data). *Cw*1701* is the first *HLA-C* length variant in the transmembrane domain, where there is an 18 nucleotide insertion relative to all other *HLA-C* alleles. This length variation is again reminiscent of *HLA-B*73*, and may represent a relic of an early stage in *HLA-B* and *-C* divergence when there was a substantial amount of length variation.

HLA-C is expressed on the cell surface at approximately 30% of the level of HLA-A and -B, primarily due to higher levels of HLA-C mRNA degradation (McCutcheon *et al.*, 1995). The patterns of expression of HLA-A, -B and -C may be a reflection of different functions *in vivo*. While the *A* and *B* loci are expressed at high levels and seem to function primarily in the presentation of peptides to cytotoxic T lymphocytes (CTL), HLA-C may function in antigen presentation to a different population of cytolytic cells which do not require a high level of cell-surface expression. The activity of natural killer (NK; Trinchieri, 1989) cells is mediated by HLA class I molecules (Storkus *et al.*, 1989). 'Public' epitopes have been mapped to residues 77 and 80 on HLA-C, and receptors specific to these epitopes are present in some populations of NK cells (Colonna *et al.*, 1993). HLA-C appears to be at least as effective at mediating NK cell lysis as HLA-A and -B; the fact that this is accomplished with a much lower level of expression than either HLA-A or -B implies that HLA-C may, in fact, be a relatively more effective NK mediator than these other class I molecules. This demonstrates that HLA class I molecules may have a larger role in the mammalian immune response than simply presenting peptides to CTL, potentially providing additional selection pressures favouring diversity.

4.2.2 HLA 'blank' alleles

Most HLA alleles have now been characterized serologically. In some cases, however, a non-homozygous individual will demonstrate only one serological allele in routine screening. The other allele is then termed a 'blank', a catch-all classification including both non-expressed and previously uncharacterized alleles. The former appear to be rare at the *HLA-A, -B* and *-C* loci, with two published examples (Balas *et al.*, 1994; Lardy *et al.*, 1992); thus, the majority of blanks are probably novel allelic variants. This has been confirmed in several recent studies of blanks at the A, B and C loci, where novel allelic sequences have been determined from the blank allele.

The most significant recent finding in the sequence analysis of blank alleles is the description of the *A*8001* sequence (Domena *et al.*, 1993; Starling *et al.*, 1994; Wagner *et al.*, 1993), a divergent allele with multiple unique substitutions which defines a sixth family of HLA-A alleles (see above). Apparently limited to Africa, this lineage demonstrates the potential diversity present in the serological blanks. Of course, not all blanks represent such divergent sequence motifs; most new *HLA-C* alleles are blanks (*Cw*12* to *Cw*17*), and one described C blank, *Cw*1401* (Takiguchi *et al.*, 1989), is closely related to *Cw*0101*, a serologically well-defined C allele. The substitutions between these two closely related alleles apparently ablate the recognition of *Cw*1401* allotype by the *Cw01* antisera. The high frequencies of HLA-C blank alleles in most populations (20–70%) are due in part to the lower level of HLA-C expression (see above), making it difficult to create monospecific antisera. These blank alleles are very likely to represent novel sequences, making the true range of allelic variation at the *HLA-C* locus unknown.

4.3 Non-classical class I diversity

In addition to the classical class I loci, there are three 'non-classical' loci found in the class I region. Designated *HLA-E, -F* and *-G*, they maintain the same exon–

intron structure seen in the classical loci, and are extremely similar at the nucleotide sequence level as well (Parham *et al.*, 1995). The pattern of mRNA expression of HLA-E and -F is similar to that of the classical genes, while HLA-G is expressed primarily in extraembryonic tissues (placenta and extravillous membrane; Wei and Orr, 1990). HLA-E and -F proteins do not appear to be expressed on the cell surface, although they are translated and are capable of complexing with β_2-microglobulin (Shimizu *et al.*, 1988). HLA-G proteins are expressed on the surface of trophoblast cells, and bind and present intracellular peptides as do other class I molecules (Lee *et al.*, 1995).

Evolutionarily, HLA-E appears to be the most divergent class I gene, having split from the lineage leading to the extant classical loci before the divergence of HLA-A, -B and -C. HLA-F and -G, on the other hand, appear to have diverged from the HLA-A lineage later than HLA-B and -C (Parham *et al.*, 1995). These genes are much less polymorphic than the classical class I genes, with few known alleles. Four *HLA-E* and four *HLA-G* alleles have been described, differing at few positions (Bodmer *et al.*, 1995). No polymorphism has been described at the *HLA-F* locus (Geraghty *et al.*, 1990).

4.3.1 Alleles at the HLA-E *and* HLA-G *loci*

HLA-E. Being polymorphic, *HLA-E* has two variable nucleotide positions in the coding region: 294 (silent) and 382 (arginine/glycine at amino acid 107; Geraghty *et al.*, 1992a; Ohya *et al.*, 1990). The A to G transition at position 532, reported by Ohya *et al.* (1990), is likely to be due to a polymerase chain reaction (PCR) misincorporation in the single clone where it was found; it was not detected in a subsequent survey of *HLA-E* polymorphism by Geraghty *et al.* (1992a). In addition, the *E*0102* allele (Bodmer *et al.*, 1995) shows two adjacent nucleotide substitutions relative to all other *HLA-E* sequences at nucleotide positions 28 and 29; as two other adjacent, unique nucleotide substitutions in this sequence (positions 310 and 311) have been shown to be due to sequencing error in the original *E*0102* sequence (Geraghty *et al.*, 1992a), it is possible that positions 28 and 29 represent sequencing errors as well. Thus, at present there are two known variable nucleotide sites in the *HLA-E* coding region, one of which results in an amino acid change.

HLA-G. Four alleles have been described for *HLA-G*, defined by polymorphisms at 11 sites (Bodmer *et al.*, 1995; Geraghty *et al.*, 1987; Morales *et al.*, 1993). The changes at positions 15 (codon -20), 36 (codon -13), 99 (codon 9), 243 (codon 57), 351 (codon 93), 354 (codon 94), 393 (codon 107), 942 (codon 290) and 999 (codon 309) are silent, while the change at position 163 results in an amino acid change at codon 31 (Thr-Ser). *G*01012* contains a three nucleotide insertion relative to the other alleles after position 741 (between codons 223 and 224). These 11 variable sites define four alleles at the HLA-G locus, designated *G*01011*, *G*01012*, *G*0102*, *G*0103* (Bodmer *et al.*, 1995). Recently, van der Ven and Ober (1994) examined *HLA-G* polymorphism in a sample of 45 African Americans. They found extensive polymorphism at this locus, although linkages among the variable sites have not been determined, which will entail sequencing complete alleles. In summary, then, *HLA-G* appears to be moderately polymorphic, although most of the variation described does not result in amino acid changes. More complete surveys of allelic variation at this locus in multiple human populations are required.

4.4 Pseudogenes

There are several pseudogenes in the class I region, most of which are clustered around the HLA-A locus (Geraghty *et al.*, 1992b). Four of these, *HLA-H*, *-J*, *-K* and *-L* have been assigned letter designations (Bodmer *et al.*, 1995), and a fifth goes by the somewhat mysterious (and unofficial) name of *HLA-X* (Chimini *et al.*, 1990). All of the pseudogenes have crippling mutations in their coding regions which ablate cell-surface expression of a class I gene product. Substantial polymorphism has been found at the HLA-H locus (formerly HLA-AR), where 48 variable positions have been described (Zemmour *et al.*, 1990). Most of these are located in exons 2 and 3 (16 and 19, respectively), providing evidence that HLA-H, like the classical class I genes, has been subject to selection for functional diversity in the recent past. HLA-J is much less polymorphic than HLA-H, with 12 variable positions scattered evenly throughout the sequence (Messer *et al.*, 1992). The lower level of variability implies either that HLA-J has been inactivated for longer than HLA-H, or that it was never subject to selection for functional diversity in the peptide binding region. No polymorphism has been described for HLA-K or -L.

4.5 Diversity-generating mechanism

The most overwhelming feature of the rapidly accumulating database of HLA allelic sequences is the extraordinary level of polymorphism exhibited by these loci. While DNA sequences from other regions of the human genome sampled from two unrelated individuals will differ at approximately one position in every 1000 bp (Li and Sadler, 1991), classical class I HLA loci are much more variable, exhibiting 20–40 times this level of nucleotide polymorphism, depending on the locus. The high level of polymorphism seen at these loci could be maintained by three mechanisms: a high rate of mutation, selection for diversity, and/or population structure.

4.5.1 Mutation

All nucleotide diversity arises through point mutation. There is no evidence, however, that the allelic diversity at the HLA locus has arisen through a high level of point mutation. If this alone were the mechanism, we would not expect to see variation maintained across species and locus boundaries, as is the case for HLA loci (Lawlor *et al.*, 1988; Mayer *et al.*, 1988); rather, the variation would be more idiosyncratic, with each species and allelic lineage displaying its own characteristic pattern of unique substitutions. There is one type of mutational event that appears to have played a significant role in HLA diversification, however: exchange (via gene conversion) among alleles. As soon as more than one variable site exists in a region of the genome, an additional level of complexity arises: the sorting of these variable sites into allelic haplotypes, resulting in a huge potential number of alleles from only a few variable positions. Much of the functional diversity at the HLA loci arises from this haplotype diversity, rather than simple mutational diversity.

The generation of allelic diversity has two components, which we will refer to as mutational and recombinational sorting. Mutational sorting is the type of allele-generating process which occurs in haploid or hemizygous systems, where mutations must occur sequentially on the same piece of DNA in order to be linked. Recombination, on the other hand, circumvents this process by allowing exchange between all alleles in the population, thereby producing chimeric alleles. The relative contribution of each type of sorting to the generation of HLA diversity is unknown, although both seem to be partially responsible for the current pattern of allelic variation. The rate of point mutation in the human genome is approximately 10^{-9} to 10^{-8} per nucleotide per generation; the HLA class I coding region encompasses slightly over 1000 nucleotides, thus giving a point mutation rate within a particular class I coding region of 10^{-6} to 10^{-5} per generation. Recombination, assuming that a physical distance of 10^6 nucleotides is equivalent to a recombination rate of 1% per generation, would tend to occur at a rate of approximately 3×10^{-5} per generation within a 3 kb HLA class I coding region. Thus, simple exchange (ignoring gene conversion) would be expected to occur 3–30 times as often as point mutation. These estimates are not particularly accurate, however, as there appears to be heterogeneity in recombination across the class I region (Crouau-Roy *et al.*, 1993; Shukla *et al.*, 1991), and the true rate of recombination within the loci has not been measured. It is worth noting that simple recombination events appear to be rare from an inspection of HLA class I sequences (Parham *et al.*, 1995).

Recent studies by Erlich and colleagues (Zangenberg *et al.*, 1995) have shown that gene conversion occurs at an extremely high frequency at the class II *DPB* locus (see Chapter 5). Using oligotyping methods to detect gene conversion events in individual sperm from heterozygous donors, the observed rate of conversion within the variable exon 2 was approximately one event in 10 000 sperm (10^{-4} per locus per generation). If the rate at the class I loci is comparable, gene conversion is expected to play the dominant role in generating new alleles, as it is at least 10 times more frequent than the estimated rate of point mutation within the HLA coding region. Moreover, the re-assortment of pre-existing variation might be expected to yield functional alleles more often than point mutation, which is presumably random with respect to functionality. Therefore, the patchwork pattern of association between blocks of variation seen among HLA alleles seems likely to be due to the action of gene conversion rather than convergent mutation.

4.5.2 The role of selection

Selection has played an important role in moulding the pattern of diversity seen among HLA alleles, as revealed by two analyses. First, there is a remarkable concordance between class I heavy chain residues involved in direct contact with peptide and T-cell receptors and those residues in the sequence which exhibit the most variability (Hughes and Nei, 1988; Nei and Hughes, 1991). *Figure 4.5* shows a sliding window plot of the number of variable nucleotide positions in a window of 50 nucleotides along the sequences of *HLA-A, -B* and *-C*. There are several peaks in each plot, corresponding to the regions of the class I genes which exhibit substantial variation. *HLA-A* has a strong peak centred around nucleotide 270 (amino acid 66), corresponding to the α_1 domain of the three-dimensional structure (see *Figure 4.2*). A second peak occurs around nucleotide 360 (amino acid 103), corresponding to a region of the β-sheet which is in contact with the bound peptide. A third peak occurs around nucleotide 510 (amino acid 146), corresponding to the α_2 domain. Finally, there is a fourth peak centred around nucleotide 620 (amino acid 182) which is not seen in the variability plots of the *B* and *C* loci (see below); this may correspond to a region of the molecule important for T-cell interactions. Since a few amino acid substitutions in the peptide-binding pockets can dramatically change the character of the peptides bound by the molecule (Barber and Parham, 1993; Falk *et al.*, 1991), the clustering of this extraordinary amount of amino acid variation in a functionally important region of the molecule provides strong evidence that selection has favoured diversity.

HLA-B has a much stronger peak in the region corresponding to the α_1 domain of the molecule, while the other peaks are lower (*Figure 4.5b*). The peaks corresponding to the α_2 and T-cell interaction domains are extremely low at the *B* locus, implying that the α_1 domain is more important for generating functional diversity. The peak centred around nucleotide 850 corresponds to the region where the *B*7301* allele (Hoffman *et al.*, 1995; Parham *et al.*, 1994; Vilches *et al.*, 1994) is maximally divergent from all other B alleles. If this low-frequency allele (<1%; Imanishi *et al.*, 1992) is removed from the analysis, a startling lack of diversity is revealed at the 3′ end of *HLA-B* compared to the *A* and *C* loci (dotted line in *Figure 4.5b*).

HLA-C shows a generally lower level of diversity compared to the other HLA loci, with no strong peaks evident in the peptide-binding region of the molecule (*Figure 4.5c*). The strong peak centred around nucleotide 980 is due to the extreme divergence of *Cw*0702* and *Cw*1701* in this region. The localized peak suggests that perhaps this divergence has been maintained by selection.

The second supporting piece of evidence for the selective importance of the variation seen at the HLA loci comes from an analysis of the types of changes that distinguish allotypes. If nucleotide variation is categorized into changes that result in amino acid differences (non-synonymous) and changes that have no effect on the amino acid sequence (synonymous), there is an excess of non-synonymous variation in the peptide-binding region of the molecule (Hughes and Nei, 1988; Nei and Hughes, 1991). This difference is statistically significant, implying that selection has tended to favour much of the non-synonymous variation which has arisen in HLA lineages, while synonymous variation (which is presumably functionally neutral) has been maintained only through genetic drift, yielding a far smaller number of variable synonymous sites.

While the pattern of diversity clearly shows that selection favours having a large number of HLA alleles, the exact means by which this diversity is maintained is unclear. Some have argued for heterotic selection (Black and Salzano, 1981; Degos *et al.*, 1974; Hedrick and Thomson, 1983; Hughes and Nei, 1988), in which heterozygotes are 'fitter' (in the Darwinian sense) than homozygotes. One prediction which follows from this scheme is that alleles (or allelic lineages) are maintained for long periods of time, and that the lineages gradually become more

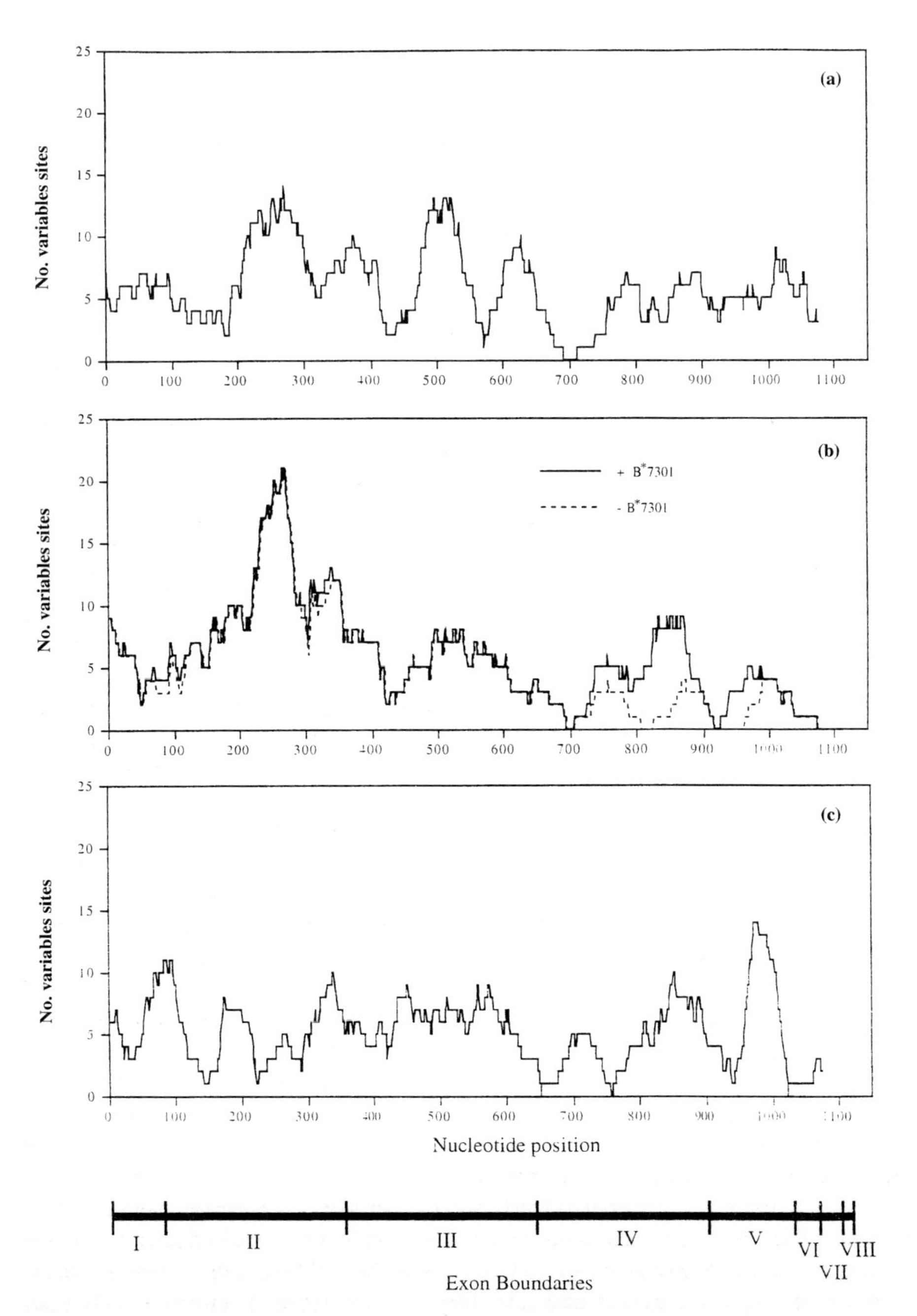

Figure 4.5. Sliding window plots of the number of variable nucleotide positions in a window of 50 nucleotides at the *HLA-A* (*A*), *HLA-B* (*B*) and *HLA-C* (*C*) loci. The alleles shown in the trees in *Figure 4.3* were used for this analysis. For HLA-B the dataset is analysed with (solid line) and without (dotted line) *B*7301*, a low-frequency divergent allele.

divergent over time. Others have preferred an alternative explanation, in which new alleles are constantly driven to moderately high frequency in the population by new pathogens (Hill *et al.*, 1991; Slade and McCallum, 1992), then subsequently replaced when novel alleles are again favoured. This frequency-dependent selection would be expected to yield a pattern of high allelic turnover, resulting in substantial population-specific variation (since not all populations have been exposed to the same diseases, and allelic diversity is presumably generated randomly). Heterozygosity at the individual level is not as important in a frequency-dependent scheme; rather, it is the possession of the 'protective' allele that is crucial to the individual's survival.

Can the pattern of variation among classical class I alleles be used to determine which of these selective regimens is more important in moulding diversity? Coalescence theory (a school of population genetics theory in which allelic diversity is modelled as divergence from a common ancestor, or coalescence point) predicts that old allelic lineages should exhibit greater diversity, since they have been accumulating mutations for a longer period of time than have new lineages. Heterotic selection, in which heterozygosity is favoured over homozygosity, tends to produce a pattern of excessive divergence between discrete allelic lineages, with some interlineage exchange (via recombination and conversion) outside of the selected region (Strobeck, 1983). This is exactly the pattern of diversity exhibited by the *HLA-A* locus. *Figure 4.3* shows a neighbour-joining tree (Saitou and Nei, 1987) of alleles representing the HLA-A, -B and -C allotypes obtained using the entire coding sequence. *Figures 4.6a* and *b* shows the effect of using only exons 1–3 or 4–8 to construct the tree (all alleles are shown on the same tree in this analysis). At the *HLA-A* locus, the same overall structure among CREGs (see above) is maintained whether the 5′ or 3′ end of the molecule is used. In other words, there has been sufficient divergence along the entire length of the allelic lineages to differentiate them from each other, and selection has apparently maintained the linkage of these substitutions over a long period of time.

This is in marked contrast to the situation seen at the *HLA-B* locus. While there is still a significant amount of structure in the tree constructed from the 5′ end of the sequences, there is almost no structure (apart from that contributed by the *B*73* lineage) in the tree constructed using the 3′ end. Why is there so little variation at the 3′ end of the sequence? McAdam *et al.* (1994) have described the decoupling of variation in exons 2 and 3 at the *HLA-B* locus due to high levels of recombination and/or conversion. If this high level of conversion has shuffled the evolution of two proximal exons, it has certainly shuffled the rest of the sequence as well. Recent studies on HLA variation in isolated human populations (Belich *et al.*, 1992; Watkins *et al.*, 1992) reveal that *HLA-B* alleles can arise and go to high frequency relatively rapidly; one can imagine the sequential evolution of alleles over time: each new allele is built from the sequence motifs of older alleles, which were built from still older alleles, and so on *ad infinitum*. In this model, selection would tend to favour diversity in exons 2 and 3, while diversity in exons 4–7 is less important for the 'success' of the allele. Other than *B*7301*, all extant *B* alleles seem to be based on nearly identical sequences at their 3′ ends. Non-human primates exhibit a much higher frequency of the divergent *B*7301*-like motif at the 3′ ends of their B alleles, so it is unlikely that no variation is tolerated in this part of the molecule. Rather, the sequential evolutionary process followed by *HLA-B* alleles in humans probably began with alleles having a non-*B*7301*-like 3′ end.

(a)

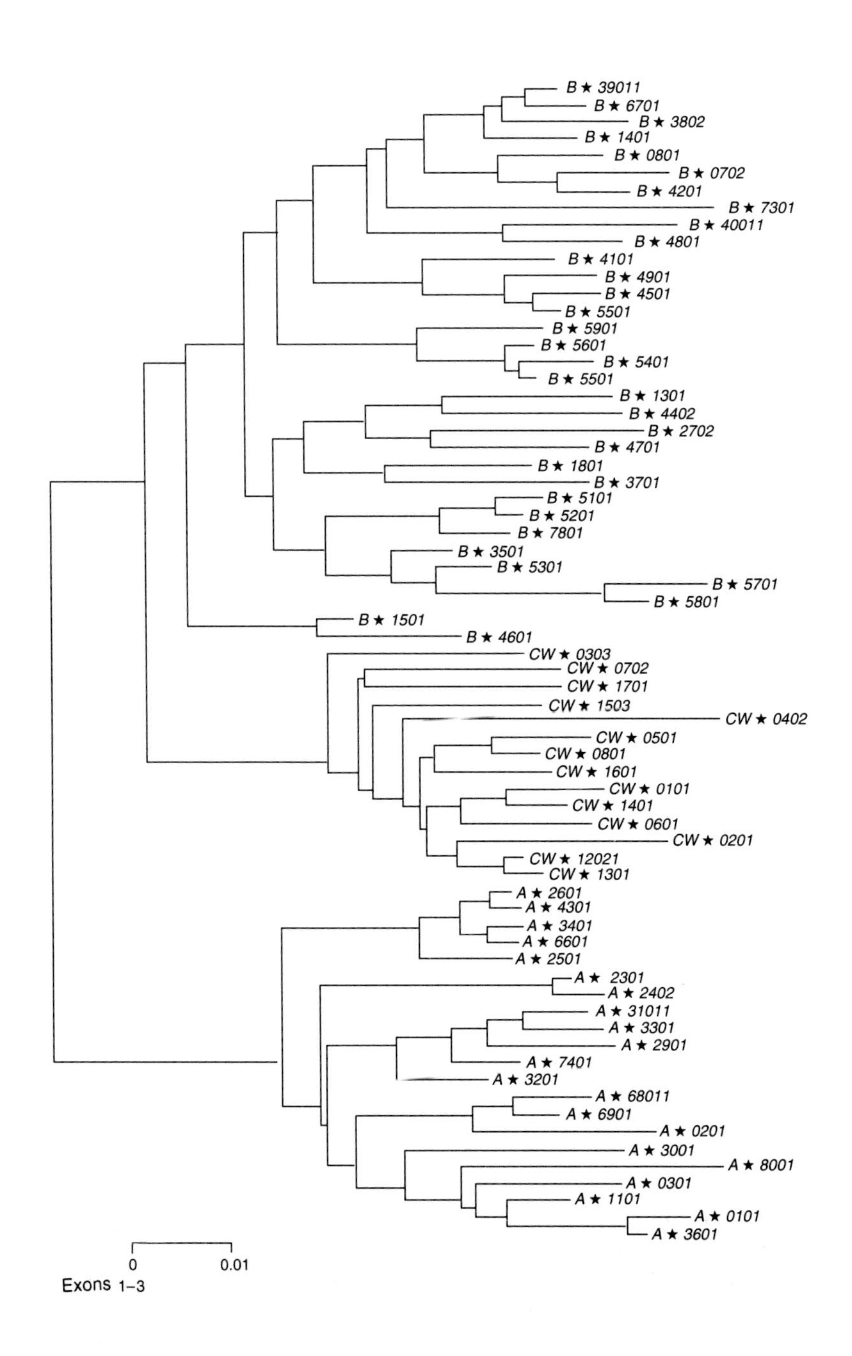

Figure 4.6. (a) A neighbour-joining tree of the combined *HLA-A*, *-B* and *-C* sequences drawn using exons 1–3 of the coding region nucleotide sequence.

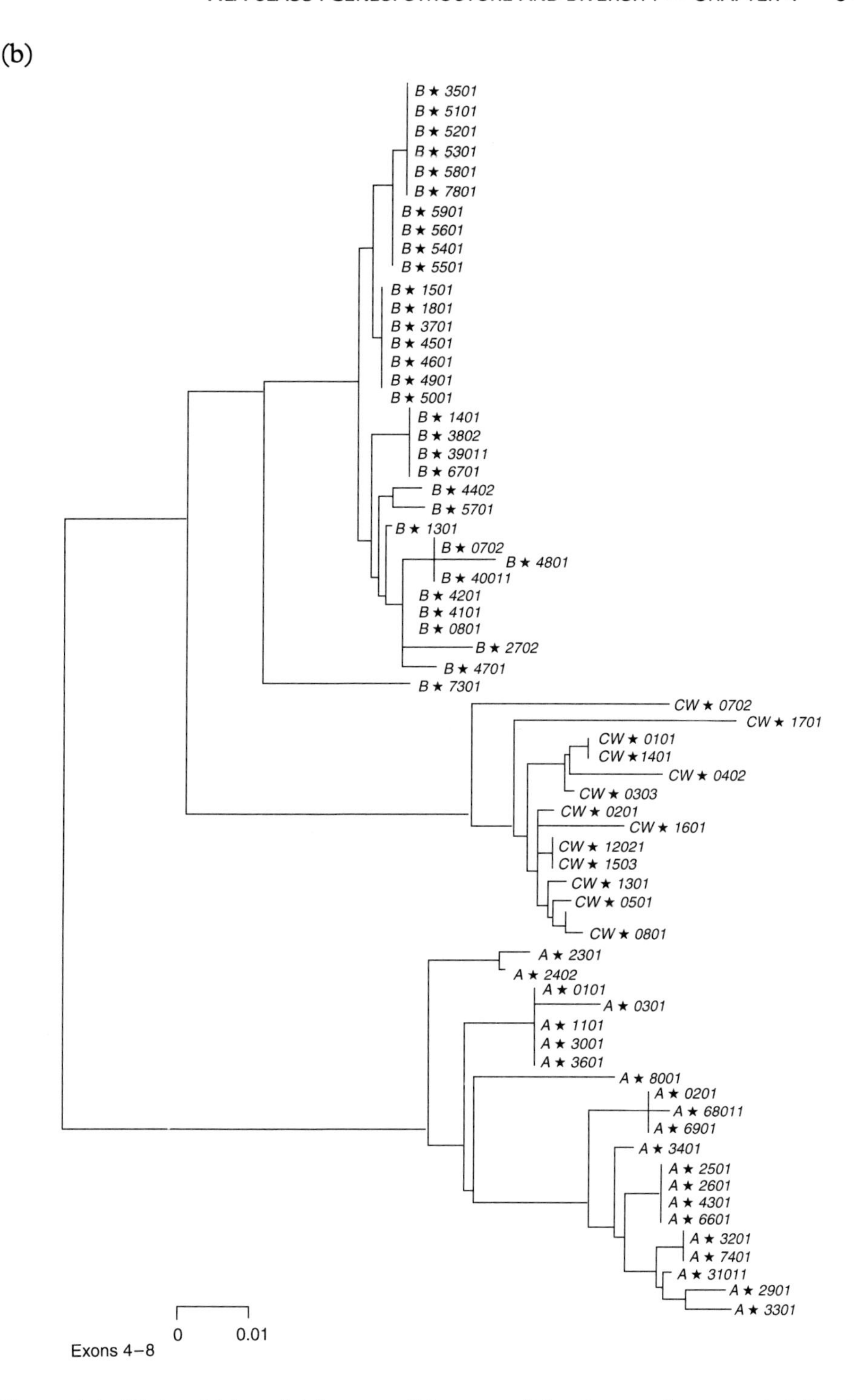

Figure 4.6. (b) A neighbour-joining tree of the same alleles drawn using exons 4–8 (4–7 in the case of *HLA-B*). Kimura 2-parameter distances (Kimura, 1980) were used for constructing the tree, and a midpoint rooting was used.

Table 4.1. *A*02* subtypes in human populations.

Allele	Relative frequency (% of *A*02*)		
	Singapore Chinese ($n = 66$)	UK Caucasian ($n = 76$)	African ($n = 37$)
*A*0201*	23	96	62
*A*0202*	–	–	22
*A*0203*	23	–	–
*A*0204*	–	–	–
*A*0205*	–	1	14
*A*0206*	8	–	–
*A*0207*	45	–	–
*A*0208*	–	–	–
*A*0209*	–	–	–
*A*0210*	2	–	–
*A*0211*	–	1	–
*A*0212*	–	–	–
*A*0213*	–	1	–
*A*0214*	–	–	3

Data reproduced from Krausa *et al.* (1995) Genetic polymorphism within HLA-A*02: significant allelic variation revealed in different populations. *Tissue Antigens* **45**: 233. © 1995 Munksgaard International Publishers Ltd, Copenhagen, Denmark.

Over time, as the other alleles have been favoured by selection, this sequential process has reduced the level of variation in the non-selected (3′) part of the allele; a 'selective sweep' (Berry *et al.*, 1991) has acted to remove diversity at the 3′ ends of the alleles. Thus, the pattern of variation seen at *HLA-B* is more consistent with that expected under a model of frequency-dependent selection due to frequent exposure to novel diseases.

To summarize, variation at HLA class I loci is distributed non-randomly along the sequence, and most of this nucleotide variation translates to protein variation. The most parsimonious explanation for such a pattern is that selection for functional diversity (balancing selection) has played a major role in HLA class I evolution. Pathogens are the probable selective factors moulding this diversity, with HLA heterozygotes better able to fight off infections (Doherty and Zinkernagel, 1975), although sexual selection (choosing mates with a different HLA type) may play a role as well (Hedrick, 1992; Wedekind *et al.*, 1995). In addition, the finding that *HLA-B*5301* appears to confer reduced morbidity in West Africans infected with *Plasmodium falciparum* malaria (Hill *et al.*, 1991) shows that disease-specific selection can play a role in moulding class I allelic diversity. HLA alleles differing in amino acid sequence can present different peptides to cytotoxic T lymphocytes, and this functional diversity has almost certainly arisen in response to pathogens with differing antigenic properties.

4.5.3 Population structure

Population structure can play a significant role in the pattern of allelic diversity of a species. If genetically distinct populations are pooled together, the apparent number of alleles is often larger than if the populations are examined separately. This could be due to different selective regimens (e.g. different parasites) in the constituent populations, or genetic drift could have randomly assorted the variation in different ways. At any rate, the level of polymorphism seen in a large, admixed population of modern humans (that found in a major city, for instance) is different from the level of HLA diversity in indigenous populations.

There is substantial variation in HLA serotype and haplotype frequencies among human populations (Imanishi *et al.*, 1992), reflecting the past history of selection, isolation and admixture in each group; for instance, the *A*4301* allele, at high frequency (11–17%) in Khoisan populations of Southern Africa, is at low frequency or absent from all other populations (Imanishi *et al.*, 1992; Madrigal *et al.*, 1993). New work at the nucleotide sequence level promises to provide a better picture of class I diversity in human populations than is possible using serological data. An example is shown in *Table 4.1*, where *HLA-A*02* subtype frequencies in three divergent populations are shown (Krausa *et al.*, 1995). The HLA-A2 antigen is at high frequency in populations worldwide, exhibiting a fairly flat geographic distribution (Imanishi *et al.*, 1992). When DNA alleles are compared among populations, however, significant differences are apparent; for instance, practically the only allele present in the Oxford population is *A*0201*, which is at much lower frequency in the sample from Singapore, while the second most common African subtype, *A*0202*, is not found in Europeans or Singaporeans (Krausa *et al.*, 1995). Different histories of selection and/or drift have yielded the divergent pattern seen among *A*02* subtypes in these modern populations.

While there are some population differences at the *HLA-A* locus, the *HLA-B* locus exhibits even more profound patterns of local diversification when examined at the sequence level. Studies on tribes of Brazilian Indians demonstrate the rapid rate of *HLA-B* evolution in isolated populations (Belich *et al.*, 1992; Watkins *et al.*, 1992). While many *A* and *C* locus alleles are identical to those found in other geographical areas, the majority of B alleles are unique to these particular populations. There is also significant divergence between populations at the *B* locus; for instance, *B*3506* is limited to the Kaingang, while *B*3505* is found in both Kaingang and Guarani (Belich *et al.*, 1992). The pattern of variation at *B*3506* suggests that it may have originated from a simple conversion event between a *B*35* allele and *B*39* allele encompassing residues 114–116 in the α_2 domain of the protein. B39 appears to be at high frequency in the Kaingang, while it is a low-frequency variant in the Guarani (20.5% and 1.5%, respectively). The high allelic turnover seen at the *HLA-B* locus, resulting in many population-specific variants, is precisely what is predicted in the frequency-dependent model of diversifying selection discussed above.

4.6 Conclusion

For the past 30 years, since the widespread application of serological typing in transplantation laboratories, studies of HLA variation have been limited by the availability of monospecific antisera to particular HLA alleles. The low resolution of this technique has now given way to sequencing studies of HLA alleles, many of which are being defined exclusively at the sequence level, with no serological specification. The patterns of variation among the class I allelic sequences will provide an indispensable database for transplantation matching, have allowed new methods of typing (using oligonucleotide hybridization and sequencing), and have yielded insights into the evolutionary mechanism of diversification at the loci. The next stage in studies of HLA polymorphism is to begin to untangle the complex pattern of interpopulation allelic variation, using the patterns as a tool to infer the history of admixture and disease pressure experienced by human populations.

References

Ayres J, Cresswell P. (1976) HLA-B specificities and w4, w6 specificities are on the same polypeptide. *Eur. J. Immunol.* **6:** 794–799.

Bahram S, Bresnahan M, Geraghty DE, Spies T. (1994) A second lineage of mammalian major histocompatibility complex class I genes. *Proc. Natl Acad. Sci. USA* **91:** 6259–6263.

Balas A, García-Sánchez F, Gómez-Reino F, Vicario JL. (1994) HLA class I allele (HLA-A2) expression defect associated with a mutation in its enhancer B inverted CAT box in two families. *Human Immunol.* **41:** 69–73.

Barber LD, Parham P. (1993) Peptide binding to major histocompatibility complex molecules. *Annu. Rev. Cell Biol.* **9:** 163–206.

Belich MP, Madrigal JA, Hildebrand WH, Zemmour J, Williams RC, Luz R, Petzl-Erler ML, Parham P. (1992) Unusual HLA-B alleles in two tribes of Brazilian Indians. *Nature* **357:** 326–329.

Berry AJ, Ajioka JW, Kreitman M. (1991) Lack of polymorphism on the *Drosophila* fourth chromosome resulting from selection. *Genetics* **129:** 1111–1117.

Black FL, Salzano FM. (1981) Evidence for heterosis in the HLA system. *Am. J. Hum. Genet.* **33:** 894–899.

Bodmer JG, Marsh SGE, Albert ED, *et al.* (1995) Nomenclature for factors of the HLA system, 1995. *Tissue Antigens* **46:** 1–18.

Campbell RD, Trowsdale J. (1993) Map of the human MHC. *Immunol. Today* **14:** 349–352.

Chimini G, Boretto J, Marguet D, Lanau F, Lauquin G, Pontarotti P. (1990) Molecular analysis of the human MHC class I region using yeast artificial chromosome clones. *Immunogenetics* **32:** 419–426.

Colonna M, Brooks EG, Falco M, Ferrara GB, Strominger JL. (1993) Generation of allospecific natural killer cells by stimulation across a polymorphism of HLA-C. *Science* **260:** 1121–1124.

Crouau-Roy B, Bouisseau C, Sommer E, Pontarotti P, Thomsen M. (1993) Analysis of HLA-A/B recombinant families with new polymorphic markers. *Hum. Immunol.* **38:** 132–136.

Dausset J. (1959) Iso-leuco-anticorps. *Acta Haematol.* **20:** 156.

Degos L, Colombani J, Chaventre A, Bengtson B, Jacquard A. (1974) Selective pressure on HLA polymorphism. *Nature* **249:** 62–63.

Doherty PC, Zinkernagel RM. (1975) Enhanced immunological surveillance in mice heterozygous at the H-2 gene complex. *Nature* **256:** 50–53.

Domena JD, Hildebrand WH, Bias WB, Parham P. (1993) A sixth family of HLA-A alleles defined by HLA-A*8001. *Tissue Antigens* **42:** 156–159.

Falk K, Rötzschke O, Stevanovic S, Jung G, Rammensee HG. (1991) Allele-specific motifs revealed by sequencing of self peptides eluted from MHC molecules. *Nature* **351:** 290–296.

Geraghty DE, Koller BH, Orr HT. (1987) A human major histocompatibility complex class I gene that encodes a protein with a shortened cytoplasmic segment. *Proc. Natl Acad. Sci. USA* **84:** 9145–9149.

Geraghty DE, Wei X, Orr HT, Koller BH. (1990) Human leukocyte antigen F (HLA-F): an expressed HLA gene composed of a class I coding sequence linked to a novel transcribed repetitive element. *J. Exp. Med.* **171:** 1–18.

Geraghty DE, Stockschleader M, Ishitani A, Hansen JA. (1992a) Polymorphism at the HLA-E locus predates most HLA-A and -B polymorphism. *Hum. Immunol.* **33:** 174–184.

Geraghty DE, Koller BH, Pei J, Hansen JA. (1992b) Examination of four class I pseudogenes. Common events in the evolution of HLA genes and pseudogenes. *J. Immunol.* **149:** 1947–1956.

Hedrick PW. (1992) Female choice and variation in the major histocompatibility complex. *Genetics* **132:** 575–581.

Hedrick PW, Thomson G. (1983) Evidence for balancing selection at HLA. *Genetics* **104:** 449–456.

Hill AVS, Allsop CE, Kwiatowski D, *et al.* (1991) Common West African HLA antigens are associated with protection from severe malaria. *Nature* **352:** 595–600.

Hoffman HJ, Kristensen TJ, Jensen TG, Graugard B, Lamm LU. (1995) Antigenic characteristics and cDNA sequences of HLA-B73. *Eur. J. Immunogenet.* **22:** 231–240.

Hughes AL, Nei M. (1988) Pattern of nucleotide substitution at major histocompatibility complex class I loci reveals overdominant selection. *Nature* **335:** 167–170.

Imanishi T, Akaza T, Kimura A, Tokunaga K, Gojobori T. (1992) W15.1: Allele and haplotype frequencies for HLA and complement loci in various ethnic groups. In: *HLA 1991, Proceedings of the Eleventh International Histocompatibility Workshop and Conference*, Vol. 1 (eds, K Tsuji, M Aizawa, T Sasasuki). Oxford University Press, Oxford, pp. 1065–1220.

Juji T, Azaka T, Tokunaga K, Miyoshi H, Kashiwase K. (1992) W2.1: The serology studies of the Eleventh Interntional Histocompatibility Workshop: an overview. In: *HLA 1991, Proceedings of the Eleventh International Histocompatibility Workshop and Conference*, Vol. 1 (eds, K Tsuji, M Aizawa, T Sasasuki). Oxford University Press, Oxford, pp. 83–108.

Kato K, Trapani JA, Allopenna J, Dupont B, Yang SY. (1989) HLA-Aw19 antigens: a genetically distinct family of HLA-A antigens comprising A29, A31, A32 and Aw33, but probably not A30. *J. Immunol.* **143:** 3371–3378.

Kimura M. (1980) A simple method for estimating evolutionary rate of base substitutions through comparative studies of nucleotide sequences. *J. Mol. Evol.* **16:** 111–120.

Krausa P, Brywka M, Savage D, *et al.* (1995) Genetic polymorphism within HLA-A*02: significant allelic variation revealed in different populations. *Tissue Antigens* **45:** 223–231.

Kumar S, Tamura K, Nei M. (1993) *MEGA: Molecular Evolutionary Genetics Analysis, version 1.01.* The Pennsylvania State University, University Park, PA.

Lee N, Malacko AR, Ishitani A, Chen M-C, Bajorath J, Marquardt H, Geraghty DE. (1995) The membrane-bound and soluble forms of HLA-G bind identical sets of endogenous peptides but differ with respect to TAP presentation. *Immunity* **3:** 591–600.

Lardy NM, Bakas RM, van der Horst AR, van Twuyver E, Bontrop RE, de Waal LP. (1992) *cis*-Acting regulatory elements abrogate allele-specific HLA class I gene expression in healthy individuals. *J. Immunol.* **148:** 2572–2577.

Lawlor DA, Ward FE, Ennis PD, Jackson AP, Parham P. (1988) HLA-A and B polymorphisms predate the divergence of humans and chimpanzees. *Nature* **335:** 268–271.

Leelayuwat C, Townsend DC, Degli-Esposti MA, Abraham LJ, Dawkins RL.(1994) A new polymorphic and multicopy MHC gene family related to non-mammalian class I. *Immunogenetics* **40:** 339–351.

Li W-H, Sadler LA. (1991) Low nucleotide diversity in man. *Genetics* **129:** 513–523.

Madrigal JA, Hildebrand WH, Belich MP, *et al.* (1993) Structural diversity in the HLA-A10 family of alleles: correlations with serology. *Tissue Antigens* **41:** 72–80.

Mayer WE, Jonker M, Klein D, Ivanyi P, van Seventer G, Klein J. (1988) Nucleotide sequences of chimpanzee MHC class I alleles: evidence for trans-species mode of evolution. *EMBO J.* **7:** 2765–2774.

McAdam SN, Boyson JE, Liu X, Garber TL, Hughes AL, Bontrop RE, Watkins DI.(1994) A uniquely high level of recombination at the HLA-B locus. *Proc. Natl Acad. Sci. USA* **91:** 5893–5897.

McCutcheon JA, Gumperz J, Smith KD, Lutz CT, Parham P. (1995) Low HLA-C expression at cell surfaces correlates with increased turnover of heavy chain mRNA. *J. Exp. Med.* **181:** 2085–2095.

Messer G, Zemmour J, Orr HT, Parham P, Weiss EH, Girdlestone J. (1992) HLA-J, a second inactivated class I HLA gene related to HLA-G and HLA-A. *J. Immunol.* **148:** 4043–4053.

Morales P, Corell A, Martinez-Laso J, Mártin-Villa JM, Varela P, Paz-Artal E, Allende L-M, Arnaiz-Villena A. (1993) Three new HLA-G alleles and their linkage disequilibria with HLA-A. *Immunogenetics* **38:** 323–331.

Muller C, Herbst H, Uchansk-Ziegler B, Ziegler A, Schunter F, Steiert I, Muller C, Wernet P. (1985) Characterization of a monoclonal anti-Bw4 antibody (Tü 109): evidence for similar epitopes on the Bw4 and Bw6 antigens. *Hum. Immunol.* **14:** 333–349.

Nei M, Hughes AL. (1991) Polymorphism and evolution of major histocompatibility complex loci in mammals. In: *Evolution at the Molecular Level* (eds RK Selander, AG Clark, TS Whittam). Sinauer, Sunderland, pp. 222–247.

Ohya K, Kondo K, Mizuno S. (1990) Polymorphism in the human class I MHC locus HLA-E in Japanese. *Immunogenetics* **32:** 205–209.

Parham P, Arnett KL, Adams EJ, Barber LD, Domena JD, Stewart D, Hildebrand WH, Little AM. (1994) The HLA-B73 antigen has a most unusual structure that defines a second lineage of HLA-B alleles. *Tissue Antigens* **43:** 302–313.

Parham P, Adams EJ, Arnett KL. (1995) The origins of HLA-A, B, C polymorphism. *Immunol. Rev.* **143:** 141–180.

Rodey GE, Fuller TC. (1987) Public epitopes and the antigenic structure of the HLA molecules. *CRC Crit. Rev. Immunol.* **7:** 229–267.

Rzhetsky A, Nei M. (1992) A simple method for estimating and testing minimum-evolution trees. *Mol. Biol. Evol.* **9:** 367–375.

Saitou N, Nei M. (1987) The neighbor-joining method: a new method for reconstructing phylogenetic trees. *Mol. Biol. Evol.* **4:** 406–425.

Shimizu Y, Geraghty DE, Koller BH, Orr HT, DeMars R. (1988) Transfer and expression of three cloned human non-HLA-A,B,C class I major histocompatibility complex genes in mutant lymphoblastoid cells. *Proc. Natl Acad. Sci. USA* **85:** 227–231.

Shukla H, Gillespie G, Srivastava R, Collins F, Chorney MA. (1991) A class I jumping clone places the HLA-G gene approximately 100 kilobases from HLA-H within the HLA-A subregion of the human *Mhc. Genomics* **10:** 905–914.

Slade RW, McCallum HI. (1992) Overdominant vs. frequency-dependent selection at MHC loci. *Genetics* **132:** 861–862.

Starling GC, Witkowski JA, Speerbacher LS, McKinney SK, Hansen JA, Choo SY. (1994) A novel HLA-A*8001 allele identified in an African-American population. *Hum. Immunol.* **39:** 163–168.

Storkus WJ, Alexander J, Payne JA, Dawson JR, Cresswell P. (1989) Reversal of natural killing susceptibility in target cells expressing transfected class I genes. *Proc. Natl Acad. Sci. USA* **86:** 2361–2364.

Strobeck C. (1983) Expected linkage disequilibrium for a neutral locus linked to a chromosomal arrangement. *Genetics* **103:** 545–555.

Takiguchi M, Nishimura I, Hayashi H, Karakl S, Kariyone A, Kano K. (1989) The structure and expression of genes encoding serologically undetected HLA-C locus antigens. *J. Immunol.* **143:** 1372–1378.

Thomsen M, Abbal M, Neugebauer M, Cambon-Thomsen A. (1989) Recombinations in the HLA system. *Tissue Antigens* **33:** 38–40.

Townsend A, Bodmer H. (1989) Antigen recognition by class I-restricted T lymphocytes. *Annu. Rev. Immunol.* **7:** 601–624.

Trinchieri G. (1989) Biology of natural killer cells. *Adv. Immunol.* **47:** 187–376.

Unanue ER, Allen PM. (1987) The basis for the immunoregulatory role of macrophages and other accessory cells. *Science* **236:** 551–557.

van der Ven K, Ober C. (1994) HLA-G polymorphism in African Americans. *J. Immunol.* **153:** 5628–5633.

Vilches C, de Pablo R, Herrero MJ, Moreno ME, Kreisler M. (1994) HLA-B73, an atypical HLA-B molecule carrying a Bw6-epitope motif variant and a B pocket identical to HLA-B27. *Immunogenetics* **40:** 166.

Wagner AG, Hughes AL, Iandoli ML, Stewart D, Herbert S, Watkins DI, Hurley CK, Rosen-Bronson S. (1993) HLA-A*8001 is a member of a newly discovered ancient family of HLA-A alleles. *Tissue Antigens* **42:** 522–529.

Watkins DI, McAdam SN, Liu X, *et al.* (1992) New recombinant HLA-B alleles in a tribe of South American Amerindians indicate rapid evolution of MHC class loci. *Nature* **357:** 329–333.

Wedekind C, Seebeck T, Bettens F, Paepke AJ. (1995) MHC-dependent mate preferences in humans. *Proc. Roy. Soc. Lond. B* **260:** 245–249.

Wei X, Orr HT. (1990) Differential expression of HLA-E, HLA-F and HLA-G transcripts in human tissue. *Hum. Immunol.* **29:** 201–240.

Zangenberg G, Huang M-M, Arnheim N, Erlich H. (1995) New HLA-DPB1 alleles generated by interallelic gene conversion detected by analysis of sperm. *Nature Genetics* **10:** 407–414.

Zemmour J, Parham P. (1992) Distinctive polymorphism at the HLA-C locus: implications for the expression of HLA-C. *J. Exp. Med.* **176:** 937–950.

Zemmour J, Koller BH, Ennis PD, Geraghty DE, Lawlor DA, Orr HT, Parham P. (1990) HLA-AR, an inactivated antigen-presenting locus related to HLA-A. *J. Immunol.* **144:** 3619–3629.

5

HLA class II genes: structure and diversity

Raymond J. Apple and Henry A. Erlich

5.1 The human major histocompatibility complex, HLA

The HLA loci on the short arm of human chromosome 6 (*Figure 5.1*) encode two distinct classes of highly polymorphic cell-surface molecules that bind and present processed antigens in the form of peptides to T lymphocytes, initiating both cellular and humoral immune responses. The class I molecules, HLA-A, -B, and -C, are found on most nucleated cells. They are cell-surface glycoproteins that bind and present processed peptides derived from endogenously synthesized proteins (e.g. viral and tumour peptides) to $CD8^+$ T cells. These heterodimers consist of an HLA-encoded α-chain associated with the non-MHC encoded monomorphic polypeptide, β_2-microglobulin. The class II molecules are encoded in the HLA-D region (*Figure 5.1*). These cell-surface glycoproteins consist of HLA-encoded α- and β-chains associated as heterodimers on the cell surface of antigen-presenting cells such as B cells and macrophages. Class II molecules serve as receptors for processed peptides; however, these peptides are derived predominantly from membrane and extracellular proteins (e.g. bacterial peptides) and are presented to $CD4^+$ T cells. The HLA-D region contains several class II genes and has three main subregions: HLA-DR, -DQ, and -DP. Both the HLA-DQ and -DP regions contain one functional gene for each of their α- and β-chains. The HLA-DR subregion contains one functional gene for the α-chain; the number of functional genes for the β-chain varies from one to two (*Figure 5.1*). All individuals express a DRB1-encoded polymorphic polypeptide that is found on the cell surface in association with the DRA-encoded polypeptide. The other functional class II DRB genes, *DRB3*, *DRB4*, and *DRB5* encode polypeptides which are found on the cell surface in association with a DRA-encoded polypeptide but at a lower level and only in certain class II haplotypes (*Figure 5.1*).

With the exception of the *DRA* and the *DPA1* loci, the genes encoding the functional class II molecules are highly polymorphic (*Table 5.1*; Bodmer *et al.*, 1995; Marsh and Bodmer, 1995). Analysis of the HLA class II crystal structure has shown that these polymorphic residues line the peptide-binding cleft and interact directly with the peptide and/or the T-cell receptor (Brown *et al.*, 1993). The extensive polymorphism at both the class I and II loci and its localization to the peptide-binding groove have led to the notion that HLA polymorphism is maintained in the population because different allelic products have the capacity to

HLA and MHC: genes, molecules and function, edited by M.J. Browning and A.J. McMichael.
© 1996 BIOS Scientific Publishers Ltd, Oxford.

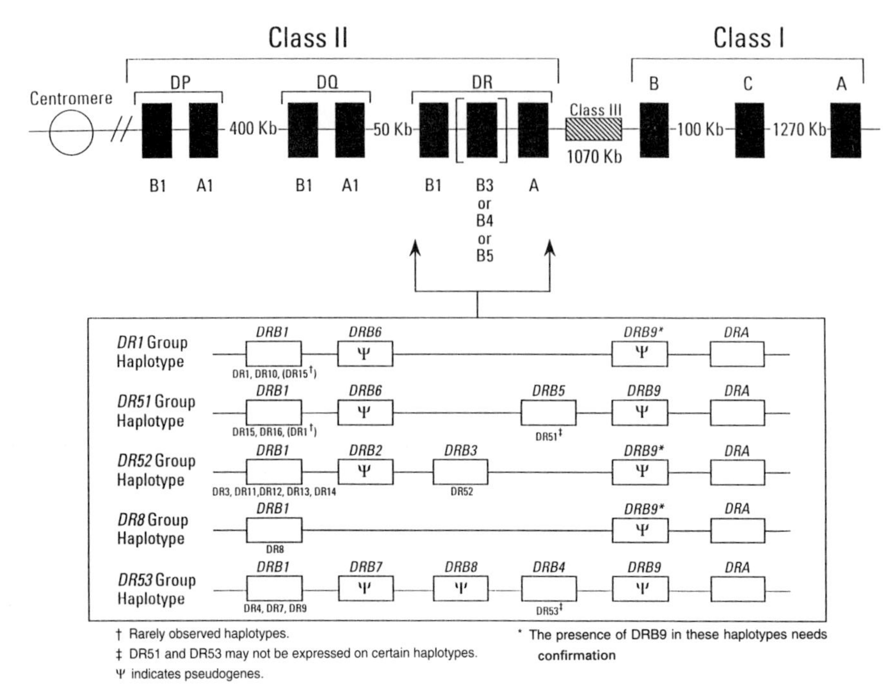

Figure 5.1. Map of the HLA complex containing the HLA class I and class II antigens.

bind different peptides; consequently, the more alleles a population has, the better it is able to cope with a large array of infectious pathogens (Doherty and Zinkernagel, 1975; Hill *et al.*, 1991).

5.2 Allelic sequence diversity

The allelic sequence diversity of the HLA class I and class II loci is the highest among mammalian coding sequence polymorphisms, with the number of alleles at some loci exceeding 100 (see *Table 5.1*). The functional significance of this extensive polymorphism, as well as the genetic mechanisms and the evolutionary forces that have generated and maintained this sequence diversity are the subject of many immunological and genetic investigations. For the HLA class II genes, the loci encoding the α- and β-chains of the DR, the DQ and the DP antigens, virtually all of the polymorphism is localized to the second exon. This exon encodes the α-helical 'walls' and the β-pleated sheet 'floor' of the peptide-binding groove formed by the α–β heterodimer. Among the α-chain loci (e.g. DRA), only the *DQA1* locus, with over 15 alleles, shows extensive polymorphism. The β-chain loci, however, are highly polymorphic. Population surveys in a variety of human populations, have identified over 65 alleles at the *DPB1* locus and over 165 at the *DRB1* locus (Marsh and Bodmer, 1995). A small number of these alleles are identical in amino acid sequence and differ only at the nucleotide level through silent substitutions; for example the 165 DNA-defined alleles for the *DRB1* locus encode 123 unique amino acid sequences (Bodmer *et al.*, 1995; Marsh and

Table 5.1. Allelic diversity at the HLA class II loci

Locus	No. of alleles
DRA1	2
DRB1	165
DRB3	5
DRB4	3
DRB5	6
DQA1	15
DQB1	25
DPA1	8
DPB1	65

The number of alleles listed here includes silent variants and is based on the 1995 WHO Nomenclature Report (Marsh and Bodmer, 1995).

Bodmer, 1995). The other DRB loci (i.e. *DRB3*) show a relatively modest number of alleles (*Table 5.1*), although some of the alleles at *DRB3* differ at many different sites.

In general, the pattern of second exon sequence diversity among the β-chain loci is a patchwork of polymorphic sequence motifs. The implication of this patchwork pattern of polymorphism for DNA-based HLA typing as well as for the evolutionary forces shaping allelic diversity will be discussed below.

5.3 HLA typing

5.3.1 Serologic and cellular methods

Historically, HLA typing has been performed using a combination of the microcytotoxicity (Amos *et al.*, 1969; Terasaki *et al.*, 1964) and mixed lymphocyte reaction (MLR) assays (Bach and Voynow, 1966). In the microcytotoxicity assay, which can be used to type both class I and II antigens, an antiserum (or monoclonal antibody) is mixed with live lymphocytes and allowed to bind to the cell surface molecules. Specific binding is then detected by the addition of complement which lyses the cells and allows the uptake of a dye. The microcytotoxicity assay requires viable cells and uses antisera obtained primarily from individuals who have been sensitized to HLA differences such as multiparous women (women who have had multiple pregnancies) or individuals who have received multiple transfusions. Consequently, these reagents are limiting in quantity and difficult to standardize. Although over 123 DRB1 allelic gene products (different amino acid sequences) have been identified (Marsh and Bodmer, 1995), serologic reagents can distinguish only 15 different groups of DR molecules encoded by these alleles; for example, the DR4 serologic specificity is found on 22 distinct *DRB1* allelic products (Marsh and Bodmer, 1995). All 22 allelic products type identically with serologic reagents but T cells can distinguish many, if not all of these 22 allelic products, suggesting that these allelic differences are functionally important in transplantation and disease.

In the mid-1960s, the mixed lymphocyte reaction was described, showing that lymphocytes from different individuals could stimulate each other to proliferate in culture (Bach and Voynow, 1966), primarily due to allelic differences in the class

II molecules of the two cell populations (Flomenberg, 1989). However, this assay only determines whether two cell types differ or are the same for HLA-D-region-encoded molecules; it does not give an actual HLA class II type. This assay was therefore used primarily to determine whether a bone marrow donor and a recipient were class II matched, although it has subsequently been shown to have little predictive value in graft-versus-host (GVH) disease (Mickelson *et al.*, 1993).

5.3.2 DNA-based techniques

Over the last decade, molecular genetic techniques have been used to isolate the genes encoding the HLA class I and class II molecules and to characterize their genomic organization (reviewed in Trowsdale *et al.*, 1985). The initial approach to HLA typing at the DNA level involved the restriction fragment length polymorphism (RFLP) method. Radioactively labelled cDNA clones or genomic fragments were used as hybridization probes for genomic DNA that had been digested with a restriction enzyme, size separated on an agarose gel, and then transferred to a membrane (Erlich *et al.*, 1986). Different alleles exhibited different banding patterns. RFLP typing required neither cell surface expression nor cell viability and, provided that several restriction enzymes were used, this technique could often subdivide HLA class II serotypes. Although an informative approach, RFLP typing was a somewhat cumbersome and time-consuming procedure that failed to distinguish much of the HLA class II sequence polymorphism.

The development of the polymerase chain reaction (PCR) in the mid-1980s (Mullis and Faloona, 1987; Saiki *et al.*, 1988) greatly facilitated the analysis of sequence polymorphism at both the class I and II loci. By generating billions of copies of the target sequence, this technique enabled the use of simple, non-radioactive methods to analyse sequence information. Based on the available database of class II allelic sequence diversity, a variety of relatively simple and rapid PCR-based methods have been developed to carry out HLA class II typing at the DNA level. The first approaches utilized labelled sequence-specific oligonucleotide (SSO) probes to hybridize to PCR products amplified from the sample and immobilized on a nylon or nitrocellulose filter, the 'dot blot' method (Saiki *et al.*, 1986). Under appropriate hybridization and wash conditions, these SSO probes would bind only to the complementary sequence in the amplified DNA and were able to distinguish single nucleotide differences. Given enough primers and probes, the SSO method is, in principle, capable of distinguishing all of the alleles at a given HLA locus. It has now been applied to typing all of the class II loci (Bugawan *et al.*, 1990; Scharf *et al.*, 1991) and, more recently, to the class I loci (Oh *et al.*, 1993; Yoshida *et al.*, 1992). Although highly informative, this method requires a separate filter and hybridization reaction for each probe; thus, for typing systems with many probes, PCR–SSO typing becomes somewhat cumbersome, particularly when typing relatively small numbers of samples.

Another PCR-based approach, based on the specificity of primer extension rather than probe hybridization, has also been applied to HLA typing. This method is known variously as allele-specific amplification (ASA), sequence-specific priming (SSP; Olerup and Zetterquist, 1992), and the amplification refractory mutation system (ARMS; Newton *et al.*, 1989). Here, a specific primer pair is designed for each polymorphic sequence motif or pair of motifs and the

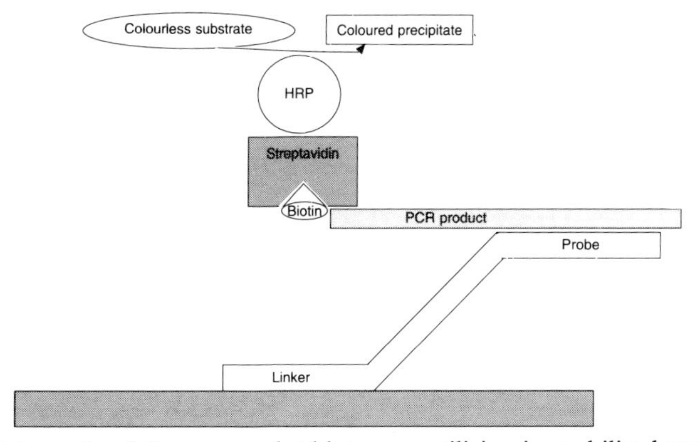

Figure 5.2. Schematic of the reverse dot blot assay utilizing immobilized probes.

presence of the targeted polymorphic sequence in a sample is detected as a positive PCR, typically identified as a band on a gel. If the PCR is negative, the sample is assumed to lack the specific motif. Since inhibition of the PCR reaction also yields a negative result, each reaction should include a positive control – an unrelated monomorphic target (Olerup and Zetterquist, 1992). Although informative and relatively fast, this approach requires many separate PCR reactions to achieve complete typing and is not suited for the rapid throughput of large sample numbers. ASA can be used in conjunction with SSO probe typing for high resolution typing. Other PCR-based methods such as PCR–RFLP (Maeda *et al.*, 1989), PCR–SSCP (single strand conformation polymorphism; Carrington *et al.*, 1992; Hoshino *et al.*, 1992), and PCR–DHA (directed heteroduplex analysis; Zimmerman *et al.*, 1993), have also been developed but are not widely used.

Recently, a new approach to SSO probe analysis of HLA polymorphism has been developed. In contrast to the conventional dot blot, which involves an immobilized PCR product that is hybridized to each of many labelled SSO probes, the 'reverse blot' method is based on the hybridization of PCR product, labelled with biotinylated primers during the amplification, to an array of immobilized probes on a nylon membrane (Saiki *et al.*, 1989; *Figure 5.2*). The presence of PCR product bound to a specific probe is detected using streptavidin-HRP (horseradish peroxidase). This enzyme can convert a soluble colourless substrate into a blue precipitate. This procedure requires only a single PCR and a single hybridization reaction to obtain information from the entire SSO probe panel; all of the probe reactivity information is contained on a single membrane, making it amenable to automated data interpretation. Recently, the hybridization, wash and colour-development steps have been automated and the probe reactivity patterns of the strips have been scanned for computer-based genotype interpretation.

The methods currently utilized to detect HLA polymorphism are described in greater detail in Chapter 6 of this volume.

5.4 Population genetics

Population genetics studies of various human populations using PCR-based typing methods have revealed the extent of allelic diversity at the HLA class II loci. In addition, these population data have been used to generate hypotheses about the

Table 5.2. Selected DRB1 alleles and DR–DQ haplotypes with unique population distributions

Class II allele	Population
*DRB1*0302*	African
*DRB1*1503*	African
*DRB1*0806*	African
*DRB1*08042*	Cayapa (Ecuador)
*DRB1*0807*	Ticuna (Brazil)
*DRB1*1105*	Filipino
Class II haplotype	
*DRB1*1502–DQB1*0502*	Filipino
*DRB1*0405–DQB1*0503*	Filipino
*DRB1*1402–DQA1*0501–DQB1*0301*	North/South American Indian/Siberian
*DRB1*1602–DQA1*0501–DQB1*0301*	North/South American Indian/Siberian

nature of the selective forces operating on these loci as well as about the pattern of human evolution and migration.

Although well over 100 *DRB1* alleles have been identified worldwide, most populations contain only 20–30 alleles. Some indigenous populations (e.g., groups of Native Americans or from Papua New Guinea) show a very restricted diversity of alleles at *DRB1* (as well as at other HLA loci), consistent with a population bottleneck in the founding of these groups (Cerna *et al.*, 1993; Imanishi *et al.*, 1992; Titus-Trachtenberg *et al.*, 1994; Trachtenberg *et al.*, 1995); for example, studies of many different Native American populations in North and South America have identified a limited set of only 4–7 *DRB1* alleles in these groups (Cerna *et al.*, 1993; Titus-Trachtenberg *et al.*, 1994; Trachtenberg *et al.*, 1995). In general, African populations contain the most diverse set of HLA class II alleles and haplotypes (Imanishi *et al.*, 1992), consistent with the hypothesis, based initially on mitochondrial DNA sequence diversity (Cann *et al.*, 1987; Vigilant *et al.*, 1991), that modern humans arose from an ancestral African population.

These population studies indicate that there are many alleles and DR–DQ haplotypes that appear to be specific for a given ethnic group. Some examples of *DRB1* alleles and DR–DQ haplotypes, unique to a particular population are shown in *Table 5.2*. Some of these population-specific alleles, such as those found uniquely in a particular indigenous group (e.g. the *DRB1*08042* allele in the Cayapa of Ecuador), may have been generated by point mutation or gene conversion from the ancestral allele after this group separated from other human groups (Titus-Trachtenberg *et al.*, 1994). In the case of population-specific alleles found in South America, these alleles may have arisen within the last 20 000–30 000 years, given the current time estimates for the colonization of North and South America. On the other hand, alleles specific to African populations could, in principle, be much older. The general issue of the 'age' of human class II alleles and the related issues of evolution rates and selective pressures are discussed below.

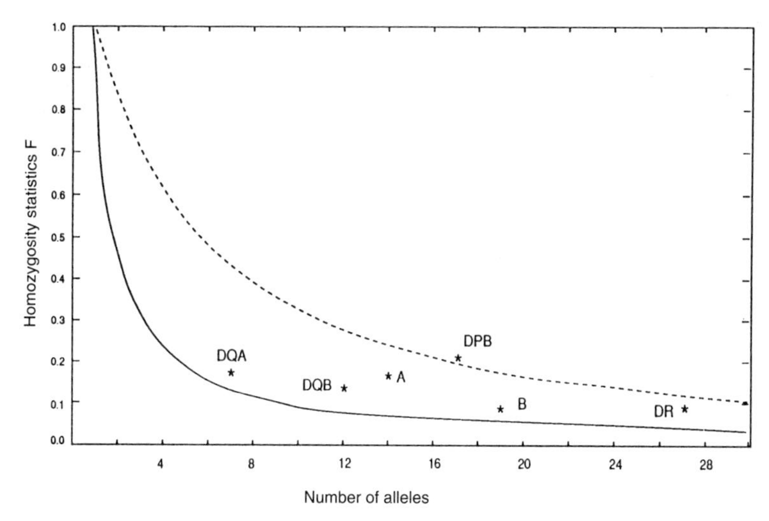

Figure 5.3. The calculated homozygosity statistic relative to the number of alleles for the HLA-A, -B, -DR, and -DP loci in the CEPH population plotted against the values predicted (a) (- - - - -) under neutrality or (b) (———) if all alleles were present at equal frequencies (minimum). HLA-A and -B typing was based on serology.

Although the frequency distributions of *DRB1*, *DQA1*, and *DQB1* alleles vary significantly among human populations, virtually all populations studied thus far have a similar property, namely a relatively even allele frequency distribution. In general, the most common allele represented has a frequency of less than 20%, and the population has several additional common alleles. This distribution, as measured by the Ewens–Watterson F-statistic (Ewens, 1972; Watterson, 1978), differs significantly from that expected for a neutral polymorphism (*Figure 5.3*) and has been interpreted as evidence of symmetric balancing selection (Begovich *et al.*, 1992; Klitz *et al.*, 1992). Balancing selection is expected to maintain multiple polymorphic alleles in a population at appreciable frequencies due to either overdominance (heterozygote advantage), frequency dependent selection, or other selective forces. (The theoretical expectation for neutral alleles is that the population contains one relatively common allele with many rare alleles, the J-shaped distribution.) One exception to the general pattern of DR balanced allele frequencies is the *DRB1* distribution found in the Philippines; in this population, one allele *DRB1*1502* (and one DR-DQ-DP haplotype, *DRB1*1502–DQB1*0502–DPB1*0101*) was found significantly more frequently than the neutral expectation (Bugawan *et al.*, 1994). This unusual distribution was attributed to strong directional selection for this frequent haplotype among the Filipino population (Bugawan *et al.*, 1994).

DQA1 and *DQB1* allele frequency distributions are, for most human populations, similar to those observed for *DRB1* and presumably reflect the same selective forces of balancing selection. The distribution of *DPB1* alleles, however, is quite different. Most human populations show a J-shaped distribution curve for *DPB1* with one common allele (frequency about 35–40%) that is not significantly different from neutrality expectations (*Figure 5.3*; Begovich *et al.*, 1992), suggesting that the selective forces shaping the *DPB1* polymorphisms have been somewhat different from the symmetrical balancing selection postulated to influence DR and DQ allele frequencies. It is unlikely, however, that the *DPB1* polymorphism is neutral. Evidence from the statistical analysis of nucleotide substitutions among the human *DPB1* sequences suggests that there is positive diversifying selection at the second exon codons that encode the amino acid residues involved in peptide binding (the antigen recognition sites or ARS). At these positions, the ratio of non-synonymous (replacement) to synonymous (silent) substitutions (dn/ds ratio) is significantly greater than 1, evidence for positive diversifying selection for functional polymorphism, whereas the ratio for codons in non-ARS sites is less than 1 (Gyllensten *et al.*, 1995). For all the class II β-chain loci, the dn/ds ratios at the ARS codons are significantly greater than 1; the dn/ds ratio for *DPB1* is even higher than the ratios calculated for either DRB1 or DQB1 (Gyllensten *et al.*, 1995).

5.5 Linkage disequilibrium

One of the characteristic properties of HLA diversity in human populations is the phenomenon of linkage disequilibrium (also known as gametic association), the non-random association of particular alleles at linked loci; for example, some *DRB1-DQA1-DQB1* haplotypes (specific combinations of linked alleles) are very much more frequent than any other combination of alleles. In any population, the number of observed haplotypes is typically a very small proportion of all possible haplotypic combinations. Within a given population, the association of particular *DRB1*, *DQA1*, and *DQB1* alleles can be almost absolute; for example, *DRB1*0301* (found in Africans, Caucasians and, more rarely, in Asians) is coupled, almost invariably, to *DQA1*0501* and *DQB1*0201*, whereas *DRB1*0302* (found only in Africans) is coupled to *DQA1*0401* and *DQB1*0402*. The set of *DRB1*0405*-containing haplotypes illustrate the population-related patterns of linkage disequilibrium. In Caucasians, this allele is coupled to *DQA1*0301* and *DQB1*0302*, among Africans *DRB1*0405* can be coupled to *DQA1*0301* and *DQB1*0201*, and in the Japanese it is coupled to *DQA1*0301* and *DQB1*0401*. In the Philippines *DRB1*0405* is coupled to *DQA1*0101* and *DQB1*0503*. *Tables 5.2* and *5.3* illustrate these patterns of DR–DQ linkage disequilibrium. Strong linkage disequilibrium between closely linked loci such as *DRB1*, *DQA1*, and *DQB1* (*DQA1* and *DQB1* arc only 12 kb apart) may reflect a lack of crossing over between the loci or, more likely, selection for particular combinations of alleles. One possible explanation for DQA1 and DQB1 linkage disequilibrium is based on preferential pairing of some α- and some β-chains. Experimental evidence for preferential pairing in the formation of DQA1/DQB1 heterodimers has been reported in the analysis of cells containing transfected *DQA1* and *DQB1* genes (Kwok *et al.*, 1989). Of course, preferential α–β-chain pairing cannot account for linkage disequilibrium between the DR and DQ loci and other selective forces may be involved. In principle, population admixture may also create linkage disequilibrium patterns but this is unlikely to account for the extensive disequilibrium found in virtually all human populations. Some HLA haplotypes reveal patterns of linkage disequilibrium that extend toward the centromere to DPB1 and toward the telomere to HLA-B and HLA-A covering almost 4 Mb (megabases) (Begovich *et al.*, 1992). These so-called ‘extended haplotypes’ (e.g.

Table 5.3 Combinations of *HLA-DQBI* and *DRBI* alleles confer IDDM susceptibility

		IDDM Risk	DQBI Codon 57
	*DRBI*0401 – DQAI*0301 – DQBI*0302*	+	Ala
	*DRBI*0401 – DQAI*0301 – DQBI*0302*	–	Asp
	*DRBI*0405 – DQAI*0301 – DQBI*0302*	+	Ala
	*DRBI*0402 – DQAI*0301 – DQBI*0302*	+	Ala
	*DRBI*0403 – DQAI*0301 – DQBI*0302*	–	Ala
	*DRBI*0405 – DQAI*0301 – DQBI*0302*	+	Ala
(Asia)	*DRBI*0405 – DQAI*0301 – DQBI*0402*	+	Asp
(Africa)	*DRBI*0405 – DQAI*0301 – DQBI*0201*	+	Ala

*A1, B8, DRB1*0301, DQB1*0201, DPB1*0101*) have been taken as evidence for strong selection for particular haplotypic combinations.

5.6 Evolution of HLA class II polymorphism

The extensive allelic sequence diversity at the HLA class II loci has elicited considerable speculation about the mechanisms that may have generated this polymorphism and the selective forces that have maintained the multiplicity of alleles in populations (Belich *et al.*, 1992; Erlich and Gyllensten, 1991; Klein and Figueroa, 1986). In general, new HLA sequence variants could arise from point mutations (nucleotide transitions or transversion), by reciprocal recombination or by segmental exchange (gene-conversion like events) within the second exon. These variants could persist in the population as a consequence of selection or, in principle, of genetic drift. Although some accounts of the evolution of HLA polymorphism assumed neutrality (Klein and Figueroa, 1986), the evidence for selective forces comes from many different analytical approaches. As noted above, the allele frequency distributions suggest balancing selection and the ratios of dn/ds in the ARS codons indicate positive diversifying selection. Comparisons of presumably newly generated alleles and their putative parental alleles suggests that selection for variants at position 86 and 57 is particularly strong (Apple *et al.*, 1992; Titus-Trachtenberg *et al.*, 1994).

Some alleles have clearly arisen by point mutation and not by a gene conversion-like event, such as *DRB1*08042* from **0802*, because the Val-86 codon in this allele (GTT) is unique among DRB1 alleles and therefore could not have been 'donated' by another allele. However, many investigators have suggested that some kind of gene conversion-like event (segmental exchange between alleles and, less frequently, among loci) could account for the patchwork patterns of polymorphism seen within the second exon of class II loci. These arguments have been based on inferences of past events from the patterns of polymorphic sequence motifs among contemporary alleles rather than on the direct detection of a gene conversion-like event. Recently, an experimental approach to detecting interallelic gene conversion and to estimating the frequency of variant class II alleles in the germline has been reported (Zangenberg *et al.*, 1995). The locus, *HLA-DPB1* was chosen because the polymorphism at this locus is clustered into six discrete regions and most of the allelic diversity looks as if it could have been generated by shuffling these discrete polymorphic sequence motifs (Bugawan *et al.*, 1990).

For this study, sperm from individuals heterozygous for DPB1 (e.g. *DPB1*0301/*0401*) were sorted into pools of 50; in each pool, one of the two alleles (and variants derived from it) was amplified for 50 cycles (sufficient to detect a single molecule) with an allele-specific primer pair. The PCR products amplified from the sperm pool were then analysed with labelled SSO probes for sequence motifs from the other, non-amplified allele, allowing the identification of novel alleles generated by interallelic segmental exchange. *DPB1* PCR products amplified from the pools containing a putative variant *DPB1* allele were cloned and the sequence of the variant *DPB1* allele was confirmed by sequencing multiple clones. The analysis of sperm pools created by mixing sperm from two homozygotes (e.g. **0301/*0301* and **0401/*0401*) served as a control for PCR artefacts (e.g. *in vitro* recombinants produced by strand-switching during PCR) as well as for variants created by point mutation rather than germline interallelic segmental exchange. The frequency of new germline *DPB1* variants, presumably generated by gene conversion-like events, was slightly less than 1/10 000. Each of the nine newly arisen *DPB1* alleles identified in this study differed from the parental allele in only one of the *DPB1* polymorphic regions, suggesting that the putative gene conversion tract was short. Because some of the sequence motifs in the polymorphic regions of the two parental alleles (e.g. *DPB1*0301* and **0401*) differed by only one nucleotide, it is formally impossible to exclude the possibility that a point mutation rather than interallelic segmental exchange was responsible for some of the new *DPB1* alleles observed in the sperm pools. No new alleles, however, were observed in the control mixture of homozygous sperm. Although the observed frequency of variant *DPB1* alleles (about 1/12 000) seems high relative to an expected mutation rate for a 300 bp segment (10^{-5}–10^{-7}), it will be important to apply this PCR-based strategy to other loci to see whether the rate of germline allelic diversification among HLA class II loci is, in fact, higher than that for other loci in the genome. Of course, only a subset of all germline variants will be favoured by selection and persist in the population. It is possible that the observation of the, presumably, new HLA class II alleles, such as those that may have been generated in the last 20 000 years, simply reflects strong selective forces rather than a high rate of germline diversification at the HLA loci. It is our view that both selective forces and a high rate of germline diversification are involved in the evolution of HLA allelic diversity, but more experiments testing the latter notion need to be performed.

Some investigators have challenged the inference that some of the HLA class II alleles are of recent origin and have suggested that most, if not all, of the alleles are millions of years old (Ayala *et al.*, 1994; Klein *et al.*, 1991). It is quite likely that, among the more than 165 *DRB1* alleles observed in human populations, some may be thousands of years old, while some may be millions of years old; preliminary data on the DR8 lineage is consistent with this view (T Bergstrom, HA Erlich, U Gyllensten, unpublished information). This situation is what might be expected from the operation of selective forces favouring diversity; a newly arisen favourable variant might be expected to co-exist with the parental allele rather than replace it. In general, most of the allelic lineages (e.g. DR3) appear to be old and predate the divergence of the hominoids (5–7 million years for chimpanzees, gorillas, and humans) but the alleles within these lineages (e.g. *DRB1*0803*, **0302*, etc.) seem to have arisen following the separation of these species and, for some alleles, after the separation of the human populations. For the class II β-chain loci, the evolution of allelic sequence diversity appears to be under selective pressure, may involve interallelic gene conversion-like events, and may be occurring on a post-speciation time scale.

5.7 Functional significance

Evidence that the extensive allelic sequence diversity found at the HLA class II loci is functionally significant comes from structural, evolutionary, and clinical, as well as from *in vitro* and *in vivo* experimental studies. As reviewed in Chapter 11, X-ray crystallographic structural studies have revealed that most of the HLA class I and class II polymorphism is restricted to amino acid residues that form the peptide-binding groove and that interact either with the peptide or with the T-cell receptor (Brown *et al.*, 1993; Bjorkman *et al.*, 1987a, 1987b; Madden *et al.*, 1992). The analysis of peptides eluted from different HLA class II molecules (reviewed in Chapter 12) shows clearly that the molecules encoded by different alleles have different inferred peptide-binding motifs. Thus, the polymorphism at the class II loci appears to determine the shape of the binding groove and, hence, the spectrum of peptides that can be bound and presented to the T cell. Furthermore, as noted above, the analysis of human and non-human primate class II alleles in terms of calculating dn/ds ratios among the ARS codons indicates that positive diversifying selection is operating on these sequences.

Most of the newly identified (and, presumably, newly arisen) alleles found in indigenous populations (e.g. South American groups) differ from the putative ancestral allele only at position 57 or position 86 residues at either end of the α-helical wall of the binding groove (Apple *et al.*, 1992; Titus-Trachtenberg *et al.*, 1994); for example, the Cayapa (Ecuador) *DRB1*08042* differs from the ancestral *DRB1*0802* only at codon 86 (Titus-Trachtenberg *et al.*, 1994), while the Ticuna (Brazil) *DRB1*0807* differs only at codon 57 (Mack *et al.*, in preparation). The analysis of intron sequences among various DR8 alleles supports the hypothesis that these population-specific alleles arose from *DRB1*0802* (T. Bergstrom *et al.*, unpublished information). The generation of *DRB1*08042* from *DRB1*0802* in South America represents an example of convergent evolution, in that *DRB1*08041* presumably arose in Africa from *DRB1*0802* by a different mutational pathway; the same protein sequence is encoded by these two alleles that differ only in the Val codon at position 86 (Apple *et al.*, 1992; Titus-Trachtenberg *et al.*, 1994). These observations suggest particularly strong selection for variants at these positions.

Clinical studies also suggest the functional importance of class II polymorphism. Susceptibility and/or protection to a large number of diseases, primarily autoimmune but also cancers and infectious diseases, have been associated with specific HLA alleles (reviewed in: Apple *et al.*, 1995; Hill *et al.*, 1991; Nepom and Erlich, 1991; Tiwari and Terasaki, 1985; see also Chapters 15 and 16). Although both linkage and association (i.e. case/control comparisons) studies have implicated the HLA region in a variety of diseases, in principle, it is possible that some of these disease associations could reflect the effects of other non-HLA genes in linkage disequilibrium with the associated HLA markers. The example of coeliac disease suggests that the HLA class II polymorphism is functionally related to disease susceptibility and not simply a linked marker (see Chapter 14). For this disease, the primary association appears to be with the genotypes DR3/X or DR5/7 (Bugawan *et al.*, 1989; Sollid *et al.*, 1989). The DR3 haplotype contains the linked alleles, *DQA1*0501* and *DQB1*0201*. The same molecule encoded *in cis* on the DR3 haplotypes is encoded *in trans* in DR5/7 heterozygotes; thus, it appears to be polymorphic residues in the DQα- and DQβ-chain that account for most, if not all, of the observed HLA associations with coeliac disease. [It is possible that, in addition to HLA class II or class I genes, other genes that map within the HLA region (e.g. *TAP-1* and *TAP-2* or *TNF-α* may also play a role in determining susceptibility or resistance to some HLA-associated diseases.)]

For several HLA-associated diseases, it appears to be specific combinations of alleles at different class II loci, rather than a single allele, that confer susceptibility or resistance. For IDDM, for example, the DR4-associated susceptibility depends on specific alleles at both the *DQB1* (i.e. *DQB1*0302* confers susceptibility) and at the *DRB1* (i.e. *DRB1*0401, *0402,*0405* confer susceptibility and *DRB1*0403* confers protection) loci (*Table 5.3*; Caillat-Zucman *et al.*, 1992; Cucca *et al.*, 1993; Erlich *et al.*, 1993). The protection from (or resistance to) IDDM conferred by certain DR–DQ haplotypes also appears to require certain *DRB1* and *DQB1* allele combinations; for example, *DRB1*1101–DQA1*0501–DQB1*0301* and *DRB1*1402–DQA1*0501–DQB1*0301* are negatively associated with IDDM (protection), whereas *DRB1*1602–DQA1*0501–DQB1*0301* is neutral (Erlich *et al.*, 1993). In addition to the well characterized DR and DQ IDDM associations, alleles at the *DPB1* locus (i.e. *DPB1*0301*) also appear to confer susceptibility to IDDM in some studies (Erlich *et al.*, 1996).

Some differences in reported HLA associations may reflect population genetics differences in the patient and control groups studied; for example, the susceptible effect of *DRB1*0405* and the protective effect for *DRB1*0403* for IDDM have generally been shown most clearly in Southern European populations (Cucca *et al.*, 1993), where these alleles are more frequent than they are in Northern Europe. Some studies have tried to interpret the HLA association patterns in terms of individual polymorphic residues, such as position 57 of the DQ-β chain (Morel *et al.*, 1988) or position 52 of the DQ-α-chain (Khalil *et al.*, 1990) for IDDM susceptibility. While there is considerable evidence that position 57 of both DQB1 and DRB1 molecules is, in fact, an important residue (see above), it is also clear that attempting to predict the disease susceptibility of an individual based on a single amino acid residue is a vastly oversimplified approach to evaluating genetic risk (Nepom and Erlich, 1991).

For those HLA-associated diseases (e.g. cervical carcinoma) in which a virus have been implicated, recent studies have shown that it is critical to type for viral genetic variation as well as that of the host (Apple *et al.*, 1994, 1995). HLA associations between human papillomavirus type 16 (HPV16)-mediated cancer cases and non-HPV16-mediated cancer cases have been shown to be different. This suggests that specific HLA class II haplotypes may influence the risk of acquiring invasive cervical carcinoma following infection with high-risk HPV viral types.

5.8 Summary

The HLA class II antigens (DR, DQ, and DP) are extremely polymorphic. These polymorphisms do not occur at random sites within the class II molecules, but are localized almost exclusively to the peptide-binding groove. Population genetics studies using PCR-based typing methods of the HLA class II allele frequency distributions suggest that the *DRB1* and *DQB1* loci are under balancing selection. However, the population distribution of *DPB1* alleles is different from that of *DRB1* and *DQB1*. It is unlikely, however, that *DPB1* polymorphism is neutral, since the ratio of synonymous to non-synonymous substitutions for codons in the antigen recognition site of *DPB1* is significantly greater than 1, as is the ratio for the *DRB1* and *DQB1* loci. Some populations, including isolated South American Indian groups, contain unique class II alleles or haplotypes, suggesting that some *DRB1* alleles may have arisen within the last 20 000–30 000 years. Some class II alleles, however, may be millions of years old. Although some alleles have clearly arisen from point mutation, the patchwork patterns of polymorphism seen within the second exon of class II loci suggest that gene conversion-like events occur between alleles, or (less frequently) between class II loci. An experimental PCR approach has been used to detect interallelic gene conversion at the DPB1 locus in sperm, and has estimated that the frequency of this event is slightly less than 1/10 000.

The polymorphism exhibited within the antigen recognition site by class II molecules affects both the specificity of bound peptide and T-cell receptor interactions. The differences in class II allele frequencies revealed by population genetics studies may influence the incidence of specific autoimmune diseases and, possibly, specific infectious diseases among different ethnic groups.

References

Amos DB, Bashir H, Boyle W, MacQueen M, Tiilikainen A. (1969) A simple micro cytotoxicity test. *Transplantation* **7:** 220–223.

Apple RJ, Begovich AB, Erlich HA. (1992) Two new HLA-DRB1 alleles found in African Americans: Implications for balancing selection at positions 57 and 86. *Tissue Antigens* **40:** 69–74.

Apple RJ, Erlich HA, Klitz W, Manos MM, Becker TM, Wheeler CM. (1994) HLA DR-DQ disease associations with cervical carcinoma show papillomavirus-type specificity. *Nature Genetics* **6:** 157–162.

Apple RJ, Becker TM, Wheeler CM, Erlich HE. (1995) Comparison of HLA DR-DQ disease associations found with cervical dysplasia and invasive cervical carcinoma. *J. Natl Cancer Inst.* **87:** 427–436.

Ayala FJ, Escalante A, O'Huigin C, Klein J. (1994) Molecular genetics of speciation and human origins. *Proc. Natl Acad. Sci. USA* **91:** 6787–6794.

Bach FH, Voynow NK. (1966) One-way stimulation in mixed leukocyte cultures. *Science* **153:** 545–547.

Begovich AB, McClure GR, Suraj VC, Helmuth RC, Fildes N, Bugawan TL, Erlich HA, Klitz W. (1992) Polymorphism, recombination and linkage disequilibrium within the HLA class II region. *J. Immunol.* **148:** 249–258.

Belich MP, Madrigal JA, Hildebrand WH, Zemmor J, Williams RC, Juz R, Petzl-Erler ML, Parham P. (1992) Unusual HLA-B alleles in two tribes of Brazilian Indians. *Nature* **357:** 326–329.

Bjorkman PJ, Saper MA, Samraoui B, Bennett WS, Strominger JL, Wiley DC. (1987a) Structure of the human class I histocompatibility antigen, HLA-A2. *Nature* **329:** 506–512.

Bjorkman PJ, Saper MA, Samraoui B, Bennett WS, Strominger JL, Wiley DC. (1987b) The foreign antigen binding site and T cell recognition regions of class I histocompatibility antigens. *Nature* **329:** 512–518.

Bodmer JG, Marsh SGE, Albert ED, et al. (1995) Nomenclature for factors of the HLA system, 1995. *Tissue Antigens* **46:** 1–18.

Brown JH, Jardetzky TS, Gorga JC, Stern LJ, Urban RG, Strominger JL, Wiley DC. (1993) Three-dimensional structure of the human class II histocompatibility antigen HLA-DR1. *Nature* **364:** 33–39.

Bugawan T, Angelini G, Larrick J, Auricchio S, Ferrara GB, Erlich HA. (1989) A combination of a particular HLA-DPβ allele and an HLA-DQ heterodimer confers susceptibility to coeliac disease. *Nature* **339:** 470–473.

Bugawan TL, Begovich AB, Erlich HA. (1990) Rapid HLA-DPβ typing using enzymatically amplified DNA and non-radioactive sequence specific oligonucleotide probes: application to tissue typing for transplantation. *Immunogenetics* **32:** 231–241.

Bugawan TL, Chang JD, Klitz W, Erlich HA. (1994) PCR/oligonucleotide probe typing of HLA class II alleles in a Filipino population reveals an unusual distribution of HLA haplotypes. *Am. J. Hum. Genet.* **54:** 331–340.

Caillat-Zucman S, Garchon HJ, Timsit J, Assan R, Boitard C, Dkilali-Saiah I, Bougueres P, Bach JF. (1992) Age dependent HLA genetic heterogeneity of type I insulin-dependent diabetes mellitus. *J. Clin. Invest.* **90:** 2242–2250.

Cann RL, Stoneking M, Wilson AC. (1987) Mitochondrial DNA and human evolution. *Nature* **325:** 31–36.

Carrington M, Miller T, White M, Gerrard B, Stewart C, Dean M, Mann D. (1992) Typing of HLA-DQA1 and DQB1 using DNA single-strand conformation polymorphism. *Hum. Immunol.* **33:** 208–212.

Cerna M, Falco M, Friedman H, Raimondi E, Maccagno A, Fernandez-Vina M, Stastny P. (1993) Differences in HLA class II alleles of isolated South American Indian populations from Brazil and Argentina. *Hum. Immunol.* **37:** 213–220.

Cucca F, Muntoni F, Lampis R, Frau F, Argiolas L, Silvetti M, Angius E, Cao A, DeVirgiliis S, Congia M. (1993) Combinations of specific DRB1, DQA1, DQB1 haplotypes are associated with insulin dependent diabetes mellitus in Sardinia. *Hum. Immunol.* **37:** 85–94.

Doherty PC, Zinkernagel RM. (1975) A biological role for the major histocompatibility antigens. *Lancet* **i:** 1406–1409.

Erlich HA, Gyllensten UG. (1991) Shared epitopes among HLA Class II alleles: Gene conversion, common ancestry, and balancing selection. *Immunol. Today* **11:** 411–414.

Erlich HA, Sheldon EL, Horn G. (1986) HLA typing using DNA probes. *Bio/Technology* **4:** 975–981.

Erlich HA, Zeidler A, Chang J, et al. (1993) HLA Class II alleles and susceptibility and resistance to insulin dependent diabetes mellitus in Mexican-American families. *Nature Genetics* **3:** 358–364.

Erlich HA, Rotter JI, Chang JD, Shaw SJ, Raffel LJ, Klitz W, Bugawan TL, Zeidler A. (1996) Association of HLA-DPB1*0301 with insulin dependent diabetes mellitus in Mexican-Americans. *Diabetes* (in press).

Ewens WJ. (1972) The sampling theory of selectively neutral alleles. *Theor. Popul. Biol.* **3:** 87–112.

Flomenberg N. (1989) Functional polymorphisms of HLA class II gene products detected by T-lymphocyte clones: summary of the Tenth International Histocompatibility Workshop Cellular Studies. In: *Immunobiology of HLA, Vol 1. Histocompatibility Testing 1987* (ed. B Dupont). Springer-Verlag, New York, pp. 532–550.

Gyllensten U, Bergstrom T, Josefsson A, Sundvall M, Erlich HA. (1995) Rapid allelic diversification of the primate class II DPB1 locus and intensified positive selection at antigen binding sites in humans. *Tissue Antigens* (in press).

Hill AVS, Allsopp CEM, Kwiatkowski D, Anstey NM, Twumasi P, Rowe PA, Bennett S, Brewster D, McMichael AJ, Greenwood BM. (1991) Common West African HLA antigens are associated with protection from severe malaria. *Nature* **352:** 595–600.

Hoshino S, Kimura A, Fukuda Y, Dohi K, Sasazuki T. (1992) Polymerase chain reaction-single-strand conformation polymorphism analysis of polymorphism in DPA1 and DPB1 genes: a simple, economical and rapid method for histocompatibility testing. *Hum. Immunol.* **33:** 98–107.

Imanishi T, Tatsuya A, Kimura A, Tokunaga K, Gojobori T. (1992) Allele and haplotype frequencies for HLA and complement loci in various ethnic groups.In: *HLA 1991, Vol. I* (eds L Tsuji, M Aizawa, T Sasazuki). Oxford University Press, Oxford, pp. 1065–1220.

Khalil I, d'Auriol L, Gobet M, Morin L, Lepage V, Deschamps I, Park MS, Degos L, Galibert F, Hors J. (1990) A combination of HLA-DQB Asp-negative and HLA-DQA Arg 52 confers susceptibility to insulin-dependent diabetes mellitus. *J. Clin. Invest.* **85:** 1315.

Klein J, Figueroa F. (1986) Evolution of the major histocompatibility complex. *CRC Crit. Rev. Immunol.* **6:** 295–386.

Klein J, O'Huigin C, Kasahara M, Vincek V, Klein D, Figueroa F. (1991) Frozen haplotypes in MHC evolution. In: *Molecular Evolution of the Major Histocompatibility Complex* (eds J Klein, D Klein). Springer-Verlag, Berlin, pp. 261–286.

Klitz W, Thomson G, Borot N, Cambon-Thomsen A. (1992) Evolutionary and population perspectives of the human HLA complex. *Evol. Biol.* **26:** 35–72.

Kwok WW, Thurtle P, Nepom GT. (1989) A genetically controlled pairing anomaly between HLA-DQα and HLA-DQβ chains. *J. Immunol.* **143:** 3598–3601.

Madden DR, Gorga JC, Strominger JL, Wiley DC. (1992) The three-dimensional structure of HLA-B27 at 2.1. A resolution suggests a general mechanism for tight peptide binding to MHC. *Cell* **70:** 1035–1048.

Maeda M, Murayama N, Ishi H, Uryu N, Ota M, Tsuji K, Inoko H. (1989) A simple and rapid method for HLA-DQA1 genotyping by digestion of PCR-amplified DNA with allele specific restriction endonucleases. *Tissue Antigens* **34:** 290–298.

Marsh SGE, Bodmer JG. (1995) HLA class II region nucleotide sequences, 1995. *Tissue Antigens* **46:** 258–280.

Mickelson EM, Bartsch GE, Hansen JA, Dupont B. (1993) The MLC assay as a test for HLA-D region compatibility between patients and unrelated donors: Results of a national marrow donor program involving multiple centers. *Tissue Antigens* **42:** 465–472.

Morel PA, Dorman JS, Todd JA, McDevitt HO, Trucco M. (1988) Aspartic acid at position 57 of the HLA-DQ beta chain protects against type I diabetes: a family study. *Proc. Natl Acad. Sci. USA* **85:** 8111–8119.

Mullis KB, Faloona F. (1987) Specific synthesis of DNA in vitro via polymerase catalyzed chain reaction. *Meth. Enzymol.* **155:** 335–350.

Nepom G, Erlich HA. (1991) MHC class II molecules and autoimmunity. *Annu. Rev. Immunol.* **9:** 493–525.

Newton CR, Graham A, Heptinstall LE, Powell SJ, Summers C, Kalsheker N, Smith JC, Markham AF. (1989) Analysis of any point mutation in DNA. The amplification refractory mutation system (ARMS). *Nucleic Acids Res.* **17:** 2503–2516.

Oh SH, Fleischhauer K, Yang SY. (1993) Isoelectric focusing subtypes of HLA-A can be defined by oligonucleotide typing. *Tissue Antigens* **41:** 135–142.

Olerup O, Zetterquist H. (1992) HLA-DR typing by PCR amplification with sequence-specific primers (PCR–SSP) in 2 hours: An alternative to serological DR typing in clinical practice including donor-recipient matching in cadaveric transplantation. *Tissue Antigens* **39:** 225–235.

Saiki RK, Bugawan TL, Horn GT, Mullis KB, Erlich HA. (1986) Analysis of enzymatically amplified β-globin and HLA-DQα DNA with allele-specific oligonucleotide probes. *Nature* **324:** 163–166.

Saiki RK, Gelfand DH, Stoffel S, Scharf S, Higuchi RH, Horn GT, Mullis KB, Erlich HA. (1988) Primer-directed enzymatic amplification of DNA with a thermostable DNA polymerase. *Science* **239:** 487–491.

Saiki RK, Walsh PS, Levenson CH, Erlich HA. (1989) Genetic analysis of amplified DNA with immobilized sequence-specific oligonucleotide probes. *Proc. Natl Acad. Sci. USA* **86:** 6230–6234.

Scharf SJ, Griffith RL, Erlich HA. (1991) Rapid typing of DNA sequence polymorphism at the HLA-DRB1 locus using the polymerase chain reaction and nonradioactive oligonucleotide probes. *Hum. Immunol.* **30:** 190–201.

Sollid LM, Markussen G, Ek J, Gjerde H, Vartdal F, Thorsby E. (1989) Evidence for a primary association of celiac disease to a particular HLA-DQ α/β heterodimer. *J. Exp. Med.* **169:** 345–350.

Terasaki PI, Mandell M, Van de Water J, Edgington TE. (1964) Human blood lymphocyte cytotoxicity reactions with allogeneic antisera. *Ann. N.Y. Acad. Sci.* **120:** 322–334.

Titus-Trachtenberg EA, Rickards O, De Stefano GF, Erlich HA. (1994) Analysis of HLA class II haplotypes in the Cayapa Indians of Ecuador: A novel DRB1 allele reveals evidence for convergent evolution and balancing selection at position 86. *Am. J. Hum. Genet.* **55:** 160–167.

Tiwari J, Terasaki P. (1985) *HLA and Disease Associations.* Springer Verlag, New York.

Trachtenberg EA, Erlich HA, Rickards O, DeStefano GF, Klitz W. (1995) HLA class II linkage disequilibrium and haplotype evolution in the Cayapa Indians of Ecuador. *Am. J. Hum. Genet.* **57:** 415–424.

Trowsdale J, Young JAT, Kelly AP, *et al.* (1985) Structure, sequence, and polymorphism in the HLA-D region. *Immunological Rev.* **85:** 5–43.

Vigilant L, Stoneking M, Harpending H, Hawkes K, Wilson AC. (1991) African populations and the evolution of human mitochondrial DNA. *Science* **253:** 1503–1507.

Watterson GA. (1977) Heterosis or neutrality? *Genetics* **85:** 789–814.

Yoshida M, Kimura A, Numano F, Sasazuki T. (1992) Polymerase-chain-reaction-based analysis of polymorphism in the HLA-B gene. *Hum. Immunol.* **34:** 257–266.

Zangenberg G, Huang M, Arnheim N, Erlich HA. (1995) New HLA-DPB1 alleles generated by interallelic gene conversion detected by analysis of sperm. *Nature Genetics* **10:** 407–414.

Zimmerman PA, Carrington MN, Nutman TB. (1993) Exploiting structural differences among heteroduplex molecules to simplify genotyping the DQA1 and DQB1 alleles in human lymphocyte typing. *Nucleic. Acids Res.* **21:** 4541–4547.

6

Detection of HLA gene polymorphism

Peter Krausa and Michael Browning

6.1 Introduction

The need for human leukocyte antigen (HLA) matching of donor and recipient for organ transplantation has driven the requirement for identification of specificities in the HLA system. As the capability for testing evolved, firstly through detection of HLA molecules at the cell surface and, more recently, by analysis of HLA genes, new specificities were uncovered and characterized. Ultimately, the complexity of HLA polymorphism has been revealed by DNA sequencing the many alleles which encode the known HLA specificities.

The genes of the class I and class II regions of the HLA gene complex represent the most highly polymorphic loci known in man. In 1987, 12 class I alleles and a class II alleles had been sequenced (Bodmer *et al.*, 1990). This has quickly grown in numbers so that in the 1995 HLA Nomenclature report (Bodmer *et al.*, 1995), some 213 class I (A, B, C) and 256 class II (DR, DP, DQ) alleles had been identified and sequenced. The revolution of DNA-based technology and its application to the identification of HLA polymorphism at the nucleotide level provided a level of resolution not possible through serological and/or biochemical means. Characterization of these HLA genes by DNA sequencing and their collection into accessible databases (Arnett and Parham, 1995; Marsh and Bodmer, 1995) provided the information for the design of DNA-based typing methods. In turn these identified new variants which were sequenced, adding to the information already held on HLA polymorphism.

Elucidation of the crystal structure of HLA class I and class II molecules (Bjorkman *et al.*, 1987; Brown *et al.*, 1993; Gorga *et al.*, 1991; Saper *et al.*, 1991) provided a physical model within which the function of HLA polymorphism could be understood (see Chapter 11). The high level of polymorphism found around the peptide-binding groove determines the nature and conformation of peptides bound by the HLA molecule (Kubo *et al.*, 1994; Saper *et al.*, 1991; Tanigaki *et al.*, 1994). HLA polymorphism which was not detectable by serology was thus shown to affect antigen presentation in the T-cell mediated immune response (Barouch *et al.*, 1995; Latron *et al.*, 1991; Rotzschke *et al.*, 1992; Tussey *et al.*, 1994; Utz *et al.*, 1992).

The polymorphic nature of the HLA system and its central role in host immunity has prompted the application of tissue typing techniques in a number of areas, including transplantation, epidemiology, disease association studies,

HLA and MHC: genes, molecules and function, edited by M.J. Browning and A.J. McMichael.
© 1996 BIOS Scientific Publishers Ltd, Oxford.

anthropology and cellular immunology. In this chapter, we review the methods which have been used to detect and characterize polymorphism within HLA.

6.2 A brief guide to HLA structure and polymorphism

The nature and effects of HLA polymorphism are dealt with in detail elsewhere in this volume, but several points require emphasis in relation to HLA typing.

HLA class I and class II antigens are highly polymorphic molecules expressed at the cell surface. HLA class I molecules comprise a major histocompatibility complex (MHC)-encoded polymorphic heavy chain associated with a non-MHC-encoded invariant light chain (β2-microglobulin) and endogenously processed peptide. HLA class II molecules comprise two MHC-encoded, non-covalently linked polypeptide chains associated with peptide. The heavy chain of HLA class I, and both α and β class II chains are type 1 transmembrane glycoproteins.

The genes that encode the polymorphic HLA class I and class II polypeptide chains cluster in the HLA region, which represents the human MHC, on the short arm of chromosome 6 (Trowsdale *et al.*, 1991). HLA class I antigens are generated from three classical loci, HLA-A, -B, -C. These classical class I molecules are expressed at the surface of most nucleated cells, and act as antigen-presenting molecules for $CD8^+$ T cells (see Chapter 13). Some non-classical HLA class I antigens also exist, namely HLA-E, -F, -G, which also associate with β2-microglobulin. These non-classical HLA molecules have restricted tissue expression, and show relatively little polymorphism as compared with the classical class I antigens (see Chapter 4). HLA-G is predominantly expressed on the trophoblast (Kovats *et al.*, 1990; Rinke *et al.*, 1990). In addition, the class I region encodes a number of pseudogenes with variable sequence homology to HLA-A, -B and -C.

Classical HLA class II antigens comprise HLA-DR, -DQ, -DP. The HLA-DR antigens consist of a polymorphic β-chain in association with an invariant α-chain. In addition to the *DRB1* locus, which codes for most of the HLA-DR polymorphism, the DR α invariant chain can combine with β-chains encoded at the *DRB3, DRB4* and *DRB5* loci to give antigens DR52, DR53 and DR51 respectively (see Chapter 5, *Figure 5.1*). HLA-DQ and DP comprise α- and β-chains, both containing polymorphism. Class II antigens are restricted in their distribution, and are constitutively expressed on B cells, macrophages, monocytes, dendritic cells, activated T cells and a few additional cell types. HLA class II molecules present antigen to $CD4^+$ T cells (see Chapter 14).

The basic intron/exon arrangements of class I and class II genes is shown in *Figure 6.1*. The great majority of sequence polymorphism is located in exons 2 and 3 of the class I heavy chain gene, and exon 2 of the class II α- and β-chain genes. These exons encode the polypeptide domains that form the peptide-binding groove of the HLA molecule. Very few sequence polymorphisms in HLA genes are silent, and the majority of nucleotide substitutions code for amino acid substitutions in or around the peptide-binding groove of the HLA molecule, potentially altering the sequence or conformation of peptide bound. These observations imply that the evolution and maintenance of HLA polymorphism is most probably mediated through selection for mutation and diversity in peptide binding, driven by pathogenic pressure (Lawlor *et al.*, 1990).

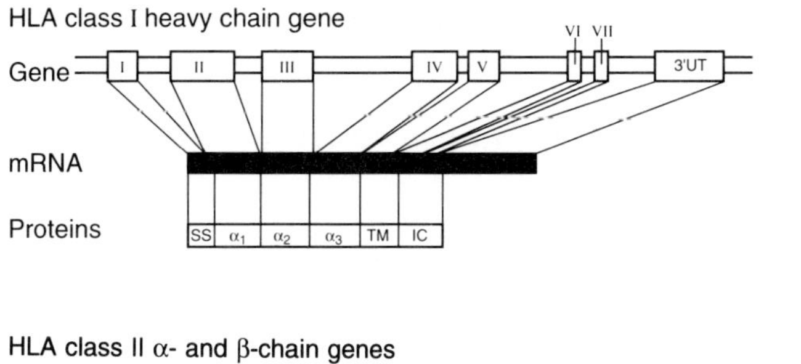

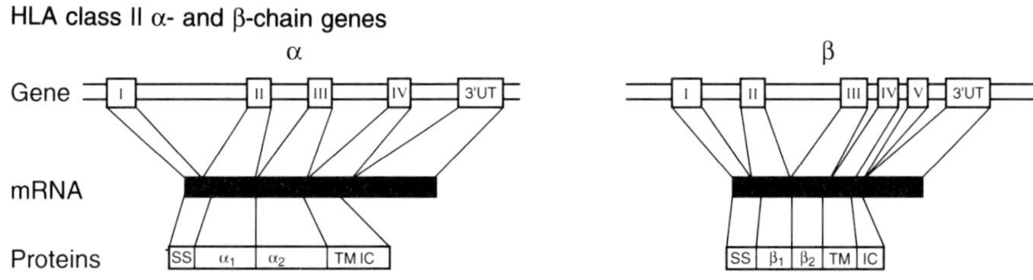

Figure 6.1. Exon/intron arrangements of HLA class I heavy chain and class II α- and β-chain genes, and the polypeptide domains of the HLA molecules that they encode. (3′UT=3′ untranslated; SS=signal sequence; TM=transmembrane region; IC=intracellular region.)

cDNA sequences of the known HLA alleles have been published (Arnett and Parham, 1995; Marsh and Bodmer, 1995). In comparing HLA cDNA sequences, the polymorphic differences are, in the most part, not represented as unique sequence motifs. Instead, most alleles gain their uniqueness through a mosaic of conserved polymorphic differences, each defined at a particular hypervariable region, resulting principally from gene conversion, recombination and exon shuffling events (Erlich *et al.*, 1991; Geraghty *et al.*, 1992; Hildebrand *et al.*, 1992, 1994; Kuhner and Peterson, 1992; Parham *et al.*., 1988; Zemmour *et al.*, 1992).

In addition to polymorphism in the coding sequences, polymorphism has been observed in the promoter regions (Cereb *et al.*, 1994; Kimura and Sasazuki, 1992a; Leen *et al.*, 1994; Louis *et al.*, 1994 and see Chapter 8) and intronic sequences of HLA genes (Cereb *et al.*, 1995), and some of these differences may also have functional implications for HLA expression. The detection and significance of promoter region polymorphism will not be discussed in this chapter, although it should be noted that DNA-based tissue typing methods can detect polymorphism in both coding sequences and promoter regions in the differentiation of HLA specificities.

6.3 Nomenclature for factors of the HLA system

A variety of techniques have been used to detect HLA polymorphism, including serological, biochemical, T-cell recognition and, most recently, molecular biological methods. These methods gauge the differences in HLA genes or their products from different perspectives. Historically, the main specificities were defined by serology, and assigned names based on the gene locus followed by a number (e.g. A10; DR1). As definition improved, many of these specificities were found to comprise several serologically definable specificities [e.g. A10 could be split into A25(10) and A26(10)]. Isoelectric focusing (IEF) was also able to identify subtypes of many serologically defined specificities, and these were named after their serological specificity followed by a number related to their isoelectric point (e.g. A2 could be defined by IEF as A2.1, A2.2, A2.3, etc.). A further level of complexity in naming HLA specificities was contributed by the use of T-cell recognition of allogeneic HLA, particularly in relation to class II specificities (e.g. the serologically defined specificity DR4 could be subdivided by T-cell clones into the T-cell-defined specificities Dw4, Dw10, Dw13, Dw14 and Dw15).

The situation was resolved by the introduction of a system of nomenclature for factors of the HLA system which was based on allelic definition of HLA gene nucleotide sequences. Under this system, alleles are defined by their gene locus and a four digit number, in which the first two digits define the specificity (largely as defined by serology) and the last two digits assign an allele number within the specificity. Thus *A*0205* defines allele 5 of HLA-A2. Similarly *DRB3*0301* defines the first (and so far only) allele of the third specificity to be defined of the *DRB3* gene. This system of nomenclature allows the unambiguous assignment of HLA specificities, within the parameters of the known HLA alleles.

6.4 Detection of polymorphism in HLA molecules

6.4.1 What the serologist saw

The HLA system was discovered through observations of agglutination of leukocytes with antisera taken from unrelated transfused individuals (Dausset, 1954). The original specificity found (Mac; Dausset, 1954) was later to be known as HLA-A2. Similarly, specificities relating to Bw4, Bw6 were identified, followed by HLA-A1 and -A3, and this marked the unfolding of the polymorphic nature of the HLA system.

The ability to detect these antigens advanced through the use of the HLA specific antisera in complement-mediated cytolysis of HLA-expressing leukocytes (Kissmeyer *et al.*, 1969; Terasaki and McClelland, 1964). This micro-lymphocytotoxicity assay (*Figure 6.2*) became the standard approach to histocompatibility testing from the 1960s through three decades. The use of polyclonal antisera from transfused individuals, or more commonly, from multiparous women (antisera against 'paternal' HLA determinants), provided the first means of elucidating the highly polymorphic nature of the HLA system. A substantive panel of well characterized antisera are necessary to detect the large number of HLA specificities, since many antisera have multiple specificities.

Serological characterization of HLA class I antigens uncovered a growing amount of polymorphism, as many of the early-defined specificities were 'split'; for example, HLA-A9 was split into A23 and A24, and HLA-B40 into B60 and B61. HLA-C specificities have been extremely difficult to define by serology, probably due to their low level of expression, and many were designated as blanks.

Serology for class II specificities (HLA-DR, -DP, -DQ) was developed at a later date (Ceppellini and van Rood, 1974; van Leeuwen *et al.*, 1973, 1982; van Rood *et al.*, 1975), following observations made in the mixed lymphocyte test (see below;

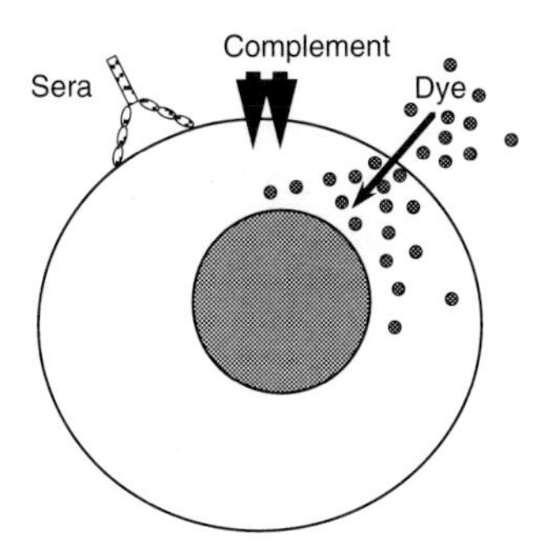

Figure 6.2. The basis of the micro-lymphocytotoxicity assay. Viable peripheral blood mononuclear cells (class I) or separated B cells (class II) are mixed with antisera of known HLA specificity. The binding of HLA-specific antibodies to the target cell activates complement, and the cell is lysed. Dye enters the dead cell, denoting a positive result, and this can be detected by microscopy. The absence of dye within the test cells indicates the HLA-specificity of the antiserum is different to that expressed by the cells.

Amos and Yunis, 1971; Mempel *et al.*, 1972; Yunis and Amos, 1971) which pointed to another component in the HLA system in addition to HLA-A, -B, -C, initially termed HLA-D. Identification of class II specificities was then achieved through serological identification, using antisera depleted of class I specificities through platelet absorption. The requirement for separation of HLA class II bearing B cells from peripheral blood mononuclear cells (PBMC), together with acquiring the appropriate antisera, adds an additional layer of complexity to serological definition of class II specificities. In addition, it quickly became apparent that serology did not adequately define the antigens encoded in the class II region. As the degree of polymorphism within HLA became apparent, it was increasingly evident that this was also the case for the antigens of the class I region.

The use of monoclonal antibodies in the field of histocompatibility testing promised the means for monospecific reagents, which would circumvent the problem of crossreactivity present in polyclonal antisera. However, as most antigenic epitopes or 'motifs' are not unique to a particular HLA type, but can be shared by several specificities, this promise was never fully realized; for instance, monoclonals have been described which have specificity for A2 (but not *A*0210*) and A69 (Brodsky *et al.*, 1979). This pattern of reactivity can be interpreted by comparing the sequences of A2 and A69 gene products, with the epitope mapping to a tryptophan at amino acid position 107, which is in an accessible position for antibody detection. Monoclonal antibodies that identify a single serologically defined HLA specificity are rare, and although more specific than polyclonal sera, typing with monoclonal reagents still relies largely on reaction patterns. Monoclonals do have the advantage that they represent a renewable source of reagent rather than using polyclonal antisera obtained from multitransfused or multiparous individuals.

Serology remains the mainstay method for many routine histocompatibility testing laboratories, particularly for typing HLA class I. However, serological typing can only define polymorphisms at the surface of the molecule, or which cause a conformational change in the molecule which is accessible to antibody. Polymorphism located within the peptide-binding groove may not be antigenic in terms of raising antibody, but its position may have great functional relevance as to the peptide-binding characteristics of the molecule (for example, see Barouch *et al.*, 1995). Characterization through biochemistry and T-cell assays made it clear that serology was not capable of identifying all clinically relevant variants. To this end, additional methods of characterization had to be considered.

6.4.2 The biochemical focus

Detection of polymorphism can also be achieved by looking at the different amino acid composition of HLA molecules through biochemical techniques such as one-dimensional IEF (Yang, 1989b). This method relies on amino acid substitutions contributing to changes in charge of the HLA molecule. The labelled HLA molecule is stripped from the cell surface and immunoprecipitated with an appropriate antibody. Electrophoresis is carried out on the immunoprecipitate through a pH gradient. The individual proteins within the immunoprecipitate will focus on a gel at points appropriate to their charge, forming discrete bands when exposed on an autoradiograph (*Figure 6.3*).

The use of IEF for tissue typing demonstrated a level of polymorphism beyond that observed in serological typing. Reports from International Histocompatibility Workshops have described many of the different IEF variants (Yang, 1989a). Hence for HLA-A2, five IEF variants were identified, four HLA-A24 variants, three HLA-A30 variants and so forth through class I and class II.

Unfortunately, IEF may not be used independently. It requires initial information from serology (or other methods) to allow interpretation of the focused bands. This is due to many of the bands of different HLA specificities focusing in close proximity to each other (*Figure 6.3*). However, IEF typing offers a level of resolution beyond serology, as variants not detected by serology can be clearly distinguished through IEF analysis. Identification of these variants has allowed their further characterization through DNA sequencing, adding to our understanding of the relevance and nature of HLA polymorphism.

6.4.3 Mixing for matching: T-cell based typing

Concurrent to observations being made using HLA-specific antisera, it was noted that lymphocytes from two unrelated sources, when mixed in culture, would proliferate (Hirschorn *et al.*, 1963). This did not occur when the lymphocytes were obtained from HLA identical sib pairs (Bach and Amos, 1967). The proliferation noted in this mixed lymphocyte reaction (MLR) could be quantified through incorporation of radioactive ^{3}H-thymidine into the dividing cells. The MLR was made easier to interpret through inactivation (using irradiation or mytomycin-c treatment) of one of the stimulator populations (Bach and Voynow, 1966). Inactivated stimulator cells that had been well characterized for HLA specificities, could be used as typing cells. The use of homozygous typing cells as stimulators (Mempel *et al.*, 1973b), delivered a more concise result since only one HLA haplotype needed to be considered (*Figure 6.4*).

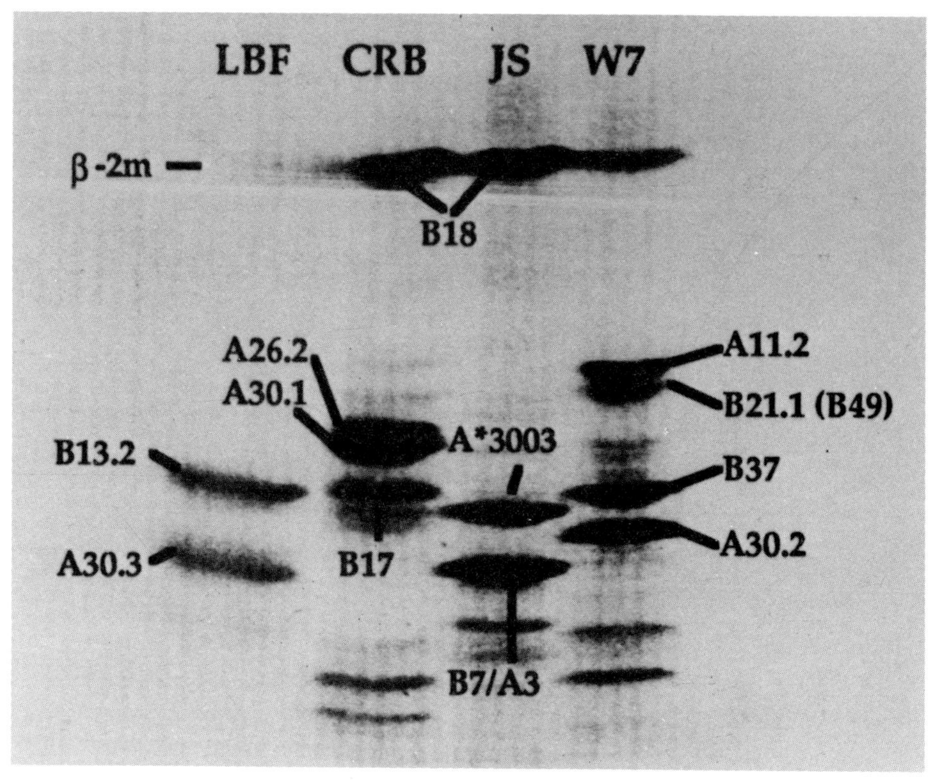

Figure 6.3. A comparison of four HLA-A30 sequenced cell lines by one-dimensional IEF. All four cell lines had been identified serologically as A30. One-dimensional IEF performed in the context of the cell lines' known serological types allows clear discrimination between A30.1 (*A*3002*), A30.2 (*A*3004*), A30.3 (*A*3001*) and *A*3003* (previously unclassified by IEF). Reprinted by permission of Elsevier Science Inc., from "Defining the allelic variants of HLA-A30 in the Sardinian population using Amplification Refractory Mutation. System-Polymerase Chain Reaction", by P. Krausa *et al.*, *Human Immunology*, **44:** pp. 35–42. Copyright 1995 by the American Society for Histocompatibility and Immunogenetics.

The MLR provided an initial technique for determination of HLA class II specificities. Specificities were denoted by the 'Dw' nomenclature (Bodmer *et al.*, 1984), which later showed relation to the serologically defined 'DR' specificities (Jaraquemada *et al.*, 1986). The MLR also provided a method for matching donor with recipient when considering transplantation. Studies have shown that serologically matched unrelated individuals can produce a positive MLR (Mempel *et al.*, 1973a). This indicates either that serology is insufficiently sensitive to perfectly match class II specificities, or the presence of mismatches amongst other 'minor' histocompatibility antigens, not considered in the serological typing. The MLR was therefore capable of detecting functional polymorphism not seen by serology. The polymorphism identified in the MLR has been shown to be significant in terms of the survival of the graft in the transplant (Morishima *et al.*, 1995). However, negative MLR does not necessarily indicate perfect matching, and these undetected mismatches may still contribute to the rejection of the graft (Eiermann *et al.*, 1992; Tiercy *et al.*, 1991).

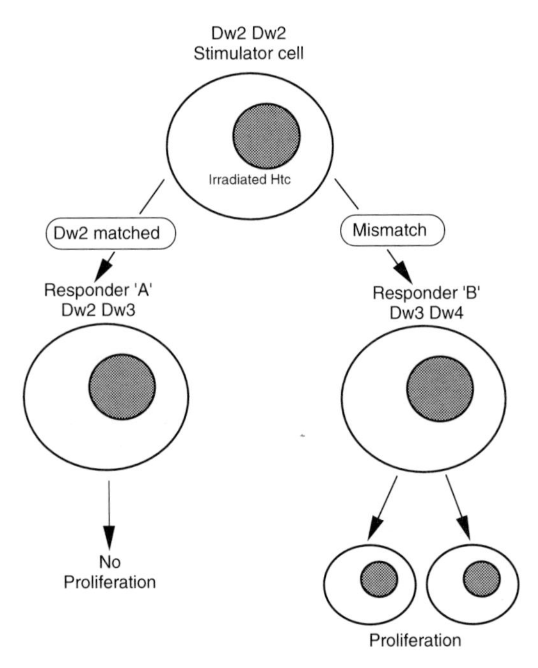

Figure 6.4. The one-way, mixed lymphocyte reaction (MLR) detects polymorphism on the basis that mixing HLA mismatched cells will cause proliferation of the 'responder' cells, detected through incorporation of ^{3}H-thymidine into the dividing cell population. Use of HLA-characterized homozygous typing cells (HTC) as the stimulator population simplifies interpretation. In the examples given, the Dw2 HTC fails to stimulate the Dw2 Dw3 responder cell 'A' through being matched at Dw2. However, responder 'B' is stimulated by the Dw2 HTC through being mismatched with Dw3 and Dw4, causing the responder cells to proliferate.

The use of T cells in HLA typing was further refined by the generation of HLA-specific T-cell clones, derived by repeated stimulation in MLR of T cells with the HLA antigenic determinant against which they were raised. This formed the basis of the primed lymphocyte test (PLT; Sheehy *et al.*, 1975), in which proliferation of the PLT clones in response to stimulation with the test cells indicated the presence of the same HLA specificity as in the original stimulating cell line. PLT typing was particularly useful for identifying and characterizing HLA-DP specificities (Morling *et al.*, 1980).

The MLR can also generate cytotoxic T lymphocytes (CTL), directed principally against allo-HLA class I molecules. The generation of such CTL clones provided another 'reagent' for demonstrating and characterizing HLA polymorphism. Alloreactive CTL clones have been used to demonstrate polymorphism within HLA specificities which were not detected through standard serological approaches (Castano *et al.*, 1991; McMichael *et al.*, 1988; Rotzschke *et al.*, 1992; Tussey *et al.*, 1994).

More recently, assays for the estimation of precursor frequencies of helper (HTLp) and cytotoxic (CTLp) T cells in the recipient capable of responding to donor allo-antigens have been introduced (Schwarer *et al.*, 1994; Sharrock *et al.*, 1987; Zhang *et al.*, 1989), and provide another index by which HLA polymorphism can be detected. Responder cells are cultured in limiting dilution with inactivated stimulator cells. Precursor frequencies for allo-reactive T cells are then estimated from the relationship between the presence or absence of a proliferative, IL-2 secreting or cytotoxic response and the responder cell concentration. These assays have been shown to be of some value in predicting graft versus host disease (GVHD) in serologically matched, unrelated-donor bone marrow transplantation (Kaminski *et al.*, 1989).

The fine specificity of allo-reactive T-cell clones demonstrates the functional significance of HLA polymorphisms not readily discriminated by serological typing. T-cell clones can cross-react, however, and they may not make ideal typing reagents for this reason (van der Poel *et al.*, 1986). However, the stimulation of allo-HLA specific T-cell responses by apparently HLA-matched stimulator cells does provide an indicator of the inadequacy of serological tissue typing in not identifying potentially significant polymorphic differences.

6.5 Molecular biological approaches to HLA typing

Analysis of HLA specificities from DNA provided a new approach to defining their polymorphic differences. Rather than looking at differences in the expressed molecule, polymorphism is characterized at the nucleotide level. DNA-based methods hold particular advantages over serological approaches to HLA typing; there is no need to rely on viable cells, reagents are renewable and there is increased potential for standardization. Recent International Histocompatibility Workshops have been instrumental in encouraging development and implementation of new typing techniques and in providing a forum for their standardization.

The development of these techniques was made possible by the sequencing of large numbers of HLA genes. In broad terms, the development of DNA typing class I alleles has encountered greater problems than those experienced for class II. The reasons for this can be explained in comparing the nature of polymorphism between class I and class II. Polymorphism within class II is contained mainly within defined hypervariable regions in exon 2, making differentiation between alleles readily achievable through hybridization with the relevant probe. This situation is more complex in class I alleles. Hypervariable regions are found at different degrees in both exons 2 and 3, which encode the peptide-binding groove of the class I molecule. Polymorphic differences in the hypervariable regions of these exons tend to be shared not only by other alleles within the same locus, but also alleles of other class I loci and pseudogenes, and there are few unique sequence motifs defining a particular allele. For these reasons, and as serology was long held to provide an adequate means for typing HLA class I, the development of DNA-based methods for typing HLA class I alleles lagged somewhat behind methods for typing HLA class II.

6.5.1 Restriction fragment length polymorphism (RFLP) analysis

Initially, there was insufficient gene sequence information to design definitive DNA typing approaches. Because only a few HLA specificities had been sequenced, analysis of DNA was an extension of other techniques. One such method was the use of restriction fragment length polymorphism (RFLP; Bell *et al.*, 1987; Bidwell *et al.*, 1987; Bodmer *et al.*, 1987; Dormoy *et al.*, 1992; Hongming *et al.*, 1986; Inoko *et al.*, 1986; Lin *et al.*, 1989; Moller *et al.*, 1985; Segurado *et al.*, 1990; Sheldon *et al.*, 1986; Tilanus *et al.*, 1986). This method relies on the use of panels of restriction enzymes, which digest DNA at specific sequence sites. The resulting digest is run on a gel to separate the different-sized DNA fragments, and this is Southern blotted on to a membrane and hybridized with an HLA class I- or class II-specific, labelled probe. HLA-specific bands are then detected on an autoradiograph. The size and pattern of banding with different restriction enzymes and different HLA probes is used to determine the HLA specificity.

RFLP analysis of DNA was useful in determining polymorphism at a time when relatively few HLA alleles had been sequenced, but did not provide unequivocal determination of HLA specificity. The size and pattern of bands identified by the HLA probes relies on polymorphic differences being at sites which can or cannot be cut by the different restriction enzymes used. Analysis of the band patterns can be difficult, especially if the probe used is not particularly specific, or the enzyme does not define any useful polymorphism. The method was mainly applied to HLA class II, with class I RFLP analysis being perceived as difficult with no advantage over serology or IEF.

The development of the polymerase chain reaction (PCR; see below) allowed the specific amplification of regions of DNA for further analysis. To this end RFLP analysis has been applied to HLA PCR products. The specificity of the PCR replaces the need for hybridization with an HLA specific probe. Because many of the sequences of HLA alleles were known by this time, enzyme restriction sites could be found within the PCR product at sites of polymorphism. This allowed clearer interpretation of PCR–RFLP typing since it was within the context of known sequence. The combination of specific PCR with RFLP has been successfully used in several studies (Hviid *et al.*, 1992; Maeda *et al.*, 1989; Medintz *et al.*, 1994; Salazar *et al.*, 1992; Tanaka *et al.*, 1992; Yunis *et al.*, 1991). However, as more HLA gene sequences became available, more definitive DNA-based HLA typing methods were developed, which have largely superceded RFLP analysis.

6.5.2 Polymerase chain reaction

PCR has been an important and powerful development in the field of molecular biology (Mullis and Faloona, 1987; Mullis *et al.*, 1986; Saiki *et al.*, 1985). PCR allows specific amplification of stretches of DNA sequence, through repeated cycles of DNA denaturation, annealing of specific primer to the DNA single strand and nucleotide extension from primer pairs using a thermally stable enzyme, to generate substantial quantities of specifically amplified regions of DNA which can then be used in further analysis (Bell, 1989). In tissue typing, PCR was used to amplify the polymorphic regions of HLA genes. This HLA PCR product could then be analysed for its polymorphic differences, to establish the tissue type. A number of such approaches have been taken, including the use of PCR in combination with sequence-specific oligonucleotide probing (PCR-SSOP; Saiki *et al.*, 1986), probing by reverse dot blot (Saiki *et al.*, 1989), sequence-based typing (SBT) of the PCR product (Santamaria *et al.*, 1992), heteroduplex analysis of PCR products (Clay *et al.*, 1994), single-stranded conformational polymorphism analysis of the PCR product (PCR-SSCP; Yoshida *et al.*, 1992) or by the use of sequence specific primers in the PCR reaction (SSP-PCR; Olerup and Zetterquist, 1991). These approaches, used singly or in combination, have all been applied as DNA-based methods for tissue-typing class I and class II HLA specificities.

6.5.3 Sequence-specific oligonucleotide probing of PCR product

One initial and popular method of DNA typing is the use of SSOP for detection of polymorphism within PCR amplified regions of HLA genes. Initial PCR amplification of hypervariable regions of the HLA genes (e.g. exons 2 and 3 of the *HLA-A, -B, -C* genes and exon 2 of the *DRB1* genes) provides a template which can be probed with labelled oligonucleotides which have specificity for particular sequence polymorphisms (*Figure 6.5*). Panels of these probes can then be used to identify which polymorphisms are present in the amplified DNA.

Tissue typing by PCR-SSOP relies on a pattern of hybridization reactivity, which decreases in complexity in relation to increasing PCR specificity. There can be some difficulty in the interpretation of the results, as the majority of the polymorphic differences detected are not unique to a particular allele, but are shared with other alleles, some of which may be found in other loci, depending on the specificity of the PCR. The complex patterns of reactivity detected may require detailed analysis to assign specificities, and certain heterozygous allele combinations can prove problematic to assign unambiguously since hybridization patterns fail to provide a definitive answer.

PCR-SSOP was first applied to HLA class II alleles, made possible through the availability of sequenced alleles. Initial studies concerned with typing HLA DQ specificities quickly spread to DR specificities as more sequence information became available. PCR-SSOP HLA class II typing was applied to studies in forensics (Higuchi *et al.*, 1988) disease association (Hill *et al.*, 1992; Wordsworth, 1991) and histocompatibility matching for transplantation (Tiercy *et al.*, 1989, 1991). Standardization of methods for PCR-SSOP typing of HLA class II happened through the 1991 International Histocompatibility Workshop (Kimura and Sasazuki, 1992b), which provided the forum for introduction of DNA-based typing to histocompatibility laboratories worldwide. The success of class II typing by this method and the experience gained, provided the incentive for further improvements, particularly to detect through non-radioactive means (Cereb *et al.*, 1995; Chen *et al.*, 1994; Erlich and Gyllensten, 1991; Giorda *et al.*, 1993) and improve sample handling.

HLA class I PCR-SSOP developed in the wake of class II, due initially to the lack of sequence data for class I alleles, but also because many workers were content with the continued use of serology. The sequencing of increasing numbers of class I alleles made clear that serology lacked the sensitivity to differentiate between certain specificities. For this reason, specificities such as the variants of HLA-A2, -A68 (Fernandez-Vina *et al.*, 1992; Tiercy *et al.*, 1994), or differentiating HLA-B35 and B53 (Allsopp *et al.*, 1991; Hill *et al.*, 1992) became the subject of

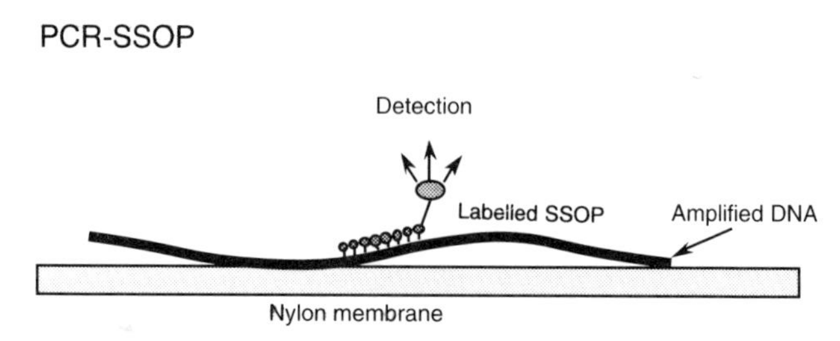

Figure 6.5. PCR-SSOP. PCR amplified DNA is fixed to a nylon membrane through UV cross-linking. A panel of labelled, specific oligonucleotide probes are then hybridized with the DNA on the membrane. The membranes are washed to an appropriate stringency, to remove any false hydridization. Detection of hybridization is through the labelled probe. Both radio-labelled and non-isotopic probes can be used in association with appropriate methods of detection.

class I typing developments. Broader PCR-SSOP methods have now been described which type the HLA-A locus (Allen *et al.*, 1994; Gao *et al.*, 1994; Oh *et al.*, 1993), the B locus (Fernandez-Vina *et al.*, 1995; Middleton *et al.*, 1995; Yoshida *et al.*, 1992) and C locus (Levine and Yang, 1994). The combination of the described methods at the different loci, has allowed for general PCR-SSOP class I DNA typing.

The reverse dot blot approach to PCR-SSOP. In parallel to development of the conventional PCR-SSOP, a variation of the method known as the 'reverse dot blot' has also been described (Saiki *et al.*, 1989). In this approach, the oligonucleotide probes are immobilized on to the membrane instead of the amplified DNA (see Chapter 5, *Figure 5.2*). The labelled PCR product can then be simultaneously hybridized against all the probes bound on a single membrane. This innovation, available as a kit, has been widely used. It means that an individual can be typed quickly through one hybridization event, rather than the lengthier conventional PCR-SSOP methods.

The raw data obtained by reverse dot blot arc similar to that of conventional PCR-SSOP, in that they may contain complex hybridization patterns which require the aid of an appropriate computer program for their interpretation. Currently, the reverse dot blot approach is limited to class II typing (Erlich *et al.*, 1991; Saiki, 1989) with class I typing only recently being attempted (Bugawan *et al.*, 1994). However, the success of conventional PCR-SSOP methods should mean, in theory, that the reverse dot blot approach should also be feasible for complete class I typing, and this is currently under development. A variation of the reverse dot blot is immobilizing probes in microtitre plates rather than nylon membranes. An advantage of using the microtitre plates is the array of dedicated equipment available for use with the format. Detection in plates therefore makes semi- or fully-automated HLA typing possible. The use of microtitre plates to perform the reverse dot blot has been reported by a number of groups for typing HLA specificities (Giorda *et al.*, 1993; Kawai *et al.*, 1994).

6.5.4 Typing using PCR-SSP

An alternative PCR-based method of typing HLA class I and class II genes is by the use of PCR-SSP. Each PCR is designed to identify a particular HLA specificity or group of specificities. In order to discriminate polymorphic differences, often of a single nucleotide, SSPs were designed on the basis of the amplification refractory mutation system (ARMS; Newton *et al.*, 1989). This entails matching the 3′ end of the primer with the required sequence polymorphism. A mismatch at the 3′ residue of the primer with the target sequence inhibits PCR amplification under appropriate conditions (*Figure 6.6*). The correct conditions rely on stringency, determined by the composition of the reaction mix and the annealing temperature in the PCR amplification program. The required specificity of the PCR is achieved through the combination of the specificity of both SSPs, defining two points of polymorphism which differentiate it from other HLA types. The advantage of knowing the priming site of each SSP is that the resulting PCR product will be of an expected size. The inclusion of an internal control amplification validates PCR conditions.

Identification of the tissue type requires a panel of specific PCR reactions which are performed simultaneously. Following the PCR, products are visualized on an agarose gel and the tissue type determined by the presence or absence of the appropriate-sized bands in each lane. PCR-SSP allows an individual sample to be typed in one step, rather than the multiple hybridizations necessary for conventional SSOP procedures. The ease of interpretation of PCR-SSP typing is a major advantage of this approach. One limitation of PCR-SSP is its non-suitability for large-scale typing as compared to PCR-SSOP. This is mainly due to restriction in sample handling and equipment, especially if PCR reactions are performed in individual tubes. However, sample throughput has improved through the use of microtitre plates and the associated technology. In addition, multiplex PCR-SSP has been described, in which several HLA-specific PCR yielding products of different band size are performed in a single reaction tube (Browning *et al.*, 1994), reducing the overall number of PCR required.

The use of PCR-SSP for HLA typing, was first described for class II specificities (Olerup and Zetterquist, 1991, 1992; Zetterquist and Olerup, 1992). HLA-DR typing was achieved through analysis of sequence polymorphism located in exon 2 of the *DRB1* alleles. Additionally, DR52 and DR53 were determined through analysis of *DRB3* and *DRB4* loci. Development of PCR-SSP spread to typing at other HLA class II loci, namely DQ (Bunce *et al.*, 1993; Olerup *et al.*, 1993) and DP (Knipper *et al.*, 1994) specificities.

Identification of class I alleles by PCR-SSP followed. Most PCR-SSP methods identify polymorphism in exons 2 and 3 of the class I gene, however, polymorphisms present in exons 1 and 4 have also been used for typing (Browning *et al.*, 1993; Bunce *et al.*, 1994, 1995a; Guttridge *et al.*, 1994; Krausa *et al.*, 1993a; 1995a,b; Sadler *et al.*, 1994). The advantages of PCR-SSP class I typing over serology were emphasized by the easy discrimination of HLA-C locus specificities (Bunce *et al.*, 1994). This has always been a weak point for serology, due to the low expression of HLA-C on the cell surface, and cross-reactivity of antisera. The ability of PCR-SSP to type *HLA-C* locus alleles, including C locus ‘blanks’, underlined the worth of DNA-based typing. Development of PCR-SSP

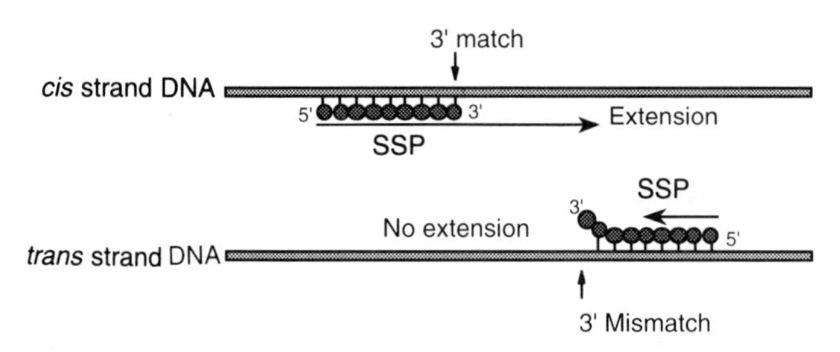

Figure 6.6. The use of ARMS in PCR-SSP. Highly specific PCR reactions are used to discriminate between different HLA types. PCR primers are designed on the principle of ARMS (Newton *et al.*, 1989), in which mismatch between template DNA and the 3′ residue of the SSP inhibits amplification. For amplification to occur under the appropriate PCR stringency, both SSPs need to be matched at their 3′ residues. In the figure, the SSP is matched to the *cis*-strand DNA, and extension proceeds on this strand. A 3′ mismatch on the SSP binding to the *trans*-strand DNA inhibits extension, so causing the PCR reaction to fail.

typing for HLA class I showed that, at each locus, DNA-based typing could rival or exceed the resolution achievable through serology.

HLA phototyping. Once PCR-SSP typing had been demonstrated at all the major HLA loci, an obvious progression was to combine all these published typing panels into a single, integrated system for typing HLA class I and class II. The use of 96- or 192-well microtitre plates for PCR provided the means for performing large numbers of PCR. This multiple locus PCR-SSP typing procedure, known as 'phototyping' (Bunce *et al.*, 1995b), provides a complete DNA-based alternative to serology. The method describes 144 PCR reactions covering HLA-A, -B, -C, DRB1, DRB3, DRB4, DRB5 and DQB1 loci (*Figure 6.7*). Tissue type is interpreted based on the pattern of reactivity with the reaction panel. As with any other DNA-based typing approach, the publication of new sequences has to be scrutinized and the change in specificity of the reaction panel noted, with new or replacement reaction combinations added as appropriate. This description of a comprehensive PCR-based system for typing HLA class I and class II has prompted the view that DNA-based typing will soon replace HLA typing by conventional methods such as serology (Dupont, 1995).

Allele-specific typing using nested PCR-SSP. The degree of specificity that can be achieved using conventional PCR-SSP is limited by the use of two polymorphic sites to define the specificity of the PCR. The resolution can be increased by the use of nested PCR, such that the specificity of the reaction is defined by a combination of four polymorphic sites. Nested PCR-SSP involves a two-step procedure. First, a specific PCR reaction is performed, which encompasses the desired allelic group and flanks the polymorphisms necessary for their discrimination. Then the first-round flanking PCR product is used as DNA template for the second round of PCR-SSP reactions. In a practical sense, nested PCR-SSP can be useful in situations where only a small amount of DNA exists, since the first round reaction generates more sample for use in typing. The real advantage of 'nesting' is that the specificity of the first round flanking reaction removes potential cross-reactivity from 'unwanted' sequences (such as pseudogenes or similar alleles in the same or other loci) from the second round PCR. This increases the availability of sequence polymorphism for definitive PCR-SSP typing reactions. The nested approach of PCR-SSP for HLA typing was first described for typing HLA-DRB1 (Bein *et al.*, 1992).

The increased power of discrimination of nested PCR-SSP was subsequently shown to provide an allelic level of definition of certain HLA 'group' specificities (Krausa *et al.*, 1995a; Krausa and Browning, 1996); for example, the frequency and diversity of *A*02* alleles was shown to vary according to the ethnic origin of a population (Krausa *et al.*, 1995a). Many of the polymorphisms that differentiate *A*02* subtypes have been shown to have functional significance in terms of T-cell recognition (Browning and Krausa, 1996). Typing at an allelic level of resolution therefore allows proper interpretation of the HLA-restricted immune response. Allele-specific typing can also uncover or strengthen associations between HLA alleles, not apparent by lower resolution methods, so providing useful information for anthropological and disease-association studies; for example, in subtyping *HLA-A*30* in a Sardinian population, linkage disequilibrium between *HLA-A*3002* and *HLA-B18* was shown (Krausa *et al.*, 1995b), and in the same population an unexpectedly high frequency of *A*0205* was found amongst *A*02* alleles (Carcassi *et al.*, 1995). It has recently been reported in the Japanese population that the autoimmune thyroid diseases, Grave's disease and Hashimoto's thyroiditis, are associated with different allelic variants of *HLA-A*02* (Sudo *et al.*, 1995).

6.5.5 Heteroduplex analysis and conformational polymorphism

Another approach to identifying HLA polymorphism, is through heteroduplex analyses (Clay *et al.*, 1994; Clay *et al.*, 1991; Oka *et al.*, 1994; Rubocki *et al.*, 1992; Sorrentino *et al.*, 1992a,b, 1993; Uhrberg *et al.*, 1994). In the PCR, it is possible that single-stranded DNA may re-anneal with an inappropriate strand to form a heterodimer, or remain single stranded. Such events produce conformationally distinct structures which can be differentiated by their electrophoretic mobilities, and may be useful in identifying HLA polymorphism.

Generation of PCR heteroduplexes have been used in matching HLA-DR (Clay *et al.*, 1991; Sorrentino *et al.*, 1992a), HLA-DQ (Rubocki *et al.*, 1992), and HLA-DP specificities (Clay *et al.*, 1994; Sorrentino *et al.*, 1992b, 1993). To enhance the discrimination offered by such analyses, an artificial single strand has been used, described as the universal heteroduplex generator (UHG; Clay *et al.*, 1994). This is a single strand of DNA containing engineered sequence polymorphisms in significant hypervariable regions. The PCR product can be denatured, re-annealed with the UHG, with the resulting heteroduplex analysed in a more defined manner.

Heteroduplex analyses have obvious applications for cross-matching in transplantation. Mismatches can be determined by re-annealing donor and recipient amplified PCR product, specific for the particular required HLA region. Mismatches can then be detected through the differential migration of

PCR-SSP Phototyping.

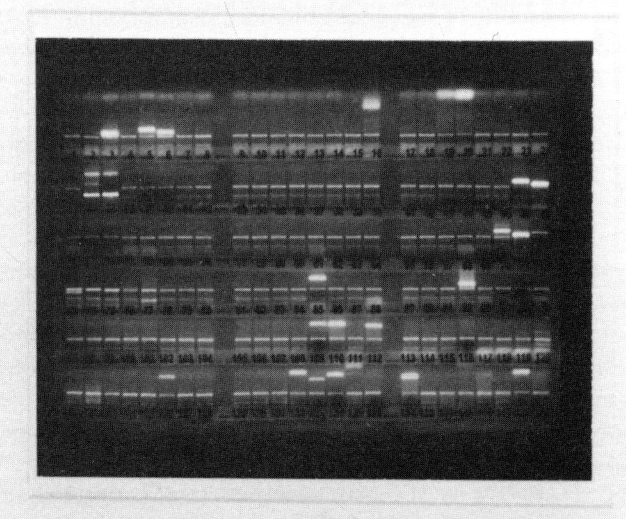

SSP type	reaction	SSP type	reaction
A*03	3	DRB1*0301/4	109,110,126
A*2301	5,6	DRB1*04	112
B*38	47,48	DRB3*0201	132,133
B*1518	70,71,72	DRB4*01	134
Bw4	26		
Bw6	27		
Cw*04	85	DQB1*02	137
Cw*0704	92	DQB1*0302	143

Figure 6.7. An example of an HLA PCR-SSP phototype. 144 PCR-SSP reactions are run simultaneously, encompassing *HLA-A, -B, -C, -DRB1, -DRB3, -DRB4, -DRB5* and *-DQB1* genes. Interpretation is reached through the presence or absence of the appropriately sized bands for each reaction following agarose gel electrophoresis. The common band present in each lane represents an internal control, which validates the PCR conditions. (Reproduced by permission of M. Bunce, Oxford.)

heteroduplex as compared to homoduplex PCR product through non-denaturing polyacrylamide gels.

The use of single-stranded conformational polymorphism of PCR product (PCR-SSCP) is a further method used in typing. The electrophoretic mobility of single-stranded DNA, under appropriate conditions, can be used to discriminate between HLA alleles containing different polymorphisms. This approach has been applied to identifying HLA class II (Shintaku *et al.*, 1993) and HLA-B specificities (Yoshida *et al.*, 1992).

6.5.6 Sequence-based typing

The full extent and complexity of polymorphism within the human MHC has only been revealed by DNA sequence analysis. In spite of their high levels of resolution, tissue typing HLA by PCR-SSOP or PCR-SSP is limited within the context of testing for known sequence polymorphism. Although many of the polymorphisms in HLA are conserved and shared amongst alleles at given hypervariable regions, unique polymorphic differences also exist, and are not always within known hypervariable regions. The use of heteroduplex or conformational polymorphism may identify such differences in sequence, but does not reveal their nature in terms of DNA sequence. Ultimately, the only definitive means of detecting such differences is through DNA sequencing the region of interest, and identifying all the relevant polymorphisms.

DNA sequencing (Sanger *et al.*, 1977) can be a lengthy and labour-intensive procedure, and, in this form, unsuitable for routine tissue-typing. Innovations in DNA sequencing, particularly with the arrival of automated sequencing, has provided the means for rapid analysis of HLA polymorphism, and so made sequence-based typing (SBT) a possibility.

Initial reports of SBT were for class II (Santamaria *et al.*, 1992), typing for DRB, DQA and DQB. Additionally, SBT has also been described for HLA-DP (Rozemuller *et al.*, 1993). A similar approach was applied to class I, in which two locus-specific sequencing reactions were performed for HLA-A, -B and -C (Santamaria *et al.*, 1993). Although this approach showed the potential of SBT for HLA class I and class II typing, it is difficult to perform routinely, due to the complexity of its performance and the requirement for expensive automated equipment. Improvements in the method have already been made, however, particularly in using fluorophor-labelled sequencing primers in conjunction with an automated sequencer (Versluis *et al.*, 1993) and the use of magnetic beads as a sequencing solid support (Spurkland *et al.*, 1993). Such approaches allow a larger throughput of samples, and potentially increase the generation of sequence information per reaction compared to previous methods.

Polymorphism not detected through methods such as PCR-SSOP and PCR-SSP, can be seen through SBT. Hence, previously unidentified alleles have been described (Santamaria *et al.*, 1992, 1993), although the possibility of sequencing errors must also be considered (Domena *et al.*, 1994). In spite of its current practical limitations, the approach of SBT does provide the most information compared to other DNA-based typing methods. As the facilities for automated sequencing improve, SBT may become a routine technique for HLA typing and other important clinical indicators. The clinical relevance of typing at this resolution in terms of organ/patient survival may well decide whether SBT is adopted as the method of choice.

6.5.7 Hybrid methods

Many of the approaches described above can be used together in a complementary fashion to elucidate the HLA type. The most common combination of methods remains the use of serology for determination of class I and a DNA-based method for class II. However, the recent and rapid development of DNA-based techniques for typing HLA class I as well as class II greatly increases the choice of methods available. Such developments reflect an awareness of the capabilities of the different methods, and how the specificity of one can provide the means for detection by another. Hence, PCR-SSP may provide the material for further analysis by SSOP, SBT, RFLP, SSCP, heteroduplex or even another round of PCR-SSP.

6.6 Conclusions and future directions

There remain many reasons for tissue-typing, including matching for organ transplantation, disease-association, forensic and anthropological studies, and studies of T-cell mediated immunity. Each of these has different requirements in terms of degree of resolution and methodology, and this has, in part, led to the variety of techniques currently available for tissue-typing HLA. The rapid progress of DNA-based HLA typing will potentially make tissue-typing available to anyone who has a reason for doing it. However, the complexity of the human MHC, and the diversity of methods available for its analysis, should ensure the need for professional histocompatibility testing laboratories for the foreseeable future.

The information generated by DNA-based methods has contributed to a more comprehensive understanding of the HLA system from a number of different perspectives. The ability to HLA type at a high resolution has uncovered previously unseen heterogeneity between population groups in terms of their HLA structure. It will hopefully improve the success of transplantation by identifying more precisely which specificities are crucial for matching to ensure graft survival. With many of the HLA polymorphisms having functional significance at the level of the T-cell mediated immune response, high-resolution typing has implications for HLA restriction of T-cell mediated responses in research as well as clinical settings.

For these reasons, the requirement for concise and dependable typing methods will continue to ensure further improvement and development in the field of HLA typing. The current advances and improvement of typing resolution have generated an extensive effort in further characterizing polymorphism within the HLA system. New alleles are still being found, and their functional and anthropological significance can subsequently be investigated. The information gained from such efforts will enhance current understanding of the structure, function and evolution of the HLA system. The diversity of methods currently available for HLA typing gives the investigator a degree of choice and flexibility, not to mention accuracy, not previously known in HLA typing.

References

Allen M, Liu L, Gyllensten U. (1994) A comprehensive polymerase chain reaction-oligonucleotide typing system for the HLA class I A locus. *Hum. Immunol.* **40:** 25–32.

Allsopp CE, Hill AV, Kwiatkowski D, Hughes A, Bunce M, Taylor CJ, Pazmany L, Brewster D, McMichael AJ, Greenwood BM. (1991) Sequence analysis of HLA-Bw53, a common West

African allele, suggests an origin by gene conversion of HLA-B35. *Hum. Immunol.* **30:** 105–109.

Amos DB, Yunis EJ. (1971) Editorial: A new interpretation of the major histocompatibility gene complexes of man and mouse. *Cell Immunol.* **2:** 517–520.

Arnett KL, Parham P. (1995) HLA Class I nucleotide sequences, 1995. *Tissue Antigens* **46:** 217–257.

Bach FH, Amos DB. (1967) Hu-1: Major histocompatibility locus in man. *Science* **156:** 1506–1508.

Bach FH, Voynow NK. (1966) One-way stimulation in mixed leukocyte cultures. *Science* **153:** 545–547.

Barouch D, Friede T, Stevanovic S, Tussey L, Rowland-Jones S, Braud V, McMichael AJ, Rammensee HG. (1995) HLA-A2 subtypes are functionally distinct in peptide binding and presentation. *J. Exp. Med.* **182:** 1847–1856.

Bein G, Glaser R, Kirchner H. (1992) Rapid HLA-DRB1 genotyping by nested PCR amplification. *Tissue Antigens* **39:** 68–73.

Bell J. (1989) The polymerase chain reaction. *Immunol. Today* **10:** 351–355.

Bell JI, Denney DJ, MacMurray A, Foster L, Watling D, McDevitt HO. (1987) Molecular mapping of class II polymorphisms in the human major histocompatibility complex. I. DR beta. *J. Immunol.* **139:** 562–573.

Bidwell JL, Bidwell EA, Laundy GJ, Klouda PT, Bradley BA. (1987) Allogenotypes defined by short DQ alpha and DQ beta cDNA probes correlate with and define splits of HLA-DQ serological specificities. *Mol. Immunol.* **24:** 513–522.

Bjorkman PJ, Saper MA, Samraoui B, Bennett WS, Strominger JL, Wiley DC. (1987) Structure of the human Class I histocompatibility antigen, HLA-A2. *Nature* **329:** 506–512.

Bodmer WF, Albert E, Bodmer JG, Dausset J, Kissmeyer NF, Mayr W, Payne R, van Rood J, Trnka Z, Walford RL. (1984) Nomenclature for factors of the HLA system 1984. *Immunogenetics* **20:** 593–601.

Bodmer J, Bodmer W, Heyes J, So A, Tonks S, Trowsdale J, Young, J. (1987) Identification of HLA-DP polymorphism with DP alpha and DP beta probes and monoclonal antibodies: correlation with primed lymphocyte typing. *Proc. Natl Acad. Sci. USA* **84:** 4596–4600.

Bodmer JG, Marsh SG, Parham P, et al. (1990) Nomenclature for factors of the HLA system, 1989. *Tissue Antigens* **35:** 1–8.

Bodmer JG, Marsh SGE, Albert ED, *et al.* (1995) Nomenclature for factors of the HLA system, 1995. *Tissue Antigens* **46:** 1–18.

Brodsky FM, Parham P, Barnstable CJ, Crumpton MJ, Bodmer WF. (1979) Monoclonal antibodies for analysis of the HLA system. *Immunol. Rev.* **47:** 3–61.

Brown JH, Jardetzky TS, Gorga JC, Stern LJ, Urban RG, Strominger JL, Wiley, DC. (1993) Three-dimensional structure of the human class II histocompatibility antigen HLA-DR1. *Nature* **364:** 33–39.

Browning M, Krausa P. (1996) Genetic diversity of HLA-A2: evolutionary and functional significance. *Immunology Today* (in press).

Browning MJ, Krausa P, Rowan A, Bicknell DC, Bodmer JG, Bodmer WF. (1993) Tissue typing the HLA-A locus from genomic DNA by sequence-specific PCR: comparison of HLA genotype and surface expression on colorectal tumor cell lines. *Proc. Natl Acad. Sci. USA* **90:** 2842–2845.

Browning MJ, Krausa P, Bodmer JG. (1994) Low resolution HLA-A locus typing by multiplex ARMS-PCR [abstract]. *Hum. Immunol.* **39:** 134.

Bugawan TL, Apple R, Erlich HA. (1994) A method for typing polymorphism at the HLA-A locus using PCR amplification and immobilized oligonucleotide probes. *Tissue Antigens* **44:** 137–147.

Bunce M, Welsh, KI. (1994) Rapid DNA typing for HLA-C using sequence-specific primers (PCR-SSP): identification of serological and non-serologically defined HLA-C alleles including several new alleles. *Tissue Antigens* **43:** 7–17.

Bunce M, Taylor CJ, Welsh KI. (1993) Rapid HLA-DQB typing by eight polymerase chain reaction amplifications with sequence-specific primers (PCR-SSP). *Hum. Immunol.* **37:** 201–206.

Bunce M, Barnardo MC, Welsh KI. (1994) Improvements in HLA-C typing using sequence-specific primers (PCR-SSR) including definition of HLA-Cw9 and Cw10 and a new allele HLA-"Cw7/8v". *Tissue Antigens* **44:** 200–203.

Bunce M, Fanning GC, Welsh KI. (1995a) Comprehensive, serologically equivalent DNA typing for HLA-B by PCR using sequence-specific primers (PCR-SSP). *Tissue Antigens* **45:** 81–90.

Bunce M, O'Neill CM, Barnardo MCNM, Krausa P, Browning MJ, Morris PJ, Welsh KI. (1995b) Phototyping: Comprehensive DNA typing for HLA-A, B, C, DRB1, DRB3, DRB4, DRB5 & DQB1 by PCR with 144 primer mixes utilising sequence-specific primers (PCR-SSP). *Tissue*

Antigens **46:** 355–367.

Carcassi C, Krausa P, Bodmer J, Contu L, Browning M. (1995) Characterization of HLA-A*02 subtypes in the Sardinian population. *Tissue Antigens* **46:** 391–393.

Castano AR, Lauzurica, P, Domenech N, Lopez de Castro J. (1991) Structural identity between HLA-A2 antigens differentially recognized by alloreactive cytotoxic T lymphocytes. *J. Immunol.* **146:** 2915–2920.

Ceppellini R, van Rood J. (1974) The HL-A system. I. Genetics and molecular biology. *Semin. Hematol.* **11:** 233–251.

Cereb N, Lee S, Maye P, Kong Y, Yang SY. (1994) Nonrandom allelic variation in the regulatory complex of HLA class I genes. *Hum. Immunol.* **41:** 46–51.

Cereb N, Maye P, Lee S, Kong Y, Yang SY. (1995) Locus-specific amplification of HLA class I genes from genomic DNA: locus-specific sequences in the first and third introns of HLA-A, -B, and -C alleles. *Tissue Antigens* **45:** 1–11.

Chen DF, Endres W, Meyer SA, Stangel W. (1994) A polymerase chain reaction-sequence-specific oligonucleotide procedure for HLA class II typing using biotin- and digoxigenin-labeled probes simultaneously in hybridization. *Hum. Immunol.* **39:** 25–30.

Clay TM, Bidwell JL, Howard MR, Bradley BA. (1991) PCR-fingerprinting for selection of HLA matched unrelated marrow donors. Collaborating Centres in the IMUST Study. *Lancet* **337:** 1049–1052.

Clay TM, Culpan D, Howell WM, Sage DA, Bradley BA, Bidwell JL. (1994) UHG crossmatching. A comparison with PCR-SSO typing in the selection of HLA-DPB1-compatible bone marrow donors. *Transplantation* **58:** 200–207.

Dausset J. (1954) Leuco-agglutinins. IV. Leuco-agglutinins and blood transfusion. *Vox Sanguis* **4:** 190–198.

Domena JD, Little AM, Arnett KL, Adams EJ, Marsh SG, Parham, P. (1994) A small test of a sequence-based typing method: definition of the B*1520 allele. *Tissue Antigens* **44:** 217–224.

Dormoy A, Urlacher A, Tongio MM. (1992) Complexity of the HLA-DP region: RFLP analysis versus PLT typing and oligotyping. *Hum. Immunol.* **34:** 39–46.

Dupont B. (1995) Editorial: "Phototyping" for HLA: the beginning of the end of HLA typing as we know it. *Tissue Antigens* **46:** 353–354.

Eiermann TH, Fakler J, Goldmann SF. (1992) The incidence of DPB1 differences between serological and mixed lymphocyte culture matched unrelated individuals: implications for selection of bone marrow donors. *Bone Marrow Transplant* **9:** 157–160.

Erlich HA, Gyllensten UB. (1991) Shared epitopes among HLA class II alleles: gene conversion, common ancestry and balancing selection. *Immunol. Today* **12:** 411–414.

Erlich H, Bugawan T, Begovich AB, Scharf S, Griffith R, Saiki R, Higuchi R, Walsh PS. (1991) HLA-DR, DQ and DP typing using PCR amplification and immobilized probes. *Eur. J. Immunogen.* **18:** 33–55.

Fernandex-Vina MA, Falco M, Sun Y, Stastny P. (1992) DNA typing for HLA class I alleles: I. Subsets of HLA-A2 and of -A28. *Hum. Immunol.* **33:** 163–173.

Fernandex-Vina M, Lazaro AM, Sun Y, Miller S, Forero L, Stastny P. (1995) Population diversity of B-locus alleles observed by high resolution DNA typing. *Tissue Antigens* **45:** 153–168.

Gao X, Jakobsen IB, Serjeantson SW. (1994) Characterization of the HLA-A polymorphism by locus-specific polymerase chain reaction amplification and oligonucleotide hybridization. *Hum. Immunol.* **41:** 267–279.

Geraghty DE, Koller BH, Pei J, Hansen JA. (1992) Examination of four HLA class I pseudogenes. Common events in the evolution of HLA genes and pseudogenes. *J. Immunol.* **149:** 1947–1956.

Giorda R, Lampasona V, Kocova M, Trucco M. (1993) Non-radioisotopic typing of human leukocyte antigen class II genes on microplates. *Biotechniques* **15:** 918–925.

Gorga JC, Brown JH, Jardetzky T, Wiley DC, Strominger JL. (1991) Crystallization of HLA-DR antigens. *Res. Immunol.* **142:** 401–407.

Guttridge MG, Burr C, Klouda PT. (1994) Identification of HLA-B35, B53, B18, B5, B78, and B17 alleles by the polymerase chain reaction using sequence-specific primers (PCR-SSP). *Tissue Antigens* **44:** 43–46.

Higuchi R, von Beroldingen C, Sensabaugh GF, Erlich HA. (1988) DNA typing from single hairs. *Nature* **332:** 543–546.

Hildebrand WH, Madrigal JA, Little AM, Parham P. (1992) HLA-Bw22: a family of molecules with identity to HLA-B7 in the alpha 1-helix. *J. Immunol.* **148:** 1155–1162.

Hildebrand WH, Domena JD, Shen SY, Lau M, Terasaki PI, Bunce M, Marsh SG, Guttridge

MG, Bias WB, Parham P. (1994) HLA-B15: a widespread and diverse family of HLA-B alleles. *Tissue Antigens* **43:** 209–218.

Hill AV, Elvin J, Willis AC, *et al.* (1992) Molecular analysis of the association of HLA-B53 and resistance to severe malaria. *Nature* **360:** 434–439.

Hirschorn K, Bach F, Kolodny RL, Firschen IL. (1963) Immune response and mitosis of human peripheral blood lymphocytes in vitro. *Science* **142:** 1185–1187.

Hongming F, Tilanus M, van Eggermond M, Giphart M. (1986) Reduced complexity of RFLP for HLA-DR typing by the use of a DR beta 3′ cDNA probe. *Tissue Antigens* **28:** 129–135.

Hviid TV, Madsen HO, Morling N. (1992) HLA-DPB1 typing with polymerase chain reaction and restriction fragment length polymorphism technique in Danes. *Tissue Antigens* **40:** 140–144.

Inoko H, Ando A, Ito M, Tsuji K. (1986) Southern hybridization analysis of DNA polymorphism in the HLA-D region. *Hum. Immunol.* **16:** 304–313.

Jaraquemada D, Navarrete C, Ollier W, Awad J, Okoye R, Festenstein H. (1986) HLA-Dw specificity assignments are independent of HLA-DQ, HLA-DR, and other class II specificities and define a biologically important segregant series which strongly activates a functionally distinct T cell subset. *Hum. Immunol.* **16:** 259–270.

Kaminski ER, Hows JM, Bridge J, Davey NJ, Brookes PA, Green JE, Goldman JM, Batchelor JR. (1991) Cytotoxic T lymphocyte precursor (CTL-p) frequency analysis in unrelated donor bone marrow transplantation: two case studies. *Bone Marrow Transplant.* **8:** 47–50.

Kawai S, Maekawajiri S, Tokunaga K, Juji T, Yamane A. (1994) A simple method of HLA-DRB typing using enzymatically amplified DNA and immobilized probes on microtiter plate. *Hum. Immunol.* **41:** 121–126.

Kimura A, Sasazuki T. (1992a) Polymorphism in the 5′ flanking region of the DQA1 gene and its relation to DR-DQ haplotype. In: *HLA 1991* (eds K Tsuji, M Aizawa, T Sasazuki). Oxford Science Publications, Oxford, pp. 382–385.

Kimura A, Sasazuki T. (1992b) Eleventh International Histocompatibility Workshop reference protocol for the HLA DNA-typing technique. In: *HLA 1991. Proceedings of the Eleventh International Histocompatibility Workshop and Conference* (eds K Tsuji, M Aizawa and T Sasazuki). Oxford University Press, Oxford, pp. 397–419.

Kissmeyer NF, Svejgaard A, Hauge M. (1969) The HL-A system defined with lymphocytotoxic and platelet antibodies in relation to kidney transplantation. *Transplant. Proc.* **1:** 357–361.

Knipper AJ, Hinney A, Schuch B, Enczmann J, Uhrberg M, Wernet P. (1994) Selection of unrelated bone marrow donors by PCR-SSP typing and subsequent nonradioactive squence-based typing for HLA DRB1/3/4/5, DQB1, and DPB1 alleles. *Tissue Antigens* **44:** 275–284.

Kovats S, Main EK, Librach C, Stubblebine M, Fisher SJ, DeMars R. (1990) A class I antigen, HLA-G, expressed in human trophoblasts. *Science* **248:** 220–223.

Krausa P, Browning MJ. (1996) A comprehensive PCR-SSP typing system for identification of HLA-A locus alleles. *Tissue Antigens* (in press).

Krausa P, Bodmer JG, Browning MJ. (1993a) Defining the common subtypes of HLA A9, A10, A28 and A19 by use of ARMS/PCR. *Tissue Antigens* **42:** 91–99.

Krausa P, Moses J, Bodmer W, Bodmer J, Browning M. (1993b) HLA-A locus alleles identified by sequence specific PCR [letter]. *Lancet* **341:** 121–122.

Krausa P, Brywka III M, Savage D, *et al.* (1995a) Genetic polymorphism within HLA-A*02: significant allelic variation revealed in different population. *Tissue Antigens* **45:** 223–231.

Krausa P, Carcassi C, Orrú S, Bodmer JG, Browning MJ, Contu L. (1995b) Defining the allelic variants of HLA-A30 in the Sardinian population using Amplification Refractory Mutation System-Polymerase Chain Reaction. *Hum. Immunol.* **44:** 35–42.

Kubo RT, Sette A, Grey HM, *et al.* (1994) Definition of specific peptide motifs for four major HLA-A alleles. *J. Immunol.* **152:** 3913–3924.

Kuhner MK, Peterson MJ. (1992) Genetic exchange in the evolution of the human MHC class II loci. *Tissue Antigens* **39:** 209–215.

Latron F, Moots R, Rothbard JB, Garrett TP, Strominger JL, McMichael A. (1991) Positioning of a peptide in the cleft of HLA-A2 by complementing amino acid changes. *Proc. Natl Acad. Sci. USA* **88:** 11325–11329.

Lawlor DA, Zemmour J, Ennis PD, Parham P. (1990) Evolution of class-I MHC genes and proteins: from natural selection to thymic selection. *Annu. Rev. Immunol.* **8:** 23–63.

Leen MPJM, Ogutu ER, Gorski J. (1994) Structural and functional analysis of DRB-promoter polymorphism and isomorphism. *Hum. Immunol.* **41:** 112–120.

Levine JE, Yang SY. (1994) SSOP typing of the Tenth International Histocompatibility Workshop reference cell lines for HLA-C alleles. *Tissue Antigens* **44:** 174–183.

Lin XY, Wang Y, Sun J, Wang FQ, Ni LQ, Chang LY, Chen SS. (1989) Association of RFLP of HLA class I genes with Chinese ankylosing spondylitis patients. *Tissue Antigens* **34:** 279–283.

Louis P, Vincent R, Cavadore P, Clot J, Eliaou JF. (1994) Differential transcriptional activities of HLA-DR genes in the various haplotypes. *J. Immunol.* **153:** 5059–5067.

Maeda M, Murayama N, Ishii H, Uryu N, Ota M, Tsuji K, Inoko H. (1989) A simple and rapid method for HLA-DQA1 genotyping by digestion of PCR-amplified DNA with allele specific restriction endonucleases. *Tissue Antigens* **34:** 290–298.

Marsh SGE, Bodmer JG. (1995) HLA Class II region nucleotide sequences 1995. *Tissue Antigens* **46:** 258–280.

McMichael AJ, Gotch FM, Santos-Aguado J, Strominger JL. (1988) Effect of mutations and variations of HLA-A2 on recognition of a virus peptide epitope by cytotoxic T lymphocytes. *Proc. Natl Acad. Sci. USA* **85:** 9194–9198.

Medintz I, Chiriboga L, McCurdy L, Kobilinsky L. (1994) Restriction fragment length polymorphism and polymerase chain reaction-HLA DQ alpha analysis of casework urine specimens. *J. Forensic Sci.* **39:** 1372–1380.

Mempel W, Albert E, Burger A. (1972) Further evidence for a separate MLC-locus. *Tissue Antigens* **2:** 250–254.

Mempel W, Grosse WH, Albert E, Thierfelder S. (1973a) Atypical MLC reactions in HL-A-typed related and unrelated pairs. *Transplant. Proc.* **5:** 401–408.

Mempel W, Grosse WH, Baumann P, Netzel B, Steinbauer RI, Scholz S, Bertrams J, Albert ED. (1973b) Population genetics of the MLC response: typing for MLC determinants using homozygous and heterozygous reference cells. *Transplant. Proc.* **5:** 1529–1534.

Middleton D, Williams F, Cullen C, Mallon E. (1995) Modification of an HLA-B PCR-SSOP typing system leading to improved allele determination. *Tissue Antigens* **45:** 232–236.

Moller E, Carlsson B, Wallin J. (1985) Implication of structural class II gene polymorphism for the concept of serologic specificities. *Immunol. Rev.* **85:** 107–128.

Morishima Y, Kodera Y, Hirabayashi N, *et al.* (1995) Low incidence of acute GVHD in patients transplated with marrow from HLA-A, B, DR-compatible unrelated donors among Japanese. *Bone Marrow Transplant* **15:** 235–239.

Morling N, Jakobsen BK, Platz P, Ryder LP, Svejgaard A, Thomsen M. (1980) A "new" primed lymphocyte typing (PLT) defined DP-antigen associated with a private HLA-DR antigen. *Tissue Antigens* **16:** 95–104.

Mullis KB, Faloona FA. (1987) Specific synthesis of DNA in vitro via a polymerase-catalyzed chain reaction. *Methods Enzymol.* **155:** 335–350.

Mullis K, Faloona F, Scharf S, Saiki R, Horn G, Erlich H. (1986) Specific enzymatic amplification of DNA in vitro: the polymerase chain reaction. *Cold Spring Harb. Symp. Quant. Biol.* **1:** 263–273.

Newton CR, Graham A, Heptinstall LE, Powell SJ, Summers C, Kalsheker N, Smith JC, Markham AF. (1989) Analysis of any point mutation in DNA. The amplification refractory mutation system (ARMS). *Nucleic Acids Res.* **17:** 2503–2516.

Oh SH, Fleischhauer K, Yang SY. (1993) Isoelectric focusing subtypes of HLA-A can be defined by oligonucleotide typing. *Tissue Antigens* **41:** 135–142.

Oka T, Matsunaga H, Tokunaga K, Mitsunaga S, Juji T, Yamane A. (1994) A simple method for detecting single base substitutions and its application to HLA-DPB1 typing. *Nucleic Acids Res.* **22:** 1541–1547.

Olerup O, Zetterquist H. (1991) HLA-DRB1*01 subtyping by allele-specific PCR amplification: a sensitive, specific and rapid technique. *Tissue Antigens* **37:** 197–204.

Olerup O, Zetterquist H. (1992) HLA-DR typing by PCR amplification with sequence-specific primers (PCR-SSP) in 2 hours: an alternative to serological DR typing in clinical practice including donor-recipient matching in cadaveric transplantation. *Tissue Antigens* **39:** 225–235.

Olerup O, Aldener A, Fogdell A. (1993) HLA-DQB1 and -DQA1 typing by PCR amplification with sequence-specific primers (PCR-SSP) in 2 hours. *Tissue Antigens* **41:** 119–134.

Parham P, Lomen CE, Lawlor DA, Ways JP, Holmes N, Coppin HL, Salter RD, Wan AM, Ennis PD. (1988) Nature of polymorphism in HLA-A, -B, and -C molecules. *Proc. Natl Acad. Sci. USA* **85:** 4005–4009.

Rinke dWT, Vloemans S, van den Elsen P, Haworth A, Stern PL. (1990) Differential expression of the HLA class I multigene family by human embryonal carcinoma and choriocarcinoma cell lines. *J. Immunol.* **144:** 1080–1087.

Rotzschke O, Falk K, Stevanovic S, Jung G, Rammensee HG. (1992) Peptide motifs of closely related HLA class I molecules encompass substantial differences. *Eur. J. Immunol.* **22:** 2453–2456.

Rozemuller EH, Bouwens AG, Bast BE, Tilanus MG. (1993) Assignment of HLA-DPB alleles by computerized matching based upon sequence data. *Hum. Immunol.* **37:** 207–212.

Rubocki RJ, Wisecarver JL, Hook DD, Cox SM, Beisel KW. (1992) Histocompatibility screening by molecular techniques: use of polymerase chain reaction products and heteroduplex formation. *J. Clin. Lab. Anal.* **6:** 337–341.

Sadler AM, Petronzelli F, Krausa P, Marsh SG, Guttridge MG, Browning MJ, Bodmer JG. (1994) Low-resolution DNA typing for HLA-B using sequence-specific primers in allele- or group-specific ARMS/PCR. *Tissue Antigens* **44:** 148–154.

Saiki RK, Scharf S, Faloona F, Mullis KB, Horn GT, Erlich HA, Arnheim N. (1985) Enzymatic amplification of beta-globin genomic sequences and restriction site analysis for diagnosis of sickle cell anemia. *Science* **230:** 1350–1354.

Saiki RK, Bugawan TL, Horn GT, Mullis KB, Erlich HA. (1986) Analysis of enzymatically amplified beta-globin and HLA-DQ alpha DNA with allele-specific oligonucleotide probes. *Nature* **324:** 163–166.

Saiki RK, Walsh PS, Levenson CH, Erlich HA. (1989) Genetic analysis of amplified DNA with immobilized sequence-specific oligonucleotide probes. *Proc. Natl Acad. Sci. USA* **86:** 6230–6234.

Salazar M, Yunis JJ, Delgado MB, Bing D, Yunis EJ. (1992) HLA-DQB1 allele typing by a new PCR-RFLP method: correlation with a PCR-SSO method. *Tissue Antigens* **40:** 116–123.

Sanger F, Nicklen S, Coulson AR. (1977) DNA sequencing with chain-terminating inhibitors. *Proc. Natl Acad. Sci. USA* **74:** 5463–5467.

Santamaria P, Boyce JM, Lindstrom AL, Barbosa JJ, Faras AJ, Rich SS. (1992) HLA class II "typing": direct sequencing of DRB, DQB, and DQA genes. *Hum. Immunol.* **33:** 69–81.

Santamaria P, Lindstrom AL, Boyce JM, Myster SH, Barbosa JJ, Faras AJ, Rich SS. (1993) HLA class I sequence-based typing. *Hum. Immunol.* **37:** 39–50.

Saper MA, Bjorkman PJ, Wiley DC. (1991) Refined structure of the human histocompatibility antigen HLA-A2 at 2.6 A resolution. *J. Mol. Biol.* **219:** 277–319.

Schwarer AP, Jiang YZ, Deacock S, Brookes PA, Barrett AJ, Goldman JM, Batchelor JR, Lechler RI. (1994) Comparison of helper and cytotoxic antirecipient T cell frequencies in unrelated bone marrow transplantation. *Transplantation* **58:** 1198–1203.

Segurado OG, Iglesias CP, Vicario JL, Corell A, Regueiro JR, Arnaiz VA. (1990) Shared SstI RFLPs by HLA-Aw19, A23/24 and A3/11 crossreacting groups. *Tissue Antigens* **35:** 206–210.

Sharrock CEM, Man S, Wanachiwanawin W, Batchelor JR. (1987) Analysis of the alloreactive T cell repertoire in man. I. Differences in precursor frequency for cytotoxic T cell responses against allogeneic MHC molecules in unrelated individuals. *Transplantation* **43:** 699–703.

Sheehy MJ, Sondel PM, Bach ML, Wank R, Bach FH. (1975) HL-A LD (lymphocyte defined) typing: a rapid assay with primed lymphocytes. *Science* **188:** 1308–1310.

Sheldon EL, Kellogg DE, Watson R, Levenson CH, Erlich HA. (1986) Use of nonisotopic M13 probes for genetic analysis: application to HLA class II loci. *Proc. Natl Acad. Sci. USA* **83:** 9085–9089.

Shintaku S, Fukuda Y, Kimura A, Hoshino S, Tashiro H, Sasazuki T, Dohi K. (1993) DNA conformation polymorphism analysis of DR52 associated HLA-DR antigens by polymerase chain reaction: a simple, economical and rapid examination for HLA matching in transplantation. *Jpn. J. Med. Sci. Biol.* **46:** 165–181.

Sorrentino R, Cascino I, Tosi R. (1992a) Subgrouping of DR4 alleles by DNA heteroduplex analysis. *Hum. Immunol.* **33:** 18–23.

Sorrentino R, Potolicchio I, Ferrara GB, Tosi R. (1992b) A new approach to HLA-DPB1 typing combining DNA heteroduplex analysis with allele-specific amplification and enzyme restriction. *Immunogenetics* **36:** 248–254.

Sorrentino R, Potolicchio I, D'Amato M, Tosi R. (1993) The HLA DP locus in bone marrow transplantation. Probeless genomic typing of the DPB1 alleles. *Bone Marrow Transplant* **1:** 17–19.

Spurkland A, Knutsen I, Markussen G, Vartdal F, Egeland T, Thorsby E. (1993) HLA matching of unrelated bone marrow transplant pairs: direct sequencing of in vitro amplified HLA-DRB1 and -DQB1 genes using magnetic beads as solid support. *Tissue Antigens* **41:** 155–164.

Sudo T, Kamikawaji N, Kimura A, Date Y, Savoie CJ, Nakashima H, Furuichi E, Kuhara S, Sasazuki T. (1995) Differences in MHC class I self peptide repertoires among HLA-A2 subtypes. *J. Immunol.* **155:** 4749–4756.

Tanaka M, Abe J, Kohsaka T, Tanae A. (1992) Analysis of HLA-DQA1 in Japanese patients with type 1 diabetes mellitus, using DNA-PCR-RFLP typing. *Acta Paediatr. Jpn* **34:** 46–51.

Tanigaki N, Fruci D, Chersi A, Falasca G, Tosi R, Butler RH. (1994) HLA-A2-binding peptides cross-react not only within the A2 subgroup but also with other HLA-A-locus allelic products. *Hum. Immunol.* **39:** 155–162.

Terasaki PI, McClelland JD. (1964) Microdoplet assay of human serum cytotoxins. *Nature* **204:** 998–1007.

Tiercy JM, Zwahlen F, Jeannet M, Mach B. (1989) Current approach to HLA typing for bone marrow transplantation: oligonucleotide typing by hybridization on DNA amplified by polymerase chain reaction. *Schweiz. Med. Wochenschr.* **119:** 1344–1346.

Tiercy JM, Morel C, Freidel AC, Zwahlen F, Gebuhrer L, Betuel H, Jeannet M, Mach B. (1991) Selection of unrelated donors for bone marrow transplantation is improved by HLA class II genotyping with oligonucleotide hybridization. *Proc. Natl Acad. Sci. USA* **88:** 7121–7125.

Tiercy JM, Djavad N, Rufer N, Speiser DE, Jeannet M, Roosnek E. (1994) Oligotyping of HLA-A2, -A3, and -B44 subtypes. Detection of subtype incompatibilities between patients and their serologically matched unrelated bone marrow donors. *Hum. Immunol.* **41:** 207–215.

Tilanus MG, Hongming F, van Eggermond MC, v.d. Bijl M, D'Amaro J, Schreuder GM, de Vries R, Giphart MJ. (1986) An overview of the restriction fragment length polymorphism of the HLA-D region: its application to individual D-, DR- typing by computerized analyses. *Tissue Antigens* **28:** 218–227.

Trowsdale J, Ragoussis J, Campbell RD. (1991) Map of the human MHC. *Immunol. Today* **12:** 443–446.

Tussey LG, Matsui M, Rowland-Jones S, Warburton R, Frelinger JA, McMichael A. (1994) Analysis of mutant HLA-A2 molecules. Differential effects on peptide binding and CTL recognition. *J. Immunol.* **152:** 1213–1221.

Uhrberg M, Hinney A, Enczmann J, Wernet P. (1994) Analysis of the HLA-DR gene locus by temperature gradient gel electrophoresis and its application for the rapid selection of unrelated bone marrow donors. *Electrophoresis* **15:** 1044–1050.

Utz U, Koenig S, Coligan JE, Biddison WE. (1992) Presentation of three different viral peptides, HTLV-1 Tax, HCMV gB, and influenza virus M1, is determined by common structural features of the HLA-A2.1 molecule. *J. Immunol.* **149:** 214–221.

van der Poel J, Pool J, Goulmy E, Giphart MJ, van Rood J. (1986) Recognition of distinct epitopes on the HLA-A2 antigen by cytotoxic T lymphocytes. *Hum. Immunol.* **16:** 247–258.

van Leeuwen A, Schuit HR, van Rood J. (1973) Typing for MLC (LD). II. The selection of nonstimulator cells by MLC inhibition tests using SD-indentical stimulator cells (MISIS) and fluorescence antibody studies. *Transplant. Proc.* **5:** 1539–1542.

van Leeuwen A, Termijtelen A, Shaw S, van Rood JJ. (1982) Recognition of a polymorphic monocyte antigen in HLA. *Nature* **298:** 565–567.

van Rood J, van Leeuwen A, Keuning JJ, van Oud Alblas A. (1975) The serological recognition of the human MLC determinants using a modified cytotoxicity technique. *Tissue Antigens* **5:** 73–79.

Versluis LF, Rozemuller E, Tonks S, Marsh SG, Bouwens AG, Bodmer JG, Tilanus MG. (1993) High-resolution HLA-DPB typing based upon computerized analysis of data obtained by fluorescent sequencing of the amplified polymorphic exon 2. *Hum. Immunol.* **38:** 277–283.

Wordsworth P. (1991) PCR-SSO typing in HLA-disease association studies. *Eur. J. Immunogen.* **18:** 139–146.

Yang SY. (1989a) Population analysis of Class I HLA antigens by one-dimensional isoelectric gel electrophoresis: Workshop summary report. In: *Immunobiology of HLA. Histocompatibility testing 1987* (ed B Dupont). Springer-Verlag, New York, pp. 309–331.

Yang SY. (1989b) A standardised method for detection of HLA-A and HLA-B alleles by one-dimensional isoelectric focusing (IEF) gel electrophoresis. In: *Immunobiology of HLA. Histocompatibility testing 1987* (ed B Dupont). Springer-Verlag, New York, pp. 332–335.

Yoshida M, Kimura A, Numano F, Sasazuki T. (1992) Polymerase-chain-reaction-based analysis of polymorphism in the HLA-B gene. *Hum. Immunol.* **34:** 257–266.

Yunis EJ, Amos DB. (1971) Three closely linked genetic systems relevant to transplantation. *Proc. Natl Acad. Sci. USA* **68:** 3031–3035.

Yunis I, Salazar M, Yunis EJ. (1991) HLA-DR generic typing by AFLP. *Tissue Antigens* **38:** 78–88.

Zemmour J, Gumperz JR, Hildebrand WH, Ward FE, Marsh SG, Williams RC, Parham P. (1992) The molecular basis for reactivity of anti-Cw1 and anti-Cw3 alloantisera with HLA-B46 haplotypes. *Tissue Antigens* **39:** 249–257.

Zetterquist H, Olerup O. (1992) Identification of the HLA-DRB1*04, -DRB1*07, and -DRB1*09 alleles by PCR amplification with sequence-specific primers (PCR-SSP) in 2 hours. *Hum. Immunol.* **34:** 64–74.

Zhang L, Li SG, Vandekerckhove B, Termijtelen A, van Rood J, Claas FH. (1989) Analysis of cytotoxic T cell precursor frequencies directed against individual HLA-A and -B alloantigens. *J. Immunol. Methods* **121:** 39–45.

7

Regulation of MHC class I gene expression

Alain Israël and Philippe Kourilsky

7.1 Introduction

The classical transplantation antigens encoded by the K, D and L loci of the H-2 complex of the mouse and HLA-A, -B and -C loci in humans are integral membrane proteins that function in the presentation of peptide antigens to T cells. The highly polymorphic major histocompatibility complex (MHC) class I heavy chains (37–45 kDa) are non-covalently associated on the cell surface with the non-polymorphic β2-microglobulin (β2-m) light chain (12 kDa). Despite large interspecies variations in the size of the class I gene family, it is possible to identify subfamilies of genes (based upon their structure and patterns of expression) that are conserved among species. One of these families contains the classical class I antigens. These molecules are responsible for allograft rejection, and are expressed on almost all tissues. Another family codes for the so-called non-classical class I antigens. In the mouse, for example, these antigens are represented by the Qa and TL molecules, which exhibit distinctive and highly regulated patterns of expression. However the function of these non-classical antigens is poorly understood. In humans non-classical class I antigens are represented by HLA-E, -F and -G. As with their murine counterpart, the function of these molecules is poorly understood.

We will review here the mechanisms that control the constitutive, as well as regulated, expression of class I genes. This review will focus mainly on classical class I genes, since the regulation of the non-classical genes has been the subject of a limited analysis.

7.2 Constitutive expression of MHC class I genes

7.2.1 Expression in the adult

Classical H-2 and HLA class I genes. Class I antigens are expressed on most somatic cells of the adult, but with varying levels in different tissues and cell types, even within a given organ (summarized in *Table 7.1*). In all species expression is highest in lymphoid cells, with lower levels in the liver, lung and heart. In the spleen, expression is highest in splenocytes, and higher in B cells than in T cells. In the murine thymus, immunoperoxidase staining for H-2 K indicates that the medulla

HLA and MHC: genes, molecules and function, edited by M.J. Browning and A.J. McMichael.
© 1996 BIOS Scientific Publishers Ltd, Oxford.

Table 7.1. Tissue distribution of MHC class I antigens

Tissue	Human[a]	Mouse[b]
Cells of the endocrine system		
Thyroid	+	
Parathyroid glands	+	
Pancreatic islets of Langerhans	+	–
Adrenals	++	
Gastrointestinal tract		
Epithelium of:		
oesophagus (basal layer)	++	I(E)
stomach	+	–
small intestine	++	++
colon		E
Pancreas:		
acinar cells	–	–
duct cells	++	E
Liver:		
Kupffer cells	++	E
sinusoids, biliary epithelium	+/–	I
hepatocytes	+/–	I
Respiratory and cardiovascular system		
Lungs:		
bronchial and alveolar epithelium, interstitial cells	++	E
bronchiolar epithelium		I
Heart:		
myocardium	+	–
endocardium		E
pericardium		I
Endothelium:		
capilliaries and large vessels in all organs		E
Nervous system		
Central		
neurons	–	–
astrocytes, microglia and oligodendrocytes		I
Peripheral	++	
Kidney		
Tubular epithelium	++	I
Glomeruli	++	E
Testis		
Germline cells, spermatozoa	+	
Epidydymis		
Spermatozoa	++	
Muscle		
Skeletal	+/–	I
Smooth	+	I
Skin		
Langerhans cells, interstitital dendritic cells (all organs)	++	I/E
Keratinocytes	+	++
Placenta	–[c]	+[d]
Lymphoid organs		
Thymus (mouse)		
medullary lymphocytes and stromal structures		+
cortical lymphocytes and epithelium reticular cells		I
Lymph nodes (all cell types)		++
Spleen (all cell types)		++

[a] From Singer and Maguire, 1990.
[b] From David-Watine *et al.*, 1987; Robinson, 1987; Singer and Maguire, 1990 (and references therein).
For the human results +/–, + and ++ refer to the expression intensity in the different organs; –: not expressed.
For the mouse results, the effect of IFN-γ treatment is denoted. I: class I genes expressed only after induction; E: enhancement; + to ++: maximal staining intensity in the unstimulated organs; –: tissue or cells which remain unstained after treatment. For placenta: [c] villous trophoblast, [d] spongiotrophoblast and labyrinthine zone.

is strongly positive. Flow cytometric analysis of thymocytes reveals the presence of two populations of class I positive cells: immature $CD4^+CD8^+$ cells are weakly stained, while single positive cells exhibit a strong class I expression. The very high expression of class I molecules on the surface of antigen-presenting dendritic cells is involved in the regulation of class I restricted cytotoxic T lymphocyte (CTL) responses (Boog *et al.*, 1988). On the other hand, class I expression is undetectable in brain cells (Berah *et al.*, 1970; Fabry *et al.*, 1994), although it can be induced by certain stimuli in astrocytes and oligodendrocytes (Bartlett *et al.*, 1989), sperm cells at certain stages of differentiation, certain cell populations of the placenta, and undifferentiated embryonal carcinoma (EC) cells (which exhibit a variety of traits characteristic of the early embryo; Gachelin, 1978). β2-m is also expressed by virtually all adult cells but not by EC cells. These data indicate that the expression of classical class I genes is highly regulated. Patterns of expression of the human classical class I molecules are, on the whole, similar to those in the mouse (*Table 7.1*).

Non-classical Qa, TL genes of the mouse. The non-classical class I genes of the Qa and TL families exhibit a much more restricted pattern of expression. Most of them are expressed only on certain subsets of lymphoid cells (e.g. the Q7 product corresponding to the Qa2 antigenic determinants, and the Q8/9 product; Robinson, 1987), whereas the Q10 gene is expressed only in liver and fetal yolk sac (Cosman *et al.*, 1982; David-Watine *et al.*, 1987). Furthermore, the expression of certain Qa/TL antigens has been documented at stages of embryonic development as early as the oocyte and two-cell stage (Warner *et al.*, 1987) and also in EC cells (Ostrand-Rosenberg *et al.*, 1989). In addition, the expression of non-classical class I genes has been detected in the brain, where classical class I genes are not normally expressed (Palmer and Frelinger, 1987; Transy *et al.*, 1987).

The expression of the human non-classical class I genes, *HLA-E*, *-F* and *-G* is dealt with in Chapter 4.

7.2.2 Expression in the embryo

During embryogenesis the expression of MHC class I genes is strictly regulated. As mentioned earlier, they are not expressed on mature sperm, and their expression becomes detectable only after the mid-somite stage of embryogenesis (Hedley *et al.*, 1989; Ozato *et al.*, 1985). In the mouse, the mRNA for H-2 K^b is undetectable at day 5.5, but readily detectable at day 11: this is true both for classical and non-classical class I genes. To circumvent the limitation of working with embryos, studies have been performed on EC cells: these cells do not express class I genes, but when induced to differentiate by treatment with retinoic acid (RA), they start to express class I antigens at the cell surface (Croce *et al.*, 1981; Morello *et al.*, 1982; Rosenthal *et al.*, 1984). Interestingly, treatment of undifferentiated EC cells with interferons induces class I expression, but no differentiation (Wan *et al.*, 1987). Similarly to class I genes, β2-m is expressed in almost all adult tissues, but not in early embryo nor in EC cells; during embryogenesis the time course of synthesis of β2-m and MHC class I genes is different, suggesting different mechanisms of regulation (Jaffe *et al.*, 1991; Morello *et al.*, 1985).

7.3 Regulated patterns of class I gene expression

Interferons α and β (IFN-α and -β) are produced by fibroblasts and other types of cell during viral infection, whereas IFN-γ is a T-cell derived lymphokine, which interacts with a separate cell-surface receptor. *In vitro,* IFNs induce expression of the classical class I genes (but not of all non-classical ones) in a wide variety of cell types of haemopoietic, lymphoid, epithelial, fibroblastic and neuronal origin (Lindahl *et al.*, 1976; for a review, see Singer and Maguire, 1990). However, the effect may be different *in vivo* as, for example, neurons are not induced to express class I upon IFN treatment in the mouse, although other types of brain cell are (Massa *et al.*, 1993). All types of IFN can elicit and increase class I expression, but IFN-γ is usually the most potent. Cells with a low basal level of expression generally respond to IFN treatment to the greatest extent and this is particularly true of the *in vivo* response to IFN-γ.

All types of IFN can also induce class I gene expression (mRNA and protein) in class I negative cells, for example in undifferentiated EC cell lines, such as F9 (Wan *et al.*, 1987). In addition cells derived from day 8.5 embryos, which do not express class I genes, can be induced to express them by treatment with IFNs, whereas day 6 cells cannot be induced (Miyazaki *et al.*, 1986; Ozato *et al.*, 1985). When U937 histiocytic lymphoma cells differentiate, their class I genes are activated by autogenous production of IFN. Whether a similar process of autogenous production of IFNs increases class I expression during various physiopathological or developmental processes remains to be assessed. The level of class I mRNA usually increases very rapidly in IFN-treated cells, indicating a transcriptional control, but post-transcriptional regulation is also important. Finally, it has been shown that IFN-γ regulates the assembly of class I heavy chain with β2-m by an unknown mechanism in certain tumour cells deficient in cell-surface class I expression (Klar and Hammerling, 1989). This may represent the induction by IFN-γ of expression of the peptide transporters, TAP, in these cells (Epperson *et al.*, 1992), as has been shown more recently for small cell lung cancer cell lines (Restifo *et al.*, 1993), as well as in other tumour cell lines deficient in MHC class I expression.

Tumour necrosis factor (TNF) is a 17 kDa protein synthesized by mononuclear phagocytes, which is cytotoxic for certain tumour cells. TNF treatment increases cell-surface expression of class I antigens in some primary and established cell lines (Collins *et al.*, 1986). In human endothelial and dermal cells, the increase is 8- to 10-fold, while the steady state mRNA levels are increased 100-fold. The effects of TNF and IFNs are different, suggesting that they function through different pathways (see below).

To escape immune detection by class I-restricted cytotoxic lymphocytes, several viruses have developed mechanisms which down-regulate class I gene expression in infected (or transformed) cells (see below). Surprisingly, Moloney murine leukaemia virus (MoMuLV) does the opposite, probably by a *trans*-acting mechanism that operates on sequences located within 1.2 kb of the class I promoter (Wilson *et al.*, 1987). After intrathymic inoculation, the radiation leukaemia virus (RadLV) also induces an increase in the expression of H-2D on the thymocytes of resistant strains, whereas leukaemic cells in susceptible strains do not express class I antigens. The mechanism(s) responsible for these modulations is still not clear.

Human adenoviruses of all serotypes can transform rodent cells in culture. Cells transformed by the oncogenic serotypes (e.g. Ad12) are highly tumorigenic in syngeneic immunocompetent rodents, whereas cells transformed by the non-oncogenic serotypes (e.g. Ad2 and Ad5) are not, but can form tumours in immunosuppressed hosts. This difference in oncogenicity has been associated with the early (E_1A) region of Ad5 and Ad12 which, early in infection and in transformed cells, produces two major transcripts, 12S and 13S, coding for nuclear phosphoproteins of 220 and 266 amino acid residues. Primary rat and mouse cells transformed by the oncogenic Ad12 virus express greatly reduced amounts of MHC class I antigens in comparison with cells transformed by the non-oncogenic Ad5 virus (Schrier *et al.*, 1983; see below).

During lytic infection of mouse embryonic cells by Ad 12 virus, the expression of the *H-2K*^{b} gene is increased. This stimulation involves the *E1A* gene products (Rosenthal *et al.*, 1985) which can apparently influence class I gene expression both positively and negatively, by mechanisms that are not fully understood but appear to involve cellular transcription factors (see below). In contrast, the non-oncogenic adenovirus Ad2 interferes with class I expression at a post-translational level, by producing a glycoprotein, designated E19, which binds to nascent class I molecules in the endoplasmic reticulum and prevents their transport to the plasma membrane (Burgert and Kvist, 1985).

7.4 Molecular mechanisms of class I gene regulation

The transcription of genes by eukaryotic RNA polymerase II is regulated by *cis*-acting DNA sequences (promoters and enhancers). The promoter region consists of the TATA box (a TA-rich region) usually located 25–30 bp 5′ to the mRNA start site, and one or more 'upstream elements' (UPE). Transcription is also regulated by enhancers, *cis*-acting elements that can act, in either orientation, over large distances. All of these sequences have been shown to bind *trans*-acting protein factors which, through their activating domain, enhance the binding or activity of one or more components of the transciption complex. Negative regulatory elements have also been described, most often by deletion mutant analyses, but considerably less is known about the way DNA-binding proteins repress transciption.

Detailed analysis of the mouse MHC class I promoters by deletion analysis and footprinting experiments (both *in vitro* and *in vivo*) has revealed several regions that are important for the expression and regulation of these genes through specific binding of proteins. These include two enhancer-like sequences: A (−200 to −158), a complex region that binds several transcription factors, and B (−120 to −61). Enhancer A is juxtaposed with an interferon response sequence, IRS (−157 to −137) (*Figure 7.1*). Both enhancers are conserved within the promoter of several genes coding for classical transplantation antigens (*H-2K*^{k}, *H-2L*^{d}) but not in the promoter of some of the *Qa-TL* region genes (such as *Q10*, see below).

7.5 Transcriptional *cis*-acting elements and *trans*-acting factors

7.5.1 Classical class I genes

Murine genes. The enhancer A region can be dissected into a series of overlapping palindromes (*Figure 7.1*). The core region of the enhancer is a perfect palindrome denoted (ab), and an imperfect copy of it, denoted (ab′), lies a few nucleotides upstream. Finally, these two palindromes are spaced by a short sequence, denoted ′d, because it is symmetrical to b′ and, therefore, somewhat analogous to a. Accordingly, a third perfect palindrome, termed (b′ ′d) overlaps (ab′). The enhancer A region can thus be described as (ab′ ′dab) (*Figure 7.1;* Israël *et al.*, 1989a; Kimura *et al.*, 1986). The division of this region into class regulatory elements CRE-I, -II and -III subregions has also been proposed (*Figure 7.1*): CRE-I corresponds to the (ab) palindrome, CRE-III to the (ab′ ′dab) region, and CRE-II to the region just upstream (Miyazaki *et al.*, 1986).

Two families of factors bind the (ab) palindrome. This is a strong binding site for the rel/NF-κB family of transcription factors (Baldwin and Sharp, 1988; Israël *et al.*, 1989a). These proteins have been the subject of numerous reviews (Baeuerle and Henkel, 1994; Siebenlist *et al.*, 1994). Five members of this family are known: p50 and p52 are synthesized under the form of cytoplasmic precursors (p105 and p100, respectively), and activate transcription weakly (they might even behave as repressors when bound to some target sites as homodimers); p65 and c-rel represent the transcriptional activators of the family, and relB is specifically expressed in the lymphoid tissues. The most frequent and most active form of NF-κB is a heterodimer between p50 and p65. The post-translational control of the activity of the *rel*/NF-κB family of transcription factors is complex. Most homo- and heterodimers of *rel*/NF-κB proteins are sequestered in the cytoplasm by a family of inhibitory proteins, collectively called IκB. Only B cells and some cells of the monocyte/macrophage lineage exhibit constitutive nuclear NF-κB activity. The most abundant of these IκBs, *IκBα/MAD-3*, was cloned as an immediate early gene of newly adherent monocytes (Haskill *et al.*, 1991). A wide variety of stimuli (phorbol esters in various cells, LPS in preB cells, cytokines like TNF or IL1) cause degradation of IκBα by the ubiquitin-proteasome pathway (following phosphorylation) and nuclear translocation of the *rel*/NF-κB dimers, where they can act on their target genes. On the other hand the p50 homodimer has a weak affinity for IκB molecules, and therefore is constitutively present in the nucleus. We originally characterized this homodimer and called it KBF1 (Kieran *et al.*, 1990; Yano *et al.*, 1987). The presence of the KBF1 binding activity seems to correlate with the basal expression of MHC class I genes. This was further demonstrated by the use of a dominant negative mutant of p50, which can form inactive heterodimers with all members of the *rel*/NF-κB family, and which inhibits

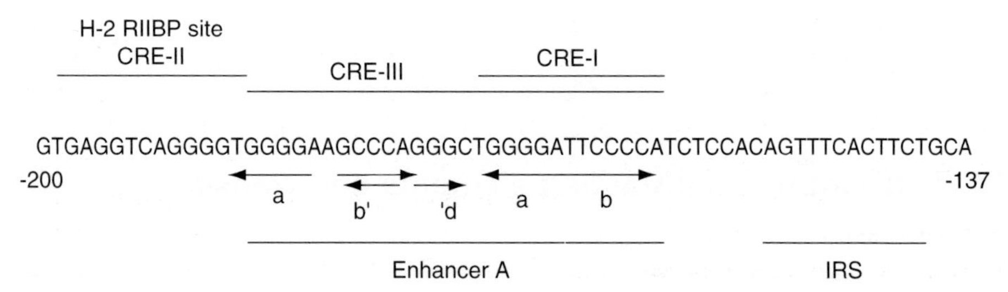

Figure 7.1. The enhancer A-IRS region of the murine $H\text{-}2K^b$ gene. The various palindromes as well as the CRE regions are described in the text.

the basal level of expression of class I genes (Logeat *et al.*, 1991). Following stimulation by inducers like TNF, more active complexes like p50/p65 heterodimers are released from their cytoplasmic inhibitors, translocate to the nucleus where they replace the KBF1 homodimer, occupying both the (ab) and the (ab′) palindromes (Israël *et al.*, 1989a). Interestingly both KBF1 and NF-κB also bind with a similar affinity to a rather divergent sequence, lying in the opposite orientation within the promoter of the mouse β2-m gene (Baldwin and Sharp, 1988; Israël *et al.*, 1987; Yano *et al.*, 1987). These transcription factors are thus expected to play a role in the coordinate regulation of class I and β2-m genes. Very recently it was shown that inactivation of the *p50* gene in mice by homologous recombination does not result in an inhibition of MHC class I gene expression (Sha *et al.*, 1995). This maybe indicates that other types of *rel*/NF-κB complexes can regulate these genes in the absence of p50.

KBF1 binding activity is present in most differentiated cells where MHC class I antigens are expressed (Burke *et al.*, 1989; Israël *et al.*, 1989b). However, it is absent in undifferentiated EC cells where class I and β2-m genes are silent, and it is induced when the cells are triggered to differentiate (Israël *et al.*, 1989). Thus, the appearance of KBF1 binding activity in EC cells parallels the onset of H-2 class I and β2-m gene expression (mRNA and proteins), suggesting that KBF1 activity is regulated during differentiation. Similarly it is absent in oligodendrocytes and neurons of the brain, which do not express MHC class I genes, but it is present in astrocytes, which do express them (Massa *et al.*, 1993). Also, in stage-specific T-cell lines, the level of class I transcripton has been shown to correlate with the amount of KBF1 binding activity.

Another factor, H-2 transcription factor 1 (H2TF1), has been shown to bind to enhancer A at exactly the same nucleotide residues as KBF1 (Baldwin and Sharp, 1988). Recent data suggests that H2TF1 is in fact the p100 precursor of the p52 subunit of NF-κB (Potter *et al.*, 1993; Scheinman *et al.*, 1993). Its potential function in the control of expression of MHC class I genes has not been firmly demonstrated.

Another family of factors, represented by MHC enhancer binding protein 1 and 2 (MBP-1 and -2); MBP-1 was cloned independently under the names of PRDII-BF1, αACRYBP1 or HIV-EP1, while MBP-2 was also cloned under the names of AGIE-BP1, AT-BP1 or HIV-EP2). MBP-1 and -2 can bind to the MHC class I enhancer [(ab) palindrome] with binding characteristics similar or identical to KBF1, H2TF1 or NF-κB, as well as to related sequences; these proteins have all been isolated by screening expression libraries using labelled oligonucleotides representing an (ab)-like sequence (Baldwin *et al.*, 1990; van't Veer *et al.*, 1992). The isolated cDNAs encode very large proteins (more than 250 kDa), containing two sets of widely separated zinc fingers and a stretch of highly acidic amino acids forming a putative transactivation domain. Although co-transfection experiments seem to indicate that MBP-2 behaves as a repressor of MHC class I gene expression, not much is known about the biological relevance of these factors.

As inferred from the known activators of the NF-κB transcription factors, the enhancer A region represents a TNF and phorbol 12-myristate 13 acetate (PMA) responsive element (Israël *et al.*, 1989). However the two adjacent palindromic sequences [(ab) and (ab′)], are required to confer TNF responsiveness, although the reason for this is not understood. The β2-m promoter contains only one and, consequently, is not stimulated by TNF.

7.5.2 The interferon-response sequence (IRS)

The IRS sequence, adjacent to the (ab) palindrome, is similar to a sequence found in several IFN-responsive genes (Israël *et al.*, 1986; Shirayoshi *et al.*, 1987; Sugita *et al.*, 1987). We found that the IRS was active with both types of IFNs only in conjunction with a specific enhancer sequence, enhancer A (Israël *et al.*, 1986), but others have observed that IRS alone could confer responsiveness to IFNs in certain cell types (Sugita *et al.*, 1987).

The analysis of proteins binding to the *H-2L* IRS following IFN-$\alpha\beta$ treatment indicates the presence of one constitutive factor and two inducible factors. One of these factors is induced in the presence of protein synthesis inhibitors and may be responsible for the rapid induction in class I transcription observed after IFN treatment, while at least one of the other two could be responsible for the return to basal level observed later (Shirayoshi *et al.*, 1988). A detailed analysis of the proteins binding to a similar sequence found in IFN-responsive genes (ISRE) has been performed. Three complexes can be observed in IFN-stimulated cells (although the pattern and kinetics of appearance of these complexes can vary from one cell type to another): ISGF-1, 2 and 3. ISGF1 is constitutive, while the two others are only observed after IFN treatment. ISGF2 has been cloned and is identical to a 59 kDa factor called IRF1; co-transfection of a vector encoding IRF1 stimulates an MHC class I promoter. ISGF1 is identical to IRF2, a gene belonging to the same family as IRF1, and behaves as a repressor which can abrogate the effect of IRF1. Another negatively acting factor, ICSBP, also belongs to the IRF1 family, but is mostly expressed in lymphocytes and macrophages, and is inducible by IFN-γ but not by IFN-$\alpha\beta$ (Bovolenta *et al.*, 1994; Nelson *et al.*, 1993; Weisz *et al.*, 1992). The ISGF3 factor can be dissociated into two elements: ISGF3-α and -γ, and is responsible for the rapid transcriptional response to IFN (Kessler *et al.*, 1990). ISGF3-γ is a 48 kDa protein which binds with a weak affinity to the IRS sequence. ISGF3-α is made of three proteins of 84, 91 and 113 kDa which cannot bind DNA and are located in the cytoplasm in untreated cells. These three proteins belong to the recently characterized family of STAT factors (signal transducers and activators of transcription) critically involved in the response to various cytokines (Darnell *et al.*, 1994): the 91 and 84 kDa proteins represent STAT1α and β, while the 113 kDa protein is STAT2. In the presence of IFN the ISGF3-α complex translocates to the nucleus and binds DNA in association with ISGF3-γ.

In conclusion, the regulation of MHC class I genes by interferons is a complex process which involves the highly regulated interplay between positive and negative factors. One of the major challenges in the future will be to determine which proteins bind to the IRS *in vivo*, and in which order.

7.5.3 Negative regulatory elements (NRE)

A region overlapping the IRS has been shown to act as a negative element in F9 EC cells, and as a positive element in 3T3 fibroblasts. The factor that binds to this sequence is not present in class I positive cells, and disappears in teratocarcinoma F9 cells when they are induced to differentiate by treatment with retinoic acid (Flanagan *et al.*, 1991). These results suggest that the expression of MHC class I genes during development involves switching from negative to positive regulation involving, at least, this sequence. Another negative regulatory element has been described in the promoter of the miniature swine MHC class I gene *PD1*. This element reduces the activity of both the homologous *PD1* promoter and a heterologous SV40 promoter, but requires the presence of a positive regulatory element to function (see below).

7.5.4 AP1 site and regulation by the jun/fos family

In vitro binding sites for the jun/fos family of transcription factors have been found around −100 and −200 in the *H-2* K^b promoter (Israël *et al.*, 1989a; Korber *et al.*, 1988). However, co-transfection experiments did not demonstrate a stimulation by the jun/fos proteins.

Independent experiments have demonstrated that co-transfection of a *c-jun* expression vector results in reduced activity of a swine *PD1* MHC class I promoter. The responsive sequence has been localized between −440 and −431, and c-jun has been shown directly to bind to this sequence (*Figure 7.2;* Howcroft *et al.*, 1993a). In an independent study, co-transfection of *c-fos* and *c-jun* expression vectors in highly metastatic cell lines derived from mouse melanoma resulted in an increased MHC class I expression and a decrease in tumorigenicity and metastatic potential of the recipient lines (Yamit-Hezi *et al.*, 1994). In contrast, transfection of the *junB* cDNA (another member of the jun/fos family) into a low metastatic cell line resulted in a decreased class I gene expression and an increased metastatic potential. The sequences responsible for these effects have not been identified.

A recent study demonstrated binding of a complex between p50 of NF-κB and fra-2 (a member of the fos family) to the (ab) palindrome (Giuliani *et al.*, 1995): hydrocortisone reduces the binding of this complex (the stoichiometry of which is unknown, but probably involves at least a p50 homodimer) as well as the level of expression of the promoter. Treatment with IFN relieves this repression.

7.5.5 The H-2 RIIBP site

Several factors bind upstream of the double palindrome of the enhancer A region, to the sequence GTGAGGTCAGGGG (*Figures 7.1* and *7.2*). Binding of factors belonging to the CREB/ATF family (involved in the response to cAMP) has been observed by bandshift experiments (Israël *et al.*, 1989a). However the MHC class I promoters respond weakly, if at all, to cAMP and therefore the biological significance of this observation is unclear. Screening of an expression library with this labelled sequence resulted in the isolation of a factor called H-2RIIBP (Hamada *et al.*, 1989). This protein is in fact a heterodimer between the RARβ retinoic acid receptor and the RXRβ retinoid X receptor. Treatment of teratocarcinoma cells with retinoic acid results in the induction of MHC class I genes through the NF-κB and H-2RIIBP binding sequences (Nagata *et al.*, 1992; Segars *et al.*, 1993; see below).

Results obtained by in vivo *footprinting.* Transgenic mice of the *H-2*^{b} haplotype carrying the HLA-B7 transgene were used for *in vivo* footprinting experiments

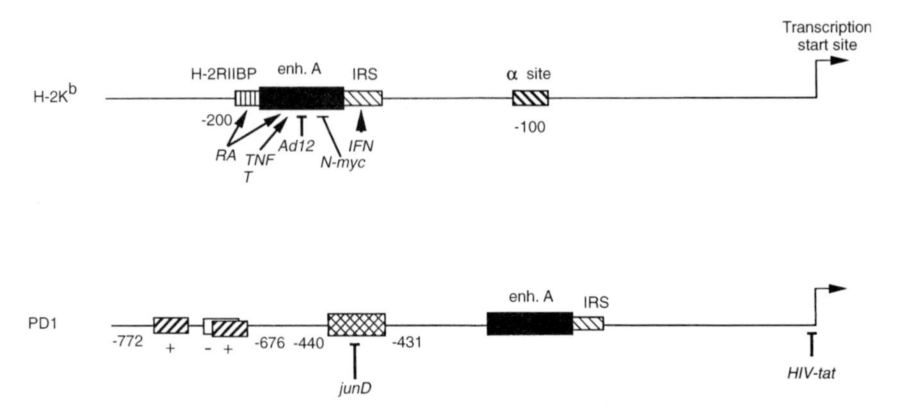

Figure 7.2. Functional map of the mouse and pig MHC class I promoters. The effectors (positive and negative) are indicated under the map in italics. + and – indicate positive and negative *cis*-acting elements. The different experimental strategies used by the different authors do not conclude that the mouse and pig MHC class I genes are regulated differently: for example the α sequence is conserved in the *PD1* promoter, but its function has not been investigated.

carried out in brain and in spleen cells. The upstream region of both the endogenous gene and the transgene are extensively occupied in spleen tissue, where MHC class I genes are expressed at high levels, but are not occupied in brain tissue, where no MHC class I expression occurs (Dey *et al.*, 1992). *In vivo* protected sites correspond to the (ab) palindrome of enhancer A, to the IRS and to a new sequence called α. This α sequence, G/TTGACGC, is located between −116 and −106 in the *HLA-B7* promoter, and between −101 and −94 in the K^b promoter (it is included in the enhancer B region, see Kimura *et al.*, 1986). This sequence is conserved among numerous class I genes of various species. It seems to exhibit a positive effect in spleen cells, but not in fibroblasts. DNA immunoprecipitation assays suggest that the RXRβ protein binds to the α sequence, possibly as a heterodimer with another nuclear hormone receptor. These results demonstrate the relevance of the results obtained by bandshift and *in vitro* footprinting, but also indicate that the mere presence of regulatory factors in a given tissue does not necessarily imply that they bind and regulate their target genes.

Similar experiments carried out on the β2-m promoter demonstrated *in vivo* occupancy of the enhancer A-IRS region, and also of another element, PAM, which is unique to the β2-m gene. These elements were occupied in spleen but not in brain cells. The PAM sequence binds nuclear factors ranging from 40 to 50 kDa in size and is capable of enhancing transcription of a reporter gene in EC and other cells (Lonergan *et al.*, 1993).

Regulatory sequences of the PDI class I gene of the miniswine. The promoter of this gene contains regions that are homologous to the enhancer A and IRS sequences (Singer and Ehrlich, 1988). Two regulatory elements, a positive and a negative, are located in the 1.1 kb region located upstream of the cap site (*Figure 7.2*; Ehrlich *et al.*, 1988). This 1.1 kb promoter region is sufficient to reproduce the normal pattern of expression of the *PD1* gene in transgenic mice (Ehrlich *et al.*, 1989). The negative region located between −771 and −676 seems to be dependent on the positive region located between −528 and −220 (Weissman and Singer, 1991). This negative region is a complex element made up of two overlapping regions, one positive and one negative; the positive region is predominantly active in lymphoid tissues, where the negative region is not active. Another regulatory element has been identified in the −489 to −395 region, which exhibits a negative effect in various cell types and seems to be involved in the transcriptional response to serum and hormones (Maguire *et al.*, 1992).

Regulation of the genes of the HLA locus. The regulatory regions of human MHC class I genes have been much less studied than their murine counterparts. The enhancer A and IRS regions are conserved and play similar roles (Ganguli *et al.*, 1989). However, some polymorphism exists in these regions among the genes of the A, B and C loci; for example, the second NF-κB binding site [palindrome (ab′) in the mouse promoters] is present in the *HLA-A* but not in the *-B* genes. This results in a weaker response of the *HLA-B* promoter to the NF-κB proteins (Girdlestone *et al.*, 1993). Similarly the (ab) palindrome of the *HLA-C* locus contains two mutations which most likely prevent binding of the NF-κB and KBF1 proteins (Park *et al.*, 1994). Sequence variations can also be detected in the IRS: for example, the IFN response of the *HLA-B7* gene is much stronger than that of the *HLA-B27* gene. This seems to be directly linked to differences in the sequence of the respective IRS (Schmidt *et al.*, 1990, 1993).

In another study, bandshift and footprint assays have revealed three distinct and independent factors that bind to the core regulatory elements of the *HLA-A* and *-B* gene loci; for example, a factor that binds to a Sp1-like sequence is crucial for normal *HLA-B* expression (Soong and Hui, 1992). Another study reports the binding of a factor that is immunologically related to the USF factor (a member of the helix-loop-helix/leucine zipper family) to a sequence that overlaps the IRS of the *B* but not of the *A* locus. This factor seems to compete with the binding of the IRF/STAT proteins (Girdlestone, 1993).

A recent study demonstrates the binding of a series of nuclear factors to positive and negative regulatory elements of the *HLA-A11* promoter. Two positive elements are located in the regions −155 to −91 and −335 to −206, and one negative element between −172 and −156. The factors binding to these sequences exhibit some tissue specificity. However, understanding the functional role of these DNA-binding activities will await a more detailed characterization of the proteins involved (Blanchet *et al.*, 1994).

Finally an example of supragenic regulation is provided by the JAR trophoblast-derived human cell line. In these cells most *HLA* genes are inactivated by CpG DNA methylation, except the *HLA-E* gene whose expression correlates with a lack of methylation (Boucraut *et al.*, 1993).

7.5.6 Promoter sequence of the non-classical Qa-TL genes

Qa-TL genes [e.g. *Q4, Q7, Q819, Q10, T3/T13* (TL)] code for class I-related molecules whose function is still unknown, but of which several are tissue-restricted (e.g. *TL* in the thymus and *Q10* in the liver and fetal yolk sac). The good overall conservation of the promoter sequences between *Qa* and classical class I genes has allowed their alignment and comparison. The promoter of the *Q10* gene has been shown to bind factors present in liver nuclear extracts (David-Watine *et al.*, 1990). Interestingly, most of the binding sequences for liver-specific factors in the *Q10* promoter do not exist in the other H-2 class I genes. Similarly, most of the regulatory elements involved in the ubiquitous expression of $H\text{-}2K^b$ or $H\text{-}2L^d$ (e.g. the enhancer A sequence) are punctually altered and not functional in *Q10*, but conserved in ubiquitously expressed genes like *Qa* or *Q8/9*. The latter all display a functional enhancer A sequence, identical to that of $H\text{-}2K^b$ in *Q8/9*, or to that of *β2-m* in *Q4*. The corresponding sequence in *Q7* is identical to the NF-κB binding site of the gene encoding the human interleukin 2 receptor (Lenardo and Baltimore, 1989). Interestingly, both genes are stimulated during the activation of T lymphocytes. Mutagenesis experiments that replaced the enhancer A sequence of *Q10* (which does not bind members of the rel/NF-κB family) by that of the L^d gene resulted in a restored expression in transfected fibroblasts (Handy *et al.*, 1989). However, the same type of experiment carried out in transgenic mice resulted in expression of the mutated construct only in the liver and thymus (Marine *et al.*, 1993). Therefore an intact (ab) palindrome is not sufficient to restore the ubiquitous pattern of expression characteristic of classical class I genes, indicating that other elements in the *Q10* promoter serve to limit ubiquitous tissue expression.

In contrast to the *Qa* genes, the sequences of the promoters from the genes located in the *TL* region are too divergent to be aligned with those of other class I genes. However, comparison of the 5′ sequences between classical class I genes and the *T18* gene reveals some homology in the IRS region. The expression of this gene is indeed stimulated by IFN-γ, but is unresponsive to IFN-αβ. By deletion analysis, promoter activity and IFN-γ responsiveness were localized to an 86 bp fragment that contains the IRS. In addition it was demonstrated that transcription of *T18* initiates much further upstream than in the classical class I genes (Horie *et al.*, 1991; Wang *et al.*, 1993).

7.6 Regulated expression

7.6.1 Regulation by retinoic acid (RA)

Retinoic acid (RA) induces differentiation in various cellular systems (especially in embryonal carcinoma cells) and acts as a morphogen at various steps of mammalian development. Treatment of human NTera-2 (NT2) EC cells induces expression of MHC class I genes (as well as β2-m) through two conserved upstream enhancers, enhancer A and the H-2RIIBP binding site (Nagata *et al.*, 1992; Segars *et al.*, 1993). NF-κB binding activity is not present in undifferentiated NT2 cells, but is induced following RA treatment. H-2RIIBP activity is present in undifferentiated cells at low levels but is greatly augmented by RA treatment because of activation of a nuclear hormone receptor heterodimer, composed of the RXRβ and the RARβ. These data show that following RA treatment, heterodimers of two transcription factor families are induced to bind to the MHC enhancers, which at least partly accounts for RA induction of MHC class I expression in NT2 EC cells.

7.6.2 Regulation by the N-myc *and* c-myc *oncogenes*

High constitutive expression of the *c-myc* oncogene in human melanoma leads to down-regulation of expression of HLA class I genes. The genes at the *HLA-B* locus are preferentially affected. This regulation takes place at the transcriptional level, but the main HLA class I enhancer (the human equivalent of the murine enhancer A sequence) is not involved in this process. Treatment with IFN-γ restores MHC class I expression (Peltenburg *et al.*, 1993; Versteeg *et al.*, 1988, 1989).

In neuroblastoma, *N-myc* suppresses the expression of MHC class I genes by reducing binding to the enhancer-A sequence, and concomitantly increases the metastatic potential of these cells (Bernards *et al.*, 1986). This reduction is due to transcriptional inhibition of the *p50* gene. Transfection of a *p50* expression vector in neuroblastoma cells that express *N-myc* at high level leads to re-expression of MHC class I antigens at the cell surface. Treatment with IFN-γ leads to the same increase. This inhibition of MHC class I gene expression cannot be observed in *N-myc*-transfected fibroblasts, indicating some cellular specificity in the effect of *N-myc* (Lenardo *et al.*, 1989; van't Veer *et al.*, 1993).

7.6.3 Regulation by the Tat protein of HIV-1

It has been observed that HIV-1 specifically and strongly decreases the activity of an MHC class I gene promoter. This repression is due to the HIV-1 Tat protein derived from a spliced viral transcript (two-exon Tat), and the target sequence is located in the proximal region of the swine PD1 promoter (Howcroft *et al.*, 1993b). This result suggests a mechanism whereby HIV-1-infected cells might be able to avoid immune surveillance, allowing the virus to persist in the infected host.

7.6.4 Regulation by adenovirus infection or transformation

A study, aimed at understanding why adenovirus 12 is oncogenic while adenovirus 5 is not, has led to the observation that Adl2-transformed primary cell lines exhibit a decreased expression of MHC class I antigens, possibly contributing to the evasion from the immune system (however, this is not the case for established cell lines; Eager *et al.*, 1985; Schrier *et al.*, 1983). This correlates with the observation that class I antigen expression is reduced in several types of human tumours (but this is not a general property of tumour cells). In Adl2-transformed cell lines, restoration of MHC class I expression reduces the oncogenicity of the cells (Tanaka *et al.*, 1985), although one report quotes the opposite (Soddu and Lewis, 1992). This inhibition of MHC class I gene expression has been ascribed to the 13S mRNA derived from the E1A region of adenovirus 12, while in transformed rat kidney cells the product of the 13S mRNA of the E1A region of Ad5 could counteract the effect of Ad12 (Bernards *et al.*, 1983). Detailed analysis of the mechanism of this repression by several groups led to different hypotheses, not necessarily exclusive. In general, there is an agreement that the repression takes place at the transcriptional level (Friedman and Ricciardi, 1988; Meijer *et al.*,

1989). One group reached the conclusion that the repression was dependent on sequences located far upstream in the H-2K^b promoter (−1725 to −1705, and −1591 to −1568; Katoh *et al.*, 1990; Ozawa *et al.*, 1993), while others pointed to the enhancer A region and demonstrated that transformation with Ad12 was accompanied by a decreased binding of members of the rel/NF-κB family to this sequence (Ackrill and Blair, 1989; Meijer *et al.*, 1992). A more detailed analysis of this phenomenon led to the conclusion that it is the processing of the cytoplasmic p105 precursor of the p50 subunit of NF-κB which is the target of Ad12 repression; a decreased processing of this molecule leads to decreased KBF1 and NF-κB binding activities (Schouten *et al.*, 1995). Reintroduction of p50 in these cells led to an increased expression of class I genes. Independent studies have demonstrated an increased binding to the H-2RIIBP site (see above) in Ad12 transformed cells, and suggested it was responsible for the decreased binding to the enhancer A region (Ackrill and Blair, 1989; Ge *et al.*, 1992; Kralli *et al.*, 1992). The activity of this negative element in Ad12-transformed cells is relieved by treatment with RA (Kralli *et al.*, 1992). In contrast, transformation of cells by the non-oncogenic adenovirus 5 results in an increased expression of MHC class I genes and concomitantly in an increased binding to the enhancer A region. The H-2RIIBP region exhibits no inhibitory effect in Ad5-transformed cells.

7.6.5 Regulation by the cytomegalovirus (CMV)

The products of the early genes of CMV induce an increase in the transcription of HLA genes (Burns *et al.*, 1993). The target sequence resides in the proximal region of the promoter, including a GCGGT sequence found in other CMV-responsive genes. Later in infection, however, synthesis of the products of the early genes results in an impaired transport of class I molecules to the cell surface (Del Val *et al.*, 1992; see also Chapter 9).

7.7 Conclusion

The picture that emerges from these *in vitro* studies is that the promoter region of class I genes can be activated through several pathways. Thus, these genes should not be considered as typical 'domestic' genes. Rather, their composite promoter might fit a variety of specific cellular situations and be able to respond to subtle changes in the cellular environment. This is also suggested by the comparative analysis of non-ubiquitously expressed class I genes, such as the *Q10* gene expressed in the liver and fetal yolk sac. The recent cloning of some of the factors responsible for the binding activities observed *in vitro* and *in vivo* has allowed confirmation of their relevance. How these different factors contribute to the fine tuning of MHC class I gene expression under physiological circumstances begins to be understood, as does the extent to which these factors may be involved in pathological processes. In the future, the use of transgenic and knockout mice will be useful in shedding some light on these areas.

References

Ackrill AM, Blair GE. (1989) Nuclear proteins binding to an enhancer element of the MHC class I promoter: differences between highly oncogenic and non-oncogenic adenovirus-transformed cells.

Virology **172:** 643–646.

Baeuerle PA, Henkel T. (1994) Function and activation of NF-κB in the immune system. *Annu. Rev. Immunol.* **12:** 141–179.

Baldwin AJ, Sharp PA. (1988) Two transcription factors, NF-κB and H2TF1, interact with a single regulatory sequence in the class I major histocompatibility complex promoter. *Proc. Natl Acad. Sci. USA* **85:** 723–727.

Baldwin ASJ, LeClair KP, Singh H, Sharp PA. (1990) A large protein containing zinc finger domains binds to related sequence elements in the enhancers of the class I major histocompatibility complex and κ immunoglobulin genes. *Mol. Cell. Biol.* **10:** 1406–1414.

Bartlett PF, Kerr RSC, Bailey KA. (1989) Expression of MHC antigens in the central nervous system. *Transplant. Proc.* **21:** 3163–3165.

Berah M, Hors J, Dausset J. (1970) A study of HLA antigens in human organs. *Transplantation* **9:** 19203–19220.

Bernards R, Schrier PI, Houweling A, Bos JL, van der Eb AJ. (1983) Tumorigenicity of cells transformed by adenovirus type 12 by evasion of T cell immunity. *Nature* **305:** 776–779.

Bernards R, Dessain SK, Weinberg RA. (1986) N-myc amplification causes down-modulation of MHC class I antigen expression in neuroblastoma. *Cell* **47:** 667–674.

Blanchet O, Gazin C, L'Haridon M, Tatari Z, Degos L, Sigaux F, Paul P. (1994) Multiple nuclear factors bind to novel positive and negative regulatory elements upstream of the human MHC class I gene HLA-A11. *Int. Immunol.* **6:** 1485–1496.

Boog CJ, Boes J. Melief CJ. (1988) Role of dendritic cells in the regulation of class I restricted cytotoxic T lymphocyte responses. *J. Immunol.* **140:** 3331–3337.

Boucraut J, Guillaudeux T, Alizadeh M, Boretto J, Chimini G, Malecze F, Semana G, Fauchet R, Pontarotti P, Le Bouteiller P. (1993) HLA-E is the only class I gene that escapes CpG methylation and is transcriptionally active in the trophoblast-derived human cell line JAR. *Immunogenetics* **38:** 117–130.

Bovolenta C, Driggers PH, Marks MS, et al. (1994) Molecular interactions between interferon consensus sequence binding protein and members of the interferon regulatory factor family. *Proc. Natl Acad. Sci. USA* **91:** 5046–5050.

Burgert HG, Kvist S. (1985) An adenovirus 2 glycoprotein blocks cell surface expression of human histocompatibility class I antigens. *Cell* **41:** 987–997.

Burke PA, Hirschfeld S, Shirayoshi Y, Kasik JW, Hamada K, Appella E, Ozato K. (1989) Developmental and tissue-specific expression of nuclear proteins that bind the regulatory element of the major histocompatibility complex class I gene. *J. Exp. Med.* **169:** 1309–1321.

Burns LJ, Waring JF, Reuter JJ, Stinski MF, Ginder GD. (1993) Only the HLA class I gene minimal promoter elements are required for transactivation by human cytomegalovirus immediate early genes. *Blood* **81:** 1558–1566.

Collins T, Lapierre L, Fiers W, Strominger J, Pober J. (1986) Recombinant human TNF increases mRNA levels and surface expression of HLA-A, -B antigens in vascular endothelial cells and dermal fibroblasts in vitro. *Proc. Natl Acad. Sci. USA* **83:** 446–450.

Cosman D, Kress M, Khoury G, Jay G. (1982) Tissue-specific expression of an unusual H-2 (class I) related gene. *Proc. Natl Acad. Sci. USA* **79:** 4947–4951.

Croce CM, Linnenbach A, Huebner K, Parnes JR, Margulies DH, Appella E, Seidman JG. (1981) Control of expression of histocompatibility antigens (H-2) and β2-microglobulin in F9 teratocarcinoma stem cells. *Proc. Natl Acad. Sci. USA* **78:** 5754–5758.

Darnell JE, Kerr IM, Stark GR. (1994) Jak-Stat pathways and transcriptional activation in response to IFNs and other extracellular signaling proteins. *Science* **264:** 1415–1421.

David-Watine B, Transy C, Gachelin G, Kourilsky P. (1987) Tissue-specific expression of the mouse Q10 H-2 class I gene during embryogenesis. *Gene* **61:** 145–154.

David-Watine B, Logeat F, Israël A, Kourilsky P. (1990) Regulatory elements involved in the liver-specific expression of the mouse MHC class I Q10 gene: characterization of a new TATA-binding factor. *Int. Immunol.* **2:** 981–993.

Del Val M, Hengel H, Hacker H, Hartlaub U, Ruppert T, Lucin P, Koszinowski UH. (1992) Cytomegalovirus prevents antigen presentation by blocking the transport of peptide-loaded major histocompatibility complex class I molecules into the medial-Golgi compartment. *J. Exp. Med.* **176:** 729–738.

Dey A, Thornton AM, Lonergan M, Weissman SM, Chamberlain JW, Ozato K. (1992) Occupancy of upstream regulatory sites in vivo coincides with major histocompatibility complex class I gene expression in mouse tissues. *Mol. Cell. Biol.* **12:** 3590–3599.

Eager K, Williams J, Brelding D, Pand S, Knowls B, Appella E, Ricciardi EP. (1985) Expression of histocompatibility antigens H-2 K, D and L is reduced in adenovirus-transformed mouse cells and is restored by interferon. *Proc. Natl Acad. Sci. USA* **82:** 5525–5529.
Ehrlich R, Maguire JE, Singer DS. (1988) Identification of negative and positive regulatory elements associated with a class I major histocompatibility complex gene. *Mol. Cell. Biol.* **8:** 695–703.
Ehrlich R, Sharrow SO, Maguire JE, Singer DS. (1989) Expression of a class I MHC transgene: effects of in vivo alpha/beta-interferon treatment. *Immunogenetics* **30:** 18–26.
Epperson DE, Arnold D, Spies T, Cresswell P, Pober JS, Johnson DR. (1992) Cytokines increase transporter in antigen processing-1 expression more rapidly than HLA class I expression in endothelial cells. *J. Immunol.* **149:** 3297–3301.
Fabry Z, Raine CS, Hart MN. (1994) Nervous tissue as an immune compartment: the dialect of the immune response in the CNS. *Immunol. Today* **15:** 218–224.
Flanagan JR, Murata M, Burke PA, Shirayoshi Y, Appella E, Sharp PA, Ozato K. (1991) Negative regulation of the major histocompatibility complex class I promoter in embryonal carcinoma cells. *Proc. Natl Acad. Sci. USA* **88:** 3145–3149.
Friedman DJ, Ricciardi RP. (1988) Adenovirus type 12 E1A represses accumulation of MHC class I mRNAs at the level of transcription. *Virology* **165:** 303–305.
Gachelin G. (1978) The cell surface antigens of mouse embryonal carcinoma cells. *Biochim. Biophys. Acta* **516:** 27–60.
Ganguli S, Vasavada HA, Weissman SM. (1989) Multiple enhancer-like sequences in the HLA-B7 gene. *Proc. Natl Acad. Sci. USA* **86:** 5247–5251.
Ge R, Kralli A, Weinmann R, Ricciardi RP. (1992) Down-regulation of the major histocompatibility complex class I enhancer in adenovirus type 12-transformed cells is accompanied by an increase in factor binding. *J. Virol.* **66:** 6969–6978.
Girdlestone J. (1993) An HLA-B regulatory element binds a factor immunologically related to the upstream stimulation factor. *Immunogenetics* **38:** 430–436.
Girdlestone J, Isamat M, Gewert D, Milstein C. (1993) Transcriptional regulation of HLA-A and -B: differential binding of members of the Rel and IRF families of transciption factors. *Proc. Natl Acad. Sci. USA* **90:** 11568–11572.
Giuliani C, Saji M, Napolitano G, Palmer LA, Taniguchi SI, Shong M, Singer DS, Kohn LD. (1995) Hormonal modulation of major histocompatibility complex class I gene expression involves an enhancer A-binding complex consisting of Fra-2 and the p50 subunit of NF-κB. *J. Biol. Chem.* **270:** 11453–11462.
Hamada K, Gleason SL, Levi BZ, Hirschfeld S, Appella E, Ozato K. (1989) H-2RIIBP, a member of the nuclear hormone receptor superfamily that binds to both the regulatory element of major histocompatibility class I genes and the estrogen response element. *Proc. Natl Acad. Sci. USA* **86:** 8289–8293.
Handy DE, Burke PA, Ozato K, Coligan JE. (1989) Site-specific mutagenesis of the class I regulatory element of the Q10 gene allows expression in non-liver tissues. *J. Immunol.* **142:** 1015–1021.
Haskill S, Beg AA, Tompkins SM, Morris JS, Yurochko AD, Sampson JA, Mondal K, Ralph P, Baldwin AJ. (1991) Characterization of an immediate-early gene induced in adherent monocytes that encodes IκB-like activity. *Cell* **65:** 1281–1289.
Hedley ML, Drake BL, Head JR, Tucker PW, Forman J. (1989) Differential expression of the class I MHC genes in the embryo and placenta during midgestational development in the mouse. *J. Immunol.* **142:** 4046–4053.
Horie M, Matsuura A, Chang KJ, Niikawa J, Shen FW. (1991) Properties of the promoter region of the T18d (T13c) Tla gene. *Immunogenetics* **33:** 171–177.
Howcroft TK, Richardson JC, Singer DS. (1993a) MHC class I gene expression is negatively regulated by the proto-oncogene, c-jun. *EMBO J.* **12:** 3163–3169.
Howcroft TK, Strebel K, Martin MA, Singer DS. (1993b) Repression of MHC class I gene promoter activity by two-exon Tat of HIV. *Science* **260:** 1320–1322.
Israël A, Kimura A, Fournier A, Fellous M, Kourilsky P. (1986) Interferon response sequence potentiates activity of an enhancer in the promoter region of a mouse H-2 gene. *Nature* **322:** 743–746.
Israël A, Kimura A, Kieran M, Yano O, Kanellopoulos J, Le Bail O, Kourilsky P. (1987) A common positive trans-acting factor binds to enhancer sequences in the promoters of mouse H-2 and beta 2-microglobulin genes. *Proc. Natl Acad. Sci. USA* **84:** 2653–2657.
Israël A, LeBail O, Hatat D, Piette J, Kieran M, Logeat F, Wallach D, Fellous M, Kourilsky P. (1989a) TNF stimulates expression of mouse MHC class I genes by inducing an NF-κB-like

enhancer binding activity which displaces constitutive factors. *EMBO J.* **8:** 3793–3800.

Israël A, Yano O, Logeat F, Kieran M, Kourilsky P. (1989b) Two purified factors bind to the same sequence in the enhancer of mouse MHC class I genes: one of them is a positive regulator induced upon differentiation of teratocarcinoma cells. *Nucleic Acids Res.* **17:** 5245–5257.

Jaffe L, Robertson EJ, Bikoff EK. (1991) Distinct patterns of expression of MHC class I and beta 2-microglobulin transcripts at early stages of mouse development. *J. Immunol.* **147:** 2740–2749.

Katoh S, Ozawa K, Kondoh S, Soeda E, Israël A, Shiroki K, Fujinaga K, Itakura K, Gachelin G, Yokoyama K. (1990) Identification of sequences responsible for positive and negative regulation by E1A in the promoter of H-2Kbm1 class I MHC gene. *EMBO J.* **9:** 127–135.

Kessler DS, Veals SA, Fu XY, Levy DE. (1990) Interferon-α regulates nuclear translocation and DNA-binding affinity of ISGF3, a multimeric transciptional activator. *Genes Dev.* **4:** 1753–1765.

Kieran M, Blank V, Logeat F, Vandekerckhove J, Lottspeich F, LeBail O, Urban MB, Kourilsky P, Baeuerle PA, Israël A. (1990) The DNA binding subunit of NF-κB is identical to factor KBF1 and homologous to the rel oncogene product. *Cell* **62:** 1007–1018.

Kimura A, Israël A, LeBail O, Kourilsky P. (1986) Detailed analysis of the mouse H-2 Kb promoter: enhancer-like sequences and their role in the regulation of class I gene expression. *Cell* **44:** 261–272.

Klar D, Hammerling GJ. (1989) Induction of assembly of MHC class I heavy chains with β2-microglobulin by interferon-γ. *EMBO J.* **8:** 475–481.

Korber B, Mermod N, Hood L, Stroynowski I. (1988) Regulation of gene expression by interferon: control of H-2 promoter responses. *Science* **239:** 1302–1306.

Kralli A, Ge R, Graeven U, Ricciardi RP, Weinmann R. (1992) Negative regulation of the major histocompatibility complex class I enhancer in adenovirus type 12-transformed cells via a retinoic acid response element. *J. Virol.* **66:** 6979–6988.

Lenardo MJ, Baltimore D. (1989) NF-κB: a pleiotropic mediator of inducible and tissue-specific gene control. *Cell* **58:** 227–229.

Lenardo M, Rustgi AK, Schievella R, Bernards R. (1989) Suppression of MHC class I gene expression by N-myc through enhancer inactivation. *EMBO J.* **8:** 3351–3355.

Lindahl P, Gresser I, Leary P, Tovey M. (1976) Interferon treatment of mice: enhanced expression of histocompatibility antigens on lymphoid cells. *Proc. Natl Acad. Sci. USA* **73:** 1284–1288.

Logeat F, Israël N, Ten R, Blank V, LeBail O, Kourilsky P, Israël A. (1991) Inhibition of transciption factors belonging to the rel/NF-κB family by a transdominant negative mutant. *EMBO J.* **10:** 1827–1832.

Lonergan M, Dey A, Becker KG, Drew PD, Ozato K. (1993) A regulatory element in the beta 2-microglobulin promoter identified by in vivo footprinting. *Mol. Cell. Biol.* **13:** 6629–6639.

Maguire JE, Frels WI, Richardson JC, Weissman JD, Singer DS. (1992) In vivo function of regulatory DNA sequence elements of a major histocompatibility complex class I gene. *Mol. Cell. Biol.* **12:** 3078–3086.

Marine JB, Shirakata Y, Wadsworth SA, Hooley JJ, Handy DE, Coligan JE. (1993) Role of the Q10 class I regulatory element region 1 in controlling tissue-specific expression in vivo. *J. Immunol.* **151:** 1989–1997.

Massa PT, Ozato K, McFarlin DE. (1993) Cell type-specific regulation of major histocompatibility complex (MHC) class I gene expression in astrocytes, oligodendrocytes, and neurons. *Glia* **8:** 201–207.

Meijer I, Jochemsen AG, De Wit CM, Bos JL, Morello D, van der Eb AJ. (1989) Adenovirus type 12 E1A down regulates expression of a transgene under control of a major histocompatibility complex class I promoter: evidence for transciptional control. *J. Virol.* **63:** 4039–4042.

Meijer I, Boot AJ, Mahabir G, Zantema A, van der Eb A. (1992) Reduced binding activity of transciption factor NF-κB accounts for MHC class I repression in adenovirus type 12 E1-transformed cells. *Cell. Immunol.* **145:** 56–65.

Miyazaki J, Appella E, Ozato K. (1986) Negative regulation of the MHC class I gene in undifferentiated embryonal carcinoma cells. *Proc. Natl Acad. Sci. USA* **83:** 9537–9541.

Morello D, Danile F, Baldacci P, Cayre Y, Gachelin G, Kourilsky P. (1982) Absence of significant H-2 and β2-m mRNA expression by mouse embryonal carcinoma cells. *Nature* **296:** 260–261.

Morello D, Duprey P, Israël A, Babinet C. (1985) Asynchronous regulation of mouse H-2D and β2-microglobulin RNA transcripts. *Immunogenetics* **22:** 441–452.

Nagata T, Segars JH, Levi BZ, Ozato K. (1992) Retinoic acid-dependent transactivation of major histocompatibility complex class I promoters by the nuclear hormone receptor H-2RIIBP in undifferentiated embryonal carcinoma cells. *Proc. Natl Acad. Sci. USA* **89:** 937–941.

Nelson N, Marks MS, Driggers PH, Ozato K. (1993) Interferon consensus sequence-binding protein, a member of the interferon regulatory factor family, suppresses interferon-induced gene transcription. *Mol. Cell. Biol.* **13:** 588–599.

Ostrand-Rosenberg S, Nickerson DA, Clements VK, Garcia EP, Lamouse-Smith E, Hood L, Stroynowski I. (1989) Embryonal carcinoma cells express Qa and TLa class I genes of the major histocompatibility complex. *Proc. Natl Acad. Sci. USA* **86:** 5084–5088.

Ozato K, Wan YJ, Orrison B. (1985) Mouse major histocompatibility class I gene expression begins at midsomite stage and is inducible at earlier-stage embryos by interferon. *Proc. Natl Acad. Sci. USA* **82:** 2427–2431.

Ozawa K, Hagiwara H, Tang X, Saka F, Kitabayashi I, Shiroki K, Fujinaga K, Israël A, Gachelin G, Yokoyama K. (1993) Negative regulation of the gene for H-2Kb class I antigen by adenovirus 12-E1A is mediated by a CAA repeated element. *J. Biol. Chem.* **268:** 27258–27268.

Palmer M, Frelinger J. (1987) Widespread expression of a Qa gene in adult mice. *J. Exp. Med.* **166:** 95–108.

Park JH, Kim JD, Kim SJ. (1994) Nuclear protein binding patterns in the 5′-upstream regulatory elements of HLA class I genes. *Yonsei Med J.* **35:** 295–307.

Peltenburg LT, Dee R, Schrier PI. (1993) Downregulation of HLA class I expression by c-myc in human melanoma is independent of enhancer A. *Nucleic Acids Res.* **21:** 1179–1185.

Potter DA, Larson CJ, Eckes P, Schmid RM, Nabel GJ, Verdine GL, Sharp PA. (1993) Purification of the major histocompatibility complex class-I transcription factor-H2TF1: the full-length product of the nfkb2 gene. *J. Biol. Chem.* **268:** 18882–18890.

Restifo NP, Esquivel F, Kawakami Y, Yewdell JW, Mulé JJ, Rosenberg SA, Bennink JR. (1993) Identification of human cancers deficient in antigen processing. *J. Exp. Med.* **177:** 265–272.

Robinson PJ. (1987) Structure and expression of polypeptides encoded in the mouse Qa region. *Immunol. Res.* **6:** 46–56.

Rosenthal A, Wright S, Cedar H, Flavell R, Grosveld F. (1984) Regulated expression of an introduced MHC H-2Kbml gene in murine embryonal carcinoma cells. *Nature* **310:** 415–416.

Rosenthal A, Wright S, Quade K, Gallimore P, Cedar H, Grosveld F. (1985) Increased MHC H-2K gene transcription in cultured mouse embryo cells after adenovirus infection. *Nature* **315:** 579–581.

Scheinman RI, Beg AA, Baldwin AS. (1993) NF-κB p100 (Lyt-10) is a component of H2TF1 and can function as an IκB-like molecule. *Mol. Cell. Biol.* **13:** 6089–6101.

Schmidt H, Gekeler V, Haas H, Engler-Blum G, Steiert I, Probst H, Muller CA. (1990) Differential regulation of HLA class I genes by interferon. *Immunogenetics* **31:** 245–252.

Schmidt H, Kellermann KE, Steiert I, Walz J, Zinser R, Muller CA. (1993) Differential regulation of human leukocyte antigen class I genes by interferon in vivo and in vitro. *J. Immunother.* **14:** 169–174.

Schouten GJ, van der Eb AJ, Zantema A. (1995) Downregulation of MHC class I expression due to interference with p105-NFκB1 processing by Ad12E1A. *EMBO J.* **14:** 1498–1507.

Schrier PI, Bernards R, Vaessen RTMJ, Houweling A, van der Eb AJ. (1983) Expression of class I major histocompatibility antigens switched off by highly oncogenic adenovirus 12 in transformed rat cells. *Nature* **305:** 771–775.

Segars JH, Nagata T, Bours V, *et al.* (1993) Retinoic acid induction of major histocompatibility complex class I genes in NTera-2 embryonal carcinoma cells involves induction of NF-κB (p50–p65) and retinoic acid recptor β-retinoid X receptor β heterodimers. *Mol. Cell. Biol.* **13:** 6157–6169.

Sha WC, Liou HC, Tuomanen EI, Baltimore D. (1995) Targeted disruption of the p50 subunit of NF-κB leads to multifocal defects in immune responses. *Cell* **80:** 321–330.

Shirayoshi Y, Miyazaki J, Burke PA, Hamada K, Appella E, Ozato K. (1987) Binding of multiple nuclear factors to the 5′ upstream regulatory element of the murine major histocompatibility class I gene. *Mol. Cell. Biol.* **7:** 4542–4548.

Shirayoshi Y, Burke PA, Appella E, Ozato K. (1988) Interferon-induced transcription of a major histocompatibility class I gene accompanies binding of inducible nuclear factors to the interferon consensus sequence. *Proc. Natl Acad. Sci. USA* **85:** 5884–5888.

Siebenlist U, Fransozo G, Brown K. (1994) Structure, regulation and function of NF-κB. *Annu. Rev. Cell. Biol.* **10:** 405–430.

Singer DS, Ehrlich R. (1988). Identification of regulatory elements associated with a class I MHC gene. *Curr. Top. Microbiol. Immunol.* **137:** 148–154.

Singer DS, Maguire JE. (1990) Regulation of the expression of class I MHC genes. *Crit. Rev.*

Immunol. **10:** 235–256.

Soddu S, Lewis AM. (1992) Driving adenovirus 12-transformed Balb/c mouse cells to express high levels of class I MHC proteins enhances, rather than abrogates, their tumorigenicity. *J. Virol.* **66:** 2875–2884.

Soong TW, Hui KM. (1992) Locus-specific transcriptional control of HLA genes. *J. Immunol.* **149:** 2008–2020.

Sugita K, Miyazaki J, Appella E, Ozato K. (1987) Interferons increase transcription of a major histocompatibility class I gene via a 5′ Interferon Consensus Sequence. *Mol. Cell. Biol.* **7:** 2625–2630.

Tanaka K, Isselbacher KJ, Khoury G, Jay G. (1985) Reversal of oncogenesis by the expression of a MHC class I gene. *Science* **228:** 23–30.

Transy C, Nash SR, David-Watine B, Cochet M, Hunt SW, Hood LE, Kourilsky P. (1987) A low polymorphic mouse H-2 class I gene from the TLa complex is expressed in a broad variety of cell types. *J. Exp. Med.* **166:** 341–361.

van't Veer LJ, Lutz PM, Isselbacher KJ, Bernards R. (1992) Structure and expression of major histocompatibility complex-binding protein 2, a 275-kDa zinc finger protein that binds to an enhancer of major histocompatibility complex class I genes. *Proc. Natl Acad. Sci. USA* **89:** 8971–8975.

van't Veer LJ, Beijersbergen RL, Bernards R. (1993) N-myc suppresses Major Histocompatibility Complex class-I gene expression through down-regulation of the p50 subunit of NF-κB. *EMBO J.* **12:** 195–200.

Versteeg R, Noordermeer IA, Kruse WM, Ruiter DJ, Schrier PI. (1988) c-myc down-regulates class I HLA expression in human melanomas. *EMBO J.* **7:** 1023–1029.

Versteeg R, Kruse-Wolters KM, Plomp AC, van Leeuwen A, Stam NJ, Ploegh HL, Ruiter DJ, Schrier PI. (1989) Suppression of class I human histocompatibility leukocyte antigen by c-myc is locus-specific. *J. Exp. Med.* **170:** 621–635.

Wan YJ, Orrison BM, Lieberman R, Lazarovici P, Ozato K. (1987) Induction of major histocompatibility class I antigens by interferons in undifferentiated F9 cells. *J. Cell Physiol.* **130:** 276–283.

Wang IM, Mehta V, Cook RG. (1993) Regulation of TL antigen expression. Analysis of the T18d promoter region and responses to IFN-gamma. *J. Immunol.* **151:** 2646–2657.

Warner CM, Gollnick SO, Flaherty L, Goldbard SB. (1987) Analysis of Qa-2 antigen expression by preimplantation mouse embryos: possible relationship to the preimplantation-embryo-development (ped) gene product. *Biol. Reprod.* **36:** 611–616.

Weissman JD, Singer DS. (1991) A complex regulatory DNA element associated with a major histocompatibility complex class I gene consists of both a silencer and an enhancer. *Mol. Cell. Biol.* **11:** 4217–4227.

Weisz A, Marx P, Sharf R, Appella E, Driggers PH, Ozato K, Levi BZ. (1992) Human interferon consensus sequence binding protein is a negative regulator of enhancer elements common to interferon-inducible genes. *J. Biol. Chem.* **267:** 25589–25596.

Wilson LD, Flyer DC, Faller DV. (1987) Murine retroviruses control MHC antigen gene expression via a trans-effect at the transcriptional level. *Mol. Cell. Biol.* **7:** 2406–2415.

Yamit-Hezi A, Plaksin D, Eisenbach L. (1994) c-fos and c-jun overexpression in malignant cells reduces their tumorigenic and metastatic potential, and affects their MHC class I gene expression. *Oncogene* **9:** 1065–1079.

Yano O, Kanellopoulos J, Kieran M, LeBail O, Israël A, Kourilsky P. (1987) Purification of KBF1, a common factor binding to both H-2 and β2-microglobulin enhancers. *EMBO J.* **6:** 3317–3324.

8

HLA/MHC class II gene regulation

R. Wassmuth

8.1 Introduction

The central task of the immune system is the discrimination between *self* and *non-self* in order to maintain self and to mount an effective immune response against non-self antigens. The immune mechanisms providing this distinction can be differentiated into those that are unspecific and those that are antigen-specific. A key feature of the antigen-specific immune response is activation of T lymphocytes. The pivotal role major histocompatibility complex (MHC) molecules play in this activation process was appreciated in the late 1970s when it was shown that antigen-specific T lymphocytes can recognize antigen only in the form of peptides bound to an MHC molecule expressed on the cell surface of an antigen-presenting cell (APC). In antigen presentation there is a functional dichotomy, as MHC class II molecules are required for the recognition of antigen by $CD4^+$, mostly helper, T lymphocytes, while $CD8^+$, mostly cytotoxic, T lymphocytes recognize antigens in the context of MHC class I molecules (Braciale *et al.*, 1987). Class II molecules are central to the initiation and propagation of an immune response. Moreover, MHC molecules play a decisive role in the generation and control of antigen-specific immune responsiveness by modulating the T-lymphocyte receptor (TCR) repertoire via positive and negative selection mechanisms in the thymus. Structurally, these immunoregulatory functions of MHC molecules are mediated by the ternary complex of the MHC molecule, the antigenic peptide and the TCR (Davis and Chien, 1993; Rammensee *et al.*, 1993).

In contrast to class I molecules, which are expressed rather ubiquitously, the expression of class II molecules is restricted to B lymphocytes, cells of myelomonocytic lineage, activated T lymphocytes and non-lymphoid cells such as endothelial cells and dendritic cells (Alonso *et al.*, 1985). MHC class II antigens are highly polymorphic molecules expressed on the cell surface as glycosylated non-covalently linked heterodimers composed of a heavy (33–35 kDa) α-chain and a light (26–29 kDa) β-chain (Kappes and Strominger, 1988; Trowsdale, 1987). So far, these molecules have only been found in vertebrates and are evolutionarily thought to have arisen from a common ancestral gene by repeated duplication and recombination events. Over the last decades, considerable knowledge has accumulated concerning the structural organization of the MHC class II genes and molecules. Being part of the immunoglobulin superfamily, the MHC genes constitute clusters of highly homologous genes. In mice, the genes for

HLA and MHC: genes, molecules and function, edited by M.J. Browning and A.J. McMichael.
© 1996 BIOS Scientific Publishers Ltd, Oxford.

the class II molecules I-E and I-A are localized on chromosome 17, while those of the human analogues HLA-DR, HLA-DQ and HLA-DP are found on chromosome 6. In addition to these characteristic class II genes other homologous genes have been found. However, their pattern of expression and function is often different or only poorly characterized. Aside from the existence of different isotypic forms and the high degree of allelic polymorphism, particularly in the region of the antigen-binding groove, diversity is generated by complementation of the α- and β-chains in the formation of the functional heterodimer. All these features contribute to a variability of the antigen-binding site and thus enhance the possibility for effective antigen presentation of a given foreign or self antigen on an individual and populationwide level. The existence of multiple class II loci invites speculation regarding differential functions of these isotypic forms, particularly since differences in expression can be observed. However, our understanding is too fragmentary to allow precise separation of isotype-specific functions. Only for some class I (e.g. *HLA-G*) and II (e.g. *HLA-DM*) genes do differential functions become evident (Morris *et al.*, 1994; Sargent, 1993; Sloan *et al.*, 1995).

An understanding of the regulation of MHC class II genes is of interest because of the central immunoregulatory function of these genes, their limited and cell-specific expression, and the necessity coordinately to express both chains of the functional heterodimer. This review will focus on the class II genes and the transcriptional mechanisms of class II gene regulation. As the subject has been extensively reviewed in the past, and due to more recent advances in the human field, the discussion will be restricted to upstream regulatory polymorphisms of HLA class II genes and some disease-relevant aspects of their regulation (Benoist and Mathis, 1990; Glimcher and Kara, 1992; Halloran and Madrenas, 1990; Ting and Baldwin, 1993). After reviewing cell type-specific patterns and key modulators of MHC class II expression, the general transcriptional mechanism of regulation will be addressed before turning to HLA class II isotype-specific features. Finally, the importance of class II regulation for human diseases will be discussed.

8.2 Biology of class II expression

Within the group of cells capable of expressing class II molecules, differences with respect to the pattern, inducibility and modulation of expression can be appreciated. Generally, these cell types can be divided into three categories:

(i) constitutive expression, which can be further augmented by induction, is seen mainly in B lymphocytes;
(ii) inducible expression, primarily in response to interferon-γ (IFN-γ), in APCs such as mononuclear phagocytes or dendritic cells and Langerhans cells which do not constitutively express class II molecules on their cell surface;
(iii) inducible expression in response to a variety of stimuli on many other non-lymphoid cell types of epithelial or endothelial origin.

Our knowledge concerning the differential roles of individual factors in class II expression is often scarce or conflicting due to the fact that there is a vast number of positive or negative and often interactive influences. Moreover, regulation is also subject to developmental and tissue-specific effects leading to opposing effects

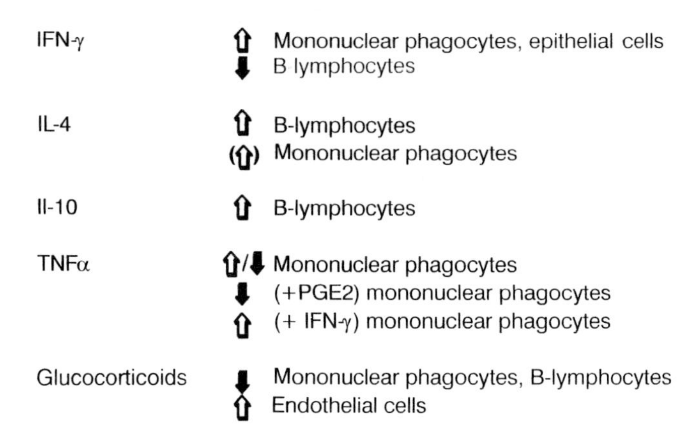

Figure 8.1. Modulation of class II expression.

of a particular factor (e.g. the effect of IFN-γ on B lymphocytes versus mononuclear phagocytes). This section will, therefore, focus on well-recognized modulators of class II expression in different cell types (*Figure 8.1*).

8.2.1 Modulators of class II expression

B lymphocytes. The level of class II expression can be influenced by a variety of external stimuli resulting in positive or negative modulation (*Figure 8.1*). In addition, class II expression is developmentally regulated since constitutive expression of MHC class II genes is acquired very early in B-lymphocyte ontogeny and occurs around the time of expression of immunoglobulin molecules on the cell surface (Lala *et al.*, 1979; Polla *et al.*, 1988). Nevertheless, these events are not coordinately regulated or interdependent (Dasch and Jones, 1986). Although expression is constitutive in B lymphocytes, a variation in the level of expression can be appreciated (Greenstein *et al.*, 1981; Mond *et al.*, 1980) and once terminal differentiation into plasma cells occurs, expression is extinguished (Dellabona *et al.*, 1989; Latron *et al.*, 1988; Venkitaraman *et al.*, 1987). A number of modulators of class II expression on B lymphocytes have been identified (reviewed in Glimcher and Kara, 1992). The most important of these is interleukin 4 (IL-4), which, on resting B lymphocytes, can up-regulate class II expression 10- to 15-fold at the transcriptional level (Boothby *et al.*, 1988; Noelle *et al.*, 1986; Rousset *et al.*, 1988). In *IL-4* transgenic mice, hyperexpression of class II molecules occurs in the absence of B lymphocyte proliferation and hyperimmunoglobulinaemia (Muller *et al.*, 1991). The effect of IL-4 can be suppressed by IFN-γ, prostaglandin E_2 (PGE_2) and corticosteroids (Dennis and Mond, 1986; McMillan *et al.*, 1988; Mond *et al.*, 1986; Polla *et al.*, 1986; Rousset *et al.*, 1988). The down-regulating effect of corticosteroids relates to inhibition of the binding of a transcription factor to the X box, a conserved regulatory *cis*-element of the class II upstream regulatory region (Celada *et al.*, 1993).

T lymphocytes. Upon activation, a considerable proportion (40–60%) of human T lymphocytes acquire class II molecules on their cell surface (Reinherz *et al.*, 1979).

In contrast to the lack of expression in the resting state in humans, however, class II molecules have been detected on quiescent canine and porcine T lymphocytes (Chang *et al.*, 1993; Doveren *et al.*, 1985). The induction of class II expression can be accomplished by exposure to mitogens (Cotner *et al.*, 1983), such as concanavalin A (ConA), phytohaemagglutinin (PHA) or pokeweed mitogen (PWM), and by soluble antigens (e.g. tetanus toxoid, TT; purified protein derivative, PPD), antibodies (e.g. HLA-DR-specific antiserum; DeWolf *et al.*, 1979) or anti-CD3 monoclonal antibodies (Robbins *et al.*, 1988), or through alloactivation in the context of a mixed lymphocyte culture (MLC; Charron *et al.*, 1980; Indiveri *et al.*, 1980). The expression in T lymphocytes can be attributed to endogenous synthesis of class II molecules and is not due to passive uptake or related to clonal expansion of a few class II-positive T lymphocytes (Singh *et al.*, 1984). Furthermore, the level of expression corresponds to the degree of proliferative T-lymphocyte response (Matis *et al.*, 1983). In addition to transcriptional activation, post-transcriptional mechanisms also have to be taken into account, as cell surface expression may occur as early as 30 minutes after mitogen stimulation (Caplen *et al.*, 1992; Zier *et al.*, 1989). Nevertheless, considerable variability with respect to the expressed isotypes, T-lymphocyte subpopulations, activational state and kinetics of expression has been appreciated (Diedrichs and Schendel, 1989; Hopkins *et al.*, 1993; Robbins *et al.*, 1988). Functionally, the expression of class II on activated T lymphocytes relates to an antigen-presenting function, as indicated by several studies, and thus includes stimulatory capacity (Suciu Foca *et al.*, 1981; Wollman *et al.*, 1980), antigen presentation function (Engleman *et al.*, 1980; Reske Kunz *et al.*, 1986; Triebel *et al.*, 1986) and effective T–B lymphocyte interaction (Ben Nun *et al.*, 1985; Reinherz *et al.*, 1981).

Antigen-presenting cells (APCs). The ability to process antigen, provide co-stimulatory signals and express peptide-loaded MHC molecules on the APC surface are prerequisites that allow effective interaction with, and modulation of, T-lymphocyte activation. Aside from B lymphocytes, well-characterized APCs include mononuclear phagocytes, dendritic cells, Langerhans cells, and vascular endothelial cells. The most important inducer of class II expression in these functionally similar cell types is IFN-γ (King and Jones, 1983; Nunez *et al.*, 1984; Zlotnik *et al.*, 1983), which acts on the transcriptional level to increase class II mRNA (Amaldi *et al.*, 1989; Rosa and Fellous, 1988). Controversial results have been obtained concerning the necessity of protein synthesis in IFN-γ-mediated class II expression as well as the importance of post-transcriptional modulation. While some studies indicate that the induction of class II expression requires protein synthesis and can be blocked by cycloheximide, a protein synthesis inhibitor (Blanar *et al.*, 1988; Böttger *et al.*, 1988; Celada and Maki, 1991), this was not observed by others (Basta *et al.*, 1988; Collins *et al.*, 1984, 1986; Fertsch *et al.*, 1987; Woodward *et al.*, 1989). However, IFN-γ also induces transcription factors involved in the expression of class II genes (Bono *et al.*, 1991; Moses *et al.*, 1992). Since, upon induction with IFN-γ, mRNA levels for class II molecules surpassed the rate expected by a transcriptional increase alone, it was speculated that a post-transcriptional mechanism may also be of importance (Amaldi *et al.*, 1989; Blanar *et al.*, 1988; Rosa and Fellous, 1988). However, it has also been noted that in a mouse macrophage cell line class II mRNAs are quite long-lived (half-life 16–20 h; Kern *et al.*, 1989). Signal transduction mechanisms in response to IFN-γ rely on phosphatidylinositol-linked activation of protein kinase C (PKC) and a consequent increase in intracellular calcium resulting in induction of c-*myc* and c-*fos* (Brenner *et al.*, 1989; Fan *et al.*, 1988; Koide *et al.*, 1988; Nezu *et al.*, 1990; Politis and Vogel, 1990; Squinto *et al.*, 1989). Nevertheless, these results have been challenged by other studies (Celada and Maki, 1991; Sasaki *et al.*, 1990). Also, upon exposure of murine macrophages to IFN-γ a rapid exchange of Na^+ and H^+ by means of the Na^+/H^+ antiporter, occurs leading to an accumulation of class II mRNA (Prpic *et al.*, 1989). Possibly, multiple pathways exist to mediate induction of class II expression, including phosphorylation of a number of intracellular substrates by tyrosine protein kinases (Lee *et al.*, 1995; Ryu *et al.*, 1992).

Synergizing with IFN-γ, tumour necrosis factor-α (TNFα) also increases expression of class II on mononuclear phagocytes (Arenzana Seisdedos *et al.*, 1988; Chang and Lee, 1986). Upon longer exposure to TNFα, however, an inhibition of class II expression has been found (Zimmer and Jones, 1990). Other cytokines capable of inducing class II expression in these APCs include IL-4 and IL-13 (Cao *et al.*, 1989; Crawford *et al.*, 1987; de Waal Malefyt *et al.*, 1993; Zlotnik *et al.*, 1987).

In contrast to its effects on B lymphocytes, IL-10 decreases class II expression in a variety of APCs. Other inhibitory agents include prostaglandins and corticosteroids (Snyder and Unanue, 1982; Snyder *et al.*, 1982). Type I interferons exert their down-regulating effect by antagonizing IFN-γ (Inaba *et al.*, 1986; Ling *et al.*, 1985).

For a variety of other cell types, mostly of endothelial or epithelial origin, expression of class II molecules as well as, at least in part, the capacity to present antigens has been demonstrated (Barclay and Mason, 1982; Burger *et al.*, 1981; Collins *et al.*, 1984; Hirschberg *et al.*, 1980, 1981; Lapierre *et al.*, 1988; Natali *et al.*, 1981; Wagner *et al.*, 1984). The modulation of expression exerted by cytokines and other stimuli corresponds, by and large, to the situation in mononuclear phagocytes. However, exceptions have been noted (Manyak *et al.*, 1988).

Other cell types. More recently, various cell types of the central nervous system have been investigated for their ability to express class II molecules and present antigen to T lymphocytes (Basta *et al.*, 1988; Lee *et al.*, 1993; Suzumura and Silberberg, 1988; Williams *et al.*, 1993; Wong *et al.*, 1984). Astrocytes and Schwann cells depend on IFN-γ to mediate expression of class II molecules. Moreover, the effect of IFN-γ can be enhanced by TNFα, but not by IL-4. In addition to these inducers, type I interferons, corticosteroids and gangliosides have been identified as inhibitory factors (Cowan *et al.*, 1991; Frei *et al.*, 1994; Massa, 1993). The ability to express MHC class II molecules in various cells may be of importance in inflammatory and possibly autoimmune diseases in the central nervous system (Kelly *et al.*, 1993; Uitdehaag *et al.*, 1993; Welsh *et al.*, 1993).

Expression of class II antigens occurs in a number of non-lymphoid cells in various organs and can, in most instances, be induced by IFN-γ (Collins *et al.*, 1984; Daar *et al.*, 1984; Groenewegen *et al.*, 1985; Houghton *et al.*, 1984; Pober *et al.*, 1983; Rosa and Fellous, 1988). Up-regulation of class II molecules has been noted in transplantation of allogenic grafts, autoimmune diseases, inflammatory reactions and malignant tumours (Carrel *et al.*, 1986; Forsum *et al.*, 1985; Halloran *et al.*, 1986). The functional significance of this in relation to antigen presentation and host defence mechanisms or tolerance remains unclear. It has been speculated that this 'aberrant expression' may be of importance in autoimmune diseases, particularly in endocrine disorders (Bottazzo *et al.*, 1985; Londei *et al.*, 1984; Sarvetnick *et al.*, 1988; Todd *et al.*, 1985; Wright *et al.*, 1986). However, the biological role of such expression and the ability of these class II molecules effectively to present antigen is a subject of debate (Lo *et al.*, 1988).

Haematopoiesis. In addition to the immunoregulatory role that class II antigens play, these molecules also function in haematopoiesis and can be considered as differentiation antigens (Broudy and Fitchen, 1986; Deeg and Huss, 1993; Falkenburg *et al.*, 1984; Fitchen *et al.*, 1982; Robinson *et al.*, 1981; Sieff *et al.*, 1982; Sparrow and Williams, 1986; Torok-Storb and Symington, 1986; Winchester *et al.*, 1977). However, a detailed discussion of this aspect is beyond the scope of this review.

8.3 Regulation of class II transcription

In eukaryotic cells, regulation of gene expression is governed by both transcriptional and post-transcriptional mechanisms (Latchmann, 1990; Roeder, 1991). The latter includes the regulation of RNA splicing, transport and stability, as well as translation (Leff *et al.*, 1986; Sharp, 1987). On the transcriptional level, chromatin structure, state of methylation, DNA sequence elements and transcription factors are of importance (Cedar, 1988; Mathis *et al.*, 1980). While changes in chromatin structure determine the general accessibility of potentially active genes, direct regulation of transcription depends on the interaction of *cis-* (regulatory DNA sequences) and *trans-* (DNA-binding transcription factors) acting elements (Mitchell and Tjian, 1989).

8.3.1 Methylation and DNase I hypersensitivity

Undermethylation of CpG sites and subsequent DNase I hypersensitivity usually indicate that a given gene is active or potentially active in that tissue. Although CpG sites can be found in class II regulatory sequences (Reith *et al.*, 1988; Sullivan and Peterlin, 1987; Wang *et al.*, 1987), no clear correlation between hypomethylation and gene expression has been observed (Gambari *et al.*, 1986; Levine and Pious, 1985; Wang and Peterlin, 1986). It is interesting, however, that the methylation pattern observed for the *DRA* gene was maintained in *DRA* transgenic mice (Feriotto *et al.*, 1991). Furthermore, in class II expressing B lymphocytes and IFN-γ-induced and -uninduced fibroblasts, two DNase I hypersensitivity sites were detected in the *DRA* gene while they were absent in class II-deficient regulatory mutants derived from patients with congenital immunodeficiency (Gonczy *et al.*, 1989).

8.3.2 Cis-*acting regulatory DNA sequences*

Regulatory DNA sequences can be divided into promoters and enhancers. While promoters are found 5′ of the gene at a short distance to the transcription initiation site, enhancers can operate at a distance and in either orientation relative to the promoter and regulate transcription in both a positive and negative manner. Promoters act by their ability to bind transcription factors, facilitating the assembly and action of a stable transcription complex. Enhancers lack promoter activity but are able considerably to increase the activity of promoters. Moreover, enhancers may convey tissue specificity of expression.

The delineation of *cis*-acting elements has become possible with the advent of DNA cloning and sequencing tools. The existence of regulatory DNA sequences was initially suspected on the basis of sequence similarity of functionally related genes and by comparing to known *cis*-acting regulatory DNA sequences or DNase I hypersensitivity. Subsequently, the importance of such sequences was confirmed by functional studies. The array of functional tools encompasses cell-free translation systems, transfection studies using reporter gene constructs and transgenic approaches. Individual regulatory sequences were pinpointed by 5′ deletion studies or by site-directed mutagenesis and linker-scanning analyses (Benoist and Mathis, 1990).

The transcriptional mechanisms of class II gene regulation are rather complex, as not only constitutive but also inducible expression has to be mediated. Moreover, developmental, cell type-specific and possibly isotype-specific regulation are required for the immunoregulatory function of class II genes. Among these genes the murine *I-A* and *I-E* α-chain genes and the human *DRA*, *DQB* and *DQA* genes have been particularly well studied. A limited number of highly conserved, critical *cis*-acting upstream regulatory elements are essential for both constitutive and inducible expression and these can be found in the region between −250 and +10 bp relative to the ATG initiation codon. In the mouse, however, an additional upstream control region could be identified in *E*α transgenic mice (Le Meur *et al.*, 1985; Pinkert *et al.*, 1985; Yamamura *et al.*, 1985). Deletion of this enhancer region extinguished expression in B lymphocytes, leaving other cell compartments intact (Burkly *et al.*, 1989; Dorn *et al.*, 1988; Le Meur *et al.*, 1989; van Ewijk *et al.*, 1988; Widera *et al.*, 1987).

As in most eukaryotic genes, a TATA box is present in some but not all class II gene proximal upstream regulatory regions (URRs). Among human class II genes, a TATA box can be found in *DRA*, *DRB*, *DPB1* and *DPB2*. Although the TATA box is thought to be important for the correct initiation of transcription, *in vitro* studies did not show any influence of the TATA box on basic transcription of class II genes (Dedrick and Jones, 1990; Tsang *et al.*, 1990; Viville *et al.*, 1991). However, transfection studies with *DRA* promoter constructs using primary human T lymphocytes indicated that transcription was lost or greatly reduced when the TATA box was deleted or mutated (Matsushima *et al.*, 1992).

As with the TATA box, a CCAAT box cannot be found in all class II promoters and the functional role of this element has still to be demonstrated. Nevertheless, it is interesting to note that in all URRs from MHC β-chain genes the CCAAT box is located at a conserved distance (17 bp) downstream of the Y box (see below), while all human and non-human primate α-chain genes lack a CCAAT box at the expected position.

Central to the constitutive and IFN-γ-induced expression are two elements, the X and Y boxes (*Figure 8.2*). The Y box is a 10 bp motif located at approximately -140 bp relative to the A of the ATG initiation codon. It contains a reverse CCAAT sequence. Moreover, in non-human primates a palindromic sequence (ATTGGCCAAT) can be found overlapping the Y box at its 3′ end due to the presence of a CCAAT sequence (Gaur *et al.*, 1992). This is not the case in the human *DRA* URR. These sequence disparities may be important for locus-specific expression and may indicate differential expression of α- and β-chain genes. This is strengthened by the observation that the 3′ end of all Y boxes, except for that of *DPA1* (ATAGGTG), reads as ATTGGCC in α-chain genes, while β-chain gene URRs are characterized by ATTGGTT in the same position. Since the transcription factor NF-Y binds specifically to the reverse CCAAT sequence contained in the Y box (Dorn *et al.*, 1987a) and since the functional significance of NF-Y binding was demonstrated for the transcription of *E*α and the murine albumin gene (Mantovani *et al.*, 1992), these chain-specific differences in Y box variability may well be of functional importance. This is also endorsed by the observation that Y box variability gave rise to differences in relative competition efficiencies in a mobility shift assay (Kimura and Sasazuki, 1992a). Moreover, in a glioblastoma cell line IFN-γ-induced expression of *DQA1* could be up-regulated by TNFα (Kimura and Sasazuki, 1992a). This *DQA1*-specific effect could be attributed to an A to G mutation in position -123 in the Y box, leading to preferential binding of NF-TRS over NF-Y as shown by gel mobility shift assays.

The X box is found 19–20 bp upstream of the Y box and covers a DNA stretch of 14 bp. Subsequently it was recognized that at the 3′ end of the X box a stretch of 7–8 bp, similar to palindromic regulatory elements known as cAMP response elements (CREs) and TPA response elements (TREs) can be found (Sloan *et al.*, 1992). This sequence has been renamed X_2, the original X box X_1. Deletions or mutations in the X or Y box reduce transcription of MHC class II genes as demonstrated by functional studies (Basta *et al.*, 1987, 1988; Boss and Strominger, 1986; Koch *et al.*, 1988; Viville *et al.*, 1991). The importance of both elements was also established *in vivo* using *E*α transgenic mice (Dorn *et al.*, 1987b). At a conserved distance upstream of the X box elements, a region termed W in *DQB* (Miwa *et al.*, 1987) and Z in *DRA* (Tsang *et al.*, 1988), has been shown to be important for IFN-γ-induced expression (Basta *et al.*, 1988; Boss and Strominger, 1986; Sakurai and Strominger, 1988; Sherman *et al.*, 1987; Thanos *et al.*, 1988). Aside from this positive regulatory function, negative or neutral effects have also been observed in *DQB* (Boss and Strominger, 1986; Miwa *et al.*, 1987; Sakurai and Strominger, 1989), *DRA* (Cogswell *et al.*, 1991; Tsang *et al.*, 1988, 1990) and the murine class II genes (Finn *et al.*, 1991; Thanos *et al.*, 1988). Within the W/Z box region, 15–17 bp upstream of the X box, a 7 bp sequence motif termed S element has been described (Servenius *et al.*, 1987). The S element has also been called heptamer, septamer or H box. Although the W/Z box is less conserved than the X and Y boxes, deletion of this element reduces constitutive and inducible expression (Dedrick and Jones, 1990; Viville *et al.*, 1991).

Aside from conservation of the individual regulatory elements W/Z, X and Y, the 18–20 bp spacing between them is also maintained in different class II genes (*Figure 8.3*). This conserved distance allows for two helical turns and thereby places these boxes on the same side of the helix. Due to their pattern of interaction

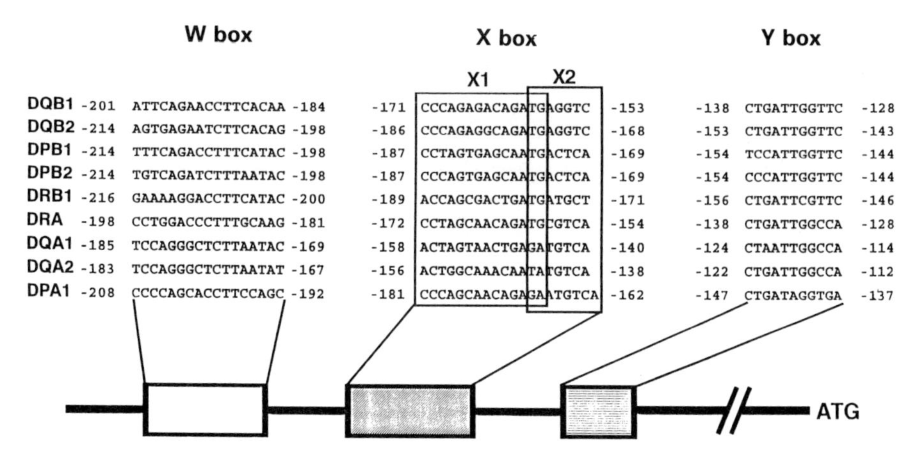

Figure 8.2. A synopsis of the *cis*-acting regulatory X, Y and W/Z boxes conserved in the proximal promoter of different class II isotypes. In addition to the individual isotypes, the positions of individual elements relative to the start site of transcription are indicated.

with transcription factors these three elements can be viewed as a transcriptional unit (*Figure 8.4* see page 256; Vilen *et al.*, 1992).

Apart from this trimeric complex, other less well-defined elements have been described: In the *DRA* and *DQB* genes a pyrimidine-rich sequence has been termed P element since expression was reduced when it was subjected to mutation (Sakurai and Strominger, 1988; Tsang *et al.*, 1990). The ability of the P element to bind transcription factors could be demonstrated for the *DQB* gene (Miwa *et al.*, 1987) and was not detected in *DRA* (Calman and Peterlin, 1988; Reith *et al.*, 1988). Possibly this pyrimidine-rich sequence stabilizes the interaction of transcription factors with the X and W/Z boxes as it is situated between them. Within the S element of murine *A*α and in humans in *DQA* only, a putative NFκB site was discovered (Dedrick and Jones, 1990). As a unique feature of the *DQA1* promoter a 14 bp DNA segment proximal to the Y box and identical in sequence to the T box was identified. In the mouse *A*α promoter and *RT1.B* α-chain genes the T box functions as a TNFα response element (Freund *et al.*, 1990). Downstream of the Y box and unique to *DRA* is the presence of an octamer motif, a regulatory element of significance in immunoglobulin regulation (Sherman *et al.*, 1989). This *DRA* octamer motif may play a role in B-lymphocyte lymphomas. Although an octamer motif was seen in the upstream region at approximately −600 bp in the *DQB* gene, it does not seem to be of functional relevance (Miwa and Strominger, 1987). Finally, the V box, which is upstream of the W/Z box and which shares sequence homology with the X box, may function as a suppressor element in *HLA-DR* (Cogswell *et al.*, 1990). Interestingly, the V box had X box-like specificity for the binding of transcription factors as seen in mobility shift studies.

Elements in the proximal promoter region are a prerequisite, but are not sufficient, for inducible and constitutive expression. Thus, intronic sequences were analysed in transient reporter gene expression systems. These studies indicated the presence of enhancer sequences contributing to lymphoid-specific expression in the *DQA*, *DQB* and *DRA* genes (Peterlin, 1991; Sullivan and

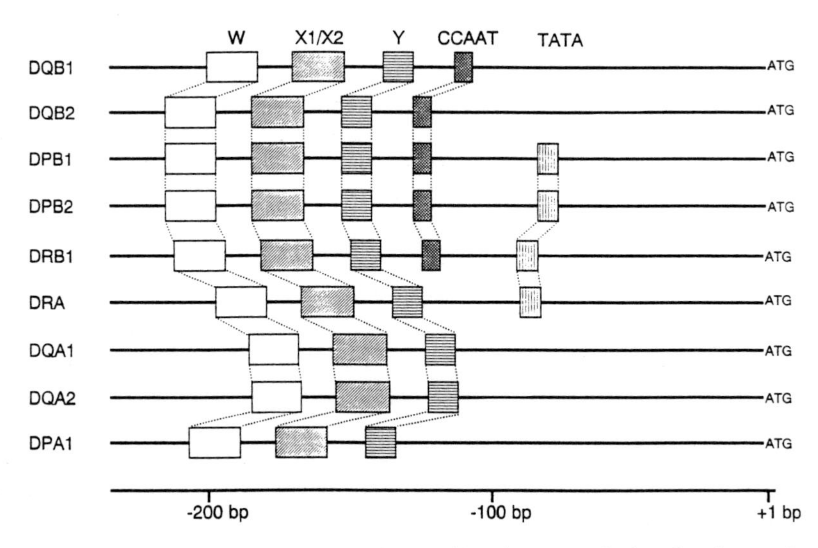

Figure 8.3. The alignment of conserved *cis*-acting elements, depicted as boxes, in the proximal promoter of different class II isotypes. To illustrate the spatial arrangement of conserved promoter elements in different isotypes, sequences have been aligned at the ATG site. Corresponding elements are shown in the same graphical layout and are connected by dotted lines.

Peterlin, 1987; Wang *et al.*, 1987). This is also supported by transfection studies in which sufficient expression of *IA* α-gene constructs was achieved by intragenic sequences alone (McCluskey *et al.*, 1988). More recently, a regulatory DNA sequence of functional importance has been identified in the first exon of the *DRA* gene (Hehlgans and Strominger, 1995). This site interacts with NF-E1 (YY1), a *trans*-activator in immunoglobulin expression.

While of the human genes *DQA*, *DQB* and *DRA* have been extensively characterized, more limited information is available on other class II genes.

HLA-DP. Proximal promoter sequences of the *HLA-DPA* and *DPB* genes also carry the well known conserved *cis*-acting elements characteristic for HLA class II genes. Functional studies confirmed the importance of the W, X, X2 and Y boxes for the constitutive and IFN-γ inducible expression of this isotype (Koide and Yoshida, 1992). Deletion analyses indicated the presence of a 15-mer sequence (5′-CTTTCCTCCGTCATC-3′) at positions -184 to -169 conferring positive regulatory effects. By investigating the DNA regulatory requirements for IFN-γ in the 5′ region of the *DPA1* gene, a 52 bp element at positions -107 to -55 could be identified (Yang *et al.*, 1990). This sequence contains the X and Y elements. Overlapping with the 5′ end of the X box a putative IFN-γ response element homologous to the S/Z element was also seen.

HLA-DM. Among the several genes newly discovered in the MHC complex two class II-like genes, designated as *HLA-DMA* and *HLA-DMB*, were identified in the region between *HLA-DNA* and *-DOB* (Kelly *et al.*, 1991). Although they share structural features with characteristic class II genes, their localization is confined to a lysosome-like compartment where peptide loading of class II molecules occurs

(Sanderson *et al.*, 1994a). Functionally, the HLA-DM heterodimer is involved in the class II pathway of antigen presentation by inducing CLIP (class II-associated invariant chain peptide) dissociation from MHC class II heterodimers and facilitating peptide loading (Denzin and Cresswell, 1995; Sloan *et al.*, 1995). In contrast to other class II genes the limited polymorphism of *HLA-DM* is not found in the antigen-binding groove (Sanderson *et al.*, 1994b). Thus, this set of genes is of particular interest in the study of class II gene regulation. Sequence analyses indicated that the URRs of the *DMA* and *DMB* genes are markedly similar to those of other class II genes (Kelly *et al.*, 1991; Radley *et al.*, 1994). Aside from the X, Y, J, CCAAT and TATA boxes, potential Sp1 and NFκB binding sites (*HLA-DMB*) were identified. These regulatory sequence similarities with other class II genes and their inducibility by IFN-γ suggest that *HLA-DM* may be coexpressed. However, this has to be shown by further analyses.

Invariant chain. The invariant chain (Ii) is closely associated with the class II pathway as it is needed to target the α/β/Ii complex to the endosomal system and to allow antigenic peptide binding upon dissociation from the class II heterodimer (reviewed in Cresswell, 1994). Despite the fact that the *Ii* gene is localized outside the MHC on chromosome 5 it is coordinately expressed with class II molecules in various cell types and can be coinduced by IFN-γ, TNFα and IL-4. It was therefore not suprising to learn that the *Ii* gene harbours homologues of the conserved DNA regulatory elements, X, Y and W boxes and S element, seen in class II genes (Brown *et al.*, 1991, 1993). Moreover, the induction of transcriptional activity is associated with the presence of *in vivo* protein/DNA interactions in Ii-expressing cells (Brown *et al.*, 1993). Similar findings were also reported for mouse *Ii* gene regulation (Eades *et al.*, 1990; Zhu and Jones, 1990). Cross-competition experiments indicated that, indeed, the *Ii* X and Y box sequences may bind *DQB* Y and *DRA* Y and X box competitor fragments in mobility shift assays (Brown *et al.*, 1991; Doyle *et al.*, 1990). However, differential uncoordinated expression of Ii and MHC antigens may also occur (Momburg *et al.*, 1986; Neiss and Reske, 1994). It is therefore of interest that the *Ii* URR contains a unique regulatory region carrying a SP1 and NFκB site and a CCAAT box proximal to the common class II elements mediating positive regulation (Brown *et al.*, 1994; Kolk and Floyd Smith, 1993; Zhu and Jones, 1990).

8.3.3 HLA class II promoters: isotypic and allelic variability

Comparisons of *cis*-acting elements of different isotypes of HLA class II genes including the invariant chain gene have indicated that there are considerable structural similarities (*Figure 8.2*) which convey similar regulatory effects when subjected to different stimuli. Similarities among corresponding URRs of HLA class II genes overall are in the range of 50%. From a teleological standpoint this is not surprising as coordinated developmental, tissue-specific expression has to be mediated, at least in part, on a transcriptional level. Nevertheless, there are distinct isotypic differences, suggestive of differential regulatory mechanisms and possibly isotype-specific functions (Ameglio *et al.*, 1983; Anichini *et al.*, 1988; Diedrichs and Schendel, 1989; Hume *et al.*, 1989; Lucey *et al.*, 1989; Manyak *et al.*, 1988; Ono and Song, 1995; Ono *et al.*, 1991a; Symington *et al.*, 1985). Despite the overall similarity, meaningful alignments of isotypic sequences are difficult to establish and only yield sensible results when expressed genes and their corresponding pseudogenes are compared. In this case, except for *DPA*, similarity ranged among different HLA class genes from 76 to 88%.

Although allelic polymorphism is a key feature of HLA class II genes, it was somewhat suprising to learn that the URRs of these genes also harbour allelic variability. The structural basis and, to some extent, the functional significance have been investigated for the *DR* and *DQ* genes.

HLA-DRA. Among different HLA class II genes *HLA-DRA* has been particularly well investigated to understand class II transcription in humans. It is a particular feature of the *HLA-DR* locus that essentially all variability is derived from multiply expressed *DRB* loci which each exhibit substantial allelic polymorphism. For *DRA*, however, there are only two alleles known (Bodmer *et al.*, 1994). Moreover, the existence of a diallelic polymorphism located outside the conserved proximal *DRA* promoter could be detected by restriction fragment length polymorphism (RFLP) and confirmed by sequence analysis (Pinet *et al.*, 1991).

HLA-DRB. Despite the fact that different *DRB* genes are closely related, the amount of *DRB1* polypeptide on the cell surface surpasses that of the other β-chain gene products by a factor of 5 to 10 (Rollini *et al.*, 1985; Sorrentino *et al.*, 1985). This difference arises at the transcriptional level as judged from mRNA studies (Stunz *et al.*, 1989). This was also true when the induction by IFN-γ was investigated (Berdoz *et al.*, 1987). Subsequent structural and functional analyses of the *DRB* proximal promoter indicated that indeed the *DRB1* promoter surpasses the *DRB3* gene promoter in transcriptional activity (Emery *et al.*, 1993). Moreover, this difference may be due to sequence variability in the X box. Compiling sequence data from several studies it seems that there are at least five URR variants at the *DRB1* locus while, except for *DRB4* (two variants), a single promoter sequence can be found for *DRB2*, *3*, *5*, and *7* (Emery *et al.*, 1993; Leen *et al.*, 1994; Louis *et al.*, 1993, 1994a, 1994b; Perfetto *et al.*, 1993; Singal *et al.*, 1993). However, the assignment of individual sequences to particular *DRB* loci may not be correct in all cases. Moreover, further structural and functional studies are necessary to appreciate the importance of upstream regulatory variability in the *DRB* genes.

HLA-DQA. Sequencing of the HLA *DQA1* URRs from different allelic *DQA1* haplotypes indicated that 12 *QAP1* variants exist in the region up to approximately 700 bp upstream of the first exon of the *DQA1* gene (Morzycka Wroblewska *et al.*, 1993; Kimura and Sasazuki, 1992b). Most of the sequence variation encoding the allelic polymorphism of the *DQA1* URRs is concentrated in a region between -250 and -300 bp upstream of the translation start site. In addition, mutations can be found in the X, X2, Y and W boxes. Population-based analysis of allele frequencies and haplotype composition indicated that the distribution of *QAP1* alleles mostly follows the distribution of *DQA1* alleles due to extensive linkage disequilibrium (Haas *et al.*, 1994a). However, unusual haplotypes were observed in 14% of the cases, indicative of recombinational events. Sequence analysis of the *DQA2* URR indicates only a very limited polymorphism compared to *DQA1* and an 85% similarity to the proximal *DQA1* promoter (Auffray *et al.*, 1987; Del Pozzo and Guardiola, 1990; Del Pozzo *et al.*, 1992; Morzycka Wroblewska *et al.*, 1993).

DQA2 is usually considered an unexpressed pseudogene due to a *cis*-acting defect in the X box of the proximal promoter (Voliva *et al.*, 1993). Nevertheless, *DQA2* mRNA has been detected in B lymphocytes (Yu and Sheehy, 1991).

HLA-DQB. The analysis of approximately 600 bp immediately upstream of the first exon of the *DQB1* gene in 14 *DQB1* alleles yielded a total of 13 different *QBP1* alleles (Andersen *et al.*, 1991; Reichstetter *et al.*, 1994). The overall similarity was in the range of 91–100%. Immediately upstream of the first exon known regulatory elements such as the W, X and Y boxes were identified. Mutations in these elements were seen for the X1, X2 and W boxes, while no variability was seen for the Y box (*Figure 8.5*). A CCAAT box located at a conserved distance of 17 bp downstream of the Y box was seen in all but two alleles (sequence CTAAT). Differences between *DQB1* URRs outside the proximal promoter region were not randomly distributed, and four clusters of highly conserved regions could be identified (*Figure 8.5*; Reichstetter *et al.*, 1994). Subjecting the *QBP1* polymorphism to a population-based study, tight linkage between the promoter region and exon 2 of *DQB1* was seen (Reichstetter *et al.*, 1995). Exceptions were mainly appreciated for *DQ5* and *DQ6* haplotypes, as single *DQB1* alleles could be linked to different, however, closely related, *QBP1* alleles, and vice versa. Possibly, the maintenance of haplotypic integrity as observed in *DQA* and *DQB* may be of functional importance.

Functional studies have indicated that this allelic polymorphism is associated with differences in transcriptional activity and affinity for DNA-binding proteins (Andersen *et al.*, 1991). Reporter gene assays indicated a fivefold difference between the *DQB1**0302* and **0301* promoters. Moreover, these differences were also associated with differential binding of transcription factors as evidenced by mobility shift studies (Andersen *et al.*, 1991; Shewey and Nepom, 1993). Studying a heterozygous melanoma cell line, it was shown that *in vitro* differences reflected the *in vivo* situation for both constitutive and IFN-γ-induced expression of class II (Nepom *et al.*, 1995). Furthermore, mutations in the W and X boxes may be responsible for the lack of *DQB2* gene expression (Shewey *et al.*, 1992; Voliva *et al.*, 1993). When comparing the *DQB1* URR with its counterpart in the coexpressed *DQA1* gene, the degree of sequence similarity was of the order of 42% only in a 535 bp overlap. Thus, the upstream regulatory DNA sequences of corresponding α- and β-chain genes of a given class II locus do not exhibit a closer resemblance when compared to other class II genes, as one might have speculated.

Taken together, the relationship of coding region (exon 2) and URR polymorphism indicates that, within a given isotype, recombinational events have occurred and that the hierarchy of similarity between alleles may be different for the coding as compared to the promoter region. This invites speculation that promoter region polymorphism is also under evolutionary constraints and URR polymorphism is not only maintained by linkage disequilibrium. Coding region polymorphism and pairing mechanisms contribute to structural differences of class II heterodimeric proteins (Charron *et al.*, 1984; Lotteau *et al.*, 1987). Allelic variability in regulatory regions may affect the overall stoichiometry of the α/β heterodimer by allele-specific differences in the rate of expression of individual α- or β-chain gene alleles. This, in turn, may affect the ability of the α/β heterodimer to bind and present peptides. These speculations on the biological implications of allelic promoter variability may be supported by the analysis of pairing

```
           -620                             -590                                         -550
       **   .    ** *   .         .           *          .   **      .          .          .    *
     TACAA AGAGAAGCCT GGGGCAAAAA TAAATTCAGT AATTTGTTGA CTCTCATAAA GCACATTAGT GGTGGAACTG CAACTCACCA

                                                          *
                                            -500                                         -460
     *     .     *  *  .   *  *   *   **     .      *   .**    **  .          .       *  .*   *
TTATTTCCTT CTAAGAACTT TGCTCTTTTC ACCAAAACTT AAGGCTCCTC AGGGTGTGTC TAAGACAACA GCAGTAAAAA TGTCTATGAC

                                                                           *
                                            -410                                         -370
.          .           .         .   *       .  **  * * *  *       .     * *  .    **    .  *
AGCAATTTTC TCTCCCCTGA AATATGATCC CCACTTAATT TGCCCTATTG AAAGAATCCC AAGTATAAGA ACAACTGGTT TTTAATCAAT

                                                                          *
                                            -320                                         -280
.     *    .           .    *    . * ***     *  *   *   .     **   .* *  ***  .          .*     *
ATTACAAAGA TGTTTACTGT TGAATCGCAT TTTT//CTTT GGCTTCTTAA AATCCCTTAG GCATTCAATC TTCAGCTCTT CCATAATTGA

                                            -230                                         -190
.   *  *   .       *   .         . *      ** .    *     .        * .   * * *  .          *       **
GAGGAAATTT TCACCTCAAA TGTTCATCCA GTGCAATTGA AAGACGTCAC AGTGCCAGGC ACTGGATTCA GAACCTTCAC AAAAAAAAAA
                                   CCAAT                                        W-Box

                                            -140                                         -100
**      *  .         *.          .           .          *     * ** .*     *   .        * . *     *
//TCTGCCCA GAGACAGATG AGGTCCTTCA GCTCCAGTGC TGATTGGTTC CTTTCCAAGG GACCATCCAA TCCTACCACG CATGGAAACA
            X1-Box   X2-Box                    Y-Box                    CCAAT

                                                                              *
                                            -50                                          -10
.      *   .           .      *  .           .     * ** .          .       ** *          .
TCCACAGATT TTTATTCTTT CTGCCAGGTA CATCAGATCC ATCAGGTCCG AGCTGTGTTG ACTACCAC// /TTTTCCCTT CGTCTCAATT
```

requirements for α/β-heterodimers, both intra- and mixed-isotype pairs, since the generation and expression of particular α/β-heterodimers depended on the level of expression of the individual chains (Germain *et al.*, 1985; Karp *et al.*, 1990; Lotteau *et al.*, 1989). This was also observed *in vivo* using a transgenic approach (Gilfillan *et al.*, 1990).

From an evolutionary standpoint it is interesting to note that interspecies relatedness for corresponding genes is higher than interlocus similarity within HLA; for example, the similarity between the bovine *DQB1* promoter and its human homologue was higher (70%) than the degree of similarity among expressed HLA class II genes (40–60%; Reichstetter *et al.*, 1994).

Disease association and promoter polymorphism. In relation to clinical medicine and human disease, class II molecules are of relevance to the field of organ transplantation and the study of autoimmune diseases. In the latter case, linkage to and association with HLA class II specificities and alleles have been demonstrated (Lepage *et al.*, 1993; Nepom, 1993; Wassmuth and Lernmark, 1989). While HLA associations may be mediated by different mechanisms, antigen presentation is one possibility. Previous studies have focused on the relevance of particular sequences or epitopes common to different alleles (Gregersen *et al.*, 1988; Todd *et al.*, 1987). However, in juvenile chronic arthritis structural evidence has been presented that allelic promoter polymorphism may be of importance for autoimmunity. Rheumatic diseases of childhood fall into several categories. One defined entity is early-onset, pauciarticular, juvenile chronic arthritis (JCA). A number of HLA associations with this disease have been described, including HLA class I (HLA-A2) and class II specificities and alleles (DR5, DR6, DR8 and DPB1*0201; Odum *et al.*, 1986; Oen *et al.*, 1982; Reekers *et al.*, 1983; Stastny and Fink, 1979; van Kerckhove *et al.*, 1990). However, the strongest association was found with a *DQA1* sequence motif on *DQA1*0401*, **0501*, **0601*, present in 89% of the patients (Haas *et al.*, 1994b). Aside from a number of differences among *QAP1* alleles in conserved regions with known regulatory functions, *QAP1-4.1* and *4.2* differ from the Y box consensus sequence by an A to G substitution at position -119. Interestingly, these variants are linked to the JCA *DQA1* susceptibility alleles and are thus highly associated with JCA (Haas *et al.*, 1995). The functional importance of this association has to be evaluated. Nevertheless, this example invites speculation that not only coding region polymorphism leading to modifications of the antigen-binding site but also differential expression of individual alleles may contribute to disease susceptibility.

Figure 8.5. Sequence information on the *QBP1-2.1* promoter linked to *DQB1*0201* is shown. Conserved regulatory regions (X, Y, W and CCAAT boxes) are indicated beneath by rectangular boxes. To illustrate the polymorphism in *DQB1* promoters, individual polymorphic positions are indicated by an asterisk (*) above the nucleotide sequence. The number of asterisks at a particular position indicates the number of different bases that could be found at that position. Thus, conserved regions can be differentiated from polymorphic stretches of sequence.

8.3.4 Transcription factors

Over 20 years ago it was suggested that coordinate expression of different genes may be mediated by common regulatory elements that are activated by the product of a specific integrator gene, in turn activated by a specific signal (Britten and Davidson, 1969). And indeed, experimental work has indicated that, as a principle, *trans*-activation of eukaryotic genes or modulation of gene activity requires the interaction of regulatory DNA sequences and transcription factors. This activation mechanism requires at least two steps:

(i) targeting of the corresponding regulatory DNA sequence and subsequent sequence-specific DNA binding of the transcription factor;
(ii) the activation of transcription (Ptashne, 1988).

Thus, transcription factors have a modular structure, separating DNA binding and activation into different domains (Latchmann, 1990). In DNA binding domains, specific basic motifs such as the helix-loop-helix and the zinc finger motif have been identified. Activation domains, often containing acidic amino acids, may act by directly contacting the RNA polymerase or by interaction with other transcription factors generating a stable transcription complex. Transcription factors have been identified by virtue of their ability to bind to DNA in a sequence-specific manner. The DNA–protein complex can be visualized by gel electrophoresis since this complex is retarded in the gel matrix leading to a slower mobility than observed with either the target DNA sequence or the transcription factor alone. This technique has therefore been termed electrophoretic retardation or electrophoretic mobility shift assay. Cloning of transcription factors was initially carried out using oligonucleotide probes which were designed according to partial amino acid sequences obtained from small amounts of protein isolates. Subsequently, expression libraries were probed with double-stranded DNA molecules containing the *cis*-acting element.

Y box binding factors. Among the Y box binding transcription factors, NF-Y has been extensively characterized. NF-Y (CP1, CBF) is a heterodimeric molecule composed of two subunits NF-YA and NF-YB (Celada *et al.*, 1988; Hooft van Huijsduijnen *et al.*, 1990; Li *et al.*, 1992a) which has similarities to the yeast transcription factors HAP2 and HAP3 (Li *et al.*, 1992b). Specific binding to the Y box is dependent on the CCAAT motif of the Y box as shown in humans and mice (Celada and Maki, 1989; Dorn *et al.*, 1987a, b; Hooft van Huijsduijnen *et al.*, 1987; Miwa *et al.*, 1987). However, this binding is not unique to class II genes as NF-Y is expressed in all tissues (Hooft van Huijsduijnen *et al.*, 1990; Mantovani *et al.*, 1992). In addition to NF-Y two other clearly distinct Y box binding factors, YEBP (Zeleznik Le *et al.*, 1991) and YB-1 (Didier *et al.*, 1988), have been described. YB-1 has been shown to be a negative regulator as its binding is negatively correlated with class II expression. Morever, the negative regulation by YB-1 affects also IFN-γ-mediated class II expression (Ting *et al.*, 1994). Binding of YB-1 may induce and stabilize single-stranded DNA thereby preventing positive regulatory effects of other transcription factors (MacDonald *et al.*, 1995).

X box binding factors. Based on the ability to bind transcription factors, the X box could be separated into an X1 and an X2 element. Among the X1 box binding elements RFX was identified by its affinity to the *DRA* X1 box (Hasegawa *et al.*, 1991; Kobr *et al.*, 1990; Reith *et al.*, 1989). RFX binds DNA in a sequence-specific manner as monomer or homodimer and thus contains both a DNA binding and a dimerization domain. Subsequent studies indicated that there is a graded affinity for human α-chain genes (Hasegawa *et al.*, 1991; Kobr *et al.*, 1990) and that RFX binds poorly to β-chain X1 boxes (Hasegawa and Boss, 1991). These binding characteristics may be of importance for isotype-specific expression. RFX has been implicated in congenital immunodeficiencies because, in some patients, there is a lack of RFX binding to *DRA* (Reith *et al.*, 1988). RFX belongs to a group of X1 box binding proteins including NF-X (Herrero Sanchez *et al.*, 1992; Kouskoff *et al.*, 1991). NF-X was originally identified in the mouse but was subsequently also seen in the human. In addition, another X1 binding factor has been identified in the analysis of IFN-γ activation in astrocytes (Moses *et al.*, 1992). Using HeLa nuclear extracts TRAX1, an X1 box binding factor different from RFX, was observed (Itoh Lindstrom *et al.*, 1995). Negative regulatory effects have been found for a factor called NF-X1 (Song *et al.*, 1994).

The X2 element is similar to a palindromic regulatory sequence resembling CREs and TREs and mediates the binding of transcription factors belonging to the leucine zipper family (e.g. hXBP-1, c-Jun, c-Fos, HB16, mXBP). As studied in the *DRA* promoter, these sequences are recognized by NF-S (Kobr *et al.*, 1990), hXBP-1 (Liou *et al.*, 1990, 1991), hXBP-1/c-Fos heterodimers (Ono *et al.*, 1991b), c-Jun homodimers and c-Jun/c-Fos heterodimers (activator protein-1; AP-1; Andersson and Peterlin, 1990). Interestingly, hXBP-1 can drive the IFN-γ-mediated expression of class II in *HLA-DR* and *-DP* but not in *HLA-DQ* (Ono *et al.*, 1991). Another set of X2 binding proteins, mXBP (mouse) and HB16 (human), show a greater than 90% similarity (Kara *et al.*, 1990; Liou *et al.*, 1988). mXBP/HB16 are leucine zipper transcription factors which can bind as homodimers and which are able to form heterodimers with c-Jun (Ivashkiv *et al.*, 1990; Kara *et al.*, 1990).

In addition to the interaction with individual factors, cooperative binding is necessary. As expected from the tightly conserved spatial arrangement of the X, Y and W/Z elements, interaction of the transcription factors RFX and NF-Y is of importance for the expression of class II genes (Reith *et al.*, 1994a, 1994b).

W/Z box binding factors. Although binding of proteins to the S element within the W/Z box has been observed, individual transcription factors have not been characterized (Cogswell *et al.*, 1991; Dedrick and Jones, 1990; Tsang *et al.*, 1990). Possibly, X box binding factors may also interact with the W/Z box (Cogswell *et al.*, 1990). For the P element, situated between the X and W/Z boxes, binding of Ets-1 and activation of the *DRA* gene have been demonstrated (Jabrane Ferrat and Peterlin, 1994). For *DPA*, a transcription factor of the zinc finger protein family has been identified requiring binding to both the S and J elements (Sugawara *et al.*, 1994, 1995).

Octamer binding factors. Unique to the *DRA* proximal promoter is an octamer motif. This element can bind two different transcription factors, Oct-1 and Oct-2 (Schreiber *et al.*, 1990; Staudt *et al.*, 1986; Zeleznik Le *et al.*, 1992). While Oct-2 is found in B lymphocytes only, Oct-1 is distributed ubiquitously. *In vitro* studies

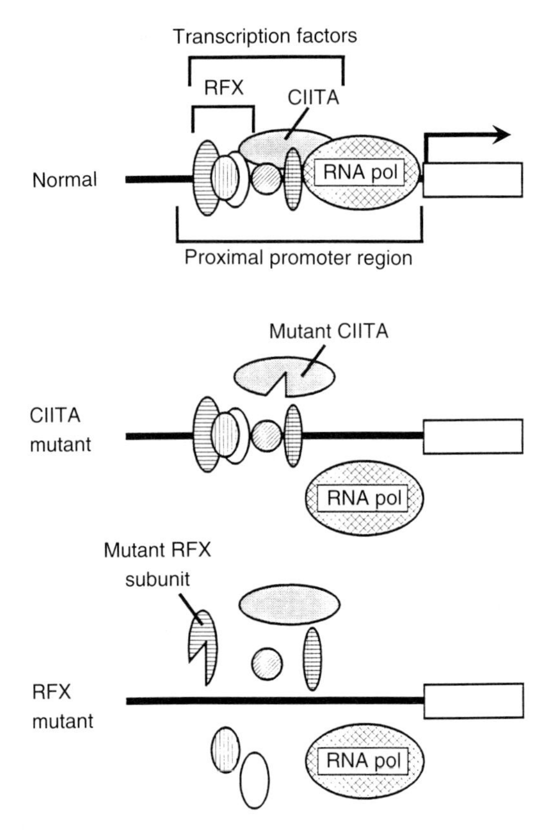

Figure 8.6. Normally, transcription of class II genes occurs when transcription factors bind to the proximal promoter region. In the case of a mutant CIITA or RFX molecule the promoter region cannot be accessed in the usual fashion and, in turn, expression of class II genes is lost. (Figure modified from Mach, 1995.)

indicated that Oct-2 rather than Oct-1 may be relevant to the transcriptional activity mediated by the octamer motif (Wright and Ting, 1992; Zeleznik Le *et al.*, 1992).

Class II trans-*activator (CIITA)*. Complementation studies in a class II mutant cell led to the identification and cloning of the *trans*-acting factor CIITA (Steimle *et al.*, 1993). Although CIITA targets particularly the X2 box and the S element, it mediates activation through interaction with other transcription factors and not by direct DNA binding (Riley *et al.*, 1995; Zhou and Glimcher, 1995). In addition to *HLA-DR*, *-DQ* and *-DP*, the invariant chain and *HLA-DM* are also regulated by CIITA (Chang and Flavell, 1995). The distribution of transcription of CIITA is tightly correlated with class II expression. Moreover, CIITA is necessary for IFN-γ induction of class II genes, and CIITA itself can be induced by IFN-γ (Chang *et al.*, 1994; Chin *et al.*, 1994; Steimle *et al.*, 1994). Also, terminal differentiation of B lymphocytes is accompanied by extinction of class II expression, a regulatory event also mediated by CIITA (Silacci *et al.*, 1994). Thus, CIITA is apparently a universal inducer of class II expression.

Congenital immunodeficiency. Much of what we know about MHC/HLA class II regulation has been learned from mutants unable to express class II molecules. Aside from deliberate generation of such class II-negative phenotypes by mutagenesis of tumour cell lines, the lack of expression of class II isotypes is also a hallmark of a genetically heterogenous group of human diseases with an autosomal recessive pattern of inheritance, also designated as type II bare lymphocyte syndrome, leading to a severe combined immunodeficiency (SCID; Lisowska Grospierre *et al.*, 1985). Clinically, patients succumb to bacterial or viral infections at a very early age due to their inability to generate an efficient humoral and cellular immune response, and bone marrow transplantation may provide the only cure (Griscelli, 1991; Klein *et al.*, 1993). The inability to express class II antigens has been mapped to the level of transcription (*Figure 8.6*; Benichou and Strominger, 1991; Bull *et al.*, 1990; de Preval *et al.*, 1985; Hume *et al.*, 1987). Moreover, complementation analyses indicated that, at least in some patients, the defect lies in the transcription factor RFX complex (Herrero Sanchez *et al.*, 1992; Reith *et al.*, 1988; Steimle *et al.*, 1995). Footprinting analyses demonstrated that the accessibility of the class II promoters in SCID was impaired leading to a bare promoter phenotype (Kara and Glimcher, 1991). In search of the underlying defect in patients with normal RFX binding, the transcription factor CIITA was identified and the defect could be attributed to a splicing mutation in the *CIITA* gene leading to a partial deletion (Steimle *et al.*, 1993).

8.4 Summary and perspective

MHC class II genes have a pivotal role in the priming of the immune system and the generation of an antigen-specific immune response by virtue of their function as antigen-presenting molecules. Expression of these molecules is subject to developmental, and tissue- and isotype-specific regulation, which occurs mainly on a transcriptional level and involves *cis*- and *trans*-acting elements. A limited number of highly conserved *cis*-acting upstream regulatory elements are critical for both constitutive and inducible expression. Most importantly, an array of regulatory sequences in the proximal promoter containing the X, Y and W/Z box elements is necessary, however, not sufficient, to explain all aspects of class II expression. In addition, less well maintained regulatory elements have been deduced in individual or particular groups of class II genes. Although regulatory sequences are conserved among different class II isotypes and across species, differences can be appreciated. More recently, allelic variability in the promoter regions in different human class II genes has been described. Since promoter variability may be associated with differences in transcription, this polymorphism may contribute to the diversity of HLA class II genes. It is tempting to speculate that these isotypic and allelic polymorphisms may be of relevance to the study of autoimmunity. Allelic variability in regulatory regions may affect the stoichiometry of class II heterodimers and, as a consequence, may have implications for the ability to express certain hybrid molecules in certain quantities; a proposition that has to be investigated in the future. The importance of class II regulatory mechanisms for human disease has been demonstrated for a group of congenital defects leading to severe combined immunodeficiencies. The defect occurs on the transcriptional level and involves the binding of transcription factors to class II promoters. Thus, transcriptional mechanisms provide powerful means positively or negatively to influence immune competence. Possibly, these mechanisms can be exploited for targeting immunomodulation. Nevertheless, much remains to be learned about the mechanisms of differential regulation of class II isotypes and alleles with respect to developmental state, cell type and the implications of these findings for function and disease.

Acknowledgements

This work was supported by a grant from the Deutsche Forschungsgemeinschaft (SFB 263). The author thanks Dr W. Ran and S. Reichstetter for critically reading this manuscript and S. Reichstetter for the preparation of the graphical work.

References

Alonso MC, Navarrete C, Solana R, Torres A, Pena J, Festenstein H. (1985) Differential expression of HLA-DR and HLA-DQ antigens on normal cells of the myelomonocytic lineage. *Tissue Antigens* **26:** 310–317.

Amaldi I, Reith W, Berte C, Mach B. (1989) Induction of HLA class II genes by IFN-gamma is transcriptional and requires a trans-acting protein. *J. Immunol.* **142:** 999–1004.

Ameglio F, Capobianchi MR, Dolei A, Tosi R. (1983) Differential effects of gamma interferon on expression of HLA class II molecules controlled by the DR and DC loci. *Infect. Immunol.* **42:** 122–125.

Andersen LC, Beaty JS, Nettles JW, Seyfried CE, Nepom GT, Nepom BS. (1991) Allelic polymorphism in transcriptional regulatory regions of HLA-DQB genes. *J. Exp. Med.* **173:** 181–192.

Andersson G, Peterlin BM. (1990) NF-X2 that binds to the DRA X2–box is activator protein 1. Expression cloning of c-Jun. *J. Immunol.* **145:** 3456–3462.

Anichini A, Castelli C, Sozzi G, Fossati G, Parmiani G. (1988) Differential susceptibility to recombinant interferon-gamma-induced HLA-DQ antigen modulation among clones from a human metastatic melanoma. *J. Immunol.* **140:** 183–191.

Arenzana Seisdedos F, Mogensen SC, Vuillier F, Fiers W, Virelizier JL. (1988) Autocrine secretion of tumor necrosis factor under the influence of interferon-gamma amplifies HLA-DR gene induction in human monocytes. *Proc. Natl Acad. Sci. USA* **85:** 6087–6091.

Auffray C, Lillie JW, Korman AJ, Boss JM, Frechin N, Guillemot F, Cooper J, Mulligan RC, Strominger JL. (1987) Structure and expression of HLA-DQ alpha and -DX alpha genes: interallelic alternate splicing of the HLA-DQ alpha gene and functional splicing of the HLA-DQ alpha gene using a retroviral vector. *Immunogenetics* **26:** 63–73.

Barclay AN, Mason DW. (1982) Induction of Ia antigen in rat epidermal cells and gut epithelium by immunological stimuli. *J. Exp. Med.* **156:** 1665–1676.

Basta PV, Sherman PA, Ting JP. (1987) Identification of an interferon-gamma response region 5' of the human histocompatibility leukocyte antigen DR alpha chain gene which is active in human glioblastoma multiforme lines. *J. Immunol.* **138:** 1275–1280.

Basta PV, Sherman PA, Ting JP. (1988) Detailed delineation of an interferon-gamma-responsive element important in human HLA-DRA gene expression in a glioblastoma multiform line. *Proc. Natl Acad. Sci. USA* **85:** 8618–8622.

Ben Nun A, Strauss W, Leeman SA, Cohn LE, Murre C, Duby A, Seidman JG, Glimcher LH. (1985) An Ia-positive mouse T-cell clone is functional in presenting antigen to other T cells. *Immunogenetics* **22:** 123–130.

Benichou B, Strominger JL. (1991) Class II-antigen-negative patient and mutant B-cell lines represent at least three, and probably four, distinct genetic defects defined by complementation analysis. *Proc. Natl Acad. Sci. USA* **88:** 4285–4288.

Benoist C, Mathis D. (1990) Regulation of major histocompatibility complex class-II genes: X, Y and other letters of the alphabet. *Annu. Rev. Immunol.* **8:** 681–715.

Berdoz J, Gorski J, Termijtelen AM, Dayer JM, Irle C, Schendel D, Mach B. (1987) Constitutive and induced expression of the individual HLA-DR beta and alpha chain loci in different cell types. *J. Immunol.* **139:** 1336–1341.

Blanar MA, Boettger EC, Flavell RA. (1988) Transcriptional activation of HLA-DR alpha by interferon gamma requires a trans-acting protein. *Proc. Natl Acad. Sci. USA* **85:** 4672–4676.

Bodmer JG, Marsh SG, Albert ED, *et al.* (1994) Nomenclature for factors of the HLA system, 1994. *Hum. Immunol.* **41:** 1–20.

Bono MR, Alcaide Loridan C, Couillin P, Letouze B, Grisard MC, Jouin H, Fellous M. (1991) Human chromosome 16 encodes a factor involved in induction of class II major histocompatibility antigens by interferon gamma. *Proc. Natl Acad. Sci. USA* **88:** 6077–6081.

Boothby M, Gravallese E, Liou HC, Glimcher LH. (1988) A DNA binding protein regulated by IL-4 and by differentiation in B cells. *Science* **242:** 1559–1562.

Boss JM, Strominger JL. (1986) Regulation of a transfected human class II major histocompatibility complex gene in human fibroblasts. *Proc. Natl Acad. Sci. USA* **83:** 9139–9143.

Bottazzo GF, Dean BM, McNally JM, MacKay EH, Swift PG, Gamble DR. (1985) In situ characterization of autoimmune phenomena and expression of HLA molecules in the pancreas in diabetic insulitis. *N. Engl. J. Med.* **313:** 353–360.

Böttger EC, Blanar MA, Flavell RA. (1988) Cycloheximide, an inhibitor of protein synthesis, prevents gamma-interferon-induced expression of class II mRNA in a macrophage cell line. *Immunogenetics* **28:** 215–220.

Braciale TJ, Morrison LA, Sweetser MT, Sambrook J, Gething MJ, Braciale VL. (1987) Antigen presentation pathways to class I and class II MHC-restricted T lymphocytes. *Immunol. Rev.* **98:** 95–114.

Brenner DA, O'Hara M, Angel P, Chojkier M, Karin M. (1989) Prolonged activation of jun and collagenase genes by tumour necrosis factor-alpha. *Nature* **337:** 661–663.

Britten RJ, Davidson EH. (1969) Gene regulation for higher cells: a theory. *Science* **165:** 349–357.

Broudy VC, Fitchen JH. (1986) Class II MHC antigens and hematpopoiesis. In: *HLA class II antigens* (eds BG Solheim, E Moller, S Ferrone). Springer Verlag, Heidelberg, pp. 386–401.

Brown AM, Barr CL, Ting JP. (1991) Sequences homologous to class II MHC W, X, and Y elements mediate constitutive and IFN-gamma-induced expression of human class II-associated invariant chain gene. *J. Immunol.* **146:** 3183–3189.

Brown AM, Wright KL, Ting JP. (1993) Human major histocompatibility complex class II-associated invariant chain gene promoter. Functional analysis and in vivo protein/DNA interactions of constitutive and IFN-gamma-induced expression. *J. Biol. Chem.* **268:** 26328–26333.

Brown AM, Linhoff MW, Stein B, Wright KL, Baldwin AS, Jr, Basta PV, Ting JP. (1994) Function of NF-kappa B/Rel binding sites in the major histocompatibility complex class II invariant chain promoter is dependent on cell-specific binding of different NF-kappa B/Rel subunits. *Mol. Cell. Biol.* **14:** 2926–2935.

Bull M, van Hoef A, Gorski J. (1990) Transcription analysis of class II human leukocyte antigen genes from normal and immunodeficient B lymphocytes, using polymerase chain reaction. *Mol. Cell. Biol.* **10:** 3792–3796.

Burger DR, Ford D, Vetto RM, Hamblin A, Goldstein A, Hubbard M, Dumonde DC. (1981) Endothelial cell presentation of antigen to human T cells. *Hum. Immunol.* **3:** 209–230.

Burkly LC, Lo D, Cowing C, Palmiter RD, Brinster RL, Flavell RA. (1989) Selective expression of class II E alpha d gene in transgenic mice. *J. Immunol.* **142:** 2081–2088.

Calman AF, Peterlin BM. (1988) Evidence for a trans-acting factor that regulates the transcription of class II major histocompatibility complex genes: genetic and functional analysis. *Proc. Natl Acad. Sci. USA* **85:** 8830–8834.

Cao H, Wolff RG, Meltzer MS, Crawford RM. (1989) Differential regulation of class II MHC determinants on macrophages by IFN-gamma and IL-4. *J. Immunol.* **143:** 3524–3531.

Caplen HS, Salvadori S, Gansbacher B, Zier KS. (1992) Post-transcriptional regulation of MHC class II expression in human T cells. *Cell Immunol.* **139:** 98–107.

Carrel S, Mach J-P, Ferremi P, Giuffré L, Accolla RS. (1986) Expression of class II MHC antigens on hematopoietic tumor cells. In: *HLA class II antigens* (eds BG Solheim, E Moller, S Ferrone). Springer Verlag, Heidelberg, pp. 412–428.

Cedar H. (1988) DNA methylation and gene activity. *Cell* **53:** 3–4.

Celada A, Maki RA. (1989) DNA binding of the mouse class II major histocompatibility CCAAT factor depends on two components. *Mol. Cell. Biol.* **9:** 3097–3100.

Celada A, Maki RA. (1991) IFN-gamma induces the expression of the genes for MHC class II I-A beta and tumor necrosis factor through a protein kinase C-independent pathway. *J. Immunol.* **146:** 114–120.

Celada A, Shiga M, Imagawa M, Kop J, Maki RA. (1988) Identification of a nuclear factor that binds to a conserved sequence of the I-A beta gene. *J. Immunol.* **140:** 3995–4002.

Celada A, McKercher S, Maki RA. (1993) Repression of major histocompatibility complex IA expression by glucocorticoids: the glucocorticoid receptor inhibits the DNA binding of the X box DNA binding protein. *J. Exp. Med.* **177:** 691–698.

Chang CH, Flavell RA. (1995) Class II transactivator regulates the expression of multiple genes involved in antigen presentation. *J. Exp. Med.* **181:** 765–767.

Chang CH, Fontes JD, Peterlin M, Flavell RA. (1994) Class II transactivator (CIITA) is sufficient for the inducible expression of major histocompatibility complex class II genes. *J. Exp. Med.* **180:** 1367–1374.

Chang HR, Smelser S, Cardon Cardo C, Houghton AN, Brennan MF. (1993) Expression and modulation of class I and class II histocompatibility leukocyte antigens on human soft tissue sarcomas. *Cancer* **72:** 2956–2962.

Chang RJ, Lee SH. (1986) Effects of interferon-gamma and tumor necrosis factor-alpha on the expression of an Ia antigen on a murine macrophage cell line. *J. Immunol.* **137:** 2853–2856.

Charron DJ, Engelman EG, Benike CJ, McDevitt HO. (1980) Ia antigens on alloreactive T cells in man detected by monoclonal antibodies. Evidence for synthesis of HLA-D/DR molecules of the responder type. *J. Exp. Med.* **152:** 127s–136s.

Charron DJ, Lotteau V, Turmel P. (1984) Hybrid HLA-DC antigens provide molecular evidence for gene trans-complementation. *Nature* **312:** 157–159.

Chin KC, Mao C, Skinner C, Riley JL, Wright KL, Moreno CS, Stark GR, Boss JM, Ting JP. (1994) Molecular analysis of G1B and G3A IFN gamma mutants reveals that defects in CIITA or RFX result in defective class II MHC and Ii gene induction. *Immunity* **1:** 687–697.

Cogswell JP, Basta PV, Ting JP. (1990) X-box-binding proteins positively and negatively regulate transcription of the HLA-DRA gene through interaction with discrete upstream W and V elements. *Proc. Natl Acad. Sci. USA* **87:** 7703–7707.

Cogswell JP, Austin J, Ting JP. (1991) The W element is a positive regulator of HLA-DRA transcription in various DR+ cell types. *J. Immunol.* **146:** 1361–1367.

Collins T, Korman AJ, Wake CT, Boss JM, Kappes DJ, Fiers W, Ault KA, Gimbrone MA, Jr, Strominger JL, Pober JS. (1984) Immune interferon activates multiple class II major histocompatibility complex genes and the associated invariant chain gene in human endothelial cells and dermal fibroblasts. *Proc. Natl Acad. Sci. USA* **81:** 4917–4921.

Collins T, Lapierre LA, Fiers W, Strominger JL, Pober JS. (1986) Recombinant human tumor necrosis factor increases mRNA levels and surface expression of HLA-A,B antigens in vascular endothelial cells and dermal fibroblasts in vitro. *Proc. Natl Acad. Sci. USA* **83:** 446–450.

Cotner T, Williams JM, Christenson L, Shapiro HM, Strom TB, Strominger J. (1983) Simultaneous flow cytometric analysis of human T cell activation antigen expression and DNA content. *J. Exp. Med.* **157:** 461–472.

Cowan EP, Pierce ML, Dhib Jalbut S. (1991) Interleukin-1 beta decreases HLA class II expression on a glioblastoma multiforme cell line. *J. Neuroimmunol.* **33:** 17–28.

Crawford RM, Finbloom DS, Ohara J, Paul WE, Meltzer MS. (1987) B cell stimulatory factor-1 (interleukin 4) activates macrophages for increased tumoricidal activity and expression of Ia antigens. *J. Immunol.* **139:** 135–141.

Cresswell P. (1994) Assembly, transport, and function of MHC class II molecules. *Annu. Rev. Immunol.* **12:** 259–293.

Daar AS, Fuggle SV, Fabre JW, Ting A, Morris PJ. (1984) The detailed distribution of MHC Class II antigens in normal human organs. *Transplantation* **38:** 293–298.

Dasch JR, Jones PP. (1986) Independent regulation of IgM, IgD, and Ia antigen expression in cultured immature B lymphocytes. *J. Exp. Med.* **163:** 938–951.

Davis MM, Chien Y. (1993) Topology and affinity of T-cell receptor mediated recognition of peptide-MHC complexes. *Curr. Opin. Immunol.* **5:** 45–49.

de Preval C, Lisowska Grospierre B, Loche M, Griscelli C, Mach B. (1985) A trans-acting class II regulatory gene unlinked to the MHC controls expression of HLA class II genes. *Nature* **318:** 291–293.

de Waal Malefyt R, Figdor CG, Huijbens R, Mohan Peterson S, Bennett B, Culpepper J, Dang W, Zurawski G, De Vries JE. (1993) Effects of IL-13 on phenotype, cytokine production, and cytotoxic function of human monocytes. Comparison with IL-4 and modulation by IFN-gamma or IL-10. *J. Immunol.* **151:** 6370–6381.

Dedrick RL, Jones PP. (1990) Sequence elements required for activity of a murine major histocompatibility complex class II promoter bind common and cell-type-specific nuclear factors. *Mol. Cell. Biol.* **10:** 593–604.

Deeg HJ, Huss R. (1993) Major histocompatibility complex class II molecules, hemopoiesis and the marrow microenvironment. *Bone Marrow Transplant* **12:** 425–430.

Del Pozzo G, Guardiola J. (1990) A SINE insertion provides information on the divergence of the HLA-DQA1 and HLA-DQA2 genes. *Immunogenetics* **31:** 229–232.

Del Pozzo G, Perfetto C, Ombra MN, Ding GZ, Guardiola J, Maffei A. (1992) DNA polymorphisms in the 5′-flanking region of the HLA-DQA1 gene. *Immunogenetics* **35:** 176–182.

Dellabona P, Latron F, Maffei A, Scarpellino L, Accolla RS. (1989) Transcriptional control of MHC class II gene expression during differentiation from B cells to plasma cells. *J. Immunol.* **142:** 2902–2910.

Dennis GJ, Mond JJ. (1986) Corticosteroid-induced suppression of murine B cell immune response antigens. *J. Immunol.* **136:** 1600–1604.

Denzin LK, Cresswell P. (1995) HLA-DM induces CLIP dissociation from MHC class II ab dimers and facilitates peptide loading. *Cell* **82:** 155–165.

DeWolf WC, Schlossman SF, Yunis EJ. (1979) DRw antisera react with activated T cells. *J. Immunol.* **122:** 1780–1784.

Didier DK, Schiffenbauer J, Woulfe SL, Zacheis M, Schwartz BD. (1988) Characterization of the cDNA encoding a protein binding to the major histocompatibility complex class II Y box. *Proc. Natl Acad. Sci. USA* **85:** 7322–7326.

Diedrichs M, Schendel DJ. (1989) Differential surface expression of class II isotypes on activated CD4 and CD8 cells correlates with levels of locus-specific mRNA. *J. Immunol.* **142:** 3275–3280.

Dorn A, Bollekens J, Staub A, Benoist C, Mathis D. (1987a) A multiplicity of CCAAT box-binding proteins. *Cell* **50:** 863–872.

Dorn A, Durand B, Marfing C, Le Meur M, Benoist C, Mathis D. (1987b) Conserved major histocompatibility complex class II boxes–X and Y–are transcriptional control elements and specifically bind nuclear proteins. *Proc. Natl Acad. Sci. USA* **84:** 6249–6253.

Dorn A, Fehling HJ, Koch W, Le Meur M, Gerlinger P, Benoist C, Mathis D. (1988) B-cell control region at the 5′ end of a major histocompatibility complex class II gene: sequences and factors. *Mol. Cell. Biol.* **8:** 3975–3987.

Doveren RF, Buurman WA, Schutte B, Groenewegen G, van der Linden CJ. (1985) Class II antigens on canine T lymphocytes. *Tissue Antigens* **25:** 255–265.

Doyle C, Ford PJ, Ponath PD, Spies T, Strominger JL. (1990) Regulation of the class II-associated invariant chain gene in normal and mutant B lymphocytes. *Proc. Natl Acad. Sci. USA* **87:** 4590–4594.

Eades AM, Litfin M, Rahmsdorf HJ. (1990) The IFN-gamma response of the murine invariant chain gene is mediated by a complex enhancer that includes several MHC class II consensus elements. *J. Immunol.* **144:** 4399–4409.

Emery P, Mach B, Reith W. (1993) The different level of expression of HLA-DRB1 and -DRB3 genes is controlled by conserved isotypic differences in promoter sequence. *Hum. Immunol.* **38:** 137–147.

Engleman EG, Benike CJ, Charron DJ. (1980) Ia antigen on peripheral blood mononuclear leukocytes in man. II. Functional studies of HLA-DR-positive T cells activated in mixed lymphocyte reactions. *J. Exp. Med.* **152:** 114s–126s.

Falkenburg JH, Jansen J, van der Vaart Duinkerken N, Veenhof WF, Blotkamp J, Goselink HM, Parlevliet J, van Rood JJ. (1984) Polymorphic and monomorphic HLA-DR determinants on human hematopoietic progenitor cells. *Blood* **63:** 1125–1132.

Fan XD, Goldberg M, Bloom BR. (1988) Interferon-gamma-induced transcriptional activation is mediated by protein kinase C. *Proc. Natl Acad. Sci. USA* **85:** 5122–5125.

Feriotto G, Pozzi L, Piva R, Deledda F, Barbieri R, Nastruzzi C, Ciucci A, Natali PG, Giacomini P, Gambari R. (1991) Transgenic mice mimic the methylation pattern of the human HLA-DR alpha gene. *Biochem. Biophys. Res. Commun.* **175:** 459–466.

Fertsch D, Schoenberg DR, Germain RN, Tou JY, Vogel SN. (1987) Induction of macrophage Ia antigen expression by rIFN-gamma and down-regulation by IFN-alpha/beta and dexamethasone are mediated by changes in steady-state levels of Ia mRNA. *J. Immunol.* **139:** 244–249.

Finn PW, Kara CJ, Grusby MJ, Folsom V, Glimcher LH. (1991) Upstream elements of the MHC class II E beta gene active in B cells. *J. Immunol.* **146:** 4011–4015.

Fitchen JH, Le Fevre C, Ferrone S, Cline MJ. (1982) Expression of Ia-like and HLA-A,B antigens on human multipotential hematopoietic progenitor cells. *Blood* **59:** 188–190.

Forsum U, Claesson K, Hjelm E, Karlsson Parra A, Klareskog L, Scheynius A, Tjernlund U. (1985) Class II transplantation antigens: distribution in tissues and involvement in disease. *Scand. J. Immunol.* **21:** 389–396.

Frei K, Lins H, Schwerdel C, Fontana A. (1994) Antigen presentation in the central nervous system. The inhibitory effect of IL-10 on MHC class II expression and production of cytokines depends on the inducing signals and the type of cell analyzed. *J. Immunol.* **152:** 2720–2728.

Freund YR, Dedrick RL, Jones PP. (1990) cis-acting sequences required for class II gene regulation by interferon gamma and tumor necrosis factor alpha in a murine macrophage cell line. *J. Exp. Med.* **171:** 1283–1299.

Gambari R, del Senno L, Barbieri R, Buzzoni D, Gustafsson K, Giacomini P, Natali PG. (1986) Lack of correlation between hypomethylation and expression of the HLA-DR alpha gene. *Eur. J. Immunol.* **16:** 365–369.

Gaur LK, Heise ER, Ting JP. (1992) Conservation of the promoter region of DRA-like genes from nonhuman primates. *Immunogenetics* **35:** 136–139.

Germain RN, Bentley DM, Quill H. (1985) Influence of allelic polymorphism on the assembly and surface expression of class II MHC (Ia) molecules. *Cell* **43:** 233–242.

Gilfillan S, Aiso S, Michie SA, McDevitt HO. (1990) The effect of excess beta-chain synthesis on cell-surface expression of allele-mismatched class II heterodimers in vivo. *Proc. Natl Acad. Sci. USA* **87:** 7314–7318.

Glimcher LH, Kara CJ. (1992) Sequences and factors: a guide to MHC class-II transcription. *Annu. Rev. Immunol.* **10:** 13–49.

Gonczy P, Reith W, Barras E, Lisowska Grospierre B, Griscelli C, Hadam MR, Mach B. (1989) Inherited immunodeficiency with a defect in a major histocompatibility complex class II promoter-binding protein differs in the chromatin structure of the HLA-DRA gene. *Mol. Cell. Biol.* **9:** 296–302.

Greenstein JL, Lord EM, Horan P, Kappler JW, Marrack P. (1981) Functional subsets of B cells defined by quantitative differences in surface I-A. *J. Immunol.* **126:** 2419–2423.

Gregersen PK, Silver J, Winchester RJ. (1988) Genetic susceptibility to rheumatoid arthritis and human leukocyte antigen class II polymorphism. The role of shared conformational determinants. *Am. J. Med.* **85:** 17–19.

Griscelli C. (1991) Combined immunodeficiency with defective expression in major histocompatibility complex class II genes. *Clin. Immunol. Immunopathol.* **61:** S106–S110.

Groenewegen G, Buurman WA, van der Linden CJ. (1985) Lymphokine dependence of in vivo expression of MHC class II antigens by endothelium. *Nature* **316:** 361–363.

Haas JP, Kimura A, Andreas A, *et al.* (1994a) Polymorphism in the upstream regulatory region of DQA1 genes and DRB1, QAP, DQA1, and DQB1 haplotypes in the German population. *Hum. Immunol.* **39:** 31–40.

Haas JP, Nevinny Stickel C, Schoenwald U, Truckenbrodt H, Suschke J, Albert ED. (1994b) Susceptible and protective major histocompatibility complex class II alleles in early-onset pauciarticular juvenile chronic arthritis. *Hum. Immunol.* **41:** 225–233.

Haas JP, Kimura A, Truckenbrodt H, Suschke J, Sasazuki T, Volgger A, Albert ED. (1995) Early-onset pauciarticular juvenile chronic arthritis is associated with a mutation in the Y-box of the HLA-DQA1 promoter. *Tissue Antigens* **45:** 317–321.

Halloran PF, Madrenas J. (1990) Regulation of MHC transcription. *Transplantation* **50:** 725–738.

Halloran PF, Wadgymar A, Autenreid P. (1986) The regulation of expression of major histocompatibility complex products. *Transplantation* **41:** 413–420.

Hasegawa SL, Boss JM. (1991) Two B cell factors bind the HLA-DRA X box region and recognize different subsets of HLA class II promoters. *Nucleic Acids Res.* **19:** 6269–6276.

Hasegawa SL, Sloan JH, Reith W, Mach B, Boss JM. (1991) Regulatory factor-X binding to mutant HLA-DRA promoter sequences. *Nucleic Acids Res.* **19:** 1243–1249.

Hehlgans T, Strominger JL. (1995) Activation of transcription by binding of NF-E1 (YY1) to a newly identified element in the first exon of the human DR alpha gene. *J. Immunol.* **154:** 5181–5187.

Herrero Sanchez C, Reith W, Silacci P, Mach B. (1992) The DNA-binding defect observed in major histocompatibility complex class II regulatory mutants concerns only one member of a family of complexes binding to the X boxes of class II promoters. *Mol. Cell. Biol.* **12:** 4076–4083.

Hirschberg H, Bergh OJ, Thorsby E. (1980) Antigen-presenting properties of human vascular endothelial cells. *J. Exp. Med.* **152:** 249s-255s.

Hirschberg H, Scott H, Thorsby E. (1981) Human endothelial cells can present antigen to sensitized T lymphocytes in vitro. *Transplant. Proc.* **13:** 100–102.

Hooft van Huijsduijnen RA, Bollekens J, Dorn A, Benoist C, Mathis D. (1987) Properties of a CCAAT box-binding protein. *Nucleic Acids Res.* **15:** 7265–7282.

Hooft van Huijsduijnen RA, Li XY, Black D, Matthes H, Benoist C, Mathis D. (1990) Co-evolution from yeast to mouse: cDNA cloning of the two NF-Y (CP-1/CBF) subunits. *EMBO J.* **9:** 3119–3127.

Hopkins J, McConnell I, Dalziel RG, Dutia BM. (1993) Patterns of major histocompatibility complex class II expression by T cell subsets in different immunological compartments. 2. Altered expression and cell function following activation in vivo. *Eur. J. Immunol.* **23:** 2889–2896.

Houghton AN, Thomson TM, Gross D, Oettgen HF, Old LJ. (1984) Surface antigens of melanoma and melanocytes. Specificity of induction of Ia antigens by human gamma-interferon. *J. Exp. Med.* **160:** 255–269.

Hume CR, Accolla RS, Lee JS. (1987) Defective HLA class II expression in a regulatory mutant is partially complemented by activated ras oncogenes. *Proc. Natl Acad. Sci. USA* **84:** 8603–8607.

Hume CR, Shookster LA, Collins N, O'Reilly R, Lee JS. (1989) Bare lymphocyte syndrome: altered HLA class II expression in B cell lines derived from two patients. *Hum. Immunol.* **25:** 1–11.

Inaba K, Kitaura M, Kato T, Watanabe Y, Kawade Y, Muramatsu S. (1986) Contrasting effect of alpha/beta- and gamma-interferons on expression of macrophage Ia antigens. *J. Exp. Med.* **163:** 1030–1035.

Indiveri F, Wilson BS, Russo C, Quaranta V, Pellegrino MA, Ferrone S. (1980) Ia-like antigens on human T lymphocytes: relationship to other surface markers, role in mixed lymphocyte reactions, and structural profile. *J. Immunol.* **125:** 2673–2678.

Itoh Lindstrom Y, Peterlin BM, Ting JP. (1995) Affinity enrichment and functional characterization of TRAX1, a novel transcription activator and X1–sequence-binding protein of HLA-DRA. *Mol. Cell. Biol.* **15:** 282–289.

Ivashkiv LB, Liou HC, Kara CJ, Lamph WW, Verma IM, Glimcher LH. (1990) mXBP/CRE-BP2 and c-Jun form a complex which binds to the cyclic AMP, but not to the 12–O-tetradecanoylphorbol-13–acetate, response element. *Mol. Cell. Biol.* **10:** 1609–1621.

Jabrane Ferrat N, Peterlin BM. (1994) Ets-1 activates the DRA promoter in B cells. *Mol. Cell. Biol.* **14:** 7314–7321.

Kappes D, Strominger JL. (1988) Human class II major histocompatibility complex genes and proteins. *Annu. Rev. Biochem.* **57:** 991–1028.

Kara CJ, Glimcher LH. (1991) In vivo footprinting of MHC class II genes: bare promoters in the bare lymphocyte syndrome. *Science* **252:** 709–712.

Kara CJ, Liou HC, Ivashkiv LB, Glimcher LH. (1990) A cDNA for a human cyclic AMP response element-binding protein which is distinct from CREB and expressed preferentially in brain. *Mol. Cell. Biol.* **10:** 1347–1357.

Karp DR, Teletski CL, Jaraquemada D, Maloy WL, Coligan JE, Long EO. (1990) Structural requirements for pairing of alpha and beta chains in HLA-DR and HLA-DP molecules. *J. Exp. Med.* **171:** 615–628.

Kelly AP, Monaco JJ, Cho SG, Trowsdale J. (1991) A new human HLA class II-related locus, DM. *Nature* **353:** 571–573.

Kelly JD, Fox LM, Lange CF, Bouchard CS, McNulty JA. (1993) Experimental autoimmune pinealitis in the rat: ultrastructure and quantitative immunocytochemical characterization of mononuclear infiltrate and MHC class II expression. *Autoimmunity* **16:** 1–11.

Kern MJ, Stuart PM, Omer KW, Woodward JG. (1989) Evidence that IFN-gamma does not affect MHC class II gene expression at the post-transcriptional level in a mouse macrophage cell line. *Immunogenetics* **30:** 258–265.

Kimura A, Sasazuki T. (1992a) HLA-DQA gene is differently regulated from other HLA class II genes. In: *Molecular Approaches to the Study and Treatment of Human Disease* (eds TO Yoshida, JM Wilson). Elsevier Science Publishers B.V. Amsterdam, pp. 97–104.

Kimura A, Sasazuki T. (1992b) Polymorphism in the 5'-flanking region of the DQA1 gene and it's relation to DR-DQ haplotype. In: *HLA 1991* (eds K Tsuji, M Aizawa, T Sasazuki). Oxford Science Publications, Oxford, pp. 382–385.

King DP, Jones PP. (1983) Induction of Ia and H-2 antigens on a macrophage cell line by immune interferon. *J. Immunol.* **131:** 315–318.

Klein C, Lisowska Grospierre B, LeDeist F, Fischer A, Griscelli C. (1993) Major histocompatibility complex class II deficiency: clinical manifestations, immunologic features, and outcome. *J. Pediatr.* **123:** 921–928.

Kobr M, Reith W, Herrero Sanchez C, Mach B. (1990) Two DNA-binding proteins discriminate between the promoters of different members of the major histocompatibility complex class II multigene family. *Mol. Cell. Biol.* **10:** 965–971.

Koch W, Candeias S, Guardiola J, Accolla R, Benoist C, Mathis D. (1988) An enhancer factor defect in a mutant Burkitt lymphoma cell line. *J. Exp. Med.* **167:** 1781–1790.

Koide Y, Yoshida TO. (1992) Enhancer-like and basal promoter elements in the upstream region of the HLA-DPB gene. In: *HLA 1991* (eds K Tsuji, M Aizawa, T Sasazuki). Oxford Science Publications, Oxford, pp. 388–390.

Koide Y, Ina Y, Nezu N, Yoshida TO. (1988) Calcium influx and the Ca2+-calmodulin complex are involved in interferon-gamma-induced expression of HLA class II molecules on HL-60 cells. *Proc. Natl Acad. Sci. USA* **85:** 3120–3124.

Kolk DP, Floyd Smith G. (1993) Induction of the murine class-II antigen-associated invariant chain by TNF-alpha is controlled by an NF-kappa B-like element. *Gene* **126:** 179–185.

Kouskoff V, Mantovani RM, Candeias SM, Dorn A, Staub A, Lisowska Grospierre B, Griscelli C, Benoist CO, Mathis DJ. (1991) NF-X, a transcription factor implicated in MHC class II gene regulation. *J. Immunol.* **146:** 3197–3204.

Lala PK, Johnson GR, Battye FL, Nossal GJ. (1979) Maturation of B lymphocytes. I. Concurrent appearance of increasing Ig, Ia, and mitogen responsiveness. *J. Immunol.* **122:** 334–341.

Lapierre LA, Fiers W, Pober JS. (1988) Three distinct classes of regulatory cytokines control endothelial cell MHC antigen expression. Interactions with immune gamma interferon differentiate the effects of tumor necrosis factor and lymphotoxin from those of leukocyte alpha and fibroblast beta interferons. *J. Exp. Med.* **167:** 794–804.

Latchmann DS, (1990) *Gene Regulation: A Eukaryotic Perspective.* Unwin Hyman, London.

Latron F, Jotterand Bellomo M, Maffei A, Scarpellino L, Bernard M, Strominger JL, Accolla RS. (1988) Active suppression of major histocompatibility complex class II gene expression during differentiation from B cells to plasma cells. *Proc. Natl Acad. Sci. USA* **85:** 2229–2233.

Le Meur M, Gerlinger P, Benoist C, Mathis D. (1985) Correcting an immune-response deficiency by creating E alpha gene transgenic mice. *Nature* **316:** 38–42.

Le Meur M, Waltzinger C, Gerlinger P, Benoist C, Mathis D. (1989) Restricted assembly of MHC class II molecules in transgenic mice. *J. Immunol.* **142:** 323–327.

Lee SC, Liu W, Roth P, Dickson DW, Berman JW, Brosnan CF. (1993) Macrophage colony-stimulating factor in human fetal astrocytes and microglia. Differential regulation by cytokines and lipopolysaccharide, and modulation of class II MHC on microglia. *J. Immunol.* **150:** 594–604.

Lee YJ, Panek RB, Huston M, Benveniste EN. (1995) Role of protein kinase C and tyrosine kinase activity in IFN-gamma-induced expression of the class II MHC gene. *Am. J. Physiol.* **268:** C127–137.

Leen MPJM, Ogutu ER, Gorski J. (1994) Structural and functional analysis of HLA-DRβ-promoter polymorphism and isomorphism. *Hum. Immunol.* **41:** 112–120.

Leff SE, Rosenfeld MG, Evans RM. (1986) Complex transcriptional units: diversity in gene expression by alternative RNA processing. *Annu. Rev. Biochem.* **55:** 1091–1117.

Lepage V, Lamm LU, Charron D. (1993) Molecular aspects of HLA class II and some autoimmune diseases. *Eur. J. Immunogen.* **20:** 153–164.

Levine F, Pious D. (1985) Different roles for cytosine methylation in HLA class II gene expression. *Immunogenetics* **22:** 427–440.

Li XY, Hofft van Huijsduijnen R, Mantovani R, Benoist C, Mathis D. (1992a) Intron-exon organization of the NF-Y genes. Tissue-specific splicing modifies an activation domain. *J. Biol. Chem.* **267:** 8984–8990.

Li XY, Mantovani R, Hooft van Huijsduijnen R, Andre I, Benoist C, Mathis D. (1992b) Evolutionary variation of the CCAAT-binding transcription factor NF-Y. *Nucleic Acids Res.* **20:** 1087–1091.

Ling PD, Warren MK, Vogel SN. (1985) Antagonistic effect of interferon-beta on the interferon-gamma-induced expression of Ia antigen in murine macrophages. *J. Immunol.* **135:** 1857–1863.

Liou HC, Boothby MR, Glimcher LH. (1988) Distinct cloned class II MHC DNA binding proteins recognize the X box transcription element. *Science* **242:** 69–71.

Liou HC, Boothby MR, Finn PW, Davidon R, Nabavi N, Zeleznik Le NJ, Ting JP, Glimcher LH. (1990) A new member of the leucine zipper class of proteins that binds to the HLA DR alpha promoter. *Science* **247:** 1581–1584.

Liou HC, Eddy R, Shows T, Lisowska Grospierre B, Griscelli C, Doyle C, Mannhalter J, Eibl M, Glimcher LH. (1991) An HLA-DR alpha promoter DNA-binding protein is expressed ubiquitously and maps to human chromosomes 22 and 5. *Immunogenetics* **34:** 286–292.

Lisowska Grospierre B, Charron DJ, de Preval C, Durandy A, Griscelli C, Mach B. (1985) A defect in the regulation of major histocompatibility complex class II gene expression in human HLA-DR negative lymphocytes from patients with combined immunodeficiency syndrome. *J. Clin. Invest.* **76:** 381–385.

Lo D, Burkly LC, Widera G, Cowing C, Flavell RA, Palmiter RD, Brinster RL. (1988) Diabetes and tolerance in transgenic mice expressing class II MHC molecules in pancreatic beta cells. *Cell* **53:** 159–168.

Londei M, Lamb JR, Bottazzo GF, Feldmann M. (1984) Epithelial cells expressing aberrant MHC class II determinants can present antigen to cloned human T cells. *Nature* **312:** 639–641.

Lotteau V, Teyton L, Burroughs D, Charron D. (1987) A novel HLA class II molecule (DR alpha-DQ beta) created by mismatched isotype pairing. *Nature* **329:** 339–341.

Lotteau V, Sands J, Teyton L, Turmel P, Charron D, Strominger JL. (1989) Modulation of HLA class II antigen expression by transfection of sense and antisense DR alpha cDNA. *J. Exp. Med.* **169:** 351–356.

Louis P, Eliaou JF, Kerlan Candon S, Pinet V, Vincent R, Clot J. (1993) Polymorphism in the regulatory region of HLA-DRB genes correlating with haplotype evolution. *Immunogenetics* **38:** 21–26.

Louis P, Pinet V, Cavadore P, Kerlan Candon S, Clot J, Eliaou JF. (1994a) Differential expression of HLA-DRB genes according to the polymorphism of their regulatory region. *C. R. Acad. Sci. III.* **317:** 161–166.

Louis P, Vincent R, Cavadore P, Clot J, Eliaou JF. (1994b) Differential transcriptional activities of HLA-DR genes in the various haplotypes. *J. Immunol.* **153:** 5059–5067.

Lucey DR, Nicholson Weller A, Weller PF. (1989) Mature human eosinophils have the capacity to express HLA-DR. *Proc. Natl Acad. Sci. USA* **86:** 1348–1351.

MacDonald GH, Itoh Lindstrom Y, Ting JP. (1995) The transcriptional regulatory protein, YB-1, promotes single-stranded regions in the DRA promoter. *J. Biol. Chem.* **270:** 3527–3533.

Mach B. (1995) MHC class II regulation–lessons from a disease. *N. Engl. J. Med.* **332:** 120–122.

Mantovani R, Pessara U, Tronche F, Li XY, Knapp AM, Pasquali JL, Benoist C, Mathis D. (1992) Monoclonal antibodies to NF-Y define its function in MHC class II and albumin gene transcription. *EMBO J.* **11:** 3315–3322.

Manyak CL, Tse H, Fischer P, Coker L, Sigal NH, Koo GC. (1988) Regulation of class II MHC molecules on human endothelial cells. Effects of IFN and dexamethasone. *J. Immunol.* **140:** 3817–3821.

Massa PT. (1993) Specific suppression of major histocompatibility complex class I and class II genes in astrocytes by brain-enriched gangliosides. *J. Exp. Med.* **178:** 1357–1363.

Mathis D, Oudet P, Chambon P. (1980) Structure of transcribing chromatin. *Prog. Nucleic Acid Res. Mol. Biol.* **24:** 1–55.

Matis LA, Glimcher LH, Paul WE, Schwartz RH. (1983) Magnitude of response of histocompatibility-restricted T-cell clones is a function of the product of the concentrations of antigen and Ia molecules. *Proc. Natl Acad. Sci. USA* **80:** 6019–6023.

Matsushima GK, Itoh Lindstrom Y, Ting JP. (1992) Activation of the HLA-DRA gene in primary human T lymphocytes: novel usage of TATA and the X and Y promoter elements. *Mol. Cell. Biol.* **12:** 5610–5619.

McCluskey J, Munitz T, Boyd L, Germain RN, Coligan JE, Singer A, Margulies DH. (1988) Cell surface expression of the amino-terminal domain of A kappa alpha. Recognition of an isolated MHC antigenic structure by allospecific T cells but not alloantibodies. *J. Immunol.* **140:** 2081–2089.

McMillan VM, Dennis GJ, Glimcher LH, Finkelman FD, Mond JJ. (1988) Corticosteroid induction of Ig+Ia- B cells in vitro is mediated via interaction with the glucocorticoid cytoplasmic receptor. *J. Immunol.* **140:** 2549–2555.

Mitchell PJ, Tjian R. (1989) Transcriptional regulation in mammalian cells by sequence-specific DNA binding proteins. *Science* **245:** 371–378.

Miwa K, Strominger JL. (1987) The HLA-DQ beta gene upstream region contains an immunoglobulin-like octamer motif that binds cell-type specific nuclear factors. *Nucleic Acids Res.* **15:** 8057–8067.

Miwa K, Doyle C, Strominger JL. (1987) Sequence-specific interactions of nuclear factors with conserved sequences of human class II major histocompatibility complex genes. *Proc. Natl Acad. Sci. USA* **84:** 4939–4943.

Momburg F, Koch N, Moller P, Moldenhauer G, Butcher GW, Hammerling GJ. (1986)

Differential expression of Ia and Ia-associated invariant chain in mouse tissues after in vivo treatment with IFN-gamma. *J. Immunol.* **136:** 940–948.

Mond JJ, Kessler S, Finkelman FD, Paul WE, Scher I. (1980) Heterogeneity of Ia expression on normal B cells, neonatal B cells, and on cells from B cell-defective CBA/N mice. *J. Immunol.* **124:** 1675–1682.

Mond JJ, Carman J, Sarma C, Ohara J, Finkelman FD. (1986) Interferon-gamma suppresses B cell stimulation factor (BSF-1) induction of class II MHC determinants on B cells. *J. Immunol.* **137:** 3534–3537.

Morris P, Shaman J, Attaya M, Amaya M, Goodman S, Bergman C, Monaco JJ, Mellins E. (1994) An essential role for HLA-DM in antigen presentation by class II major histocompatibility molecules. *Nature* **368:** 551–554.

Morzycka Wroblewska E, Harwood JI, Smith JR, Kagnoff MF. (1993) Structure and evolution of the promoter regions of the DQA genes. *Immunogenetics* **37:** 364–372.

Moses H, Panek RB, Benveniste EN, Ting JP. (1992) Usage of primary cells to delineate IFN-gamma-responsive DNA elements in the HLA-DRA promoter and to identify a novel IFN-gamma-enhanced nuclear factor. *J. Immunol.* **148:** 3643–3651.

Muller W, Kuhn R, Rajewsky K. (1991) Major histocompatibility complex class II hyperexpression on B cells in interleukin 4–transgenic mice does not lead to B cell proliferation and hypergammaglobulinemia. *Eur. J. Immunol.* **21:** 921–925.

Natali PG, De Martino C, Quaranta V, Nicotra MR, Frezza F, Pellegrino MA, Ferrone S. (1981) Expression of Ia-like antigens in normal human nonlymphoid tissues. *Transplantation* **31:** 75–78.

Neiss U, Reske K. (1994) Non-coordinate synthesis of MHC class II proteins and invariant chains by epidermal Langerhans cells derived from short-term in vitro culture. *Int. Immunol.* **6:** 61–71.

Nepom BS. (1993) The role of the major histocompatibility complex in autoimmunity. *Clin. Immunol. Immunopathol.* **67:** S50–55.

Nepom GT, Chung J, West KA. (1995) Differential expression of HLA-DQB1 alleles in a heterozygous cell line. *Immunogenetics* **42:** 143–148.

Nezu N, Ryu K, Koide Y, Yoshida TO. (1990) Regulation of HLA class II molecule expressions by IFN-gamma. The signal transduction mechanism in glioblastoma cell lines. *J. Immunol.* **145:** 3126–3135.

Noelle RJ, Kuziel WA, Maliszewski CR, McAdams E, Vitetta ES, Tucker PW. (1986) Regulation of the expression of multiple class II genes in murine B cells by B cell stimulatory factor-1 (BSF-1). *J. Immunol.* **137:** 1718–1723.

Nunez G, Giles RC, Ball EJ, Hurley CK, Capra JD, Stastny P. (1984) Expression of HLA-DR, MB, MT and SB antigens on human mononuclear cells: identification of two phenotypically distinct monocyte populations. *J. Immunol.* **133:** 1300–1306.

Odum N, Morling N, Friis J, Heilmann C, Hyldig Nielsen JJ, Jakobsen BK, Pedersen FK, Platz P, Ryder LP, Svejgaard A. (1986) Increased frequency of HLA-DPw2 in pauciarticular onset juvenile chronic arthritis. *Tissue Antigens* **28:** 245–250.

Oen K, Petty RE, Schroeder ML. (1982) An association between HLA-A2 and juvenile rheumatoid arthritis in girls. *J. Rheumatol.* **9:** 916–920.

Ono SJ, Song Z. (1995) Mapping of the interaction site of the defective transcription factor in the class II major histocompatibility complex mutant cell line clone-13 to the divergent X2–box. *J. Biol. Chem.* **270:** 6396–6402.

Ono SJ, Bazil V, Sugawara M, Strominger JL. (1991a) An isotype-specific trans-acting factor is defective in a mutant B cell line that expresses HLA-DQ, but not -DR or -DP. *J. Exp. Med.* **173:** 629–637.

Ono SJ, Liou HC, Davidon R, Strominger JL, Glimcher LH. (1991b) Human X-box-binding protein 1 is required for the transcription of a subset of human class II major histocompatibility genes and forms a heterodimer with c-fos. *Proc. Natl Acad. Sci. USA* **88:** 4309–4312.

Perfetto C, Zacheis M, McDaid D, Meador JW, Schwartz BD. (1993) Polymorphism in the promoter region of HLA-DRB genes. *Hum. Immunol.* **36:** 27–33.

Peterlin BM. (1991) Transcriptional regulation of HLA-DRA gene. *Res. Immunol.* **142:** 393–399.

Pinet V, Eliaou JF, Clot J. (1991) Description of a polymorphism in the regulatory region of the HLA-DRA gene. *Hum. Immunol.* **32:** 162–169.

Pinkert CA, Widera G, Cowing C, Heber Katz E, Palmiter RD, Flavell RA, Brinster RL. (1985) Tissue-specific, inducible and functional expression of the E alpha d MHC class II gene in transgenic mice. *EMBO J.* **4:** 2225–2230.

Pober JS, Collins T, Gimbrone MA, Jr, Cotran RS, Gitlin JD, Fiers W, Clayberger C, Krensky AM, Burakoff SJ, Reiss CS. (1983) Lymphocytes recognize human vascular endothelial and dermal fibroblast Ia antigens induced by recombinant immune interferon. *Nature* **305:** 726–729.

Politis AD, Vogel SN. (1990) Pharmacologic evidence for the requirement of protein kinase C in IFN-induced macrophage Fc gamma receptor and Ia antigen expression. *J. Immunol.* **145:** 3788–3795.

Polla BS, Poljak A, Ohara J, Paul WE, Glimcher LH. (1986) Regulation of class II gene expression: analysis in B cell stimulatory factor 1–inducible murine pre-B cell lines. *J. Immunol.* **137:** 3332–3337.

Polla BS, Ohara J, Paul WE, *et al.* (1988) Differential induction of class II gene expression in murine pre-B-cell lines by B-cell stimulatory factor-1 and by antibodies to B-cell surface antigens. *J. Mol. Cell. Immunol.* **3:** 363–373.

Prpic V, Yu SF, Figueiredo F, Hollenbach PW, Gawdi G, Herman B, Uhing RJ, Adams DO. (1989) Role of Na+/H+ exchange by interferon-gamma in enhanced expression of JE and I-A beta genes. *Science* **244:** 469–471.

Ptashne M. (1988) How eukaryotic transcriptional activators work. *Nature* **335:** 683–689.

Radley E, Alderton RP, Kelly A, Trowsdale J, Beck S. (1994) Genomic organization of HLA-DMA and HLA-DMB. Comparison of the gene organization of all six class II families in the human major histocompatibility complex. *J. Biol. Chem.* **269:** 18834–18838.

Rammensee HG, Falk K, Rotzschke O. (1993) MHC molecules as peptide receptors. *Curr. Opin. Immunol.* **5:** 35–44.

Reekers P, Schretlen ED, van de Putte LB. (1983) Increase of HLA-"DRw6" in patients with juvenile chronic arthritis. *Tissue Antigens* **22:** 283–288.

Reichstetter S, Krellner PH, Meenzen CM, Kalden JR, Wassmuth R. (1994) Comparative analysis of sequence variability in the upstream regulatory region of the HLA-DQB1 gene. *Immunogenetics* **39:** 207–212.

Reichstetter S, Brünnler G, Kalden JR, Wassmuth R. (1996) DQB1 promoter sequence variability and linkage in Caucasians. *Hum. Immunol.* (in press).

Reinherz EL, Kung PC, Pesando JM, Ritz J, Goldstein G, Schlossman SF. (1979) Ia determinants on human T-cell subsets defined by monoclonal antibody. Activation stimuli required for expression. *J. Exp. Med.* **150:** 1472–1482.

Reinherz EL, Morimoto C, Penta AC, Schlossman SF. (1981) Subpopulations of the T4+ inducer T cell subset in man: evidence for an amplifier population preferentially expressing Ia antigen upon activation. *J. Immunol.* **126:** 67–70.

Reith W, Satola S, Sanchez CH, Amaldi I, Lisowska Grospierre B, Griscelli C, Hadam MR, Mach B. (1988) Congenital immunodeficiency with a regulatory defect in MHC class II gene expression lacks a specific HLA-DR promoter binding protein, RF-X. *Cell* **53:** 897–906.

Reith W, Barras E, Satola S, Kobr M, Reinhart D, Sanchez CH, Mach B. (1989) Cloning of the major histocompatibility complex class II promoter binding protein affected in a hereditary defect in class II gene regulation. *Proc. Natl Acad. Sci. USA* **86:** 4200–4204.

Reith W, Kobr M, Emery P, Durand B, Siegrist CA, Mach B. (1994a) Cooperative binding between factors RFX and X2bp to the X and X2 boxes of MHC class II promoters. *J. Biol. Chem.* **269:** 20020–20025.

Reith W, Siegrist CA, Durand B, Barras E, Mach B. (1994b) Function of major histocompatibility complex class II promoters requires cooperative binding between factors RFX and NF-Y. *Proc. Natl Acad. Sci. USA* **91:** 554–558.

Reske Kunz AB, Reske K, Rude E. (1986) Cloned murine Ia+ BK-BI-2.6.C6 T cells function as accessory cells presenting protein antigens to long-term-cultured antigen-specific T cell lines. *J. Immunol.* **136:** 2033–2040.

Riley JL, Westerheide SD, Price JA, Brown JA, Boss JM. (1995) Activation of class II MHC genes requires both the X box region and the class II transactivator (CIITA). *Immunity* **2:** 533–543.

Robbins PA, Maino VC, Warner NL, Brodsky FM. (1988) Activated T cells and monocytes have characteristic patterns of class II antigen expression. *J. Immunol.* **141:** 1281–1287.

Robinson J, Sieff C, Delia D, Edwards PA, Greaves M. (1981) Expression of cell-surface HLA-DR, HLA-ABC and glycophorin during erythroid differentiation. *Nature* **289:** 68–71.

Roeder RG. (1991) The complexities of eukaryotic transcription initiation: regulation of preinitiation complex assembly. *Trends Biochem. Sci.* **16:** 402–408.

Rollini P, Mach B, Gorski J. (1985) Linkage map of three HLA-DR beta-chain genes: evidence for a recent duplication event. *Proc. Natl Acad. Sci. USA* **82:** 7197–7201.

Rosa FM, Fellous M. (1988) Regulation of HLA-DR gene by IFN-gamma. Transcriptional and post-transcriptional control. *J. Immunol.* **140:** 1660–1664.

Rousset F, Malefijt RW, Slierendregt B, Aubry JP, Bonnefoy JY, Defrance T, Banchereau J, De Vries JE. (1988) Regulation of Fc receptor for IgE (CD23) and class II MHC antigen expression on Burkitt's lymphoma cell lines by human IL-4 and IFN-gamma. *J. Immunol.* **140:** 2625–2632.

Ryu K, Koide Y, Yamashita Y, Yoshida TO. (1992) Signal transduction mechanisms of HLA class II molecule expression by IFN-gamma: tyrosine protein kinase is involved before protein kinase C activation in a glioblastoma cell line. In: *HLA 1991* (eds K Tsuji, M Aizawa, T Sasazuki). Oxford Science Publications, Oxford, pp. 379–382.

Sakurai M, Strominger JL. (1988) B-cell-specific enhancer activity of conserved upstream elements of the class II major histocompatibility complex DQB gene. *Proc. Natl Acad. Sci. USA* **85:** 6909–6913.

Sakurai M, Strominger JL. (1989) Studies of expression of the DQ beta promoter and its 5′ deletion derivatives in normal and mutant human B cell lines. *Tissue Antigens* **34:** 64–77.

Sanderson F, Kleijmeer MJ, Kelly A, Verwoerd D, Tulp A, Neefjes JJ, Geuze HJ, Trowsdale J. (1994a) Accumulation of HLA-DM, a regulator of antigen presentation, in MHC class II compartments. *Science* **266:** 1566–1569.

Sanderson F, Powis SH, Kelly AP, Trowsdale J. (1994b) Limited polymorphism in HLA-DM does not involve the peptide binding groove. *Immunogenetics* **39:** 56–58.

Sargent IL. (1993) Maternal and fetal immune responses during pregnancy. *Exp. Clin. Immunogenet.* **10:** 85–102.

Sarvetnick N, Liggitt D, Pitts SL, Hansen SE, Stewart TA. (1988) Insulin-dependent diabetes mellitus induced in transgenic mice by ectopic expression of class II MHC and interferon-gamma. *Cell* **52:** 773–782.

Sasaki A, Levison SW, Ting JP. (1990) Differential suppression of interferon-gamma-induced Ia antigen expression on cultured rat astroglia and microglia by second messengers. *J. Neuroimmunol.* **29:** 213–222.

Schreiber E, Harshman K, Kemler I, Malipiero U, Schaffner W, Fontana A. (1990) Astrocytes and glioblastoma cells express novel octamer-DNA binding proteins distinct from the ubiquitous Oct-1 and B cell type Oct-2 proteins. *Nucleic Acids Res.* **18:** 5495–5503.

Servenius B, Rask L, Peterson PA. (1987) Class II genes of the human major histocompatibility complex. The DO beta gene is a divergent member of the class II beta gene family. *J. Biol. Chem.* **262:** 8759–8766.

Sharp PA. (1987) Splicing of messenger RNA precursors. *Science* **235:** 766–771.

Sherman PA, Basta PV, Ting JP. (1987) Upstream DNA sequences required for tissue-specific expression of the HLA-DR alpha gene. *Proc. Natl Acad. Sci. USA* **84:** 4254–4258.

Sherman PA, Basta PV, Heguy A, Wloch MK, Roeder RG, Ting JP. (1989) The octamer motif is a B-lymphocyte-specific regulatory element of the HLA-DR alpha gene promoter. *Proc. Natl Acad. Sci. USA* **86:** 6739–6743.

Shewey LM, Nepom GT. (1993) Allele-specific DNA-protein interactions associated with the X-box regulatory region of the DQB1*0302 gene. *Autoimmunity* **15** (Suppl.): 8–11.

Shewey LM, Beaty JS, Andersen LC, Nepom GT. (1992) Differential expression of related HLA class II DQ beta genes caused by nucleotide variation in transcriptional regulatory elements. *J. Immunol.* **148:** 1265–1273.

Sieff C, Bicknell D, Caine G, Robinson J, Lam G, Greaves MF. (1982) Changes in cell surface antigen expression during hemopoietic differentiation. *Blood* **60:** 703–713.

Silacci P, Mottet A, Steimle V, Reith W, Mach B. (1994) Developmental extinction of major histocompatibility complex class II gene expression in plasmocytes is mediated by silencing of the transactivator gene CIITA. *J. Exp. Med.* **180:** 1329–1336.

Singal DP, Qiu X, D'Souza M, Sood SK. (1993) Polymorphism in the upstream regulatory regions of HLA-DRB genes. *Immunogenetics* **37:** 143–147.

Singh SK, Abramson EJ, Krco CJ, David CS. (1984) Endogenous synthesis of Ia antigens by a cloned T cell line. *J. Mol. Cell. Immunol.* **1:** 147–156.

Sloan JH, Hasegawa SL, Boss JM. (1992) Single base pair substitutions within the HLA-DRA gene promoter separate the functions of the X1 and X2 boxes. *J. Immunol.* **148:** 2591–2599.

Sloan VS, Cameron P, Porter G, Gammon M, Amaya M, Mellins E, Zaller DM. (1995) Mediation by HLA-DM of dissociation of peptides from HLA-DR. *Nature* **375:** 802–806.

Snyder DS, Unanue ER. (1982) Corticosteroids inhibit murine macrophage Ia expression and interleukin 1 production. *J. Immunol.* **129:** 1803–1805.

Snyder DS, Beller DI, Unanue ER. (1982) Prostaglandins modulate macrophage Ia expression. *Nature* **299:** 163–165.

Song Z, Krishna S, Thanos D, Strominger JL, Ono SJ. (1994) A novel cysteine-rich sequence-specific DNA-binding protein interacts with the conserved X-box motif of the human major histocompatibility complex class II genes via a repeated Cys-His domain and functions as a transcriptional repressor. *J. Exp. Med.* **180:** 1763–1774.

Sorrentino R, Lillie J, Strominger JL. (1985) Molecular characterization of MT3 antigens by two-dimensional gel electrophoresis, NH2-terminal amino acid sequence analysis, and southern blot analysis. *Proc. Natl Acad. Sci. USA* **82:** 3794–3798.

Sparrow RL, Williams N. (1986) The pattern of HLA-DR and HLA-DQ antigen expression on clonable subpopulations of human myeloid progenitor cells. *Blood* **67:** 379–384.

Squinto SP, Doucet JP, Block AL, Morrow SL, Davenport WD. (1989) Induction of macrophage-like differentiation of HL-60 leukemia cells by tumor necrosis factor-alpha: potential role of fos expression. *Mol. Endocrinol.* **3:** 409–419.

Stastny P, Fink CW. (1979) Different HLA-D associations in adult and juvenile rheumatoid arthritis. *J. Clin. Invest.* **63:** 124–130.

Staudt LM, Singh H, Sen R, Wirth T, Sharp PA, Baltimore D. (1986) A lymphoid-specific protein binding to the octamer motif of immunoglobulin genes. *Nature* **323:** 640–643.

Steimle V, Otten LA, Zufferey M, Mach B. (1993) Complementation cloning of an MHC class II transactivator mutated in hereditary MHC class II deficiency (or bare lymphocyte syndrome). *Cell* **75:** 135–146.

Steimle V, Siegrist CA, Mottet A, Lisowska Grospierre B, Mach B. (1994) Regulation of MHC class II expression by interferon-gamma mediated by the transactivator gene CIITA. *Science* **265:** 106–109.

Steimle V, Durand B, Barras E, Zufferey M, Hadam MR, Mach B, Reith W. (1995) A novel DNA-binding regulatory factor is mutated in primary MHC class II deficiency (bare lymphocyte syndrome). *Genes Dev.* **9:** 1021–1032.

Stunz LL, Karr RW, Anderson RA. (1989) HLA-DRB1 and -DRB4 genes are differentially regulated at the transcriptional level. *J. Immunol.* **143:** 3081–3086.

Suciu Foca N, Khan R, Hardy M, Godfrey M, Susinno E, Reed E, Woodward K, Reemtsma K. (1981) Expression of HLA-D and DR gene products on in vitro and in vivo primed human T cells. *Transplant. Proc.* **13:** 1020–1025.

Sugawara M, Scholl T, Ponath PD, Strominger JL. (1994) A factor that regulates the class II major histocompatibility complex gene DPA is a member of a subfamily of zinc finger proteins that includes a Drosophila developmental control protein. *Mol. Cell. Biol.* **14:** 8438–8450.

Sugawara M, Scholl T, Mahanta SK, Ponath PD, Strominger JL. (1995) Cooperativity between the J and S elements of class II major histocompatibility complex genes as enhancers in normal and class II-negative patient and mutant B cell lines. *J. Exp. Med.* **182:** 175–184.

Sullivan KE, Peterlin BM. (1987) Transcriptional enhancers in the HLA-DQ subregion. *Mol. Cell. Biol.* **7:** 3315–3319.

Suzumura A, Silberberg DH. (1988) MHC antigen expression on glial cells. *Ann. N. Y. Acad. Sci.* **540:** 495–497.

Symington FW, Levine F, Braun M, Brown SL, Erlich HA, Torok Storb B. (1985) Differential Ia antigen expression by autologous human erythroid and B lymphoblastoid cell lines. *J. Immunol.* **135:** 1026–1032.

Thanos D, Mavrothalassitis G, Papamatheakis J. (1988) Multiple regulatory regions on the 5' side of the mouse E alpha gene. *Proc. Natl Acad. Sci. USA* **85:** 3075–3079.

Ting JP, Baldwin AS. (1993) Regulation of MHC gene expression. *Curr. Opin. Immunol.* **5:** 8–16.

Ting JP, Painter A, Zeleznik Le NJ, MacDonald G, Moore TM, Brown A, Schwartz BD. (1994) YB-1 DNA-binding protein represses interferon gamma activation of class II major histocompatibility complex genes. *J. Exp. Med.* **179:** 1605–1611.

Todd I, Pujol Borrell R, Hammond LJ, Bottazzo GF, Feldmann M. (1985) Interferon-gamma induces HLA-DR expression by thyroid epithelium. *Clin. Exp. Immunol.* **61:** 265–273.

Todd JA, Bell JI, McDevitt HO. (1987) HLA-DQ beta gene contributes to susceptibility and resistance to insulin-dependent diabetes mellitus. *Nature* **329:** 599–604.

Torok-Storb B, Symington FW. (1986) Class II MHC antigens and erythropoiesis. In: *HLA Class II Antigens* (eds BG Solheim, E Moller, S Ferrone). Springer Verlag, Heidelberg, pp. 302–411.

Triebel F, de Roquefeuil S, Blanc C, Charron DJ, Debre P. (1986) Expression of MHC class II and Tac antigens on IL2-activated human T cell clones that can stimulate in MLR, AMLR, PLT and can present antigen. *Hum. Immunol.* **15:** 302–315.

Trowsdale J. (1987) Genetics and polymorphism: class II antigens. *Br. Med. Bull.* **43:** 15–36.

Tsang SY, Nakanishi M, Peterlin BM. (1988) B-cell-specific and interferon-gamma-inducible regulation of the HLA-DR alpha gene. *Proc. Natl Acad. Sci. USA* **85:** 8598–8602.

Tsang SY, Nakanishi M, Peterlin BM. (1990) Mutational analysis of the DRA promoter: cis-acting sequences and trans-acting factors. *Mol. Cell. Biol.* **10:** 711–719.

Uitdehaag BM, de Groot CJ, Kreike A, Van der Meide PH, Polman CH, Dijkstra CD. (1993) The significance of in-situ Ia antigen expression in the pathogenesis of autoimmune central nervous system disease. *J. Autoimmun.* **6:** 323–335.

van Ewijk W, Ron Y, Monaco J, Kappler J, Marrack P, Le Meur M, Gerlinger P, Durand B, Benoist C, Mathis D. (1988) Compartmentalization of MHC class II gene expression in transgenic mice. *Cell* **53:** 357–370.

van Kerckhove C, Luyrink L, Elma MS, Maksymowych WP, Levinson JE, Larson MG, Choi E, Glass DN. (1990) HLA-DP/DR interaction in children with juvenile rheumatoid arthritis. *Immunogenetics* **32:** 364–368.

Venkitaraman AR, Culbert EJ, Feldmann M. (1987) A phenotypically dominant regulatory mechanism suppresses major histocompatibility complex class II gene expression in a murine plasmacytoma. *Eur. J. Immunol.* **17:** 1441–1446.

Vilen BJ, Penta JF, Ting JP. (1992) Structural constraints within a trimeric transcriptional regulatory region. Constitutive and interferon-gamma-inducible expression of the HLA-DRA gene. *J. Biol. Chem.* **267:** 23728–23734.

Viville S, Jongeneel V, Koch W, Mantovani R, Benoist C, Mathis D. (1991) The E alpha promoter: a linker-scanning analysis. *J. Immunol.* **146:** 3211–3217.

Voliva CF, Tsang S, Peterlin BM. (1993) Mapping cis-acting defects in promoters of transcriptionally silent DQA2, DQB2, and DOB genes. *Proc. Natl Acad. Sci. USA* **90:** 3408–3412.

Wagner CR, Vetto RM, Burger DR. (1984) The mechanism of antigen presentation by endothelial cells. *Immunobiology* **168:** 453–469.

Wang Y, Peterlin BM. (1986) Methylation patterns of HLA-DR alpha genes in six mononuclear cell lines. *Immunogenetics* **24:** 298–303.

Wang Y, Larsen AS, Peterlin BM. (1987) A tissue-specific transcriptional enhancer is found in the body of the HLA-DR alpha gene. *J. Exp. Med.* **166:** 625–636.

Wassmuth R, Lernmark A. (1989) The genetics of susceptibility to diabetes. *Clin. Immunol. Immunopathol.* **53:** 358–399.

Welsh J, Sapatino B, Rosenbaum B, Smith R, Linthicum S. (1993) Correlation between susceptibility to demyelination and interferon-gamma induction of major histocompatibility complex class II antigens on murine cerebrovascular endothelial cells. *J. Neuroimmunol.* **48:** 91–97.

Widera G, Burkly LC, Pinkert CA, Bottger EC, Cowing C, Palmiter RD, Brinster RL, Flavell RA. (1987) Transgenic mice selectively lacking MHC class II (I-E) antigen expression on B cells: an in vivo approach to investigate Ia gene function. *Cell* **51:** 175–187.

Williams K, Jr, Ulvestad E, Cragg L, Blain M, Antel JP. (1993) Induction of primary T cell responses by human glial cells. *J. Neurosci. Res.* **36:** 382–390.

Winchester RJ, Ross GD, Jarowski CI, Wang CY, Halper J, Broxmeyer HE. (1977) Expression of Ia-like antigen molecules on human granulocytes during early phases of differentiation. *Proc. Natl Acad. Sci. USA* **74:** 4012–4016.

Wollman EE, Cohen D, Fradelizi D, Sasportes M, Dausset J. (1980) Different stimulating capacity of B and T lymphocytes in primary and secondary allogeneic reactions: cellular detection of HLA-D products on T lymphocytes. *J. Immunol.* **125:** 2039–2043.

Wong GH, Bartlett PF, Clark Lewis I, Battye F, Schrader JW. (1984) Inducible expression of H-2 and Ia antigens on brain cells. *Nature* **310:** 688–691.

Woodward JG, Omer KW, Stuart PM. (1989) MHC class II transcription in different mouse cell types. Differential requirement for protein synthesis between B cells and macrophages. *J. Immunol.* **142:** 4062–4069.

Wright JR, Jr, Lacy PE, Unanue ER, Muszynski C, Hauptfeld V. (1986) Interferon-mediated induction of Ia antigen expression on isolated murine whole islets and dispersed islet cells. *Diabetes* **35:** 1174–1177.

Wright KL, Ting JP. (1992) In vivo footprint analysis of the HLA-DRA gene promoter: cell-specific interaction at the octamer site and up-regulation of X box binding by interferon gamma. *Proc. Natl Acad. Sci. USA* **89:** 7601–7605.

Yamamura K, Kikutani H, Folsom V, Clayton LK, Kimoto M, Akira S, Kashiwamura S,

Tonegawa S, Kishimoto T. (1985) Functional expression of a microinjected Ed alpha gene in C57BL/6 transgenic mice. *Nature* **316:** 67–69.

Yang Z, Sugawara M, Ponath PD, Wessendorf L, Banerji J, Li Y, Strominger JL. (1990) Interferon gamma response region in the promoter of the human DPA gene. *Proc. Natl Acad. Sci. USA* **87:** 9226–9230.

Yu LP, Sheehy MJ. (1991) The cryptic HLA-DQA2 ("DX alpha") gene is expressed in human B cell lines. *J. Immunol.* **147:** 4393–4397.

Zeleznik Le NJ, Azizkhan JC, Ting JP. (1991) Affinity-purified CCAAT-box-binding protein (YEBP) functionally regulates expression of a human class II major histocompatibility complex gene and the herpes simplex virus thymidine kinase gene. *Proc. Natl Acad. Sci. USA* **88:** 1873–1877.

Zeleznik Le NJ, Itoh Lindstrom Y, Clarke JB, Moore TL, Ting JP. (1992) The B cell-specific nuclear factor OTF-2 positively regulates transcription of the human class II transplantation gene, DRA. *J. Biol. Chem.* **267:** 7677–7682.

Zhou H, Glimcher LH. (1995) Human MHC class II gene transcription directed by the carboxyl terminus of CIITA, one of the defective genes in type II MHC combined immune deficiency. *Immunity* **2:** 545–553.

Zhu L, Jones PP. (1990) Transcriptional control of the invariant chain gene involves promoter and enhancer elements common to and distinct from major histocompatibility complex class II genes. *Mol. Cell. Biol.* **10:** 3906–3916.

Zier K, Gansbacher B, Ikegaki N, Kennett R, Polakova K. (1989) Expression of HLA-DR mRNA in T cells following activation is early and can precede DNA synthesis. *Autoimmunity* **5:** 59–70.

Zimmer T, Jones PP. (1990) Combined effects of tumor necrosis factor-alpha, prostaglandin E2, and corticosterone on induced Ia expression on murine macrophages. *J. Immunol.* **145:** 1167–1175.

Zlotnik A, Shimonkevitz RP, Gefter ML, Kappler J, Marrack P. (1983) Characterization of the gamma-interferon-mediated induction of antigen-presenting ability in P388D1 cells. *J. Immunol.* **131:** 2814–2820.

Zlotnik A, Fischer M, Roehm N, Zipori D. (1987) Evidence for effects of interleukin 4 (B cell stimulatory factor 1) on macrophages: enhancement of antigen presenting ability of bone marrow-derived macrophages. *J. Immunol.* **138:** 4275–4279.

9

Cell biology of MHC class I molecules

Vincenzo Cerundolo and Veronique Braud

9.1 Introduction

Analysis of the peptides eluted from major histocompatibility complex (MHC) class I and class II molecules showed that there are two different intracellular pools from which MHC class I and class II molecules can draw peptides for binding. The majority of MHC class II molecules are loaded with peptides derived from glycoproteins or secreted proteins, whereas the majority of MHC class I molecules draw peptides from cytosolic proteins (Chicz *et al.*, 1993; Jardetzky *et al.*, 1991; Hunt *et al.*, 1992). Although several exceptions have been described, this dichotomy of antigen presentation by MHC class I and class II molecules is central to the current understanding of T-cell immune response. The aim of this review is to explain this general rule in terms of the cell biology of MHC class I molecules. In the first part of the chapter we shall describe the mechanisms by which peptides are generated and transported into the endoplasmic reticulum (ER). We shall then analyse the folding and assembly of the class I/β2-microglobulin (β2-m) complex, and describe the interaction of class I molecules with other proteins within the lumen of the ER. Finally, we shall describe how several viruses may escape cytotoxic T-cell (CTL) recognition by interfering with the class I antigen presentation pathway.

9.2 Generation of MHC class I bound peptides

9.2.1 Cytosol as the primary site for antigen processing

Class I heavy chains are highly polymorphic transmembrane glycoproteins of 45 kDa encoded in the MHC. During biosynthesis, class I heavy chains associate with β2-m in the ER and the heterodimer is transported through the Golgi to the cell surface. Zinkernagel and Doherty (1974) demonstrated that virus-specific CTL kill virus-infected cells only if they share the same MHC class I allele with target cells. The nature of this association, defined as MHC class I-restricted killing, remained obscure for several years. The generally held view was that virus-specific CTL could recognize a binary complex made of MHC class I molecules and a viral protein on the surface of target cells. This model implied that only surface-expressed viral proteins could be seen by CTL. Further experiments did not

HLA and MHC: genes, molecules and function, edited by M.J. Browning and A.J. McMichael.
© 1996 BIOS Scientific Publishers Ltd, Oxford.

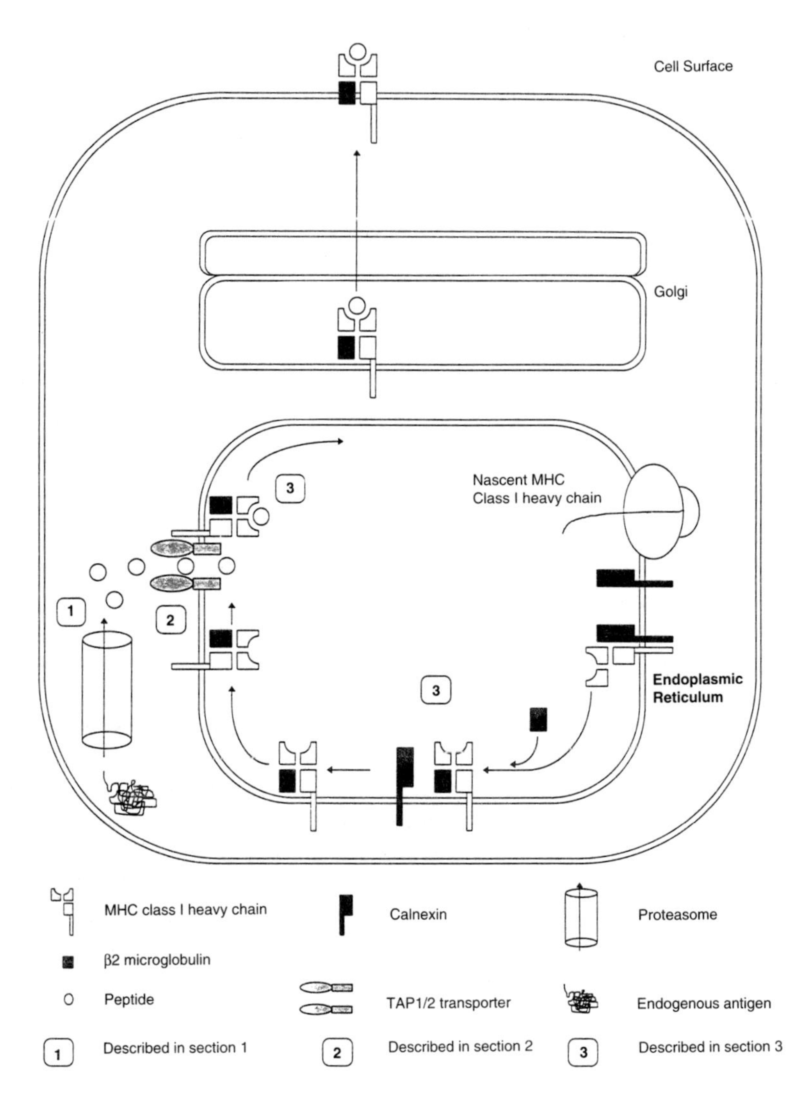

Figure 9.1. MHC class I assembly in human cells.

confirm this hypothesis. In 1985, Townsend and colleagues demonstrated that cytosolic fragments of the influenza nucleoprotein (NP), lacking known signal sequences to cross the ER membrane, sensitized target cells for lysis by influenza specific CTL (Townsend *et al.*, 1985). These results were extended in 1986 by showing that synthetic peptides, corresponding to the influenza NP sequence, could sensitize target cells for lysis by NP-specific CTL (Townsend *et al.*, 1986a). Similar results were obtained with glycoproteins, lacking the N-terminal hydrophobic signal sequence normally required for co-translational transport into the ER lumen (Townsend *et al.*, 1986b). Cells infected with a recombinant vaccinia virus coding for a signal sequence-deleted haemagglutinin (HA) were very efficiently recognized as targets by HA-specific CTL. In these experiments, the signal-sequence deleted HA was rapidly degraded in the cytosol, with a half-life of less than 30 minutes, compared with greater than 4 hours for wild-type HA. The findings that killed virus particles or purified viral proteins did not sensitize target cells for class I-restricted mediated lysis, although they did sensitize target cells for class II-restricted recognition (Morrison *et al.*, 1986) suggested that two pathways for presenting antigens to T cells existed: one dealing with newly synthesized protein of virus-infected cells, the other concerned with endocytosed soluble proteins. Although several exceptions were described (Kovacsovics-Bankowski and Rock, 1995; Stearz *et al.*, 1988; Tevethia *et al.*, 1980; Wraith and Vessey, 1986; Yide *et al.*, 1988), the majority of the results obtained to date are consistent with this model.

These results led to the hypothesis that the majority of class I-restricted CTL recognize degraded fragments of proteins that have been exposed to the cytosol of target cells. This presentation pathway, which is quite distinct from the pH-dependent presentation of exogenous proteins (see Chapter 10), requires three components which will be analysed in the next sections (*Figure 9.1*):

(i) a degradation system that generates peptides able to bind to MHC class I molecules;
(ii) a transport system that translocates peptides from the cytosol to the ER;
(iii) binding of peptides to MHC class I molecules.

9.2.2 Degradation of CTL target proteins

The role of the cytosol. The nature of the cytosolic protease(s) that generates CTL epitopes is not known. Unlike lysosomal degradation, where selective degradation is controlled by sorting of proteins to the lysosomes, the mechanisms that control protein half-life in the cytosol, and generation of T-cell epitopes are still ill-defined events. The hallmark of class I-restricted presentation is its resistance to pharmacological agents that inhibit lysosomal degradation by raising lysosome pH. A series of peptide aldehydes were recently described that were able to block both cytosolic and lysosomal degradation. Experiments carried out using these inhibitors confirmed that proteolysis is required for the generation of class I-bound peptides (Rock *et al.*, 1994). However, as these protease inhibitors lack specificity, it is very difficult to draw any conclusion about the nature of the protease(s) involved in the cytosolic generation of T-cell epitopes. Attention has recently been focused on a large multi-subunit cytosolic endo-peptidase, called the proteasome. Proteasomes have a broad proteolytic activity, degrading bonds on the carboxyl side of basic, hydrophobic and acidic residues (Goldberg and Rock, 1992). Two proteasome subunits, called LMP2 and LMP7, are encoded in the MHC, closely linked to the peptide transporter (*TAP*) genes (Glynne *et al.*, 1991; Kelly *et al.*, 1991). The close association of *LMP2* and *LMP7* with *TAP1* and *TAP2* genes, and the finding that all four genes can be up-regulated by gamma interferon (IFNγ), suggest that products of these genes might play a role in the antigen presentation pathway. Binding of LMP2 and LMP7 to the proteasome displaces two housekeeping proteasome subunits called MB1 and δ in human cells (Belich *et al.*, 1994), and subunit 2 and 10 in mouse cells (Fruh *et al.*, 1994). It was shown that displacement of these housekeeping proteasome subunits by LMP2 and LMP7 alters the cleavage specificity of purified proteasome. The rate of cleavage after hydrophobic and basic residues is increased, whereas the rate of cleavage after acidic residues is reduced (Driscoll *et al.*, 1993; Gaczynska *et al.*, 1993).

Although a role of LMP2 and LMP7 in antigen presentation was initially called into question (Arnold *et al.*, 1992; Momburg *et al.*, 1992; Yewdell *et al.*, 1994), further experiments carried out, both with LMP knockout mice (Fehling *et al.*, 1994; van Kaer *et al.*, 1994) and with human (Cerundolo *et al.*, 1995) and mouse (Sibille *et al.*, 1995) cells bearing a deletion of *LMP2* and/or *LMP7* genes, demonstrated that cells lacking LMP2 and/or LMP7 have a defect in antigen presentation. This defect is selective for certain epitopes, whereas other epitopes are efficiently generated. The mechanisms by which LMP products control the generation of certain CTL epitopes is not known. It is of interest that the lack of presentation of the 'LMP dependent' epitopes cannot be overcome by increasing the rate of degradation of the target protein (Cerundolo *et al.*, 1995). This finding suggests that LMP-deficient cells have a qualitative change in cleavage specificities that interferes with presentation of certain epitopes. This possibility might be addressed by site-directed mutagenesis of CTL flanking regions.

One mechanism known to regulate protein half-life, both in prokaryotic and eukaryotic cells, was initially described by Varshavsky and colleagues (Bachmair *et al.*, 1986). The results of these studies demonstrated that the amino terminus of a protein may play an important role in determining its metabolic stability. This amino-terminal degradation signal, defined as N-end rule, comprises two distinct determinants, each of which is necessary but by itself not sufficient to render the protein metabolically unstable. One determinant is the protein amino-terminal residue. The second determinant identified is a specific internal lysine residue, which is covalently bound to a 7 kDa protein called ubiquitin (Hershko and Ciechanover, 1992). Ubiquitinated proteins are destined for rapid degradation, and are degraded by the proteasome (Goldberg and Rock, 1992). Whether ubiquitination plays a role in antigen presentation remains to be demonstrated. In 1988, Townsend and his colleagues showed that the half-life of influenza NP is controlled by the amino-terminal residue (Townsend *et al.*, 1988). Ubiquitin was linked to influenza NP via either an arginine (which is an N-end rule 'de-stabilizing' residue) or via a methionine (which is an N-end rule 'stabilizing' residue). Cleavage of ubiquitin by the ubiquitin C-terminal hydrolase exposes the two different N-terminal residues. Analysis of the half-life of the two constructs showed that influenza NP conjugated to ubiquitin via an arginine at position 1 has a half-life shorter than 30 minutes, whereas NP preceded by a methionine has a half-life longer than 4 hours. These results demonstrate that a single amino acid change (Met to Arg), linking ubiquitin to NP, is sufficient to destabilize NP. Although these results are consistent with ubiquitin-dependent degradation playing a role in antigen presentation, further evidence needs to be obtained to prove that the enhanced NP degradation is actually ubiquitin-mediated. To address a role of ubiquitination in antigen presentation, Rock and colleagues (Michalek *et al.*, 1993) carried out antigen presentation experiments using the temperature-sensitive mutant cell line E36 ts20 (Kulka *et al.*, 1988). This cell line has a mutant thermolabile ubiquitin-activating enzyme E1, which fails to degrade short-lived proteins at the restrictive temperature. Incubation of E36 ts20 cells at the non-permissive temperature (43°C), for more than 1 hour, inactivates E1 and reduces the efficiency of ubiquitination and rate of degradation of protein substrates. Rock and colleagues showed that, after treating E36 ts20 cells at 43°C, micro-injected ovalbumin was not able to sensitize the target cells for lysis by ovalbumin-specific CTL (Michalek *et al.*, 1993). As a control, the authors showed that presentation of the vaccinia-encoded peptide epitope was not impaired at the non-permissive temperature. These results are consistent with ubiquitination of ovalbumin being involved in the generation of the CTL epitope. The significance of these results, however, was recently called into question by further experiments carried out with the same cells by Cox and colleagues (Cox *et al.*, 1995), who found that presentation of endogenous antigens was not affected by inactivation of E1 ubiquitin-activating enzyme.

The role of the ER in the processing of CTL epitopes. It is still not clear whether antigen presentation is completed in the cytosol or whether peptides longer than the optimum length can be transported into the lumen of the ER, and then trimmed by ER-resident peptidases. Indeed, *in vitro* transport assays have shown that the peptide transporter can transport peptides up to 40 amino acids long. These results suggest that class I molecules are offered a mixture of different length peptides from which they select the peptides with the highest binding affinity. Consistent with a role of the ER in antigen presentation, it was previously shown that peptides derived from the signal sequence can bind to HLA-A2 class I molecules (Henderson *et al.*, 1992; Wei and Cresswell, 1992). These results suggest that the signal peptidase is capable of generating peptides that bind to MHC class I molecules.

The role of the ER in the processing of CTL epitopes was directly addressed by two recent papers. Snyder and colleagues (1994) made a recombinant vaccinia virus encoding a tandem of two CTL epitopes preceded by a signal sequence to introduce them into the ER of TAP negative cells. The results of these experiments demonstrated that only the C-terminal epitope was presented. Although it is possible that the N-terminal epitope is destroyed by a 'nibbling' effect of the signal peptidase, these results are consistent with an ER aminopeptidase which can efficiently generate optimal length peptides in the lumen of the ER. Similar conclusions were derived from the results obtained by Elliott and colleagues (1995). Peptides extended at the N-terminus were co-translationally synthesized in the lumen of the ER of TAP-negative cells by tagging them to the haemagglutinin signal sequence (Elliott *et al.*, 1995). Combined high-pressure liquid chromatography (HPLC) purification of peptides eluted from class I molecules and CTL recognition showed that only the optimal length peptides were bound to class I molecules. The targeting in the ER of a much longer fragment of the influenza NP (170 amino acids) gave similar results. These findings are consistent with an endopeptidase, as well as an aminopeptidase, involved in the generation of T cell epitopes within the lumen of ER. It is of interest that expression in the ER of full-length NP failed to sensitize TAP-negative cells for lysis by H-2 D^b and K^k restricted CTL, whereas the NP epitope presented through HLA-B8 was efficiently generated. Lack of killing was also observed after expressing influenza haemagglutinin in TAP-negative cells. These findings demonstrate that the generation in the lumen of the ER of CTL epitopes contained within full-length proteins is not very efficient, and might be selective for certain epitopes. The nature of the ER resident peptidase remains to be defined. It will be informative to study whether the peptide aldehydes, shown to block cytosolic degradation, may also interfere with the proteolytic activity of ER-resident proteases.

9.3 Transport of peptides from the cytosol to the ER

9.3.1 Peptide transporter-deficient cells

Until recently, association of class I heavy chains with β2-m was thought to be independent of peptide binding to class I molecules. The generally held view was that folding of class I heavy chains, assembly with β2-m, and transport of the class I–β2-m complex to the cell surface were not linked to processing of intracellular proteins and generation of peptides. Over the last five years it has become clear that class I assembly and peptide binding to class I molecules are coupled phenomena. These findings were made possible by the use of a series of processing mutant cells, which have deletions or mutations of the genes encoding for the peptide transporter proteins (TAP1/2 complex). Mouse (Townsend *et al.*, 1989) and human cells (Cerundolo *et al.*, 1990; Kelly *et al.*, 1992; Spies *et al.*, 1992) lacking either or both TAP1 and TAP2 molecules have been described (*Figure 9.2*). Both mouse and human TAP-deficient cells have a similar phenotype:

(i) they are not able to present intracellular antigens to class I-restricted CTL, although they are able to present exogenously added synthetic peptides;
(ii) newly synthesized class I molecules are unstable, and after cell lysis they are not detectable by conformation-specific antibodies in the absence of peptide. Addition of MHC class I peptide ligands to whole cells or to cell lysates stabilizes the folded conformation of class I molecules.

Two conclusions were drawn from the analysis of the phenotype of these cells:

(i) the results led to the hypothesis that the stability of class I–β2-m complex depends on the occupation of the binding site by a peptide ligand. This hypothesis was consistent with the three-dimensional crystal structure of class I molecules, showing that peptide ligands are deeply embedded in the antigen-binding groove, and that the class I α1 and α2 domains make contacts with several residues of the β2-m;
(ii) the phenotype of these cells was consistent with a shortage of class I-binding peptides in the lumen of the ER. This phenotype could be accounted for by a loss of a transport mechanism required to translocate peptides from the cytosol into the lumen of the ER. The analysis of the human cell lines LBL 721.174 and Cresswell's derivative T2 gave the opportunity to demonstrate that the gene(s) controlling this phenotype mapped in the class II region of the MHC (Cerundolo *et al.*, 1990).

Further evidence that genes encoded in the MHC control presentation of intracellular antigens has come from studies with rat class I molecules. Livingstone and colleagues described a rat MHC-linked locus (named *cim* for class I modifier), mapped to the rat MHC class II region, which controls the alloantigenic structure and rate of intracellular transport of the rat class I molecule RT1A.[a] (Livingstone *et al.*, 1989). Different alleles of the *cim* locus (*cim a* and *cim b*) regulate the specificity of class I antigen presentation and the rate of transport out of the ER:

(i) RT1A.[a]-specific CTL can distinguish between RT1A.[a] synthesized in a strain coding for the *cim a* or *cim b* allele;

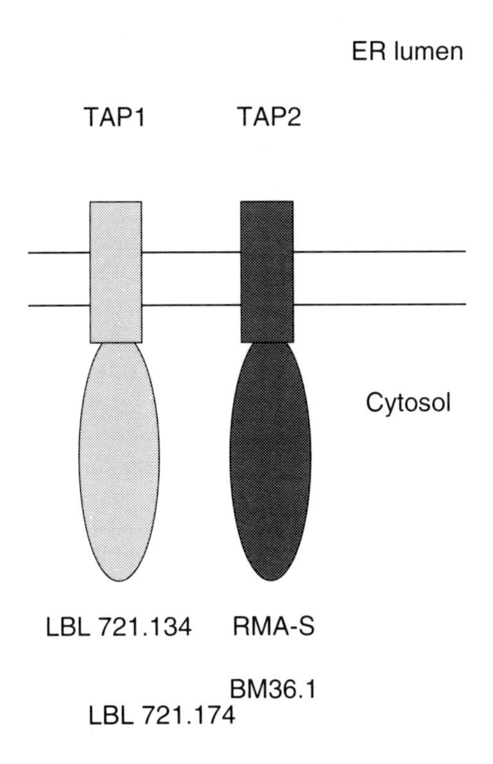

Figure 9.2. Peptide-transporter-deficient cells. Processing mutant cells lacking either TAP1 (LBL 721.134) or TAP2 molecules (RMA-S, BM36.1) or both (LBL 721.174, T2).

(ii) RT1A.[a] shows fast transport kinetics out of the ER in a *cim a* strain, whereas the same class I molecule shows a much slower rate of transport in a *cim b* strain;
(iii) F1 hybrids (*cim a* strain crossed with *cim b* strain) show that *cim a* is dominant over *cim b*.

The findings that genes in the rat and human MHC control antigen processing and presentation led one year later to the identification and cloning of a cluster of genes in the MHC of human, rat and mouse that are structurally related to the ATP-binding cassette (ABC) superfamily of transporters (Deverson *et al.*, 1990; Monaco *et al.*, 1990; Spies *et al.*, 1990; Trowsdale *et al.*, 1990). The product of these genes was named TAP for peptide Transporter Associated with Antigen Processing. Further experiments demonstrated that transfection of these genes restored the normal phenotype in the mutant cells. Similar to the phenotype described in the TAP-deficient cells, mice deficient in TAP1 alone (Van Kaer *et al.*, 1992) lack stable surface expression of H-2$K^b$and H-2D^b molecules. Addition of peptide to splenocytes from these mice stabilizes class I molecules at the cell surface. These results indicate that TAP1 is essential *in vivo* for expression of stable class I molecules on the surface of cells. The finding that TAP1-deficient mice have a severe reduction in the number of $CD4^-8^+$ lymphocytes in both central and peripheral lymphoid organs, suggests a role of the TAP complex in the generation of class I-restricted T-cell repertoire, probably through its function in peptide supply within the thymus during T-cell ontogeny. The role of the TAP transporter in the immune response has recently been confirmed by the report of a case of primary immunodeficiency (De la Salle *et al.*, 1994) due to a single mutation, leading to a premature stop codon in the *TAP2* gene.

9.3.2 Structure and polymorphism of the peptide transporter

The TAP complex is made of two proteins (TAP1 and TAP2) both of which are encoded in the MHC. The product of these genes belongs to the ABC superfamily. The ABC transport superfamily includes over 30 members. All of them are closely related in terms of domain organization, structure, ATP hydrolysis and probably evolutionary origin. Immunoprecipitation of the TAP1 protein, using an antibody raised to a peptide from the C-terminus of the TAP1 ATP-binding domain, demonstrated that TAP1 associates with TAP2 protein forming a complex (Kelly *et al.*, 1992; Spies *et al.*, 1992). These results strongly support the notion that, like other members of the ATP superfamily, the functional TAP protein is a heterodimer with two transmembrane and two ATP-binding domains.

TAP1 and TAP2 each consists of one half of a typical member of the ABC transporter protein: one hydrophobic domain and one ATP-binding domain. The human *TAP1* gene encodes a protein of 808 amino acids The known *TAP1* alleles differ at two coding positions (333 I/V and 637 D/G). Four allelic forms of the *TAP2* gene have been identified with substitutions at position 379 V/I, 565 A/T, 665 T/A and 687 Q/STOP. The ATP-binding domain of TAP1 and TAP2 share 61% homology and the hydrophobic domains share 30% homology. No functional differences have been associated with any of the allelic forms of TAP1 and TAP2.

The genetic analysis of the processing mutant cells gave the opportunity to demonstrate that deletion or mutation of either TAP1 and/or TAP2 proteins can determine loss of presentation of viral antigens and unstable class I–β2-m complexes. Although some exceptions were described (Esquivel *et al.*, 1992; Hosken and Bevan, 1992; Zhou *et al.*, 1992), the majority of viral epitopes analysed to date appear to require a functional TAP complex to be presented at the cell surface.

9.3.3 Analysis of TAP-dependent peptide translocation

Definitive evidence that the TAP complex translocates peptides from the cytosol to the ER was provided by experiments carried out using two *in vitro* transport assays.

In the first assay permeabilized cells with the bacterial toxin streptolysin-O were used. This toxin makes holes in the cytosolic membrane, thus allowing access to the cytosol without penetrating the ER membrane (Androlewicz *et al.*, 1993; Neefjes *et al.*, 1993a). This property of streptolysin O gives the opportunity to add iodinated peptides into the cytosol and then measure their transport into the ER. In order to assess transport in the ER, peptides containing an N-linked glycosylation site (NST) were used (Neefjes *et al.*, 1993a). The efficiency of peptide translocation in the ER was measured by purifying the radiolabelled peptides with Con-A sepharose beads. The second transport assay is based on the translocation of labelled peptide through microsome preparations from livers of TAP-positive mice (Shepherd *et al.*, 1993). Livers from TAP-negative mice were used as a control.

Experiments carried out using these two assays showed that TAP transport is substrate-specific and ATP-dependent (Androlewicz *et al.*, 1993; Momburg *et al.*, 1994a, 1994b; Neefjes *et al.*, 1993a; Shepherd *et al.*, 1993).

ATP-dependence of peptide translocation. Circumstantial evidence supporting a role for the ATP-binding domain in the TAP complex was previously provided by the analysis of the antigen-processing mutant lymphoblastoid cell line BM36.1 (Kelly *et al.*, 1992). The TAP2 protein in BM36.1 cells was shown to be larger than the parental TAP2 by approximately 6 kDa, suggesting that it may be a mutant form. Sequence analysis of *TAP2* cDNA from BM36.1 revealed a deletion of 2 bp 3′ to the Walker B ATP-binding domain, causing a frameshift with replacement of 52 amino acids in the ATP-binding domain, and a 51 amino acid extension of the protein. Transfection of the normal *TAP2* gene into BM36.1 reversed the mutant phenotype, and the wild-type protein competed with the elongated mutant form for binding to the TAP1 protein. Presentation of viral antigens and proportion of stable class I molecules (Kelly *et al.*, 1992) were restored to levels comparable to the parental line BM28.1. These results showed that the mutation in the ATP-binding domain of the TAP2 protein is responsible for the mutant phenotype of BM36.1 cells, which might result from loss of ATP hydrolysis required for its function *in vivo*.

Experiments carried out using *in vitro* peptide transport assays demonstrated that hydrolysis of ATP is required for TAP-dependent peptide translocation across the ER membrane (Androlewicz *et al.*, 1993; Neefjes *et al.*, 1993a; Shepherd *et al.*, 1993). It was shown that 1 mM ATP, but not ATPγ S, drives accumulation of peptide in microsomes from TAP1+/+ mice. In the presence of 50 μM ATP and an ATP-regenerating system, microsomes from TAP1+/+ mice translocated fourfold more peptide than microsomes from TAP−/− mice. Depletion of endogenous ATP by addition of apyrase confirmed a role of ATP hydrolysis in TAP-dependent peptide translocation (Androlewicz *et al.*, 1993).

To study whether ATP hydrolysis is required both for the binding of peptides to the TAP complex and for the peptide translocation across the ER membrane, experiments were carried out using cross-linkable peptides. These experiments demonstrated that, in the absence of ATP, radiolabelled peptides can be cross-linked to the TAP complex. Analysis of binding of photolabelled peptide to the TAP complex on TAP-transfected cell lines demonstrated that the formation of a peptide-binding site occurs only in the presence of both TAP1 and TAP2 molecules. It is of interest that TAP1 and TAP2 are differentially labelled, depending upon the photolabelled peptide used (Androlewicz *et al.*, 1994). Similar results were obtained by Van Endert and colleagues by expressing individual TAP1 and TAP2 chains in insect cells, and comparing peptide translocation at 4°C and 37°C (Van Endert *et al.*, 1994). Scatchard analysis of a saturation-binding experiment using the peptide RRYNASTEL gave a straight line, corresponding to a binding affinity of 4.1×10^{-7} M. These results argue that a functional TAP is composed of a TAP1 and TAP2 heterodimer, and that expression of either TAP1 or TAP2 is non-functional. These conclusions are consistent with the results obtained from the analysis of TAP-deficient processing mutant cells.

Selectivity of TAP-dependent peptide translocation. The first evidence that allelic forms of the peptide transporter translocate a different pool of peptides was provided by experiments carried out using rat cells with different alleles of the rat peptide transporters. It was shown that rat *TAP2* alleles (*a* allele and *b* allele) control the HPLC profiles of peptides eluted from the rat class I molecules RT1.A^{a} (Powis *et al.*, 1992). This finding, which may explain the previously described *cim* phenomenon (see above), was consistent with a functional polymorphism of the rat transporter at the level of the peptide transport in the ER.

Further experiments carried out using *in vitro* transport assays confirmed these results. The specificity of the peptide transporter for different peptides was assessed by studying the translocation of an index peptide, and comparing the amount of transported peptide in the presence of competitor peptides. These experiments showed that the peptide transporter has a substrate specificity, and has a preference for peptides of 8–11 amino acids. The efficiency of peptide translocation is significantly reduced with peptides shorter than eight residues and peptides longer than 16 amino acids (Androlewicz and Cresswell, 1994; Momburg *et al.*, 1994a, b; Shepherd *et al.*, 1993). The length preference of translocated peptides may differ for allelic variants of TAP complex. It was shown that the rat *TAP^{b}* allele is more permissive than the *TAP^{a}* allele, as it can efficiently translocate peptides up to 12–14 amino acids (Heemels and Ploegh, 1993).

Efficiency of peptide translocation is also controlled by the amino- and carboxyl-terminal amino acid residues. Peptides ending with an acidic residue are not efficiently transported by human, rat and mouse *TAP* alleles. Human *TAP* alleles and the rat *TAP^{a}* allele translocate more efficiently peptides with a hydrophobic and basic carboxyl-terminus, whereas mouse *TAP* alleles and the rat *TAP^{b}* allele translocate more efficiently peptides with a hydrophobic C-terminus (Heemels *et al.*, 1993; Momburg *et al.*, 1994a; Schuhmacher *et al.*, 1994). Peptide transport of optimal length peptides requires free NH_2- and COOH- peptide termini. Indeed, acetylation and amidation of the peptide N- and C-termini, respectively, reduces the efficiency of peptide translocation (Momburg *et al.*, 1994a). It has been shown, however, that N-formylated peptides can be efficiently presented by H-2M3 class I molecules in a TAP-dependent fashion (Attaya *et al.*, 1992; Fischer *et al.*, 1991). It is becoming clear that peptides with an optimal C-terminal residue are not necessarily efficiently transported. Indeed, residues other than the C- and N-termini may contribute to the overall specificity of peptide translocation. It was shown that mouse transporters do not efficiently translocate 9mer peptides which have a proline at position 3 (Deres *et al.*, 1992; Heemels and Ploegh, 1994; Neisig *et al.*, 1995). As several CTL epitopes have a proline at position 3, it is possible that transport of longer peptides might be a way to overcome the less efficient transport of peptide with a proline at position 3. Indeed, it was shown that addition of natural flanking sequences to poorly transported peptides, but immunodominant epitopes, might increase the efficiency of translocation (Neisig *et al.*, 1995). These results are consistent with the hypothesis that peptides longer than the optimal length might be transported and then trimmed in the ER to generate optimal length peptides.

Although the significance of these results has not been confirmed with antigen presentation experiments, the substrate specificity of human and mouse *TAP* alleles appears to correlate with peptide-binding motifs and length requirement of human and mouse class I molecules. Indeed, mouse class I molecules bind peptides which have hydrophobic residues at position 9, whereas human class I molecules can bind peptides ending with hydrophobic or basic residues. This correlation fuelled the hypothesis that the C-terminus of class I binding peptides is optimized in the cytosol, whereas the N-terminus is trimmed in the lumen of the ER. No definitive evidence, however, is available to confirm this hypothesis.

9.4 Folding and assembly of MHC class I molecules

9.4.1 Interactions between free heavy chain, β2-m and peptide

We have discussed up to this point the mechanisms by which peptides might be generated in the cytosol and transported into the ER. It is not known, however, whether peptides interact with the α-1 and α-2 domain of class I molecules first, thus enabling β2-m to bind and stabilize the complex, or whether β2-m and heavy chain may associate without peptide and are then stabilized by the binding of those peptides that will fit into the groove (*Figure 9.3*).

Analysis of class I assembly in mutant cells (Cerundolo *et al.*, 1990; Townsend *et al.*, 1989, 1990) and in cells lacking β2-m (Elliott *et al.*, 1991) showed that both routes are possible *in vitro*. The results of these experiments demonstrated that:

(i) *association of heavy chain, β2-m and peptide is reversible and obeys the law of mass action.* The formation of the trimolecular complex (heavy chain + β2-m + peptide) is a multi-equilibrium reaction. Each equilibrium can be driven by increasing the concentration of each ligand *in vitro* (Elliott *et al.*, 1991; Townsend *et al.*, 1990);
(ii) *both longer peptides and optimal length peptides can bind to heavy chain–β2-m complexes* (Cerundolo *et al.*, 1991). Only optimal length peptides, however are able to bind to unfolded free heavy chain and trigger their conformational change (Elliott *et al.*, 1991) (equilibrium K3 of *Figure 9.3*);
(iii) *peptide and β2-m bind cooperatively to class I heavy chains.* Intermediate peptide-class I complexes and β2-m-class I complexes (*Figure 9.3*: equilibria K3 and K1 respectively) are unstable. The stability of each intermediate complex is increased by the association of the other ligand (*Figure 9.3*: equilibria K2 and K4). This property of class I–β2-m–peptide complex is consistent with the finding that, in cells lacking either a supply of peptide ligands in the lumen of the ER (TAP-negative cells) or β2-m, the majority of class I molecules are unstable, and after lysis rapidly lose a folded conformation. Addition of peptide ligands or β2-m into the lysate stabilizes the complex (Elliott *et al.*, 1991; Townsend *et al.*, 1990).

These results, initially obtained with class I molecules purified from detergent-lysed cells, were subsequently confirmed with molecules purified in detergent-free buffer using secreted molecules from *Drosophila* cells (Matsumura *et al.*, 1992), CHO cells (Fahnestock *et al.*, 1992, 1994) and *in vitro* translated class I molecules (Kvist and Hamann, 1990; Levy *et al.*, 1991). Thermal stability profiles of secreted

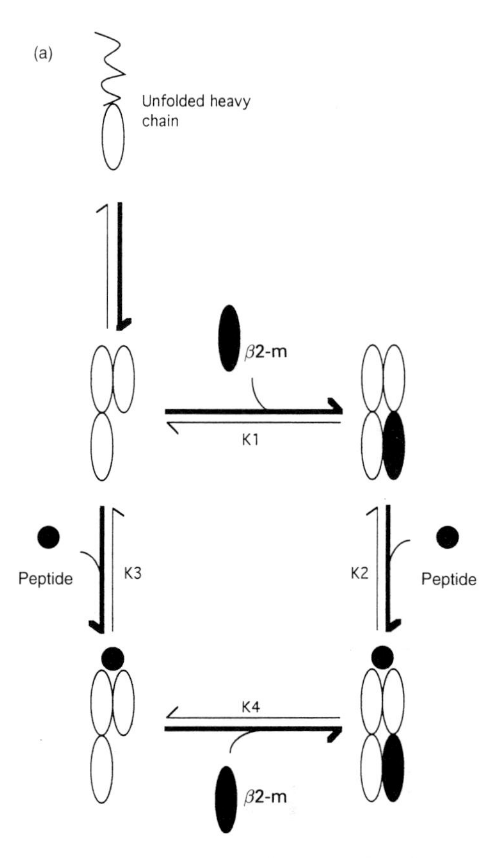

Figure 9.3. (a) Association of heavy chain, β2-microglobulin and peptide in normal cells.

empty K^d molecules were compared to the profiles of peptide-loaded K^d molecules. The results of these experiments demonstrated that occupation of K^d binding site by a peptide ligand increases the transition midpoint of the melting curve from 45°C to 57°C (Fahnestock *et al.*, 1992).

It is not known which route of assembly is followed *in vivo*. Association of MHC class I molecules was studied in intact cells with or without peptide transporter (Neefjes *et al.*, 1993b). Newly synthesized class I molecules were precipitated at different time points with the conformation-specific antibody W6/32. Free class I molecules were precipitated with the free heavy-chain-specific antibody HC10. The results of these experiments suggest that association of heavy chains with β2-m precedes binding of the peptides. This interpretation is consistent with the finding that empty class I–β2-m complexes are associated with the TAP complex (see *Figure 9.1*). It is not known, however, whether the conformation-specific antibody W6/32 can precipitate peptide-folded free heavy chains in the absence of β2-m. In contrast to the latter hypothesis, experiments carried out using β2-m-deficient mice demonstrated that functional D^b molecules can be detected on the surface of β2-m-negative cells (Bix and Raulet, 1992). Furthermore, β2-m negative cells can be lysed by alloreactive CTL (Bix and Raulet, 1992) and NP-specific CTL (Vitiello *et al.*, 1990) even in serum-free conditions. These results

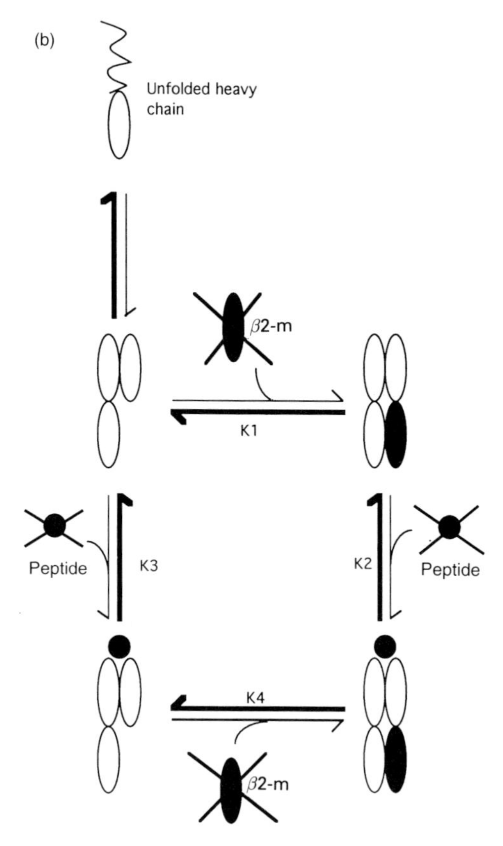

Figure 9.3. (b) Association of heavy chain, β2-m microglobulin and peptide in cells lacking either β2-m or TAP.

suggest that binding of peptide to free heavy chains can occur *in vivo* in the absence of β2-m. Further experiments supported the latter hypothesis through analysis of Con A-stimulated splenocytes from TAP-negative, β2-m-negative and double mutant TAP-negative and β2-m-negative mice (Machold *et al.*, 1995).

9.4.2 Kinetics of peptide binding to MHC class I molecules

Naturally bound peptides eluted from MHC class I molecules share similar properties:

(i) homogenous length (8–10 residues);
(ii) allele-specific binding motifs.

The majority of class I molecules purified from TAP-positive cells are occupied by endogenous peptides. The presence of naturally bound peptides associated to class I molecules prevents binding of exogenous peptides. Thus, the percentage of class I molecules purified from normal cells that can bind peptides is very low (Chen and Parham, 1989). The identification of TAP-negative cells gave the opportunity to purify a larger proportion of class I molecules able to bind exogenous peptides (Cerundolo *et al.*, 1991; Townsend *et al.*, 1989). The analysis of the peptide-binding kinetics to MHC class I molecules purified from the TAP-negative cells T2 has shown that both peptide length and binding motifs are important requirements for the binding affinity of the peptide–class I complex. Different length peptides can bind to class I–β2-m complexes, but only optimal length peptides form stable complexes with class I molecules. Addition or deletion of a single residue at the C- and N-terminus of an optimum length peptide greatly reduces both the binding affinity of the peptide and its capacity to form a stable complex with class I molecules (Cerundolo *et al.*, 1991). The rate of dissociation of peptides from the H-2 D^b class I molecule increases by 100-fold if the 9mer is extended by one residue on the C-terminus (Cerundolo *et al.*, 1991). Additions to the N-terminus appear to be tolerated slightly better. The rate of dissociation of the D^b-binding peptide NP366-374 extended by one or two residues at the N-terminus is approximately fivefold longer than extensions at the C-terminus. This observation was confirmed by measuring the dissociation rate from HLA-Aw68 molecules (data not shown), and from K^d class I molecules purified from CHO cells (Fahnestock *et al.*, 1994). Different results, however, were obtained by lengthening the N- and C-termini of K^b-binding peptides (Matsumura *et al.*, 1992). Two residue extension at the N-terminus of the VSV peptide (VSV 52–59) reduced the binding affinity by a factor of approximately 100, whereas two residue extensions at the C-terminus only reduced the affinity by a factor of approximately five. Similar results were obtained with the K^b-binding ovalbumin peptide 257–264 (Matsumura *et al.*, 1992). These findings indicate that the contribution of the precise positioning of a peptide N- and C-terminus for the formation of a stable class I–peptide complex may vary for different peptides and for different class I molecules. Measurements of binding kinetics of optimal length peptides to class I molecules purified from detergent-lysed TAP-negative cells gave a K_d value in the nM range at 4°C, and extremely slow off rates, corresponding to a half-life longer than 100 hours at 4°C, and of about 3 to 7 hours at 37°C. These values fell in the middle range of an extensive series of measurements done by other groups (Boyd *et al.*, 1992; Christinck *et al.*, 1991; Fahnestock *et al.*, 1994; Matsumura *et al.*, 1992; Ojcius *et al.*, 1993; Schumacher *et al.*, 1990). The finding that only optimal length peptides can form long-lived complexes with MHC class I molecules is consistent with the peptide amino- and carboxyl-termini forming stabilizing contacts with the class I-binding site. Indeed, analysis of the crystal structure of several mouse and human class I molecules has shown the presence of two pockets at the ends of the groove (A and F), where the cleft narrows, interacting with the charged N- and C-termini of a bound peptide (Madden *et al.*, 1991). In addition, we have demonstrated that the α amino and carbonyl groups at the N- and C-termini play a major role in both inducing the conformational change in free heavy chain and formation of a stable class I–peptide complex (Elliott *et al.*, 1992a, 1992b). The longer peptides can be anchored in the specificity pockets without forming stabilizing contacts at ends of the cleft. Thus, long and short peptides may both bind in the ER, but long peptides are likely to have dissociated by the time the class I molecules arrive at the cell surface. The rapid off rate of longer peptides would therefore lead to enrichment of class I molecules containing 8–9 residue peptides. The relative contribution of binding motifs to the kinetics of peptide binding to class I molecules was investigated by replacing anchor residues with alanine. These substitutions showed a reduction in the binding affinity by 100- to 500-fold (Elliott *et al.*, 1992b).

The basis for the formation of long-lived peptide class I complexes *in vivo* is not known. The findings that peptides are deeply embedded in the class I groove (Bjorkman *et al.*, 1987; Garrett *et al.*, 1989; Madden *et al.*, 1991), and that optimum length peptides induce a conformational change of free class I heavy chains (Elliott *et al.*, 1991) suggest that the formation of long-lived peptide–class I molecule complexes is based on a 'peptide trapping' mechanism, analogous to that proposed for class II–peptide interaction (Sadegh-Nasseri and Germain, 1991). Indeed, the comparison of the association-binding kinetics of long and short peptides to D^b and Aw68 class I molecules demonstrated that long peptides bind with a simple kinetics and rapid dissociation rates.

Results obtained with D^b and Aw68 molecules, as obtained from the mutant T2 cells, however, showed that optimum length peptides can interact with class I molecules with two sets of kinetics (Cerundolo *et al.*, manuscript in preparation). At the beginning of the reaction, both the forward- and backward-binding rate constants appear comparably rapid (100- to 480-fold faster than predicted from measured binding affinity constant and off rate), but, after contact of class I molecules with the peptides, the kinetics change and give rise to stable complexes with very slow off rates. These results suggest that either the class I binding site or the peptide ligand may change conformation during the binding reaction. The fact that this effect correlates with the ability of optimal length peptides to induce a conformational change in the free heavy chain (Elliott *et al.*, 1991), suggests that it is a change in the class I binding site that is responsible. Peptide binding to class I molecules may trigger a conformational change of the antigen-binding site, which results in a shift from rapid to slow peptide-binding kinetics. The complex of heavy chain and β2-m formed in the absence of peptide may have a more open conformation that allows rapid access of the anchor residues of the peptide to the specificity pockets in the antigen-binding site. The conformational change induced by these interactions may then trap the peptide and allow the formation of multiple stabilizing hydrogen bonds with the peptide's N- and C-termini (Madden *et al.*, 1991).

The result of this effect is that peptides with relatively low binding affinities compared to many antibodies (Mason and Williams, 1980) can form extremely stable complexes rapidly, via the formation of an unstable intermediate. A related phenomenon was described for peptides binding to class II molecules (Sadegh-Nasseri and Germain, 1991; Sadegh-Nasseri and McConnell, 1989). This mechanism may have evolved to cope with the requirement for each class I and class II molecule to form stable complexes with many different peptide ligands. Class I and class II molecules may employ analogous conformational transitions for 'trapping' peptides with inherently low binding affinities.

9.4.3 Cellular compartment in which MHC class I molecules bind peptides

Analysis of the effects of the antibiotic brefeldin A (BFA) (Lippincott-Schwartz *et al.*, 1989) and of the adenovirus E3/19K glycoprotein (Anderson *et al.*, 1985; Cox *et al.*, 1991) (see Section 9.5.1) on antigen processing and presentation shows that class I molecules bind peptides in a pre-Golgi compartment. Although several studies suggest that the ER is a primary site for loading peptides into class I molecules, definitive evidence is still missing. BFA blocks transport of newly synthesized proteins from the ER to the Golgi apparatus, but it also has an effect on the medial- and trans-Golgi, endosomes, and lysosososomes (Lippincott-Schwartz *et al.*, 1990, 1991). After BFA treatment, proteins and lipids from the Golgi complex redistribute into the ER. Thus, MHC class I molecules restricted to the ER by BFA treatment, have been exposed to proteins localized in the Golgi compartment. BFA-treated cells are not able to present viral antigens to class I-restricted CTL, whereas the capacity to present exogenous peptide epitopes is retained (Nutchern *et al.*, 1989; Yewdell and Bennick, 1989). Furthermore, it was recently shown that BFA has no effect on class I assembly and stability, and that antigenic peptides can be recovered from BFA-treated cells (Lapham *et al.*, 1993). These results indicate that class I molecules associate with peptides prior to transport out of the ER. It is not possible to rule out, however, that BFA-treated cells rapidly redistribute their Golgi to the ER, thus exposing ER-retained class I molecules to a non-ER environment. Another agent that blocks antigen presentation is the adenovirus protein E3/19K. This protein associates in the ER with some class I alleles (H-2D^b, H-2K^b, H-2K^d and H-2L^d but not H-2D^d or H-2K^k) and retains them in the ER (Anderson *et al.*, 1985; Cox *et al.*, 1991). However, retention of such complexes in the ER is via retrieval from post-ER compartments (Jackson *et al.*, 1993). After infection, target cells are not able to present viral antigens in association with K^d molecules, but there is no effect on D^d-restricted killing (Cox *et al.*, 1990). Similarly to BFA treated cells, class I molecules retained in the ER by the E3/19K protein are stable and associated with β2-m. Furthermore, class I molecules, in which the cytoplasmic domain was substituted with an E3/19K cytoplasmic domain sequence, are retained within the lumen of the ER , and are loaded with intracellular peptides (Lapham *et al.*, 1993). Thus, class I molecules bind peptides in the compartment in which they are retained by the E3/19K protein. Further evidence that class I molecules bind peptides within the lumen of the ER was provided by the analysis of the cellular distribution of TAP1, by light microscopy and immunoelectron microscopy (Kleijmeer *et al.*, 1992). These studies showed that the TAP1/2 complex is localized in the ER and *cis*-Golgi, and is oriented with its ATP-binding domain in the cytosol. To identify the subcellular compartment required for class I loading and assembly, Salter and colleagues recently compared the effect of several inhibitors of intracellular transport for their effect on the biosynthetic pathway of class I molecules (Tector *et al.*, 1994). The results of their studies demonstrated that treatment with the phosphatase inhibitor okadaic acid (OKA) blocks protein transport out of the ER. It is of interest, however, that unlike the block obtained with BFA, class I molecules retained in the ER by OKA treatment are not loaded with intracellular peptides, remain associated with calnexin, and the maturation of glycans on class I molecules is completely blocked. These results demonstrate that OKA blocks assembly of class I complexes, but it does not interfere with synthesis of heavy chains. Although the authors did not analyse the effect of OKA on the TAP-dependent peptide translocation, these results suggest that class I molecules acquire stability in a location beyond the block induced by the phosphatase inhibitor. This may correspond to an intermediate compartment between the Golgi and the ER.

9.4.4 Maturation of MHC class I molecules in vivo

Analysis of the biosynthetic pathway of class I molecules in normal cells demonstrated that the majority of class I heavy chains fold and associate with β2-m in approximately 5 minutes (Neefjes *et al.*, 1993b). However, marked differences between several class I alleles were described. Pulse-chase experiments of mouse class I molecules showed that H-2K^k molecules mature at a faster rate than H-2D^k molecules (Williams *et al.*, 1985). Transport of L^d is particularly slow, and L^d molecules at the cell surface are not associated with β2-m (Beck *et al.*, 1986; Weis and Murre, 1985). H-2L^d, as well as H-2D^b, are able to reach the cell surface in β2-m-negative cells (Allen *et al.*, 1988). Differences in the rate of assembly and transport of human class I molecules were also observed. Analysis of naturally occurring variants related to the fast assembler HLA-B35 and the slow assembler HLA-B51 demonstrated that the assembly efficiency is controlled by the α2 domain in which these molecules differ at eight amino acid residues (Hill *et al.*, 1993). The mechanisms and biological significance of these differences are not known.

The analysis of the maturation rate of class I molecules in TAP-negative cells has shown some interesting differences in the maturation rate of human and mouse alleles expressed in human cells. Cresswell and his colleagues showed that the majority of human class I molecules expressed in the human cell line T2 are retained in the lumen of the ER. In contrast, mouse class I molecules leave the ER and reach the cell surface (Alexander *et al.*, 1989). These results led to the hypothesis that the transport of class I molecules in the human cell line T2 is controlled and regulated in the lumen of the ER by accessory molecules which can distinguish between human and mouse class I alleles.

Evidence supporting a role of accessory molecules on the biosynthetic pathway of class I molecules will be discussed in the next sections.

9.4.5 TAP and MHC class I assembly

It has recently been shown by two groups that both in human and mouse cells, the TAP1/2 complex can associate with newly synthesized class I–β2-m complexes (Ortmann *et al.*, 1994; Suh *et al.*, 1994). Association of TAP with class I molecules can only be observed if class I molecules are correctly folded and associated with β2-m, whereas no binding of free class I molecules was detected. The mechanisms that control this interaction are not known. However, it was shown that TAP1 alone is able to associate with class I molecules (Androlewicz *et al.*, 1994). The finding that dissociation of class I molecules from the TAP complex is increased by the addition in the lysate of class I-binding peptide ligands, suggests that empty class I–β2-m complexes bind to TAP. Binding of peptides increases the dissociation rate of class I molecules from TAP, thus allowing their exit from the ER. The possibility that association of class I molecules with TAP facilitates the channelling of cytosolic peptides into the class I-binding groove is intriguing.

Although the significance of TAP/class I association remains to be shown, the hypothesis that translocated peptides can be sampled almost immediately by class I–β2-m complexes, and selected on the basis of the correct binding motif is intriguing. It is clear, however, that association of class I molecules with TAP is not necessary for antigen presentation and for the folding and maturation of class I molecules. It was shown that newly synthesized class I molecules in TAP-negative cells are correctly folded and associate *in vivo* with β2-m. Precipitation of class I molecules, immediately after lysis, shows that class I folding and assembly are not impaired in TAP-negative cells (Elliott *et al.*, 1991). Indeed, folded D^b molecules can be captured at the cell surface of the processing mutant RMA-S cells by adding to the medium a conformation specific monoclonal antibody (Ortiz and Hammerling, 1991). Furthermore, the lack of the TAP complex does not prevent peptide loading of class I molecules. Peptide epitopes, co-translationally synthesized in the ER of the TAP-negative cells T2, are efficiently presented to class I-restricted CTL (Anderson *et al.*, 1991; Elliott *et al.*, 1995; Hammond *et al.*, 1993; Snyder *et al.*, 1994).

Evidence that TAP/class I association may facilitate peptide loading *in vivo* has come from the characterization of a novel genetic defect in the human cell line LBL 721.220 (.220). The human B cell line .220 has a deletion of both copies of *HLA-A* and *B* class I alleles. Transfection of class I genes in .220 cells has revealed a reduction in stability and surface expression of class I molecules. It is of interest that this phenotype is specific for particular class I alleles: HLA-A1 and B8 were shown to be unstable, whereas surface expression of HLA-A2, -A3 and -B7 was comparable to the level of surface expression in the parental line (Greenwood *et al.*, 1994). Restoration of the normal phenotype after fusion with the β2-m negative Daudi cell line and the TAP-negative cell line .174 indicates that .220 cells express functional β2-m and TAP (Grandea *et al.*, 1995). The authors showed that association of class I molecules with calnexin is normal, whereas association of class I molecules with TAP complex is severely compromised. As the normal phenotype is restored by stable transfer, by microcell fusion, of a human chromosome 6, these results are consistent with a loss of function of an unidentified gene or genes, possibly linked to the MHC, which facilitate association of class I molecules with the TAP complex, and the capture by class I molecules of translocated cytosolic peptides.

9.4.6 Calnexin and MHC class I assembly

Calnexin, previously named p88 in mouse cells (Ahluwalia *et al.*, 1992) and IP90 in human cells (Galvin *et al.*, 1992), is a Ca^{2+}-binding trans-membrane protein. Several reports demonstrated that calnexin associates transiently with many different newly synthesized proteins, suggesting that it can act in assisting protein assembly and/or in the retention within the ER of incompletely folded molecules.

It was suggested that oligosaccharide residues may serve as substrates for calnexin binding. This hypothesis is consistent with the finding that polypeptides bound to calnexin are retained on Concanavalin-A-sepharose (Ou *et al.*, 1993). It is therefore possible that calnexin acts as a lectin that facilitates folding of glycoproteins in the lumen of the ER by binding to their N-linked oligosaccharide side-chains. Treatment of metabolically labelled cells with the glucosidase inhibitor castanospermine prevents the binding of glycoproteins to calnexin, and results in an increased rate of degradation of nascent T-cell receptor (TCR) α subunits (Kearse *et al.*, 1994). It was shown, however, that calnexin can also bind to unglycosylated TCR subunits and unglycosylated class I molecules (Carreno *et al.*, 1995; Rajagopalan *et al.*, 1994).

Degen and Williams (1991) demonstrated that calnexin can interact with newly synthesized class I heavy chains in mouse cells. Binding of calnexin to class I molecules occurs within 5 minutes of synthesis. This association is transient, and by approximately 30 minutes the complex dissociates. Dissociation of class I–β2-m complex from calnexin correlates with its exit from the ER, and acquisition of resistance to Endo H digestion. The identification of the class I domain which determines the interaction with calnexin, remains controversial. Deletion of the transmembrane region of the class I heavy chain and replacement with the region of the $Q7^b$ molecule, encoding the glycosylphosphatidylinositol (GPI) anchor site, resulted in a negligible detection of class I–calnexin complexes by using cross-linking techniques (Margolese *et al.*, 1993). These results indicated that the class I transmembrane region might be the site of interaction with calnexin. These conclusions, however, were called into question by a recent report by Hansen and colleagues (Carreno *et al.*, 1995). Immunoprecipitation experiments and Western blot analysis showed that GPI-linked $Q7^b$ class I molecules were associated with calnexin, whereas a soluble $Q7^b$ isoform was not calnexin associated. These results demonstrate that the transmembrane region is not the site of interaction with calnexin, and implicate the 9 amino acid fragment, connecting the $\alpha3$ domain with the transmembrane region. The discrepancy between the two studies could be due to the different sensitivities of the techniques used.

Experiments carried out in β2-m-deficient cells and TAP-negative cells gave the opportunity to study the effect of β2-m binding and peptide loading on the kinetics of calnexin/free heavy chain association. The results of these experiments demonstrated that, both in human (Daudi) (Hochstenbach *et al.*, 1992) and mouse (R1E) (Degen *et al.*, 1992) β2-m-deficient cells, the majority of class I molecules remain associated with calnexin throughout their lifetime. An increased association between calnexin and D^b–β2-m complexes was also observed in the murine TAP-deficient RMA-S cells (Degen *et al.*, 1992). The association of heavy chain–β2-m complex to calnexin was not confirmed in human TAP-deficient cells. Sugita and Brenner showed that, in human cells, binding of β2-m to free heavy chains triggers the dissociation of free class I molecules from calnexin (Sugita and Brenner, 1994). This unstable intermediate associates with the TAP complex, and it is subsequently stabilized by the binding of cytosolic peptides. These results suggest that calnexin may retain empty or unfolded class I molecules in the lumen of the ER, thus preventing or reducing the amount of empty class I molecules on the surface. This hypothesis was recently confirmed by two papers both showing that the level of surface expression of MHC class I molecules can be regulated by transfection of calnexin. In the first paper, *Drosophila melanogaster* cells, which lack the TAP complex, were used as a model to study the rate of maturation of class I molecules (Jackson *et al.*, 1994). Previous experiments demonstrated that transfection of murine class I heavy chain and β2-m in *Drosophila* cells generates 'empty' class I complexes, which leave the ER and are expressed on the cell surface (Jackson *et al.*, 1992). The rate of maturation of class I–β2-m complexes in *Drosophila* cells appears to be faster than the rate observed in murine TAP-defective RMA-S cells. Transfection of calnexin in *Drosophila* cells retarded the intracellular transport of both free heavy chains and empty class I–β2-m complexes (Jackson *et al.*, 1994), at a rate comparable to the rate observed in RMA-S cells. In a second paper, it was shown that retention of free class I molecules in the ER is mediated by a motif contained within the cytosolic tail of calnexin (Rajagopalan and Brenner, 1994). Deletion of calnexin C-terminal motif (RKPRRE) results in the exit of calnexin from the ER, and its accumulation on the surface and in lysosomes. Transfection of a cytoplasmic tail-deleted mutant calnexin in β2-m-deficient cells increases the level of surface expression of free class I molecules.

Although these results are consistent with the hypothesis that calnexin plays a role in the maturation of class I molecules, direct evidence is still missing. It is of interest that a recent paper showed that class I molecules synthesized in a calnexin-deficient human cell line mature at a normal rate, and are expressed on the surface at levels comparable to the level of expression in the parental line (Scott and Dawson, 1995). No detectable difference has been found in the HPLC profiles eluted from both cell lines. Further experiments need to be done formally to address the significance of association of class I molecules with calnexin and TAP molecules.

9.5 Viral mechanisms preventing class I-restricted antigen presentation

Intracellular parasites and viruses have developed several mechanisms to escape CTL recognition. RNA and DNA viruses appear to use different strategies to evade presentation through MHC class I molecules. RNA viruses, as a result of the low fidelity of the reverse transcriptase, have a high frequency of single amino acid mutations. Amino acid substitutions, which decrease CTL recognition, thus conferring a growth advantage for the virus, are positively selected. Mutations within CTL epitopes of HIV (Phillips *et al.*, 1991) were described. It was shown that the majority of mutations are clustered within CTL epitopes which are presented by the class I molecules expressed by the same patients. Virus-specific CTL response selects for amino acid substitutions which decrease both the binding affinity of peptide epitopes to the class I molecules (Phillips *et al.*, 1991), and alter the ability of the peptides to trigger a CTL response (Bertoletti *et al.*, 1994; Klenerman *et al.*, 1994; Meier *et al.*, 1995), antagonizing recognition of wild-type epitopes.

Over the last few years, several viral proteins encoded by DNA viruses have been characterized which interfere with the biosynthetic pathway of MHC class I molecules and the processing and presentation of intracellular antigens. A summary of the immune evasion strategies used by DNA viruses is reported below.

9.5.1 Adenoviruses

Adenoviruses can affect surface expression of class I molecules at two distinct levels.

(i) *The oncogenic adenovirus 12 (Ad12)* synthesizes a viral protein E1A which was shown selectively to inhibit transcription of MHC class I genes (Bernards *et al.*, 1983; Schrier *et al.*, 1983). Decrease of class I mRNA in Ad12-transformed cells is due to an inhibition of post-transcriptional processing of MHC class I mRNA in the nucleus (Vaessen *et al.*, 1987). In addition, mRNA levels of both TAP1 and TAP2 are significantly reduced in Ad12-

transformed cells in comparison to levels of expression in the parental line or Ad5-transformed cells (Rotem-Yehudar *et al.*, 1994). A reduction of the amount of peptides available within the lumen of the ER also contributes to the reduction of surface-expressed class I molecules.

(ii) *The adenovirus 2 (Ad2)* encodes a glycoprotein E19 from the early region E3 (E3/19K). It was shown that this protein binds to MHC class I molecules and retains them in the ER (Anderson *et al.*, 1985; Cox *et al.*, 1991) (see Section 9.4.3). This retention mechanism appears to be specific for class I molecules, as no defect in the maturation of other glycoproteins was observed. E3/19K glycoprotein contains an ER retention sequence in its cytosolic domain (DEKKMP). Deletion of this motif alters the capacity of E3/19K to retain class I molecules in the lumen of the ER, but it has no effect on the ability of E3/19K to bind to class I molecules. Cells infected with E3/19K deleted of the cytosolic tail have a normal surface expression of MHC class I molecules, and are able to present processed antigen to $CD8^+$ T cells.

9.5.2 Mouse and human cytomegaloviruses (MCMV and HCMV)

Mouse (MCMV) and human (HCMV) cytomegaloviruses down-regulate the expression of MHC class I molecules using different mechanisms.

MCMV. Del Val and colleagues showed that after infection of mouse cells with MCMV, presentation of an L^d epitope contained within the pp89 protein is abolished after expression of the early phase (E) genes (Del Val *et al.*, 1989). Further experiments showed that this antigen-presentation defect is due to retention of class I molecules in the lumen of the ER (Del Val *et al.*, 1992). This effect appears to be selective for MHC class I molecules. The mechanisms by which MCMV blocks MHC class I transport are not known. Koszinowski and colleagues (personal communication), using MCMV deletion mutants, were recently able to isolate a candidate gene coding for a small glycoprotein, which might be involved in the down-regulation of MHC class I molecules.

HCMV. Down-regulation of HLA class I surface expression is observed after infection with HCMV. A homologue of human class I heavy chains is encoded by the *UL18* gene of the HCMV. It was shown that the product of this gene is able to bind to β2-m (Browne *et al.*, 1990). Using recombinant vaccinia viruses, Browne and colleagues showed that the *UL18* gene product can be co-precipitated with β2-m, and the UL18–β2-m complex can be detected at the cell surface of infected cells using an anti-β2-m antibody. These results were interpreted by saying that high expression of UL18 binds to and saturates the endogenous β2-m, thus preventing presentation of viral epitopes through endogenous class I molecules. This hypothesis was later called into question because it was shown that *UL18*-deleted virus does not abolish the down-regulation of MHC class I molecules (Browne *et al.*, 1992). This result suggests that UL18 might not be responsible for the down-regulation of MHC class I molecules. Indeed, it was recently shown that down-regulation of class I molecules is due to a reduced half-life of class I molecules (Beersma *et al.*, 1993). The results of these studies showed that free heavy chains as well as class I–β2-m complexes are rapidly degraded after HCMV infection. The increased rate of class I degradation is not prevented by the binding of peptide ligands to class I molecules.

9.5.3 Herpes simplex virus (HSV)

HSV type 1 and type 2 interfere with MHC class I antigen presentation by altering the TAP-dependent peptide translocation.

It was previously shown that cells infected by HSV have a low level of expression of MHC class I molecules at the cell surface (Jennings *et al.*, 1985; Koelle *et al.*, 1993). Also, after HSV infection, MHC class I molecules remain unstable and are retained in the ER (Hill *et al.*, 1994; York *et al.*, 1994). These results suggested that this phenotype could be due either to a block at the level of transport of peptides from the cytosol to the ER or to a block in the loading of peptides on to class I molecules. The use of virus deletion mutants gave the opportunity to identify a 9 kDa protein, called ICP47, which controls the phenotype of HSV-infected cells. Indeed, recombinant adenovirus-encoded ICP47 prevents MHC class I transport to the surface and T cell recognition of HSV CTL epitopes. The mechanisms by which ICP47 inhibit antigen presentation were recently described by two separate groups. It was shown that ICP47 binds to the TAP complex and blocks transport of peptides into the ER (Fruh *et al.*, 1995; Hill *et al.*, 1995). Association of ICP47 with the TAP complex was demonstrated by co-precipitating the complex with ICP47 and TAP1-specific antibodies. Using streptolysin O semipermeabilized cells (see Section 9.3.3), it was shown that TAP-dependent transport of peptide from the cytosol to the ER was inhibited after infection with HSV1 and adenovirus encoded ICP47. Virus deletion mutants, lacking ICP47, lost the ability to inhibit peptide translocation across the ER membrane.

9.5.4 Epstein-Barr virus (EBV)

One of the EBV proteins, the nuclear antigen EBNA1, was recently shown to be able to escape CTL recognition. This protein is responsible for the maintenance of the EBV episomes in infected cells. EBNA1 is the only EBV protein that is expressed in malignancies associated with EBV, such as Burkitt's lymphomas (Klein, 1994). Despite the high level of expression of this protein in EBV-infected cells, EBNA1-specific CTL were not previously described. In contrast, other EBV nuclear antigens (EBNA2, 3, 4, 5) or latent membrane proteins (LMP1, 2A, and B) are highly immunogenic, and several CTL epitopes were characterized.

It has recently been shown that a unique amino acid motif, contained within the sequence of EBNA1, controls its lack of CTL recognition. EBNA1 contains an approximately 230 amino acid Gly/Ala repetitive sequence, the function of which is not defined. Masucci and her colleagues showed that presentation of EBNA4 CTL epitopes cloned upstream of the Gly/Ala motif of EBNA1 was severely impaired. Deletion of the Gly/Ala repeat was sufficient to relieve the block in antigen presentation (Levitskaya *et al.*, 1995). Inhibition of presentation of EBNA4 CTL epitopes was also observed after inserting the EBNA1 Gly/Ala repeat in the EBNA4 sequence.

These results were recently confirmed by Townsend and colleagues (Mukherjee *et al.*, manuscript in preparation) by using a different strategy. A K^d-binding peptide was identified downstream of the EBNA1 Gly/Ala repeat region. A recombinant vaccinia virus was then generated, encoding a sequence of approximately 50 residues containing the K^d-binding peptide, and used to raise CTL in Balb/c mice. It was shown that those CTL were able to lyse cells transfected or infected with vaccinia viruses expressing Gly/Ala deleted EBNA1, whereas no killing was observed when full length EBNA1 protein was expressed. The mechanisms by which the EBNA1 Gly/Ala repeat blocks antigen presentation are not known. The finding that overexpression of EBNA1 does not prevent recognition of EBNA4 vaccinia-infected fibroblasts by EBNA4-specific CTL suggests that the presence of the Gly/Ala repeat interferes only with cytosolic degradation of EBNA1, thus inhibiting generation of CTL epitopes.

9.5.5 Vaccinia virus

Vaccinia viruses are often used for the expression of recombinant proteins to sensitize target cells for CTL lysis. Synthesis of vaccinia-encoded recombinant proteins can be driven either by an early promoter or by a late promoter. The early promoter is active before viral DNA replication, whereas the late promoter is active after DNA replication. Two papers showed that vaccinia virus interferes with the presentation of influenza HA and NP to class I-restricted CTL (Coupar *et al.*, 1986; Townsend *et al.*, 1988). The inhibitory effect is selective for certain epitopes, and is more profound during the late phase of infection. The presentation of HA and NP is restored by increasing the rate of degradation of recombinant proteins. The mechanisms by which vaccinia virus blocks presentation of certain epitopes are not known. A recent report has shown that this block in antigen presentation is not controlled by two serine protease inhibitors (the serpin proteins B13R and B22R), which are encoded by the vaccinia virus (Blake *et al.*, 1995).

9.6 Conclusions

The analysis of the events leading to the presentation of intracellular antigens to CTL indicates three levels at which selection of the peptide repertoire presented to the immune system could occur:

(i) polymorphisms of class I molecules;
(ii) peptide transport specificity;
(iii) degradation of intracellular proteins.

9.6.1 Polymorphisms of class I molecules

The very presence of class I allele-specific motifs implies a selection operated by class I molecules in the peptide repertoire. Indeed, Falk and colleagues (Falk *et al.*, 1990) have shown that the peptide defining any particular antigen can be isolated only from cells that express the class I molecule known to function as a restriction element for that antigen. The class I molecules in a cell, therefore, define the set of peptides that can be isolated by acid extraction. This finding implies that class I molecules themselves are responsible for this selective effect and that unbound peptides might have an extremely short half-life. This concept has recently been challenged by a series of papers published by Srivastava and colleagues (Srivastava *et al.*, 1986). The results of these studies demonstrated that injection into mice of the ER-resident protein gp96/grp94, purified from tumour cells generates tumour-specific CTL. Similar results were obtained using gp96/grp94 purified from virus-infected cells (Suto and Srivastava, 1995). These findings imply that gp96/grp94 is capable of binding polypeptides containing CTL epitopes, and able to trigger a specific CTL response *in vivo* (Li and Srivastava, 1993).

Epitope selection can also be influenced by competition for binding to class I molecules in the lumen of the ER. Tussey and colleagues (Tussey *et al.*, 1995) have analysed the immunogenicity of two overlapping CTL epitopes. Influenza NP B8 CTL epitope (NP 380–388) overlaps the NP HLA-B27.02 CTL epitope (NP 381–388). The authors demonstrated that co-expression of HLA-B8 and HLA-B27.02 in the same cell line leads to a suboptimal loading of the B8 epitope, possibly due to competition for binding. Mutation of the B27.02 epitope anchor residue (Arg to Lys at position 382) reduces the binding of the B27.02 epitope, and restores NP presentation through B8.

9.6.2 Peptide transport specificity

The rat *cim* phenomenon is possibly the only reported example of functional polymorphism of the peptide transporter into the ER. Several examples of mouse CTL that can kill human target cells transfected with the appropriate mouse class I molecules have been described and, vice versa, of human CTL that kill mouse target cells transfected with human class I molecules. Indeed, it was shown that mouse HLA-A2 and HLA-B27 transgenic mice are able to generate a CTL response that is specific for the same epitopes generated in human cells. Furthermore, the human *TAP2* allele was shown to restore the normal phenotype in the murine RMA-S cells (Yewdell *et al.*, 1993). Thus, murine and human processing pathways appear to be interchangeable without any effect on the presentation of individual epitopes. It remains to be proven whether the TAP substrate specificity, which has been well documented *in vitro*, has any relevance *in vivo*.

9.6.3 Degradation of intracellular proteins

Rules and motifs that control proteolysis in the cytosol and in the ER are still poorly understood. The recent results obtained with LMP-deficient cells demonstrate that proteasome subunit composition may regulate the proteolytic cleavage of target proteins. Polymorphisms at the level of cytosolic degradation of self and non-self proteins in different tissues or during infections might play a central role in the development of the T-cell repertoire and autoimmune diseases.

References

Ahluwalia N, Bergeron J, Wada I, Degen E, Williams D. (1992) The p88 molecular chaperone is identical to the endoplasmic reticulum membrane protein, calnexin. *J. Biol. Chem.* **267:** 10914–10918.

Alexander J, Payne JA, Murray R, Frelinger J, Cresswell P. (1989) Differential transport requirements of HLA and H-2 class I glycoproteins. *Immunogenetics* **29:** 380–388.

Allen H, Fraser J, Flyer S, Calvin S, Flavell R. (1988) β-2 microglobulin is not required for cell surface expression of the murine class I histocompatibility antigen H-2Db or of a truncated H-2Db. *Proc. Natl Acad. Sci. USA* **79:** 7447–7451.

Anderson K, Cresswell P, Gammon M, Hermes J, Williamson A, Zweerink H. (1991) Endogenously synthesized peptide with an endoplasmic reticulum signal sequence sensitizes antigen processing mutant cells to class-I restricted cell-mediated lysis. *J. Exp. Med.* **174:** 489-492.

Anderson M, Paabo S, Nilsson T, Peterson P. (1985) Impaired intracellular transport of class I MHC antigens as a possible means for adenoviruses to evade immune surveillance. *Cell* **43:** 215–222.

Androlewicz M, Cresswell P. (1994) Human transporters associated with antigen processing possess a promiscuous peptide-binding site. *Immunity* **1:** 7–14.

Androlewicz M, Anderson K, Cresswell P. (1993) Evidence that transporters associated with antigen processing translocate a major histocompatibility complex class I-binding peptide into the endoplasmic reticulum in an ATP-dependent manner. *Proc. Natl Acad. Sci. USA* **90:** 9130–9134.

Androlewicz M, Ortmann B, Van Endert P, Spies T, Cresswell P. (1994) Characteristics of peptide and MHC class I β-2 microglobulin binding to the transporters associated with antigen processing (TAP1 and TAP2). *Proc. Natl Acad. Sci. USA* **91:** 12716–12720.

Arnold D, Driscoll J, Androlewicz M, Hughes E, Cresswell P, Spies T. (1992) Proteasome subunits encoded in the MHC are not generally required for the processing of peptides bound by MHC class I molecules. *Nature* **360:** 171–174.

Attaya M, Jameson S, Martinez CK, Hermel E, Aldrich A, Forman J, Fischer-Lindahl K, Bevan M, Monaco JJ. (1992) Ham-2 corrects the class I antigen-processing defect in RMA-S cells. *Nature* **355:** 647–649.

Bachmair A, Finley D, Varshavsky A. (1986) In vivo half-life of a protein is a function of its amino-terminal residue. *Science* **234:** 179–186.

Beck J, Hansen T, Cullen S, Lee D. (1986) Slower processing, weaker β2-m association, and lower surface expression of H-2Ld are influenced by its amino terminus. *J. Immunol.* **13:** 916–923.

Beersma M, Bijlmakers M, Pleogh H. (1993) Human cytomegalovirus down-regulates HLA class I expression by reducing the stability of class I H chains. *J. Immunol.* **151:** 4455–4464.

Belich M, Glynne R, Senger G, Sheer D, Trowsdale J. (1994) Proteasome components with reciprocal expression to that of the MHC-encoded LMP proteins. *Curr. Biol.* **4:** 769–776.

Bernards R, Schrier P, Houweling A, Bos J, van der Eb A, Zijlstra M, Melief C. (1983) Tumorigenicity of cells transformed by adenoviruses type 12 by evasion of T-cell immunity. *Nature* **305:** 776–779.

Bertoletti A, Sette A, Chisari FV, Penna A, Levrero M, De CM, Fiaccadori F, Ferrari C. (1994) Natural variants of cytotoxic epitopes are T-cell receptor antagonists for antiviral cytotoxic T cells. *Nature* **369:** 407–410.

Bix M, Raulet D. (1992) Functionally conformed free class I heavy chains exist on the surface of beta-2 microglobulin negative cells. *J. Exp. Med.* **176:** 829–834.

Bjorkman P, Saper M, Samraoui B, Bennett W, Strominger J, Wiley D. (1987) Structure of human class I histocompatibility antigen, HLA-A2. *Nature* **329:** 506–511.

Blake N, Kettle S, Law K, Gould K, Bastin J, Townsend A, Smith G. (1995) Vaccinia virus serpins B13R and B22R do not inhibit antigen presentation to class I-restricted cytotoxic T lymphocytes. *J. Gen. Virol.* **76:** 2393–2398.

Boyd L, Kozlowski S, Margulies D. (1992) Solution binding of an antigenic peptide to a major histocompatibility complex class I molecule and the role of β-2 microglobulin. *Proc. Natl Acad. Sci. USA* **89:** 2242–2246.

Browne H, Smith G, Beck S, Minson T. (1990) A complex between the MHC class I homologue encoded by human cytomegalovirus and β-2 microglobulin. *Nature* **347:** 770–772.

Browne H, Churcher M, Minson T. (1992) Construction and characterization of a human cytomegalovirus mutant with the UL18 (class I homologue) gene deleted. *J. Virol.* **66:** 6784–6790.

Carreno B, Schreiber K, McKean D, Stroynowski I, Hansen T. (1995) Aglycosylated and phosphatidylinositol-anchored MHC class I molecules are associated with calnexin. *J. Immunol.* **154:** 5173–5180.

Cerundolo V, Alexander J, Anderson K, Lamb C, Cresswell P, McMichael A, Gotch F, Townsend A. (1990) Presentation of viral antigen controlled by a gene in the major histocompatibility complex. *Nature* **345:** 449–452.

Cerundolo V, Elliott T, Elvin J, Bastin J, Rammensee H-G, Townsend A. (1991) The binding affinity and dissociation rates of peptides for class I histocompatibility complex molecules. *Eur. J. Immunol.* **21:** 2069–2075.

Cerundolo V, Kelly A, Elliott T, Trowsdale J, Townsend A. (1995) Genes encoded in the MHC affecting the generation of peptides for TAP transport. *Eur. J. Immunol.* **25:** 554–562.

Chen B, Parham P. (1989) Direct binding of influenza peptides to class I HLA molecules. *Nature* **337:** 743–745.

Chicz RM, Urban RG, Gorga JC, Vignali DA, Lane WS, Strominger JL. (1993) Specificity and promiscuity among naturally processed peptides bound to HLA-DR alleles. *J. Exp. Med.* **178:** 27–47.

Christinck E, Luscher M, Barber B, Williams D. (1991) Peptide binding to class I MHC on living cells and quantitation of complexes required for CTL lysis. *Nature* **352:** 67–70.

Coupar B, Andrew M, Both G, Boyle D. (1986) Temporal regulation of influenza hemagglutinin expression in vaccinia virus recombinant and effects on the immune response. *Eur. J. Immunol.* **16:** 1479–1487.

Cox J, Yewdell J, Eisenlohr L, Johnson P, Bennink J. (1990) Antigen presentation requires transport of MHC class I molecules from the endoplasmic reticulum. *Science* **247:** 715–718.

Cox JH, Bennink JR, Yewdell JW. (1991) Retention of adenovirus E19 glycoprotein in the endoplasmic reticulum is essential to its ability to block antigen presentation. *J. Exp. Med.* **174:** 1629–1637.

Cox JH, Galardy P, Bennink JR, Yewdell JW. (1995) Presentation of endogenous and exogenous antigens is not affected by inactivation of E1 ubiquitin-activating enzyme in temperature-sensitive cell lines. *J. Immunol.* **154:** 511–519.

De la Salle H, Hanau D, Fricker D. *et al.* (1994) Homozygous human TAP peptide transporter mutation in HLA class I deficiency. *Science* **265:** 237–241.

Degen E, Williams D. (1991) Participation of a novel 88-Kd protein in the biogenesis of murine class I histocompatibility molecules. *J. Cell Biol.* **112:** 1099–1115.

Degen E, Cohen-Doyle M, Williams D. (1992) Efficient dissociation of the p88 chaperone from major histocompatibility complex class I molecules requires both β-2 microglobulin and peptide. *J. Exp. Med.* **175:** 1653–1661.

Del Val M, Münch K, Reddehase M, Koszinowski U. (1989) Presentation of CMV immediate-early antigen to cytolytic T lymphocytes is selectively prevented by viral genes expressed in the early phase. *Cell* **58:** 305–315.

Del Val M, Hengel H, Hacker H, Hartlaub U, Ruppert T, Lucin P, Koszinowski U. (1992) Cytomegalovirus prevents antigen presentation by blocking the transport of peptide-loaded MHC class I molecules into the medial Golgi compartment. *J. Exp. Med.* **176:** 729–738.

Deres K, Schumacher T, Wiesmuller K, Stevanovic S, Greiner G, Jung G, Ploegh H. (1992) Preferred size of peptides that bind to H-2Kb is sequence dependent. *Eur. J. Immunol.* **22:** 1603–1608.

Deverson EV, Gow IR, Coadwell WJ, Monaco JJ, Butcher GW, Howard JC. (1990) MHC class II region encoding proteins related to the multidrug resistance family of transmembrane transporters. *Nature* **348:** 738–741.

Driscoll J, Brown MG, Finley D, Monaco JJ. (1993) MHC-linked LMP gene products specifically alter peptidase activities of the proteasome. *Nature* **365:** 262–264.

Elliott T, Cerundolo V, Elvin J, Townsend A. (1991) Peptide-induced conformational change of the class I heavy chain. *Nature* **351:** 402–406.

Elliott T, Elvin J, Cerundolo V, Allen H, Townsend A. (1992a) Structural requirements for the peptide induced conformational change of free MHC class I heavy chains. *Eur. J. Immunol.* **22:** 2085–2091.

Elliott T, Cerundolo V, Townsend A. (1992b) Short peptides assist the folding of free class I heavy chains in solution. *Eur. J. Immunol.* **22:** 3121–3125.

Elliott T, Willis A, Cerundolo V, Townsend A. (1995) Processing of major histocompatibility class I-restricted antigens in the endoplasmic reticulum. *J. Exp. Med.* **181:** 1481–1491.

Esquivel F, Yewdell J, Bennink J. (1992) RMA/S cells present endogenously synthesized cytosolic proteins to class I-restricted cytotoxic T lymphocytes. *J. Exp. Med.* **175:** 163–168.

Fahnestock M, Tamir I, Narhi L, Bjorkman P. (1992) Thermal stability comparison of purified empty and peptide-filled forms of a class I MHC molecule. *Science* **258:** 1658–1662.

Fahnestock ML, Johnson JL, Feldman RM, Tsomides TJ, Mayer J, Narhi LO, Bjorkman PJ. (1994) Effects of peptide length and composition on binding to an empty class I MHC heterodimer. *Biochemistry* **33:** 8149–8158.

Falk K, Rotzschke O, Rammensee HG. (1990) Cellular peptide composition governed by major histocompatibility complex class I molecules. *Nature* **348:** 248–251.

Fehling HJ, Swat W, Laplace C, Kuhn R, Rajewsky K, Muller U, von Boehmer H. (1994) MHC class I expression in mice lacking the proteasome subunits LMP-7. *Science* **265:** 1234–1237.

Fischer LK, Hermel E, Loveland BE, Wang CR. (1991) Maternally transmitted antigen of mice: a model transplantation antigen. *Annu. Rev. Immunol.* **9:** 351–372.

Fruh K, Gossen M, Wang K, Bujard H, Peterson P, Yang Y. (1994) Displacement of housekeeping proteasome subunits by MHC-encoded LMPs: a newly discovered mechanism for modulating the multicatalytic proteinase complex. *EMBO J.* **13:** 3236–3244.

Fruh K, Ahn K, Djaballah HPS, van Endert P, Tampe R, Peterson P. (1995) A viral inhibitor of peptide transporters for antigen presentation. *Nature* **375:** 415–418.

Gaczynska M, Rock KL, Goldberg AL. (1993) Gamma-interferon and expression of MHC genes regulate peptide hydrolysis by proteasomes. *Nature* **365:** 264–267.

Galvin K, Krishna S, Ponchel F, Frohlich M, Cummings DE, Carlson R, Wands JR, Isselbacher KJ, Pillai S, Ozturk M. (1992) The major histocompatibility complex class I antigen-binding protein p88 in the product of the calnexin gene. *Proc. Natl Acad. Sci. USA* **89:** 8452–8456.

Garrett T, Saper M, Bjorkman P, Strominger J, Wiley D. (1989) Specificity pockets for the side chains of peptide antigens in HLA-Aw68. *Nature* **342:** 692–696.

Glynne R, Powis S, Beck S, Kelly A, Kerr L, Trowsdale J. (1991) A proteasome-related gene between the two ABC transporter loci in the class II region of the human MHC. *Nature* **353:** 357–360.

Goldberg AL, Rock KL. (1992) Proteolysis, proteasomes and antigen presentation. *Nature* **357:** 375–380.

Grandea A, Androlewicz M, Athwal R, Geraghty D, Spies T. (1995) Dependence of peptide binding by MHC class I molecules on their interaction with TAP. *Science* **270:** 105–108.

Greenwood R, Shimizu Y, Sekhon G, DeMars R. (1994) Novel allele-specific, post-translational reduction in HLA class I surface expression in a mutant human B cell line. *J. Immunol.* **153:** 5525–5536.

Hammond S, Bollinger R, Tobery T, Silliciano R. (1993) Transporter-independent processing of HIV-1 envelope protein for recognition by $CD8^+$ T cells. *Nature* **364:** 158–161.

Heemels M, Ploegh H. (1994) Substrate specificity of allelic variants of the TAP peptide transporter. *Immunity* **1:** 775–784.

Heemels M, Schumacher K, Wonigeit K, Ploegh H. (1993) Peptide translocation by variants of the transporter associated with antigen processing. *Science* **262:** 2059–2063.

Henderson RA, Michel H, Sakaguchi K, Shabanowitz J, Appella E, Hunt DF, Engelhard VH. (1992) HLA-A2.1-associated peptides from a mutant cell line: A second pathway of antigen presentation. *Science* **255:** 1264–1266.

Hershko A, Ciechanover A. (1992) The ubiquitin system for protein degradation. *Annu. Rev. Biochem.* **61:** 761–807.

Hill A, Takiguchi M, McMichael A. (1993) Different rates of HLA class I molecule assembly which are determined by amino acid sequence in the alpha 2 domain. *Immunogenetics* **37:** 95–101.

Hill A, Barnett B, McMichael A, McGeoch D. (1994) HLA class I molecules are not transported to the cell surface in cell infected with herpes simplex virus types 1 and 2. *J. Immunol.* **152:** 2736–2741.

Hill A, Jugovic P, York I, Russ G, Bennink J, Yewdell J, Ploegh H, Johnson D. (1995) Herpes simplex virus turns off the TAP to evade host immunity. *Nature* **375:** 411–415.

Höchstenbach F, David V, Watkins S, Brenner M. (1992) Endoplasmic reticulum resident protein of 90 kilodaltons associates with the T- and B-cell antigen receptors and major histocompatibility complex antigens during their assembly. *Proc. Natl Acad. Sci. USA* **89:** 4734–4738.

Hosken NA, Bevan MJ. (1992) An endogenous antigenic peptide bypasses the class I antigen presentation defect in RMA-S. *J. Exp. Med.* **175:** 719–729.

Hunt DF, Henderson RA, Shabanowitz J, Sakaguchi K, Michel H, Sevilir N, Cox AA, Appella E, Engelhard VH. (1992) Characterization of peptides bound to class I MHC molecule HLA-A2.1 by mass spectrometry. *Science* **255:** 1261–1263.

Jackson M, Song E, Yang Y, Peterson P. (1992) Empty and peptide containing conformers of class I major histocompatibility complex molecules expressed in *Drosophila melanogaster* cells. *Proc. Natl Acad. Sci. USA* **89:** 12117–12121.

Jackson M, Nilsson T, Peterson P. (1993) Retrieval of transmembrane proteins to the endoplasmic reticulum. *J. Cell. Biol.* **121:** 317-333.

Jackson MR, Cohen DMF, Peterson PA, Williams DB. (1994) Regulation of MHC class I transport by the molecular chaperone, calnexin (p88, IP90). *Science* **263:** 384–387.

Jardetzky T, Lan W, Robinson R, Madden D, Wiley D. (1991) Identification of self-peptides bound to purified HLA-B27. *Nature* **353:** 326–329.

Jennings S, Rice P, Kloszewski E, Anderson R, Thompson D, Tevethia S. (1985) Effect of herpes simplex virus type 1 and 2 on surface expression of class I MHC antigens on infected cells. *J. Virol.* **56:** 757–761.

Kearse KP, Williams DB, Singer A. (1994) Persistence of glucose residues on core oligosaccharides prevents association of TCR-alpha and TCR-beta proteins with calnexin and results specifically in accelerated degradation of nascent TCR-alpha proteins within the endoplasmic reticulum. *EMBO J.* **13:** 3678–3686.

Kelly A, Powis S, Glynne R, Radley E, Beck S, Trowsdale J. (1991) A second proteasome-related gene in the human MHC class II region. *Nature* **353:** 667–668.

Kelly A, Powis SH, Kerr LA, Mockridge I, Elliott T, Bastin J, Uchanska-Ziegler B, Ziegler A, Trowsdale J, Townsend A. (1992) Assembly and function of the two ABC transporter proteins encoded in the human major histocompatibility complex. *Nature* **355:** 641–664.

Kleijmeer M, Kelly A, Geuze H, Slot J, Townsend A, Trowsdale J. (1992) MHC-encoded transporters are located in the ER and cis-Golgi. *Nature* **357:** 342–344.

Klein G. (1994) Epstein–Barr virus strategy in normal and neoplastic B cells. *Cell* **77:** 791–793.

Klenerman P, Rowland JS, McAdam S, *et al.* (1994) Cytotoxic T-cell activity antagonized by naturally occurring HIV-1 Gag variants. *Nature* **369:** 403–407.

Koelle D, Tiggers M, Burke R, Symington F, Riddel S, Abbo H, Corey L. (1993) Herpes simplex virus infection of human fibroblasts and keratinocytes inhibits recognition by cloned $CD8^+$ cytotoxic T lymphocytes. *J. Clin. Invest.* **91:** 961–969.

Kovacsovics-Bankowski M, Rock K. (1995) A phagosome-to-cytosol pathway for exogenous antigens presented on MHC class I molecules. *Science* **267:** 243–246.

Kulka RG, Raboy B, Schuster R, Parag HA, Diamond G, Ciechanover A, Marcus M. (1988) A Chinese hamster cell cycle mutant arrested at G2 phase has a temperature-sensitive ubiquitin-activating enzyme, E1. *J. Biol. Chem.* **263:** 15726–15731.

Kvist S, Hamann U. (1990) A nucleoprotein peptide of influenza A virus stimulates assembly of HLA-B27 class I heavy chains and beta 2-microglobulin translated in vitro. *Nature* **348:** 446–448.

Lapham C, Bacik I, Yewdell J, Kane K, Bennink J. (1993) Class I molecules retained in the endoplasmic reticulum bind antigenic peptides. *J. Exp. Med.* **177:** 1633–1641.

Levitskaya J, Coram M, Levitsky V, Imreh S, Steigerwald-Muller P, Klein G, Kurilla M, Masucci M. (1995) Inhibition of antigen processing by the internal repeat region of the Epstein–Barr virus nuclear antigen-1. *Nature* **357:** 685–688.

Levy F, Larsson R, Kvist S. (1991) Translocation of peptides through microsomal membranes is a rapid process and promotes assembly of HLA-B27 heavy chain and beta 2-microglobulin translated in vitro. *J. Cell. Biol.* **115:** 959–970.

Li Z, Srivastava P. (1993) Tumor rejection antigen gp96/grp94 is an ATPase: implications for protein folding and antigen presentation. *EMBO J.* **12:** 3143–3151.

Lippincott-Schawarts J, Yaun LC, Bonifacino JS, Klausner RD. (1989) Rapid distribution of Golgi proteins in the ER in cell trated with brefeldin A: Evidence for membrane cycling from Golgi to the ER. *Cell* **56:** 801–831.

Lippincott-Schwartz J, Donaldson JG, Schweizer A, Berger EG, Hauri HD, Yuan LC, Klausner R. (1990) Microtubule dependent retrograde transport of proteins into the ER in the presence of brefeldin A suggests an ER recycling pathway. *Cell* **60:** 821–836.

Lippincott-Schawarts J, Yaun L, Tipper C, Amherdt M, Orci L, Klausner R. (1991) Brefeldin A effects on endosomes, lysosomes, and the TGN suggest a general mechanism for regulating organelle structure and membrane traffic. *Cell* **67:** 601–616.

Livingstone AM, Powis SJ, Diamond AG, Butcher GW, Howard JC. (1989) A trans-acting major histocompatibility complex-linked gene whose alleles determine gain and loss changes in the antigenic structure of a classical class I molecule. *J. Exp. Med.* **170:** 777–795.

Machold R, Andree S, Van Kaer L, Ljunggren H-G, Ploegh H. (1995) Peptide influences the folding and intracellular transport of free major histocompatibility complex class I heavy chains. *J. Exp. Med.* **181:** 1111–1122.

Madden D, Gorga J, Strominger J, Wiley D. (1991) The structure of HLA-B27 reveals nonamer self-peptides bound in an extended conformation. *Nature* **353:** 321–325.

Margolese L, Waneck GL, Suzuki CK, Degen E, Flavell RA, Williams DB. (1993) Identification of the region on the class I histocompatibility molecule that interacts with the molecular chaperone, p88 (Calnexin, IP90). *J. Biol. Chem.* **268:** 17959–17966.

Mason D, Williams A. (1980) The kinetics of antibody binding in solution and at the cell surface. *Biochem. J.* **87:** 1–20.

Matsumura M, Saito Y, Jackson M, Song E, Peterson P. (1992) In vitro peptide binding to soluble empty class I major histocompatibility complex molecules isolated from transfected *Drosophila melanogaster* cells. *J. Biol. Chem.* **267:** 23589–23595.

Meier UC, Klenerman P, Griffin P, James W, Koppe B, Larder B, McMichael A, Phillips R. (1995) Cytotoxic T lymphocyte lysis inhibited by viable HIV mutants. *Science* **270:** 1360–1362.

Michalek MT, Grant EP, Gramm C, Goldberg AL, Rock KL. (1993) A role for the ubiquitin-dependent proteolytic pathway in MHC class I-restricted antigen presentation. *Nature* **363:** 552.

Momburg F, Ortiz NV, Neefjes J, *et al.* (1992) Proteasome subunits encoded by the major histocompatibility complex are not essential for antigen presentation. *Nature* **360:** 174–177.

Momburg F, Roelse J, Howard J, Butcher G, Hammerling G, Neefjes J. (1994a) Selectivity of MHC-encoded peptide transporters from human, mouse and rat. *Nature* **367:** 648–651.

Momburg F, Roelse J, Hammerling G, Neefjes J. (1994b) Peptide size selection by the major histocompatibility complex-encoded peptide transducer. *J. Exp. Med.* **179:** 1613–1623.

Monaco JJ, Cho S, Attaya M. (1990) Transport protein genes in the murine MHC: possible implications for antigen processing. *Science* **250:** 1723–1726.

Morrison L, Lukacker A, Braciale V, Fan D, Braciale T. (1986) Differences in antigen presentation to MHC class I and class II restricted influenza specific cytotoxic T lymphocyte clones. *J. Exp. Med.* **163:** 903–910.

Neefjes J, Momburg F, Hammerling G. (1993a) Selective and ATP-dependent translocation of peptides by the MHC-encoded transporter. *Science* **261:** 769–771.

Neefjes J, Hammerling G, Momburg F. (1993b) Folding and assembly of major histocompatibility complex class I heterodimers in the endoplasmic reticulum of intact cells precedes the binding of peptide. *J. Exp. Med.* **178:** 1971–1980.

Neisig A, Roelse J, Sijts A, Ossendorp F, Feltkamp M, Kast W, Melief C, Neefjes J. (1995) Major differences in transporter associated with antigen presentation (TAP)-dependent translocation of MHC class I-presentable peptides and the effect of flanking sequences. *J. Immunol.* **154:** 1273–1279.

Nuchtern J, Bonifacino J, Biddison W, Klausner R. (1989) Brefeldin A implicates egress from the endoplasmic reticulum in class I restricted antigen presentation. *Nature* **339:** 223–226.

Ojcius D, Godeau F, Abastado J-P, Casanova J-L, Kourilsky P. (1993) Real-time measurement of antigenic peptide binding to empty and preloaded single-chain major histocompatibility complex class I molecules. *Eur. J. Immunol.* **23:** 1118–1124.

Ortiz NV, Hammerling GJ. (1991) Surface appearance and instability of empty H-2 class I molecules under physiological conditions. *Proc. Natl Acad. Sci. USA* **88:** 3594–3597.

Ortmann B, Androlewicz MJ, Cresswell P. (1994) MHC class I/beta 2-microglobulin complexes associate with TAP transporters before peptide binding. *Nature* **368:** 864–867.

Ou WJ, Cameron PH, Thomas DY, Bergeron JJM. (1993) Association of folding intermediates of glycoproteins with calnexin during protein maturation. *Nature* **364:** 771–776.

Phillips RE, Rowland-Jones S, Nixon D, *et al.* (1991) Human immunodeficiency virus genetic variation that can escape cytotoxic T cell recognition. *Nature* **354:** 453–459.

Powis SJ, Deverson E, Coadwell W, Ciruela A, Huskisson N, Smith H, Butcher G, Howard J. (1992) Effect of polymorphism of an MHC-linked transporter on the peptides assembled in a class I molecule. *Nature* **357:** 211–215.

Rajagopalan S, Brenner MB. (1994) Calnexin retains unassembled major histocompatibility complex class I free heavy chains in the endoplasmic reticulum. *J. Exp. Med.* **180:** 407–412.

Rajagopalan S, Xu Y, Brenner MB. (1994) Retention of unassembled components of integral membrane proteins by calnexin. *Science* **263:** 387–390.

Rock KL, Gramm C, Rothstein L, Clark K, Stein R, Dick L, Hwang D, Goldberg AL. (1994) Inhibitors of the proteasome block the degradation of most cell proteins and the generation of peptides presented on MHC class I molecules. *Cell* **78:** 761–771.

Rotem-Yehudar R, Winogråd H, Sela S, Coligan J, Ehrlich R. (1994) Downregulation of peptide transporter genes in cell lines transformed with the high oncogenic adenovirus 12. *J. Exp. Med.* **180:** 477–488.

Sadegh-Nasseri S, Germain R. (1991) A role for peptide in determining MHC class II structure. *Nature* **353:** 167–169.

Sadegh-Nasseri S, McConnell H. (1989) A kinetic intermediate in the reaction of an antigenic peptide and I-Ek. *Nature* **337:** 274–276.

Schrier P, Bernards RRV, Houweling A, van der Eb A. (1983) Expression of class I MHC antigens switched off by highly oncogenic adenovirus 12 in transformed rat cells. *Nature* **305:** 771–775.

Schumacher T, Heemels M, Neefjes J, Kast W, Melief C, Ploegh H. (1990) Direct binding of peptide to empty MHC class I molecules on intact cells and *in vitro*. *Cell* **62:** 563–567.

Schumacher T, Kantesaria D, Hemels M, Ashton-Rickardt P, Shepherd J, Fruth K, Yang Y, Peterson P, Tonegawa S, Ploegh H. (1994) Peptide length and sequence specificity of the mouse TAP1/TAP2 translocator. *J. Exp. Med.* **179:** 533–540.

Scott J, Dawson J. (1995) MHC class I expression and transport in a calnexin deficient cell line. *J. Immunol.* **155:** 143–148.

Sibille C, Gould K, Willard-gallo K, Thomson S, Rivett J, Powis S, Butcher G, Baetselier P. (1995) LMP2+ proteasome are required for the presentation of specific antigens to cytotoxic T lymphocytes. *Curr. Biol.* **5:** 923–930.

Shepherd J, Schumacher T, Ashton-Rickardt P, Imaeda S, Ploegh H, Janeway C, Tonegawa S. (1993) TAP1-dependent peptide translocation in vitro is ATP dependent and peptide selective. *Cell* **74:** 577–580.

Snyder HL, Yewdell JW, Bennink JR. (1994) Trimming of antigenic peptides in an early secretory compartment. *J. Exp. Med.* **180:** 2389–2394.

Spies T, Bresnahan M, Bahram S, Arnold D, Blanck G, Mellins E, Pious D, DeMars R. (1990) A gene in the human major histocompatibility complex class II region controlling the class I antigen presentation pathway. *Nature* **348:** 744–747.

Spies T, Cerundolo V, Colonna M, Cresswell P, Townsend A, DeMars R. (1992) Presentation of viral antigen by MHC class I molecules is dependent on a putative peptide transporter heterodimer. *Nature* **355:** 644–646.

Srivastava P, Deleo A, Old L. (1986) Tumour rejection antigens of chemically induced sarcomas of inbred mice. *Proc. Natl Acad. Sci. USA* **83:** 3407–3411.

Stearz U, Karasuyama H, Garner A. (1988) Cytotoxic T lymphocytes against a soluble protein. *Nature* **329:** 449–450.

Sugita M, Brenner MB. (1994) An unstable beta-2-microglobulin: Major histocompatibility complex class I heavy chain intermediate dissociates from calnexin and then is stabilized by binding peptide. *J. Exp. Med.* **180:** 2163–2171.

Suh W, Cohen D, Fruh K, Wang K, Peterson P, Williams D. (1994) Interaction of MHC class I molecules with the transporter associated with antigen processing. *Science* **264:** 1322–1326.

Suto R, Srivastava P. (1995) A mechanism for the specific immunogenicity of heat shock protein-chaperoned peptides. *Science* **269:** 1585–1588.

Tector M, Zhang Q, Salter R. (1994) Phosphatase inhibitors block in vivo binding of peptides to class I major histocompatibility complex molecules. *J. Biol. Chem.* **269:** 25816–25822.

Tevethia S, Flyer D, Tjian R. (1980) Biology of simian virus 40 (SV40) transplantation antigen (TrAg) VI. Mechanism of induction of SV40 transplantation immunity in mice by purified SV40 T antigen (D2 protein). *Virology* **107:** 13–18.

Townsend ARM, Gotch FM, Davey J. (1985) Cytotoxic T cells recognise fragments of influenza nucleoprotein. *Cell* **42:** 457–467.

Townsend A, Rothbard J, Gotch F, Bahadur B, Wraith D, McMichael A. (1986a) The epitopes of influenza nucleoprotein recognized by cytotoxic T lymphocytes can be defined with short synthetic peptides. *Cell* **44:** 959–968.

Townsend A, Bastin J, Gould K, Brownlee G. (1986b) Cytotoxic T lymphocytes recognize influenza haemagglutinin that lacks a signal sequence. *Nature* **234:** 575–577.

Townsend A, Ohlen C, Bastin J, Ljunggren HG, Foster L, Karre K. (1989) Association of class I major histocompatibility heavy and light chains induced by viral peptides. *Nature* **340:** 443–448.

Townsend A, Bastin J, Gould K, Brownlee G, Andrew M, Coupar B, Boyle D, Chan S, Smith G. (1988) Defective presentation to class I-restricted cytotoxic T lymphocytes in vaccinia-infected cells is overcome by enhanced degradation of antigen. *J. Exp. Med.* **168:** 1211–1224.

Towsend A, Elliott T, Cerundolo V, Foster L, Barber B, Tse A. (1990) Assembly of MHC class I molecules analyzed in vitro. *Cell* **62:** 285–295.

Trowsdale J, Hanson I, Mockridge I, Beck S, Townsend A, Kelly A. (1990) Sequences encoded in the class II region of the MHC related to the 'ABC' superfamily of transporters. *Nature* **348:** 741–744.

Tussey L, Rowland-Jones S, Zheng T, Androlewicz M, Cresswell P, Frelinger J, McMichael A. (1995) Different MHC class I alleles compete for presentation of overlapping viral epitopes. *Immunity*

3: 65–77.

Vaessen R, Houweling A, van der Eb A. (1987) Post-transcriptional control of class I MHC mRNA expression in adenovirus 12-transformed cells. *Science* **235:** 1486–1488.

Van Endert P, Tampe R, Meyer T, Tisch R, Bach J-F, McDevitt H. (1994) A sequential model for peptide binding and transport by the transporters associated with antigen processing. *Immunity* **1:** 491–500.

Van Kaer L, Ashton-Rickardt P, Ploegh H, Tonegawa S. (1992) TAP1 mutant mice are deficient in antigen presentation, surface class I molecules and CD4–8+ T cells. *Cell* **71:** 1205–1214.

Van Kaer L, Aston-Rickardt P, Eichelberger M, Gaczynska M, Nagashima K, Rock K, Goldberg A, Doherty P, Tonegawa S. (1994) Altered peptidase and viral specific T cell response in LMP2 mutant mice. *Immunity* **1:** 533–541.

Vitiello A, Potter TA, Sherman LA. (1990) The role of beta 2-microglobulin in peptide binding by class I molecules. *Science* **250:** 1423–1426.

Wei ML, Cresswell P. (1992) HLA-A2 molecules in an antigen processing mutant cell contain signal sequence-derived peptides. *Nature* **356:** 443–446.

Weis J, Murre C. (1985) Differential expression of H-2Dd and H-2Ld histocompatibility antigens. *J. Exp. Med.* **161:** 356–365.

Williams D, Swiedler S, Hart G. (1985) Intracellular transport of membrane glycoproteins: two closely related histocompatibility antigens differ in their rates of transmit to the cell surface. *J. Cell Biol.* **101:** 725–734.

Wraith D, Vessey A. (1986) Influenza virus-specific cytotoxic T-cell recognition: simulation of nucleoprotein-specific clones with intact antigen. *Immunology* **59:** 173–177.

Yewdell J, Bennink J. (1989) Brefeldin A specifically inhibits presentation of protein antigens to cytotoxic T lymphocytes. *Science* **244:** 1072–1075.

Yewdell J, Esquivel F, Arnold D, Spies T, Eisenlohr L, Bennink J. (1993) Presentation of numerous viral peptides to mouse major histocompatibility complex (MHC) class I-restricted T lymphocytes is mediated by the human MHC-encoded transporter or by a hybrid mouse–human transporter. *J. Exp. Med.* **177:** 1785–1790.

Yewdell J, Lapham C, Bacik I, Spies T, Bennink J. (1994) MHC-encoded proteasome subunits LMP2 and LMP7 are not required for efficient antigen presentation. *J. Immunol.* **152:** 1163–1170.

Yide J, Wai-Kuo S, Berkower I. (1988) Human T cell response to the surface antigen of hepatitis B (HBsAg). Endosomal and non-endosomal processing pathways are accessible to both endogenous and exogenous antigen. *J. Exp. Med.* **168:** 293–297.

York I, Roop C, Andrews D, Riddel S, Graham F, Johnson D. (1994) A cytosolic herpes simplex virus protein inhibits antigen presentation to CD8+ T lymphocytes. *Cell* **77:** 525–535.

Zhou X, Glas R, Liu T, Ljunggren HG, Jondal M. (1993) Antigen processing mutant T2 cells present viral antigen restricted through H-2Kb. *Eur. J. Immunol.* **23:** 1802–1808.

Zinkernagel R, Doherty P. (1974) Immunological surveillance against altered self components by sensitized T lymphocytes in lymphocytic choriomeningitis. *Nature* **251:** 547–549.

10

MHC class II assembly and transport

Richard W. Wubbolts, Jacques J. Neefjes and
Mar Fernandez-Borja

10.1 Introduction

The immune system can eradicate pathogens from the body by several regulated destruction mechanisms. These include removal of intact antigen by antibodies from the body fluids (i.e. the humoral immune response). Pathogens that have invaded host cells will not be accessible to this defence mechanism. The cellular immune response, however, can recognize cells that are infected or altered.

The main cellular immune response is elicited by thymocytes that recognize pathogen protein fragments in the context of major histocompatibility complex (MHC) molecules. These highly polymorphic peptide-binding molecules can be subdivided into two main groups: MHC class I and MHC class II. Thymocytes bearing the CD8 co-receptor for the T-cell receptor (TCR) complex primarily recognize antigenic fragments in the context of MHC class I molecules (Rammensee *et al.*, 1993). CD4 co-stimulation of the TCR complex confers immune reactivity towards fragments in the context of MHC class II molecules (reviewed in Brodsky and Guagliardi, 1991; Cresswell, 1994; Germain, 1994; Germain and Margulies, 1993; Neefjes *et al.*, 1991 and see Chapter 14).

Peptide extraction studies have shown that MHC class I molecules mainly associate with protein fragments (of 8–9 amino acids in length) of nuclear or cytosolic origin (Rammensee *et al.*, 1993; Townsend and Bodmer, 1989). MHC class II molecules primarily bind peptides (of 7–25 amino acids) derived from proteins that can be degraded in endosomal compartments (Chicz *et al.*, 1992; Hunt *et al.*, 1992; Mircheff *et al.*, 1994; Newcomb and Cresswell, 1993; Rotzschke and Falk, 1994; Rudensky *et al.*, 1992; Unanue, 1984). Why MHC class I and II molecules bind peptides of different length could be explained by the differences in their three-dimensional structures (Bjorkman and Burmeister, 1994; Brown *et al.*, 1988, 1993; Jardetzky *et al.*, 1994; Stern *et al.*, 1994). This is discussed elsewhere in this book (see Chapter 11).

In general, MHC class II molecules are expressed on so-called professional antigen-presenting cells (APCs), that is, macrophages, monocytes, dendritic cells and B cells (Knight and Stagg, 1993). Pathogenic fragments presented in the context of MHC class II molecules usually do not induce T-cell-mediated cytotoxicity, but locally modulate the immune response by, for example, the production of cytokines (reviewed in Paul and Seder, 1994). This, in contrast to

HLA and MHC: genes, molecules and function, edited by M.J. Browning and A.J. McMichael.
© 1996 BIOS Scientific Publishers Ltd, Oxford.

MHC class I-mediated reactivity, will not kill the cells but allows the APC to boost a local immune response when allogenic proteins are present in the immediate environment. It should be noted, however, that a reasonable portion of the CD4 positive T-cell population does show cytotoxic activity. Migration of the APCs or activated T or B cells through the body can transpose the local response to a more generalized immune response.

Elimination of the genes encoding the MHC class II molecules in mice have established that MHC class II molecules are essential for the selection of $CD4^+$ T cells (Cosgrove *et al.*, 1991). $CD4^+$ T cells are necessary for mounting an antibody response, which illustrates the strong connections between humoral and cellular immune responses.

Specific TCR/CD4 recognition of MHC class II/peptide complex can occur in a transient fashion (Valitutti *et al.*, 1995). Binding of peptide to MHC class II molecules, therefore, can induce a prolonged immune response stimulating many T cells during its existence. In order to do so, peptide/MHC class II complexes are very stable and invariably expressed at the plasma membrane (Buus *et al.*, 1986; Lanzavecchia and Watts, 1994; Lanzavecchia *et al.*, 1992; Nelson *et al.*, 1994).

In this chapter we will describe the biosynthesis of MHC class II molecules and how this results in selection of peptides in endosomal/lysosomal compartments prior to cell-surface expression.

10.2 Biosynthesis of MHC class II molecules

10.2.1 Assembly of MHC class II molecules

The peptide-binding moiety of MHC class II molecules is formed by two transmembrane glycoproteins, the α-chain and the β-chain. The $\alpha 1$ and $\beta 1$ domains of both subunits cooperate to form a peptide-binding groove facing away from the plasma membrane. The $\alpha 2$ and $\beta 2$ domains form membrane proximal immunoglobulin-like domains that support the peptide-binding portion of the class II molecule. Intrachain disulphide bonds are present within the $\beta 1$, $\alpha 2$ and $\beta 2$ domains, while both chains associate non-covalently (Brown *et al.*, 1988, 1993).

Polymorphism of the MHC class II complex is the result of genetic variations in the β-chain (Bell *et al.*, 1985) resulting mainly in alteration of the peptide-binding site, that is the region that is seen by the TCR complex (Tate *et al.*, 1995). The human genes are encoded by the *DR, DP, DQ* loci in the MHC II region, in mice by the *I-A* and *I-E* loci (Monaco, 1993). As a result, humans heterozygous for all three alleles will express six different MHC class II proteins at the cell surface.

The α-chain has an apparent molecular weight (MW) of 33–35 kDa while the MW of the β-chain ranges from 25 to 30 kDa. Signal sequences for endoplasmic reticulum (ER) targeting and membrane translocation are present on α- and β-chain amino termini. The transmembrane regions (20–25 amino acids) result in a type I orientation leaving a short carboxy terminal tail (12–15 amino acids) cytoplasmically exposed.

During translation and insertion in the ER membrane, N-linked glycans are added to both chains, one to the β-chain and two to the α-chain (*Figure 10.1*; Cresswell *et al.*, 1987). The α-chain is also reported to be O-glycosylated and its

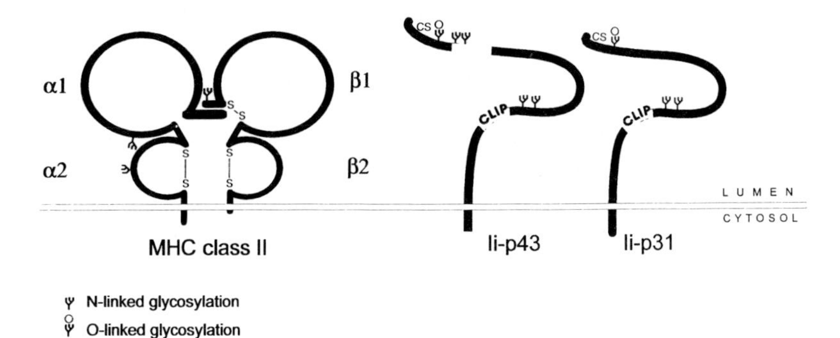

Figure 10.1. An MHC class II heterodimer and invariant chain (Ii). N-linked and O-linked glycosylation sites are indicated. Shaded areas represent regions that vary among the different forms of Ii as a result of alternative splicing or alternate insertion. (CS, chondroitin sulphate.)

cytoplasmic tail can be modified by palmitoylation and phosphorylation (Kaufman *et al.*, 1984).

In vitro expression studies showed that both chains rapidly associate after biosynthesis (Bijlmakers *et al.*, 1994b; Hedley *et al.*, 1994; Miller and Germain, 1986; Nijenhuis *et al.*, 1994; Sekaly *et al.*, 1986). Co-expression of soluble forms of class II subunits in insect cells (Kozono *et al.*, 1994; Sloan *et al.*, 1995; Stern and Wiley, 1992) or *Escherichia coli* (Kozono *et al.*, 1994) renders peptide-binding-competent molecules. Transfection experiments also showed that intra-isotypic mismatched α- and β-chains did assemble but were not transported out of the ER as assessed by continued endoglycosidase H sensitivity (Sant *et al.*, 1991). Apparently, there is a preference for MHC class II association of subunits encoded by the same locus. Thus, association of α- and β-chain occurs without requirement of additional factors to form the peptide-binding moiety. Apparently, the cytoplasmic tail and transmembrane region of the proteins are dispensable for the assembly process. However, they may affect subtle changes in structure causing loss of conformational-dependent epitopes and a block in transport (Cosson and Bonifacino, 1992; Wade *et al.*, 1995).

10.2.2 Invariant chain, an MHC class II specific chaperonin

In vivo, heterodimers in the ER rapidly associate with a third polypeptide, the invariant chain (Ii or CD74) (Claesson *et al.*, 1983; Kvist *et al.*, 1982; Machamer and Cresswell, 1982; Strubin *et al.*, 1986; Sung and Jones, 1981). This association greatly affects MHC class II function (reviewed in Cresswell, 1994, 1992; Sant and Miller, 1994). It is shown to affect MHC class II conformation (Anderson and Cresswell, 1994; Melnick and Argon, 1995; Peterson and Miller, 1990), assembly (Bikoff *et al.*, 1995; Teyton *et al.*, 1990), and transport (reviewed in Cresswell, 1992; Humbert *et al.*, 1993; see below), inhibiting premature peptide binding

(Teyton *et al.*, 1990) but promoting high affinity peptide binding (Sherman *et al.*, 1995; Sloan *et al.*, 1995) and affecting antigen presentation (Bodmer *et al.*, 1994; Naujokas *et al.*, 1993; Peterson and Miller, 1992).

Studies on Ii-deficient mice showed that MHC class II molecules are present at the cell surface of APCs in lower amounts (four- to eight-fold); transport of the complexes is impeded, while assembly seems affected in an allele-specific manner. The antigen-presenting capacity of splenocytes from these mice was reduced for most but not all antigens. Moreover, antigens offered to APCs that needed antigenic processing were not presented, while peptides were (Bikoff *et al.*, 1993, 1995; Elliott *et al.*, 1994; Schaiff *et al.*, 1992; Viville *et al.*, 1993). The MHC class II complexes that did mature showed reduced stability in sodium dodecyl sulphate (SDS) (Viville *et al.*, 1993), a biochemical feature previously shown for MHC class II molecules tightly bound to peptides (Germain and Hendrix, 1991; Sadegh-Nasseri and Germain, 1991; see below).

10.2.3 Structural features of the invariant chain

Ii is a non-polymorphic transmembrane glycoprotein (Koch *et al.*, 1987; McKnight *et al.*, 1989; O'Sullivan *et al.*, 1987; Strubin *et al.*, 1986). Its transmembrane (TM) region serves as a non-cleavable ER membrane insertion signal, as is often observed for type II-oriented TM proteins (Lipp and Dobberstein, 1988). Hence, Ii exposes its amino terminus at the cytosolic face of the ER membrane.

Different forms of the non-polymorphic Ii protein exist. The majority of Ii has a MW of 31–33 kDa (p33). Sequence data given in this chapter are of the human Ii-p33 which is 216 amino acids long. In humans, usage of an alternative start codon (at amino acid −16 relative to the p33 start methionine) generates a 35 kDa protein (O'Sullivan *et al.*, 1987; Strubin *et al.*, 1986). Humans, mice and rats contain an intron splice variant that contains 64 additional amino acids near the carboxy terminus of the protein, inserted at amino acid 193 (Koch, 1988; Koch *et al.*, 1987; O'Sullivan *et al.*, 1987; Strubin *et al.*, 1986). The insert contains five cysteine residues and shows homology with repetitive sequences in thyroglobin type I (McKnight *et al.*, 1989) and with a region within entactin, a basement membrane protein (Cresswell, 1994). Thus, mice and rats express two Ii forms, p33 and p41, while humans express the two additional Ii variants, p35 and p43.

Extensive post-translational modifications further increase the variation of Ii (reviewed in Cresswell, 1992; Sant and Miller, 1994). The p33 and p35 variants contain two consensus sites for N-linked glycosylation (at Asn 114 and 120) both of which are used (Claesson *et al.*, 1983; Machamer and Cresswell, 1982). The splice variant introduces two additional N-linked glycosylation sites (Koch, 1988; Machamer and Cresswell, 1984). O-linked glycosylation, sialylation on both N- and O-linked glycans (Machamer and Cresswell, 1984), phosphorylation (Spiro and Quaranta, 1989), palmitoylation of a cysteine residue cytoplasmic of the TM region (Cys 44) (Koch and Hammerling, 1986; Spiro and Quaranta, 1989) increase heterogeneity. Additionally, for a minor portion of Ii, heterogeneity is increased further by addition of chondroitin sulphates to a serine residue 202 which forms the attachment site for proteoglycan formation on the Ii chain (Giacoletto *et al.*, 1986; Sant *et al.*, 1985).

10.2.4 Retention and sorting signals in MHC class II subunits

Ii variants were shown to form trimers rapidly after translation (Lamb and Cresswell, 1992; Markes *et al.*, 1990), even when the cytoplasmic and TM regions were removed by proteinase K digestion (Marks *et al.*, 1990). Transfecting cDNAs of the respective Ii variants showed that these trimers incorporated different forms of the Ii chains (Arunachalam *et al.*, 1994). Trimers containing p35/p43 were efficiently retained in the ER (Arunachalam *et al.*, 1994; Lotteau *et al.*, 1990; Marks *et al.*, 1990; Roche *et al.*, 1991). A fraction of the complexes formed with p33 and/or p41 ended up in endosomal or lysosomal compartments (Bakke and Dobberstein, 1990; Lamb and Cresswell, 1992; Simonis *et al.*, 1989) upon overexpression, thereby causing a swollen morphology of endosomal structures (Pieters *et al.*, 1993; Romagnoli *et al.*, 1993; Schutze *et al.*, 1994). Mutational analysis of the 16 amino acids upstream of the p33 start codon revealed an ER retention signal which was attributed to a doublet of basic amino acids present in human p35 and p43 (Schutze *et al.*, 1994). It is thought that association of MHC class II with Ii conceals this retention signal, thereby releasing both Ii and MHC class II from the ER.

MHC class II associated with Ii, as trimers either with or without p35/p43 (Arunachalam *et al.*, 1994), are transported to an endosomal compartment (Lamb *et al.*, 1991; Loss and Sant, 1993; Lotteau *et al.*, 1990; Neefjes *et al.*, 1990). It was shown that preformed Ii trimers gained MW by three sequential additions of MHC class II heterodimers (Lamb and Cresswell, 1992; Roche *et al.*, 1991). Nonamer formation ($Ii_3(\alpha\beta)_3$) caused increase in MW and sedimentation coefficient of the Ii trimer, yet it only marginally altered its Stokes radius (Roche *et al.*, 1991). Thus, Ii trimers capture MHC class II molecules in the ER into their extended structures, masking ER retention signals in the Ii chains. The nonameric complex is then transported through the Golgi, allowing conversion of high-mannose to complex oligosaccharides and sialic acid modification. A proteolytic fragment of Ii, p25, that lacks its TM region is found associated with MHC class II molecules early during biosynthesis but remains largely endoglycosidase-H (endo-H) sensitive. This indicates that it is generated and remains present in a location or conformation inaccessible to the glycosylation enzymes in the Golgi that forms complex sugars (Pieters *et al.*, 1991; Schutze *et al.*, 1994). The function of this ER proteolytic processing of Ii is unknown.

One group reported that transported 'empty' Ii trimers largely remained endo H sensitive, which was suggested to be the result of ER egress by the formation of autophagocytic vesicles that fuse with lysosomal-like structures. Inhibiting the autophagocytic pathway by 3-methyl-adenine abrogated transport from the ER (Chervonsky and Sant, 1995). Other groups showed that Ii chains did acquire complex type carbohydrates upon transport to endosomes and identified endosomal localization signals in the amino terminal region at Leu-Ile (amino acids 7, 8; Bremnes *et al.*, 1994; Pieters *et al.*, 1993) and Pro-Met-Leu (amino acids 15–17; Bakke and Dobberstein, 1990; Bremnes *et al.*, 1994; Odorizzi *et al.*, 1994). These signals are essential for the delivery of MHC class II dimers (Lotteau *et al.*, 1990) or Ii-transferrin receptor chimeras (Odorizzi *et al.*, 1994) to a late endosomal/lysosomal compartment. The same regions mediate localization to clathrin-coated pits and promote rapid internalization, although only 19% of the Ii-transferrin receptor followed this pathway (Odorizzi *et al.*, 1994). The TM region of Ii was shown to affect endosomal localization as well (Odorizzi *et al.*, 1994). Moreover, when chimeric Ii molecules were constructed that contained only ten amino acids of the trans-Golgi network (TGN) resident protein 1,4-β galactosyltransferase at the luminal portion in the transmembrane region, they were retained in the trans-Golgi (Nilsson *et al.*, 1991). This study indicated that either introduction of the galactosyltransferase motif or deletion of an Ii transmembrane signal was dominant over the Ii sorting signals present in the cytoplasmic tail of Ii.

Other signals affecting MHC class II transport are present in MHC class II chains themselves; for example, amino acid sequences in the luminal portions of MHC class II molecules (at amino acids 80–82; Chervonsky *et al.*, 1994) have been claimed to contribute to the efficacy of transport. Additionally, cytoplasmic tail deletions of both MHC class II α- and β-chain reduced the internalization efficiency from the cell surface (Pinet *et al.*, 1995).

Finally, signals in the Ii luminal portion can modulate multimerization as tested by *in vitro* translation (amino acids 163–183; Bijlmakers *et al.*, 1994a) or by transfection studies (mouse amino acids 153–215; Bertolino *et al.*, 1995). Trimerization of Ii may have a function in forming the endosome targetting signal. Co-expressing wild type Ii with Ii lacking N-terminal sorting signals showed increased surface expression of the Ii trimers. Only one in three Ii at the cell surface was wild-type Ii, although equal amounts of the proteins were present in the stable cell lines. This suggested that a single Ii endosomal localization signal present in the $Ii_3(\alpha\beta)_3$ nonameric complex is not sufficient for efficient endosomal localization. However, presence of one Ii cytoplasmic tail did increase the internalization rate of surface oligomeric Ii–MHC class II complexes (Arneson and Miller, 1995). This suggests that the Ii tail contains an additional cell-surface internalization signal, which is in agreement with other studies (Henne *et al.*, 1995; Odorizzi *et al.*, 1994; Roche *et al.*, 1993).

This clearly illustrated how multimerization may modulate sorting of the Ii–MHC class II complexes, possibly varying at different locations in the cell. Thus, MHC class II molecules are transported by direct sorting to endosomal structures from the TGN (Loss and Sant, 1993; Neefjes and Ploegh, 1992; Peters *et al.*, 1991b) or, alternatively, transport to the cell surface is followed by rapid internalization (Henne *et al.*, 1995; Reid and Watts, 1990; Roche *et al.*, 1992). In either case, MHC class II complexes are delayed in transport from endosomal compartments by retention mechanisms allowing peptide binding to occur in these compartments.

10.2.5 MHC class II association with ER resident proteins

Folding intermediates of the MHC class II oligomeric complex interact with various ER chaperonins. Many of these proteins have promiscuous protein-binding activity and facilitate protein folding by, for example, preventing aggregation. ER chaperonins contain ER retention signals, such as the KDEL (i.e. Lys, Asp, Glu, Leu) motif (or a different cytoplasmic tail motif for p88), restricting localization of the protein and its associated folding intermediate (reviewed in Bergeron *et al.*, 1994; Gething and Sambrook, 1992; Melnick and Argon, 1995). Three different ER chaperonins have been reported to associate with MHC class II. Calnexin (or p88), a transmembrane molecule, has been shown to preferentially bind monoglycosylated proteins, however the non-glucosylated protein CD3ϵ also binds p88. Immunoglobulin binding protein (BIP) (or GP78) is a soluble ER protein that binds relatively hydrophobic peptide stretches. GRP94 exists in a soluble and transmembrane form, its substrate specificity is not known but it has been shown to bind Ig chains in a ternary complex with BIP or after BIP association.

Calnexin only binds intermediate complexes in the assembly process of Ii and MHC class II molecules (Anderson and Cresswell, 1994; Schreiber *et al.*, 1994). Association of BIP was seen for free MHC class II subunits or heterodimers only in cells lacking Ii expression (Bonnerot *et al.*, 1994; Nijenhuis and Neefjes, 1994). Furthermore, BIP was shown to associate with a slowly forming high molecular 'aggregate' of the β-chain (in presence of Ii but absence of α-chain). In time, the ratio BIP: β-chain (from 1:1 to 2:1) increased and correlated with maturation of the β-chain in the aggregate. This maturation was proposed to be a prerequisite for pre-Golgi degradation of the β-chain (Cotner and Pious, 1995). Again in the absence of Ii, MHC class II heterodimers co-precipitated with GRP94 (Schaiff *et al.*, 1992). Also a protein shown to have protein disulphide isomerase (PDI) activity, ERp72, was co-purified in this study (Schaiff *et al.*, 1992). PDI activity breaks and reforms disulphide bridges between cysteine residues allowing disulphide bond reshuffling within a protein. Other unidentified ER resident proteins of 200 kDa (Anderson and Miller, 1992), 74 kDa (Schaiff *et al.*, 1992) and 62–64 kDa (named gp62; Bonnerot *et al.*, 1994) have been isolated with MHC class II from the Ii-deficient cells.

10.2.6 Functional implications of Ii association

MHC class II complexes are inefficiently folded in the absence of Ii, this causes a delay in the release from ER chaperonins and thus, reduces transport rates to the cell surface. Moreover, it was recently shown that Ii may reverse a transient aggregation state in which the MHC II molecules are present shortly after translation (Marks *et al.*, 1995). Not only the folding process but also the MHC class II molecules that are eventually cell-surface exposed are altered, since conformation-specific epitope recognition by monoclonal antibodies (Anderson and Miller, 1992; Naujokas *et al.*, 1993; Peterson and Miller, 1990) as well as certain T-cell responses (Naujokas *et al.*, 1993; Rath *et al.*, 1992) were different in Ii-deficient cells.

An important function suggested for Ii association was to prevent premature peptide binding to MHC class II molecules in the ER. Indeed, no peptides are isolated from Ii-associated MHC class II complexes (Newcomb and Cresswell, 1993) and isolated Ii prevents MHC class II molecules from binding peptides *in vitro* (Bijlmakers *et al.*, 1994; Teyton *et al.*, 1990). In support of this, it was shown that mild denaturation of the $Ii_3(\alpha\beta)_3$ complexes or *in vitro* proteolysis of associated Ii by the cysteine protease cathepsin B generated peptide-binding-competent MHC class II molecules (Roche *et al.*, 1993). However, stably transfected cell lines (Busch *et al.*, 1995) and analyses of the Ii deficient mice (Bikoff *et al.*, 1993; Elliott *et al.*, 1994; Viville *et al.*, 1993) showed that introduction of presentable peptides into the ER, even in the absence of Ii, did not lead to MHC class II binding. This illustrates that peptide binding by MHC class II molecules in the ER will be inefficient due to Ii association and an unfavourable environment for peptide binding (e.g. neutral pH, redox potential or the absence of co-factors like HLA-DM, see below).

Ii-mediated transport of MHC class II molecules to a lysosomal environment that favours peptide binding generally results in a profound effect on antigen presentation (Roche *et al.*, 1992; Stockinger *et al.*, 1989). In some studies, this seemed to be independent of the N-terminal endosomal localization signals showing that MHC class II internalized via the cell surface is equally capable of presenting peptides (Anderson *et al.*, 1993). Interestingly, the p41 variant of Ii appears to be superior in enhancing the MHC class II antigen presentation (Peterson and Miller, 1992). The chondroitin sulphate modification of a small Ii subpopulation (2–5%) markedly enhances antigen-presenting capacity to primary T cells as well. This was shown to be the result of the Ii-chondroitin sulphate interaction with CD44 on the T cells which then could provide co-stimulation of the TCR (Naujokas *et al.*, 1993).

Another function of Ii was suggested by its structural homology to cystatins. Cystatins are endogenous cathepsin inhibitors. *In vitro*, cathepsin L and H activities were inhibited by Ii in the low micromolar range. On the other hand, cathepsin B activity seemed unaffected (Katunuma *et al.*, 1994). This suggests that dissociated Ii from MHC class II molecules may inhibit specific proteolytic activity in the endosomal compartment and thus influence antigenic peptide generation. Furthermore, Ii may prolong residence of molecules in endosomes by a non-specific mechanism. Transfer from early to late endosomes of the endocytic tracer horseradish peroxidase is significantly delayed in transfected cells expressing wild-type Ii with respect to cells that express Ii lacking its first 20 amino acids (Gorvel *et al.*, 1995). Ii has been shown to associate with some MHC class I molecules in the ER and mediate sorting of these molecules to endosomal structures, indicating that Ii itself may be able to mediate transport of associated molecules (Sugita and Brenner, 1995). The function of this interaction is unclear at this moment.

10.3 Intracellular transport of class II molecules

The transport of MHC class II to the plasma membrane diverts in the TGN from the constitutive secretory pathway. After passage through the TGN, class II molecules remain intracellular for 1–3 hours before being expressed at the cell surface (Neefjes *et al.*, 1990). During this period, class II molecules are sorted to an acidic endosomal/lysosomal compartment where class II-associated Ii is degraded (*Figure 10.2*). Class II molecules are now able to acquire peptides derived from antigens that have reached the same compartment either by phagocytosis, pinocytosis or by receptor-mediated endocytosis [e.g. via the cell-surface immunoglobulins on B cells, Fc receptors or the recently described DEC-205 receptor in dendritic cells (Jiang *et al.*, 1995)]. The newly formed class II-peptide complexes are then transported to the cell surface along an as yet unknown pathway.

Peptide association confers an unusual biochemical feature on MHC class II molecules: part of the MHC class II molecules become resistant to incubation of SDS at temperatures lower than 50°C and consequently migrate at approximately

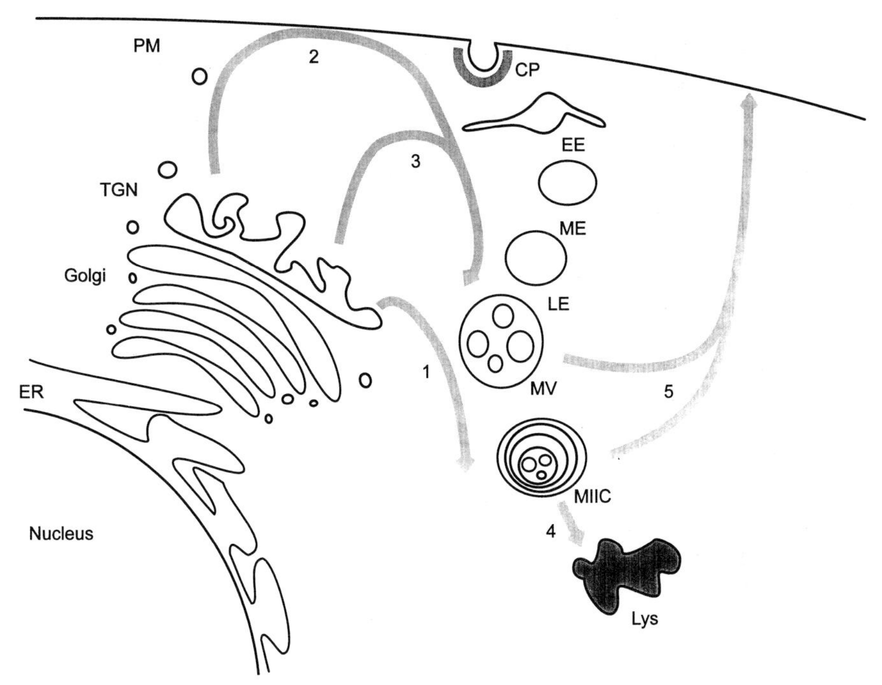

Figure 10.2. Intracellular transport pathways of MHC class II molecules. (1) Direct sorting to MIIC; (2) sorting via the cell surface followed by rapid internalization; (3) direct sorting to early endosomal structures; (4) some MHC class II molecules may eventually end up in lysosomes; (5) unknown transport route for peptide-loaded class II molecules to the cell surface. (CP, coated pit; EE, early endosome; ME, middle endosome; LE, late endosome; MV, multivesicular structures; MIIC, MHC class II compartment; Lys, lysosome; PM, plasma membrane; TGN, trans-Golgi network; ER, endoplasmic reticulum.)

50–55 kDa upon SDS polyacrylamide gel electrophoresis (SDS-PAGE) under non-reducing conditions (Dornmair *et al.*, 1989). *In vitro* peptide-loading experiments show that these SDS stable 'compact' complexes are slowly formed after peptide loading (Buus *et al.*, 1986). Interestingly, the loading is more efficient at acidic pH and is stabilized by subsequent neutral pH treatment (Harding *et al.*, 1991b; Jensen, 1990; Sadegh-Nasseri and Germain, 1991). *In vitro*, the half-life of these SDS stable class II-peptide complexes is long only when they are formed slowly (Sadegh-Nasseri and McConnell, 1989; Sadegh-Nasseri *et al.*, 1994). It should be mentioned that SDS-stability is merely an easy readout for peptide loading of MHC class II molecules, but it is unclear how it translates into a physiological function. It has been claimed that peptides that induce SDS-stability have a longer half-life and will be present in the cell surface MHC class II complex in higher concentrations than other peptides (Lanzavecchia *et al.*, 1992; Nelson *et al.*, 1994). Furthermore, longer presence of these MHC-associated antigenic peptides at the cell surface will allow more TCR to be triggered (Valitutti *et al.*, 1995).

10.3.1 Where do class II molecules intersect the endocytic pathway?

An essential event in the intracellular transport of class II molecules is their sorting in the TGN to endosomal compartments where they can interact with peptides. The intracellular site where newly synthesized class II molecules intersect the endocytic pathway, as well as the compartments involved in peptide loading are still a matter of debate. Three main routes have been suggested:

(i) direct sorting of the newly synthesized class II molecules from the TGN to a specialized compartment for peptide loading (Calafat *et al.*, 1994; Neefjes *et al.*, 1990; Peters *et al.*, 1991b; Qiu *et al.*, 1994; Tulp *et al.*, 1994; West *et al.*, 1994);
(ii) sorting of the class II–Ii complexes to early endosomes (EE) followed by transport to later endocytic compartments or to the cell surface as class II-peptide complexes (Castellino and Germain, 1995; Cresswell, 1985; Guagliardi *et al.*, 1990; Roche *et al.*, 1993; Romagnoli *et al.*, 1993);
(iii) sorting of the class II–Ii complexes to the cell surface followed by internalization (Roche *et al.*, 1993).

Subcellular fractionation studies of B lymphocytes combined with the detection of SDS-stable class II molecules (Castellino and Germain, 1995), as well as immunofluorescence staining studies of class II transfected cells (Romagnoli *et al.*, 1993), support the model of targeting of class II molecules from the TGN to EE and subsequent transport to later endocytic compartments. It is proposed that class II molecules associate with peptides in several compartments, depending on the conditions of pH and proteolytic activity required for the breakdown of the antigen (Castellino and Germain, 1995). Alternatively, targeting to EE by rapid internalization of plasma membrane class II–Ii complexes has been proposed (Roche *et al.*, 1993). However, a number of immunoelectron microscopy and subcellular fractionation studies failed to detect any class II molecules in EE, suggesting that the major route followed by MHC class II–Ii complexes is the direct targeting from the TGN to a specialized peptide-loading compartment without passing along the cell surface or EE (Amigorena *et al.*, 1994; Benaroch *et al.*, 1995; Calafat *et al.*, 1994; Neefjes *et al.*, 1990; Peters *et al.*, 1991b; Qiu *et al.*, 1994; Tulp *et al.*, 1994; West *et al.*, 1994). It should be mentioned, however, that the transport route taken by MHC class II–Ii complexes is not of essential importance for antigen presentation as long as MHC class II molecules end up in the compartment where peptide loading takes place (Nijenhuis *et al.*, 1994).

10.3.2 What are the requirements for peptide loading of class II molecules?

Ii-chain degradation and its subsequent dissociation from class II molecules are prerequisites for peptide loading and surface expression of class II molecules. Lysosomotropic agents that increase the vacuolar pH, such as monensin, chloroquine, primaquine or ammonium chloride, or protease inhibitors like leupeptin, inhibit the degradation of Ii resulting in accumulation of class II–Ii complexes and reduced cell surface expression of class II molecules (Blum and Cresswell, 1988; Germain and Hendrix, 1991; Loss and Sant, 1993; Maric *et al.*, 1994; Neefjes and Ploegh, 1992; Pieters *et al.*, 1991; Roche and Cresswell, 1991).

This indicates that Ii degradation requires an acidic environment and the participation of proteases. Furthermore, Ii acts as an endosomal stop-transport signal since its degradation is required for plasma membrane expression of class II. Two distinct types of proteases are involved in Ii degradation, aspartic and cysteine proteases, which sequentially cleave the luminal region of the Ii polypeptide (Maric *et al.*, 1994). The action of aspartic proteases such as cathepsin D and E leads to the generation of two fragments of 21 kDa and 11–14 kDa, named LIP (for leupeptin-induced peptide) and SLIP (for small leupeptin-induced peptide), respectively, which includes the 15 amino-terminal amino acids that contain the endosomal targeting signal. Treatment with leupeptin, a specific inhibitor of cysteine proteases, such as cathepsin B, causes the intracellular accumulation of these intermediates, indicating that a cysteine protease is responsible for the final steps of Ii degradation and release from class II molecules (Neefjes and Ploegh, 1992; Roche and Cresswell, 1991; Stebbins *et al.*, 1995).

The smallest detected fragments associated with MHC class II molecules are termed CLIP (for class II-associated invariant chain peptides). CLIP consists of a nested set of peptides that are derived from amino acids 83–109 of Ii, and which occupy the peptide-binding groove of class II molecules, preventing the acquisition of immunogenic peptides (Riberdy *et al.*, 1992; Romagnoli and Germain, 1994; Sette *et al.*, 1992).

The mechanism of exchange of CLIP for an immunogenic peptide was studied in mutant cell lines deficient in antigen presentation but expressing normal levels of *HLA-DR3*. The mutant cell lines were isolated by negative selection for a conformational-dependent *DR3* epitope recognized by the monoclonal antibody 16.23 (Mellins *et al.*, 1990) and they were found to express class II at the cell surface in association with CLIP peptides (Mellins *et al.*, 1994; Riberdy *et al.*, 1992; Sette *et al.*, 1992). Class II molecules isolated from the mutant cells are not SDS-stable. Further analysis of these mutant cell lines revealed that the defect was due to the deletion or mutation of genes that map to the class II region of the MHC: the *HLA-DMA* and *HLA-DMB* genes (Fling *et al.*, 1994; Mellins *et al.*, 1991; Morris *et al.*, 1994). Furthermore, the expression of *HLA-DM* cDNA in these mutant cell lines restored the wild-type phenotype in the mutant cell lines (Fling *et al.*, 1994; Morris *et al.*, 1994). HLA-DM is a heterodimer, highly homologous to MHC class I and class II molecules (Cho *et al.*, 1991; Denzin *et al.*, 1994), consisting of a 33–35 kDa α-chain and a 30–31 kDa β-chain. *In vitro* experiments demonstrate that under acidic conditions (pH 5.0), soluble HLA-DM induced the dissociation of a subset of peptides, including CLIP, bound to isolated class II molecules and facilitated the subsequent binding of exogenous peptides (Denzin and Cresswell, 1995; Sherman *et al.*, 1995; Sloan *et al.*, 1995). The observation that HLA-DM also increases the off-rate of some peptides, including CLIP, but not all peptides, suggests that DM is involved in selection of class II-binding peptides (Sloan *et al.*, 1995). The current model proposes that HLA-DM has a catalytic role in peptide loading of class II molecules: one HLA-DM molecule would sequentially interact with several class II–CLIP complexes, thereby mediating the release of CLIP and the incorporation of new peptide (Denzin and Cresswell, 1995). This model is based on different experimental data. First, HLA-DM is present in low amounts relative to HLA-DR (Denzin and Cresswell, 1995) and, furthermore, it has a long half-life *in vivo* (Denzin *et al.*, 1994). The mechanism of action of HLA-DM at a molecular level is still unknown, although antibody blocking experiments suggest that the interaction between HLA-DM and class II molecules is required (Denzin and Cresswell, 1995). On the other hand, although HLA-DM is essential for peptide loading of *HLA-DR3* alleles, antigen presentation by other human class II alleles or class II from different species may not be equally dependent on HLA-DM (Stebbins *et al.*, 1995).

Another event preceding peptide loading of class II molecules is obviously the proteolytic breakdown of exogenous antigens. A family of acidic proteases, cathepsins, are known to participate in antigen processing. The aspartyl protease cathepsin D is able to generate T-cell-stimulatory peptides from endocytosed antigens like sperm whale myoglobin, lysozyme and ovalbumin (Rodriguez and Diment, 1995; Van Noort and Jacobs, 1994; Van Noort and van der Drift, 1989; Van Noort *et al.*, 1991). In the case of the processing of ovalbumin, Rodriguez and Diment (1995) have shown that cathepsin D is able to generate a fragment that is recognized by ovalbumin-specific T helper cells. In contrast, cathepsins B and L do not generate T-cell-stimulatory fragments and, in addition, destroy the ovalbumin epitope generated by cathepsin D. This epitope, when first loaded on to fixed macrophages, could be protected from degradation by cathepsin L suggesting that binding to class II molecules prevents further degradation of the peptide. Cathepsins, possible peptidases and other lysosomal proteases may also have a role in trimming the peptide ends once the fragment is bound to the class II molecule (Rodriguez and Diment, 1995; Van Noort and Jacobs, 1994; Van Noort and van der Drift, 1989; Van Noort *et al.*, 1991).

10.3.3 In which endocytic compartment does peptide loading occur?

To identify the intracellular compartments where peptide loading of class II molecules occurs, a number of studies have taken advantage of what is known about the events and molecules involved in this process. Both immunoelectron microscopy and subcellular fractionation have been used to address this question.

One immunoelectron microscopy study has shown co-localization of internalized surface immunoglobulin, class II molecules, Ii and cathepsins B and D, in EE of a human B-lymphoblastoid cell line (Guagliardi *et al.*, 1990). However, in a recent work, Peters *et al.* (1995) failed to detect cathepsin D and MHC class II molecules in EE. After subcellular fractionation, Castellino and Germain (1995) detected SDS-stable class II molecules in early endosomal fractions, suggesting that peptide loading may occur in this compartment. However, peptide loading of newly synthesized class II molecules is unlikely to occur in an early endocytic compartment, since EE lack HLA-DM (Denzin *et al.*, 1994; Sanderson *et al.*, 1994).

Many data support the existence of a specialized peptide-loading compartment in APCs. Using immunoelectron microscopy, Peters *et al.* (1991b) first described a new compartment highly enriched in class II molecules in a human lymphoblastoid B-cell line. This compartment, named MHC class II-containing compartment or MIIC, consists of vesicles of 200–300 nm in diameter which display a heterogeneous morphology that varies from multivesicular, multilamellar or tubulovesicular, depending on the cell type (*Figure 10.3*; Harding and Geuze, 1992; Peters *et al.*, 1991b). Subsequent immunoelectron-microscopic studies have

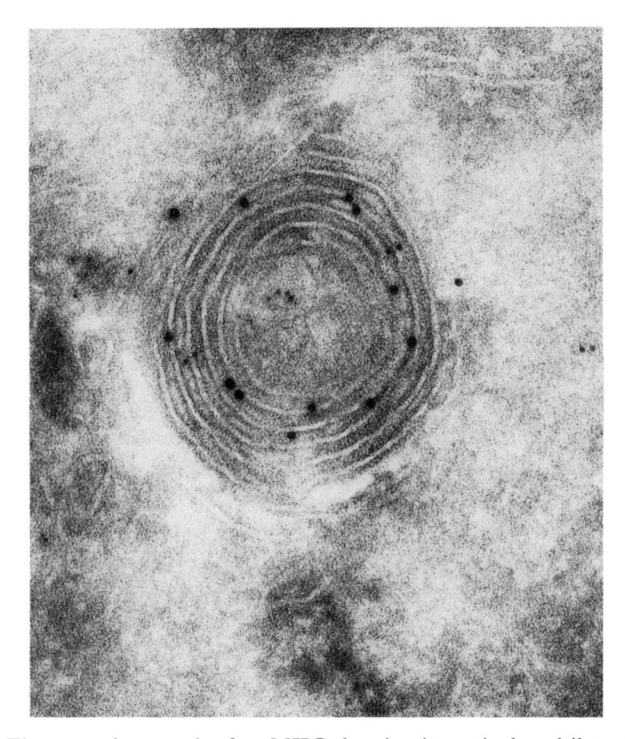

Figure 10.3. Electronmicrograph of an MIIC showing its typical multilamellar morphology and colocalization of immunodetected CD63 (small gold particles) and MHC class II molecules (large gold particles).

confirmed the presence of MIIC in human B cells (Calafat *et al.*, 1994; Peters *et al.*, 1995; West *et al.*, 1994) as well as in other APCs like macrophages (Harding and Geuze, 1992, 1993), dendritic cells (Arkema *et al.*, 1991, 1993; Kleijmeer *et al.*, 1995; Nijman *et al.*, 1995), Langerhans cells (Kleijmeer *et al.*, 1994) and also in the human melanoma cell line Mel JuSo (Sanderson *et al.*, 1994). In addition to class II molecules, MIIC contains the lysosomal enzyme β-hexosaminidase and the lysosomal-associated membrane proteins lamp-1 and CD63, but it is negative for typical endosomal markers like the mannose-6-phosphate receptor (late endosomes) and the transferrin receptor (EE). Furthermore, both cathepsin D (Harding and Geuze, 1992, 1993; Peters *et al.*, 1995) and HLA-DM (Karlsson *et al.*, 1994; Nijman *et al.*, 1995; Riberdy *et al.*, 1994; Sanderson *et al.*, 1994) have been localized in MIIC structures. In HLA-DM-deficient cell lines, MHC class II–CLIP complexes, immediate precursors of peptide-loaded class II molecules, accumulate in vesicles which contain the MIIC markers. MIICs are acidic structures linked to the endocytic pathway and accessible to endocytic tracers after long incubation [($\sim$)30 minutes]. Due to the characteristics mentioned above, MIICs are considered a subpopulation of lysosomes (Harding and Geuze, 1992, 1993; Peters *et al.*, 1991b, 1995) and, given the importance of cathepsin D and HLA-DM in class II-antigen presentation, MIICs are proposed to be the site for the peptide-loading of MHC class II molecules.

Studies based on different subcellular fractionation techniques support the existence of a specialized compartment where class II accumulate and associate with peptide. Class II-enriched vesicles distinct from EE, late endosomes and dense lysosomes have been isolated by Percoll gradient (Harding and Geuze, 1993; Qiu *et al.*, 1994; West *et al.*, 1994), free flow electrophoresis (Amigorena *et al.*, 1994), density gradient electrophoresis (Sanderson *et al.*, 1994; Tulp *et al.*, 1994) or by a combination of sucrose density gradients and free flow electrophoresis (Amigorena *et al.*, 1995). This compartment presented pre- or light-lysosomal features in human B lymphoblasts (Qiu *et al.*, 1994; West *et al.*, 1994), in a human melanoma cell line (Sanderson *et al.*, 1994) and in mouse macrophages (Harding and Geuze, 1993) and actually was identified as MIIC by immunoelectron microscopy both by morphology and presence of markers like lamp-1 and cathepsin D (Harding and Geuze, 1993; Sanderson *et al.*, 1994; West *et al.*, 1994). In contrast, in the mouse A20 B cell line, a specialized compartment was identified as a subpopulation of recycling endosomes and named CIIV for class II vesicles (Amigorena *et al.*, 1994, 1995). By immunoelectron microscopy, CIIV were shown to consist of vesicles of 300–500 nm in diameter with membrane infoldings or internal vesicles (Amigorena *et al.*, 1994, 1995). Similarly to MIIC, CIIV are negative for invariant chain, contain SDS-stable compact MHC class II compact molecules and are linked to the endocytic pathway. However, CIIV lack lysosomal markers present in MIIC and contain markers of the early endosomal recycling pathway like the transferrin receptor.

MIIC is not a pre-existing intracellular compartment. Transfection experiments showed that expression of class II molecules in non-APCs suffices to induce the formation of MIIC-like structures independently of the co-expression of Ii. In these cells, MIIC have the same multilamellar morphology as MIIC described in B cells, they contain the lysosomal markers lamp-1 and CD63 and are reached by endocytic tracers after relatively long periods of uptake (Calafat *et al.*, 1994). Furthermore, MIIC in B cells are not stable compartments since their maintenance requires the continuous synthesis of proteins (Calafat *et al.*, 1994). Thus, the existence of MIIC depends upon the expression of class II molecules which, when accumulated in an endosomal compartment, may induce the formation of internal membrane sheets typical for the multilamellar morphology of MIIC.

MIIC show morphological heterogeneity which varies from multivesicular, multilamellar to tubulovesicular structures (Calafat *et al.*, 1994; Harding and Geuze, 1993; West *et al.*, 1994). These different morphologies could be either cell type-specific, for instance tubulovesicular MIIC are only found in macrophages (Harding and Geuze, 1993), or correspond to different maturation states of the compartment (Nijman *et al.*, 1995; Peters *et al.*, 1995). In fact, both multivesicular and multilamellar structures have been found to coexist in dendritic cells (Nijman *et al.*, 1995) and in the A20 murine B cell line (Amigorena *et al.*, 1994). Currently it is proposed that the multivesicular MIIC could actually correspond to the CIIV compartment described in mouse B lymphocytes (Mellman *et al.*, 1995; Peters *et al.*, 1995). Two possibilities arise:

(i) MIIC and CIIV are two separate discrete compartments both displaying antigen processing and peptide loading activities;
(ii) MIIC are originated from CIIV or vice versa.

The latter possibility has been suggested by Peters *et al.* (1995) who have described by immunohistochemically three different maturation states for MIIC compartments on the basis of the presence of the luminal portion of Ii, which is the part that is first lost during Ii proteolysis. Thus, MIICs are observed that contain either complete Ii, or exclusively the cytoplasmic tail of Ii; other MIICs show no detectable Ii. Interestingly, this progressive loss of Ii epitopes correlates with the transition from multivesicular to multilamellar MIIC. In addition, multivesicular structures are first accessible to endocytic tracers. The authors propose that multivesicular TGN-associated structures could correspond to the CIIV compartment described previously, and that they represent early MIIC. In fact, the heterogeneity in MIIC morphology has been explained in terms of different maturation states where the multivesicular structure is a precursor of the multilamellar one. First of all, multivesicular MIIC contain relatively low amounts of MHC class II molecules when compared to multilamellar MIIC (Riberdy *et al.*, 1994). Moreover, multivesicular MIIC contain Ii, while multilamellar structures are Ii-negative (Kleijmeer *et al.*, 1994; Nijman *et al.*, 1995; Peters *et al.*, 1995) and, as mentioned above, the alteration in morphology coincides with the progressive degradation of the amino-terminal fragment of Ii, indicative of a maturation process (Peters *et al.*, 1995). Accordingly, leupeptin (an inhibitor of Ii degradation) induces the accumulation of the Ii intermediate forms p22 and p10 in multivesicular compartments (Amigorena *et al.*, 1995; Zachgo *et al.*, 1992). Early multivesicular MIIC are reached by gold-albumin prior to the late multilamellar MIIC. In the same line of evidence, Qiu *et al.* (1994) showed by subcellular fractionation that floppy and compact forms of class II molecules reside in distinct subcellular compartments. The floppy forms are considered to be conformational intermediates in the formation of the compact SDS-stable MHC class II molecules (Sadegh-Nasseri and Germain, 1992). The compact forms were found in denser vesicles that co-sedimented with lamp-1 and β-hexosaminidase-positive lysosomes and were negative for Ii, indicating that they could actually correspond to late MIICs. Furthermore, these vesicles, and not the class II floppy form-containing vesicles, contained T-cell stimulatory peptides, suggesting that the final steps in antigen processing take place in late MIIC (Qiu *et al.*, 1994).

The relative abundance of CIIV (or multivesicular MIIC) and MIIC may depend on the cell type and the differentiation state of the cell; for instance, MIIC are barely present in the A20 murine B-cell line when compared to the human B-cell lines IM9 and JY (Amigorena *et al.*, 1994; Peters *et al.*, 1995). By contrast, A20 cells apparently contain abundant CIIV. MIIC levels vary among different types of human APCs, which correlates with the antigen-presenting capacity of the different types of human cells. MIIC are more abundant in dendritic cells than in peripheral blood B cells, while monocytes show very little MIIC (Nijman *et al.*, 1995). On the other hand, freshly isolated dendritic cells contain higher numbers of MIIC than differentiated cultured dendritic cells (Nijman *et al.*, 1995).

Both compartments, CIIV (or early MIIC) and late MIIC, seem to be able to process antigen: T-cell-stimulatory peptides have been found in subcellular fractions, which were identified as CIIV in a murine B-cell line (Barnes and Mitchell, 1995) and as MIIC in human B lymphocytes (Qiu *et al.*, 1994), and mouse macrophages (Harding and Geuze, 1993), although different antigens were used in every study. In fact, the association of processed antigen to class II molecules could take place in different endocytic compartments depending on the sensitivity of the antigen to proteolysis (Escola *et al.*, 1995). Class II molecules in CIIV may become loaded with peptides derived from antigens more labile to proteolysis, such as ovalbumin, while class II in mature MIIC bind peptides from more proteolysis-resistant antigens. However, late lysosomal-like MIIC structures probably play a prominent role in antigen processing due to the acidic pH and reducing environment required for the unfolding and denaturation of antigens. In macrophages, the hen egg lysozyme encapsulated in liposomes that are able to release their contents either to endosomes or lysosomes, is more efficiently processed in lysosomes (Harding *et al.*, 1991a).

10.3.4 How do peptide-loaded class II molecules reach the cell surface?

The final question on how class II-peptide complexes reach the cell surface for antigen presentation has not been addressed so far. It is known, however, that the egress of class II from the peptide-loading compartment requires the binding of peptide (Germain and Rinker, 1993; Neefjes and Ploegh, 1992). The association of antigenic peptides promotes stability and prevents aggregation of class II molecules (Germain and Rinker, 1993). Thus, peptide would play a similar role in the peptide-loading compartment as Ii plays in the ER. However, the pathway followed by class II molecules from MIIC to the cell surface is unclear. Possibly, multivesicular MIICs are exocytotic compartments which fuse with the plasma membrane in a similar way to the cytotoxic T-cell granules (which are multivesicular structures containing perforin and granzymes that are exocytosed upon specific interaction between the T cell and the target cell). These cytotoxic granules are acidic endocytic vesicles, positive for the lysosomal markers lamp-1 and CD63 and negative for the mannose-6-phosphate receptor (Peters *et al.*, 1991a), characteristics also shown by MIIC. There is some evidence that the exocytosis of peptide-loaded class II molecules is a regulated event as well, because the location of class II molecules shifts from MIIC to the cell surface upon culturing of freshly isolated dendritic cells, which suggests exocytosis of class II in activated differentiated cells (Nijman *et al.*, 1995). Similarly, dendritic cells which remain in the undifferentiated state by culturing in the presence of granulocyte/ macrophage colony-stimulating factor and interleukin-4, undergo differentiation upon the addition of tumour necrosis factor-α (Sallusto and Lanzavecchia, 1994). This cytokine induces an increased cell surface expression of class II molecules.

References

Amigorena S, Drake JR, Webster P, Mellman I. (1994) Transient accumulation of new class II MHC molecules in a novel endocytic compartment in B lymphocytes. *Nature* **369:** 113–120.

Amigorena S, Webster P, Drake J, Newcomb J, Cresswell P, Mellman I. (1995). Invariant chain cleavage and peptide loading in major histocompatibility complex class II vesicles. *J. Exp. Med.* **181:** 1729–1741.

Anderson KS, Cresswell P. (1994) A role for calnexin (IP90) in the assembly of class II MHC molecules. *EMBO J.* **13:** 675–682.

Anderson MS, Miller J. (1992) Invariant chain can function as a chaperone protein for class II major histocompatibility complex molecules. *Proc. Natl Acad. Sci. USA* **89:** 2282–2286.

Anderson MS, Swier K, Arneson L, Miller J. (1993) Enhanced antigen presentation in the absence of the invariant chain endosomal localization signal. *J. Exp. Med.* **178:** 1959–1969.

Arkema JM, Schadee-Eestermans IL, Broekhuis-Fluitsma DM, Hoefsmit EC. (1991) Localization of class II molecules in storage vesicles, endosomes and lysosomes in human dendritic cells. *Immunobiology* **183:** 396–407.

Arkema JM, Schadee-Eestermans IL, Broekhuis-Fluitsma DM, Hoefsmit EC. (1993) Immunocytochemical characterization of dendritic cells. *Adv. Exp. Med. Biol.* **329:** 17–21.

Arneson LS, Miller J. (1995) Efficient endosomal localization of major histocompatibility complex class II-invariant chain complexes requires multimerization of the invariant chain targeting sequence. *J. Cell Biol.* **129:** 1217-1228.

Arunachalam B, Lamb CA, Cresswell P. (1994) Transport properties of free and MHC class II-associated oligomers containing different isoforms of human invariant chain. *Int. Immunol.* **6:** 439–451.

Bakke O, Dobberstein B. (1990) MHC class II-associated invariant chain contains a sorting signal for endosomal compartments. *Cell* **63:** 707–716.

Barnes KA, Mitchell RN. (1995) Detection of functional class II-associated antigen: role of a low density endosomal compartment in antigen processing. *J. Exp. Med.* **181:** 1715–1727.

Bell JI, Denny DW Jr, McDevitt HO. (1985) Structure and polymorphism of murine and human class II major histocompatibility antigens. *Immunol. Rev.* **84:** 51–71.

Benaroch P, Yilla M, Raposo G, Ito K, Miwa K, Geuze HJ, Ploegh HL. (1995) How MHC class II molecules reach the endocytic pathway. *EMBO J.* **14:** 37–49.

Bergeron JJ, Brenner MB, Thomas DY, Williams DB. (1994) Calnexin: a membrane-bound chaperone of the endoplasmic reticulum. *Trends Biochem. Sci.* **19:** 124–128.

Bertolino P, Staschewski M, Trecol-Biemont M-C, Freisewinkel IM, Schenck K, Chretien I, Forquet F, Gerlier D, Rabourdin-Combe C, Koch N. (1995) Deletion of a C-terminal sequence of the class II-associated invariant chain abrogates invariant chains oligomer formation and class II antigen presentation. *J. Immunol.* **154:** 5620–5629.

Bijlmakers MJ, Benaroch P, Ploegh HL. (1994a) Mapping functional regions in the lumenal domain of the class II-associated invariant chain. *J. Exp. Med.* **180:** 623–629.

Bijlmakers MJ, Benaroch P, Ploegh HL. (1994b) Assembly of HLA DRI molecules translated in vitro: binding of peptide in the endoplasmic reticulum precludes association with invariant chain. *EMBO J.* **13:** 2699–2707.

Bikoff EK, Huang LY, Episkopou V, van Meerwijk J, Germain RN, Robertson EJ. (1993) Defective major histocompatibility complex class II assembly, transport, peptide acquisition, and CD4+ T cell selection in mice lacking invariant chain expression. *J. Exp. Med.* **177:** 1699–1712.

Bikoff EK, Germain RN, Robertson EJ. (1995) Allelic differences affecting invariant chain dependency of MHC class II subunit assembly. *Immunity* **2:** 301–310.

Bjorkman PJ, Burmeister WP. (1994) Structures of two classes of MHC molecules elucidated: crucial differences and similarities. *Curr. Opin. Struct. Biol.* **4:** 852–856.

Blum JS, Cresswell P. (1988) Role for intracellular proteases in the processing and transport of class II HLA antigens. *Proc. Natl Acad. Sci. USA* **85:** 3975–3979.

Bodmer H, Viville S, Benoist C, Mathis D. (1994) Diversity of endogenous epitopes bound to MHC class II molecules limited by invariant chain. *Science* **263:** 1284–1286.

Bonnerot C, Marks MS, Cosson P, Robertson EJ, Bikoff EK, Germain RN, Bonifacino JS. (1994) Association with BiP and aggregation of class II MHC molecules synthesized in the absence of invariant chain. *EMBO J.* **13:** 934–944.

Bremnes B, Madsen T, Gedde-Dahl M, Bakke O. (1994) An LI and ML motif in the cytoplasmic tail of the MHC-associated invariant chain mediate rapid internalization. *J. Cell Sci.* **107:** 2021–2032.

Brodsky FM, Guagliardi LE. (1991) The cell biology of antigen processing and presentation. *Annu. Rev. Immunol.* **9:** 707–744.

Brown JH, Jardetzky T, Saper MA, Samraoui B, Bjorkman PJ, Wiley DC. (1988) A hypothetical model of the foreign antigen binding site of class II histocompatibility molecules. *Nature* **332:** 845–850.

Brown JH, Jardetzky TS, Gorga JC, Stern LJ, Urban RG, Strominger JL, Wiley DC. (1993) Three-dimensional structure of the human class II histocompatibility antigen HLA-DR1. *Nature* **364:** 33–39.

Busch R, Vturina IY, Drexler J, Momburg F, Hammerling GJ. (1995) Poor loading of major histocompatibility complex class II molecules with endogenously synthesized short peptides in the absence of invariant chain. *Eur. J. Immunol.* **25:** 48–53.

Buus S, Sette A, Colon SM, Jenis DM, Grey HM. (1986) Isolation and characterization of antigen-la complexes involved in T cell recognition. *Cell* **47:** 1071–1077.

Calafat J, Nijenhuis M, Janssen H, Tulp A, Dusseljee S, Wubbolts R, Neefjes J. (1994) Major histocompatibility complex class II molecules induce the formation of endocytic MIIC-like structures. *J. Cell Biol.* **126:** 967–977.

Castellino F, Germain RN. (1995) Extensive trafficking of MHC class II-invariant chain complexes in the endocytic pathway and appearance of peptide-loaded class II in multiple compartments. *Immunity* **2:** 73–88.

Chervonsky A, Sant AJ. (1995) In the absence of major histocompatibility complex class II molecules, invariant chain is translocated to late endocytic compartments by autophagy. *J. Eur. Immunol.* **25:** 911–918.

Chervonsky AV, Gordon L, Sant AJ. (1994) A segment of the MHC class II beta chain plays a critical role in targeting class II molecules to the endocytic pathway. *Int. Immunol.* **6:** 973–982.

Chicz RM, Urban RG, Lane WS, Gorga JC, Stern LJ, Vignali DA, Strominger JL. (1992) Predominant naturally processed peptides bound to HLA-DR1 are derived from MHC-related molecules and are heterogeneous in size. *Nature* **358:** 764–768.

Cho SG, Attaya M, Monaco JJ. (1991) New class II-like genes in the murine MHC. *Nature* **353:** 573–576.

Claesson L, Larhammar D, Rask L, Peterson PA. (1983) cDNA clone for the human invariant gamma chain of class II histocompatibility antigens and its implications for the protein structure. *Proc. Natl Acad. Sci. USA* **80:** 7395–7399.

Cosgrove D, Gray D, Dierich A, Kaufman J, Lemeur M, Benoist C, Mathis D. (1991) Mice lacking MHC class II molecules. *Cell* **66:** 1051–1066.

Cosson P, Bonifacino JS. (1992) Role of transmembrane domain interactions in the assembly of class II MHC molecules. *Science* **258:** 659–662.

Cotner T, Pious D. (1995) HLA-DR beta chains enter into an aggregated complex containing GRP-78/BiP prior to their degradation by the pre-Golgi degradative pathway. *J. Biol. Chem.* **270:** 2379–2386.

Cresswell P. (1985) Intracellular class II HLA antigens are accessible to transferrin-neuraminidase conjugates internalized by receptor-mediated endocytosis. *Proc. Natl Acad. Sci. USA* **82:** 8188–8192.

Cresswell P. (1992) Chemistry and functional role of the invariant chain. *Curr. Opin. Immunol.* **4:** 87–92.

Cresswell P. (1994) Assembly, transport, and function of MHC class II molecules. *Annu. Rev. Immunol.* **12:** 259–293.

Cresswell P, Blum JS, Kelner DN, Marks MS. (1987) Biosynthesis and processing of class II histocompatibility antigens. *Crit. Rev. Immunol.* **7:** 31–53.

Denzin LK, Cresswell P. (1995) HLA-DM induces CLIP dissociation from MHC class II $\alpha\beta$ dimers and facilitates peptide loading. *Cell* **82:** 155–156.

Denzin LK, Robbins NF, Carboy-Newcomb C, Cresswell P. (1994) Assembly and intracellular transport of HLA-DM and correction of the class II antigen-processing defect in T2 cells. *Immunity* **1:** 595–606.

Dornmair K, Rothenhausler B, McConnell HM. (1989) Structural intermediates in the reactions of antigenic peptides with MHC molecules. *Cold Spring Harb. Symp. Quant. Biol.* **54:** 409–416.

Elliott EA, Drake JR, Amigorena S, Elsemore J, Webster P, Mellman I, Flavell RA. (1994) The invariant chain is required for intracellular transport and function of major histocompatibility complex class II molecules. *J. Exp. Med.* **179:** 681–694.

Escola J-M, Grivel J-C, Chavrier P, Gorvel J-P. (1995) Different endocytic compartments are involved in the tight association of class II molecules with processed hen egg lysozyme and ribonuclease A in B cells. *J. Cell Sci.* **108:** 2337–2345.

Fling SP, Arp B, Pious D. (1994) HLA-DMA and -DMB genes are both required for MHC class II/ peptide complex formation in antigen-presenting cells. *Nature* **368:** 554–558.

Germain RN. (1994) MHC-dependent antigen processing and peptide presentation: providing ligands for T lymphocyte activation. *Cell* **76:** 287–299.

Germain RN, Hendrix LR. (1991) MHC class II structure, occupancy, and surface expression determined by post-endoplasmic reticulum antigen binding. *Nature* **353:** 134–139.

Germain RN. Margulies DH. (1993) The biochemistry and cell biology of antigen processing and presentation. *Annu. Rev. Immunol.* **11:** 403–450.

Germain RN, Rinker AG, Jr. (1993) Peptide binding inhibits protein aggregation of invariant-chain free class II dimers and promotes surface expression of occupied molecules. *Nature* **363:** 725-728.

Gething MJ, Sambrook J. (1992) Protein folding in the cell. *Nature* **355:** 33–45.

Giacoletto KS, Sant AJ, Bono C, Gorka J, O'Sullivan DM, Quaranta V, Schwartz BD. (1986) The human invariant chain is the core protein of the human class II-associated protcoglycan. *J. Exp. Med.* **164:** 1422–1439.

Gorvel J-P, Escola J-M, Stang E, Bakke O. (1995) Invariant chain induces a delayed transport from early to late endosomes. *J. Biol. Chem.* **270:** 2741–2746.

Guagliardi LE, Koppelman B, Blum JS, Marks MS, Cresswell P, Brodsky FM. (1990) Co-localization of molecules involved in antigen processing and presentation in an early endocytic compartment. *Nature* **343:** 133–139.

Harding CV, Geuze HJ. (1992) Class II MHC molecules are present in macrophage lysosomes and phagolysosomes that function in the phagocytic processing of Listeria monocytogenes for presentation to T cells. *J. Cell Biol.* **119:** 531–542.

Harding CV, Geuze HJ. (1993) Immunogenic peptides bind to class II MHC molecules in an early lysosomal compartment. *J. Immunol.* **151:** 3988–3998.

Harding CV, Collins DS, Slot JW, Geuze HJ, Unanue ER. (1991a) Liposome-encapsulated antigens are processed in lysosomes, recycled, and presented to T cells. *Cell* **64:** 393–401.

Harding CV, Roof RW, Allen PM, Unanue ER. (1991b) Effects of pH and polysaccharides on peptide binding to class II major histocompatibility complex molecules. *Proc. Natl Acad. Sci. USA* **88:** 2740–2744.

Hedley ML, Urban RG, Strominger JL. (1994) Assembly and peptide binding of major histocompatibility complex class II heterodimers in an in vitro translation system. *Proc. Natl Acad. Sci. USA* **91:** 10479–10483.

Henne C, Schwenk F, Koch N, Moller P. (1995) Surface expression of the invariant chain (CD74) is independent of concomitant expression of major histocompatibility complex class II antigens. *Immunology* **84:** 177–182.

Humbert M, Raposo G, Cosson P, Reggio H, Davoust J, Salamero J. (1993) The invariant chain induces compact forms of class II molecules localized in late endosomal compartments. *Eur. J. Immunol.* **23:** 3158–3166.

Hunt DF, Michel H, Dickinson TA, Shabanowitz J, Cox AL, Sakaguchi K, Appella E, Grey HM, Sette A. (1992) Peptides presented to the immune system by the murine class II major histocompatibility complex molecule I-Ad. *Science* **256:** 1817–1820.

Jardetzky TS, Brown JH, Gorga JC, Stern LJ, Urban RG, Chi YI, Stauffacher C, Strominger JL, Wiley DC. (1994) Three-dimensional structure of a human class II histocompatibility molecule complexed with superantigen. *Nature* **368:** 711–718.

Jensen PE. (1990) Regulation of antigen presentation by acidic pH. *J. Exp. Med.* **171:** 1779–1784.

Jiang W, Swiggard WJ, Heufler C, Peng M, Mirza A, Steinman RM, Nussenzweig MC. (1995) The receptor DEC-205 expressed by dendritic cells and thymic epithelial cells is involved in antigen processing. *Nature* **375:** 151–155.

Karlsson L, Peleraux A, Lindstedt R, Liljedahl M, Peterson PA. (1994) Reconstitution of an operational MHC class II compartment in nonantigen-presenting cells. *Science* **266:** 1569–1573.

Katunuma N, Kakegawa H, Matsunaga Y, Saibara T. (1994) Immunological significances of invariant chain from the aspect of its structural homology with the cystatin family. *FEBS Lett.* **349:** 265–269.

Kaufman JF, Krangel MS, Strominger JL. (1984) Cysteines in the transmembrane region of major histocompatibility complex antigens are fatty acylated via thioester bonds. *J. Biol. Chem.* **259:** 7230–7238.

Kleijmeer MJ, Oorschot VM, Geuze HJ. (1994) Human resident Langerhans cells display a lysosomal compartment enriched in MHC class II. *J. Invest. Dermatol.* **103:** 516–523.

Kleijmeer MJ, Ossevoort MA, van Veen CJH, van Hellemond JJ, Neefjes JJ, Kast WM, Melief CJM, Geuze HJ. (1995) MHC class II compartments and the kinetics of antigen presentation in activated mouse spleen dendritic cells. *J. Immunol.* **154:** 5715–5724.

Knight SC, Stagg AJ. (1993) Antigen-presenting cell types. *Curr. Opin. Immunol.* **5:** 374–382.

Koch N. (1988) Postranslational modifications of the Ia-associated invariant protein p41 after gene transfer. *Biochemistry* **27:** 4097–4102.

Koch N, Hammerling GJ. (1986) The HLA-D-associated invariant chain binds palmitic acid at the cysteine adjacent to the membrane segment. *J. Biol. Chem.* **261:** 3434–3440.

Koch N, Lauer W, Habicht J, Dobberstein B. (1987) Primary structure of the gene for the murine Ia antigen-associated invariant chains (Ii). An alternatively spliced exon encodes a cysteine-rich domain highly homologous to a repetitive sequence of thyroglobulin. *EMBO J.* **6:** 1677–1683.

Kozono H, White J, Clements J, Marrack P, Kappler J. (1994) Production of soluble MHC class II proteins with covalently bound single peptides. *Nature* **369:** 151–154.

Kvist S, Wiman K, Claesson L, Peterson PA, Dobberstein B. (1982) Membrane insertion and oligomeric assembly of HLA-DR histocompatibility antigens. *Cell* **29:** 61–69.

Lamb CA, Cresswell P. (1992) Assembly and transport properties of invariant chain trimers and HLA-DR-invariant chain complexes. *J. Immunol.* **148:** 3478–3482.

Lamb CA, Yewdell JW, Bennink JR, Cresswell P. (1991) Invariant chain targets HLA class II molecules to acidic endosomes containing internalized influenza virus. *Proc. Natl Acad. Sci. USA* **88:** 5998–6002.

Lanzavecchia A, Watts C. (1994) Peptide partners call the tune. *Nature* **371:** 198–199.

Lanzavecchia A, Reid PA, Watts C. (1992) Irreversible association of peptides with class II MHC molecules in living cells. *Nature* **357:** 249–252.

Lipp J, Dobberstein B. (1988) Signal and membrane anchor functions overlap in the type II membrane protein I gamma CAT. *J. Cell Biol.* **106:** 1813–1820.

Loss GE, Jr, Sant AJ. (1993) Invariant chain retains MHC class II molecules in the endocytic pathway. *J. Immunol.* **150:** 3187-3197.

Lotteau V, Teyton L, Peleraux A, Nilsson T, Karlsson L, Schmid SL, Quaranta V, Peterson PA. (1990) Intracellular transport of class II MHC molecules directed by invariant chain. *Nature* **348:** 600–605.

Machamer CE, Cresswell P. (1982) Biosynthesis and glycosylation of the invariant chain associated with HLA-DR antigens. *J. Immunol.* **129:** 2564–2569.

Machamer CE, Cresswell P. (1984) Monensin prevents terminal glycosylation of the N- and O-linked oligosaccharides of the HLA-DR-associated invariant chain and inhibits its dissociation from the alpha-beta chain complex. *Proc. Natl Acad. Sci. USA* **81:** 1287–1291.

Maric MA, Taylor MD, Blum JS. (1994) Endosomal aspartic proteinases are required for invariant-chain processing. *Proc. Natl Acad. Sci. USA* **91:** 2171–2175.

Marks MS, Blum JS, Cresswell P. (1990) Invariant chain trimers are sequestered in the rough endoplasmic reticulum in the absence of association with HLA class II antigens. *J. Cell Biol.* **111:** 839–855.

Marks MS, Germain RN, Bonifacino JS. (1995) Transient aggregation of major histocompatibility complex class II chains during assembly in normal spleen cells. *J. Biol. Chem.* **270:** 10475–10481.

McKnight AJ, Mason DW, Barclay AN. (1989) Sequence of a rat MHC class II-associated invariant chain cDNA clone containing a 64 amino acid thyroglobulin-like domain. *Nucleic Acids Res.* **17:** 3983–3984.

Mellins E, Smith L, Arp B, Cotner T, Celis E, Pious D. (1990) Defective processing and presentation of exogenous antigens in mutants with normal HLA class II genes. *Nature* **343:** 71–76.

Mellins E, Kempin S, Smith L, Monji T, Pious D. (1991) A gene required for class II-restricted antigen presentation maps to the major histocompatibility complex. *J. Exp. Med.* **174:** 1607–1615.

Mellins E, Cameron P, Amaya M, Goodman S, Pious D, Smith L, Arp B. (1994) A mutant human histocompatibility leukocyte antigen DR molecule associated with invariant chain peptides. *J. Exp. Med.* **179:** 541–549.

Mellman I, Pierre P, Amigorena S. (1995) Lonely MHC molecules seeking immunogenic peptides for meaningful relationships. *Curr. Opin. Cell Biol.* **7:** 564–572.

Melnick J, Argon Y. (1995) Molecular chaperones and the biosynthesis of antigen receptors. *Immunol. Today* **16:** 243–250.

Miller J, Germain RN. (1986) Efficient cell surface expression of class II MHC molecules in the absence of associated invariant chain. *J. Exp. Med.* **164:** 1478–1489.

Mircheff AK, Gierow JP, Wood RL. (1994) Traffic of major histocompatibility complex class II molecules in rabbit lacrimal gland acinar cells. *Invest. Ophthalmol. Vis. Sci.* **35:** 3943–3951.

Monaco JJ. (1993) Structure and function of genes in the MHC class II region. *Curr. Opin. Immunol.* **5:** 17–20.

Morris P, Shaman J, Attaya M, Amaya M, Goodman S, Bergman C, Monaco JJ, Mellins E. (1994) An essential role for HLA-DM in antigen presentation by class II major histocompatibility molecules. *Nature* **368:** 551–554.

Naujokas MF, Morin M, Anderson MS, Peterson M, Miller J. (1993) The chondroitin sulfate form of invariant chain can enhance stimulation of T cell responses through interaction with CD44. *Cell* **74:** 257–268.

Neefjes JJ, Ploegh HL. (1992) Inhibition of endosomal proteolytic activity by leupeptin blocks surface expression of MHC class II molecules and their conversion to SDS resistance alpha-beta heterodimers in endosomes. *EMBO J.* **11:** 411–416.

Neefjes JJ, Stollorz V, Peters PJ, Geuze HJ, Ploegh HL. (1990) The biosynthetic pathway of MHC class II but not class I molecules intersects the endocytic route. *Cell* **61:** 171–183.

Neefjes JJ, Schumacher TN, Ploegh HL. (1991) Assembly and intracellular transport of major

histocompatibility complex molecules. *Curr. Opin. Cell Biol.* **3:** 601–609.
Nelson CA, Petzold SJ, Unanue ER. (1994) Peptides determine the lifespan of MHC class II molecules in the antigen-presenting cell. *Nature* **371:** 250–252.
Newcomb JR, Cresswell P. (1993) Characterization of endogenous peptides bound to purified HLA-DR molecules and their absence from invariant chain-associated alpha beta dimers. *J. Immunol.* **150:** 499–507.
Nijenhuis M, Neefjes J. (1994) Early events in the assembly of major histocompatibility complex class II heterotrimers from their free subunits. *Eur. J. Immunol.* **24:** 247–256.
Nijenhuis M, Calafat J, Kuijpers KC, Janssen H, de Haas M, Nordeng TW, Bakke O, Neefjes JJ. (1994) Targeting major histocompatibility complex class II molecules to the cell surface by invariant chain allows antigen presentation upon recycling. *Eur. J. Immunol.* **24:** 873–883.
Nijman HW, Kleijmeer MJ, Ossevoort MA, Oorschot VMJ, Vierboom MPM, van de Keur M, Kenemans P, Kast WM, Geuze HJ, Melief CJM. (1995) Antigen capture and major histocompatibility class II compartments of freshly isolated and cultured human blood dendritic cells. *J. Exp. Med.* **182:** 163–174.
Nilsson T, Lucocq JM, Mackay D, Warren G. (1991) The membrane spanning domain of beta-1,4-galactosyltransferase specifies trans Golgi localization. *EMBO J.* **10:** 3567–3575.
O'Sullivan DM, Noonan D, Quaranta V. (1987) Four Ia invariant chain forms derive from a single gene by alternate splicing and alternate initiation of transcription/translation. *J. Exp. Med.* **166:** 444–460.
Odorizzi CG, Trowbridge IS, Xue L, Hopkins CR, Davis CD, Collawn JF. (1994) Sorting signals in the MHC class II invariant chain cytoplasmic tail and transmembrane region determine trafficking to an endocytic processing compartment. *J. Cell Biol.* **126:** 317–330.
Paul WE, Seder RA. (1994) Lymphocyte responses and cytokines. *Cell* **76:** 241–251.
Peters PJ, Borst J, Oorschot V, Fukuda M, Krahenbuhl O, Tschopp J, Slot JW, Geuze HJ. (1991a) Cytotoxic T lymphocyte granules are secretory lysosomes, containing both perforin and granzymes. *J. Exp. Med.* **173:** 1099–1109.
Peters PJ, Neefjes JJ, Oorschot V, Ploegh HL, Geuze HJ. (1991b) Segregation of MHC class II molecules from MHC class I molecules in the Golgi complex for transport to lysosomal compartments. *Nature* **349:** 669–676.
Peters PJ, Raposo G, Neefjes JJ, Oorschot V, Leijendekker RL, Geuze HJ, Ploegh HL. (1995) Major histocompatibility complex class II compartments in human B lymphoblastoid cells are distinct from early endosomes. *J. Exp. Med.* **182:** 325–334.
Peterson M, Miller J. (1990) Invariant chain influences the immunological recognition of MHC class II molecules. *Nature* **345:** 172–174.
Peterson M, Miller J. (1992) Antigen presentation enhanced by the alternatively spliced invariant chain gene product p4l. *Nature* **357:** 596–598.
Pieters J, Horstmann H, Bakke O, Griffiths G, Lipp J. (1991) Intracellular transport and localization of major histocompatibility complex class II molecules and associated invariant chain. *J. Cell Biol.* **115:** 1213–1223.
Pieters J, Bakke O, Dobberstein B. (1993) The MHC class II-associated invariant chain contains two endosomal targeting signals within its cytoplasmic tail. *J. Cell Sci.* **106:** 831–846.
Pinet V, Vergelli M, Martin R, Bakke O, Long EO. (1995) Antigen presentation mediated by recycling of surface HLA-DR molecules. *Nature* **375:** 603–606.
Qiu Y, Xu X, Wandinger-Ness A, Dalke DP, Pierce SK. (1994) Separation of subcellular compartments containing distinct functional forms of MHC class II. *J. Cell Biol.* **125:** 595–605.
Rammensee HG, Falk K, Rotzschke O. (1993) MHC molecules as peptide receptors. *Curr. Opin. Immunol.* **5:** 35–44.
Rath S, Lin RH, Rudensky A, Janeway CA, Jr. (1992) T and B cell receptors discriminate major histocompatibility complex class II conformations influenced by the invariant chain. *Eur. J. Immunol.* **22:** 2121–2127.
Reid PA, Watts C. (1990) Cycling of cell-surface MHC glycoproteins through primaquine-sensitive intracellular compartments. *Nature* **346:** 655–657.
Riberdy JM, Newcomb JR, Surman MJ, Barbosa JA, Cresswell P. (1992) HLA-DR molecules from an antigen-processing mutant cell line are associated with invariant chain peptides. *Nature* **360:** 474–477.
Riberdy JM, Avva RR, Geuze HJ, Cresswell P. (1994) Transport and intracellular distribution of MHC class II molecules and associated invariant chain in normal and antigen-processing mutant cell lines. *J. Cell Biol.* **125:** 1225–1237.

Roche PA, Cresswell P. (1991) Proteolysis of the class II-associated invariant chain generates a peptide binding site in intracellular HLA-DR molecules. *Proc. Natl Acad. Sci. USA* **88:** 3150–3154.

Roche PA, Marks MS, Cresswell P. (1991) Formation of a nine-subunit complex by HLA class II glycoproteins and the invariant chain. *Nature* **354:** 392–394.

Roche PA, Teletski CL, Karp DR, Pinet V, Bakke O, Long EO. (1992) Stable surface expression of invariant chain prevents peptide presentation by HLA-DR. *EMBO J.* **11:** 2841–2847.

Roche PA, Teletski CL, Stang E, Bakke O, Long EO. (1993) Cell surface HLA-DR-invariant chain complexes are targeted to endosomes by rapid internalization. *Proc. Natl Acad. Sci. USA* **90:** 8581–8585.

Rodriguez GM, Diment S. (1995) Destructive proteolysis by cysteine proteases in antigen presentation of ovalbumin. *Eur. J. Immunol.* **25:** 1823–1827.

Romagnoli P, Germain RN. (1994) The CLIP region of invariant chain plays a critical role in regulating major histocompatibility complex class II folding, transport, and peptide occupancy. *J. Exp. Med.* **180:** 1107–1113.

Romagnoli P, Layet C, Yewdell J, Bakke O, Germain RN. (1993) Relationship between invariant chain expression and major histocompatibility complex class II transport into early and late endocytic compartments. *J. Exp. Med.* **177:** 583–596.

Rotzschke O, Falk K (1994) Origin, structure and motifs of naturally processed MHC class II ligands. *Curr. Opin. Immunol.* **6:** 45–51.

Rudensky AY, Preston-Hurlburt P, al-Ramadi BK, Rothbard J, Janeway CA, Jr. (1992) Truncation variants of peptides isolated from MHC class II molecules suggest sequence motifs. *Nature* **359:** 429–431.

Sadegh-Nasseri S, McConnell HM. (1989) A kinetic intermediate in the reaction of an antigenic peptide and I-Ek. *Nature* **337:** 274–276.

Sadegh-Nasseri S, Germain RN. (1991) A role for peptide in determining MHC class II structure. *Nature* **353:** 167–170.

Sadegh-Nasseri S, Germain RN. (1992) How MHC class II molecules work: peptide-dependent completion of protein folding. *Immunol. Today* **13:** 43–46.

Sadegh-Nasseri S, Stern LJ, Wiley DC, Germain RN. (1994) MHC class II function preserved by low-affinity peptide interactions preceding stable binding. *Nature* **370:** 647–650.

Sallusto F, Lanzavecchia A. (1994) Efficient presentation of soluble antigen by cultured human dendritic cells is maintained by granulocyte/macrophage colony-stimulating factor plus interleukin 4 and downregulated by tumor necrosis factor α. *J. Exp. Med.* **179:** 1109–1118.

Sanderson F, Kleijmeer MJ, Kelly A, Verwoerd D, Tulp A, Neefjes JJ, Geuze HJ, Trowsdale J. (1994) Accumulation of HLA-DM, a regulator of antigen presentation, in MHC class II compartments. *Science* **266:** 1566–1569.

Sant AJ, Miller J. (1994) MHC class II antigen processing: biology of invariant chain. *Curr. Opin. Immunol.* **6:** 57–63.

Sant AJ, Cullen SE, Giacoletto KS, Schwartz BD. (1985) Invariant chain is the core protein of the Ia-associated chondroitin sulfate proteoglycan. *J. Exp. Med.* **162:** 1916–1934.

Sant AJ, Hendrix LR, Coligan JE, Maloy WL, Germain RN. (1991) Defective intracellular transport as a common mechanism limiting expression of inappropriately paired class II major histocompatibility complex alpha/beta chains. *J. Exp. Med.* **174:** 799–808.

Schaiff WT, Hruska KA, Jr, McCourt DW, Green M, Schwartz BD. (1992) HLA-DR associates with specific stress proteins and is retained in the endoplasmic reticulum in invariant chain negative cells. *J. Exp. Med.* **176:** 657–666.

Schreiber KL, Bell MP, Huntoon CJ, Rajagopalan S, Brenner MB, McKean DJ. (1994) Class II histocompatibility molecules associate with calnexin during assembly in the endoplasmic reticulum. *Int. Immunol.* **6:** 101–111.

Schutze MP, Peterson PA, Jackson MR. (1994) An N-terminal double-arginine motif maintains type II membrane proteins in the endoplasmic reticulum. *EMBO J.* **13:** 1696–1705.

Sekaly RP, Tonnelle C, Strubin M, Mach B, Long EO. (1986) Cell surface expression of class II histocompatibility antigens occurs in the absence of the invariant chain. *J. Exp. Med.* **164:** 1490–1504.

Sette A, Ceman S, Kubo RT, *et al.* (1992) Invariant chain peptides in most HLA-DR molecules of an antigen-processing mutant. *Science* **258:** 1801–1804.

Sherman MA, Weber DA, Jensen PE. (1995) DM enhances peptide binding to class II MHC by release of invariant chain-derived peptide. *Immunity* **3:** 197–205.

Simonis S, Miller J, Cullen SE. (1989) The role of the la-invariant chain complex in the posttranslational processing and transport of Ia and invariant chain glycoproteins. *J. Immunol.* **143:** 3619–3625.

Sloan VS, Cameron P, Porter G, Gammon M, Amaya M, Mellins E, Zaller DM. (1995) Mediation by HLA-DM of dissociation of peptides from HLA-DR. *Nature* **375:** 802–806.

Spiro RC, Quaranta V. (1989) The invariant chain is a phosphorylated subunit of class II molecules. *J. Immunol.* **143:** 2589–2594.

Stebbins CC, Loss GE, Jr, Elias CG, Chervonsky A, Sant AJ. (1995) The requirement for DM in class II-restricted antigen presentation and SDS-stable dimer formation is allele and species dependent. *J. Exp. Med.* **181:** 223–234.

Stern LJ, Wiley DC. (1992) The human class II MHC protein HLA-DR1 assembles as empty alpha beta heterodimers in the absence of antigenic peptide. *Cell* **68:** 465–477.

Stern LJ, Brown JH, Jardetzky TS, Gorga JC, Urban RG, Strominger JL, Wiley DC. (1994) Crystal structure of the human class II MHC protein HLA-DR1 complexed with an influenza virus peptide. *Nature* **368:** 215–221.

Stockinger B, Pessara U, Lin RH, Habicht J, Grez M, Koch N. (1989) A role of Ia-associated invariant chains in antigen processing and presentation. *Cell* **56:** 683–689.

Strubin M, Berte C, Mach B. (1986) Alternative splicing and alternative initiation of translation explain the four forms of the Ia antigen-associated invariant chain. *EMBO J.* **5:** 3483–3488.

Sugita M, Brenner MB. (1995) Association of the invariant chain with major histocompatibility complex class I molecules directs trafficking to endocytic compartments. *J. Biol. Chem.* **270:** 1443–1448.

Sung E, Jones PP. (1981) The invariant chain of murine Ia antigens: its glycosylation, abundance and subcellular localization. *Mol. Immunol.* **18:** 899–913.

Tate K, Lee C, Edelman S, Carswell-Crumpton C, Liblau R, Jones PP. (1995) Interactions among polymorphic and conserved residues in MHC class II proteins affect MHC-peptide conformation and T cell recognition. *Int. Immunol.* **7:** 747–761.

Teyton L, O'Sullivan D, Dickson PW, Lotteau V, Sette A, Fink P, Peterson PA. (1990) Invariant chain distinguishes between the exogenous and endogenous antigen presentation pathways. *Nature* **348:** 39–44.

Townsend A, Bodmer H. (1989) Antigen recognition by class I-restricted T lymphocytes. *Annu. Rev. Immunol.* **7:** 601–624.

Tulp A, Verwoerd D, Dobberstein B, Ploegh HL, Pieters J. (1994) Isolation and characterization of the intracellular MHC class II compartment. *Nature* **369:** 120–126.

Unanue ER. (1984) Antigen-presenting function of the macrophage. *Annu. Rev. Immunol.* **2:** 395–428.

Valitutti S, Muller S, Cella M, Padovan E, Lanzavecchia A. (1995) Serial triggering of many T-cell receptors by a few peptide-MHC complexes. *Nature* **375:** 148–151.

Van Noort JM, Jacobs MJM. (1994) Cathepsin D, but not cathepsin B, releases T cell stimulator fragments from lysozyme that are functional in the context of multiple murine class II MHC molecules. *Eur. J. Immunol.* **24:** 2175–2180.

Van Noort JM, van der Drift ACM. (1989) The selectivity of cathepsin D suggests an involvement of the enzyme in the generation of T-cell epitopes. *J. Biol. Chem.* **264:** 14159–14164.

Van Noort JM, Boon J, van der Drift AC, Wagenaar JP, Boots AM, Boog CJ. (1991) Antigen processing by endosomal proteases determines which sites of sperm-whale myoglobin are eventually recognized by T cells. *Eur. J. Immunol.* **21:** 1989–1996.

Viville S, Neefjes J, Lotteau V, Dierich A, Lemeur M, Ploegh H, Benoist C, Mathis D. (1993) Mice lacking the MHC class II-associated invariant chain. *Cell* **72:** 635–648.

Wade WF, Khrebtukova I, Schreiber KL, McKean DJ, Wade TK. (1995) Truncated MHC class II cytoplasmic and transmembrane domains: effect on plasma membrane expression. *Mol. Immunol.* **32:** 433–436.

West MA, Lucocq JM, Watts C. (1994) Antigen processing and class II MHC peptide-loading compartments in human B-lymphoblastoid cells. *Nature* **369:** 147–151.

Zachgo S, Dobberstein B, Griffiths G. (1992) A block in degradation of MHC class II-associated invariant chain correlates with a reduction in transport from endosome carrier vesicles to the prelysosome compartment. *J. Cell Sci.* **103:** 811–822.

11

Crystal structures of MHC class I and class II molecules

Ted Jardetzky

11.1 Introduction

The immune response to pathogens requires the selective recognition and elimination of foreign antigens. The major histocompatibility complex (MHC) glycoproteins are involved in the regulation of this response, through the presentation of peptide antigens to the T cell receptor (TCR). It is the complex of one peptide with one MHC molecule on an antigen-presenting cell (APC), that is the target of a T cell, in a process known as MHC-restricted recognition of antigen (Germain, 1994). T cell responses can be cytotoxic, for example MHC class I-restricted killing of infected cells (see Chapter 13), or lead to the release of activating lymphokines, as in MHC class II-restricted responses (see Chapter 14). This MHC-restricted recognition allows the immune system to discriminate between self and non-self. Although self-peptides are bound to MHC molecules similarly to antigenic peptides, many T cells that react against self-peptides die during maturation in the thymus (Nossal, 1994), while others are held at bay by additional regulatory mechanisms in peripheral tissues. Yet in order to guarantee an immune response, MHC molecules are designed to bind peptides of great sequence diversity. In contrast to both antibodies and TCRs, MHC molecules do not undergo genetic recombination to produce new binding sites for antigens. Instead, one MHC protein is able to bind a vast number of peptides of different sequence, and MHC molecules do not themselves discriminate between self and non-self-peptides. Crystallographic studies of MHC molecules have provided detailed molecular explanations for how this peptide recognition occurs and how it differs between MHC class I and MHC class II proteins. In the process, these structural studies have helped clarify not only mechanisms but also important concepts of self and immunity.

The resolution of the crystal structure of the human MHC molecule HLA-A2 (Bjorkman *et al.*, 1987a, 1987b) provided the first structural insights into this family of proteins. Since that time the application of X-ray crystallographic techniques has continued to provide in-depth information on the mechanisms by which MHC proteins function. The crystal structures of many different MHC class I proteins, of both mouse and human origin, have provided key observations

HLA and MHC: genes, molecules and function, edited by M.J. Browning and A.J. McMichael.
© 1996 BIOS Scientific Publishers Ltd, Oxford.

that help to explain the specificity and stability of peptide binding. In addition, the elucidation of the structure of MHC class II molecules has demonstrated how the same MHC peptide binding-site fold has been adapted to recognize peptides in a distinct manner (Brown *et al.*, 1993). The evolution of MHC class I and class II molecules that recognize distinct cellular populations of antigenic peptides is also reflected in their different atomic interactions with antigenic peptides. Finally, the interaction of bacterial superantigens with MHC class II molecules has also yielded to crystallographic investigation, with the structures of two complexes recently solved (Jardetzky *et al.*, 1994; Kim *et al.*, 1994). Given the many functional interactions that MHC proteins have with other molecules of the antigen presentation system, such as the transporters of antigenic peptides (TAPs), the invariant chain, HLA-DM molecules, in addition to the T-cell receptor, the structural investigation of MHC function promises to continue to yield significant new insights in the future.

11.2 MHC class I crystal structures

11.2.1 Structure of the MHC molecule

The overall structure of MHC class I molecules was revealed by the solution of the crystal structure of the HLA-A2 molecule, and consists of a peptide-binding super-domain supported by two immunoglobulin-like domains (Bjorkman *et al.*, 1987a, b). *Figure 11.1* shows a view of the extracellular portion of the HLA-A2 molecule. Three 'domains' ($\alpha1$, $\alpha2$, and $\alpha3$) of the protein are formed by the heavy, or α, chain of the class I molecule, which is anchored to the cell by a single transmembrane segment. The heavy chains associate non-covalently with the $\beta2$-microglobulin molecule, to form a heterodimer as shown in *Figure 11.1a*. The $\alpha1$ and $\alpha2$ domains of the heavy chain fold to form the peptide-binding site of the MHC molecule (see below), while the immunoglobulin-like domains of $\alpha3$- and β-microglobulin lie below the plane of the peptide-binding super-domain. MHC class I heavy chains are extremely polymorphic, and the crystal structure of HLA-A2 revealed that this polymorphism clusters in space within the peptide-binding site of the MHC molecule. Heavy-chain polymorphism directly changes the chemical nature of the peptide-binding site of the MHC molecule and therefore influences the selection of antigenic peptides. Peptide residues which point out of the binding site, and MHC residues along the tops of the $\alpha1$ and $\alpha2$ domain helices, are thought to form the direct set of interactions with TCRs (Bjorkman *et al.*, 1987b). While peptide interactions are restricted to residues of the class I $\alpha1$ and $\alpha2$ domains, the $\alpha3$ domain has been shown to be important for the binding of the T-cell accessory cell molecule CD8 (Giblin *et al.*, 1994; Salter *et al.*, 1989,

Figure 11.1. (a) Structure of the extracellular domains of the MHC class I molecule. A polymorphic heavy chain ($\alpha1$, $\alpha2$, and $\alpha3$ shown in light grey) binds non-covalently to $\beta2$-microglobulin (black), forming a binding site for peptide. Note the $\alpha1$ and $\alpha2$ domains form one super-domain which binds peptide in a groove formed by two α helices above a β-sheet. The lower domains, $\alpha3$ and $\beta2$-microglobulin, are immunoglobulin-like domains. The $\alpha3$ domain is implicated in interactions with the co-receptor molecule CD8 on T cells. (b) Top view of the MHC class I peptide-binding site. The $\alpha1$ domain is above and shown in

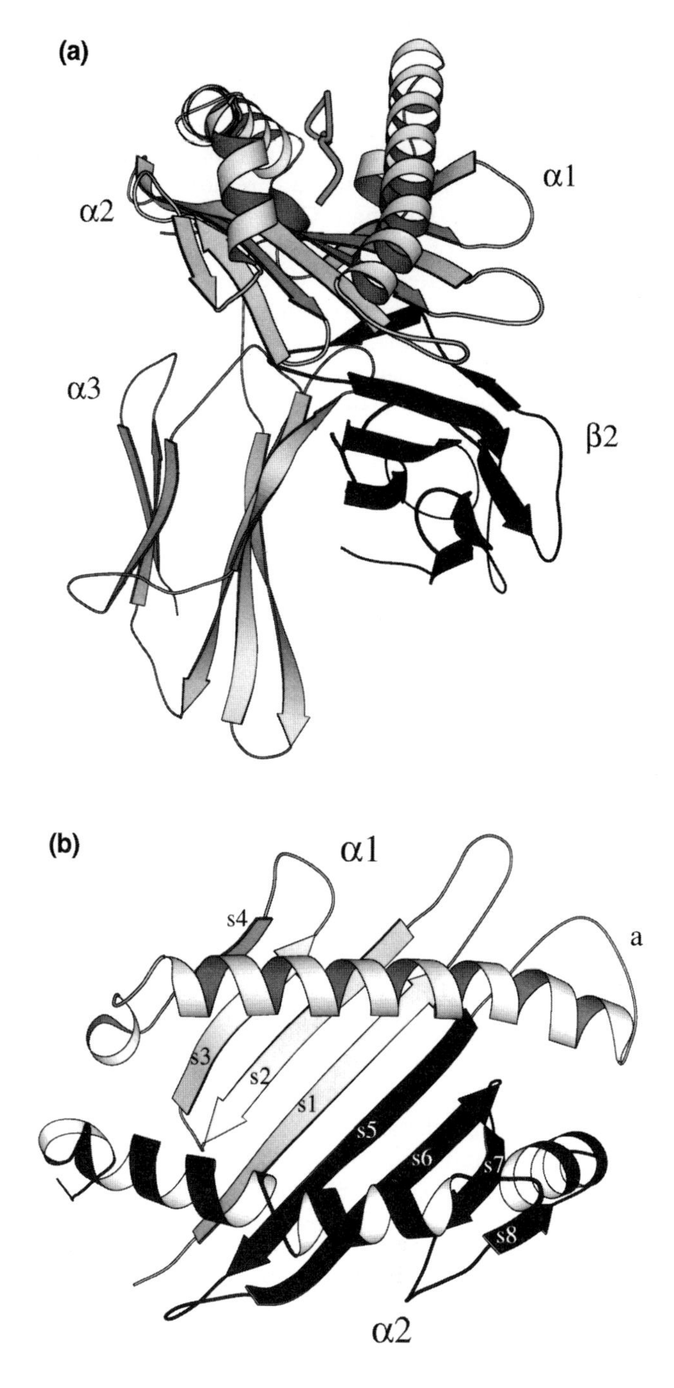

light grey, while $\alpha2$ is below and is coloured black. The connecting peptide from the C-terminus of the $\alpha1$ helix to the first strand of the $\alpha2$ domain is labelled *a* for comparison with *Figure 11.5b*. The peptide-binding site strands are labelled from s1 to s8 to clarify the connectivity of the domains. Compare this connectivity to the corresponding view of the MHC class II peptide-binding site shown in *Figure 11.5*. Peptide is not shown, but is bound between the α helices in an extended conformation. (The figure was made with the program MOLSCRIPT; Kraulis, 1991.)

1990). A ridge of residues of the immunoglobulin-like domain, including residue 245, has been implicated by mutagenesis studies to interact directly with CD8, although efforts to elucidate these interactions by crystallographic techniques have so far failed to yield results.

The overall folded structure of the HLA-A2 molecule is observed in many other MHC class I molecules from both mouse (Fremont *et al.*, 1992, 1995; Young *et al.*, 1994; Zhang *et al.*, 1992) and man (Garrett *et al.*, 1989; Madden *et al.*, 1992), as well as in the structures of MHC class II molecules (see below; Brown *et al.*, 1993). The details of the interactions of each of these variant MHC class I molecules with peptides of varying sequence depends upon the polymorphic residues of the peptide-binding site, and studies of the interaction of peptides with different MHC class I molecules have revealed numerous variations on how peptides interact with the binding site. However, the conservation of functional and structural features among these many MHC class I molecules provides a unifying view of peptide–MHC interactions.

11.2.2 Architecture of the peptide-binding site

The MHC peptide-binding-site fold is composed of a pseudosymmetrical super-domain of eight anti-parallel β-strands supporting two anti-parallel α-helices (*Figure 11.1b*). The helices are separated by a varying distance along their length, forming a large groove that closes at both ends of the helices (Garrett *et al.*, 1989; Saper *et al.*, 1991). This structural motif has been observed in all MHC proteins solved to date, including both mouse and human MHC class II molecules, which bear little sequence homology to class I molecules. In addition, the neonatal Fc receptor (Burmeister *et al.*, 1994) has been shown to have a similar fold to MHC molecules, although the Fc receptor does not bind peptides, and structural rearrangements of the $\alpha 2$ α-helix occur that contribute to the loss of an open binding site.

The peptide-binding site of the MHC class I molecule has a prominent interhelical groove. Peptide binding has been shown to be important for the stable folding of MHC class I molecules, and, in the original crystal structures of molecules isolated from B-lymphoblastoid cell lines, electron density corresponding to a mixture of co-purifying self-peptides was observed in this groove. The comparison of the structures of HLA-A2 and HLA-Aw68 clarified how MHC polymorphism could change the shapes and specificities of subsite pockets within the peptide-binding site, and a number of pockets that interact with peptide side-chains could be identified, even prior to the elucidation of the conformation of bound peptides (Garrett *et al.*, 1989; Saper *et al.*, 1991).

11.2.3 The conformation of bound peptides

MHC class I molecules bind peptides in an extended conformation (Fremont *et al.*, 1992; Madden *et al.*, 1991; Matsumura *et al.*, 1992; Zhang *et al.*, 1992), leaving the main chain atoms and termini of the peptides open for direct interactions with residues of the MHC molecule (*Figures 11.2* and *11.3* see pages 257 and 258). The first interpretation of the conformation of bound peptides could be made with the elucidation of the structure of the human MHC class I molecule HLA-B27 (Madden *et al.*, 1991), in which the electron density of a mixture of endogenous peptides allowed the tracing of a nonameric peptide backbone. The combination of crystallographic data and sequence information on the pool of bound peptides (Jardetzky *et al.*, 1991) combined to provide a first interpretation of the key MHC interactions with bound peptides. Subsequent high resolution structures of single peptide–MHC class I complexes have greatly clarified the interactions between MHC molecules, peptides and conserved water molecules within the peptide-binding site (Fremont *et al.*, 1992, 1995; Madden *et al.*, 1993; Silver *et al.*, 1992).

An extended conformation of bound peptides has been observed in all peptide complexes solved to date (*Figure 11.2* see page 257). The amino-terminus of the peptide lies to the left in *Figure 11.2*, near the N-terminal end of the $\alpha1$ helix and the C-terminal end of the $\alpha2$ helix. This orientation of bound peptides seems to be the same for different MHC class I molecules and is probably dictated by the specific interactions at both the N- and C-termini of the peptide with the MHC residues. Three residues at the amino-terminus and two residues at the C-terminus of bound peptides adopt an essentially invariant conformation in the MHC class I binding site (Madden *et al.*, 1992; Matsumura *et al.*, 1992), while the central residues of the peptide can be more flexibly accommodated within the peptide-binding site (*Figure 11.2a* and *b*); for instance, the structure of the HLA-Aw68 molecule (Guo *et al.*, 1992) refined to high resolution shows density only for the N-terminal and C-terminal residues of the peptide, suggesting that, in the mixture of peptides, the central residues adopt a variety of different conformations. This was also directly observed for the structures of five individual peptide–HLA-A2 structures (Madden *et al.*, 1993) and four of these superimposed peptides are shown in *Figure 11.2* (see page 257). The N- and C-terminal residues of these peptides show a common conformation, while the central residues adopt different extended main-chain and side-chain conformations in the binding site. For both HLA-A2 and HLA-Aw68, the most important anchoring peptide side-chain residues that are important for peptide binding and stability are found at positions P2 and P9 (Falk *et al.*, 1991), and residues between positions P3 and P7 are less critical for binding. In contrast, for the H2-K^b molecule, the presence of an anchor residue at position P5 of the peptide is important for peptide binding, restricting the variability in peptide conformation compared to HLA-A2 or HLA-Aw68 (Fremont *et al.*, 1992, 1995; Zhang *et al.*, 1992). Therefore the position of side-chain anchor residues in MHC class I binding motifs can reduce the variability in conformation that peptides can adopt within the binding site.

11.2.4 Peptide interactions and the length of bound peptides

A set of MHC residues are involved in interactions with the peptide main-chain atoms at the two ends of the peptide-binding site, forming networks of hydrogen bonds between MHC, peptide and water molecules (*Figure 11.3*, page 258). The conformation of the first three amino-terminal peptide residues and the two C-terminal peptide residues is very nearly identical in the many different peptide–MHC structures solved so far (*Figure 11.2*, page 257). The hydrogen bond networks observed at the ends of bound peptides are important in determining the stability of the peptide–MHC complexes and comprise an invariant group of atomic interactions among MHC and peptide molecules of variable sequence. These conserved hydrogen bond interactions are therefore of particular interest because they define a common mechanism of peptide recognition.

The overall shape of the peptide-binding site of the MHC class I molecule is designed to accommodate peptides typically varying in length between 8 and 11 amino acids, although a subset of longer peptides is found associated with MHC class I molecules. The structure of a decamer peptide bound to HLA-Aw68 shows how longer peptides can extend out of the binding groove at their C-terminus (Collins *et al.*, 1994). However, the majority of peptide–MHC class I structures show that the interaction of pockets at either end of the peptide-binding site with the peptide N- and C-termini is a conserved aspect of the binding of peptides by MHC class I molecules (Fremont *et al.*, 1992, 1995; Guo *et al.*, 1992; Madden *et al.*, 1992, 1993; Silver *et al.*, 1992; Young *et al.*, 1994; Zhang *et al.*, 1992). One result of the recognition of the peptide N-terminus is that the sequences of bound peptides are in register relative to each other. Thus sequencing of complex, eluted endogenous peptide pools has provided binding motif information, although individual peptide sequences are not resolved (Falk *et al.*, 1991).

At the amino terminus of bound peptides four MHC tyrosines (Tyr7, Tyr59, Tyr159 and Tyr171) form a network of hydrogen bonds as shown in *Figure 11.3*. These tyrosines are mostly conserved in MHC sequences and interact with both the amino and carbonyl groups of the first amino acid of the bound peptide. Tyrosines 7, 59, and 171 form a pentagonal hydrogen-bonding network with the N-terminus and a conserved water molecule in the high resolution MHC class I structures determined so far (Fremont *et al.*, 1992, 1995; Guo *et al.*, 1992; Madden *et al.*, 1992; Silver *et al.*, 1992; Young *et al.*, 1994; Zhang *et al.*, 1992). At the C-terminus of bound peptides, MHC residues Lys146, Thr143, Tyr84 interact with the carboxyl group of the last amino acid (*Figure 11.3*, page 258). A water molecule forms an additional hydrogen bond to the carboxyl group, while interacting with both Thr80 and Asp77. The amino group of the last peptide residue is also involved in a hydrogen bond to Asp77. The carbonyl of the second-to-last peptide residue is hydrogen bonded to a conserved tryptophan (Trp147), forming an interaction that is also found in MHC class II interactions with peptides.

Two exceptions to these observations have also been studied crystallographically. One involves a decamer peptide which extends out of the C-terminal end of the HLA-A2 peptide-binding site, while the second is observed in the structure of the mouse H2-M3 molecule, which recognizes N-formylated peptide antigens (Wang *et al.*, 1995). In the structure of HLA-A2 with a decamer of the calreticulin signal sequence (Collins *et al.*, 1994), the last peptide residue forms a different set of hydrogen bond interactions to extend out of the peptide-binding site. In this case, only one of the conventional hydrogen bonds observed in other peptide–MHC class I structures is conserved. The carbonyl of the ninth peptide amino acid interacts with Thr143, but hydrogen bonds with Tyr84 and Thr80 are absent. However, Lys146 is found to rearrange and maintain an interaction with the C-terminal carboxylate that has exited the conventional C-terminal pocket. Rearrangements of both Lys146 and Thr80 allow the C-terminal glycine to exit the site. However, removal of the C-terminal glycine residue from this 10-mer generates a 9-mer peptide which has a significantly higher thermostability (T=65.8°C vs. 51.7°C), suggesting that the lost hydrogen bonds and rearrangements of the MHC molecule are made at a significant cost to the stability of the complex. In the H2-M3 molecule, leucine 163 points into the peptide-binding site, decreasing the size of the N-terminal binding pocket, and tryptophan 167 and

continues on p. 265 ▶

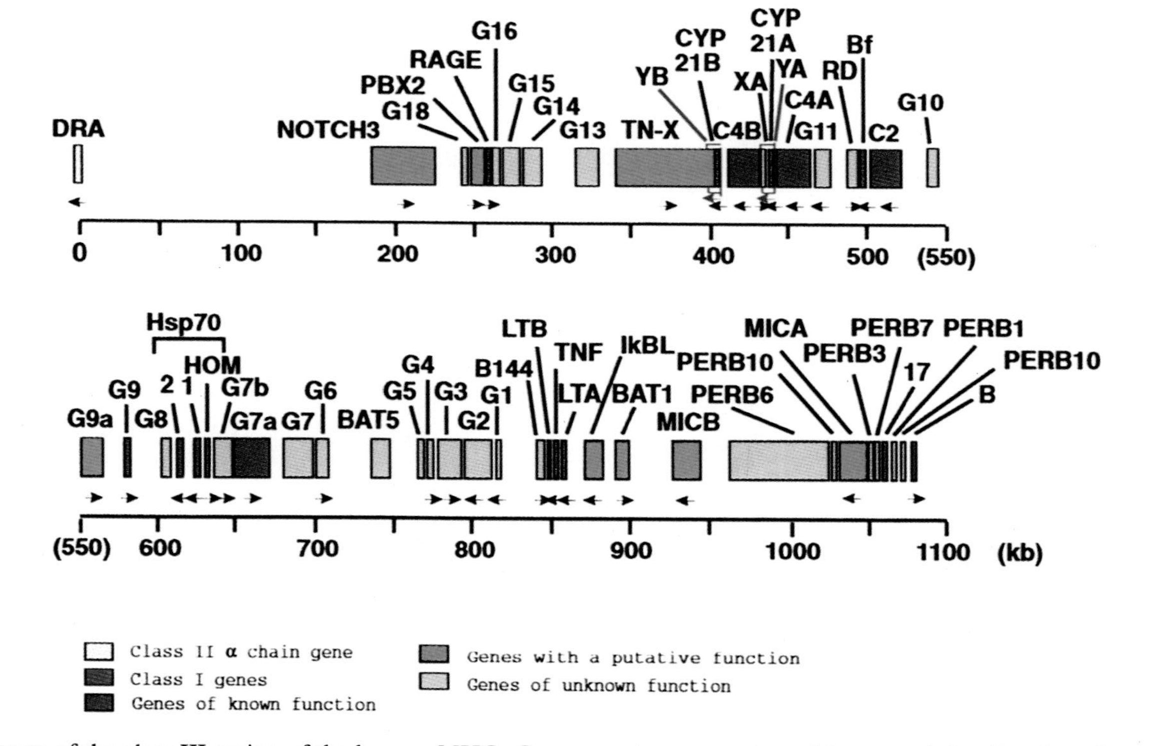

Figure 3.1. Molecular map of the class III region of the human MHC. Genes are shown as coloured boxes and the direction of transcription, where known, is given by an arrow. The scale is in kb. A key to the different coloured boxes is given below the map.

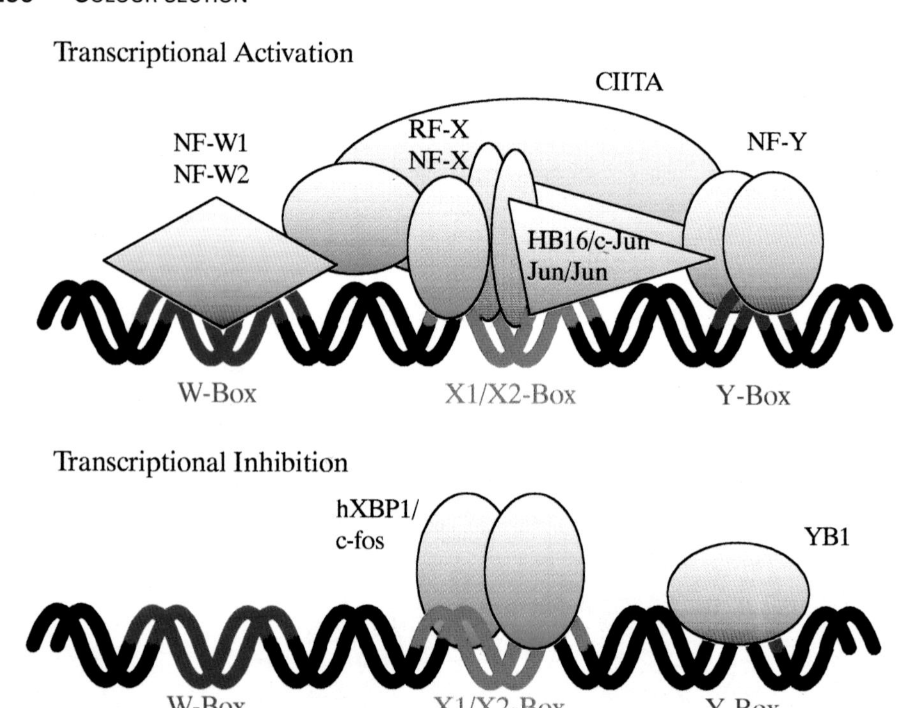

Figure 8.4. *Cis-* and *trans*-acting factors in transcriptional regulation of class II expression.

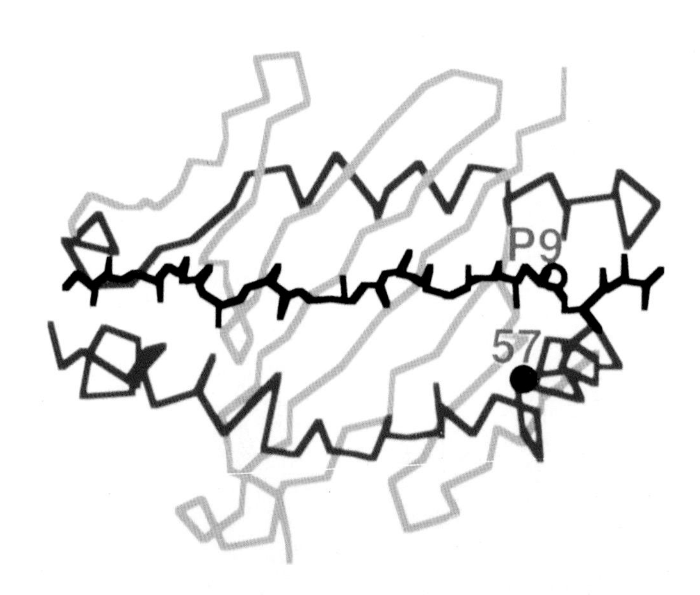

Figure 16.2. A role for HLA-DQβ57 in peptide binding in type 1 diabetes. Biochemical and structural studies have shown that the charge at DQβ57 in the DQ DQβ-chain α helix (red) is important for peptide (black) binding through interaction with the anchor residue P9 (Kwok *et al.*, 1995; Marshall *et al.*, 1994; Reich *et al.*, 1994; Wucherpfennig and Strominger, 1995). (Figure reproduced from *The Journal of Experimental Medicine* (1995) **181:** 1597–1601 by copyright permission of The Rockefeller University Press.

(a)

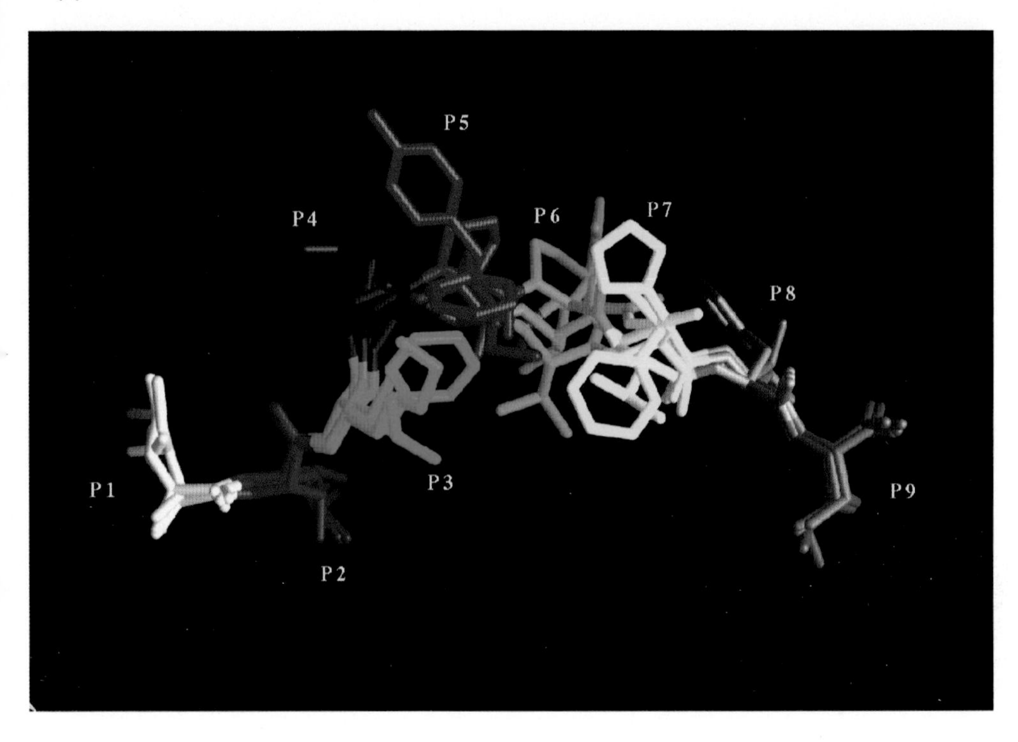

(b)

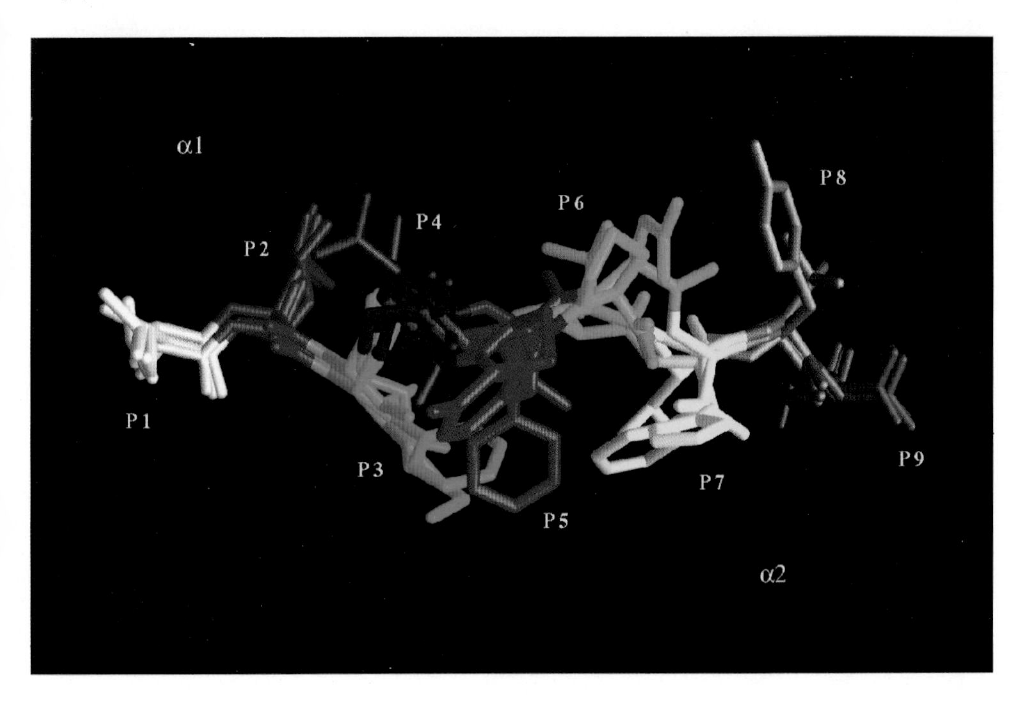

Figure 11.2. For full figure legend see p. 263.

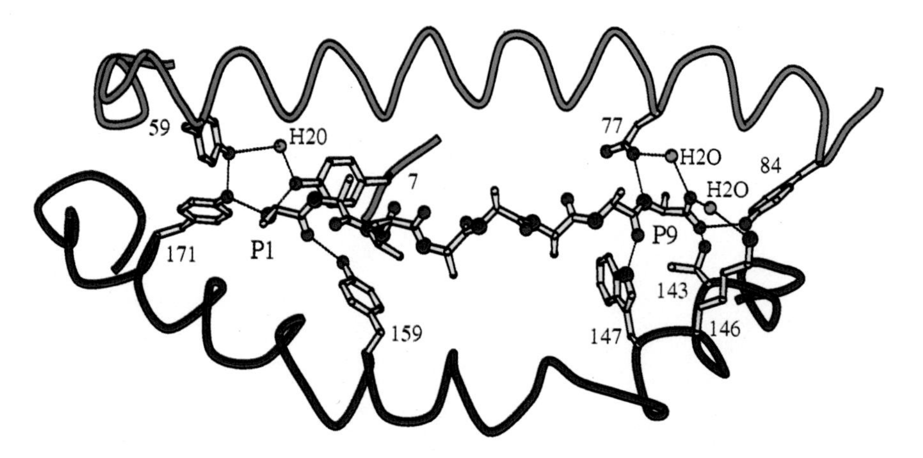

Figure 11.3. For full figure legend see p. 263.

Figure 11.4. For full figure legend see p. 263.

Figure 11.6. For full figure legend see p. 263.

(a)

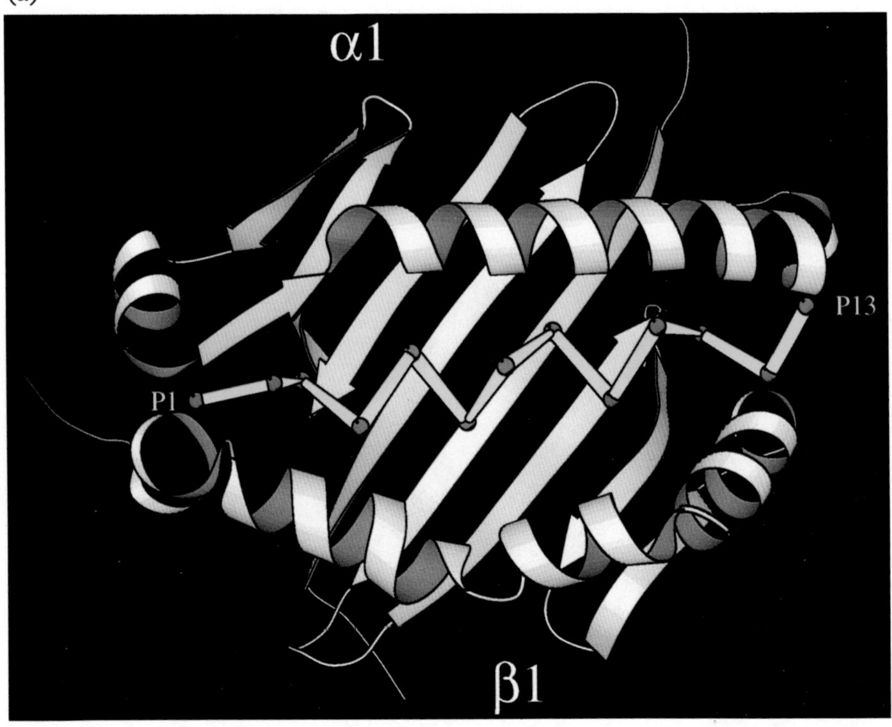

(b)

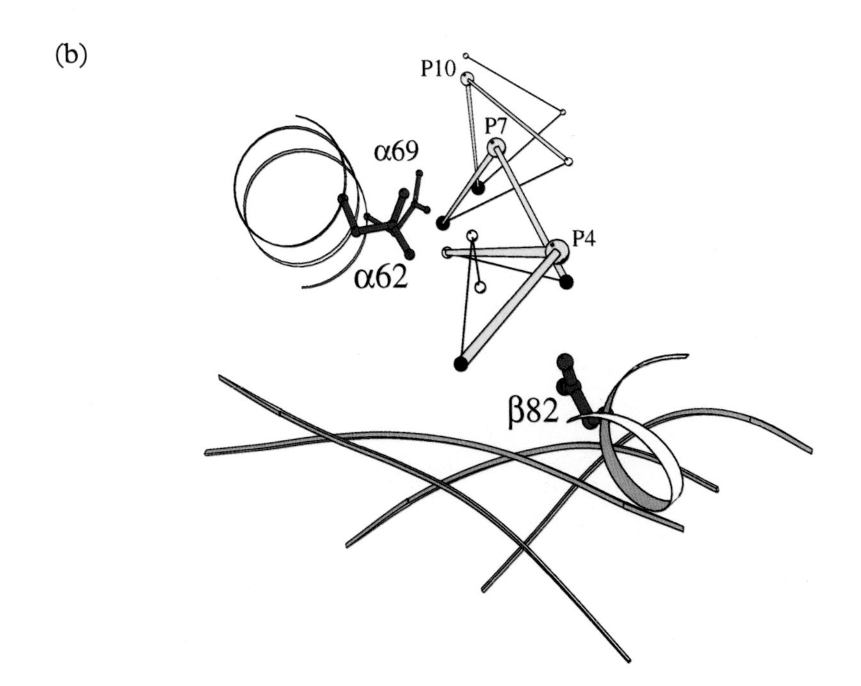

Figure 11.7. For full figure legend see p. 263.

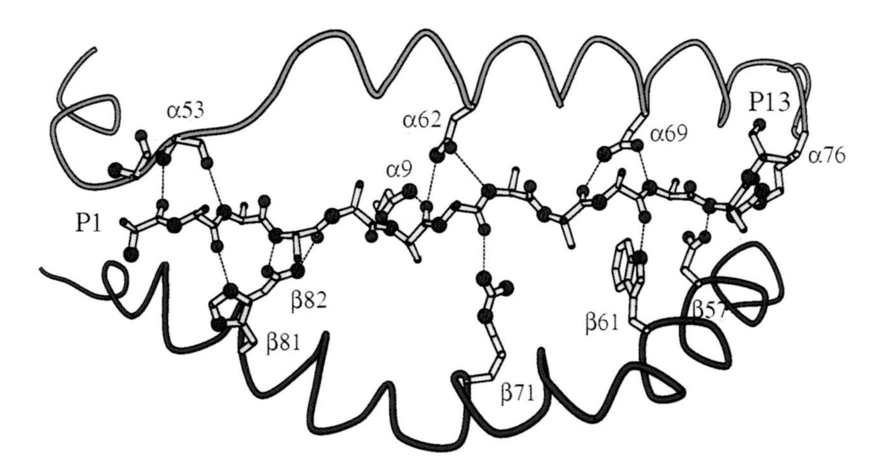

Figure 11.8. For full figure legend see p. 264.

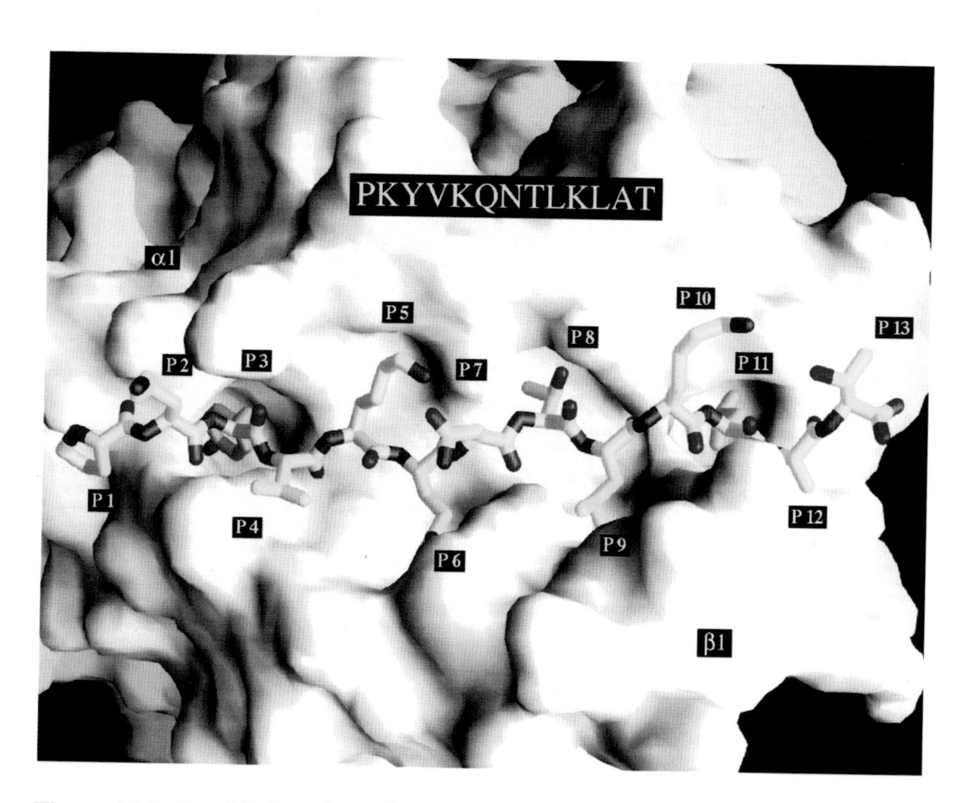

Figure 11.9. For full figure legend see p. 264.

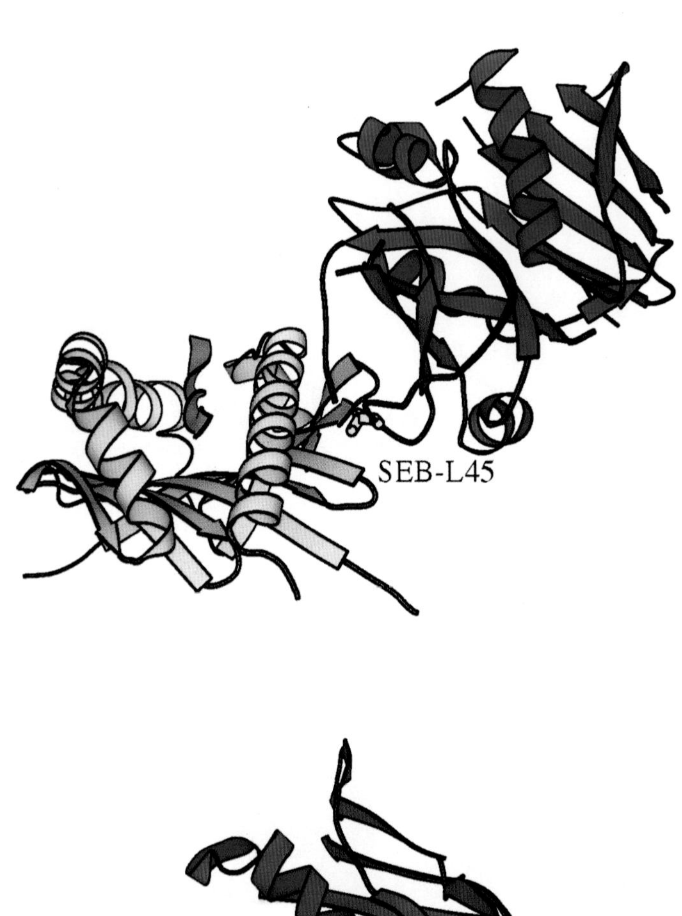

Figure 11.10. For full figure legend see p. 264.

Figure 11.2. Conformational variation of four nonameric peptides bound to HLA-A2. (a) Side view of bound peptides. The peptides shown are the HIV-1 reverse transcriptase peptide, the HTLV-1 tax peptide, the Influenza A matrix peptide, and the HIV-1 gp 120 peptide (described in Madden *et al.*, 1993). Peptide residues are coloured by position in the sequence and labelled accordingly, with P1 at the N-terminus of the peptide and P9 at the C-terminus. All peptides are in an extended conformation, allowing access to the peptide main-chain. Note the good superposition of positions, P1, P2, P3, P8 and P9. In contrast, positions P4–P7 show much greater variability in side-chain and main-chain positioning. (b) Top view of the same HLA-A2 bound peptides. The MHC class I helical regions are indicated by $\alpha 1$ and $\alpha 2$. Compare with *Figure 11.4*, which shows the surface of the MHC class I molecule with one of these peptides bound. (The figure was made with the program GRASP; Honig and Nicholls, 1995.)

Figure 11.3. Conserved hydrogen bonds in the formation of peptide–MHC class I complexes. The $\alpha 1$ helix is shown in light blue (top) and the $\alpha 2$ helix in dark blue (bottom). A nonameric peptide bound in the binding site of HLA-B27 is shown with conserved hydrogen bond interactions. Note how the hydrogen bonds cluster at either the N-terminus (P1) or the C-terminus (P9) of the bound peptide. Three water molecules participate in these interactions and are shown in green. MHC residues are shown with carbon atoms in grey, while peptide residues are shown with carbon atoms coloured yellow. Compare this distribution of conserved hydrogen bond interactions with those observed for peptides bound to MHC class II molecules shown in *Figure 11.7.* (The figure was produced with the program MOLSCRIPT; Kraulis, 1991.)

Figure 11.4. HLA-A2 in a complex with a single peptide derived from Influenza A matrix protein. The $\alpha 1$ and $\alpha 2$ helices of the binding site are labelled (same orientation as in *Figure 11.1b*, and each peptide residue position is indicated, from P1 to P9. Peptide atoms are coloured blue for nitrogen, red for oxygen, and yellow for carbon. The HLA-A2 molecular surface is shown in white. Note how peptide side-chains bind into distinct subsite pockets of the HLA-A2 binding site, especially at the anchor residue positions P2 and P9. Also note how the N- and C-termini of the bound peptide are buried in the binding site. Compare this with MHC class II in *Figure 11.8.* (The figure was produced with the program GRASP; Honig and Nicholls, 1995.)

Figure 11.6 Side view of the MHC class II super-dimer observed in different crystal forms. Two HLA-DR1 heterodimers are shown in the parallel packing orientation observed in different crystals. The α-chain is coloured light red (right dimer) or dark red (left dimer) and the β-chain is coloured light blue (right dimer) or dark blue (left dimer). The two peptide-binding sites are oriented in a parallel direction with bound peptide shown in yellow. The two HLA-DR molecules are related by an approximate two-fold axis near the N-terminus of the $\beta 1$ α-helix. The dimer may play a role in the aggregation of the T-cell receptor, but this remains speculative. (The figure was made with the program MOLSCRIPT; Kraulis, 1991.)

Figure 11.7. Conformation of peptides bound to MHC class II molecules. (a) Top view of the HLA-DR1 peptide-binding site, showing the left-handed helical twist of bound peptide that is similar to polyproline type II helix (ppII). The peptide C-β atoms are shown in light blue and are connected by yellow rods. This conformation is observed for the DR1-HA complex, in a mixture of endogenous-peptides bound to HLA-DR1, and in a complex of invariant chain peptides bound to HLA-DR3. The conformation of bound peptides may be more constrained due to the formation of hydrogen bonds between the MHC molecule and the peptide main-chain throughout the peptide-binding site (see *Figure 11.7*). In addition, bidentate hydrogen bonds between asparagine residues and the peptide main-chain may induce a preference for this conformation. (b) End view of the DR1 peptide-binding site, showing the ppII conformation of bound peptide. Asparagine residues α62, α69 and β82, which form bidentate hydrogen bond rings with bound peptides, are shown. The peptide bonds that interact with these asparagines are shown as yellow rods connecting peptide Cβ positions P3–P5, P6–P8 and P9–P11. Peptide residues that interact with the four most important HLA-DR1 pockets are shown as black Cβ atoms. Peptide positions that point out of the HLA-DR1 peptide-binding site are shown as blue Cβ atoms and are labelled (P4, P7 and P10). (The figure was drawn with the program MOLSCRIPT; Kraulis, 1991.)

Figure 11.8. Hydrogen bond interactions between MHC class II and bound peptides. The HLA-DR1 peptide-binding site is shown, with the α1 helix in light blue (top) and the β1 helix in red (bottom). Conserved hydrogen bond interactions are shown as black lines between HLA-DR1 residues and a bound peptide. These interactions have been observed in two HLA-DR1 structures. In a complex of invariant chain peptides bound to HLA-DR3, one of these hydrogen bonds is broken, due to a proline residue near DRα62. Non-conserved hydrogen bonds are shown in dark blue. Note the distribution of these hydrogen bonds throughout the peptide-binding site, in comparison to those observed for MHC class I complexes (see *Figure 11.3*). Also note the bidentate hydrogen bonds formed between asparagine residues α62, α69 and β82 with the peptide main-chain.

Figure 11.9. HLA-DR1 in a complex with a single peptide derived from Influenza A haemagglutinin. The α1 and β1 helices of the binding site are labelled (same orientation as in *Figure 11.5*), and each peptide residue position is indicated from P1 to P13. Peptide atoms are coloured blue for nitrogen, red for oxygen and yellow for carbon. The HLA-DR1 molecular surface is shown in white. Note how peptide side-chains bind into distinct subsite pockets of the HLA-DR1 binding site, especially at the anchor residue positions P3 and P11. Also note how the N- and C-termini of the bound peptide extend out of the binding site. Compare this with MHC class I in *Figure 11.4*. (The figure was produced with the program GRASP; Honig and Nicholls, 1995.)

Figure 11.10. HLA-DR1 in complexes with two bacterial superantigens, SEB and TSST-1. An end view of the MHC class II peptide-binding site is shown, looking from the C-terminal end of the bound peptides. The β1 helix is to the left in the figure and the α1 helix is to the right. The DR1–SEB complex is shown in the top panel. SEB interacts exclusively with the α1 domain of the DR1 molecule. The DR1–TSST-1 complex is shown in the bottom panel. TSST-1 shares an overlapping binding site on the α1 domain of the DR1 molecule with SEB, but the TSST-1 superantigen is reoriented and binds across the top of the DR1 peptide-binding site. Therefore, TSST-1 interacts with the class II α-chain and β-chain, and potentially with MHC-bound peptide. Both toxins use a leucine residue to interact with a hydrophobic pocket of the α-chain. The leucines are shown in yellow and labelled in each panel.

tyrosine 171 are replaced by leucine and phenylalanine, respectively. The N-formylated peptide does not extend into this region of the binding site, but starts at a position corresponding to P2 in other peptide–MHC class I complexes. In addition to the loss of tyrosine 171, tryptophan 163 is also replaced by leucine. The hydrogen bonds between peptide and H2-M3 differ significantly from other MHC class I complexes and demonstrate that the MHC peptide-binding site fold is consistent with other solutions to the problem of binding peptides.

However, in most MHC class I complexes, the amino and carbonyl groups of the first and last peptide residues, as well as the carbonyl group of the second-to-last residue, are typically completely engaged by conserved MHC class I amino acids. The networks of hydrogen bonds at these positions involve both α-helices and strands of the β-sheet that form the binding site, and therefore the formation of these hydrogen bonds ties together disparate parts of the binding site and probably contributes significantly to the stabilization of the MHC fold. Substitution of either the peptide amino or carboxyl group with a methyl group destabilizes the complex of the influenza matrix peptide bound to HLA-A2, while the peptide substituted at both ends can no longer promote the stable folding of the MHC class I molecule *in vitro* (Bouvier and Wiley, 1994). In addition, the mutation of conserved tyrosines in HLA-A2 that interact with the N-terminus of peptides had significant effects on the presentation of peptides to T cells (Latron *et al.*, 1992).

The binding of the ends of peptides by MHC class I molecules also requires that the central region of the peptide-binding site accommodate peptides of different lengths. A lack of density in the central region of the peptide-binding site observed in the mixture of self-peptides bound to HLA-Aw68 (Guo *et al.*, 1992) suggested that different peptides could be bound in distinct conformations within this site. The longer peptides identified by sequencing of the endogenous peptide pool might 'bulge' out of the binding site. This interpretation may depend on the MHC allele and location of anchor residue positions. However, the variability in peptide conformation seen in five different peptide–HLA-A2 structures (Madden *et al.*, 1993) further suggests the ability of the binding site to accommodate different peptide main-chain conformations and side-chain types (*Figure 11.2*, page 257). Longer peptides may zigzag, back and forth rather than bulge out of the MHC binding site. The length of peptides that can bind to any MHC molecule also depends on the structural details of the binding site; for example, it has been suggested that alleles that bind octameric peptides, such as H2-K^b, have a deep recess in the middle of the binding site that allows shorter peptides to span the distance between the N- and C-terminal pockets (Young *et al.*, 1994). Recent work has shown that tricyclic compounds such as phenanthridine can be used as a molecular spacer for the key N- and C-terminal residues of peptides, generating a new class of hybrid compounds that bind to MHC class I molecules with high affinity and specificity (Weiss *et al.*, 1995).

11.2.5 The interaction of side-chains with subsite pockets and the origins of allele-specific motifs

The identification of variable peptide side-chain binding pockets of MHC molecules provides a ready explanation for the selection of distinct peptide sequence motifs by different MHC alleles (Bjorkman *et al.*, 1987b; Garrett *et al.*, 1989; Matsumura *et al.*, 1992; Saper *et al.*, 1991). The initial analysis of these pockets in *HLA-A*0201* has identified six distinct pockets, A–F (Saper *et al.*, 1991). Two of these (A and F) interact with the amino and carboxy termini of bound peptides, while the remaining B–E pockets interact with peptide side-chains. The side-chains that are most critical for the high affinity binding of peptides have been termed anchor residues (Falk *et al.*, 1991) and the structures of MHC class I molecules have shown how these anchor residues interact with defined subsite pockets of the MHC class I molecules (*Figure 11.4*). The pockets and their relative importance in peptide-binding affinity can vary among alleles, and are sometimes referred to by the peptide residue position relative to the N-terminus (e.g. the P2 pocket is equivalent to the B pocket in HLA-A2). While the P2 pocket (*Figure 11.4*, see page 258) defines a primary anchor residue for HLA-B27, HLA-A2 and HLA-Aw68 peptide-binding motifs, the P5 pocket is a primary anchor site for H2-K^b molecules.

Understanding how MHC pockets selectively discriminate between peptide side-chains will require additional structural investigations, although in some cases simple explanations for specificity are evident; for example, the specificity of the *HLA-B*2705* molecule for peptides containing arginine at position 2 is due to the formation of a buried salt bridge between the peptide arginine and residue 45 of the MHC molecule (Madden *et al.*, 1992). In other alleles, the residues that form this pocket are changed, resulting in distinct specificities. *HLA-A*0201* prefers peptides with leucine at position 2 and *HLA-A*6801* prefers valine. The three refined structures of *HLA-B*2705, HLA-A*0201*, and *HLA-A*6801* has allowed an analysis of the P2 specificity differences among these alleles (Guo *et al.*, 1993). The structures suggest that subtle structural differences in the MHC molecules around the pocket residues can contribute to the selective side-chain differences. The complexity of understanding the determinants of specificity of the MHC molecules for peptides is nearly as complex as the protein-folding problem itself. This has also been recently demonstrated for the H2-K^b molecule, where the interplay of peptide residues at P2 and P5 can be an important determinant of peptide-sequence specificity (Fremont *et al.*, 1995). These structures have shown that, at least in some cases, peptide residue interactions with the MHC molecule are not independent of the remaining peptide sequence and that groups of distinct peptide-sequence sub-motifs may be identifiable for each MHC allele. In addition to the primary anchor positions, the majority of peptide side-chains interact with the MHC peptide binding site (*Figure 11.4*, page 258) and therefore can contribute to secondary specificities within allele-specific motifs.

11.2.6 Conformational changes in the MHC class I molecule upon binding peptides

The recognition of peptide–MHC complexes by the TCR probably occurs via direct interactions with both peptides and MHC molecules at the unique molecular surface generated by each peptide–MHC complex. However, the possibility that peptides might induce MHC conformational changes that are recognized by the TCR has been raised by both structural and functional studies of these molecules. These structural changes can be segregated into two categories:

(i) a global shift in α-helical secondary structures;
(ii) the rearrangement of specific MHC side-chains.

A concerted and peptide-dependent change in MHC structure could be a novel mechanism for generating significantly altered target surfaces for recognition by the TCR, and follows from classic ideas of altered self as the explanation for MHC restricted immune responses. However, for such a mechanism to be more significant than the changes in the chemical nature of different bound peptides, these MHC conformational differences would have to be peptide-specific and also demonstrate a variability as great as that of the many different peptide sequences that can be bound. This does not seem to be the case for the peptide–MHC structures that have been solved to date. The structures of H2-K^b with the SEV-9 and VSV-8 peptides showed a concerted movement of the first α-helical segment of the $\alpha 2$ helix (residues 144–151) that forms part of the C-terminal region of the peptide-binding site (Fremont *et al.*, 1992). In the structures of five HLA-A2 complexes, this concerted shift in the $\alpha 2$ α-helix has also been observed, and the structures were grouped into two categories in which the main-chain positions of this region were essentially identical (Madden *et al.*, 1993). Group I included three of the structures, while group II included the other two. Within each group the coordinate errors were within experimental error and did not correlate with the length of bound peptides, as both nonameric and decameric peptide complexes were found in the same group. The possibility that these conformational differences may be due to differences in crystal contacts for each of these structures has been raised (Madden *et al.*, 1993). However, even if this MHC structural change is indeed peptide-induced, it is clearly not unique for every peptide and would therefore simply define groups of MHC–peptide complexes with a similar $\alpha 2$ helix position. Direct interactions between the TCR and the peptide–MHC surface would still be the primary driving force for specificity.

In addition to the observed shifts in secondary structure positions, MHC amino acid side-chains have been observed to adopt distinct conformations that are peptide-dependent and might thereby have consequences for T-cell recognition of these complexes. In comparing three H2-K^b structures, four MHC side-chains (Lys66, Glu152, Arg155 and Trp167) have been shown to take on peptide-specific conformations (Fremont *et al.*, 1995). Similarly, in the comparison of five HLA-A2 complexes, three residues (Arg97, Tyr116 and Trp167) also change conformation in response to peptide binding (Madden *et al.*, 1993). The binding of peptides that extend out of the C-terminal region of the peptide-binding site also requires the conformational adjustments of Tyr84 and Lys146 (Collins *et al.*, 1994). Residues Arg97 and Tyr116 lie on the β strands beneath bound peptides and have been shown to undergo a concerted conformational change upon the binding of different subsets of peptides (Madden *et al.*, 1993). This MHC conformational change seems to depend on the size of the peptide C-terminal side-chain as well as the packing of the peptide side-chain two residues before the C-terminus. The other conformational changes occur at, or near, the surface of the MHC molecule (Lys66, Tyr84, Lys146, Glu152, Arg155 and Trp167), where they could be directly recognized by TCR, contributing to the unique identity of each peptide–MHC complex.

11.3 MHC class II crystal structures

11.3.1 Structural comparison between MHC class I and class II

Although MHC class I and class II molecules have little sequence homology, these proteins were predicted to have similar three-dimensional structures, based on structural and functional observations (Brown *et al.*, 1988). The determination of the crystal structure of the human MHC class II molecule HLA-DR1 has validated these earlier predictions and allowed a molecular description of the important differences between MHC class I and class II proteins that leads to distinct functional attributes for each class (Brown *et al.*, 1993; Stern and Wiley, 1994). As shown in *Figure 11.5*, the overall structure of HLA-DR1 is very similar to that of MHC class I molecules. MHC class II molecules are heterodimers consisting of two chains (α and β) of approximately equal size, in contrast to the disparate sizes of the class I heavy and light chains. In class II molecules, each chain consists of two extracellular domains and a transmembrane anchoring sequence. In class I only the heavy chain is anchored to the membrane, while β2-microglobulin is a soluble protein. The α- and β-chains of class II both contribute to the formation of the peptide-binding site with their N-terminal domains (α1 and β1). As seen in *Figure 11.5*, α1 and β1 form a pseudosymmetric super-domain very reminiscent of the class I peptide-binding fold, although, on closer inspection, details of the peptide-binding site and some regions of secondary structure differ between these molecules. The correspondence in domain nomenclature between class I and class II molecules is as follows: class I α1 corresponds to class II α1; class I α2 corresponds to class II β1; class I α3 corresponds to class II β2; class I β2-microglobulin corresponds to class II α2.

Although the structures of MHC class I and class II molecules appear nearly identical at first, there are readily identifiable differences. The class I and class II folds can be easily distinguished by features of the α1 domain helix. At the N-terminus of the class II α1 helix, there is a region of extended chain, which is helical in class I molecules (*Figure 11.5b*). This region interacts directly with the main-chain of bound peptides in class II molecules as described below (Stern and Wiley, 1994). Another region of clear difference is the C-terminal region of the α1 helix. In class II molecules the helix ends and a connecting peptide extends downward to the immunoglobulin-like domain below the peptide-binding site (class II α2, *Figure 11.5b*), while in class I molecules the connecting peptide extends around and back towards the middle of the peptide-binding site to connect to the first strand of the class I α2 domain (*Figure 11.1b*), which is in the centre of the peptide-binding site. In addition to these differences, the class II β2 domain is positioned differently relative to the peptide-binding site compared to

Figure 11.5. Structure of the extracellular domains of the human MHC class II molecule, HLA-DR1. (a) Side view of the HLA-DR1 molecule. The protein consists of two chains. The α-chain is shown in light grey and the β-chain is shown in black. Each chain contributes two domains to the heterodimer, forming a structure that is similar to that observed for MHC class I (see *Figure 11.1*). The α1 and β1 domains form a super-domain (analogous to MHC class I α1 and α2), that binds peptide. The β2 domain has been implicated in direct interactions with the co-receptor molecule CD4 found on the surface of

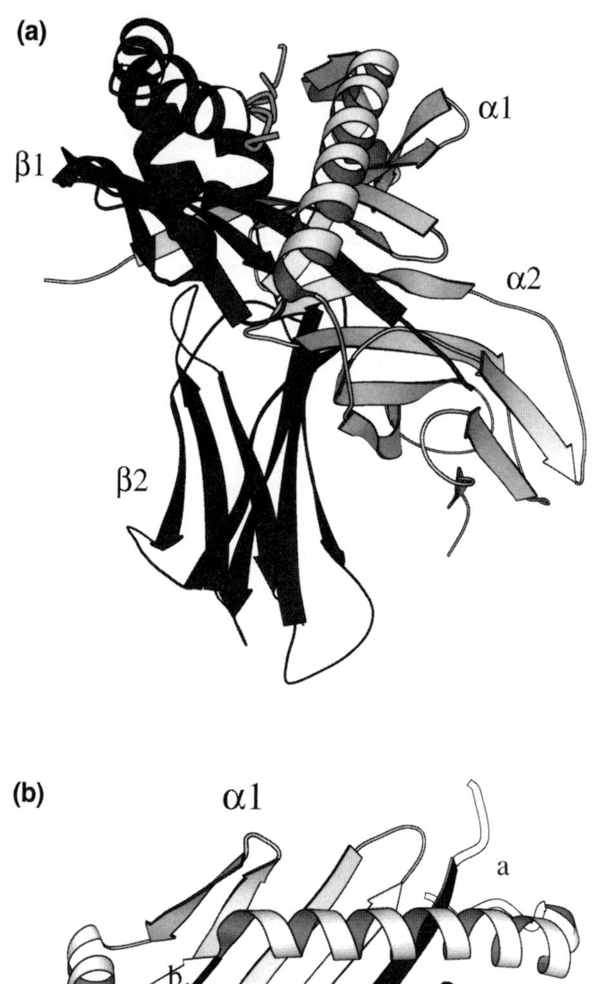

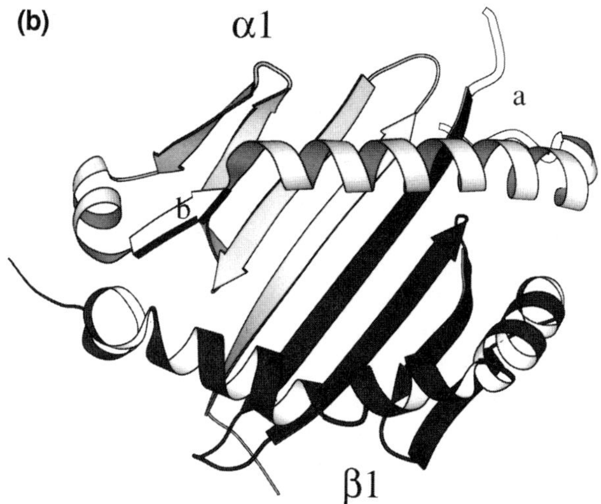

T cells. (b) Top view of the MHC class II peptide-binding site. The $\alpha 1$-domain is light grey (top) an $\beta 1$ is black (bottom). Note the similarity with MHC class I (*Figure 11.1b*). Obvious structural differences between MHC class I and class II molecules are highlighted with the letters *a* and *b*. First, the connectivity between the C-terminal helix of the $\alpha 1$ domain is different between MHC class I and class II (labelled *a*, compare with *Figure 11.1b*). In addition, the N-terminal region of the $\alpha 1$ α-helix in class I corresponds to an extended conformation in MHC class II, shown as a strand in white in the figure (labelled *b*). This extended region can interact with bound peptide to form a short segment of parallel β-sheet. The figure was drawn with the program MOLSCRIPT.

the class I $\alpha3$ domain, swinging further out from underneath the β-sheet of the peptide-binding site and making minimal contacts with the peptide-binding site. The CD4 co-receptor on the T cell is thought to interact directly with the class II $\beta2$ domain in a region that is structurally homologous to the CD8 interaction site defined on MHC class I molecules (Cammarota *et al.*, 1992; Konig *et al.*, 1992). Other changes in the peptide-binding site directly affect the interactions of peptides with the class II molecule and these will be described below.

Many different crystal structures of MHC class II molecules show the formation of a dimer of the class II dimer, sometimes referred to as a super-dimer, which has been the object of some speculation regarding a possible role in T-cell stimulation (Brown *et al.*, 1993; Shafer *et al.*, 1995). The relative orientation of these MHC class II molecules is shown in *Figure 11.6*, page 259. The parallel orientation of the two peptide-binding sites suggests the possibility that these dimers could be involved in the cross-linking of TCR and the subsequent T-cell activation that ensues (Brown *et al.*, 1993). However, this model still requires validation from functional studies which could show that the crystallographically observed complex is important to T-cell activation. Although many different crystal forms have been described for both HLA-DR1 and HLA-DR3 that show this super-dimer, these crystals are all grown under similar conditions which could favour particular packing interactions (Shafer *et al.*, 1995). The recent determination of a mouse MHC class II structure shows a different mode of dimer formation (D. Fremont and W. Hendrickson, personal communication). In the absence of clear functional data, the question of the importance of the super-dimer remains open.

11.3.2 The conformation of bound peptides

HLA-DR molecules bind peptides in an extended conformation that is distinct from that found for MHC class I-bound peptides (Brown *et al.*, 1993; Jardetzky *et al.*, 1996; Stern *et al.*, 1994). Three structures have been determined which allow interpretation of the conformation of bound peptides. These include the HLA-DR1 molecule complexed with an antigenic peptide from influenza virus (Stern *et al.*, 1994), HLA-DR1 with a mixture of endogenous peptides (Jardetzky *et al.*, 1996) and the structure of the HLA-DR3 molecule complexed with naturally processed peptides of the invariant chain (P. Ghosh and D.C. Wiley, personal communication). In all three structures, the peptide is seen to adopt a left-handed helical conformation with ϕ/ψ angles that are similar to that of the polyproline II (ppII) helix. *Figure 11.7a*, page 260 shows a diagram of peptide in the HLA-DR1 binding site, demonstrating the helical twist of the peptide extending throughout the site. The peptide ϕ/ψ angles for most of the amino acids cluster around $-80°$ and $130°$, respectively, which results in a regular projection of side-chains from the peptide main-chain path with a nearly threefold regularity. *Figure 11.7b* (see page 260) shows the regularity of the main-chain conformation and the left-handed helical nature of the peptide conformation, looking from the N- to C-terminus of a bound peptide. A similar conformation of peptides has been observed for proline-containing peptides, which bind to SH3 domains (Feng *et al.*, 1994). Some adjustment of the main-chain conformation is likely to be necessary for the accommodation of different peptide side-chains into the MHC pockets. However, the current structural data suggest that a common core length of approximately 13 peptide residues bound in the MHC class II binding site will adopt very similar overall conformations.

11.3.3 Conserved MHC class II peptide interactions

Peptides that bind to MHC class II molecules are typically longer (~13–25 amino acids) and more variable in length than the majority of peptides that bind to class I molecules (~8–11 amino acids). This difference in length preference is reflected in the differences in how these two classes of MHC molecules interact with peptides. In contrast to MHC class I molecules, where conserved interactions are concentrated at the N- and C-termini of the bound peptides (see *Figure 11.3*, page 258), MHC class II molecules do not have similar interactions with the peptide termini. Conserved MHC class II residues interact with the peptide main-chain at relatively regular intervals throughout the length of the peptide, as shown in *Figure 11.8* on page 261 (Jardetzky *et al.*, 1996; Stern *et al.*, 1994). Consequently, the termini of peptides bound to MHC class II molecules are free to extend out of the peptide-binding site (*Figures 11.8* and *11.9*, page 261). There is little length dependence on the binding of longer peptides to MHC class II molecules and therefore potentially very long antigen-processing intermediates could be associated with the MHC class II molecules and subsequently trimmed by proteases (Deng *et al.*, 1993).

In the DR1–influenza A haemagglutinin (HA) and DR1–endogenous peptide structures there are 12 potential hydrogen bonds between conserved MHC residues and the main-chain of the bound peptides (*Figure 11.8*). Other residues that are not conserved also form hydrogen bond interactions with the peptide in the two structures and these are also shown in *Figure 11.8*, on page 261. In the case of the HLA-DR3-invariant chain (Ii) peptide structure, one of the hydrogen bonds (from DRα62N) is lost due to the presence of a proline in the peptide sequence, but it is formed instead with the hydroxyl group of a peptide threonine side-chain (Ghosh *et al.*, 1995). The conserved hydrogen-bond interactions include the main-chain of DRα53, and the side-chains of DRα62, DRα69, DRα76, DRβ61, DRβ81 and DRβ82. Three of these HLA-DR residues are asparagines that form bidentate hydrogen bonds to the peptide main-chain (*Figures 11.7b* and *11.8*, asnα62, asnα69, and asnβ82). These ring structures have been observed in other protein structures (Le Questel *et al.*, 1993). Although such amide side-chain interactions with peptide main-chain are, at first glance, analogous to hydrogen bonding in β-sheets (e.g. residue DRα53 in *Figure 11.8*), an amide side-chain does not have adjustable ϕ/ψ angles to accommodate different β strand conformations. However, amide side-chains, such as the asparagines in MHC class II, can interact with a ppII peptide conformation. The distribution of asparagines in the binding site (*Figures 11.7a* and *11.8*) and their potential preference for extended, ppII-like peptide conformations, may be important in restricting possible peptide conformations when bound to MHC class II (Jardetzky *et al.*, 1996).

11.3.4 The interaction of peptide side-chains with subsite pockets

Similar to MHC class I, MHC class II molecules have regions of the peptide-binding site which are designed to accommodate peptide side-chains (Jardetzky *et al.*, 1996; Stern *et al.*, 1994). These can be seen in the surface representation of the HLA-DR1 molecule shown in *Figure 11.9*, see page 261. For HLA-DR1, two subsite pockets point down towards the β-sheet at either end of the peptide-binding site (P3 and P11 in *Figure 11.9*). The larger of these pockets (P3) accommodates the large hydrophobic side-chains of Tyr, Phe, Trp and Met, which has been shown to be the most critical peptide side-chain for high-affinity binding (Jardetzky *et al.*, 1990). The C-terminal pocket can also accommodate hydrophobic side-chains, although preferences at this site tend to be for the non-aromatic side-chains such as leucine, methionine or alanine (Hammer *et al.*, 1992, 1993). In the DR1–influenza haemagglutin (DR1-HA) structure there are five peptide positions (P3, P6, P8, P9 and P11) for which more than 90% of the side-chain is buried (Stern *et al.*, 1994). One of these positions, corresponding to five amino acids after the N-terminal tyrosine (P8 in *Figure 11.9*, page 261), defines an MHC subsite that shows preference for alanines in HLA-DR1, but differs in other MHC class II alleles (Hammer *et al.*, 1993). In spite of a low specificity for peptide sequences, the majority of the peptide side-chains do interact with the MHC molecule binding site; for example, in the DR1–HA structure, if one excludes the first and last HA amino acids that extend out of the binding site, only three of eleven side-chains (P2, P7 and P10) have more surface exposed to the solvent than is buried by the MHC molecule (Stern and Wiley, 1994). Relatively shallow side-chain pockets extend to either side of the peptide-binding site and these tend to have less specificity for peptide amino acids than the deeper P3 and P11 pockets, although the P8 pocket is important in distinguishing allele-specific motifs (Hammer *et al.*, 1993). In general, allelic differences between HLA-DR molecules can change the chemical nature of these subsite pockets and thereby lead to changes in preferences for particular peptide-sequence motifs.

11.3.5 Conformational changes in the MHC class II molecule upon binding peptides

There are two issues of interest regarding peptide-induced conformations of the MHC class II molecule. The first concerns the possibility that conformational changes occur upon peptide binding and that these conformational changes reflect a common mechanism in peptide binding. The second concerns the possibility that individual peptides may induce alternative MHC conformations, similar to the hypothesis discussed earlier for MHC class I molecules. Unfortunately there is still a paucity of structural data with which to address both of these points. There are too few single peptide–MHC class II structures to address the possibility of peptide-dependent conformational changes directly. In addition, higher resolution structures, comparable to those obtained for MHC class I, would be desirable for the detection of any small conformational effects.

Both of these possible structural changes in MHC class II molecules are of great interest. The production of HLA-DR in insect cell systems produces protein which can be loaded efficiently with added peptides (Kozono *et al.*, 1994; Sloan *et al.*, 1995; Stern and Wiley, 1992; Wallny *et al.*, 1995). In the absence of peptides, these DR molecules tend to aggregate, suggesting the possibility that the proteins are misfolded in the absence of peptide, or that empty peptide binding sites can interact weakly with other regions of the MHC structure. Kinetic observations indicate that relatively fast peptide-binding is followed by a slow conformational change of the complex to the slow-dissociating complex (Sadegh-Nasseri *et al.*, 1994). As the pre-complex has a fast dissociation rate, it may be that the final clarification of the MHC class II peptide-binding mechanism will come from a structural understanding of the difference between a pre-complex and a final slow-dissociating complex.

The possibility that peptide-binding may induce unique MHC conformations that could then be detected by the TCR is similar to the possibilities discussed above with MHC class I. Although there is limited structural data, it seems likely that there may well be some unique aspects of MHC conformation associated with peptide binding. However, as with class I, these may be associated with groups of peptides of different sequence. Perhaps such peptide-associated structural changes could explain puzzling inhibition results obtained with bacterial superantigens (see below), but this remains to be determined experimentally.

11.3.6 The binding of superantigens to MHC class II molecules

MHC class II molecules present another class of antigens to T cells (Kotzin *et al.*, 1993; Thibodeau and Sékaly, 1995). These antigens are called superantigens, because of their ability to stimulate massive T-cell responses. Both bacterial and viral superantigens have been identified and these share common features in their ability to stimulate the immune system. Superantigens differ fundamentally in their interactions with both MHC class II and TCR from conventional peptide antigens, acting on the T-cell system as intact proteins rather than as peptide fragments. Superantigens are basically bifunctional proteins, with a MHC-binding and a TCR-binding function (Jardetzky, 1995; Seth *et al.*, 1994).

Two structures of bacterial superantigens bound to HLA-DR1 have been solved, highlighting the differences, not only between superantigen and conventional antigen binding to MHC molecules, but also demonstrating variability amongst the binding of the superantigens to MHC. The two complexes are of HLA-DR1 bound to the *Staphylococcus* enterotoxin B (SEB) (Jardetzky *et al.*, 1994) and the toxic shock syndrome toxin (TSST-1; Kim *et al.*, 1994). These two complexes are shown in *Figure 11.9*. Both toxins bind outside of the conventional antigen-binding site. For SEB, the binding occurs only to the $\alpha 1$ domain of the α-chain, with interactions along loops of the β-sheet and along the side and top of part of the α-helix. For TSST-1, the toxin is positioned very differently relative to the MHC peptide binding site, forming interactions with the α-chain that overlap the SEB binding site and extending over the top of the peptide-binding site, over the peptide and across to the β-chain α-helix. The TSST-1 toxin shares a leucine in common with SEB (*Figure 11.10*, see page 262), which inserts into a hydrophobic pocket formed by two β-sheet loops and the helix of the α-chain. This interaction serves as a pivot point around which the two toxins are rotated relative to each other in the different complexes.

The two toxin structures help to explain some confusing observations about superantigen-binding sites on the MHC class II molecules. SEB and TSST-1 bind to overlapping regions of the MHC molecule, and both superantigens interact with MHC class II residues such as Lysα39 and Metα36 (Thibodeau *et al.*, 1994). Mutations of these residues can affect the binding of both toxins to the MHC class

II molecules, consistent with both crystal structures. However, many binding studies have shown that SEB and TSST-1 do not compete for binding to MHC molecules at the cell surface (Jardetzky, 1995), and that cellular factors may modulate the binding of superantigens to MHC class II (Thibodeau *et al.*, 1994; Yagi *et al.*, 1994). In fact, peptides may be the cellular factors that affect superantigen binding and thereby create subsets of MHC class II at the cell surface which can interact with different superantigens. Preliminary data support this hypothesis for TSST-1 and SEA. For TSST-1, a direct interaction with peptides in the MHC antigen-binding site is consistent with the crystallographic complex. For SEB, peptide-dependent binding to MHC may also occur, although functional data is lacking and the crystallographic data would suggest that peptide effects on SEB binding would be transmitted through some conformational change in the MHC molecule. Why superantigens have evolved to bind subsets of MHC molecules at the cell surface remains to be determined, but this selective binding could have significant implications for the pathogenicity of particular superantigens as well as for their potential association with autoimmune disease induction. The structure of a superantigen–MHC complex bound to a TCR has not been solved, but the two current structures of SEB and TSST-1 with HLA-DR1 indicate that the ternary complexes will differ from conventional antigen–MHC–TCR complexes.

References

Bjorkman PJ, Saper MA, Samraoui B, Bennett WS, Strominger JL, Wiley DC. (1987a) The foreign antigen binding site and T-cell recognition regions of class I histocompatibility antigens. *Nature* **329:** 512–518.

Bjorkman, PJ, Saper MA, Samraoui B, Bennett WS, Strominger JL, Wiley DC. (1987b) Structure of the human class I histocompatibility antigen, HLA-A2. *Nature* **329:** 506–512.

Bouvier M, Wiley DC. (1994) Importance of peptide amino and carboxyl termini to the stability of MHC class I molecules. *Science* **265:** 398–402.

Brown JH, Jardetzky T, Saper MA, Samraoui B, Bjorkman PJ, Wiley DC. (1988) A hypothetical model of the foreign antigen binding site of class II histocompatibility molecules. *Nature* **332:** 845–850.

Brown JH, Jardetzky TS, Gorga JC, Stern LJ, Urban RG, Strominger JL, Wiley DC. (1993) Three-dimensional structure of the human class II histocompability antigen HLA-DR1. *Nature* **364:** 33–39.

Burmeister WP, Gastinel LN, Simister NE, Blum ML, Bjorkman PJ. (1994) Crystal structure at 2.2 A resolution of the MHC-related neonatal Fc receptor. *Nature* **372:** 336–343.

Cammarota G, Scheirle, A, Takacs B, Doran DM, Knorr R, Bannwarth W, Guardiola J, Sinigaglia F. (1992) Identification of a CD4 binding site on the beta 2 domain of HLA-DR molecules. *Nature* **356:** 799–801.

Collins EJ, Garboczi DN, Wiley DC. (1994) Three-dimensional structure of a peptide extending from one end of a class I MHC binding site. *Nature* **371:** 626–629.

Deng H, Apple R, Clare-Salzler M, Trembleau S, Mathis D, Adorini L, Sercarz E. (1993) Determinant capture as a possible mechanism of protection afforded by major histocompatibility complex class II molecules in autoimmune disease. *J. Exp. Med.* **178:** 1675–1680.

Falk K, Rotzschke O, Stevanovic S, Jung G, Rammensee HG. (1991) Allele-specific motifs revealed by sequencing of self-peptides eluted from MHC molecules. *Nature* **351:** 290–296.

Feng S, Chen JK, Yu H, Simon JA, Schreiber SL. (1994) Two binding orientations for peptides to the Src SH3 domain: development of a general model for SH3-ligand interactions. *Science* **266:** 1241–1247.

Fremont DH, Matsumura M, Stura EA, Peterson PA, Wilson IA. (1992) Crystal structures of two viral peptides in complex with murine MHC class I H-2Kb. *Science* **257:** 919–927.

Fremont DH, Stura EA, Matsumura M, Peterson PA, Wilson IA. (1995) Crystal structure of an H-2Kb-ovalbumin peptide complex reveals the interplay of primary and secondary anchor positions in the major histocompatibility complex binding groove. *Proc. Natl Acad. Sci. USA* **92:** 2479–2483.

Garrett TP, Saper MA, Bjorkman PJ, Strominger JL, Wiley DC. (1989) Specificity pockets for the side chains of peptide antigens in HLA-Aw68. *Nature* **342:** 692–696.

Germain RN. (1994) MHC-dependent antigen processing and peptide presentation: providing ligands for T lymphocyte activation. *Cell* **76:** 287–299.

Ghosh P, Amaya M, Mellins E, Wiley DC. (1995) *Nature* **378:** 457–462.

Giblin PA, Leahy DJ, Mennone J, Kavathas PB. (1994) The role of charge and multiple faces of the CD8 alpha/alpha homodimer in binding to major histocompatibility complex class I molecules: support for a bivalent model. *Proc. Natl Acad. Sci. USA* **91:** 1716–1720.

Guo HC, Jardetzky TS, Garrett TP, Lane WS, Strominger JL, Wiley DC. (1992) Different length peptides bind to HLA-Aw68 similarly at their ends but bulge out in the middle. *Nature* **360:** 364–366.

Guo HC, Madden DR, Silver ML, Jardetzky TS, Gorga JC, Strominger JL, Wiley DC. (1993) Comparison of the P2 specificity pocket in three human histocompatibility antigens: HLA-A*6801, HLA-A*0201, and HLA-B*2705. *Proc. Natl Acad. Sci USA* **90:** 8053–8057.

Hammer J, Takacs B, Sinigaglia F. (1992) Identification of a motif for HLA-DR1 binding peptides using M13 display libraries. *J. Exp. Med.* **176:** 1007–1013.

Hammer J, Valsasnini P, Tolba K, Bolin D, Higelin J, Takacs B, Sinigaglia F. (1993) Promiscuous and allele-specific anchors in HLA-DR-binding peptides. *Cell* **74:** 197–203.

Honig B, Nicholls A. (1995) Classical electrostatics in biology and chemistry. *Science* **268:** 1144–1149.

Jardetzky TS. (1995) Structural studies of the interaction of superantigens with class II major histocompatibility complex molecules. *Bacterial Superantigens: Structure, Function and Therapeutic Potential* (eds J Thibodeau, R-P Sékaly). Austin, RG. Landes, pp. 67–82.

Jardetzky TS, Gorga JC, Busch R, Rothbard J, Strominger JL, Wiley DC. (1990) Peptide binding to HLA-DR1: a peptide with most residues substituted to alanine retains MHC binding. *EMBO J.* **9:** 1797–1803.

Jardetzky TS, Lane WS, Robinson RA, Madden DR, Wiley DC. (1991) Identification of self peptides bound to purified HLA-B27. *Nature* **353:** 326–329.

Jardetzky TS, Brown JH, Gorga JC, Stern LJ, Urban RG, Chi YI, Stauffacher C, Strominger JL, Wiley DC. (1994) Three-dimensional structure of a human class II histocompatibility molecule complexed with superantigen. *Nature* **368:** 711–718.

Jardetzky TS, Brown JH, Gorga JC, Stern LJ, Urban RG, Strominger JL, Wiley DC. (1995) Crystallographic analysis of endogenous peptides associated with HLA-DR1 suggests a common, polyproline II-like conformation for bound peptides. *Proc. Natl Acad. Sci. USA* **93:** 734–738.

Kim J, Urban RG, Strominger JL, Wiley DC. (1994) Toxic shock syndrome toxin-1 complexed with a class II major histocompatibility molecule HLA-DR1. *Science* **266:** 1870–1874.

Konig R, Huang LY, Germain RN. (1992) MHC class II interaction with CD4 mediated by a region analogous to the MHC class I binding site for CD8. *Nature* **356:** 797–798.

Kotzin BL, Leung DY, Kappler J, Marrack P. (1993) Superantigens and their potential role in human disease. *Adv. Immunol.* **54:** 99–166.

Kozono H, White J, Clements J, Marrack P, Kappler J. (1994) Production of soluble MHC class II proteins with covalently bound single peptides. *Nature* **369:** 151–154.

Kraulis PJ. (1991) MOLSCRIPT: a program to produce both detailed and schematic plots of protein structures. *J. Appl. Cryst.* **24:** 945–949.

Latron F, Pazmany L, Morrison J, Moots R, Saper MA, McMichael A, Strominger JL. (1992) A critical role for conserved residues in the cleft of HLA-A2 in presentation of a nonapeptide to T cells. *Science* **257:** 964–967.

Le Questel JY, Morris DG, Maccallum PH, Poet R, Milner-White EJ. (1993) Common ring motifs in proteins involving asparagine or glutamine amide groups hydrogen-bonded to main-chain atoms. *J. Mol. Biol.* **231:** 888–896.

Madden DR, Gorga JC, Strominger JL, Wiley DC. (1991) The structure of HLA-B27 reveals nonamer self-peptides bound in an extended conformation. *Nature* **353:** 321–325.

Madden DR, Gorga JC, Strominger JL, Wiley DC. (1992) The three-dimensional structure of HLA-B27 at 2.1 A resolution suggests a general mechanism for tight peptide binding to MHC. *Cell* **70:** 1035–1048.

Madden DR, Garboczi DN, Wiley DC. (1993) The antigenic identity of peptide-MHC complexes: a comparison of the conformations of five viral peptides presented by HLA-A2. *Cell* **75:** 693–708.

Matsumura M, Fremont DH, Peterson PA, Wilson IA. (1992) Emerging principles for the recognition of peptide antigens by MHC class I molecules. *Science* **257:** 927–934.

Nossal GJ. (1994) Negative selection of lymphocytes. *Cell* **76:** 229–239.

Sadegh-Nasseri S, Stern LJ, Wiley DC, Germain RN. (1994) MHC class II function preserved by low-affinity peptide interactions preceding stable binding. *Nature* **370:** 647–650.

Salter RD, Norment AM, Chen BP, Clayberger C, Krensky AM, Littman DR, Parham P. (1989) Polymorphism in the alpha 3 domain of HLA-A molecules affects binding to CD8. *Nature* **338:** 345–347.

Salter RD, Benjamin RJ, Wesley PK, Buxton SE, Garrett TP, Clayberger C, Krensky AM, Norment AM, Littman DR, Parham P. (1990) A binding site for the T-cell co-receptor CD8 on the alpha 3 domain of HLA-A2. *Nature* **345:** 41–46.

Saper MA, Bjorkman PJ, Wiley DC. (1991) Refined structure of the human histocompatibility antigen HLA-A2 at 2.6 A resolution. *J. Mol. Biol.* **219:** 277–319.

Seth A, Stern LJ, Ottenhoff TH, Engel I, Owen MJ, Lamb JR, Klausner RD, Wiley DC. (1994) Binary and ternary complexes between T-cell receptor, class II MHC and superantigen in vitro. *Nature* **369:** 324–327.

Shafer PH, Pierce SK, Jardetzky TS. (1995) The structure of MHC class II: a role for dimers of dimers. *Semin. Immunol.* **7:** 389–398.

Silver ML, Guo HC, Strominger JL, Wiley DC. (1992) Atomic structure of a human MHC molecule presenting an influenza virus peptide. *Nature* **360:** 367–369.

Sloan VS, Cameron P, Porter G, Gammon M, Amaya M, Mellins E, Zaller D. (1995) Mediation by HLA-DM of dissociation of peptides from HLA-DR. *Nature* **375:** 802–806.

Stern LJ, Wiley DC. (1992) The human class II MHC protein HLA-DR1 assembles as empty alpha beta heterodimers in the absence of antigenic peptide. *Cell* **68:** 465–477.

Stern LJ, Wiley DC. (1994) Antigenic peptide binding by class I and class II histocompatibility proteins. *Structure* **2:** 245–251.

Stern LJ, Brown JH, Jardetzky TS, Gorga JC, Urban RG, Strominger JL, Wiley DC. (1994) Crystal structure of the human class II MHC protein HLA-DR1 complexed with an influenza virus peptide. *Nature* **368:** 215–221.

Thibodeau J, Sékaly R.-P. (Eds) (1995) *Bacterial Superantigens: Structure, Function and Therapeutic Potential.* Austin, R.G. Landes.

Thibodeau J, Cloutier I, Labreque N, Mourad W, Jardetzky T, Sékaly R.-P. (1994) Subsets of HLA-DR1 molecules defined by SEB and TSST-1 binding. *Science* **266:** 1874–1878.

Wallny HJ, Sollami G, Karjalainen K. (1995) Soluble mouse major histocompatibility complex class II molecules produced in Drosophila cells. *Eur. J. Immunol.* **25:** 1262–1266.

Wang C.-R, Castano AR, Peterson PA, Slaughter C, Fischer-Lindahl K, Deisenhofer J. (1995) Nonclassical binding of formylated peptide in a crystal structure of the MHC class Ib molecule H2-M3. *Cell* **82:** 655–664.

Weiss GA, Collins EJ, Garboczi DN, Wiley DC, Schreiber SL. (1995) A tricyclic ring system replaces the variable regions of peptides presented by three alleles of human MHC class I molecules. *Chem. Biol.* **2:** 401–407.

Yagi J, Uchiyama T, Janeway C Jr. (1994) Stimulator cell type influences the response of T cells to staphylococcal enterotoxins. *J. Immunol.* **152:** 1154–1162.

Young AC, Zhang W, Sacchettini JC, Nathenson SG. (1994) The three-dimensional structure of H-2Db at 2.4 A resolution: implications for antigen-determinant selection. *Cell* **76:** 39–50.

Zhang W, Young AC, Imarai M, Nathenson SG, Sacchettini JC. (1992) Crystal structure of the major histocompatibility complex class I H-2Kb molecule containing a single viral peptide: implications for peptide binding and T-cell receptor recognition. *Proc. Natl Acad. Sci. USA* **89:** 8403–8470.

12

Peptides associated with MHC class I and class II molecules

Miles P. Davenport and Adrian V.S. Hill

12.1 Importance of specific peptide-binding motifs for MHC alleles

T cells recognize their peptide ligands bound to major histocompatibility complex (MHC) molecules on the surface of cells. The MHC molecule plays an important role in determining which peptide antigens are 'visible' to T cells. The antigen processing pathways of the MHC class I and class II molecules (see Chapters 9 and 10) determine to a large degree the source proteins from which peptides may be derived. In the case of MHC class I these are largely intracellular, and for class II largely extracellular or cell surface in origin. However, beyond these broad categories of source protein, the MHC molecule itself is important in determining which peptides from this pool will be bound. An MHC molecule will bind only to peptides conforming to certain structural requirements, and thus is only able to present a subset of the available peptides to T cells. Importantly, these structural requirements for peptide binding vary between the family of polymorphic MHC alleles of a given locus, resulting in the presentation of different peptides by each. The high degree of polymorphism at MHC loci is thought to have evolved as a result of some selective advantage associated with the presence of multiple MHC alleles with different peptide-binding specificities. The presence of MHC alleles with different binding specificities may make it less likely that a pathogen can 'escape' from binding of its peptides to a given MHC molecule. Therefore, it is of both evolutionary and functional interest to deduce the peptide-binding specificity of individual MHC alleles, so as to understand how the structural diversity of the different MHC alleles affects their abilities to interact with antigenic peptides.

The lack of a simple optimal method for identifying the rules of peptide binding to MHC molecules is indicated by the large number of methods that have been applied to study this question. Many MHC alleles have now been analysed using various techniques. However, we are still far from understanding the more subtle complexities of MHC–peptide interactions. The broad picture, however, seems clear. Peptides bound to MHC class I are of a fairly uniform length of between 8 and 10 amino acids (Hunt *et al.*, 1992a). The length of peptides bound to MHC

HLA and MHC: genes, molecules and function, edited by M.J. Browning and A.J. McMichael.
© 1996 BIOS Scientific Publishers Ltd, Oxford.

class II is variable, but usually of between 12 and 20 amino acids (Chicz *et al.*, 1993; Rudensky *et al.*, 1992). In both cases much longer peptides are occasionally found (Chicz *et al.*, 1994; Urban *et al.*, 1994).

Analysis of the peptide-binding specificity of MHC molecules has focused on peptide 'anchor residues' fitting into 'pockets' of the MHC molecule (Bjorkman *et al.*, 1987; Stern *et al.*, 1994). This has proved useful in the prediction of ligands for MHC class I antigens (Hill *et al.*, 1992; Pamer *et al.*, 1991; Rötzschke *et al.*, 1991), although there is good evidence that, even for class I molecules, binding is much more complex than this simple scheme would imply (Chen *et al.*, 1994; Ruppert *et al.*, 1993). Motifs for MHC class II have yet to prove their utility in the prediction of class II epitopes. This may well be because the rules for binding to MHC class II are more complex than for class I, and remain to be established. Nonetheless, an impressive amount of information on the peptide-binding specificity of MHC molecules has been accumulated over the last 5 years.

12.2 Methods of motif identification

The methods for analysis of MHC class I and class II motifs are broadly similar and fall generally into three categories:

(i) those that utilize measurements of peptide binding to MHC;
(ii) those that involve analysis of the natural ligands bound to MHC molecules;
(iii) those that rely on theoretical analyses of MHC–peptide interactions.

The first two of these methods have been more extensively used than the third.

The existence of multiple methods for the analysis of MHC–peptide interactions has sometimes proved problematic, since the results reported in different studies are not always consistent. This may be due to differences in the techniques used to analyse the peptide specificity of MHC molecules. One explanation for such inconsistency is comparison of results derived from peptide-binding studies with those derived from analysis of eluted peptides. The latter observes which peptides have been naturally processed and presented on the MHC molecule. The effects of processing are, therefore, integral to the selection of these peptide ligands. Peptide-binding studies, on the other hand, measure the MHC–peptide interaction in isolation. Since, most often, motifs are used to try to define peptides which will be processed and presented on MHC, peptide–binding studies may be misleading, simply because they do not incorporate the constraints of processing. Peptide-binding studies may also lead to different results from those observed by analysis of eluted peptides because the MHC binding occurs under different conditions. This is particularly relevant in the case of MHC class II molecules, where both pH and the molecular environment of the class II molecules are thought to affect peptide–MHC interactions (Buelow *et al.*, 1994; Sette *et al.*, 1992; Sherman *et al.*, 1994).

Therefore, at present, there is no perfect method for analysis of MHC–peptide interactions. Frequently motifs must be generated from the synthesis of several lines of conflicting evidence. Therefore, we present a description of the common methods of analysis in use, and try to outline their utility.

12.2.1 Peptide binding studies

Methods of analysing peptide binding differ greatly between MHC class I and class II molecules. Whereas MHC class II molecules readily undergo peptide exchange, both on the cell surface and *in vitro*, MHC class I molecules do not. For this reason, very different methods of analysis have evolved. In order to study peptide binding to MHC class I molecules it is necessary to generate 'empty' class I molecules. The crucial advance in this area has been the generation of mutant cell lines which are defective in the transport and assembly of class I molecules and therefore generate large amounts of 'empty' molecules (see Chapter 9). These cell lines are defective in the 'transporter associated with antigen processing' (TAP) molecule(s) and are, therefore, unable to load their class I molecules with peptides. These 'empty' class I molecules are unstable unless they are stabilized by an appropriate peptide. Thus, assays have developed which measure the ability of different peptides to stabilize these empty molecules. These defective cell lines can be transfected with the class I molecule of interest, then aliquots of these cells can be metabolically labelled and lysed in detergent in the presence of different peptides. Class I molecules are subsequently immunoprecipitated from the lysate using conformation-specific antibodies, which bind only to correctly assembled class I molecule. The amount of class I heavy chain immunoprecipitated by this method is therefore a measure of the ability of the added peptide to stabilize class I molecules and compares well to measurements using a labelled peptide and direct binding to MHC (Elvin *et al.*, 1993). Since these defective cell lines contain their own endogenous class I molecules, it is usually necessary to separate these from the transfected class I variant. This can be done by using allele-specific antibodies, if these are available, or, more commonly, by separating the different immunoprecipitated class I variants using isoelectric focusing and analysing only the class I type of interest.

The assembly of MHC class I molecules has also been measured using fluorescence-activated cell sorting (FACS) analysis of the amount of MHC present at the cell surface. In the absence of peptide, the mutant cell lines are largely devoid of cell-surface class I molecules. The addition of peptide leads to the presence of stable class I molecules at the cell surface, which can be detected by fluorescently labelled antibody and FACS analysis. Due to the presence of the endogenous class I molecules, allele-specific antibodies may be required to differentiate endogenous from transfected class I molecules. Methods that directly measure the binding of peptides to class I molecules have also been developed. In these assays, peptides are either radioactively labelled or biotinylated for later detection (Cerundolo *et al.*, 1991; Lopez *et al.*, 1992). These assays can avoid the problem of the presence of endogenous class I alleles within the mutant cell lines, because a labelled peptide, which binds specifically to the allele of interest, can be used as the indicator, and competition experiments performed against 'cold' test peptides.

The binding specificity of MHC class I may be probed by the use of any of the above methods of analysis. One approach is the use of a known epitope which has all residues sequentially altered to all naturally occurring amino acids. The effect of these changes is to provide an indication of the peptide-binding specificity of the class I molecules. However, the results of these studies do not always correlate with the results from analysis of naturally occurring peptide ligands.

The ability of MHC class II to exchange peptide has allowed for extensive use of *in vitro* binding assays for the analysis of MHC–peptide interaction. In their simplest form, these assays rely on the co-incubation of affinity-purified MHC molecules and labelled peptide. The details of these assays vary somewhat, but most commonly include incubation at 37°C and low pH (pH 5.0) for periods of 24–72 hours (Harding *et al.*, 1991; Jensen, 1991). MHC-associated peptides may be then separated from free peptides by the use of size exclusion chromatography (Buus *et al.*, 1987) or capture by immobilized antibody and then quantified. The low pH may facilitate peptide binding, probably due to increased exchange of the class II-associated invariant chain peptide (CLIP). However, pH may have unpredictable effects on the binding of individual peptides (Jensen, 1991; Sette *et al.*, 1992). Peptides used in these assays may be labelled by a variety of methods. The use of iodine-125(^{125}I)-labelled peptides has largely been replaced by the use of biotinylated peptides and an enzymatic (horseradish peroxidase or alkaline phosphatase) detection system. Since the labelling of peptides by either of these methods may alter their affinity for the MHC molecule, competition assays between a labelled peptide and cold competitor peptides are the most reliable method of comparing peptide affinities and, in addition, do not require the labelling of multiple peptides. One problem with these methods of analysis, paradoxically, is their extreme sensitivity. Many peptides appear to bind at some level to MHC class II molecules. Only by comparing the affinity of a peptide with that of known T-cell epitopes or self-peptides is it possible to assess whether it is likely to be a good T-cell epitope.

Assays that utilize direct binding of labelled peptides to live antigen presenting cells (APCs) have also been used. This involves the incubation of APCs with labelled peptides. ^{125}I labelling of peptides containing a tyrosine has been used, and binding detected by radiometry (Ceppellini *et al.*, 1989). More recently, biotin-labelled peptides have been used and detected through incubation with streptavidin-FITC and FACS analysis (Busch *et al.*, 1990). One advantage of these systems is that they do not require the purification of MHC molecules and peptide binding thus occurs in a relatively natural environment.

The peptide-binding motifs of MHC class II molecules have been studied by measuring the effects of different modifications on peptide binding using peptide analogues with multiple single amino acid substitutions or truncations of a known epitope. The peptides that have been most commonly studied in this way have been those known to bind promiscuously to multiple MHC alleles. The use of these peptides raises a question as to how well allele-specific differences in peptide binding can be determined by using them.

One modification of this approach is the use of polyalanine peptides, with substitution of different amino acid residues at various positions (Hammer *et al.*, 1994a; Jardetzky *et al.*, 1990). This method can be used to study both the nature and the spacing of anchors required for interaction with the MHC molecule. More recently, the measurement of the binding affinities of peptide analogues with sequential substitution of all naturally occurring amino acids at all positions of a peptide has been coupled with a computer algorithm to calculate how well each position of a peptide is likely to bind to a given MHC molecule (Hammer *et al.*, 1994b, 1995; Reay *et al.*, 1994). Therefore, this method analyses the 'fit' of a peptide at all amino acid positions, and takes into account positive and negative effects of residues on binding. This analysis can be used to scan whole proteins to identify likely MHC binding peptides. Approaches such as this rely on the assumption that every amino acid of a peptide contributes independently to MHC binding. In addition, since this method depends on binding measurements alone, it does not incorporate any possible restrictions of peptide processing on the repertoire of peptides able to bind.

Another approach to measuring the peptide-binding specificity of class II molecules has been the use of phage display libraries (Hammer *et al.*, 1992, 1993). In this system, synthetic oligonucleotides encoding a random nine amino acid stretch flanked by four glycines at either end are inserted into the pIII phage coat protein of the M13 bacteriophage. In this way a large library of phage is created, each bearing a different random nonamer on its surface. Large numbers of phage ($\approx 10^{10}$) are then incubated with biotinylated MHC molecules. Those that encode a nonamer with high affinity will preferentially associate with the class II molecules. The MHC molecules and any bound phage are then removed using streptavidin-coated beads. After extensive washing, the class II-associated phage are eluted from the class II molecule and expanded in *Escherichia coli*. After several rounds of sequential purification and expansion in this manner, the inserts from the MHC-associated pool and the initial pool of phage are sequenced and compared. The MHC-associated pool of peptides shows increases in the frequency of some amino acids at particular positions relative to the unselected pool. Again, analysis of these enrichments can be used to identify potential anchor residues for MHC–peptide interactions. Although this is, in principle, a useful means of analysis, the method has not been extensively used.

12.2.2 Analysis of natural peptide ligands

The aim of most studies of MHC–peptide specificity is the prediction of epitopes within either pathogen-derived proteins or autoantigens. The most direct route to this would be the analysis of the peptides from these proteins that are presented on MHC molecules. This is made difficult by the low numbers of such peptides bound to MHC molecules on the cell surface. In most cases it has only been possible to sequence the most abundant self-peptides present. In a few cases the analysis of antigenic peptides has been possible through the use of T-cell assays to identify an immunostimulatory high pressure liquid chromatography (HPLC) fraction and subsequent mass spectrometry to sequence the peptide of interest (Cox *et al.*, 1994). However, most work still revolves around the analysis of 'self-peptides' associated with MHC molecules and attempts to use this information to predict which peptides may be processed from a given antigen.

The purification of peptides bound to MHC molecules has become a relatively standard procedure. Antibody affinity chromatography is used to purify MHC molecules from detergent lysates of cells (usually aliquots of 10^{10} cells, equivalent to about 10 ml of cell pellet), and the peptides are eluted from the MHC molecules by 0.1% trifluoroacetic acid or 10% acetic acid. The peptides are then separated from the MHC molecules using 3 kDa or 10 kDa cut-off ultrafiltration, and then analysed by HPLC.

The simplest analysis of these peptides involves pool sequencing of all the eluted peptides by Edman degradation (Falk *et al.*, 1991b), sometimes after removal of any dominant peptide peaks seen on HPLC to prevent their biasing the pool (Falk *et al.*, 1994b). Analysis of these results shows that the concentrations of some amino acids are greatly increased at particular cycles, indicating that these amino acids are enriched at these positions. In the case of MHC class I molecules, where peptides are short and relatively uniform in length, these 'peaks' of enrichment tend to be fairly sharply defined (over one sequencing cycle; *Figure 12.1*), clearly demonstrating the positions at which particular amino acids are favoured. In the case of MHC class II molecules, peptides are longer and the distance from their amino termini to the first hydrophobic 'anchor' position may vary between about three and six amino acids. Therefore, any enrichments tend to be seen over broader peaks (three sequencing cycles) and it is more difficult to interpret the relative alignment of anchors. In either case, comparison of the pool-sequencing data with the sequences of individual peptides generally assists in the analysis of which anchors are important for peptide binding.

We have recently developed an empirical method for predicting MHC ligands based on the analysis of pool sequence data (Davenport *et al.*, 1995a). This relies on the use of a computer program to score how well an individual peptide conforms to the patterns of amino acid enrichment seen in pool sequence data. Thus this method should incorporate information on processing requirements as well as MHC-binding affinities during processing of natural ligands.

The sequencing of individual peptide ligands eluted from MHC molecules may be performed by a variety of methods. One option is Edman sequencing of an individual peptide peak from HPLC. From these data an unambiguous sequence may be immediately obvious, but more frequently, a clear stretch of sequence may contain points of ambiguity. In this case, the ambiguity is frequently resolved by comparison of the sequence with the computer database of known proteins. This approach is not without its hazards, however, as there may be a temptation to overinterpret poor sequencing data. This difficulty may be overcome in several ways. First, binding studies with synthetic analogues of the eluted peptides offer a simple way to assess whether these may be natural ligands for the MHC molecule in question. These self-peptides should have a high affinity for MHC, being among the most prominent species present on the cell surface. Second, the HPLC elution time of these synthetic analogues may also be compared with that of the HPLC fraction from which the sequence was derived. Finally, analysis of the eluted peptides using mass spectrometry (MS) may be of assistance. A simple time of flight (TOF) mass spectrum is usually sufficient to identify all the major peptides within an HPLC fraction and their masses. Putative peptide sequences from the fraction can then be analysed to see if the predicted masses of these peptides coincides with the observed masses (Chicz *et al.*, 1992). If the match is good, this is excellent evidence that the sequence is correct. Additionally, in the case of MHC class II molecules, where peptides are long and frequently the Edman sequencing provides little information on the C-terminus, the MS data may help to identify the exact C-terminus of the peptide. Again, since there are only a small number of possible predicted masses representing different C-terminal truncations of the peptide, a good match with the observed mass is confirmation of the sequence.

On its own, MS has also been used to sequence MHC-derived peptides. Tandem mass spectrometry, essentially two mass analysers in series, is required for the sequencing of these peptides. As with the TOF instrument, peptides are first ionized and then subjected to mass analysis. Whereas TOF

(a) Experiment 1

	Amino-acid residues (pmol)																		
	A	R	N	D	E	Q	G	H	I	L	K	M	F	P	S	T	Y	V
Cycle	Ala	Arg	Asn	Asp	Glu	Gln	Gly	His	Ile	Leu	Lys	Met	Phe	Pro	Ser	Thr	Tyr	Val
1	172.6	0.0	31.9	25.7	44.8	125.9	112.4	2.8	144.4	123.8	60.0	30.7	63.3	117.9	75.9	49.0	50.3	104.9
2	42.5	0.0	16.2	14.1	25.6	53.1	44.7	1.6	69.6	511.0	15.5	71.0	10.5	38.7	16.2	16.1	12.2	86.5
3	99.8	0.0	9.5	18.3	12.3	20.4	31.8	11.1	51.5	110.8	5.8	55.7	19.4	30.4	12.0	8.7	20.9	46.0
4	36.0	0.6	12.7	26.4	59.5	21.7	56.2	1.3	10.4	22.7	24.6	5.2	5.2	52.4	10.9	14.0	5.2	28.8
5	35.1	0.1	13.4	18.6	28.1	19.8	55.6	2.8	21.4	23.9	47.2	4.1	6.2	39.1	7.5	10.5	11.6	29.0
6	30.3	0.9	16.8	14.1	21.4	17.3	28.5	1.4	68.1	43.4	14.7	4.4	5.8	40.8	9.2	20.3	5.0	106.2
7	42.1	0.0	11.7	9.5	27.2	21.8	19.0	3.2	36.3	27.3	7.9	5.7	8.0	54.1	5.4	13.6	14.0	62.8
8	37.9	0.3	13.4	8.1	37.3	24.3	21.1	1.8	11.6	15.1	33.8	3.4	5.1	22.3	8.8	17.9	10.2	22.4
9	23.3	0.0	5.1	6.0	15.7	10.5	14.8	0.7	11.5	27.5	8.7	3.1	2.7	11.9	5.6	6.7	5.1	60.2
10	12.0	0.7	2.6	4.4	6.5	5.2	10.2	0.4	4.5	12.1	4.5	1.0	1.8	7.1	2.7	3.2	2.3	20.4

(b) Experiment 2

Cycle	Ala	Arg	Asn	Asp	Glu	Gln	Gly	His	Ile	Leu	Lys	Met	Phe	Pro	Ser	Thr	Tyr	Val
1	110.8	10.8	4.0	3.1	10.0	14.5	55.7	0.2	60.3	44.4	10.8	8.2	37.5	20.3	27.4	14.6	19.8	48.0
2	13.4	1.6	2.0	1.9	6.8	11.0	9.0	0.0	37.9	302.7	0.0	26.2	5.0	6.3	4.4	4.5	3.3	26.5
3	62.4	3.5	5.0	9.1	4.9	10.0	12.6	0.1	35.7	71.5	0.0	24.5	13.8	13.4	8.9	4.8	17.9	19.6
4	16.9	2.2	4.5	8.9	25.3	7.9	24.5	0.1	6.2	10.3	2.8	1.3	2.0	22.1	4.9	5.0	1.8	9.3
5	22.3	1.6	6.8	8.6	14.3	9.9	31.8	0.0	16.6	15.1	8.2	1.9	4.0	16.3	4.5	4.6	5.7	18.3
6	10.6	1.3	6.6	3.6	6.4	6.2	10.1	0.1	38.7	27.1	0.0	1.4	2.7	12.6	3.2	6.1	1.3	39.2
7	19.2	1.0	4.7	2.5	7.2	9.0	5.6	0.2	22.3	16.1	0.0	1.9	3.9	17.4	1.9	3.5	3.6	27.2
8	13.4	1.2	3.1	1.3	7.9	6.3	6.9	0.3	4.7	6.7	3.0	0.6	2.0	5.1	2.2	4.9	1.6	5.3
9	5.7	0.5	0.9	0.8	2.9	2.0	2.7	0.2	3.8	11.5	0.4	0.3	0.6	2.0	1.0	1.1	0.4	10.8
10	2.9	0.6	0.5	0.5	1.0	0.9	1.8	0.3	1.6	5.1	0.4	0.3	0.3	0.8	0.4	0.3	0.2	3.6

(c) The HLA-A2-restricted peptide motif

	Position									
	1	2	3	4	5	6	7	8	9[†]	
Dominant anchor residue		**L**							**V**	
Strong		**M**		**E**		**V**		**K**		
				K						
Weak	I		A	G	I	I	A	E	L	
	L		Y	P	K	L	Y	S		
	F		F	D	Y	T	H			
	K		P	T	N					
	M		M		G					
	Y		S		F					
	V		R		V					
					H					
										Protein source
Reported epitopes, aligned	I	L	K	E	P	V	H	G	V	HIV reverse transcriptase 461–485
	G	I	L	G	F	V	F	T	L	Influenza matrix protein 57–68
	I	L	G	F	V	F	T	L	T	Influenza matrix protein 57–68
	F	L	Q	S	R	P	E	P	T	HIV Gag protein 446–460
	A	M	Q	M	L	K	E			HIV Gag protein 193–203
	P	I	A	P	G	Q	M	R	E	HIV Gag protein 219–233
	Q	M	K	D	C	T	E	R	Q	HIV Gag protein 418–443

Figure 12.1. Analysis of pool sequencing data derived from peptides eluted from HLA-A2.1 (Falk *et al.*, 1991b). Sequencing cycle is shown down the left, and the yield of each amino acid residue in picomoles (pmol) is shown. Underlined positions show an increase in yield of an amino acid consistent with this being an anchor residue for the peptides. The derived motif is shown below, and compared with the sequences of known T-cell epitopes. (Reproduced, with permission, from *Nature* **351**, 290–296. Copyright (1991) Macmillan Magazines Limited.)

instruments most frequently use laser desorption (vaporization of the sample by firing a laser at a solid target) to ionize peptides, electrospray ionization (creating a vapour from a fine steam of solvent) is usually used in tandem MS. In the first mass separator (MS-1) the masses of ions from all the peptides within the HPLC fraction are seen (there are frequently many, even in an apparently homogeneous HPLC peak). One of these ions is then selected and allowed into a collision chamber, where it collides with atoms of an inert gas (helium or argon usually) and is broken up by collisions with these. The fragments from these collisions are then allowed into the second mass analyser (MS-2). Thus, the fragments generated from the breakdown of an individual peptide are seen. Since peptides tend to fragment in characteristic patterns, it is usually possible to see fragments differing from each other by one amino acid (Hines *et al.*, 1992). The mass difference between these fragments corresponds to the mass of the amino acid that was lost. Thus the sequence can be pieced together (*Figure 12.2*).

In the case of peptides eluted from MHC class I molecules, tandem MS has been very successful and widely used, and may in some instances be capable of sequencing femtomole amounts of peptide (Edman chemistry, in comparison, is limited to picomole amounts). On the other hand, sequencing of MHC class II-derived peptides does not seem to have advanced so rapidly. Although the first reports of MS sequencing of class I (Hunt *et al.*, 1992a) and class II- (Hunt *et al.*, 1992b) derived peptides were in the same year, there remain only a few reports of sequencing of class II-derived peptides by MS alone (Davenport *et al.*, 1995b; Hunt *et al.*, 1992b). This is probably due to the increased length of class II peptides, which results in greater difficulty sequencing them.

12.2.3 Theoretical methods

Whilst the analysis of peptide binding and natural peptide ligands have both proved useful tools, theoretical models for analysis of MHC–peptide interactions have also been proposed. Early attempts at prediction were based upon the analysis of only a few known T-cell epitopes and the suggestion that peptides lie in a helical conformation within the peptide-binding groove of the MHC molecule (DeLisi and Berzofsky, 1985; Rothbard and Taylor, 1988). More recent analyses have benefited from new structural analyses. In particular, understanding of the crystal structure of both class I (Bjorkman *et al.*, 1987) and class II (Brown *et al.*, 1993) molecules (see Chapter 11) and the identification of an increasing number of MHC ligands (Rammensee *et al.*, 1995) have assisted these analyses. Attempts have been made to analyse peptide-binding motifs based on analysis of a few known epitopes (Hobohm and Meyerhans, 1993), or by using molecular models to try to identify key anchor positions within the MHC structure (Thorpe and Travers, 1994; Thorpe *et al.*, 1995). These methods have been of limited value in predicting epitopes, but when used in combination with analysis of eluted peptides or peptide binding, they may provide some useful insights (Corr *et al.*, 1992; Huczko *et al.*, 1993).

12.3 Allele-specific motifs

A large number of allele-specific motifs have now been identified using the methods described above. Many of these motifs have been defined using only one

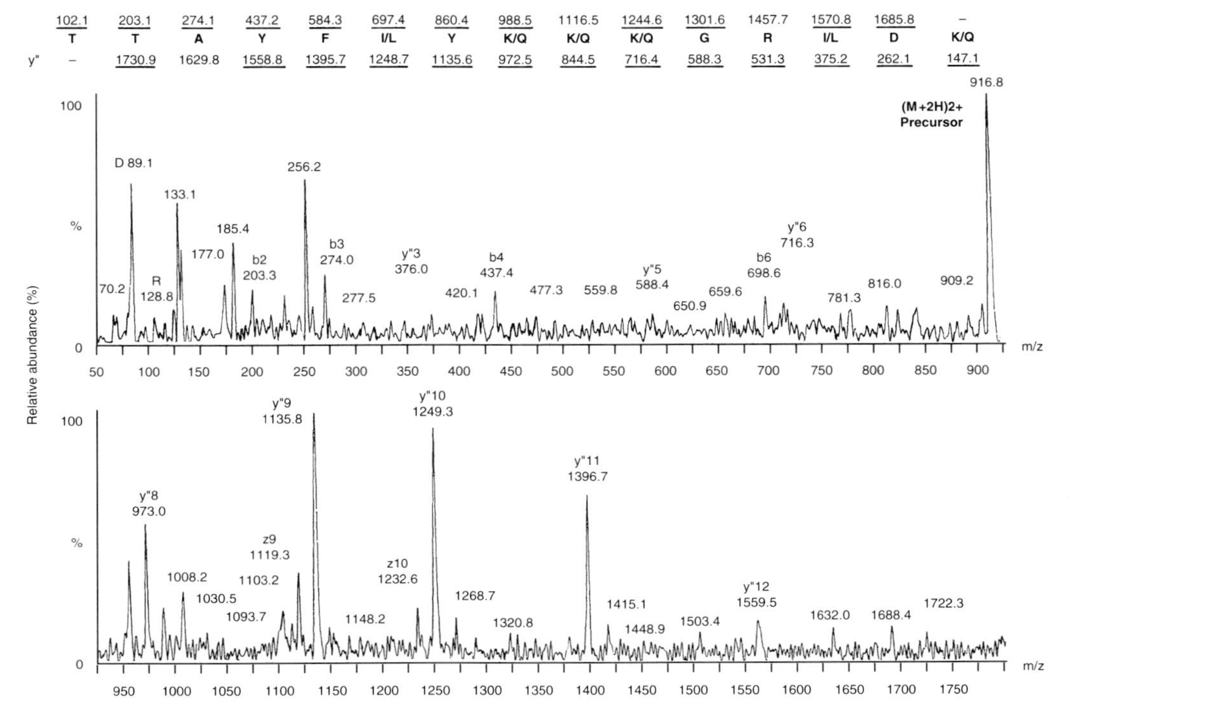

Figure 12.2. Sequencing of a self peptide eluted from *HLA-DRB1*1302* (Davenport *et al.*, 1995b). The 'precursor ion' is doubly charged with a mass-to-charge (*m/z*) ratio of 916.8. The true molecular mass of the peptide is thus 1831.6 Da. This precursor ion was subjected to collision-induced dissociation with argon gas, generating the observed product ions. The most obvious series here is a singly charged ion series marked y″. The difference between y″12 (molecular mass 1559.5 Da) and y″11 (1396.7 Da) is 162.8 Da, corresponding to the mass of a tyrosine residue. Similarly, the difference between y″11 and y″10 corresponds to a phenylalanine residue, y″10 and y″9 a leucine or isoleucine residue, and so on. The deduced peptide sequence [corresponding to invariant chain (66–80)] is shown above, as are the predicted product ions from this peptide (underlined masses indicate those ions that were observed in the spectrum). (Reproduced from Davenport *et al.* (1995b), with permission, from *Proceedings of the National Academy of Sciences USA* 1995.)

of the methods described, and may have to be modified in future to incorporate new data derived from other methods of analysis. We have attempted here to provide a current reference list (*Table 12.1*) for those wishing to examine motifs for different alleles, rather than a definitive listing of final motifs. We have therefore attempted to include data for all reported motifs, even in some cases where these may appear incorrect. Readers are advised to refer to the original references to inspect the primary data and compare the methods used before conducting experiments based on this summary. In particular, it is important to appreciate that, whereas anchor residues may be important in focusing attention on likely candidate peptides, they offer only a rough guide. Non-anchor residues of the peptide also play an important role in peptide binding (Chen *et al.*, 1994; Ruppert *et al.*, 1993). Analysis of anchor residues generally provides information on the positive contribution of particular amino acids to binding. Some residues may also have dominant negative effects and thus prevent peptides from binding.

12.4 Utility of allele-specific motifs

The ultimate test of the usefulness of MHC motifs is a purely pragmatic one: do these motifs allow the prediction of T-cell epitopes? In the case of MHC class I molecules, simple motifs such as those outlined in *Table 12.1* have been used extensively in this way. The general method employed is to scan a candidate protein for possible nonamer, octamer and decamer peptides, which conform to a simple two- or three-position motif. These peptides can then be synthesized and tested in an MHC class I-binding assay. Peptides that appear to have a significant affinity for MHC class I are then used in cytotoxic T-cell (CTL) lysis assays to determine whether they are CTL epitopes. These motifs have allowed precise prediction of epitopes in both murine and human systems (Hill *et al.*, 1992; Pamer *et al.*, 1991; Rötzschke *et al.*, 1991). However, the sensitivity and specificity of such predictions are probably quite low. Commonly only 10–50% of peptides synthesized as conforming to a given motif will bind (Hill *et al.*, 1992; Ruppert *et al.*, 1993). More sophisticated methods of analysis promise more precise prediction of peptide epitopes (Rovero *et al.*, 1994), although these are yet to be tested.

MHC class II motifs have yet to prove their value in the prediction of T-cell epitopes. This may be in part because conventional methods for epitope identification are somewhat easier than for class I, and thus there has been less incentive to develop these assays. In addition, the motifs for MHC class II tend to be less specific than for class I (*Table 12.1*), and thus generate many more predicted peptide ligands. Prediction of T-cell epitopes for MHC class II molecules may well require more sophisticated methods of analysis which incorporate information on the contribution of all residues to the binding energy of a given peptide (Davenport *et al.*, 1995a; Hammer *et al.*, 1994b).

Progress in the direct analysis of antigenic peptides bound to MHC molecules may well overtake the use of motifs within the next few years. The use of T-cell identification of peptides and sequencing by MS has allowed the identification of peptide targets for CTL lysis in peptides involved in the immune response to melanoma (Cox *et al.*, 1994) and graft versus host disease (den Haan *et al.*, 1995). Sequencing of antigenic peptides eluted from an MHC class II molecule (H2-A^k)

continues on p. 302 ▶

Table 12.1(a). HLA class I motifs. Amino acids are indicated by single letter code. The relative importance of different residues is indicated by their size: **Bold** > plain > (small). Molecular designations for naming of alleles are given where known, as are previous equivalents (Bodmer *et al.*, 1994). We have summarized here the data available through different methods to generate a single motif. In the cases where several credible motifs exist, we have attempted to include only those features that are consistent in the different reported motifs. Consult original references in order to assess how these may differ. The table extends to position 9 in all cases. In fact, it appears that some MHC molecules may be more or less tolerant of longer or shorter peptides. In particular, peptides of eight or ten amino acids are common, and generally have their C-terminal anchor in the last position (position 8 or 10). Therefore position 9 in the table should perhaps read 'C-terminal anchor'

Allele	Position									References
	1	2	3	4	5	6	7	8	9	
HLA-A										
HLA-A★0101			**E**						**Y**	Pool sequencing (Falk *et al.*, 1994c)
HLA-A1			**D**							Individual peptide sequencing (DiBrino *et al.*, 1994)
		T S		P			L			Pool plus individual peptide sequencing, and peptide binding with polyalanine analogues (Kubo *et al.*, 1994)
		(I/M/L)		(G/I)	(G/N/Y)	(G/V/I)	(M/I)			Predictive value of motif (DiBrino *et al.*, 1993b)
HLA-A★0201		**L**							**V**	Pool sequencing (Falk *et al.*, 1991b)
HLA-A2.1		**M**								Mass spectrometry sequencing of self peptides (Hunt *et al.*, 1992a)
				E K		V		K	L	Pool sequencing and peptide binding with polyalanine analogues (Kubo *et al.*, 1994)
										Predictive value of motif (Tanigaki *et al.*, 1994a)
										Promiscuity of binding among HLA-A2 subtypes (Barouch *et al.*, 1995; del Guercio *et al.*)
HLA-A★0202		L							**L**	Pool sequencing (Barouch *et al.*, 1995)
HLA-A2.2F	(A/F/G/V/I)	(A)	(F/A)	(E/D)	(I)	(I/L/V/T)	(H/Q)		(V)	
HLA-A★0205		**V**							**L**	Pool sequencing (Barouch *et al.*, 1995; Rötzschke *et al.*, 1992)
HLA-A2.2Y		L/I Q	(P/I)	(G/E/D)		I		K		

Table 12.1(a). Continued

Allele	Position 1	2	3	4	5	6	7	8	9	References
*HLA-A*0214*		**V/Q**							**L**	Pool and individual peptide sequencing (Barouch *et al.*, 1995)
HLA-A2		L				I/L				
		(A)				V/F			(V/M)	
*HLA-A*0301*		**V**							**K**	Individual peptide sequencing and peptide binding with polyglycine peptides (DiBrino *et al.*, 1993d)
HLA-A3.1		**L**	F			I/M			**Y**	
		M	Y			V/L				Pool sequencing (Maier *et al.*, 1994)
						F				Pool sequencing and peptide binding with polyalanine analogues (Kubo *et al.*, 1994)
		(I/A/ S/T)							(R/F)	Predictive value of motif (DiBrino *et al.*, 1993b; Kast *et al.*, 1994)
*HLA-A*1101*									**K**	Binding studies with multiple substitutions of an EBV epitope (Zhang *et al.*, 1993)
HLA-A11.1		V/I								
		T	(M/L/ F/I/Y A)				(L/I/Y/ V/F)		(R)	Pool sequencing and individual peptide sequencing (by MS-MS), plus binding with polyalanine peptides (Kubo *et al.*, 1994)
		(F/Y/ M/A)								Pool sequencing (Falk *et al.*, 1994c)
										Predictive value (Kast *et al.*, 1994)
*HLA-A*2401*		**Y**							**I**	Pool sequencing (Maier *et al.*, 1994)
HLA-A24(9)					I	F			**L**	Pool sequencing, individual peptide sequencing (by MS-MS), and binding with alanine-substituted analogues (Kubo *et al.*, 1994)
					V				**F**	
									(W)	Predictive value (Kast *et al.*, 1994)
*HLA-A*3101*									**R**	Pool sequencing (Falk *et al.*, 1994c)
HLA-A31(19)	L/V	F/L				L/F/				
	Y/F	Y/W				V/I				
*HLA-A*3301*									**R**	Pool sequencing (Falk *et al.*, 1994)
HLA-A33(19)		A/I								
		L/F								
	(D/E)	Y/V	(L/K)	(P)	(P)	(I/L/ F)				

HLA-A★6801 HLA-A68(28)		V T							K R	Individual peptide sequencing (and discussion of crystal structure and peptide length variation) (Guo *et al.*, 1992)
HLA-A★6901 HLA-A69(28)		**V/T** A	I/F L/M			I/F L			**V/L**	Pool sequencing (Barouch *et al.*, 1995)
HLA-B										
HLA-B★0701 HLA-B7.1	A	**P**	R						L	Individual peptide sequencing by mass spectrometry, plus structural modelling (Huczko *et al.*, 1993) Pool sequencing (Maier *et al.*, 1994)
HLA-B★0702 HLA-B7.2	A	**P**	M/R K/Q F (L/Y/H)	(L/E/G/T)	(V)	(R/T)	(L)		(L)	Pool and individual peptide sequencing (Barber *et al.*, 1995)
HLA-B★0801 HLA-B8		I/L (P)	**K** **R**		**K** **R**				I/L	Pool sequencing (Malcherek *et al.*, 1993; Sutton *et al.*, 1993) Individual peptide sequencing (DiBrino *et al.*, 1994)
HLA-B★1501 HLA-B62(15)		**Q** **L**			I/V				**F** **Y**	Pool and individual peptide sequencing (Falk *et al.*, 1995c) Predictive value of motif (Tanigaki *et al.*, 1994a)
HLA-B★2702 HLA-B27		**R**							**F/Y** **I/L** **W**	Pool and individual peptide sequencing (Rötzschke *et al.*, 1994) Peptide binding with multiple single amino acid substitutions (with B★2701/3/4/5/6) (Tanigaki *et al.*, 1994b)
HLA-B★2705 HLA-B27	(K)	**R**	(L/I/F)	(K)			(I/T)	(K)	K R (L/F)	Individual peptide sequencing (Jardetzky *et al.*, 1991) Pool and individual peptide sequencing (Rötzschke *et al.*, 1994) Peptide binding with multiple single amino acid substitutions (Fukazawa *et al.*, 1994; Tanigaki *et al.*, 1994b; Wen *et al.*, 1994) Predictive value of motif (Tanigaki *et al.*, 1994a)
HLA-B★3501 HLA-B35	(M)	**P** (A/V)	(F/E/I/N/Y)	(E)					**Y** F/M L/I	Pool sequencing (Falk *et al.*, 1993b, 1994a; Hill *et al.*, 1992) Predictive value of motif (Tanigaki *et al.*, 1994a)

Table 12.1(a). Continued

Allele	Position									References
	1	2	3	4	5	6	7	8	9	
HLA-B★3701		**D**								Pool sequencing (Falk *et al.*, 1993b)
HLA-B37		E			V			F/M	I	
					I			L	L	
HLA-B★3801									**F**	Pool sequencing (Falk *et al.*, 1995a)
HLA-B38(16)		H	D/E						**L**	
HLA-B★3901		**R**				I/V			**L**	Pool sequencing (Falk *et al.*, 1995a)
HLA-B39.1		**K**				L				
HLA-B★3902		**K**			I/L				**L**	Pool sequencing (Falk *et al.*, 1995a)
HLA-B39.2		**Q**			F/V					
HLA-B★4001		**E**							**L**	Pool and individual peptide sequencing (Falk *et al.*, 1995c)
HLA-B60(40)							I/V			
HLA-B★4006		**E**	F/I			I			**V**	Pool and individual peptide sequencing (Falk *et al.*, 1995c)
HLA-B61(40)			L/V							
			Y/N							
HLA-B★4402		**E**	M/I						**Y**	Pool sequencing (Fleischhauer *et al.*, 1994)
HLA-B44(12)			L/D						**F**	Prediction based on modelling (Thorpe and Travers, 1994)
HLA-B★4403		**E**	M/I						**Y**	Pool sequencing (Fleischhauer *et al.*, 1994)
HLA-B44(12)			L/D						**F**	
			V							

HLA-B★5101		**A**							**F**	Pool and individual peptide sequencing (Falk *et al.*, 1995b)
HLA-B51(5)		**P**							**I**	Peptide binding (Connan *et al.*, 1994)
		G								Predictive value of motif (Tanigaki *et al.*, 1994a)
HLA-B★5102		**P**							**I**	Pool and individual peptide sequencing (Falk *et al.*, 1995b)
HLA-B51(5)		**A**							**V**	
		G	Y							
HLA-B★5103		**A**								Pool and individual peptide sequencing (Falk *et al.*, 1995b)
HLA-B51(3)		**P**	Y						V/I	
		G							F	
HLA-B★5201								**I**	**I**	Pool and individual peptide sequences (Falk *et al.*, 1995b)
HLA-B52(5)		Q	F/Y		L/I			**V**	**V**	
			W		V					
HLA-B★5301		**P**								Pool sequencing (Hill *et al.*, 1992)
HLA-B53				E	I					
HLA-B★5401		**P**	F/M							Pool sequencing (Barber *et al.*, 1995)
HLA-B54(22)			R/Y							
			N							
	(M/F)		(K/L/ Q)	(G)	(V)				(A)	
HLA-B★5501		**P**								Pool and individual peptide sequencing (Barber *et al.*, 1995)
HLA-B55(22)	A		R/M	T	V	F/G		M		
			Y/K							
HLA-B★5502		**P**								Pool and individual peptide sequencing (Barber *et al.*, 1995)
HLA-B55(22)	A		R/M							
			Y/F							
			H/K	H				M		
			L							

Table 12.1(a). Continued

Allele	Position 1	2	3	4	5	6	7	8	9	References
HLA-B★5601		**P**								Pool and individual peptide sequencing (Barber *et al.*, 1995)
HLA-B56(22)			Y/N							
	A		R/F						A	
			Q/K							
			H/L							
HLA-B★5801		**A**							**I**	Pool and individual peptide sequencing (Falk *et al.*, 1995c)
HLA-B58(17)		**S**		P/E	V/I				**W**	
		T		K	L/M					
					F					
HLA-B★6701		**P**								Pool sequencing (Barber *et al.*, 1995)
HLA-B67	A		F/M	G						
			Y/I							
			A							
HLA-B★7801		**P**							**?**	Pool sequencing (C-terminal not identified; Falk *et al.*, 1995b)
HLA-B7801		**A**				I/L		A		
		G				F/V				
HLA-C										
HLA-C★0301			V/I	P		F			**L**	Pool sequencing (Falk *et al.*, 1993a)
HLA-Cw3			Y/L			Y			F/M	
			M						I	
HLA-C★0401		**Y**								Pool sequencing (Falk *et al.*, 1993a)
HLA-Cw4		**P**	D	D/E	A	V/I	A	K	L/F	
		F		P		L			M	
HLA-C★0602									**L**	Pool sequencing (Falk *et al.*, 1993a)
HLA-Cw6					I/L	V/I			I/V	
					F/M	L			Y	
HLA-C★0701					**V/Y**	**V**			**Y**	Pool sequencing (Falk *et al.*, 1993a)
HLA-Cw7		Y	P/G	D	I/L	**I**				
		P	A	E	F/M	L/M			F/L	

Table 12.1(b). HLA class II. The relative importance of different residues is indicated by their size: **Bold** > plain > (small). This table differs from that for class I because, rather than integrate the results arrived at by different methods, we have listed them separately. This is to indicate how much these may differ and allow an independent appraisal of how useful (or not) a given motif may be (discussed also in Section 12.4). Additional references that may be helpful in assessing the value of a motif are also included. In addition, whereas for class I peptides the amino acid numbering is from the first residue, in the case of class II the numbering is from the primary hydrophobic anchor residue (which is usually at absolute position 3–6 in naturally processed peptides). The presence of proline in position 2 of pool sequences and individual peptides has been noted previously (Falk *et al.*, 1994; Kropshofer *et al.*, 1993), and may be a useful addition to these motifs in the assessment of the likelihood that a given peptide is an epitope. Where both the DRB1 and DRB3/4/5 allelic products have not been separated and thus are analysed together, we have made no comment on the minor component. This component seems unlikely to contribute greatly to the observed motif, since it represents only a minor fraction of the total HLA-DR (Berdoz *et al.*, 1987), and analysis of individual peptides indicates that they bind predominantly to the *DRB1* gene product (Davenport *et al.*, 1995b; Kinouchi *et al.*, 1994). The reliability of these motifs may vary greatly. In particular it is important to note that, whereas alignment of individual peptide ligands is useful in predicting binding motifs for MHC class I molecules, this method may be unreliable as a sole method for identifying motifs for MHC class II molecules (e.g. see comparison of motifs for *DRB1*★0101 derived by different methods)

Allele	Position									Method	Additional references
	1	2	3	4	5	6	7	8	9		
HLA-DR											
DRB1★0101 DR1 (Dw1)	K R						H-bond donor		Hø	Individual peptide sequencing (Edman) (Chicz *et al.*, 1992)	Binding studies with single point mutations and truncations of promiscuous peptides (O'Sullivan *et al.*, 1991a, 1991b)
										Phage display library (Hammer *et al.*, 1992)	Pool sequencing (Kropshofer *et al.*, 1993) Binding studies with different anchor combinations and positions in polyalaninepeptides (Hammer *et al.*, 1994a, Hill *et al.*, 1994)
	Y F W (L)			M L (I)		G/A (S)			L (M/A or Hø)		Predictive algorithm for peptide binding based on analysis of pepdite-binding studies performed with peptide analogues substituted with all amino acids at all positions (Hammer *et al.*, 1994b)

Table 12.1(b). Continued

Allele	Position 1	2	3	4	5	6	7	8	9	Method	Additional references
	Hø					S/T A/V I/L P/C			V/A I/L F (Hø)	Pool/individual peptide sequencing Peptide binding with alanine-substituted analogues (Kropshofer *et al.*, 1992)	Peptide binding and T-cell recognition studies with multiply substituted peptides (Geluk *et al.*, 1992) Pool sequencing (Kropshofer *et al.*, 1993)
	Y/V L/F I (A/M/W)			L/A (I/V/M/N)					L/A I/V N/F Y	Pool sequencing (Falk *et al.*,1994b)	Compare match with several motifs with measured binding to overlapping peptides (Geluk *et al.*, 1994) Peptide binding studies with substituted and truncated peptide analogues (Malcherek *et al.*, 1994)
DRB1★0301 DR17(DR3/Dw3)	Hø		Hø	D/N Q/R K/E S		R K H				Peptide binding (Sidney *et al.*, 1992)	
	F/I L/V Y			D/N Q/T						Individual peptide sequencing (Chicz *et al.*, 1993)	
	L I F			D		K/R (E/Q/N)		L	Y L	Individual peptide and pool sequencing (Malcherek *et al.*, 1993)	
	I/L V/M Y/F I/L V/M Y/F A			D D/N Q/E S/T		 K R H				Peptide binding and analysis of previous results to conclude that there are two 'submotifs' for peptide binding to this allele (Geluk *et al.*, 1994)	

*DRB1*0401* DR4(Dw4)	F/L V				N/Q S/T	Individual peptide sequencing (Chicz *et al.*, 1993)	Binding studies with different anchor combinations and positions in poly-alanine peptides (Hammer *et al.*, 1994a) (Hill *et al.*, 1994)
	F/W Y/L I/V M	(no R or K)	S/T L/I V/M	(no R or K)	(no R or K, no D or E)	Peptide binding with many unrelated peptides, truncation analysis and single amino acid substitutions (Sette *et al.*, 1993)	Predictive algorithm for peptide binding based on analysis of peptide binding studies performed with peptide analogues substituted with all amino acids at all positions (Hammer *et al.*, 1995)
	W Y	M (A/V/L)	T (S)	L/Q (M/N)		Phage display library (Hammer *et al.*, 1993)	
	Y/W I/V	D/A (no K/R)	S/T (no I/L)	I/L (no K/R)	(no R or K, no D or E)	Individual peptide sequencing and binding studies with single amino acid substituted analogues (Max *et al.*, 1994)	
*DRB1*0402* DR4(Dw10)	V I L M	Y/F W/I L/M R/N (no D or E)	N/Q S/T K	R/K H/N Q/P (D and E rare)	Polar aliphatic M	Pool and individual peptide sequencing (Rammensee *et al.*, 1995)	Predictive algorithm for peptide binding based on analysis of peptide binding studies performed with peptide analogues substituted with all amino acids at all positions (Hammer *et al.*, 1995)
*DRB1*0404* DR4(Dw14)	V I L M	F/Y W/I L/V M/A D/E (no R or K)	N T S Q R	polar charged aliphatic	polar aliphatic K	Pool and individual peptide sequencing (Rammensee *et al.*, 1995)	Predictive algorithm for peptide binding based on analysis of peptide binding studies performed with peptide analogues substituted with all amino acids at all positions (Hammer *et al.*, 1995)

Table 12.1(b). Continued

Allele	Position 1	2	3	4	5	6	7	8	9	Method	Additional references
*DRB1*0405* DR4(Dw15)	Y					T/V	T/V		D	Individual peptide sequencing (TV=>D spacing is 3 residues) (Kinouchi *et al.*, 1994)	
	F (I/L V/M W/Y)			L I F		N				Individual peptide sequencing, binding studies with alanine substituted analogues (Matsushita *et al.*, 1994)	
	F/Y (W/ V/I L/M)			V/I L/M D/E		N/S T/Q K/D	polar charged aliphatic		D (E/Q)	Pool and individual peptide sequencing (Rammensee *et al.*, 1995)	
*DRB1*0701* DR7(Dw17)	F I L V Y					N S T				Individual peptide sequences (Chicz *et al.*, 1993)	
*DRB1*0801* DR8(Dw8.1)	F/I L/V Y				H K R					Individual peptide sequencing (Chicz *et al.*, 1993)	
*DRB1*0901* DR9	Y/F W/L			A S					I/L V	Individual peptide sequencing (Futaki *et al.*, 1995)	
*DRB1*1101* DR11(DR5/Dw5)	Y F					R K		R K		Individual peptide sequencing (Newcomb and Cresswell, 1993)	Peptide binding with singple point mutations and truncations of promiscuous peptides (O'Sullivan *et al.*, 1991a, 1991b)

	W			M/L (V)		R (K)				Phage display library (Hammer *et al.*, 1993)	Pool sequencing (Kropshofer *et al.*, 1993). Binding studies with different anchor combinations and positions in polyalanine peptides (Hammer *et al.*, 1994a)
DRB1★1201 DR12(DR5/'DwDB6')	I/L F/Y (V)		L/N M (V/A)			V/Y (F/I N/A)			Y/F M (I/V)	Pool and individual peptide sequencing (Falk *et al.*, 1994b)	
DRB1★1301 DR13(DR6/Dw18)	L/I V					+'ve or polar			Y	Individual peptide (by Edman or tandem MS and pool sequence. Peptide binding with alanine substituted analogues (Davenport *et al.*, 1995b)	Peptide binding and T-cell responses to single amino acid-substituted analogues of tetanus toxoid (830–843) (Boitel *et al.*, 1995)
DRB1★1302 DR13(DR6/Dw19)	I/L F/Y					+'ve or polar			Y	Individual peptide (by Edman or tandem MS and pool sequence. Peptide binding with alanine substituted analogues (Davenport *et al.*, 1995b)	
DRB1★1501 DR15(DR2/Dw2)	I L V									Individual peptide sequencing (Chicz *et al.*, 1993)	Peptide-binding studies with multiple single amino acid substitutions and truncations of myelin basic protein (84–102) peptide (Wucherpfenning *et al.*, 1994)
	L I V			F Y			M I V F			Individual peptide sequencing and peptide binding with alanine substitution and truncation of myelin basic protein peptide (Vogt *et al.*, 1994)	Peptide-binding studies with multiple single amino acid substitutions and truncations of myelin basic protein (84–102) peptide (Wucherpfennig *et al.*, 1994)
DRB5★0101 DR51(Dw2)	F Y	+'ve	(G/A S/P/T N)	Q/V I/M		(G/A S/P/T N)		R/K	R/K	Individual peptide sequencing and peptide binding with single amino acid substituted and truncation analogues of a myelin basic protein peptide (Vogt *et al.*, 1994)	

Table 12.1(b). Continued

Allele	Position 1	2	3	4	5	6	7	8	9	Method	Additional references
HLA-DQ											
DQA1★0501 *DQB1★0201* DQ2	K			I	(A)				F	Pool sequencing (Verreck *et al.*, 1994)	Peptide binding assays (Johansen *et al.*, 1994) Pool sequence (identifies only an N-terminal anchor) (Kropshofer *et al.*, 1993)
DQA1★0101 *DQB1★0501* DQ5(1)	L				Y F W					Individual peptide sequencing (author comments 'one of several possible alignments') (Chicz *et al.*, 1994)	
DQA1★0301 *DQB1★0301* DQ7(DQ3.1)	no charged or polar	no charged	A/G S/T	no D or E	A/V L/I					Peptide binding (Sidney *et al.*, 1994)	
DQA1★0501 *DQB1★0301* DQ7(DQ3.1)	F/Y I/M (L/V)	(A)	(A)	(A)	V/L (I/M/Y)		Y/F M/L V/I			Pool sequencing (Falk *et al.*, 1994b)	
DQA1★0301 *DQB1★0302* DQ8(DQ3.2)	R K					A G			N E D	Individual peptide sequencing (Chicz *et al.*, 1994)	
HLA-DP											
DPA1★0102 *DPB1★0201* DPw2	F/L M (V/W/Y)				F/L (M/Y)			I/A (M/V)		Pool sequencing (Rötzechke and Falk, 1994)	Comparison of known epitopes (Hammond *et al.*, 1991)
DPA1★0201 *DPB1★0401* DPw4	F/L Y/M (I/V/A)						F/L Y (M/V/I/A)			Pool sequencing (Falk *et al.*, 1994)	
	I	G/L K/Y D (E)	R	L	V (F)	T/E (I)			(A)	Pool sequencing of a mixture of DPw3 and DPw4 bound peptides. (non-standard alignment of anchors) (Verreck *et al.*, 1994)	

Table 12.1(c). Murine MHC. The relative importance of different residues is indicated by their size: **Bold** > plain > (small). Motifs for MHC class Ia, class Ib, class II, and CD1 are provided. Motifs for MHC class II should be interpreted carefully, since in some cases the data on which these are based rely on molecular modelling alone (see comments above)

Allele	Position 1	2	3	4	5	6	7	8	9	(10)	12/13	References
Class I												
H-2D^b					**N**				**M**			Pool sequencing (Falk *et al.*, 1991)
		(M)	(I/L/F/V)	(K/E/Q/V)		(L/F)			I/L			Peptide-binding studies with substituted peptides (Deres *et al.*, 1993)
H-2D^d		G	P		+'ve				L/I/F			Analysis of self peptides and molecular modelling (Corr *et al.* 1993)
H-2K^b					**F/Y**			**L**				Pool sequencing (Falk *et al.*, 1991)
			Y					M/I				Binding studies (Deres *et al.*, 1993; Jameson and Bevan, 1992)
								V				Epitope prediction (Lipford *et al.*, 1993; Rötzschke *et al.*, 1991)
												Effects of non-dominant anchor residues and their interaction (Chen *et al.*, 1994; Fremont *et al.*, 1995)
H-2K^d		**Y**							**I/L**			Alignment of known epitopes (Falk *et al.*, 1991)
			(N/I/L)	(P)	(M)	(K/F)	(T/N)		V			Pool sequencing (Falk *et al.*, 1991)
												Binding with peptide analogues (Eberl *et al.*, 1993; Romero *et al.*, 1991)
												Epitope prediction (Pamer *et al.*, 1991)
H2-K^k		**E**							**I**			Comparison of known epitopes (Gould *et al.*, 1991)
		D										Pool (Norda *et al.*, 1993) and individual peptide sequencing with peptide binding (Brown *et al.*, 1994)
												Epitope prediction (Cossins *et al.*, 1993)
H-2K^{kml}		E							I			Pool sequencing (Norda *et al.*, 1993)
H-2L^d		**P**							**F/L**			Individual peptide (Corr *et al.*, 1992) and pool sequencing (Rammensee *et al.*, 1995)
		(S)	Q					R	**M**			

Table 12.1(c). Continued.

Allele	Position 1	2	3	4	5	6	7	8	9	(10)	12/13	References
Class Ib												
H-2M3[a]	fM											Binds N-formyl peptides as short as two amino acids (Vyas *et al.*, 1995). May bind unformylated peptides with lower affinity (Smith *et al.*, 1994)
H-2Qa-2							**H**					Pool sequencing (Rötzschke *et al.*, 1993)
	K	Q/L	L/N	P	I/V	K/M I/L		E/Q N	L/I F			Individual peptide sequencing (Joyce *et al.*, 1994)
Class II												
H-2A[b]	**N** Q/D						**P** I/S					Alignment of individual peptides (Rudensky *et al.*, 1991, 1992)
H-2A[g7]	**K**	A		A			L I	(R/K)		V	D/E	Individual peptide sequencing and peptide binding (Reich *et al.*, 1994)
H-2A[q]	Hø				R/K							Peptide binding with alanine-substituted analogues and comparison with known epitopes (Brand *et al.*, 1994)
H-2A[s]	**I** V/F T	**T** A/S					**H** R					Individual peptide sequencing (Rudensky *et al.*, 1992)
H-2E[b]	W/F Y			L/I F/V		Q/N A			K R			Individual peptide sequencing (Rudensky *et al.*, 1991, 1992) Modelling of structure (Rammensee *et al.*, 1995)
H-2E[d]	**W/Y** **F** I/L V			**K/R** **I**		**I/L** **V/G**			**K/R**			Pool and individual peptide sequencing (Rammensee *et al.*, 1995)
H-2E[k]	**I/L** **V**					**Q/N**			**R/K**			Individual peptide sequencing (Marrack *et al.*, 1993) Peptide binding (Leighton *et al.*, 1991), with substitution of all amino acids at all positions (Reay *et al.*, 1994)
H-2E[s]	**I/L** **V**			**I/L** **V**		**Q/N**			**K/R**			Individual peptide sequencing (Marrack *et al.*, 1993) Modelling of structure (Rammensee *et al.*, 1995)

CD1	**F** **W**	(H/N)	**I/L** **M**	**W**	Phage display library and peptide binding (Castano *et al.*, 1995)

Hø, hydrophobic amino acid (I/L/V/F/W/Y). +'ve, positively charged amino acids (R/K). fM, formylated methionine.

Table 12.1(d). Other known motifs. The relative importance of different residues is indicated by their size: **Bold** > plain > (small)

Allele	Position 1	2	3	4	5	6	7	8	9	References
Class I										
BoLA-A11 (Bovine)		**P** Q	(A)		(P)				**I/V** L/Y	Pool sequencing (Hegde *et al.*, 1995)
BoLA-A20 (Bovine)	A/I	**K**	K	D/P	D/P	K/I P/V D	I/V L	R/Q	**R**	Pool sequencing (Bamford *et al.*, 1995)
BoLA-HD1 (Bovine)		(E/Q/T)	**I** L (M/F)	P H	N					Pool sequencing (Gaddum *et al.*, submitted)
BoLA-HD6 (Bovine)		**Q** I/L	(K/Y H)	P	(H/K)		(H)		(R/H)	Pool sequencing (Gaddum *et al.*, submitted)
BoLA-HD7 (Bovine)		**V** I/A Q	P I	P	F		**I**			Pool sequencing (Gaddum *et al.*, submitted)
RT1.A[a] (Rat)		**L/M** **V/I**							**R** **K**	Molecular modelling (Thorpe *et al.*, 1995)
Class II										
RT1.B[1] (Rat)	S						E			Peptide binding and alignment of known epitopes (Hickey, 1994; Wegmann *et al.*, 1994)

has also been attempted in the case of T-cell recognition of hen egg lysozyme (Nelson *et al.*, 1992; Vignali *et al.*, 1993). However, since both the detection of antigen by $CD4^{+}$ T cells and the sequencing of MHC class II-derived peptides are less sensitive than for class I peptides, it may be some time before this method can be applied to disease-related epitopes.

The identification of associations between HLA and disease offers an excellent opportunity to utilize the knowledge gained from analysis of peptide-binding motifs to predict the epitopes responsible for the disease association. This methodology has been applied to the analysis of the association between HLA-B53 and protection from severe malaria (Hill *et al.*, 1991, 1992), leading to the identification of a CTL epitope for HLA-B53. The use of these techniques in the identification of autoantigens responsible for the associations between HLA molecules and autoimmune diseases is more challenging. Whereas it may be expected that the main epitopes involved in the immune response to antigens from infectious agents will be those that are processed and presented with high efficiency (i.e. the immunodominant epitopes), this may well not be the case for autoimmune targets. Those self-epitopes that are presented with high efficiency are probably those that induce self-tolerance best. It is perhaps more likely that poorly presented epitopes, which have failed to induce sufficient self tolerance, may be the targets of autoimmune responses. Since the present methods of motif analysis are designed to identify peptides that are processed efficiently and bind to MHC with high affinity, they may well be biased away from the detection of autoantigenic peptides.

Our understanding of the rules for peptide binding to MHC remain incomplete. In addition, although the application of these rules to the prediction of T-cell epitopes in infectious disease has yielded very promising results, it is unclear how well these rules may apply in the detection of epitopes in autoimmune disease. In the future, however, the limitations to the use of epitope prediction methods may well be overcome by rapidly developing new techniques for the direct sequencing of T-cell epitopes from eluted peptides.

Acknowledgements

We would like to thank Tim Elliott and Brian Green for helpful comments and careful reading of the manuscript. Miles Davenport received support from the Lionel Murphy Foundation and The Wellcome Trust. Adrian Hill is a Wellcome Trust Principal Research Fellow.

References

Bamford AI, Douglas A, Friede T, Stevanovic S, Rammensee HG, Adair BM. (1995) Peptide motif of a cattle MHC class I molecule. *Immunol. Lett.* **45:** 129–136.

Barber LD, Gillece-Castro B, Percival L, Li X, Clayberger C, Parham P. (1995) Overlap in the repertoires of peptides bound in vivo by a group of related HLA-B allotypes. *Curr. Biol.* **5:** 179–190.

Barouch D, Friede T, Stevanovic S, Tussey L, Smith K, Rowland-Jones S, Braud V, McMichael A, Rammensee H-G. (1995) HLA-A2 subtypes are functionally distinct in peptide binding and presentation. *J. Exp. Med.* **182:** 1847–1856.

Berdoz J, Gorski J, Termijtelen A-M, Dayer J-M, Irlé C, Schendel D, Mach B. (1987)

Constitutive and induced expression of the individual HLA-DR β and α chain loci in different cell types. *J. Immunol.* **139:** 1336–1341.

Bjorkman PJ, Saper MA, Samraoui B, Bennet WS, Strominger JL, Wiley DC. (1987) Structure of human class I histocompatibility antigen, HLA-A2. *Nature* **329:** 506–512.

Bodmer JG, Marsh SGE, Albert ED, *et al.* (1994) Nomenclature for factors of the HLA system, 1994. *Hum. Immunol.* **41:** 1–10.

Boitel B, Blank U, Mege D, Corradin G, Sidney J, Sette A, Acuto O. (1995) Strong similarities in fine specificity among DRB1*1302-restricted tetanus toxin tt830-843-specific TCRs in spite of highly heterogeneous CDR3. *J. Immunol.* **154:** 3245–3255.

Brand DD, Myers LK, Terako K, Whittington KB, Stuart JM, Kang AH, Rosloniec EF. (1994) Characterization of the T cell determinants in the induction of autoimmune arthritis by bovine alpha 1(II)-CB11 in h-2q mice. *J. Immunol.* **152:** 3088–3097.

Brown EL, Wooters JL, Ferenz CR, O'Brien CM, Hewick RM, Herrmann SH. (1994) Characterisation of peptide binding to the murine MHC class I H-2K^k molecule: sequencing of the bound peptides and direct binding of synthetic peptides to isolated class I molecules. *J. Immunol.* **153:** 3079–3092.

Brown JH, Jardetzky TS, Gorga JC, Stern LJ, Urban RG, Strominger JL, Wiley DC. (1993) Three-dimensional structure of the human class II histocompatibility antigen HLA-DR1. *Nature* **364:** 33–39.

Buelow R, Kuo S, Paborsky L, Wilson KJ, Rothbard JB. (1994) Detergent-enhanced dissociation of endogenous peptides from PI-DRB1*0401. *Eur. J. Immunol.* **24:** 2182–2185.

Busch R, Strang G, Howland K, Rothbard JB. (1990) Degenerate binding of immunogenic peptides to HLA-DR proteins on B cell surfaces. *Int. Immunol.* **2:** 443–450.

Buus S, Sette A, Colon SM, Miles C, Grey HM. (1987) The relationship between major histocompatibility complex (MHC) restriction and the capacity of Ia to bind immunogenic peptides. *Science* **235:** 1353–1358.

Castano AR, Tangri S, Miller JEW, Holcombe HR, Jackson MR, Huse WD, Kronenberg M, Peterson PA. (1995) Peptide binding and presentation by mouse CD1. *Science* **269:** 223–226.

Ceppellini R, Frumento G, Ferrara GB, Tosi R, Chersi A, Pernis B. (1989) Binding of labelled influenza matrix peptide to HLA DR in living B lymphoid cells. *Nature* **339:** 392–394.

Cerundolo V, Elliott T, Elvin J, Bastin J, Rammensee H-G, Townsend A. (1991) The binding affinity and dissociation rates of peptides for class I major histocompatibility complex molecules. *Eur. J. Immunol.* **21:** 2069–2075.

Chen W, Khilko S, Fecondo J, Margulies DH, McCluskey J. (1994) Determinant selection of major histocompatibility complex class I-restricted antigenic peptides is explained by class I-peptide affinity and is strongly influenced by nondominant anchor residues. *J. Exp. Med.* **180:** 1471–1483.

Chicz RM, Urban RG, Lane WS, Gorga JC, Stern LJ, Vignali DAA, Strominger JL. (1992) Predominant naturally processed peptides bound to HLA-DR1 are derived from MHC-related molecules and are heterogeneous in size. *Nature* **358:** 764–768.

Chicz RM, Urban RG, Gorga JC, Vignali DAA, Lane WS, Strominger JL. (1993) Specificity and promiscuity among naturally processed peptides bound to HLA-DR alleles. *J. Exp. Med.* **178:** 27–47.

Chicz RM, Lane WS, Robinson RA, Trucco M, Strominger JL, Gorga JC. (1994) Self-peptides bound to the type I diabetes associated class II MHC molecules HLA-DQ1 and HLA-DQ8. *Int. Immunol.* **6:** 1639–1649.

Connan F, Hlavac F, Hoebeke J, Guillet J-G, Choppin J. (1994) A simple assay for the detection of peptides promoting the assembly of HLA class I molecules. *Eur. J. Immunol.* **24:** 777–780.

Corr M, Boyd LF, Frankel SR, Kozlowski S, Padlan EA, Margulies DH. (1992) Endogenous peptides of a soluble major histocompatibility complex class I molecule, H-2L^d: sequence motif, quantitative binding, and molecular modeling of the complex. *J. Exp. Med.* **176:** 1681–1692.

Corr M, Boyd LF, Padlan EA, Margulies DH. (1993) H-2D^d exploits a four residue peptide binding motif. *J. Exp. Med.* **178:** 1877–1892.

Cossins J, Gould KG, Smith M, Driscoll P, Brownlee GG. (1993) Precise prediction of a K^k-restricted T-cell epitope in the NS1 protein of influenza virus using an MHC allele-specific motif. *Virology* **193:** 289–295.

Cox AL, Skipper J, Chen Y, Henderson RA, Darrow TL, Shabanowitz J, Engelhard VH, Hunt DF, Slingluff CL. (1994) Identification of a peptide recognized by five melanoma-specific human cytotoxic T cell lines. *Science* **264:** 716–719.

Davenport MP, Ho Shon I, Hill AVS. (1995a) An empirical method for the prediction of T-cell epitopes. *Immunogenetics* **42:** 392–397.

Davenport MP, Quinn CL, Chicz RM, Green BN, Willis AC, Lane WS, Bell JI, Hill AVS. (1995b) Naturally processed peptides from two disease-resistance associated HLA-DR13 alleles show related sequence motifs and the effects of the dimorphism at position 86 of the HLA-DRβ chain. *Proc. Natl Acad. Sci. USA* **92:** 6567–6571.

del Guercio M-F, Sidney J, Hermanson G, Perez C, Grey HM, Kubo RT, Sette A. (1995) Binding of a peptide antigen to multiple HLA alleles allows definition of an A2-like supertype. *J. Immunol.* **154:** 685–693.

DeLisi C, Berzofsky JA. (1985) T cell antigenic sites tend to be amphipathic structures. *Proc. Natl Acad. Sci. USA* **82:** 7048–7052.

den Haan JMM, Sherman NE, Blokland E, *et al.* (1995) Identification of a graft versus host disease-associated human minor histocompatibility antigen. *Science* **268:** 1476–1480.

Deres K, Beck W, Faath S, Jung G, Rammensee HG. (1993) MHC/peptide binding studies indicate hierarchy of anchor residues. *Cell. Immunol.* **151:** 158–167.

DiBrino M, Parker KC, Shiloac J, Knierman M, Lukszo J, Turner RV, Biddison WE, Coligan JE. (1993a) Endogenous peptides bound to HLA-A3 possess a specific combination of anchor residues that permit identification of potential antigenic peptides. *Proc. Natl Acad. Sci. USA* **90:** 1508–1512.

Dibrino M, Tsuchida T, Turner RV, Parker KC, Coligan JE, Biddison WE. (1993b) HLA-A1 and HLA-A3 T cell epitopes derived from influenza virus proteins predicted from peptide binding motifs. *J. Immunol.* **151:** 5930–5935.

DiBrino M, Parker KC, Shiloach J, Turner RV, Tsuchida T, Garfield M, Biddison WE, Coligan JE. (1994) Endogenous peptides with distinct amino acid anchor residue motifs bind to HLA-A1 and HLA-B8. *J. Immunol.* **152:** 620–631.

Eberl G, Sabbatini A, Servis C, Romero P, Maryanski JL, Corradin G. (1993) MHC class I H-2Kd restricted antigenic peptides: additional constraints for the binding motif. *Int. Immunol.* **5:** 1489–1492.

Elvin J, Potter C, Elliott T, Cerundolo V, Townsend A. (1993) A method to quantify binding of unlabelled peptides to class I MHC molecules and detect their allele specificity. *J. Immunol. Methods* **158:** 161–171.

Falk K, Rötzschke O, Deres K, Metzger J, Jung G, Rammensee H-G. (1991a) Identification of naturally processed nonapeptides allows their quantification in infected cells and suggests an allele-specific T cell epitope forecast. *J. Exp. Med.* **174:** 425–434.

Falk K, Rötzschke O, Stevanovic S, Jung G, Rammensee H-G. (1991b) Allele-specific motifs revealed by sequencing of self-peptides eluted from MHC molecules. *Nature* **351:** 290–296.

Falk K, Rötzschke O, Grahovac B, Schendel D, Stevanovic S, Gnau V, Jung G, Strominger JL, Rammensee H-G. (1993a) Allele-specific peptide ligand motifs of HLA-C molecules. *Proc. Natl Acad. Sci. USA* **90:** 12005–12009.

Falk K, Rötzschke O, Grahovac B, Schendel DJ, Stevanovic S, Jung G, Rammensee HG. (1993b) Peptide motifs of HLA-B35 and -B37 molecules. *Immunogenetics* **38:** 161–162.

Falk K, Rötzschke O, Grahovac B, Schendel DJ, Stevanovic S, Jung G, Rammensee HG. (1994a) Peptide motifs of HLA-B35 and -B37 molecules (erratum). *Immunogenetics* **39:** 379.

Falk K, Rötzschke O, Stevanovic S, Jung G, Rammensee H-G. (1994b) Pool sequencing of natural HLA-DR, DQ, and DP ligands reveals detailed peptide motifs, constraints of processing, and general rules. *Immunogenetics* **39:** 230–242.

Falk K, Rötzschke O, Takiguchi M, Grahovac B, Gnau V, Stevanovic S, Jung G, Rammensee H-G. (1994c) Peptide motifs of HLA-A1,-A11,-A31, and -A33 molecules. *Immunogenetics* **40:** 238–241.

Falk K, Rötzschke O, Takiguchi M, Gnau V, Stevanovic S, Jung G, Rammensee H-G. (1995a) Peptide motifs of HLA-B38 and B39 molecules. *Immunogenetics* **41:** 162–164.

Falk K, Rötzschke O, Takiguchi M, Gnau V, Stevanovic S, Jung G, Rammensee H-G. (1995b) Peptide motifs of HLA-B51, -B52 and -B78 molecules and implications for Behçet's disease. *Int. Immunol.* **7:** 223–228.

Falk K, Rötzschke O, Takiguchi M, Gnau V, Stevanovic S, Jung G, Rammensee H-G. (1995c) Peptide motifs of HLA-B58, B60, B61 and B62 molecules. *Immunogenetics* **41:** 165–168.

Fleischhauer K, Avila D, Vilbois F, Traversari C, Bordignon C, Wallny H-J. (1994) Characterization of natural peptide ligands for HLA-B*4402 and -B*4403: implications for peptide involvement in allorecognition of a single amino acid change in the HLA-B44 heavy chain. *Tissue Antigens* **44:** 311–317.

Fremont DH, Stura EA, Matsumura M, Peterson P, Wilson IA. (1995) Crystal structure of an H-

2K^b-ovalbumin peptide complex reveals the interplay of primary and secondary anchor positions in the major histocompatibility complex binding groove. *Proc. Natl Acad. Sci. USA* **92:** 2479–2483.

Fukazawa T, Wang J, Huang F, Wen J, Tyan D, Williams KM, Raybourne RB, Yu DTY. (1994) Testing the importance of each residue in a HLA-B27–binding peptide using monoclonal antibodies. *J. Immunol.* **152:** 1190–1196.

Futaki G, Kobayashi H, Sato K, Taneichi M, Katagiri M. (1995) Naturally processed HLA-DR9/DR53 (DRB1*0901/DRB4*0101)-bound peptides. *Immunogenetics* **42:** 299–301.

Gaddum RM, Willis AC, Ellis SA. (1996) Peptide motifs for three cattle (BoLA) class I antigens. (Submitted.)

Geluk A, Van-Meijgaarden KE, Janson AA, Drijfhout JW, Meloen RH, De Vries RR, Ottenhoff TH. (1992) Functional analysis of DR17(DR3)-restricted mycobacterial T cell epitopes reveals DR17–binding motif and enables the design of allele-specific competitor peptides. *J. Immunol.* **149:** 2864–2871.

Geluk A, van-Meijgaarden KE, Southwood S, Oseroff C, Drijfhout JW, de Vries RR, Ottenhoff TH, Sette A. (1994) HLA-DR3 molecules can bind peptides carrying two alternative specific submotifs. *J. Immunol.* **152:** 5742–5748.

Gould KG, Scotney H, Brownlee GG. (1991) Characterization of two distinct major histocompatibility complex class I K^k-restricted T-cell epitopes within the influenza A/PR/8/34 virus haemagglutinin. *J. Virol.* **65:** 5401–5409.

Guo H-C, Jardetzky TS, Garret TPJ, Lane WS, Strominger JL, Wiley DC. (1992) Different length peptides bind to HLA-Aw68 similarly at their ends but bulge out in the middle. *Nature* **360:** 364–366.

Hammer J, Takacs B, Sinigaglia F. (1992) Identification of a motif for HLA-DR1 binding peptides using M13 display libraries. *J. Exp. Med.* **176:** 1007–1013.

Hammer J, Valsasnini P, Tolba K, Bolin D, Higelin J, Takacs B, Sinigaglia F. (1993) Promiscuous and allele-specific anchors in HLA-DR-binding peptides. *Cell* **74:** 197–203.

Hammer J, Belunis C, Bolin D, Papadopoulos J, Walsky R, Higelin J, Danho W, Sinigaglia F, Nagy ZA. (1994a) High-affinity binding of short peptides to major histocompatibility complex class II molecules by short anchor combinations. *Proc. Natl Acad. Sci. USA* **91:** 4456–4460.

Hammer J, Bono E, Gallazzi F, Belunis C, Nagy Z, Sinigaglia F. (1994b) Precise prediction of major histocompatibility complex class II-peptide interaction based on peptide side chain scanning. *J. Exp. Med.* **180:** 2353–2358.

Hammer J, Gallazzi F, Bono E, Karr RW, Guenot J, Valsasnini P, Nagy ZA, Sinigaglia F. (1995) Peptide binding specificity of HLA-DR4 molecules: correlation with rheumatoid arthritis association. *J. Exp. Med.* **181:** 1847–1855.

Hammond SA, Obah E, Stanhope P, Monell CR, Strand M, Robbins FM, Bias WB, Karr RW, Koenig S, Siliciano RF. (1991) Characterization of a conserved T cell epitope in HIV-1 gp41 recognized by vaccine induced human cytolytic T cells. *J. Immunol.* **146:** 1470–1477.

Harding CW, Roof RW, Allen PM, Unanue ER. (1991) Effects of pH and polysaccharides on peptide binding to class II major histocompatibility complex molecules. *Proc. Natl Acad. Sci. USA* **88:** 2740–2744.

Hegde NR, Ellis SA, Gaddum RM, Tregaskes CA, Sarath G, Srikumaran S. (1995) Peptide motif for cattle MHC class I antigen BoLA-A11. *Immunogenetics* **42:** 302–303.

Hickey WF. (1994) An MHC binding motif for the Lewis rat and its relationship to autoimmune reactions. *Neuropathol. Appl. Neurobiol.* **20:** 198–200.

Hill AVS, Allsopp CEM, Kwiatkowski D, *et al.* (1991) Common West African HLA antigens are associated with protection from severe malaria. *Nature* **352:** 595–600.

Hill AVS, Elvin J, Willis AC, *et al.* (1992) Molecular analysis of the association of HLA-B53 and resistance to severe malaria. *Nature* **360:** 434–439.

Hill CM, Liu A, Marshall KW, Mayer J, Jorgensen B, Yuan B, Cubbon RM, Nichols EA, Wickers LS, Rothbard JB. (1994) Exploration of the requirements for peptide binding to HLA-DRB1*0101 and DRB1*0401. *J. Immunol.* **152:** 2890–2898.

Hines WM, Falick AM, Burlingame AL, Gibson BW. (1992) Pattern-based algorithm for peptide sequencing from tandem high energy collision-induced dissociation mass spectra. *J. Am. Soc. Mass Spectrom.* **3:** 326–336.

Hobohm U, Meyerhans A. (1993) A pattern search method for putative anchor residues in T cell epitopes. *Eur. J. Immunol.* **23:** 1271–1276.

Huczko EL, Bodnar WM, Benjamin D, Sakaguchi K, Zhou Zhu N, Shabanovitz J, Henderson RA, Appella E, Hunt DF, Engelhard VH. (1993) Characteristics of endogenous peptides eluted from the class I molecule HLA-B7 determined by mass spectrometry and computer modelling. *J. Immunol.* **151:** 2572–2587.

Hunt DF, Henderson RA, Shabanowitz J, Sakaguchi K, Michel H, Sevilir N, Cox AL, Appella E, Engelhard VH. (1992a) Characterization of peptides bound to class I MHC molecule HLA-A2.1 by mass spectrometry. *Science* **255:** 1261–1263.

Hunt DF, Michel H, Dickinson TA, Shabanowitz J, Cox AL, Sakaguchi K, Appella E, Grey HM, Sette A. (1992b) Peptides presented to the immune system by the murine class II major histocompatability complex molecule I-A^d. *Science* **256:** 1817–1820.

Jameson SC, Bevan MJ. (1992) Dissection of major histocompatibility complex (MHC) and T cell receptor contact residues in a Kb-restricted ovalbumin peptide and an assessment of the predictive power of MHC-binding motifs. *Eur. J. Immunol.* **22:** 2663–2667.

Jardetzky TS, Gorga JC, Busch R, Rothbard J, Strominger JL, Wiley DC. (1990) Peptide binding to HLA-DR1: A peptide with most residues substituted to alanine retains MHC binding. *EMBO J.* **9:** 1797–1803.

Jardetzky TS, Lane WS, Robinson RA, Madden DR, Wiley DC. (1991) Identification of self peptides bound to purified HLA-B27. *Nature* **353:** 326–329.

Jensen PE. (1991) Enhanced binding of peptide antigen to purified class II major histocompatibility glycoproteins at acidic pH. *J. Exp. Med.* **174:** 1111–1120.

Johansen BH, Buus S, Vartdal H, Eriksen JA, Thorsby E, Sollid L. (1994) Binding of peptides to HLA-DQ molecules: peptide binding properties of the disease-associated HLA-DQ(α1*0501, β1*0201) molecule. *Int. Immunol.* **6:** 453–461.

Joyce S, Tabaczewski P, Angeletti RH, Nathenson SG, Stroynowski I. (1994) A nonpolymorphic major histocompatibility complex class Ib molecule binds a large array of diverse self-peptides. *J. Exp. Med.* **179:** 579–588.

Kast WM, Brandt RMP, Sidney J, Drjfhout J-W, Kubo RT, Grey HM, Melief CJM, Sette A. (1994) Role of HLA-A motifs in identification of potential epitopes in human papillomavirus type 16 E6 and E7 proteins. *J. Immunol.* **152:** 3904–3912.

Kinouchi R, Kobayashi H, Sato K, Kimura S, Katagiri M. (1994) Peptide motifs of HLA-DR4/DR53 (DRB1*0405/DRB4*0101) molecules. *Immunogenetics* **40:** 376–378.

Kropshofer H, Max H, Müller CA, Hesse F, Stevanovic S, Jung G, Kalbacher H. (1992) Self-peptide released from class II HLA-DR1 exhibits a hydrophobic two-residue contact motif. *J. Exp. Med.* **175:** 1799–1803.

Kropshofer H, Max H, Halder T, Kalbus M, Muller CA, Kalbacher H. (1993) Self-peptides eluted from four HLA-DR alleles share hydrophobic anchor residues near the NH_2-terminal including proline as a stop signal for trimming. *J. Immunol.* **151:** 4732–4742.

Kubo RT, Sette A, Grey HM, *et al.* (1994) Definition of specific peptide motifs for four major HLA-A alleles. *J. Immunol.* **152:** 3913–3924.

Leighton J, Sette A, Sidney J, Appella E, Ehrhardt C, Fuchs S, Adorini L. (1991) Comparison of the structural requirements for interaction of the same peptide with I-E^k and I-E^d molecules in the activation of MHC class II restricted T cells. *J. Immunol.* **147:** 198–204.

Lipford GB, Hoffman M, Wagner H, Heeg K. (1993) Primary in vivo responses to ovalbumin. Probing the predictive value of the K^b binding motif. *J. Immunol.* **150:** 1212–1222.

Lopez JA, Luescher IF, Cerottini JC. (1992) Direct binding of peptides to MHC class I molecules on living cells. Analysis at the single cell level. *J. Immunol.* **149:** 3827–3835.

Maier R, Falk K, Rötzschke O, Maier B, Gnau V, Stevanovic S, Jung G, Rammensee H-G, Meyerhans A. (1994) Peptide motifs of HLA-A3, -A24, and -B7 molecules as determined by pool sequencing. *Immunogenetics* **40:** 306–308.

Malcherek G, Falk K, Rötzschke O, Rammensee H-G, Stevanovic S, Gnau V, Jung G, Melms A. (1993) Natural peptide ligand motifs of two HLA molecules associated with myasthenia gravis. *Int. Immunol.* **5:** 1229–1237.

Malcherek G, Gnau V, Stevanovic S, Rammensee H-G, Jung G, Melms A. (1994) Analysis of allele specific contact sites of natural HLA-DR17 ligands. *J. Immunol.* **153:** 1141–1149.

Marrack P, Ignatowicz L, Kappler JW, Boymel J, Freed JH. (1993) Comparison of peptides bound to spleen and thymus class II. *J. Exp. Med.* **178:** 2173–2183.

Matsushita S, Takahashi K, Motoki M, Komoriya K, Ikagawa S, Nishimura Y. (1994) Allele specificity of structural requirement for peptides bound to HLA-DRB1*0405 and -DRB1*0406 complexes: implications for the HLA-associated susceptibility to methimazole-induced insulin autoimmune syndrome. *J. Exp. Med.* **180:** 873–883.

Max H, Halder T, Kalbus M, Gnau V, Jung G, Kalbacher H. (1994) A 16mer peptide of the human autoantigen calreticulin is a most prominent HLA-DR4w4–associated self-peptide. *Hum. Immunol.* **41:** 39–45.

Nelson CA, Roof RW, McCourt DW, Unanue ER. (1992) Identification of the naturally processed form of hen egg white lysozyme bound to the murine major histocompatibility complex class II molecule I-A^k. *Proc. Natl Acad. Sci. USA* **89:** 7380–7383.

Newcomb JR, Cresswell P. (1993) Characterization of endogenous peptides bound to purified HLA-DR molecules and their absence from invariant chain-associated α/β dimers. *J. Immunol.* **150:** 499–507.

Norda M, Falk K, Rötzschke O, Stevanovic S, Jung G, Rammensee H-G. (1993) Comparison of the H-2^k and H-$2K^{kml}$ restricted peptide motifs. *J. Immunother.* **14:** 144–149.

O'Sullivan D, Arrhenius T, Sidney J, *et al.* (1991a) On the interaction of promiscuous antigenic peptides with different DR alleles. *J. Immunol.* **147:** 2663–2669.

O'Sullivan D, Sidney J, Del-Guercio MF, Colon SM, Sette A. (1991b) Truncation analysis of several DR binding epitopes. *J. Immunol.* **146:** 1240–1246.

Pamer EG, Harty JT, Bevan MJ. (1991) Precise prediction of a dominant class I MHC-restricted epitope of *Listeria monocytogenes*. *Nature* **353:** 852–855.

Rammensee H-G, Friede T, Stevanovic S. (1995) MHC ligands and peptide motifs: first listing. *Immunogenetics* **41:** 178–228.

Reay PA, Kantor RM, Davis MM. (1994) Use of global amino acid replacements to define the requirements for MHC binding and T cell recognition of moth cytochrome c (93–103). *J. Immunol.* **152:** 3946–3957.

Reich EP, Von Grafenstein H, Barlow A, Swenson KE, Williams K, Janeway CA. (1994) Self peptides isolated from MHC glycoproteins of non-obese diabetic mice. *J. Immunol.* **152:** 2279–2288.

Romero P, Corradin G, Luescher IF, Maryanski Jl. (1991) H-$2K^d$-restricted antigenic peptides share a simple binding motif. *J. Exp. Med.* **174:** 603–612.

Rothbard JB, Taylor WR. (1988) A sequence pattern common to T cell epitopes. *EMBO J.* **7:** 93–100.

Rötzschke O, Falk K. (1994) Origin, structure and motifs of naturally processed MHC class II molecules. *Curr. Opin. Immunol.* **6:** 45–51.

Rötzschke O, Falk K, Stevanovic S, Jung G, Walden P, Rammensee HG. (1991) Exact prediction of a natural T cell epitope. *Eur. J. Immunol.* **21:** 2891–2894.

Rötzschke O, Falk K, Stevanovic S, Jung G, Rammensee HG. (1992) Peptide motifs of closely related HLA class I molecules encompass substantial differences. *Eur. J. Immunol.* **22:** 2453–2456.

Rötzschke O, Falk K, Stevanovic S, Grahovac B, Soloski MJ, Jung G, Rammensee HG. (1993) Qa-2 molecules are peptide receptors of higher stringency than ordinary class I molecules. *Nature* **361:** 642–644.

Rötzschke O, Falk K, Stevanovic S, Gnau V, Jung G, Rammensee H-G. (1994) Dominant aromatic/aliphatic C-terminal anchor in HLA-B*2702 and B*2705 peptide motifs. *Immunogenetics* **39:** 74–77.

Rovero P, Riganelli D, Fruci D, Vigano S, Pegoraro R, Greco G, Butler R, Clementi S, Tanigaki N. (1994) The importance of secondary anchor residue motifs of HLA class I proteins: a chemometric approach. *Mol. Immunol.* **31:** 549–554.

Rudensky AY, Preston-Hurlburt P, Hong S-C, Barlow A, Janeway CA. (1991) Sequence analysis of peptides bound to MHC class II molecules. *Nature* **353:** 622–627.

Rudensky AY, Preston-Hurlburt P, Al-Ramadi BK, Rothbard J, Janeway CA. (1992) Truncation variants of peptides isolated from MHC class II molecules suggest sequence motifs. *Nature* **359:** 429–431.

Ruppert J, Sidney J, Celis E, Kubo RT, Grey HM, Sette A. (1993) Prominent role of secondary anchor residues in peptide binding to HLA-A2.1 molecules. *Cell* **74:** 929–937.

Sette A, Southwood S, O'Sullivan D, Gaeta FCA, Sidney J, Grey HM. (1992) Effect of pH on MHC class II-peptide interactions. *J. Immunol.* **148:** 844–851.

Sette A, Sidney J, Oseroff C, Delguercio MF, Southwood S, Arrhenius T, Powell ME, Colon SM, Gaeta FCA, Grey HM. (1993) HLA DRw4–binding motifs illustrate the biochemical basis of degeneracy and specificity in peptide-DR interactions. *J. Immunol.* **151:** 3163–3170.

Sherman MA, Runnels HA, Moore JC, Stern LJ, Jensen PE. (1994) Membrane interactions influence the peptide binding behaviour of DR1. *J. Exp. Med.* **179:** 229–234.

Sidney J, Oseroff C, Southwood S, Wall M, Ishioka G, Koning F, Sette A. (1992) DRB1*0301 molecules recognize a structural motif distinct from the one recognized by most DRβ1 alleles. *J. Immunol.* **149:** 2634–2640.

Sidney J, Oseroff C, del Guercio M-F, Southwood S, Krieger JI, Ishioka GY, Sakaguchi K, Appella E, Sette A. (1994) Definition of a DQ3.1-specific binding motif. *J. Immunol.* **152:** 4516–4525.

Smith GP, Dabhi VM, Pamer EG, Fischer Lindahl K. (1994) Peptide presentation by the MHC class Ib molecule, H2–M3. *Int. Immunol.* **6:** 1917–1926.

Stern LJ, Brown JH, Jardetzky TS, Gorga JC, Urban RG, Strominger JL, Wiley DC. (1994) Crystal structure of the human class II MHC protein HLA-DR1 complexed with an influenza virus peptide. *Nature* **368:** 215–221.

Sutton J, Rowland-Jones S, Rosenberg W, *et al.* (1993) A sequence pattern for peptides presented to cytotoxic T lymphocytes by HLA B8 revealed by analysis of epitopes and eluted peptides. *Eur. J. Immunol.* **23:** 447–453.

Tanigaki N, Fruci D, Groome N, Butler RH, Londei M, Tosi R. (1994a) Exploring myelin basic protein for HLA class I-binding sequences. *Eur. J. Immunol.* **24:** 2196–2202.

Tanigaki N, Fruci D, Vigneti E, Starace G, Rovero P, Londei M, Butler RH, Tosi R. (1994b) The peptide binding specificity of HLA-B27 subtypes. *Immunogenetics* **40:** 192–198.

Thorpe CJ, Travers PJ. (1994) Prediction of an HLA-B44 binding motif by the alignment of known epitopes and molecular modeling of the antigen binding cleft. *Immunogenetics* **40:** 303–305.

Thorpe CJ, Moss DS, Powis SJ, Howard JC, Butcher GW, Travers PJ. (1995) An analysis of the antigen binding site of RT1.A^{a} suggests an allele specific motif. *Immunogenetics* **41:** 329–331.

Urban RG, Chicz RM, Lane WS, Strominger JL, Rehm A, Kenter MJH, Uytdehaag FGCM, Ploegh H, Uchanska-Ziegler B, Ziegler A. (1994) A subset of HLA-B27 molecules contains peptides much longer than nonamers. *Proc. Natl Acad. Sci. USA* **91:** 1534–1538.

Verreck FAW, van de Poel A, Termijtelen A, Amons R, Drijfhout J-W, Koning F. (1994) Identification of an HLA-DQ2 peptide binding motif and HLA-DPw3–bound self peptide by pool sequencing. *Eur. J. Immunol.* **24:** 375–379.

Vignali DAA, Urban RG, Chicz RM, Strominger JL. (1993) Minute quantities of a single immunodominant foreign epitope are presented as large nested sets by major histocompatability complex class II molecules. *Eur. J. Immunol.* **23:** 1602–1607.

Vogt AB, Kropshofer H, Kalbacher H, Kalbus M, Rammensee H-G, Coligan JE, Martin R. (1994) Ligand motifs of HLA-DRB5*0101 and DRB1*1501 molecules delineated from self peptides. *J. Immunol.* **153:** 1665–1673.

Vyas JM, Rodgers JR, Rich RR. (1995) H-2M3a violates the paradigm for major histocompitibility complex class I peptide binding. *J. Exp. Med.* **181:** 1817–1825.

Wegmann KW, Zhao W, Griffin AC, Hickey WF. (1994) Identification of myocarditogenic peptides derived from cardiac myosin capable of inducing experimental allergic myocarditis in the Lewis rat. The utility of a class II binding motif in selecting self-reactive peptides. *J. Immunol.* **153:** 892–900.

Wen J, Wang J, Kuipers JG, Huang F, Williams KM, Raybourne RB, Yu DT. (1994) Analysis of HLA-B*2705 peptide motif using T2 cells and monoclonal antibody. *Immunogenetics* **39:** 444–446.

Wucherpfennig KW, Sette A, Southwood S, Oseroff C, Matsui M, Strominger JL, Hafler DA. (1994) Structural requirements for binding of an immunodominant myelin basic protein peptide to DR2 isotypes and for its recognition by human T cell clones. *J. Exp. Med.* **179:** 279–290.

Zhang Q-J, Gavioli R, Klein G, Masucci MG. (1993) An HLA-A11-specific motif in nonamer peptides derived from viral and cellular proteins. *Proc. Natl Acad. Sci. USA* **90:** 2217–2221.

13

Function of HLA class I restricted T cells

Andrew McMichael

13.1 Introduction

Although cytotoxic T lymphocytes (CTL) were first identified in the context of alloreactivity (Cerotinni and Brunner, 1974), their natural role only became apparent when Blanden and colleagues started to examine their role in antiviral immunity (Blanden, 1974). Shortly thereafter Zinkernagel and Doherty made their classic observation that recognition of virus-infected cells was restricted by the class I molecules of the polymorphic major histocompatibility complex (MHC; Zinkernagel and Doherty, 1975). They showed that lymphocytic choriomeningitis virus (LCMV) specific CTL from $H2^d$ mice only lysed infected target cells that shared $H\text{-}2^d$ class I molecules. This finding was soon extended as a general principle for other virus infections in mice (Zinkernagel and Doherty, 1979) and humans (McMichael, 1980).

The CTL that were MHC class I-restricted in this way were subsequently demonstrated to carry the CD8 surface molecule (Biddison *et al.*, 1981) and this was shown to bind to MHC class I molecules (Norment *et al.*, 1988). The explanation for MHC restriction, however, remained elusive until Townsend and colleagues showed that the virus antigen recognized by CTL could be artificially fragmented and represented by short synthetic peptides added to uninfected cells (Townsend *et al.*, 1985, 1986). Different class I molecules presented different peptide epitopes (McMichael *et al.*, 1986; Townsend *et al.*, 1986). A year later, the structure of HLA-A2 was determined by Bjorkman and colleagues and found to contain peptide fragments within a groove on the outer surface (Bjorkman *et al.*, 1987a, 1987b). Rammensee and colleagues then showed that peptides could be eluted from class I MHC molecules. These peptides were short, usually nonamers, and that peptides eluted from each class I allelic product showed common residues at particular positions, often at position 2 and always at the carboxy terminus (Falk *et al.*, 1991; Rötzschke *et al.*, 1990).

The last piece of the jigsaw was to determine how the peptides are derived from virus, and other intracellular, proteins and how they gain access to the MHC class I molecules. This is reviewed elsewhere in this volume (see Chapter 9), but in essence intracellular proteins are degraded in the cytosol, probably by the proteasome complex (Driscoll *et al.*, 1993), and then transported into the lumen of the endoplasmic reticulum where they stabilize the folding of newly synthesized class I molecules (Townsend and Trowsdale, 1993; Townsend *et al.*, 1990). The

HLA and MHC: genes, molecules and function, edited by M.J. Browning and A.J. McMichael.
© 1996 BIOS Scientific Publishers Ltd, Oxford.

proteasome generates peptides with appropriate carboxy termini and the transporter selects for size and, to a limited extent, C-terminus (Elliott *et al.*, 1993; Neefjes *et al.*, 1993). Some final trimming of peptides occurs in the endoplasmic reticulum (Elliott *et al.*, 1995).

A key finding to emerge from all of these studies was that MHC type determined the epitope peptides for an intracellular pathogen. Thus there are important differences in the CTL response to an intracellular pathogen determined by HLA type. This review examines the function of class I-restricted T cells with this point in view. The implication is that the subtle differences in specificity of these T cells imposed by variable HLA structures selects the HLA polymorphism.

13.2 Virus-specific CTL

Viruses are arguably the most dangerous intracellular pathogens and the HLA class I system may have evolved primarily to deal with them. Most of this review will, therefore, concern virus infections. Other important intracellular pathogens will be considered in subsequent sections.

13.2.1 Viruses that stimulate CTL

Early CTL studies were carried out using LCMV (Zinkernagel and Doherty, 1975, 1979). Soon after, influenza-specific CTL were demonstrated in mice (Zweerink *et al.*, 1977) and humans (McMichael *et al.*, 1977). Other viruses for which human CTL have been demonstrated are listed in *Table 13.1*. For some viruses it has been difficult, but not impossible, to demonstrate MHC class I-restricted CTL. Measles-specific $CB8^+$ CTL were hard to find (Fleischer and Kreth, 1983), although $CD4^+$ CTL were more easily identified (Jacobson *et al.*, 1984; van Binnendijk *et al.*, 1993). Herpes simplex virus-specific CTL were also difficult to demonstrate in humans (see below; Koelle *et al.*, 1993). Mice with the $H2^{bm\text{-}1}$ MHC antigen do not make a Sendai virus-specific CTL response, for genetic reasons (Kast *et al.*, 1986), but human non-responders are less clearly genetically defined. Many humans appear not to respond to influenza virus but probably because T cell memory has faded in the absence of antigen since their last natural infection (McMichael *et al.*, 1983a). A few adult patients infected with human immunodeficiency virus (HIV) fail to make any CTL response for reasons that are not clear; one possibility is an abortive response with exhaustion of reactive T cells (Moskophidis *et al.*, 1993; Pantaleo *et al.*, 1994).

Where virus antigens that stimulate CTL have been determined, most CTL responses are directed towards internal virus proteins (Bennink *et al.*, 1986; Gotch *et al.*, 1987; Townsend *et al.*, 1985b; Yewdell *et al.*, 1985). Although surface glycoproteins are recognized, these responses tend to be subdominant (Townsend *et al.*, 1985b). Often CTL specificity is dominated by a few epitopes, sometimes only by one (Gavioli *et al.*, 1993; Gotch *et al.*, 1988; Martinon *et al.*, 1990; Townsend *et al.*, 1986). This is remarkable because each virus protein contains many potential epitopes with the right sequence motifs to bind to the class I molecules present. The reasons for this are not clear. One possibility is that the restriction occurs at the level of antigen processing so that only very few epitopes

Table 13.1. HLA class I-restricted virus-specific CTL

Virus	Reference
Influenza	McMichael *et al.*, 1977
Epstein–Barr virus	Rickinson *et al.*, 1981
Mumps	Kress and Kreth, 1982
Measles	Fleischer and Kreth, 1983
Cytomegalovirus	Borysiewicz *et al.*, 1983
Respiratory syncitial virus	Bangham and McMichael, 1986
HTLV1	Kannagi *et al.*, 1983
HIV-1	Walker *et al.*, 1987
HIV-2	Gotch *et al.*, 1993
Herpes simplex	Koelle *et al.*, 1993
Hepatitis A virus	Vallbracht *et al.*, 1989
Hepatitis B virus	Pignatelli *et al.*, 1987
Hepatitis C virus	Koziel *et al.*, 1993

are exposed at the cell surface. Alternatively, there may be competition between responding CTL for diminishing antigen so that there will be a tendency for the responses to focus on one dominant epitope and, ultimately, one clone (Nowak and McMichael, 1995; Nowak *et al.*, 1995). This extreme is sometimes seen (Argaet *et al.*, 1994; Moss *et al.*, 1995). This tendency may accentuate the effect of the MHC type on the anti-virus response, for instance making the response less cross-reactive against different virus strains in some individuals.

13.2.2 The role of CTL in virus infections

CTL can have both beneficial and harmful roles in virus infections. Zinkernagel and Hengartner have argued that viruses may be divided into those that are cytopathic and those that are non-cytopathic, and that CTL are always beneficial for the former and harmful for the latter (Zinkernagel and Hengartner, 1994). It seems more likely that for many viruses these two roles may alternate or coexist and that the balance may tip from one to the other.

Beneficial effects in acute virus infections. The prototype virus for these studies has been influenza in the mouse model. Certain strains of influenza virus cause a dose-dependent fatal pneumonitis in mice after intranasal administration. At lower doses the mice recover. Ada and co-workers showed that recovery coincided with the appearance of CTL in the lungs and then the spleens of these mice and that this occurred before the appearance of neutralizing antibodies (Yap and Ada, 1978a). His group went on to show that irradiated, infected mice given $CD8^+$ spleen cells from immune animals recovered (Yap and Ada, 1978b); $CD4^+$ T cells did not show this effect. Influenza-specific CTL clones mediated this recovery (Lin and Askonas, 1981; Lukacher *et al.*, 1984). Similar effects were shown following adoptive transfer of immune $CD8^+$ T cells in nude mice, and antibody had no effect on recovery, though it was very effective in prophylaxis (Wells *et al.*, 1981). Not all CTL clones mediated recovery, however, and some $CD4^+$ clones also had this effect, possibly Th1 clones (Graham *et al.*, 1994; Taylor *et al.*, 1990). CTL clones that produced interferon-gamma were most effective (Morris *et al.*,

1982). Similar results have been obtained for respiratory syncitial virus (RSV; Cannon *et al.*, 1988) and herpes simplex virus (Bonneau and Jennings, 1990). The CTL clone-transfer experiments offer the most compelling evidence of the role of CTL in recovery from acute viral infection. However, β2-microglobulin knockout mice do recover from influenza and Sendai virus infection, though this may be delayed; they probably utilize alternative CD4$^+$ CTL (Eichelberger *et al.*, 1991; Hou *et al.*, 1992).

In humans it has not been possible to carry out such cell-transfer experiments in acute virus infections, although recent results in patients with Epstein–Barr virus (EBV) lymphoma and cytomegalovirus suggest this might be feasible in some situations (see below). A correlation has been shown between memory CTL levels and reduction in the titre of virus shed in nasal washings of volunteers infected with influenza virus (McMichael *et al.*, 1983b). This finding is in line with evidence that live virus infection (which stimulates CTL) offers longer term protection than killed or subunit vaccines (which only weakly stimulate CTL; Hoskins *et al.*, 1979). Once influenza virus is cleared, memory gradually fades so reinfection can and does occur (McMichael et al., 1983a). It remains to be determined whether the response to reinfection, which ranges from subclinical infection to death from acute pneumonitis, is influenced by the level of CTL memory as well as antibody.

For humans there is also evidence that CTL are important in the recovery from acute measles (Sissons *et al.* 1985) and acute hepatitis A virus (HAV) infection (Vallbracht *et al.* 1989). In the latter case, CTL are found in the liver as jaundice develops; recovery thereafter is normally rapid (Fleischer *et al.*, 1990).

Beneficial effects in chronic virus infections. Epstein–Barr virus (EBV) infection is a good example of a persistent virus infection where CTL control but do not eliminate the infection. More than 85% of adult humans are infected with EBV and the vast majority harbour the virus, which persistently infects B lymphocytes and epithelial cells of the oropharynx without any harm. However the virus does contribute to four human malignancies, Burkitts lymphoma, some B cell lymphomas, Hodgkins disease and nasopharyngeal cancer. In addition it readily transforms B lymphocytes *in vitro* (reviewed in Rickinson *et al.*, 1992).

There is considerable evidence that EBV is controlled by CTL. CD8$^+$ T cells inhibit the outgrowth of transformed B lymphoblastoid lines in culture, and CTL are maintained at a high level in all infected individuals (Moss *et al.*, 1978; Rickinson *et al.*, 1981). Immunosuppression to the point where the EBV-specific CTL response is impaired is associated with the development of B cell lymphomas with the phenotype of transformed B-LCL (Cohen, 1991; Crawford *et al.*, 1980). Burkitts lymphomas reduce expression of all virus proteins except EBNA-1 (Rowe *et al.*, 1987), which has a mechanism for evading the CTL response (see below). Human lymphocytes, if infused into mice with severe combined immunodeficiency, spontaneously develop B cell lymphomas, which are more likely in the absence of EBV-specific CD8$^+$ T cells (Rowe *et al.*, 1991). Most convincingly, patients with B cell lymphomas after bone marrow transplantation can be treated by infusion of CD8$^+$ T cells from the EBV-positive donor (Papadopoulos *et al.*, 1995) or, better still, with EBV-specific CTL lines (Rooney *et al.*, 1995). In the acute phase of EBV infection there is a strong CTL response which results in control of this phase of the infection but leaves some persistently infected cells expressing only EBNA-1. Thereafter there is a continuing breakthrough of cells expressing other latent EBV antigens, but these are eliminated by the CTL response; at the same time the continual antigen stimulation maintains the strong CTL response. Other herpes virus infections show similarities. CMV also generates a CTL response and appears to be controlled by this (Borysiewicz *et al.*, 1983), although much less detail is known compared to EBV. Down-regulation of class I HLA molecules seems to be part of the strategy used by this virus, implying that $CD8^+$ CTL have strong anti-viral effects (Beersma *et al.*, 1993). Specific CTL in bone marrow transplant recipients protect against cytomegalovirus (CMV) disease (Li *et al.*, 1994; Quinnan *et al.*, 1982), and CMV infection may be preventable in immunodeficient bone marrow transplant recipients by infusion of CMV-specific CTL clones (Riddell *et al.*, 1994). Likewise herpes simplex virus also down-regulates expression of class I HLA by interfering with the ability of the transporters to translocate peptides into the lumen of the endoplasmic reticulum (ER) (Hill *et al.*, 1995; York *et al.*, 1994). Also this virus can hide in neurons which normally express very little class I HLA antigen.

Human immunodeficiency virus (HIV) is another persisting virus infection where CTL responses are probably important, although ultimately they fail to control the infection (McMichael and Walker, 1994). The initial CTL response becomes detectable during the acute phase of infection when virus replication is at a very high level (Koup *et al.*, 1994). CTL are detected before any neutralizing antibody and their appearance is associated with a large (10–100-fold) reduction in plasma viraemia. As for EBV, CTL regulate virus replication *in vitro*, although by means of cytokines or chemokines, presently unidentified (Kurth *et al.*, 1995; Lusso *et al.*, 1995), rather than, or as well as, killing of infected cells (Walker *et al.*, 1986). During the asymptomatic phase, virus is kept at relatively low levels, but later the CTL response fails (Carmichael *et al.*, 1993; Klein *et al.*, 1995), along with the Th1 response (Clerici and Shearer, 1993) and virus clearly escapes (Nowak and McMichael, 1995; Nowak *et al.*, 1995). Some evidence indicates that CTL induction by vaccines can lower virus loads after challenge for simian immunodeficiency virus (SIV) rather than HIV; Gallimore *et al.*, 1995). Several exposed but uninfected individuals have been shown to have CTL responses, suggesting that it might be possible to clear infection in some individuals by CTL (Rowland-Jones *et al.*, 1995). Escape mutation may be an important way by which virus can evade CTL responses (see below; Nowak and McMichael, 1995). Other probable means by which HIV escapes from the CTL are by integrating provirus without expression of virus proteins and by directly (by infection) impairing T cell help that is necessary to maintain the CTL response (Walter *et al.*, 1995).

Harmful antiviral CTL responses. LCMV infection in mice is the classic example of immunopathology mediated by CTL (Buchmeier *et al.*, 1980). When mice are infected intracerebrally at birth, they make no response to this virus, showing that the virus infection, which is not cytopathic, causes no real harm. If they are infected as adults, a strong CTL response is made and they develop a fatal choriomeningitis mediated by $CD8^+$ CTL. Zinkernagel and Hengartner have argued that CTL are likely to cause immunopathology for all non-cytopathic viruses (Zinkernagel and Hengartner, 1994).

There are probable similar examples in human disease; one of the clearest is hepatitis B virus infection, where the virus can infect hepatocytes in a non-cytopathic way resulting in a carrier state (Missale *et al.*, 1993; Pignatelli *et al.*, 1987; Rehermann *et al.*, 1995). Activation of a CTL response will cause an acute hepatitis, although it may be possible to eliminate the virus. However, the virus infection predisposes to liver cancer so that chronic infection, while not cytopathic, is not harmless. Therefore elimination of the virus by CTL is beneficial even at the cost of acute hepatitis.

There has been some controversy as to whether the strong anti-HIV response could be harmful. CTL might contribute to the reduction in $CD4^+$ T cells. Ho and Wei and colleagues (Ho *et al.*, 1995; Wei *et al.*, 1995) have calculated virus turnover rates of around 2 days in infected patients; this is accompanied by an equivalent loss of $CD4^+$ T cells, implying that the virus is cytopathic *in vivo*. A strong CTL response will tend to reduce virus replication by killing virus infected $CD4^+$ T cells before they release virus particles, but thereby reducing $CD4^+$ T cell levels (Cheynier *et al.*, 1994). A weak CTL response would fail to kill $CD4^+$ T cells before virus is released and so $CD4^+$ T cells would be killed by virus, while the total virus load may increase rapidly. Both of these events may occur simultaneously if $CD8^+$ CTL are not evenly distributed, as is likely to be the case in lymph nodes (Cheynier *et al.*, 1994). Also at the same time CTL activity may contribute to neurological disease (Sethi *et al.*, 1988) and to bronchoalveolitis (Plata *et al.*, 1987), where the targets may be infected glial cells and macrophages. Other examples where CTL may contribute to immunopathology are mumps, meningitis (Kreth *et al.*, 1982) and a number of rashes associated with acute virus infections (Sissons *et al.*, 1985).

13.2.3 Evasion of CTL responses by virus infections

Probably the greatest compliment that a virus can pay to the CTL response is the evolution of a specific evasion strategy. A number have been identified. The first was the identification of the adenovirus E19 protein which retains HLA class I molecules in the ER, so reducing surface expression of viral antigenic peptides (Andersson *et al.*, 1985). Herpes simplex virus expresses a small protein ICP47 which binds to the cytosolic part of the transporters associated with antigen processing (TAP) and greatly reduces their capacity to move peptides into the ER, and thereby reduces class I HLA-peptide surface expression (Hill *et al.*, 1995). The EBV protein EBNA-1 contains a long glycine-alanine repeat sequence which impairs the processing pathway for EBNA-1, although not for other virus proteins; the virus only expresses this antigen in latency (Levitskaya, 1995). EBV can also reduce expression of the adhesion molecules ICAM-1 and LFA-3, impairing CTL–target cell interaction (Gregory *et al.*, 1988). CMV reduces class I HLA antigen expression by yet other mechanisms (Beersma *et al.*, 1993). Some of these, and other viruses, may also evade the CTL response by infecting cells such as neurons and hepatocytes with low expression of class I MHC.

An important factor in the evasion of the immune response by HIV and HTLV-1 is their natural variability (Niewiesk *et al.*, 1995; Nowak and McMichael, 1995; Nowak et al., 1995; Phillips *et al.*, 1991). The high and continuous virus replication rate, together with error-prone reverse transcription, results in huge numbers of mutant viruses, many defective (Coffin, 1990). The fact that infected cells integrate a single virus genome means that all virus proteins generated will carry the same mutations and the whole cell can escape a CTL response. Similarly, the tendency of the CTL response to focus on a small number of epitopes makes escape feasible, particularly in the response dominant at that time (Nowak and McMichael, 1995). Even where there is a parallel CTL response to two or more epitopes, a small advantage could be significant, reducing the rate of CTL killing relative to virus replication. A further advantage might be gained if the altered epitope peptide antagonizes the CTL response to the wild-type epitope, as has been clearly observed *in vitro* (Bertoletti *et al.*, 1994; Klenerman *et al.*, 1994; Meier *et al.*, 1995).

A clear but artificial example of HIV escape from CTL was found in a patient treated by infusion of a CTL clone specific for an epitope in the virus nef protein (Koenig, 1995). A mutant virus with a deletion in the epitope that removed the epitope occurred. A similar observation of a point mutation destroying an epitope for LCMV was observed in mice (Pircher *et al.*, 1990). At a population level, EBV in Papua New Guinea has an altered amino acid sequence in an HLA-A11-presented epitope; HLA-A11 is common in this population and evokes strong EBV-specific CTL (De Campos Lima *et al.*, 1993).

13.2.4 HLA polymorphism and virus infections

It is clear that HLA type selects the epitopes that elicit the anti-viral CTL response. In general, individuals with the same HLA type respond to the same epitopes, but there are exceptions. In one example there was competition between overlapping epitopes, one presented by HLA-B8 and the other by HLA-B2702, so that when both HLA types were present only B2702 presented the peptide (Tussey, 1995). In another example, differences in the dominant T-cell receptors resulted from a cross-reaction between an EBV EBNA-3 peptide plus HLA-B8 and HLA-B44 (Argaet *et al.*, 1994; Burrows *et al.*, 1994). For HIV-specific CTL, differences in the dominant epitopes recognized by CTL can result from different patterns of mutation in the virus (Nowak *et al.*, 1995; Phillips *et al.*, 1991).

Do these differences in selected epitopes result in different outcomes to the virus infection? The most extreme example must be the $H2/^{bm-1}$ mice, a laboratory strain with a recent mutation in the K^b molecule. These mice fail to make any CTL response to Sendai virus and die when infected with the more virulent version of this virus (Kast *et al.*, 1986). In contrast, the HLA types we find in the population are old, as are most of the viruses, so total non-responders have probably been removed. HIV is new and more striking effects may be seen. Some patients fail to make any CTL response (and do very badly) but it is not clear whether this has anything to do with their HLA type. Different epitopes differ in the amount of variation that can occur, and different HLA molecules may be more or less susceptible to the effects of mutation in the epitopes. In our studies, HLA B8 seems particularly prone to these effects compared to HLA-A2 (Nowak and McMichael, 1995), possibly because of the positioning of the three amino acids of the peptides that bind in the groove (S. Reid *et al.*, submitted for publication). HLA-B27 almost always elicits a strong response to a conserved epitope (Nowak and McMichael, 1995). There is evidence that patients with HLA-B8 and B35 tend to do badly (Itescu *et al.*, 1992; Steel *et al.*, 1988). The strength of particular responses may well reflect the number of CTL precursors in the primary infection and this may influence the virus load and hence ultimate outcome. In acute virus infections, such as influenza, the strength of the response may influence the duration of CTL memory. This, in turn, may affect the outcome in subsequent exposures to the virus; influenza is particularly interesting in this context because, while the surface glycoproteins can mutate to evade antibody memory completely, T cell responses are much more focused on the much less variable internal proteins such as nucleoprotein or matrix protein (Askonas *et al.*, 1982). Thus when new pandemic strains of virus appear, immune protection can only be mediated by T cell memory and this may determine the severity of the infection.

Overall, apparently subtle differences in the CTL response influenced by HLA type could have profound effects at a population level. It is likely that these differences contribute significantly to the continuing selection of the HLA class I polymorphism.

13.3 Bacteria-specific CTL

CTL can also recognize bacterially infected cells (*Table 13.2*). This was first shown in mice where H-2 restricted T cells could transfer immunity to *Listeria monocytogenes* (Zinkernagel, 1974). Bacterially infected cells are lysed by CTL and peptide epitopes have been defined, restricted by H-2 K^d (Pamer *et al.*, 1991).

In humans it has been harder to show bacteria-specific CTL for pathogenic intracellular bacteria. Many attempts have been made to demonstrate mycobacteria-specific CTL. First, $CD8^+$ T cells that proliferated in response to bacterially infected cells were described (Rees *et al.*, 1988). More recently, heat-shock protein specific $CD8^+$ T cells have been found capable of lysing infected macrophages (Schoel *et al.*, 1994). Recombinant mycobacteria have also been developed for use as vaccines and can induce class I MHC-restricted responses *in vivo* (Yasutomi *et al.*, 1995). In tuberculosis it is not clear whether CTL-mediated lysis of macrophages infected with mycobacteria might be counterproductive and actually facilitate spread of the infection. On the other hand, secretion of interferon gamma by specific $CD8^+$ T cells as well as by Th1 cells, should activate bacteria-containing macrophages and would thus have an anti-mycobacterial effect. The increased incidence of tuberculosis associated with HIV infection is testimony to the importance of the Th1 component (Wallis *et al.*, 1992; Zhang *et al.*, 1994), although, as indicated above, CTL are themselves T-helper cell (probably Th1) -dependent (Walter *et al.*, 1995).

Salmonella-specific CTL have been sought, initially because of the possibility of using recombinant salmonella as a vaccine. Normally salmonella infects enterocytes and macrophages and grows in intracellular vesicles. A recombinant salmonella-expressing plasmodium circumsporozoite (CSP) antigen was first shown to stimulate $CD8^+$ CTL responses in mice (Aggarwal *et al.*, 1990). The CTL-lysed cells pulsed with a known CSP epitope peptide. Gao and colleagues (Gao *et al.*, 1992) attempted to demonstrate lysis of cells infected with recombinant salmonella that expressed influenza nucleoprotein, but were not successful, despite demonstrating that the target cells expressed the nucleoprotein as inclusion bodies in large quantity. Similar findings were reported by Verjans and colleagues (Verjans *et al.*, 1995). Gao and co-workers went on to show that the same recombinants failed to induce CTL responses in mice *in vivo*, but

Table 13.2. MHC class I-restricted bacteria-specific CTL

Bacteria	Reference
Listeria (mice)	Pamer *et al.*, 1991; Zinkernagel, 1974
Mycobacterium tuberculosis	Schoel *et al.*, 1994
Chlamydia	Hermann *et al.*, 1993
Salmonella	Aggarwal *et al.*, 1990
Yersinia	Hermann *et al.*, 1993

recombinant salmonella that expressed fragments of influenza nucleoprotein in the periplasmic space could induce NP-specific CTL. However, even these bacteria could not sensitize cells for lysis. It was thought that the reasons were low levels of expression and poor access of bacterial proteins to the host cell cytosol. More recently, Bowness and colleagues (Bowness *et al.*, in preparation) have demonstrated that $CD8^+$ CTL specific for a heat shock protein 70 peptide could be demonstrated in a patient with post-salmonella-reactive arthritis.

Reactive arthritis (ReA), which occurs in some HLA-$B27^+$ individuals after infection with intracellular bacterial infections such as *Salmonella, Shigella, Yersinia* and *Chlamydia*, has stimulated a search for HLA-B27-restricted CTL specific for these bacteria. H-$2D^b$ restricted CTL that lyse *Y. enterocolitica*-infected cells have been demonstrated in mice and a peptide epitope defined (Starnbach and Bevan, 1994). Hermann and colleagues (Hermann *et al.*, 1993) have been able to stimulate CTL from synovial cells of patients with post-*Yersinia* ReA using HLA-B27-transfected L cells infected with *Y. enterocolitica*. *Chlamydia*-specific CTL were obtained in a similar way. *Chlamydia*-specific CTL have also been obtained by restimulation of peripheral blood lymphocytes of patients with peptides representing chlamydial sequences. The CTL appeared with kinetics suggestive of a secondary, rather than a primary, CTL response (Bowness *et al.*, in preparation).

These experiments demonstrate that CTL specific for bacterial antigens can be found when appropriate methods are used. It is not clear, however, how effective they are at lysing bacterially infected cells *in vivo*. Intracellular bacteria may reside in vesicles in relatively small numbers within the cell. Bacterial products would have to traverse both bacterial and cell membranes to gain access to the MHC class I processing pathway. It is possible that this might occur in some specialized APCs *in vivo*, but other infected cells may not be lysed. For models of cross-reactive autoimmune reactions (Benjamin and Parham, 1990), induction of CTL may be sufficient, but in general, it is less clear what role CTL might play in antibacterial immunity. It is possible that by lysing infected cells, CTL might actually facilitate spread of the bacteria.

13.4 Parasite-specific CTL

The first clear demonstration of parasite-specific CTL was for the bovine parasite *Theileria*, which infects lymphocytes (Goddeeris *et al.*, 1986). Specific CTL may be important in the control of this infection. More detailed studies have been made of the malaria parasite *Plasmodium falciparum* in humans and *P. yoeli* and *berghei* in

Table 13.3. MHC class I restricted parasite specific CTL

Parasite	Reference
Theileria (cattle)	Goddeeris *et al.*, 1986
Plasmodium yoeli (mice)	Hoffman *et al.*, 1990; Rodrigues *et al.*, 1992; Weiss *et al.*, 1990
Plasmodium berghei (mice)	Romero *et al.*, 1990
Plasmodium falciparum	Hill *et al.*, 1992; Aidoo *et al.*, 1995

mice (*Table 13.3*; Aggarwal *et al.*, 1990; Hoffman *et al.*, 1990; Rodrigues *et al.*, 1992; Romero *et al.*, 1990; Weiss *et al.*, 1990). In mouse models of infection it is clear that CTL against sporozoite antigens such as CSP play an important defensive role, and vaccine induction of such CTL can prevent infection (Hoffman *et al.*, 1990; Romero *et al.*, 1990; Weiss *et al.*, 1990). In addition, mice can be protected by transfer of $CD8^+$ T cell clones (Rodrigues *et al.*, 1992). These results have stimulated an intensive effort to identify *Plasmodium*-specific CTL in humans and design vaccines that stimulate them.

A further stimulus was the finding that HLA-B53 was decreased in frequency, by about half, in children in West Africa with severe life-threatening malaria (Hill *et al.*, 1991). HLA-B53, while common in the Gambia (25% antigen frequency), is rare in Europe (< 1%), suggesting that it might have been selected because it offers this resistance. In total, the resistance effect was of the same order of magnitude as that offered by the much rarer haemoglobin S, which has the severe disadvantage of lethality when homozygous. In more recent studies the HLA-B53 association was not found in East Africa (Kenya); a possible reason is the different level of continuous exposure in the latter country (Hill *et al.*, 1994). Hill and colleagues showed that HLA-B53 presents a peptide derived from a liver stage antigen of *Plasmodium* to CTL that were detectable in immune adults (Hill *et al.*, 1992), so it is reasonable to argue that the protection might be mediated by CTL. A CSP peptide epitope has also been described in humans immunized with irradiated sporozoites, and Aidoo and colleagues have found other epitopes presented by other HLA molecules in exposed Gambians (Aidoo *et al.*, 1995; Malik *et al.*, 1991). However, levels of CTL may naturally be quite low (Doolan *et al.*, 1993). A hypothesis that would encompass all of the above findings might be that the annual seasonal infection in the Gambia leads to a gradual build up of CTL-mediated immunity; this might occur more rapidly in the children with HLA-B53 who are thereby less likely to suffer severe infection in childhood. In parts of Kenya where the parasite is endemic, immunity might be acquired much more rapidly, so that this HLA-mediated effect might be less pronounced (Hill *et al.*, 1994).

The protective effect of HLA-B53 against malaria in West Africa could account for the increased frequency of this allele in that population. This study is the best example of an HLA association with protection against an infectious disease in humans. The observation that this effect is not apparent in a population where the infection is different does not weaken the original observation, which is well founded statistically, but instead emphasizes the complex nature of the selective forces that are probably at work. The prevalence of other present and past infective threats must also be significant factors.

13.5 Tumour-specific CTL

There is now abundant evidence that CTL can be raised against a variety of experimental and natural tumours (De Plaen *et al.*, 1988; Maryanski *et al.*, 1982, reviewed in McMichael and Bodmer, 1992). In humans there have been detailed studies of melanoma-specific CTL (Anichini *et al.*, 1986; Degiovanni *et al.*, 1988; Knuth *et al.*, 1989; Mukherji and McAlister, 1983) with identification of a number of peptide epitopes, derived from MAGE-1 (Traversari *et al.*, 1993), tyrosinase and other proteins (Boel *et al.*, 1995; Zakut *et al.*, 1993). Some of these CTL have come from long-term survivors and they may have played some role in the remission, although this is not completely clear. Despite repeated demonstrations that melanoma-specific CTL can be found, this remains a particularly malignant cancer with a very poor prognosis for the vast majority of patients.

Specific CTL have been described in other tumours, such as ovarian cancer (Ferrini *et al.*, 1985) and cervical (human papilloma virus-positive) tumours, as well as some leukaemias (reviewed in Melief and Kast, 1991). However, they have been much harder to demonstrate against common and devastating solid tumours such as lung and colon cancers (*Table 13.4*). Therefore their role *in vivo* is hard to assess, although they may well be valuable in therapy.

There has been much discussion as to whether the often-seen down-regulation of class I molecule expression on the surface of these tumours represents escape from CTL-mediated attack (Momberg *et al.*, 1989; Smith *et al.*, 1989). A strong argument in favour of this hypothesis is that down-regulation of class I HLA requires two chromosomes to be affected. A common mechanism is loss of expression of β2-microglobulin on *both* chromosomes. It has been argued that this would be extremely unlikely to happen by chance without selection by CTL (Bodmer and McMichael, 1992). However, tumour cells with the mutator phenotype, often found in colon cancers (Fishel *et al.*, 1993), show some predilection for mutations in a particular region of the β2-microglobulin gene (Bicknell *et al.*, 1994). Given this phenotype, it might not be so improbable that double mutations could occur, without selection or with only minimal selection. Similarly, the K-ras oncogene has been shown to affect MHC class I expression (Alon *et al.*, 1987). Also against the CTL hypothesis is the repeated failure to generate secondary CTL specific for tumours such as colon cancer that show loss of HLA class I expression, the patchy down-regulation of the HLA molecules, and the lack of any cellular infiltrate histologically. On the other hand, loss of the CTL could occur long before the tumour became clinically apparent.

The issue is of general as well as clinical importance. If CTL are commonly active against tumours, they may have a primary immunosurveillance role, with the

Table 13.4. HLA class I-restricted CTL against solid tumours

Cancer	Reference
Melanoma	Anichini *et al.*, 1986; Degiovanni *et al.*, 1988
Small cell lung cancer	Yoshino *et al.*, 1994
Ovarian cancer	Yoshino *et al.*, 1994
Squamous cell carcinoma	Slingluff *et al.*, 1994; Yasumura *et al.*, 1994
Pancreas	Wahab *et al.*, 1991
Cervical carcinoma	Kast *et al.*, 1993
Lymphomas	Rickinson *et al.*, 1992 (review)

CTL continuously destroying abnormal cells as they arise. Alternatively, this could be much less important than their antimicrobial role. In the latter case, the microbial target for the CTL would be the strong selective force that selects the polymorphism. There is some common ground in that there are major cancers caused by viruses: nasopharangeal cancer, liver cancer and cervical cancer. For the present the issue remains unresolved and somewhat controversial. Stronger support for the anti-tumour effects of CTL may come from clinical trials of immunotherapy based on CTL or from attempts to vaccinate against cancers.

13.6 Conclusions

The function of class I molecules of the MHC is to present antigenic peptides to $CD8^+$ CTL. The primary role of these cells is to eliminate intracellular pathogens, particularly viruses but also some bacteria and parasites. The importance of these functions is underlined by the enormous polymorphism of the HLA class I system, which is consistent with small but significant advantages being offered by each HLA molecule against infection with some pathogens. The studies on malaria indicate that it can be possible to measure these advantages. A major challenge now in this field will be to extend these observations to other pathogens.

Acknowledgements

I am indebted to many colleagues for discussion of the ideas detailed here, particularly, Frances Gotch, Sarah Rowland-Jones, Paul Bowness, Rodney Phillips, Martin Nowak and Adrian Hill. I am also grateful to the Medical Research Council for support.

References

Aggarwal A, Kumar S, Jaffe R, Hone D, Gross M, Sadoff J. (1990) Oral salmonella: malaria circumsporozoite recombinants induce CD8+ cytotoxic T cells. *J. Exp. Med.* **172:** 1083–1090.

Aidoo M, Lalvani A, Allsopp CE, *et al.* (1995) Identification of conserved antigenic components for a cytotoxic T lymphocyte-inducing vaccine against malaria. *Lancet* **345:** 1003–1007.

Alon Y, Hammerling GJ, Segal S, Bar-Eli M. (1987) Association in the expression of Kirsten-ras oncogene and the major histocompatibility complex class I antigens in fibrosarcoma tumor cell variants exhibiting different metastatic capabilities. *Cancer Res.* **47:** 2553–2557.

Andersson M, Paabo S, Nilsson T, Peterson PA. (1985) Impaired intracellular transport of class I MHC antigens as a possible means for adenovirus to evade immune surveillance. *Cell* **43:** 215–222.

Anichini A, Fossati G, Parmiani G. (1986) Heterogeneity of clones from a human metastatic melanoma detected by autologous cytotoxic T lymphocyte clones. *J. Exp. Med.* **163:** 215–220.

Argaet VP, Schmidt CW, Burrows SR, *et al.* (1994) Dominant selection of an invariant T cell antigen receptor in response to persistent infection by Epstein-Barr virus. *J. Exp. Med.* **180:** 2335–2340.

Askonas BA, McMichael AJ, Webster RG. (1982) The immune response to influenza virus and the problem of protection against infection. In: *Basic and Applied Influenza Research* (eds AS Beare). CRC Press, Boca Raton, Florida, pp. 157–188.

Baier M, Werner A, Bannert N, Metzner K, Kurth R. (1995) HIV suppression by interleukin-16. *Nature* **378:** 563.

Bangham CR, McMichael AJ. (1986) Specific human cytotoxic T cells recognize B-cell lines persistently infected with respiratory syncytial virus. *Proc. Natl Acad. Sci. USA* **83:** 9183–9187.

Beersma MF, Bijlmakers MJ, Ploegh HL. (1993) Human cytomegalovirus down-regulates HLA class I expression by reducing the stability of class I H chains. *J. Immunol.* **151:** 4455–4464.

Benjamin R, Parham P. (1990) Guilt by association: HLA B27 and ankylosing spondylitis. *Immunology Today* **11:** 137–142.

Bennink JR, Yewdell JW, Smith GL, Moss B. (1986) Anti-influenza cytotoxic T lymphocytes recognise the three viral polymerases and a non-structural protein: responsiveness to individual viral antigens is MHC controlled. *J. Virol.* **61:** 1098–1102.

Bertoletti A, Sette A, Chisari FV, Penna A, Levrero M, De Carli M, Fiaccadori F, Ferrari C. (1994) Natural variants of cytotoxic epitopes are T-cell receptor antagonists for antiviral cytotoxic T cells [see comments]. *Nature* **369:** 407–410.

Bicknell DC, Rowan A, Bodmer WF. (1994) Beta 2-microglobulin gene mutations: a study of established colorectal cell lines and fresh tumors. *Proc. Natl Acad. Sci. USA* **91:** 4751–4756.

Biddison WE, Shearer GM, Chang TW. (1981) Regulation of influenza virus specific cytotoxic T cell responses by monoclonal antibodies. *J. Immunol.* **127:** 487–491.

Bjorkman P, Saper M, Samraoui B, Bennett W, Strominger J, Wiley D. (1987a) The foreign antigen binding site and T cell recognition regions of class I histocompatibility antigens. *Nature* 329: 512–519.

Bjorkman P, Saper M, Samraoui B, Bennett W, Strominger J, Wiley D. (1987b) Structure of human class I histocompatibility antigen, HLA-A2. *Nature* **329:** 506–511.

Blanden RV. (1974) T cell response to viral and bacterial infection. *Transplant. Rev.* **19:** 56–88.

Bodmer WF, McMichael AJ. (1992) Introduction: A new look at cancer immunology. *Cancer Surveys* **13:** 1–4.

Boel P, Wildmann C, Sensi ML, Brasseur R, Renauld JC, Coulie P, Boon T, van der Bruggen P. (1995) BAGE: a new gene encoding an antigen recognized on human melanomas by cytolytic T lymphocytes. *Immunity* **2:** 167–175.

Bonneau RH, Jennings SR. (1990) Herpes simplex virus-specific cytolytic T lymphocytes restricted to a normally low responder H-2 allele are protective in vivo. *Virology* **174:** 599–604.

Borysiewicz LK, Morris SM, Page J, Sissons JG. (1983) Human cytomegalovirus-specific cytotoxic T-lymphocytes: requirements for in vitro generation and specificity. *Eur. J. Immunol.* **13:** 804–809.

Buchmeier MJ, Welsh RM, Dutko FJ, Oldstone MBA. (1980) The virology and immunology of lymphocytic choriomeningitis virus infection. *Adv. Immunol.* **30:** 275–331.

Burrows SR, Khanna R, Burrows JM, Moss DJ. (1994) An alloresponse in humans is dominated by cytotoxic T lymphocytes (CTL) cross-reactive with a single Epstein-Barr virus CTL epitope: implications for graft-versus-host disease. *J. Exp. Med.* **179:** 1155–1161.

Cannon MJ, Openshaw PJM, Askonas B. (1988) Cytotoxic cells clear virus but augment lung pathology in mice infected with respiratory syncytial virus. *J. Exp. Med.* **168:** 1163–1168.

Carmichael A, Jin X, Sissons P, Borysiewicz L. (1993) Quantitative analysis of the human HIV-1 specific cytotoxic T lymphocyte (CTL) response at different stages of infection: Differential CTL responses to HIV-1 and Epstein-Barr virus in late disease. *J. Exp. Med.* **177:** 249–256.

Cerottini J-C, Brunner KT. (1974) Cell mediated cytotoxicity, allograft rejection and tumour immunity. *Adv. Immunol.* **18:** 67–132.

Cheynier R, Henrichwark S, Hadida F, Pelletier E, Oksenhendler E, Autran B, Wain Hobson S. (1994) HIV and T cell expansion in splenic white pulps is accompanied by infiltration of HIV-specific cytotoxic T lymphocytes. *Cell* **78:** 373–387.

Clerici M, Shearer GM. (1993) A TH1 to Th2 switch is a critical step in the etiology of HIV infection. *Immunol. Today* **14:** 107–110.

Cocchi F, DeVico AL, Garzino-Demo A, Arya SK, Gallo RC, Lusso P. (1995) Identification of RANTES, MIP-1α and MIP-1β as the major HIV-suppressive factors produced by CD8$^+$ T cells. *Science* **270:** 1811–1815.

Coffin JM. (1990) Genetic variation in retroviruses. In: *Virus Variability Epidemiology, and Control* (eds E Kurstak, RG Marusyk, FA Murphy, MHV VanRegenmortel). Plenum Publishing Corporation, New York, pp. 11–33.

Cohen JI. (1991) Epstein-Barr virus lymphoproliferative disease associated with acquired immunodeficiency. *Medicine* **70:** 137–160.

Crawford DH, Thomas JA, Janossy G, Sweny P, Fernando ON, Moorgead JF, Thompson JH. (1980) Epstein-Barr virus nuclear antigen positive lymphoma after cyclosporin A treatment in patients with renal allograft. *Lancet* **1:** 1355–1356.

De Campos Lima P, Gavioli R, Zhang QJ, Wallace LE, Dolcetti R, Rowe M, Rickinson AB,

Masucci MG. (1993) HLA-A11 epitope loss isolates of Epstein-Barr virus from a highly A11+ population. *Science* **260:** 98–100.

De Plaen E, Lurquin C, Van Pel A, Mariame B, Szikora JP, Wolfel T, Sibille C, Chomez P, Boon T. (1988) Immunogenic (tum-) variants of mouse tumor P815: cloning of the gene of tum-antigen P91A and identification of the tum- mutation. *Proc. Natl Acad. Sci. USA* **85:** 2274–2227.

Degiovanni G, Lahaye T, Herin M, Hainaut P, Boon T. (1988) Antigenic heterogeneity of human melanoma tumour detected by autologous CTL clones. *Eur. J. Immunol.* **18:** 671–676.

Doolan DL, Khamboonruang C, Beck HP, Houghten RA, Good MF. (1993) Cytotoxic T lymphocyte (CTL) low-responsiveness to the Plasmodium falciparum circumsporozoite protein in naturally-exposed endemic populations: analysis of human CTL response to most known variants. *Int. Immunol.* **5:** 37–46.

Driscoll J, Brown MG, Finlay D, Monaco JJ. (1993) MHC-linked LMP gene products specifically alter peptidase activities of the proteasome. *Nature* **365:** 262–264.

Eichelberger M, Allan W, Zijlstra M, Jaenisch R, Doherty PC. (1991) Clearance of influenza virus respiratory infection in mice lacking class I major histocompatibility complex-restricted CD8+ T cells. *J. Exp. Med.* **174:** 875–880.

Elliott T, Driscoll P, Smith M, McMichael AJ. (1993) Peptide epitope selection by class I MHC molecules. *Current Biology* **3:** 854–866.

Elliott T, Willis A, Cerundolo V, Townsend A. (1995) Processing of major histocompatibility class I-restricted antigens in the endoplasmic reticulum. *J. Exp. Med.* **181:** 1481–1491.

Falk K, Rötzschke O, Stevanovic S, Jung G, Rammensee H-G. (1991) Allele-specific motifs revealed by sequencing of self-peptides eluted from MHC molecules. *Nature* **351:** 290–296.

Ferrini S, Biassoni R, Moretta A, Buzzone M, Nicolin A, Moretta L. (1985) Clonal analysis of T lymphocytes isolated from ovarian carcinoma ascitic fluid. Phenotype and functional characterization of T cell clones capable of lysing autologous carcinoma cells. *Int. J. Cancer* **36:** 337–343.

Fishel R, Lescoe MK, Rao MR, Copeland NG, Jenkins NA, Garber J, Kane M, Kolodner R. (1993) The human mutator gene homolog MSH2 and its association with hereditary nonpolyposis colon cancer. *Cell* **75:** 1027–1038.

Fleischer B, Kreth HW. (1983) Clonal expansion and functional analysis of virus-specific T lymphocytes from cerebrospinal fluid in measles encephalitis. *Hum. Immunol.* **7:** 239–248.

Fleischer B, Fleischer S, Maier K, Wiedmann KH, Sacher M, Thaler H, Vallbracht A. (1990) Clonal analysis of infiltrating T lymphocytes in liver tissue in viral hepatitis A. *Immunology* **69:** 14–19.

Gallimore A, Cranage N, Cook N, *et al.* (1995) Early suppression of SIV replication by CD8+ nef specific cytotoxic T lymphocytes in infected macaques. *Nature Medicine* **1:** 1167–1173.

Gao XM, Tite JP, Lipscombe M, Rowland JS, Ferguson DJ, McMichael AJ. (1992) Recombinant Salmonella typhimurium strains that invade nonphagocytic cells are resistant to recognition by antigen-specific cytotoxic T lymphocytes. *Infect. Immun.* **60:** 3780–3789.

Gavioli R, Kurilla MG, de Campos Lima PO, Wallace LE, Dolcetti R, Murray RJ, Rickinson AB, Masucci MG. (1993) Multiple HLA A11-restricted cytotoxic T-lymphocyte epitopes of different immunogenicities in the Epstein-Barr virus-encoded nuclear antigen 4. *J. Virol.* **67:** 1572–1578.

Goddeeris BM, Morrison WI, Teale AJ, Bensaid A, Baldwin CL. (1986) Bovine cytotoxic T-cell clones specific for cells infected with the protozoan parasite Theileria parva: parasite strain specificity and class I major histocompatibility complex restriction. *Proc. Natl Acad. Sci. USA* **83:** 5238–5242.

Gotch FM, McMichael AJ, Smith GL, Moss B. (1987) Identification of the virus molecules recognised by influenza specific cytotoxic T lymphocytes. *J. Exp. Med.* **165:** 408–416.

Gotch FM, McMichael AJ, Rothbard J. (1988) Recognition of influenza A matrix protein by HLA-A2 restricted cyotoxic T lymphocytes. Use of analogues to orientate the matrix peptide in the HLA A2 binding site. *J. Exp. Med.* **168:** 2045–2058.

Gotch FM, McAdam S, Allsopp C, Gallimore A, Elvin J, Kieny M-P, Hill A, McMichael AJ, Whittle H. (1993) Cytotoxic T cells in HIV.2 seropositive Gambians - Identification of a virus specific MHC restricted peptide epitope. *J. Immunol.* **151:** 3361–3369.

Graham MB, Braciale VL, Braciale TJ. (1994) Influenza virus-specific CD4+ T helper type 2 T lymphocytes do not promote recovery from experimental virus infection. *J. Exp. Med.* **180:** 1273–1282.

Gregory CD, Murray RJ, Edwards CF, Rickinson AB. (1988) Downregulation of cell adhesion molecules LFA-3 and ICAM-1 in Epstein-Barr virus-positive Burkitts lymphoma underlies tumor cell escape from virus-specific T cell surveillance. *J. Exp. Med.* **167:** 1811–1824.

Hermann E, Yu DT, Meyer zum Buschenfelde KH, Fleischer B. (1993) HLA-B27-restricted

CD8 T cells derived from synovial fluids of patients with reactive arthritis and ankylosing spondylitis. *Lancet* **342:** 646–650.

Hill AVS, Allsopp CEM, Kwiatkowski D, *et al.* (1991) Common West African HLA antigens are associated with protection from severe malaria. *Nature* **352:** 595–600.

Hill AV, Elvin J, Willis AC, *et al.* (1992) Molecular analysis of the association of HLA-B53 and resistance to severe malaria. *Nature* **360:** 434–439.

Hill AV, Yates SN, Allsopp CE, Gupta S, Gilbert SC, Lalvani A, Aidoo M, Davenport M, Plebanski M. (1994) Human leukocyte antigens and natural selection by malaria. *Phil. Trans. R. Soc. Lond. B. Biol. Sci.* **346:** 379–385.

Hill A, Jugovic P, York I, Russ G, Bennink J, Yewdell J, Ploegh H, Johnson D. (1995) Herpes simplex virus turns off the TAP to evade host immunity. *Nature* **375:** 411–415.

Ho DD, Neumann AU, Perelson AS, Chen W, Leonard JM, Markowitz M. (1995) Rapid turnover of plasma virions and CD4 lymphocytes in HIV-1 infection. *Nature* **373:** 123–126.

Hoffman SL, Weiss W, Mellouk S, Sedegah M. (1990) Irradiated sporozoite vaccine induces cytotoxic T lymphocytes that recognize malaria antigens on the surface of infected hepatocytes. *Immunol. Lett.* **25:** 33–38.

Hoskins TW, Davies JR, Smith AJ, Miller CL, Allchin A. (1979) Assessment of inactivated influenza A vaccine after three outbreaks of influenza A at Christs Hospital. *Lancet* **i:** 33–35.

Hou S, Doherty PC, Zijlstra M, Jaenisch R, Katz JM. (1992) Delayed clearance of Sendai virus in mice lacking class I MHC-restricted CD8+ T cells. *J. Immunol.* **149:** 1319–1325.

Itescu S, Mathur Wagh U, Skovron ML, Brancato LJ, Marmor M, Zeleniuch Jacquotte A, Winchester R. (1992) HLA-B35 is associated with accelerated progression to AIDS. *J. Acquir. Immune Defic. Syndr.* **5:** 37–45.

Jacobson S, Richert JR, Biddison WE, Satinsky A, Hartzman RJ, McFarland HF. (1984) Measles virus-specific T4+ human cytotoxic T cell clones are restricted by class II HLA antigens. *J. Immunol.* **133:** 754–757.

Kannagi M, Sugamura K, Sato H, Okochi K, Uchino H, Hinuma Y. (1983) Establishment of human cytotoxic T cell lines specific for human adult T cell leukaemia virus-bearing cells. *J. Immunol.* **130:** 2942–2946.

Kast WM, Bronkhorst AM, de Waal LP, Melief CJ. (1986) Cooperation between cytotoxic and helper T lymphocytes is protective against lethal Sendai virus infection. *J. Exp. Med.* **164:** 723–738.

Kast WM, Brandt RM, Drijfhout JW, Melief CJ. (1993). Human leukocyte antigen-A2.1 restricted candidate cytotoxic T lymphocyte epitopes of human papillomavirus type 16 E6 and E7 proteins identified by using the processing-defective human cell line T2. *J. Immunother.* **14:** 115–120.

Klein MR, van Baalen CA, Holwerda AM, Kerkhof Garde SR, Bende RJ, Keet IP, Eeftinck Schattenkerk JK, Osterhaus AD, Schuitemaker H, Miedema F. (1995) Kinetics of Gag-specific cytotoxic T lymphocyte responses during the clinical course of HIV-1 infection: a longitudinal analysis of rapid progressors and long-term asymptomatics. *J. Exp. Med.* **181:** 1365–1372.

Klenerman P, Rowland-Jones S, McAdam S, *et al.* (1994) Naturally occurring HIV-1 gag variants antagonise cytotoxic T cell activity. *Nature* **369:** 403–407.

Knuth A, Wolfel T, Klehmann E, Boon T, Meyer-zum-Buschenfeld KD. (1989) Cytolytic T-cell clones against an autologous human melanoma: Specificity study and definition of three antigens by immunoselection. *Proc. Natl Acad. Sci. USA* **86:** 2804–2808.

Koelle DM, Tigges MA, Burke RL, Symington FW, Riddell SR, Abbo H, Corey L. (1993) Herpes simplex virus infection of human fibroblasts and keratinocytes inhibits recognition by cloned CD8+ cytotoxic T lymphocytes. *J. Clin. Invest.* **91:** 961–968.

Koenig S, Conley AJ, Brewan YA, *et al.* (1995) Transfer of HIV-1 specific cytotoxic T lymphocytes to an AIDS patient leads to selection for mutant HLA variants and subsequent disease progression. *Nature Medicine* **1:** 330–336.

Koup R, Safrit JT, Cao Y, Andrews CA, McLeod G, Borkowsky W, Farthing C, Ho DD. (1994) Temporal association of cellular immune responses with the initial control of viraemia in primary human immunodeficiency virus type 1 syndrome. *J. Virol.* **68:** 4650–4655.

Koziel MJ, Dudley D, Afdhal N, Choo QL, Houghton M, Ralston R, Walker BD. (1993) Hepatitis C virus (HCV)-specific cytotoxic T lymphocytes recognize epitopes in the core and envelope proteins of HCV. *J. Virol.* **67:** 7522–7532.

Kress HG, Kreth HW. (1982) HLA restriction of secondary mumps-specific cytotoxic T lymphocytes. *J. Immunol.* **129:** 844–849.

Kreth HW, Kress L, Kress HG, Ott HF, Eckert G. (1982) Demonstration of primary cytotoxic T cells in venous blood and cerebrospinal fluid in children with mumps meningitis. *J. Immunol.* **128:** 2411–2415.

Li CR, Greenberg PD, Gilbert MJ, Goodrich JM, Riddell SR. (1994) Recovery of HLA-restricted cytomegalovirus (CMV)-specific T-cell responses after allogeneic bone marrow transplant: correlation with CMV disease and effect of ganciclovir prophylaxis. *Blood* **83:** 1971–1979.

Lin Y, Askonas BA. (1981) Biological properties of an influenza A virus specific killer T cell clone. *J. Exp. Med.* **154:** 225–234.

Levitskaya J, Coram M, Levitsky V, Imreh S, Steigerwald Mullen PM, Klein G, Kurilla MG, Masucci MG. (1995) Inhibition of antigen processing by the internal repeat region of the Epstein-Barr virus nuclear antigen-1. *Nature* **375:** 685–688.

Lukacher AE, Braciale VL, Braciale TJ. (1984) In vivo effector function of influenza virus specific cytotoxic T lymphocyte clones is highly specific. *J. Exp. Med.* **160:** 814–826.

Malik A, Egan JE, Houghten RA, Sadoff JC, Hoffman SL. (1991) Human cytotoxic T lymphocytes against the Plasmodium falciparum circumsporozoite protein. *Proc. Natl Acad. Sci. USA* **88:** 3300–3304.

Martinon F, Gomard E, Hannoun C, Levy JP. (1990) In vitro human cytotoxic T cell responses against influenza A virus can be induced and selected by synthetic peptides. *Eur. J. Immunol.* **20:** 2171–2176.

Maryanski JL, Van Snick J, Cerottini J-C, Boon T. (1982) Immunogenic variants obtained by mutagenesis of mouse mastocytoma P815. III. Clonal analysis of the syngeneic cytolytic T lymphocyte response. *Eur. J. Immunol.* **12:** 401–406.

McMichael AJ. (1980) HLA restriction of human cytotoxic T-cells. *Springer Sem. Immunopathol.* **3:** 3–22.

McMichael AJ, Walker B. (1994) Cytotoxic T lymphocyte epitopes: implications for HIV vaccines. *AIDS* **8:** S155–S173.

McMichael AJ, Ting A, Zweerink HJ, Askonas BA. (1977) HLA restriction of cell mediated lysis of influenza virus infected human cells. *Nature* **270:** 524–526.

McMichael AJ, Gotch FM, Dongworth DW, Clark A, Potter CW. (1983a) Declining T cell immunity to influenza 1977-82. *Lancet* **ii:** 762–764.

McMichael AJ, Gotch FM, Noble GR, Beare PAS. (1983b) Cytotoxic T-cell immunity to influenza. *N. Engl. J. Med.* **309:** 13–17.

McMichael AJ, Gotch F, Rothbard J. (1986) HLA B37 determines an influenza A virus nucleoprotein epitope recognized by cytotoxic T lymphocytes. *J. Exp. Med.* **164:** 1397–1406.

McMichael AJ, Bodmer WF. (eds) (1992) A new look at cancer immunology. *Cancer Surveys* **13.**

Melief CJ, Kast WM. (1991) Cytotoxic T lymphocyte therapy of cancer and tumor escape mechanisms. *Semin. Cancer Biol.* **2:** 347–354.

Missale G, Redeker A, Person J, Fowler P, Guilhot S, Schlicht H. J, Ferrari C, Chisari FV. (1993) HLA-A31- and HLA-Aw68-restricted cytotoxic T cell responses to a single hepatitis B virus nucleocapsid epitope during acute viral hepatitis. *J. Exp. Med.* **177:** 751–762.

Momberg F, Ziegler A, Harpprecht J, Moller P, Moldenhauer G, Hammerling GJ. (1989) Selective loss of HLA-A or HLA-B antigen expression in colon carcinoma. *J. Immunol.* **142:** 352–358.

Morris AG, Lin Y-L, Askonas BA. (1982) Immune interferon release when a cloned cytotoxic T cell meets its correct influenza-infected target cell. *Nature* **295:** 150–152.

Moskophidis D, Laine E, Zinkernagel RM. (1993) Peripheral clonal deletion of antiviral memory CD8+ T cells. *Eur. J. Immunol.* **23:** 3306–3311.

Moss DJ, Rickinson AB, Pope JH. (1978) Long term T cell mediated immunity to Epstein-Barr virus in man. I. Complete regression of virus-induced transformation on cultures of seropositive donor leukocytes. *Int. J. Cancer* **22:** 662–668.

Moss PAH, Rowland-Jones SL, Frodsham P, McAdam S, Giangrande P, McMichael AJ, Bell JI. (1995) Persistent high frequency of human immunodeficiency virus specific cytotoxic T cells in peripheral blood of infected donors. *Proc. Natl Acad. Sci USA* **92:** 5773–5777.

Mukherji B, McAlister TJ. (1983) Clonal analysis of cytotoxic T cell response against human melanoma. *J. Exp. Med.* **158:** 240–245.

Neefjes J, Momberg F, Hammerling G. (1993) Selective and ATP-dependent translocation of peptides by the MHC-encoded transporter. *Science* **261:** 769–771.

Niewiesk S, Daenke S, Parker CE, Taylor G, Weber J, Nightingale S, Bangham CR. (1995) Naturally occurring variants of human T-cell leukemia virus type I Tax protein impair its recognition by cytotoxic T lymphocytes and the transactivation function of Tax. *J. Virol.* **69:** 2649–2653.

Norment AM, Salter RD, Parham P, Engelhard VH, Littman DR. (1988) Cell-cell adhesion mediated by CD8 and MHC class I molecules. *Nature* **336:** 79–82.

Nowak M, McMichael AJ. (1995) How HIV defeats the immune system. *Sci. Am.* **273:** 42–49.

Nowak M, May RM, Phillips RE, *et al.* (1995) Antigenic oscillations and shifting immunodominance in HIV-1 infections. *Nature* **375:** 606–611.

Pamer EG, Harty JT, Bevan MJ. (1991) Precise prediction of a dominant class I MHC-restricted epitope of Listeria monocytogenes. *Nature* **353:** 852–855.

Pantaleo G, Demarest JF, Soudeyns H, *et al.* (1994) Major expansion of CD8+ T cells with a predominant V beta usage during the primary immune response to HIV. *Nature* **370:** 463–467.

Papadopoulos EB, Ladanyi M, Emanuel D, *et al.* (1994) Infusions of donor leukocytes to treat Epstein–Barr virus-associated lymphoproliferative disorders after allogeneic bone marrow transplantation. *N. Engl. J. Med.* **330:** 1185–1191.

Phillips RE, Rowland-Jones S, Nixon DF, *et al.* (1991) Human immunodeficiency virus genetic variation that can escape cytotoxic T cell recognition. *Nature* **354:** 453–459.

Pignatelli M, Waters J, Lever A, Iwarson S, Gerety R, Thomas HC. (1987) Cytotoxic T cell responses to the nucleocapsid proteins of HBV in chronic hepatitis. Evidence that antibody modulation may cause protracted infection. *J. Hepatol.* **4:** 15–21.

Pircher H, Moskphidis A, Rohrer U, Burki K, Hengartner H, Zinkernagel RM. (1990) Viral escape by selection of cytotoxic T cell-resistant variants in vivo. *Nature* **346:** 629–633.

Plata F, Autran B, Martins LP, Wain-Hobson S, Raphael M, Mayaud C, Denis M, Guillon JM, Debre P. (1987) AIDS virus specific cytotoxic T lymphocytes in lung disorders. *Nature* **328:** 348–351.

Quinnan GV, Kirmani N, Rook AH, Manischewitz JF, Jackson L, Moreschi G, Santos GW, Saral R, Burns WH. (1982). Cytotoxic T cell in cytomegalovirus infection: HLA restricted T lymphocyte and non-T lymphocyte cytotoxic responses correlate with recovery from cytomegalovirus infection on bone-marrow transplant recipients. *N. Engl. J. Med.* **307:** 7–13.

Recs A, Scoging A, Mehlert A, Young DB, Ivanyi J. (1988) Specificity of proliferative response of human CD8 clones to mycobacterial antigens. *Eur. J. Immunol.* **18:** 1881–1887.

Rehermann B, Fowler P, Sidney J, Person J, Redeker A, Brown M, Moss B, Sette A, Chisari FV. (1995) The cytotoxic T lymphocyte response to multiple hepatitis B virus polymerase epitopes during and after acute viral hepatitis. *J. Exp. Med.* **181:** 1047–1058.

Rickinson AB, Moss DJ, Allen DJ, Wallace LE, Rowe M, Epstein MA. (1981) Reactivation of Epstein-Barr virus-specific cytotoxic T cells by in vitro stimulation with autologous lymphoblastoid cell line. *Int. J. Cancer* **27:** 593–601.

Rickinson AB, Murray RJ, Brooks J, Griffin H, Moss DJ, Masucci M. (1992) T cell recognition of Epstein Barr virus associated lymphomas. *Cancer Surveys* **13:** 53–80.

Riddell SR, Walter BA, Gilbert MJ, Greenberg PD. (1994) Selective reconstitution of CD8+ cytotoxic T lymphocyte responses in immunodeficient bone marrow transplant recipients by the adoptive transfer of T cell clones. *Bone Marrow Transplant* **14:** 578–584.

Rodrigues M, Nussenzweig RS, Romero P, Zavala F. (1992) The in vivo cytotoxic activity of CD8+ T cell clones correlates with their levels of expression of adhesion molecules. *J. Exp. Med.* **175:** 895–905.

Romero P, Maryanski JL, Cordey AS, Corradin G, Nussenzweig RS, Zavala F. (1990) Isolation and characterization of protective cytolytic T cells in a rodent malaria model system. *Immunol. Lett.* **25:** 27–31.

Rooney CM, Smith CA, Ng CY, Loftin S, Li C, Krance RA, Brenner MK, Heslop HE. (1995) Use of gene-modified virus-specific T lymphocytes to control Epstein-Barr-virus-related lymphoproliferation. *Lancet* **345:** 9–13.

Rötzschke O, Falk K, Deres K, Schild H, Norda M, Metzger J, Jung G, Rammensee H-G. (1990) Isolation and analysis of naturally processed viral peptides as recognised by cytotoxic T cells. *Nature* **348:** 252–254.

Rowe M, Rowe DT, Gregory CD, Young LS, Farrelli PJ, Rupani H, Rickinson AB. (1987) Differences in B cell growth phenotype reflect novel patterns of Epstein-Barr virus latent gene expression in Burkitt's lymphoma. *EMBO J.* **6:** 2743.

Rowe M, Young LS, Crocker J, Stokes H, Henderson S, Rickinson AB. (1991) Epstein-Barr virus (EBV)-associated lymphoproliferative disease in the SCID mouse model: implications for the pathogenesis of EBV-positive lymphomas in man. *J. Exp. Med.* **173:** 147–158.

Rowland-Jones SL, Sutton J, Ariyoshi K, *et al.* (1995) HIV-specific cytotoxic T cells in HIV-exposed but uninfected Gambian women. *Nature Medicine* **1:** 59–64.

Schoel B, Zugel U, Ruppert T, Kaufmann SH. (1994) Elongated peptides, not the predicted nonapeptide stimulate a major histocompatibility complex class I-restricted cytotoxic T lymphocyte clone with specificity for a bacterial heat shock protein. *Eur. J. Immunol.* **24:** 3161–3169.

Sethi KK, Naher H, Stroehmann I. (1988) Phenotypic heterogeneity of cerebrospinal fluid-derived cytotoxic T cell clones. *Nature* **335:** 178–180.

Sissons JG, Colby SD, Harrison WO, Oldstone MB. (1985) Cytotoxic lymphocytes generated in vivo with acute measles virus infection. *Clin. Immunol. Immunopathol.* **34:** 60–68.

Slinghuff CL, Cox AL, Stover JM, Moore MM, Hunt DF, Englehard VH. (1994) Cytotoxic T-lymphocyte response to autologous human squamous cell cancer of the lung: epitope reconstitution with peptides extracted from HLA Aw68. *Cancer Res.* **54:** 2731–2737.

Smith ME, Marsh SGE, Bodmer JG, Gelsthorpe K, Bodmer WF. (1989) Loss of HLA-A,B,C, allele products and lymphocyte function associated antigen-3 in colorectal neoplasia. *Proc. Natl Acad. Sci. USA* **86:** 5557–5561.

Starnbach MN, Bevan MJ. (1994) Cells infected with Yersinia present an epitope to class I MHC-restricted CTL. *J. Immunol.* **153:** 1603–1612.

Steel CM, Ludlum C, Beatson D, Peutherer JF, Cuthbert RJG, Simmonds P, Morrison H, Jones M. (1988) HLA haplotype A1 B8 DR3 as a risk factor for HIV-related disease. *Lancet* **i:** 1185–1188.

Taylor PM, Esquivel F, Askonas BA. (1990) Murine CD4+ T cell clones vary in function in vitro and in influenza infection in vivo. *Int. Immunol.* **2:** 323-328.

Townsend A, Trowsdale J. (1993) The transporters associated with antigen presentation. *Semin. Cell. Biol.* **4:** 53–61.

Townsend A, Rothbard J, Gotch F, Bahadur B, Wraith D, McMichael A. (1986) The epitopes of influenza nucleoprotein recognized by cytotoxic T lymphocytes can be defined with short synthetic peptides. *Cell* **44:** 959–968.

Townsend A, Elliott T, Cerundolo V, Foster L, Barber B, Tse A. (1990) Assembly of MHC class I molecules analyzed in vitro. *Cell* **62:** 285–295.

Townsend ARM, Gotch FM, Davey J. (1985) Cytotoxic T cells recognize fragments of the influenza nucleoprotein. *Cell* **42:** 457–467.

Traversari C, Van der Bruggen P, Luescher IF, Lurquin C, Van Pel A, De Plaen E, Amar-Costasec A, Boon T. (1993) A nonapeptide encoded by human MAGE-1 is recognized on HLA-A1 by CTL directed against tumor antigen MZ2-E. *J. Exp. Med.* **176:** 1453–1457.

Tussey LG, Rowland Jones S, Zheng TS, Androlewicz MJ, Cresswell P, Frelinger JA, McMichael AJ. (1995) Different MHC class I alleles compete for presentation of overlapping viral epitopes. *Immunity* **3:** 65–77

Vallbracht A, Maier K, Stierhof YD, Wiedmann KH, Flehmig B, Fleischer B. (1989) Liver-derived cytotoxic T cells in hepatitis A virus infection. *J. Infect. Dis.* **160:** 209–217.

van Binnendijk RS, Versteeg van Oosten JP, Poelen MC, Brugghe HF, Hoogerhout P, Osterhaus AD, Uytdehaag FG. (1993) Human HLA class I- and HLA class II-restricted cloned cytotoxic T lymphocytes identify a cluster of epitopes on the measles virus fusion protein. *J. Virol.* **67:** 2276–2284.

Verjans GM, Janssen R, UytdeHaag FG, van Doornik CE, Tommassen J. (1995) Intracellular processing and presentation of T cell epitopes, expressed by recombinant Escherichia coli and Salmonella typhimurium, to human T cells. *Eur. J. Immunol.* **25:** 405–410.

Wahab ZA, Metzgar RS. (1991) Human cytotoxic lymphocytes reactive with pancreatic adenocarcinoma cells. *Pancreas* **6:** 307–317.

Walker CM, Moody DJ, Stites DP, Levy JA. (1986) CD8+ lymphocytes can control HIV infection in vitro by suppressing virus replication. *Science* **234:** 1563–1566.

Walker DB, Chakrabati S, Moss B, et al. (1987) HIV specific cytotoxic T lymphocytes in seropositive individuals. *Nature* **328:** 345–348.

Wallis RS, Ellner JJ, Shiratsuchi H. (1992) Macrophages, mycobacteria and HIV: the role of cytokines in determining mycobacterial virulence and regulating viral replication. *Res. Microbiol.* **143:** 398–405.

Walter EA, Greenberg PD, Gilbert MJ, Finch RJ, Watanabe KS, Thomas ED, Riddell SR. (1995) Reconstitution of cellular immunity against cytomegalovirus in recipients of allogeneic bone marrow by transfer of T-cell clones from the donor. *N. Engl. J. Med.* **333:** 1038–1044.

Wei X, Ghosh SK, Taylor ME, et al. (1995) Viral dynamics in human immunodeficiency virus type 1 infection. *Nature* **373:** 117–122.

Weiss WR, Mellouk S, Houghten RA, Sedegah M, Kumar S, Good MF, Berzofsky JA, Miller

LH, Hoffman SL. (1990) Cytotoxic T cells recognize a peptide from the circumsporozoite protein on malaria-infected hepatocytes. *J. Exp. Med.* **171:** 763–773.

Wells MA, Ennis FA, Albrecht P. (1981) Recovery from a viral respiratory infection. II. Passive transfer of immune spleen cells to mice with influenza pneumonia. *J. Immunol.* **126:** 1042–1046.

Yap KL, Ada GL. (1978a) Cytotoxic T cells in the lungs of mice infected with influenza A virus. *Scand. J. Immunol.* **7:** 73–80.

Yap KL, Ada GL. (1978b) Transfer of specific cytotoxic T lymphocytes protects mice inoculated with influenza virus. *Nature* **273:** 238–240.

Yasumura S, Weidmann E, Hirabayashi H, Johnson JT, Herberman RB, Whiteside TL. (1994) HLA restriction and T-cell-receptor V beta gene expression of cytotoxic T lymphocytes reactive with human squamous-cell carcinoma of the head and neck. *Int. J. Cancer* **57:** 297–305.

Yasutomi Y, Koenig S, Woods RM, *et al.* (1995) A vaccine-elicited, single viral epitope-specific cytotoxic T lymphocyte response does not protect against intravenous, cell-free simian immunodeficiency virus challenge. *J. Virol.* **69:** 2279–2284.

Yewdell JW, Bennink JR, Smith GL, Moss B. (1985) Influenza A virus nucleoprotein is a major target for cross-reactive anti-influenza A virus specific cytotoxic T lymphocytes. *Proc. Natl Acad. Sci. USA* **82:** 1785–1789.

York IA, Roop C, Andrews DW, Riddell SR, Graham FL, Johnson DC. (1994) A cytosolic herpes simplex virus protein inhibits antigen presentation to CD8+ T lymphocytes. *Cell* **77:** 525–535.

Yoshino I, Goedegebuure PS, Peoples GE, Parikh AS, DiMaio JM, Lyerly HK, Gazdar AF, Eberlein TJ. (1994) HER2/neu-derived peptides are shared antigens among human non-small cell lung cancer and ovarian cancer. *Cancer Res.* **54:** 3387–3390.

Zakut R, Topalian SL, Kawakami Y, Mancini M, Eliyahu S, Rosenberg SA. (1993) Differential expression of MAGE-1, -2, and -3 messenger RNA in transformed and normal human cell lines. *Cancer Res.* **53:** 5–8.

Zhang M, Gong J, Iyer DV, Jones BE, Modlin RL, Barnes PF. (1994). T cell cytokine responses in persons with tuberculosis and human immunodeficiency virus infection. *J. Clin. Invest.* **94:** 2435–2442.

Zinkernagel RM. (1974) Restriction by H-2 gene complex of transfer of cell-mediated immunity to Listeria monocytogenes. *Nature* **251:** 230–233.

Zinkernagel RM, Doherty PC. (1975) H-2 compatibility requirement for T-cell mediated lysis of target cells infected with lymphocytic choriomeningitis virus. *J. Exp. Med.* **141:** 1427–1436.

Zinkernagel RM, Doherty PC. (1979) MHC-restricted cytotoxic T cells: studies on the biological role of polymorphic major transplantation antigens determining T cell restriction-specificity, function and responsiveness. *Adv. Immunol.* **27:** 51–177.

Zinkernagel RM, Hengartner H. (1994). T-cell-mediated immunopathology versus direct cytolysis by virus: implications for HIV and AIDS. *Immunol. Today* **15:** 262–268.

Zweerink HJ, Askonas BA, Millican D, Courtneidge SA, Skehel JJ. (1977) Cytotoxic T lymphocytes to type A influenza virus: viral haemagglutinin induces strain A specificity while infected cells confer cross-reactive cytotoxicity. *Eur. J. Immunol.* **7:** 630–635.

14

Class II-restricted T cell function

Knut E.A. Lundin, Ludvig M. Sollid and Erik Thorsby

14.1 Introduction

T cells are said to be restricted by major histocompatibility complex [MHC; in man human leukocyte antigen (HLA)] molecules in their recognition of antigen. The reason is that T cells can only recognize antigen, or rather peptide-fragments of antigen, when presented by MHC molecules. T cells restricted by MHC class II molecules mainly belong to the $CD4^+$, T cell receptor (TCR) α/β^+ T cell subset, although it has been shown that some $CD8^+$ T cells and some $TCR\gamma/\delta^+$ T cells may also be restricted by class II molecules. Here we discuss $CD4^+$ T cell recognition of peptides presented by class II molecules and the function of such class II-restricted T cells. Emphasis will be put on HLA class II-restricted human T cells and some aspects of their function that we have been particularly interested in. Furthermore, we discuss the role of class II-restricted T cells in HLA-associated diseases and how they may be manipulated as part of immunological intervention.

14.2 Polymorphism of MHC class II molecules

An important feature of the MHC class II (as well as class I) molecules is their genetic diversity. In humans there exist three polymorphic series of class II molecules, which are termed HLA-DR, -DQ and -DP. In the mouse, only two series of class II molecules exist, I-A and I-E. The genetic organization and diversity of the MHC class II region and the three-dimensional structure of class II molecules is dealt with in detail elsewhere in this book (see Chapters 2, 5 and 11). It is important to note, however, that most of the polymorphic residues of class II molecules are situated in or around the peptide-binding cleft, where they may interact with a bound peptide or with the TCR (Madden, 1995; Stern *et al.*, 1994).

14.3 Expression of class II molecules

Class II molecules are constitutively expressed on cells which may serve as antigen presenting cells (APCs) for $CD4^+$ T cells, such as macrophages, monocytes, dendritic cells and B cells (Daar *et al.*, 1984). However, class II molecules may be induced on many other cell types which normally are class II negative, by, for instance, interferon-γ (IFN-γ) and tumour necrosis factor (TNF). Control of MHC class II gene expression is discussed elsewhere in this book (see Chapter 8),

HLA and MHC: genes, molecules and function, edited by M.J. Browning and A.J. McMichael.
© 1996 BIOS Scientific Publishers Ltd, Oxford.

and involves genomic elements upstream of the structural MHC genes (Woolfrey and Nepom, 1995).

Concomitant expression of all three classes of HLA class II molecules is often found. Exceptions to this exist, however; for instance, some B cells express HLA-DQ, but not -DR and -DP molecules (Ono *et al.*, 1991), or -DR but not -DQ molecules (Woolfrey and Nepom, 1995). This is also the case for other cells such as some peripheral blood mononuclear cells (Freudenthal and Steinman, 1990; Gonwa *et al.*, 1986) and some epithelial cells of the small intestinal mucosa (Scott *et al.*, 1987).

14.4 CD4$^+$ T cells recognize peptide–MHC class II complexes

During the early 1970s, several animal studies demonstrated that antigen-specific T cell help of B cells (Kindred and Shreffler, 1972), activation of helper T cells by APC (Rosenthal and Shevach, 1973) and killing of target cells by antigen-specific T cells (Zinkernagel and Doherty, 1974) required histocompatibility between the T cells and the interacting cells. Sharing of the MHC class II region was found to be important for CD4$^+$ (helper) T cell interactions, and sharing of the MHC class I region to be important for CD8$^+$ (cytotoxic) T cell interactions. This phenomenon was called MHC restriction of antigen-specific T cell responses.

The mechanism responsible for MHC restriction remained an enigma until X-ray crystallography studies of the HLA-A2 class I molecule revealed that it had a peptide-binding cleft (Bjorkman *et al.*, 1987). Thus MHC molecules are peptide-binding receptors that collect peptides inside the cell and transport them to the cell surface, where the complex of a peptide and an MHC molecule may be recognized by the TCR. Corresponding X-ray crystallography studies of HLA-DR1 later showed that the MHC class II peptide-binding cleft is similar to that of class I molecules, with the exception that it is more open at both ends allowing the bound peptide to protrude out of the cleft (Stern *et al.*, 1994). Class II binding peptides are bound in an extended conformation. The core region of bound peptide that directly interacts with the class II molecule comprises about 9–10 amino acids. Studies of natural T cell epitopes, by the use of both proteolytic fragments of antigens and synthetic peptides, indicate that peptides recognized by class II-restricted T cells are usually 10–17 amino acids long, and this fits well with the size of peptides that can be eluted from class II molecules (Chicz *et al.*, 1993). The specificity in the binding of peptides is determined by interactions between polymorphic amino acids of the MHC molecule that form binding pockets in its peptide-binding cleft, and side-chains of so-called anchor residues in the peptide. The characteristics, the number and the spacing of the anchor residues of peptides bound to a given MHC molecule constitute a peptide-binding motif which is unique for each MHC variant molecule (Hammer, 1995; Rammensee, 1995 and see Chapter 12).

When a T cell recognizes a peptide–HLA complex by its TCR, this normally will result in a signal for T-cell activation. How the TCR recognizes the peptide–MHC complex and how this again results in T-cell activation is dealt with later in this chapter.

14.5 The origin of peptides bound by class II molecules

Generation of class II-bound peptides is dependent on digestion and processing of the antigen by the APC (Ziegler and Unanue, 1982). This is discussed in detail by Neefjes and colleagues elsewhere in this book (see Chapter 10). Briefly, antigen processing takes place in endosomal and prelysosomal compartments. Peptide binding mainly occurs in specialized organelles, which are termed the MHC class II compartment. The transport of newly synthesized class II molecules from the endoplasmic reticulum (ER) to this compartment is aided by the invariant chain. In the MHC class II compartment, HLA-DM molecules participate in unloading invariant chain-derived fragments (so-called CLIP peptides) and loading of the class II molecules with peptides generated by the enzymatic processing of proteins (Denzin and Cresswell, 1995). A surprisingly large fraction of the peptides bound by class II molecules is derived from the APC itself (Chicz *et al.*, 1993). Only in some very few cases has it been possible to identify peptides from an externally added antigen among the peptides that were eluted from the class II molecules (Nelson *et al.*, 1992). Notably, peptides derived from the HLA molecules of the APC themselves are commonly bound by their class II molecules (Chicz *et al.*, 1993), and peptides derived from one HLA molecule of an APC can be presented by another HLA molecule of the same cell (de Koster *et al.*, 1989). $CD4^+$ T cell-recognition of peptides derived from allogeneic APC themselves may thus be important for allorecognition (see Chapter 17).

14.6 Protein antigens as immunogens for class II-restricted T cells

In protein antigens there may exist dominant and cryptic T cell epitopes (Sercarz *et al.*, 1993), the former being mainly recognized after immunization with intact protein antigens. It is probably the APC that determines whether epitopes are dominant or cryptic, but this phenomenon has a major impact on T cell function. Epitope dominance and crypticity has been demonstrated most convincingly in mice; for instance, after immunization with intact hen egg lysozyme, the T cell response in H-2^a mice (expressing both the I-A^a and I-E^a class II molecules) focuses on one particular fragment of hen egg lysozyme presented by I-A^a, although other epitopes may be recognized if other fragments are used as immunogens. There are several possible mechanisms leading to epitope dominance (Perkins *et al.*, 1991). High-affinity binding peptides may be protected against further proteolytic digestion (Mouritsen *et al.*, 1992), and capture of one determinant could lead to increased digestion of other epitopes (Deng *et al.*, 1993).

Human T cell responses towards foreign (or self) antigens appear also to be influenced by dominance and crypticity of epitopes. Primary stimulation *in vitro* with short peptides induces a broader epitope reactivity pattern than when T cells are first stimulated with an intact protein antigen (Manca *et al.*, 1991; Markovic-Plese *et al.*, 1995; Matsuo *et al.*, 1995). Stimulation of human T cells with short peptides to define T cell epitopes within a given antigen may thus not be the ideal strategy for definition of dominant epitopes. Crypticity of a given epitope is probably not a static phenomenon; for example, HIV gp120 down-regulates cell-surface expression of intact CD4 molecules but, at the same time, increases the presentation of processed CD4 epitopes by the same APC to specific T cells (Salemi *et al.*, 1995).

14.7 Role of different cells as APC

Different cell types expressing class II molecules may not always be able to induce the same function of T cells. At least three different levels, all involving peptide–MHC class II complexes, must be considered during the activation of $CD4^+$ T cells:

(i) developing thymocytes interact with peptide–MHC class II complexes during their thymic development and education;
(ii) naive T cells that have entered circulation are activated by recognition of peptide–MHC class II complexes on specialized APC (which are able also to deliver additional signals necessary for activation of naive T cells), and may thereby mature to effector and, in some cases, to memory T cells;
(iii) the effector T cells may interact with peptide–MHC class II complexes on any class II-positive target cells.

During thymic education, the thymocytes are subject to positive and negative selection (Davis and Bjorkman, 1988; Marrack and Kappler, 1988). T cells with TCR that recognize self-peptide–MHC complexes with high affinity are clonally deleted (i.e. negative selection), whereas the positive selection ensures that T cells in the periphery may recognize antigenic peptides bound to self MHC molecules. These processes probably take place mainly in distinct anatomical compartments in the thymus and are exerted by different cell types.

Studies with TCR transgenic mice indicate that dendritic cells are most efficient at activating naive T cells (Croft *et al.*, 1992). Dendritic cells appear also to have some degree of antigen-specificity, since they can use the mannose receptor for macropinocytosis and can therefore efficiently take up infectious non-self antigens (Sallusto *et al.*, 1995). Activated B cells are somewhat less efficient, whereas resting B cells and both spleen-derived and bone-marrow-derived macrophages are poor APC for naive T cells. These findings correlate at least partly with data from B-cell deficient mice, where it has been found that B cells are not needed for priming of naive T cells (Epstein *et al.*, 1995). These differences seem, to a large extent, to be caused by differences among APC in their expression of accessory molecules such as the CD28 ligand, B7, which are essential for delivery of co-stimulatory signals (see below). In humans, dendritic cells, but not adherent monocytes, can induce T cell responses to human immunodeficiency virus (HIV) gp160 and keyhold limpet cyanin (KLH) in individuals who have not previously been immunized with these antigens (Fagnoni *et al.*, 1995).

The APC for already activated, that is effector or memory, T cells constitute a broader repertoire of cell types, as depicted in *Figure 14.1*. Already activated T cells generally require less co-stimulatory signals from the APC to exert their effects, as discussed later. B cells carrying membrane-bound immunoglobulin (Ig) require minute amounts of corresponding antigen to serve as APC of peptides derived from this antigen (Lanzavecchia, 1990; Simitsek *et al.*, 1995). Also, for

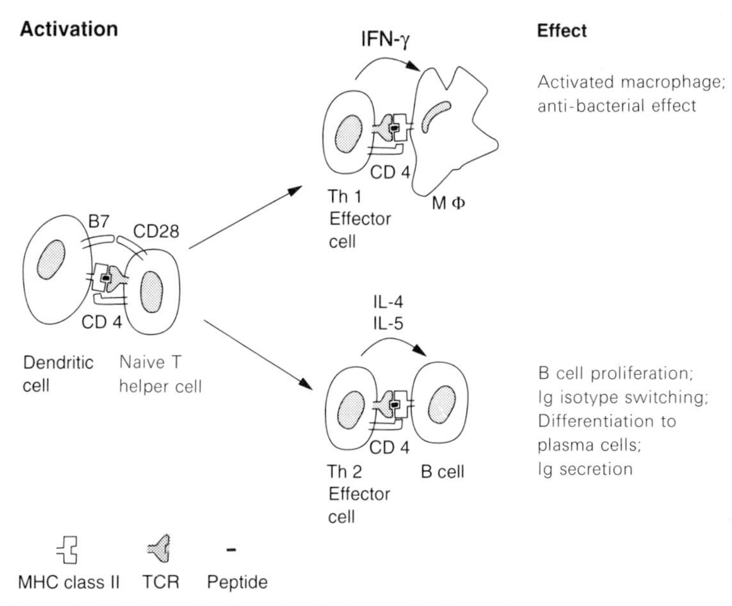

Figure 14.1. Activation of naive T cells by professional APC (usually dendritic cells) in lymph nodes, spleen and mucosa-associated lymphoid tissue leads to functional maturation of the responding T cells. As effector T cells they have, amongst other functions, the ability to interact with and activate macrophages and thereby enhance an antibactericidal effect, or interact with B cells and thereby enhance antibody production. The most crucial step in all these interactions is the interaction between the TCR and the HLA class II-peptide complexes, as indicated.

effector T cells, dendritic cells are important APC *in vivo* (Steinman, 1991). On the other hand, many other class II-expressing cells, which most probably cannot act as APC for naive T cells, may also present antigen to activated $CD4^+$ T cells. Endothelial cells, once class II is induced with IFN-γ, for example are one such cell type. They present antigen less efficiently than peripheral blood APC, possibly due to low levels of B7 expression (Savage *et al.*, 1995). Similarly, class II-expressing epithelial cells from the thyroid can stimulate already activated T cells (Grubeck-Loebenstein *et al.*, 1988). In contrast, some class II-positive non-lymphoid cells may fail to stimulate activated $CD4^+$ T cells (Eckels *et al.*, 1988). These differences are most probably caused by different co-stimulatory properties of various cell types.

It is important that T cell activation may result in different T cell responses depending on the nature of the stimulating cell type. The colon cancer epithelial cell line HT29, which expresses HLA-DQ molecules after treatment with IFN-γ, is very sensitive to killing by alloreactive and antigen-specific T cell clones, but it does not induce detectable T cell proliferation *in vitro* (Gedde-Dahl *et al.*, 1994; Lundin *et al.*, 1990a). This supports the notion that non-bone-marrow-derived cells, including malignant cells, are not efficient in activating primary T cell responses, but may be targets for activated T cells.

Induction of class II molecules in cells which normally do not express class II molecules has been suggested to be a mechanism for initiation of autoimmune diseases (Bottazzo *et al.*, 1983). However, the evidence that naive T cells require more signals for activation than those generated by recognition of peptide–MHC class II complexes makes this hypothesis less attractive.

14.8 Are different T cell functions associated with recognition of peptides bound by different class II molecules?

An important question is whether the different class II isotypes (i.e. DR, DQ or DP; I-A or I-E) exert distinct T cell functions or whether they merely enlarge the peptide-binding repertoire. In the mouse (at least for those mouse strains which express both I-A and I-E) fairly balanced proportions of T cells seem to be restricted by I-A versus I-E. The previous proposals that peptides bound to I-A molecules preferentially stimulate helper T cells, whereas those bound to I-E molecules in addition also stimulate suppressor T cells (Baxevanis *et al.*, 1982) have been difficult to substantiate. Moreover, transgenic and 'knock-out' mice expressing only I-A or only I-E show fairly equal properties in terms of maturation of $CD4^+$, $CD8^-$ T cells in the thymus, and generation of CTL responses and antibody production (Cosgrove *et al.*, 1992). With respect to the origin of peptides occupying the various class II molecules in mouse and humans major differences are not seen, which argues against specialized roles in immune regulation (Chicz *et al.*, 1993).

In humans a similar relationship between HLA-DQ as a 'suppressor' class II molecule and DR as a 'helper' class II molecule has been suggested (Hirayama *et al.*, 1987). DQ-restricted, antigen-specific T cells were found in individuals with a low, but not a high response to streptococcal antigen (Kamikawaji *et al.*, 1991). The presence of antibodies towards DQ but not DR molecules present during primary culture could abrogate non-responsiveness against *Mycobacterium leprae* in leprosy patients (Ottenhoff *et al.*, 1990). The findings of DQ-restricted, $CD8^+$ suppressor T cells in leprosy (Salgame *et al.*, 1991) seemed very unusual, but have recently received some support, since I-E class II molecules in TCR transgenic mice may cause positive selection of $CD8^+$ T cells (Kirberg *et al.*, 1994).

Most human antigen-specific T cell clones generated from peripheral blood mononuclear cells have been found to be DR-restricted. This led to the proposal that HLA-DQ is not important for presentation of antigen in the periphery but is instead mainly involved in thymic education of T cells (Altmann *et al.*, 1991; Möller *et al.*, 1990). We believe that this assumption is unlikely, since in many experimental systems antigen-specific DQ-restricted T cells can be found. This includes T cell clones directed at a variety of bacterial, viral and self antigens. Some examples are given in *Table 14.1*.

Using mycobacteria, DQ-restricted T cells seem to be rare (Mustafa *et al.*, 1993; Ottenhoff *et al.*, 1986). On the other hand, DQ-restricted T cells may constitute a large part of the T cell repertoire against influenza virus matrix protein (Adler *et al.*, 1994) or gluten proteins (Gjertsen *et al.*, 1994). Others have found preferential restriction by DP molecules of peripheral blood T cells specific for streptococcal antigen (Dong *et al.*, 1995), or mycobacterial antigen (Gaston *et al.*, 1991), or synovial T cells specific for chlamydial antigen (Hermann *et al.*, 1992). Although the list in *Table 14.1* is not complete, it indicates that, for a large variety of foreign antigens, T cells restricted by class II molecules other than HLA-DR can be found. On the other hand a quantitative imbalance seems to exist, in that antigen-

Table 14.1. Antigen-specific HLA-DQ restricted CD4$^+$ T cells found in the blood or in tissues

Antigen	T cell source	Reference
Hepatitis B virus	Blood	Ferrari *et al.*, 1992; Serra *et al.*, 1993
Myelin basic protein	Blood	Chou *et al.*, 1989
ras oncogene products	Blood	Gedde-Dahl *et al.*, 1992a, b; Fossum *et al.*, 1993
Malaria antigens	Blood	Guttinger *et al.*, 1991
Cytomegalovirus	Blood	Gehrz *et al.*, 1987
Herpes simplex virus	Blood	Yasukawa and Zarling 1984; Lundin *et al.*, 1990b
Herpes simplex virus	Tissue	Koelle *et al.*, 1994
Borrelia burgdorferi	Blood	Shanafeldt *et al.*, 1991
Thyroid antigen	Tissue	Dayan *et al.*, 1991
Heat shock protein	Tissue	Quayle *et al.*, 1992
Wheat gluten	Tissue	Lundin *et al.*, 1993, 1994
Wheat gluten	Blood	Gjertsen *et al.*, 1994; Jensen *et al.*, 1995
Influenza virus	Blood	Adler *et al.*, 1994
M. leprae	Blood	Ottenhoff *et al.*, 1986

specific HLA-DR-restricted T cells predominate over DQ- or DP-restricted T cells in many of the experimental systems tested so far.

The apparent difference between mouse T cell responses (quantitatively balanced between I-A and I-E restriction) and human T cell responses (skewed towards DR restriction) deserves an explanation. One major difference between studies of mice and humans is that in the mouse T cells and APC are usually derived from spleen and lymph nodes, whereas in humans most studies have used T cells and APC from peripheral blood, where some APC may not express all three series of class II molecules (see Section 14.3). In contrast, when antigen-specific T cells are harvested from the tissues, more non-DR-restricted T cells may be found. Several examples are mentioned above, and illustrated in *Table 14.1*. In our own studies of coeliac disease, we found an almost exclusive HLA-DQ2 or -DQ8 restriction of gliadin-specific T cells from the small intestinal mucosa of DQ2- or DQ8-positive patients (*Figure 14.2*), whereas in peripheral blood gliadin-specific T cells restricted both by HLA-DR or -DQ molecules could be found (Gjertsen *et al.*, 1994; Lundin *et al.*, 1993). One explanation of these findings could be that APC from tissues, in particular inflamed tissues, often express HLA-DQ molecules at a higher concentration than their counterparts in peripheral blood. Thus, it is possible that we will have to redefine our conceptions on quantitative aspects of HLA-DR versus HLA-DQ restriction as more T cells from inflamed tissues are studied.

14.9 TCR usage in antigen recognition

Data have just begun to emerge on the three-dimensional structure of the TCRβ chain (Bentley *et al.*, 1995). The similarity between Ig domains and TCR is confirmed, and very close contacts between the variable and constant domains of TCR are found. So far, theoretical models based on sequence homology with Ig have been presented (Chothia *et al.*, 1988; Davis and Bjorkman, 1988). The TCR

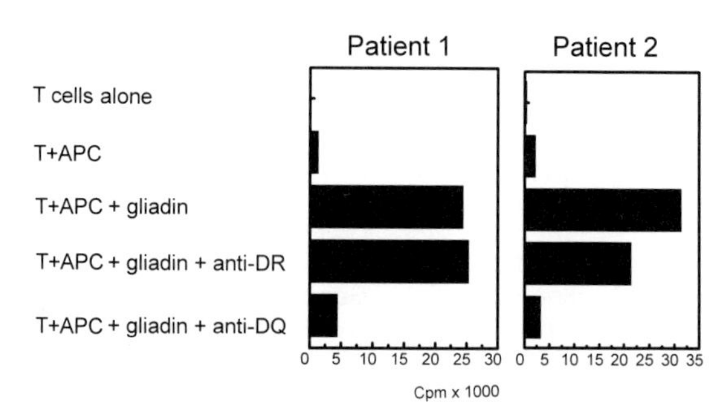

Figure 14.2. Inhibition with anti-HLA antibodies of polyclonal, gliadin-specific T-cell lines from the small intestinal mucosa of coeliac-disease patients. Shown here are two polyclonal T cell lines from two different patients, Patient 1 being DQ2 positive and Patient 2 being DQ8 positive. As APC we used HLA-matched Epstein-Barr virus-transformed and B lymphoblastoid cell lines, respectively. The typical patterns for such polyclonal T cell lines have been shown to be preferential recognition of gliadin when presented by DQ2 or DQ8, and *not* by the DR molecules also carried by the patients (Lundin *et al.*, 1993, 1994).

α- and β-chains probably carry complementarity-determining-regions (CDR) with analogy to Ig. The CDRs determined by the VJ junction of the α chain and the VDJ junction of the β chain presumably interact with the antigenic peptide, whereas the CDRs of Vα and Vβ segments, respectively, interact with the α-helices of the MHC molecules (Jorgensen *et al.*, 1992; Katayama *et al.*, 1995). Some data from animal models suggest very close correlations between TCR usage and specificity for particular peptide–MHC class II complexes (Zamvil and Steinman, 1990). Several groups have searched for similar findings for MHC class II in humans, but, on balance, close correlations are hardly the general rule (Obata and Kashiwagi, 1992); for instance, in our studies of a panel of five different HLA-DQ6-restricted T cell clones specific for a particular 16 amino-acid-long peptide from a mutated *ras* oncogene protein, we found that they carried highly diverse TCRα/β combinations (Spurkland *et al.*, 1994). In fact, similar low levels of correlations were foreseen earlier (Davis and Bjorkman, 1988). Several possible explanations may be found and it may be relevant that individual T cells may not only 'see' the peptide–MHC class II complex in one particular orientation (Janeway *et al.*, 1995).

14.10 T cell activation

The TCR is associated with the CD3 complex in the cell membrane. The CD3 complex consists of five invariant polypeptide chains, designated γ, δ, ϵ, ξ and η, respectively. They deliver signals to the T cell during TCR recognition of peptide–MHC complexes (Irwing and Weiss, 1991). The affinity of the TCR for the peptide–MHC complex is generally low (Karjalainen, 1994). However, several co-receptors increase the avidity of the T cell for the APC, which may be particularly important for activation of naive T cells. The CD4 and CD8 molecules interact with more constant parts of MHC class II and class I molecules, respectively, CD2 interacts with LFA-3 (CD58), while LFA-1 interacts with ICAM-1. The interaction between B7 (on the APC) and CD28 and CTLA-4 (on the T cell) may be one of the most important for activating naive T cells, but may play a lesser role in reactivation of already activated T cells. B7 is found in at least three variants (B7-1: CD80, B7-2: CD86 and B7-3; Boussiotis *et al.*, 1993). These interactions are dynamically regulated (Dustin and Springer, 1989).

Another molecule of T cells of importance for T cell activation is CD45, which exists in several isoforms due to alternative splicing of the same gene. In general, 'naive' T cells which have not yet encountered antigenic stimulation after migration from the thymus express high levels of CD45RA and low levels of CD45RO, whereas 'memory' T cells display an inverse expression of these molecules.

TCR stimulation induces activation of tyrosine kinases, including the ξ chain-associated protein tyrosine kinase (PTK) ZAP-70 and $p59^{fyn}$ and the CD4/8-associated PTK $p56^{lck}$ (Robey and Allison, 1995). The function of these kinases is also influenced by the CD45-associated tyrosine phosphatase activity, leading to complex intracellular events in the responding T cells. Some of these mechanisms have received further attention since they can be manipulated during immunotherapeutic interventions of class II-restricted T cells.

Importantly, recent reports indicate that the TCR interaction with the peptide–MHC class II complex involves more than a simple occupancy of the TCR interaction site (Evavold *et al.*, 1993; Germain *et al.*, 1995; Janeway 1995). Subtle differences in the peptide recognized may have dramatic effects on the effector functions of the T cell, from full activation, partial activation (e.g. cytokine production in the absence of proliferation), to anergy induction (leaving the cell unresponsive to concomitant signals). Exactly how the altered peptide ligands exert these effects is currently being debated, and it is puzzling how a low-affinity interaction between the TCR and the peptide–MHC complex can lead to such dramatic differences.

14.11 No T cell response to many peptide–HLA class II complexes

Peptide fragments may bind to a class II molecule without evoking a T cell response, because of negative selection of T cells carrying the corresponding TCR in the thymus. In addition, peptide–HLA class II complexes may induce anergy in the corresponding T cells in 'the periphery', or they may induce regulatory ('suppressor') T cells. T cells specific for particular peptide–HLA complexes may thus be absent functionally from the T cell repertoire. This is the rule for T cells carrying TCR specific for self-peptide–HLA complexes, but may also involve peptides derived from foreign proteins bound to HLA molecules, because of structural similarity to self-peptide–HLA complexes. Evidence to support the existence of 'holes' in the T cell repertoire exists in mice, as several peptides from a foreign antigen that bind with high affinity to the corresponding H-2 molecules have been found to be non-immunogenic (Schaeffer *et al.*, 1989). Holes in the T cell repertoire probably also exist in man. Following vaccination with recombinant hepatitis B virus surface antigen (HBsAg), it has been found that some individuals fail to give an antibody response and that this trait is associated with the HLA-B8, -DR3, -DQ2 haplotype. The defect is apparently caused by absence of responding

T cells and not by defective antigen presentation or suppressor T cells (Egea *et al.*, 1991; Salazar *et al.*, 1995).

14.12 Effector functions of activated CD4$^+$ T cells

There are several distinct effector functions of class II-restricted T cells: B cell help, induction and regulation of inflammatory responses and cytotoxicity. The T cells have the capability of exerting their function in cell-to-cell contact with their target cells, where they may secrete lytic substances and thereby kill the target cells. A more important function of class II-restricted T cells may be their secretion of cytokines with both local and systemic functions.

14.12.1 Effector functions: cytokines

The secretion of soluble factors (cytokines) is possibly the most important effector function of class II-restricted T cells. At least in the mouse the secretion pattern of T cells seems to be polarized and not random (Mosmann and Coffman, 1989; Parronchi *et al.*, 1991). CD4$^+$ T cells have been classified as Th1, Th2 and Th0 cells, respectively, as depicted in *Figure 14.3*. The Th1 subset is predominantly 'inflammatory' and secretes, among other cytokines, IFN-γ. This subset also secretes interleukin (IL)-2, which helps in expansion and functional maturation of CD8$^+$, class I-restricted T cells. The Th2 subset is mainly responsible for helping the B cell system by secretion of IL-4, IL-5 and other cytokines. The Th0 subset may be the progenitor of both Th1 and Th2 cells. Some reports have shown that T cells in different inflamed tissues secrete polarized patterns of cytokines compatible with a Th1/Th2 dichotomy (Maggi *et al.*, 1991; Yamamura *et al.*, 1993). More recently, however, data have been presented using tissue-infiltrating T cells or T cells from peripheral blood showing either a broad cytokine pattern or patterns which do not fit the previous classifications (Howe *et al.*, 1995; Nilsen *et al.*, 1995). Therefore, more studies are needed and the concept of distinct patterns of cytokine secretion by two different subsets of T cells (Th1 and Th2) has been questioned (Kelsoe, 1995).

It is not known what dictates the polarization of the T cell lymphokine patterns. The T cell response of an individual towards one antigen (e.g. *M. tuberculosis*) may be dominated by a Th1 response, whereas their response towards another antigen (e.g. *Lolium perrenne*) may be dominated by a Th2 response (Parronchi *et al.*, 1991). However, the antigen is not the only factor, since some patients with *M. leprae* infection have a predominantly Th2 response (Yamamura *et al.*, 1993). Another factor involved is perpetuation of the polarized T cells by their own cytokines, since IFN-γ and IL-4 support a Th1 and a Th2 response, respectively (Maggi *et al.*, 1992). The initial stimulus for T cells to enter the different pathways could be macrophage-derived IL-12 and NK cell-derived IFN-γ on the one side and mast cell-derived IL-4 on the other (Garside and Mowat, 1995). Most recently, evidence has been presented that the expression of co-stimulatory molecules, especially the B7 variants, on the APC is important (Thompson, 1995).

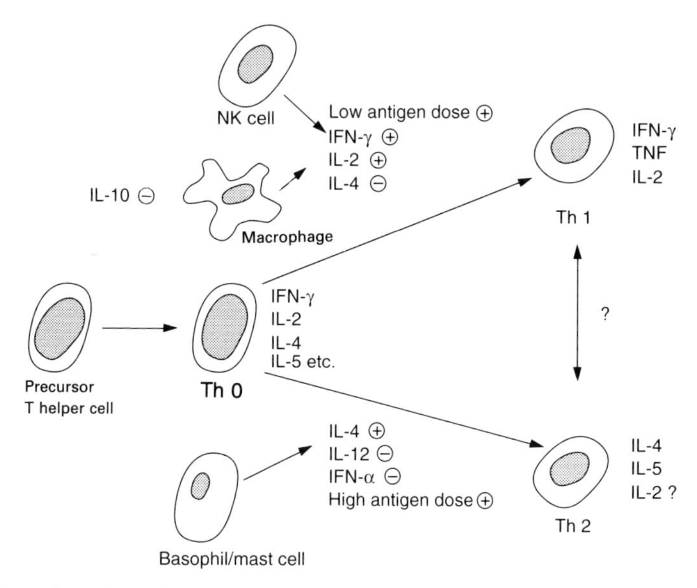

Figure 14.3. Overview of some of the known cytokine interactions. Class II-restricted T cells probably undergo a functional maturation from a naive state, where they are said to belong to a Th0 subset of cells, and then enter either a Th1 or Th2 differentiation pathway. The cytokines in the local milieu where this differentiation takes place greatly influence the final patterns. Some of these cytokines may be derived from already differentiated T cells, others may be derived from other cells like natural killer (NK) cells, macrophages and basophils/mast cells. Importantly, the rigid polarization patterns indicated here may not always be found in T cells, since in many cases intermediate patterns are found, as discussed in the text. There may be a role for transition of T cells from one type of polarized pattern in the direction of the other, but this is uncertain.

Another important factor seems to be the antigen concentration. High ligand density favours Th1 activation, while low ligand density favours Th2 activation (Janeway and Bottomly, 1994; Kumar *et al.*, 1995).

Polarized cytokine secretion may be important for autoimmune diseases (Romagnani, 1994). In experimental allergic encephalomyelitis of mice, Th1, but not Th2 cells, specific for myelin basic protein (MBP) or proteolipid protein (PLP) can transfer disease (Baron *et al.*, 1993). Insulin-dependent diabetes mellitus (IDDM) can be transferred in non-obese diabetic (NOD) mice by Th1 but not by Th2 cells specific for membrane proteins from an insulinoma cell line (Healey *et al.*, 1995). Similarly, insulin-specific Th1 and Th0 cell clones (secreting both IFN-γ and IL-4) from the pancreatic islets of pre-diabetic NOD mice accelerate the development of diabetes in young mice, or transfer disease in NOD/SCID (severe combined immunodeficiency) mice (Daniel *et al.*, 1995). Not all reports fit, however, into this simple picture. Some Th1-like T cell clones do not accelerate disease but instead suppress progression of IDDM (Akthar *et al.*, 1995). The findings of a polarized class II-restricted T cell response in the induction phase of immunopathological conditions are interesting, as they open the possibility to redirect such T cell responses away from tissue-damaging responses towards protective responses, as will be discussed later.

14.12.2 Effector functions: cytotoxicity

Class I-restricted T cells are the main cytotoxic effector cells in most kinds of viral infections (Zinkernagel and Doherty, 1987, and see Chapter 13). Some evidence suggests, however, that class II-restricted T cells may also be important cytotoxic cells in the response against infectious agents (Mahon *et al.*, 1995; Muller and Louis, 1989). The defence against herpes simplex virus (HSV) infection serves as an interesting model for the importance of class II-restricted T cells in viral infection. In humans, the T cell response towards HSV is predominantly mediated by $CD4^+$, HLA-DR-restricted T cells (Schmid, 1988; Yusukawa and Zarling, 1984). Similar findings have been made in mice, where protection against HSV infection is adoptively transferred by $CD4^+$ T cells, but not $CD8^+$ T cells (Manickan *et al.*, 1995). A possible explanation for the seemingly unusual preferential recognition of infectious virus by $CD4^+$ cells may be that HSV turns off the class I-associated transporter mechanisms (Hill *et al.*, 1995). Thus, class II-restricted recognition of HSV can serve as a second line of anti-viral defence mechanism.

Cytotoxic activity of class II-restricted T cells has been a controversial subject. It has long been known that $CD4^+$ T cells can kill other cells in a class II-restricted fashion (Meuer *et al.*, 1982). Human $CD4^+$ T cells cultured with IL-2 acquire cytotoxic activity (Fleischer, 1984), and a similar maturation process is also possible *in vivo*. Two distinct pathways of cytotoxic activity exist, granule exocytosis on the one side, and Fas–Fas ligand interaction on the other. The relative contribution of each of these pathways in cytotoxicity induced by $CD4^+$ and $CD8^+$ T cells is still uncertain (Berke, 1995). It is possible that $CD4^+$ cytotoxic T cells are predominantly dependent on the Fas–Fas ligand interaction (Stalder *et al.*, 1994). At any rate, both we and others have found a remarkable cytotoxic, class II-restricted potential of T cell clones against, for instance, epithelial cells expressing relevant class II molecules (Bogen *et al.*, 1986; Lundin *et al.*, 1990a). This indicates that cytotoxic activity by $CD4^+$ T cells against various cell types *in vivo* may be more important than previously considered.

14.12.3 Effector functions: regulation

T cells may suppress activation of other T cells; for instance, after adoptive transfer of T cells from immunologically manipulated animals to naive animals. Whether suppressor T cells exist as a separate lineage has been controversial and has received less and less support (Bloom *et al.*, 1992; Möller, 1988). Lanzavecchia has suggested that specific suppression could be a function of class II-restricted cytotoxic T cells (Lanzavecchia, 1989). Such T cells could abrogate a response by killing the APC or by killing activated T cells, which, in man, express class II molecules (Ottenhoff and Mutis, 1990; Siliciano *et al.*, 1988). Another mechanism by which T cells could specifically suppress an immune response would be to compete for growth factors like IL-2, after they have themselves become anergic (Lombardi *et al.*, 1994). A third mechanism that also may lead to suppression is secretion of suppressive cytokines, in particular TGF-β, which may be important for induction of oral tolerance (Garside *et al.*, 1995). The relative contribution of each of these mechanisms is unclear, but the findings have induced a wave of enthusiasm for the ability to down-regulate unwanted immune reactions.

14.13 Where do class II-restricted T cells exert their function?

T cells do not migrate randomly but are, instead, under continuous influence of their homing receptors (Mackay, 1993). Importantly, homing is not a static phenomenon for each individual T cell, but is influenced by the T cell's maturation level. Naive T cells home to peripheral lymphoid organs (lymph nodes, spleen, mucosa-associated lymphoid tissues) due to their expression of L-selectin which bind to endothelial cells in a variety of secondary lymphoid organs (other homing receptors may also be involved). Later, when such naive T cells have been activated and matured to effector T cells, L-selectin is down-regulated and the T cells instead start to display homing receptors. Typically, gut wall grafts from immunocompetent mice can repopulate the gut-associated lymphoid tissue of immunodeficient SCID mice but not the peripheral lymph nodes (Rudolphi *et al.*, 1994). The migration patterns of antigen-specific T cells from TCR-transgenic animals is dependent on injection site and antigen stimulation (Kearny *et al.*, 1994). In man, at least one case of a pronounced compartmentalization of antigen-specific T cells (in this case to the liver in hepatitis C patients) has been found (Minutello *et al.*, 1993). Thus, it is anticipated that the T cell repertoire in various tissues is quantitatively and qualitatively different and may deviate from that found in peripheral blood.

14.14 T cell recognition in HLA class II-associated diseases

A number of human diseases show striking HLA class II associations (Thorsby, 1995). Some of these diseases (e.g. rheumatoid arthritis, Grave's disease, pemphigus vulgaris, multiple sclerosis and others) appear to be associated with particular HLA-DR variants, whereas other diseases are associated with particular HLA-DQ variants (coeliac disease, narcolepsy and others). For still other diseases, evidence for a primary involvement of some HLA-DQ molecules has been given, but HLA-DR molecules seem to contribute in determining susceptibility (e.g. type I diabetes). This strong HLA class II association suggests in itself that class II-restricted T cells specific for putative (auto)antigens are crucial in the pathogenesis of HLA-associated diseases.

There are a number of human autoimmune diseases where class II-restricted T cells specific for the putative autoantigen have been described. In multiple sclerosis, several investigators have found T cell clones specific for MBP (Pette *et al.*, 1990; Wucherpfennig and Strominger, 1995). This has been received with great interest, since certain animals develop a multiple sclerosis-like disease (experimental allergic encephalomyelitis) when they are immunized with MBP. However, since in several of these studies it has been shown that healthy individuals may also have MBP-specific T cells in their peripheral blood, the relevance of such T cells in patients has been questioned.

It may be of greater interest to study tissue-infiltrating T cells in autoimmune diseases, since such T cells may have unique properties that are not always mirrored by peripheral blood T cells. In type 1 diabetes the target organ is almost inaccessible in man, but in the mouse diabetogenic T cell clones are found in the pancreas of NOD mice (Daniel *et al.*, 1994) and the same could be the case in humans. In the chronic liver disease primary biliary cirrhosis antibodies against the pyruvate dehydrogenase complex of mitochondria are used as a diagnostic marker, and many T cells in the lesions recognize this antigen (Löhr *et al.*, 1993; van de Water *et al.*, 1995).

We have chosen coeliac disease as a model for HLA class II-associated diseases for several reasons:

(i) the disease shows a very strong association to one particular HLA-DQ heterodimer DQ(α1*0501, β1*02) (i.e. DQ2, encoded in either *cis* or *trans* position), with a weaker association to another DQ heterodimer DQ(α1*0301, β1*0302), i.e. DQ8; Sollid and Thorsby, 1993; Sollid *et al.*, 1989);
(ii) the triggering agent (the gliadin moiety of wheat gluten) is known and patients are treated by eliminating gluten from the diet (Trier, 1991);
(iii) an *in vivo* or *in vitro* challenge of the small intestinal mucosa of patients on a gluten-free diet by gluten mirrors the start of the disease;
(iv) the target organ, the small intestine, is easily accessible for taking biopsies.

The association of coeliac disease to two particular HLA class II heterodimers suggests that presentation of gliadin peptides by these heterodimers to $CD4^+$ T cells in the intestinal mucosa is involved in the disease immunopathology. To note, lamina propria $CD4^+$ T cells become activated and express IL-2 receptor (CD25) when the patients reintroduce gluten to their diet or when their small intestinal biopsies are challenged with gluten in an organ culture system (Halstensen *et al.*, 1993). We isolated such CD25-expressing T cells and demonstrated their gluten-reactivity (Lundin *et al.*, 1993, 1994). Importantly, the vast majority of gluten-specific T cells (both polyclonal T cell lines and T cell clones) were restricted by the disease-associated HLA-DQ molecules, that is DQ(α1*0501, β1*0201) in the patients who were DQ2, and DQ(α1*0301, β1*0302) in patients who were DQ8 (*Figure 14.4*). Thus, neither the other DQ molecules carried by the patients, nor any of their DR molecules seemed to be important in the presentation of gluten antigens to small intestinal T cells. Using a large panel of B cells carrying the DQ(α1*0501, β1*02) heterodimer where amino acid residues of the DQβ chain had been systematically changed by site-directed mutagenesis in the direction of DQ-β*0301 (which does not confer susceptibility to disease), evidence for a pronounced heterogeneity of T cell recognition patterns was obtained, since each single T cell clone tested had a unique pattern of susceptibility for substitutions (Paulsen *et al.*, 1995). Finally, although all the T cell clones recognized the gliadin moiety of gluten (the harmful fraction), testing of a large panel of small intestinal T-cell clones against three different purified gliadins again demonstrated heterogeneity in reactivity patterns.

The gliadin-reactive T cells were the very first antigen-specific T cells isolated from the human small intestinal mucosa and offered a unique possibility to investigate possible effector-function mechanisms of a T cell-mediated disease. Stimulating them with gliadin (presented by DQ2-expressing APC), we found a broad cytokine secretion pattern with high levels of IFN-γ (Nilsen *et al.*, 1995), which can kill epithelial cells and induce HLA class II expression, both typical findings of the coeliac disease lesion (Brandtzaeg, 1991). Soluble factors from the T cell clones can induce typical morphological changes in small intestinal biopsies, presumably caused by IFN-γ (Przemioslo *et al.*, 1995). Many T cell clones also secreted IL-4 and IL-5, which could increase local antibody production.

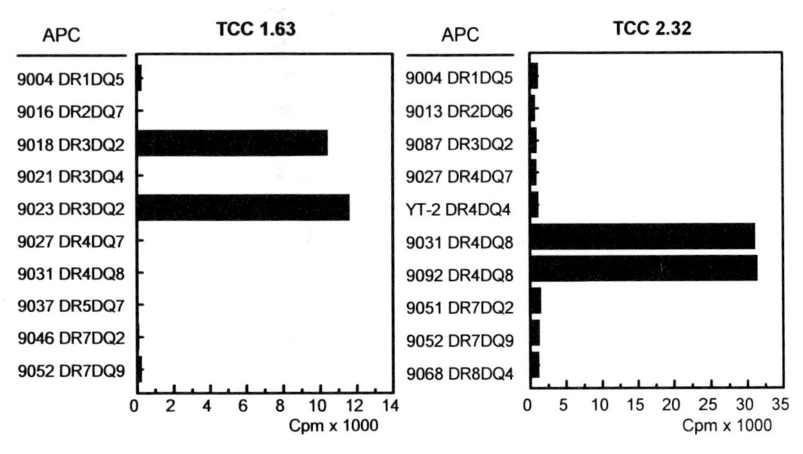

Figure 14.4. HLA restriction patterns of gliadin-specific T cell clones (TCC) from the small intestinal mucosa of coeliac-disease patients. T cell clones were isolated from polyclonal T cell lines shown in *Figure 14.2*, that is from a DR3,DQ2-positive patient (1.63) and a DR4,DQ8-positive patient (2.32), respectively. Both clones were inhibited by anti-HLA-DQ monoclonal antibodies only. These two T cell clones are representative of large panels of clones (Lundin *et al.*, 1993, 1994).

Together, our studies emphasize the importance of using tissue-infiltrating T cells in studies of HLA-associated diseases.

14.15 Class II-restricted T cells as targets for immunological intervention

Given the instrumental role of class II-restricted T cells in the induction and effector phases of any T cell responses, they are particularly attractive for manipulations with the goal of:

(i) inducing protective immunity against infectious agents;
(ii) preventing immunopathological reactions to self antigens.

For vaccine purposes, intact, but attenuated bacteria or viruses have generally been used. Recently, however, some recombinant vaccines have been tested; for example, HBsAg and HIV antigens. Normally, the use of such recombinant vaccines relies on the definition of important cell epitopes. There are several unresolved questions, including delivery of the antigen, use of adjuvants to induce a T cell response and a truly safe way of vaccination to avoid potential immunopathological reactions. Vaccination against HIV and hepatitis C particularly is problematical due to the rapid mutation rate of antigenic epitopes of these viruses. One of the diseases where vaccination is really needed is leprosy, where we will need to define carefully ways of antigen delivery to avoid the state of non-responsiveness seen in lepromatous leprosy patients.

Inhibition of class II-restricted T cells is possible at several levels. An agent like cyclosporin A, which acts by inhibiting IL-2 production, has become the fundamental immunosuppressive drug after organ transplantation. The effect on autoimmune conditions in humans has been investigated, and there is no doubt that development of type 1 diabetes can be halted (Feutren *et al.*, 1986). However, due to serious side-effects caused by this agent, it is not attractive for life-long treatment. The same arguments hold true for more radical immunosuppressive regimens, like induction of pronounced T cell depletion or MHC class II blockade by monoclonal antibodies, although such methods have taught us much about autoimmunity in animal models.

Strategies have been developed for selective immune suppression by interfering with various facets of the trimolecular complex consisting of peptide, MHC class II and the TCR (Gaur and Fathman, 1994; Guéry and Adorini, 1993; Lanzavecchia, 1993). One obvious possibility is to prevent binding of peptides that are the targets for autopathogenic T cells, by other peptides with high affinity for the same MHC class II molecule, but which are not recognized by the same T cells. The phenomenon that different peptides compete for MHC class II binding and thereby interfere with T cell activation has long been known (Rock and Benacerraf, 1984). Employing this strategy in animal models has shown that it works *in vivo* as well (Hurtenbach *et al.*, 1993). The strategy has the obvious disadvantage that displacing already bound peptides can be very difficult, since a complex of a given peptide and an MHC class II molecule usually remains stable for the lifespan of the class II molecule (Lanzavecchia *et al.*, 1992), and soluble short peptides have very short half-lives, making it difficult to maintain high tissue-levels of a putative blocking peptide. On the other hand, it has been shown in the experimental allergic encephalomyelitis model that at least one autoantigenic peptide from MBP binds with very low affinity to the relevant H-2 molecule (Fairchild *et al.*, 1993), in contrast to some foreign peptides which bind with much higher affinity (Buus *et al.*, 1987). The low-affinity-binding of the MBP peptide could be involved in escape from tolerance induction, and may facilitate the use of other blocking peptides for disease inhibition (Wraith *et al.*, 1989). Whether non-peptidic agents can be used for class II blocking remains to be seen. Some data indicate that carbohydrates and carbopeptides have poor capacity in this respect (Ishioka *et al.*, 1992).

A more attractive possibility would be selectively to delete or anergize T cells that are involved in the autoimmune process. In theory, this could be achieved by deleting them from the repertoire by, for example, using monoclonal antibodies against their TCR. This requires a non-random TCR usage among pathogenic T cells, a phenomenon that has been observed in some, but not in all, animal models and only rarely in humans (see above). A further possibility has recently been described, namely the use of altered peptide ligands which do not fully activate the T cell via TCR, but rather anergize or induce a qualitatively different signal to the responding T cell (de Magistris *et al.*, 1992; Racioppi *et al.*, 1993; Sloan-Lancaster and Allen, 1995). Some evidence suggests that it is possible to design peptides with the ability to antagonize different T cells with TCR that are specific for the same antigen (Kuchroo *et al.*, 1994; Snoke *et al.*, 1993). Although very attractive as a model, such an approach is seriously hindered if the T cell response is highly diverse in autoimmune diseases, as may be the case particularly late in the course of the disease. Also, the use of altered peptide ligands to induce a redirected cytokine profile in a human Th0-like T cell clone has so far not been successful (Lamb *et al.*, 1995).

Induction of oral tolerance is also a very attractive possibility, which takes direct advantage of the ability of the gut-associated lymphoid tissue to induce toleration in T cells against the enormous burden of potentially antigenic peptides. In the

Lewis rat experimental allergic encephalomyelitis model, feeding with the encephalitogenic antigen MBP prior to immunization with MBP leads to a distinct up-regulation of inhibitory cytokines like TGF-β, and down-regulation of all inflammatory cytokines (Khoury *et al.*, 1992). Importantly, $CD4^+$ T cells may be very important for induction of this oral tolerant state (Chen *et al.*, 1995a; Garside *et al.*, 1995). In a TCR transgenic mouse model, feeding with the antigen at higher doses has been shown to induce clonal deletion by apoptosis in gut-associated lymphoid tissues (Peyer's patches), whereas at lower doses induction of TGF-β, IL-4 and IL-10 was observed (Chen *et al.*, 1995b).

These are only some of the strategies currently being investigated. We will surely see a further development in this field and await the results of clinical testing along these lines (Trentham *et al.*, 1993).

14.16 Conclusion

During the last few years our knowledge of the involvement of class II molecules in activation of $CD4^+$ T cells, as well as the function of class II-restricted T cells, has greatly increased. The molecular interaction between the antigenic peptides and the MHC class II molecules has been unveiled, and we have much information concerning the way the TCR interacts with the peptide–MHC class II complexes. Furthermore, knowledge concerning how this interaction is translated into various signals in the responding T cell, and prospects of how such interactions can be manipulated on the level of peptide substitutions are at hand. We are also beginning to understand how class II-restricted T cells isolated from the site of action of the tissues react. Evidence has been presented that, in at least some HLA-associated diseases, preferential presentation of immunogenic peptides to T cells by the disease-associated HLA molecules may take place. Further work in this field should enable us specifically to prevent development of the HLA-associated diseases in susceptible individuals or halt the progression, if the disease has already developed.

Acknowledgements

We thank Drs Frode Vartdal and Jan E. Brinchmann for helpful advice in the preparation of this manuscript. Knut E.A. Lundin was supported by Pronova, Ludvig M. Sollid by the Norwegian Medical Research Council. Due to limited space, only a small fraction of relevant references could be mentioned. We apologise to those who have not been mentioned.

References

Adler S, Frank R, Lanzavecchia A, Weiss S. (1994) T cell epitope analysis with peptides simultaneously synthesized on celullose membranes: fine mapping of two DQ dependent epitopes. *FEBS Lett.* **352:** 167–170.

Akthar I, Gold JP, Pan LY, Ferrara JLM, Yang XD, Kim JI, Tan KN. (1995) CD4+β islet cell-reactive T cell clones that suppress autoimmune diabetes in nonobese diabetic mice. *J. Exp. Med.* **182:** 87–97.

Altmann DM, Sansom D, Marsh SGE. (1991) What is the basis for HLA–DQ associations with autoimmune disease? *Immunol. Today* **12:** 267–270.

Baron JL, Madri JA, Ruddle NH, Hashim G, Janeway CA. (1993) Surface expression of $\alpha 4$ integrin by CD4 T cells is required for their entry into brain parenchyma. *J. Exp. Med.* **177:** 57–68.

Baxevanis CN, Ishii N, Nagy ZA, Klein J. (1982) Role of the E^k molecule in the generation of suppressor T cells in the response to LDHb. *Scand. J. Immunol.* **16:** 25–31.

Bentley GA, Boulot G, Karjalainen K, Mariuzza RA. (1995) Crystal structure of the β chain of a T cell antigen receptor. *Science* **267:** 1984–1986.

Berke G. (1995) The CTL's kiss of death. *Cell* **81:** 9–12.

Bjorkman PJ, Saper MA, Samrauoi B, Bennett WS, Strominger JL, Wiley DC. (1987) The foreign antigen binding site and T cell recognition regions of class I histocompatibility antigens. *Nature* **329:** 512–518.

Bloom BR, Salgame P, Diamond B. (1992) Revisiting and revising suppressor T cells. *Immunol Today* **13:** 131–136.

Bogen B, Malissen B, Haas W. (1986) Idiotope-specific T cell clones that recognize syngeneic immunoglobulin fragments in the context of class II molecules. *Eur. J. Immunol.* **16:** 1373–1378.

Bottazzo GF, Pujol–Borell R, Hanafusa T. (1983) Role of aberrant HLA–DR expression and antigen presentation in induction of endocrine autoimmunity. *The Lancet* **12:** 1115–1118.

Boussiotis VA, Freeman GJ, Gribben JG, Daley J, Gray G, Nadler LM. (1993) Activated human B lymphocytes express three CTLA-4 counterreceptors that costimulate T-cell activation. *Proc. Natl Acad. Sci. USA* **90:** 11059–11063.

Brandtzaeg P. (1991) Immunologic basis for celiac disease, inflammatory bowel disease, and type B chronic gastritis. *Curr. Opin. Gastroenterol.* **7:** 450–462.

Buus S, Sette A, Colon SM, Miles C, Grey HM. (1987) The relation between major histocompatibility complex (MHC) restriction and the capacity of Ia to bind immunogenic peptides. *Science* **235:** 1353–1358.

Chen Y, Inobe J, Weiner HL. (1995a) Induction of oral tolerance to myelin basic protein in CD8–depleted mice: Both CD4+ and CD8+ cells mediate active suppression. *J. Immunol.* **155:** 910–916.

Chen Y, Inobe J, Marks R, Gonnella P, Kuchroo VK, Weiner HL. (1995b) Peripheral deletion of antigen–reactive T cells in oral tolerance. *Nature* **376:** 177–180.

Chicz RM, Urban RG, Gorga JC, Vignali DAA, Lane WS, Strominger JL. (1993) Specificity and promiscuity among naturally processed peptides bound to HLA–DR alleleles. *J. Exp. Med.* **178:** 27–47.

Chothia C, Boswell DR, Lesk AM. (1988) The outline structure of the T-cell $\alpha\beta$ receptor. *EMBO* **7:** 3745–3755.

Chou YK, Vainiene M, Whitman R, Bourdette D, Chou CHJ, Hashim G, Offner H, Vandenbark AA. (1989) Response of human T lymphocyte lines to myelin basic protein: Association of dominant epitopes with HLA class II restriction molecules. *J. Neurosci. Res.* **23:** 207–216.

Cosgrove D, Bodmer H, Bogue M, Benoist C, Mathis D. (1992) Evaluation of the functional equivalence of major histocompatibility complex class II A and E complexes. *J. Exp. Med.* **176:** 629–634.

Croft M, Duncan DD, Swain SL. (1992) Response of naive antigen-specific CD4+ T cells in vitro: Characteristics and antigen-presenting cell requirements. *J. Exp. Med.* **176:** 1431–1437.

Daar AS, Fuggle SV, Fabre JW, Ting A, Morris PJ. (1984) The detailed distribution of MHC class II antigens in normal human organs. *Transplantation* **38:** 293–298.

Daniel D, Gill RG, Schloot N, Wegmann D. (1995) Epitope specificity, cytokine production profile and diabetogenic activity of insulin-specific T cell clones isolated from NOD mice. *Eur. J. Immunol.* **25:** 1056–1062.

Davis MM, Bjorkman PJ. (1988) T-cell antigen receptor genes and T-cell recognition. *Nature* **334:** 395–402.

Dayan CM, Londei M, Corcoran AE, Grubeck–Loebenstein B, James RFL, Rapoport B, Feldmann M. (1991) Autoantigen recognition by thyroid-infiltrating T cells in Graves disease. *Proc. Natl Acad. Sci. USA* **88:** 7415–7419.

de Koster HS, Anderson DC, Termijtelen A. (1989) T cells sensitized to synthetic HLA–DR3 peptide give evidence of continous presentation of denaturated HLA–DR3 molecules by HLA–DP. *J. Exp. Med.* **169:** 1191–1196.

de Magistris MT, Alexander J, Coggeshall M, Altman A, Gaeta FCA, Grey HM, Sette A. (1992) Antigen analog-major histocompatibility complexes act as antagonists of the T cell receptor. *Cell* **68:** 625–634.

Deng H, Apple R, Clare-Salzler M, Trembleau S, Mathis D, Adorini L, Sercarz E. (1993)

Determinant capture as a possible mechanism of protection afforded by major histocompatibility complex class II molecules in autoimmune disease. *J. Exp. Med.* **178:** 1675–1680.

Denzin LK, Cresswell P. (1995) HLA–DM induces CLIP disassociation from MHC class II alphabeta dimers and facilitates peptide loading. *Cell* **82:** 155–165.

Dong RP, Kamikawaji N, Toida N, Fujita Y, Kimura A, Sasazuki T. (1995) Characterization of T cell epitopes restricted by HLA–DP9 in streptococcal M12 protein. *J. Immunol.* **154:** 4356–4545.

Dustin ML, Springer TA. (1989) T-cell receptor cross-linking transiently stimulates adhesiveness through LFA-1. *Nature* **341:** 619–624.

Eckels DD, Sell TW, Long EO, Sekaly RP. (1988) Presentation of influenza hemagglutinin peptide in the presence of limited allostimulation by HLA-DR1 transfected human fibroblasts. *Hum. Immunol.* **21:** 173–181.

Egea E, Iglesias A, Salazar M, Morimoto C, Kruskall MS, Awdeh Z, Schlossman SF, Alper CA, Yunis EJ. (1991) The cellular basis for lack of antibody response to hepatitis B vaccine in humans. *J. Exp. Med.* **173:** 531–538.

Epstein MM, Di Rosa F, Jankovic D, Sher A, Matzinger P. (1995) Successful T cell priming in B cell deficient mice. *J. Exp. Med.* **182:** 915–922.

Evavold BD, Sloan–Lancaster J, Allen PM. (1993) Tickling the TCR: selective T-cell functions stimulated by altered peptide ligands. *Immunol. Today* **14:** 602–609.

Fagnoni FF, Takamizawa M, Godfrey WR, Rivas A, Azuma M, Okumura K, Engleman EG. (1995) Role of B70/B7-2 in CD4+ T-cell immune responses induced by dendritic cells. *Immunology* **85:** 467–474.

Fairchild PJ, Wildgoose R, Atherton E, Webb S, Wraith DC. (1993) An autoantigenic T cell epitope forms unstable complexes with class II MHC: a novel route for escape from tolerance induction. *Int. Immunol.* **5:** 1151–1158.

Ferrari C, Cavalli A, Penna A, *et al.* (1992) Fine specificity of the human T-cell response to the Hepatitis B virus preS1 antigen. *Gastroenterology* **103:** 255–263.

Feutren G, Assan R, Karsenty G, *et al.* (1986) Cyclosporin increases the rate and length of remissions in insulin–dependent diabetes of recent onset. *Lancet* **2:** 119–123.

Fleischer B. (1984) Acquisition of specific cytotoxic activity by human T4+ T lymphocytes in culture. *Nature* **308:** 365–367.

Fossum B, Gedde–Dahl III T, Hansen T, Eriksen JA, Thorsby E, Gaudernack G. (1993) Overlapping epitopes encompassing a point mutation (12 Gly→Arg) in p21 ras can be recognized by HLA–DR, –DP and –DQ restricted T cells. *Eur. J. Immunol.* **23:** 2687–2691.

Freudenthal PS, Steinman RM. (1990) The distinct surface of human blood dendritic cells, as observed after an improved isolation method. *Proc. Natl Acad. Sci. USA* **87:** 7698–7702.

Garside P, Mowat AM. (1995) Polarization of Th–cell responses: a phylogenetic consequence of nonspecific immune defence? *Immunol. Today* **16:** 220–223.

Garside P, Steel M, Liew FY, Mowat AM. (1995) CD4+ but not CD8+ T cells are required for the induction of oral tolerance. *Int. Immunol.* **7:** 501–504.

Gaston JSH, Life PF, van der Zee R, Jenner PJ, Colston MJ, Tonks S, Bacon PA. (1991) Epitope specificity and MHC restriction of rheumatoid arthritis synovial T cell clones which recognize a mycobacterial 65 kDa heat shock protein. *Int. Immunol.* **3:** 965–972.

Gaur A, Fathman CG. (1994) Immunotherapeutic strategies directed at the trimolecular complex. *Adv. Immunol.* **56:** 219–265.

Gedde–Dahl III T, Eriksen JA, Thorsby E, Gaudernack G. (1992a) T cell responses against products of oncogenes. Generation and characterization of human T cell clones specific for p21 ras derived synthetic peptides. *Hum. Immunol.* **33:** 266–274.

Gedde–Dahl T III, Spurkland A, Eriksen JA, Thorsby E, Gaudernack G. (1992b) Memory T cells of a patient with follicular thyroid carcinoma recognize peptides derived from mutated p21 ras (Gln–Leu61). *Int. Immunol.* **4:** 1331–1337.

Gedde–Dahl III T, Nilsen E, Thorsby E, Gaudernack G. (1994) Growth inhibition of a colonic adenocarcinoma cell line (HT29) by T cells specific for mutant p21 ras. *Cancer Immunol. Immunother.* **38:** 127–134.

Gehrz RC, Fuad S, Liu YNC, Bach FH. (1987) HLA class II restriction of T helper cell response to cytomegalovirus (CMV) I. Immunogenetic control of restriction. *J. Immunol.* **138:** 3145–3151.

Germain RN, Levine EH, Madrenas J. (1995) The T-cell receptor as a diverse signal transduction machine. *Immunologist* **3:** 113–121.

Gjertsen HA, Sollid LM, Ek J, Thorsby E, Lundin KEA. (1994) T cells from the peripheral blood of coeliac disease patients recognize gluten antigens when presented by HLA-DR, -DQ, or -DP molecules. *Scand. J. Immunol.* **39:** 567–574.

Gonwa TA, Frost JP, Karr RW. (1986) All human monocytes have the capability of expressing HLA-DQ and HLA-DP molecules upon stimulation with interferon-γ. *J. Immunol.* **137:** 519–524.

Grubeck-Loebenstein B, Londei M, Greenall C, Pirich K, Kassal H, Waldhausl W, Feldmann M. (1988) Pathogenetic relevance of HLA class II expressing thyroid follicular cells in non-toxic goiter and in Graves' disease. *J. Clin. Invest.* **81:** 1608–1614.

Guery JC, Adorini L. (1993) Selective immunosuppression of class II-restricted T cells by MHC-binding peptides. *Crit. Rev. Immunol.* **13:** 195–206.

Guttinger M, Romagnoli P, Vandel L, Meloen R, Takacs B, Pink JRL, Sinigaglia F. (1991) HLA polymorphism and T cell recognition of a conserved region of p190, a malaria vaccine candidate. *Int. Immunol.* **3:** 899–906.

Halstensen TS, Scott H, Fausa O, Brandtzaeg P. (1993) Gluten stimulation of coeliac mucosa in vitro induces activation (CD25) of lamina propria CD4+ T cells and macrophages but no crypt-cell hyperplasia. *Scand. J. Immunol.* **38:** 581–590.

Hammer J. (1995) New methods to predict MHC–binding sequences within protein antigens. *Curr. Opin. Immunol.* **7:** 263–269.

Healey D, Ozegbe P, Arden S, Chandler P, Hutton J, Cooke A. (1995) In vivo activity and in vitro specificity of CD4+ Th1 and Th2 cells derived from the spleens of diabetic NOD mice. *J. Clin. Invest.* **95:** 2979–2985.

Hermann E, Mayet WJ, Thomssen H, Sieper J, Poralla T, Meyer zum Buschenfelde KH, Fleischer B. (1992) HLA–DP restricted Clamydia trachomatis specific synovial fluid T cell clones in Chlamydia–induced Reiter's disease. *J. Rheumatol.* **19:** 1243–1246.

Hill A, Jugovic P, York I, Russ G, Bennink J, Yewdell J, Ploegh H, Johnson D. (1995) Herpes simplex virus turns off the TAP to evade host immunity. *Nature* **375:** 411–414.

Hirayama K, Matsushita S, Kikuchi I, Iuchi M, Ohta N, Sasazuki T. (1987) HLA- DQ is epistatic to HLA-DR in controlling the immune response to schistosomal antigen in humans. *Nature* **327:** 426–430.

Howe RC, Wondimu A, Demissee A, Frommel D. (1995) Functional heterogeneity among CD4+ T-cell clones from blood and skin lesions of leprosy patients. Identification of T-cell clones distinct from Th0, Th1 and Th2. *Immunology* **84:** 585–594.

Hurtenbach U, Lier E, Adorini L, Nagy ZA. (1993) Prevention of autoimmune diabetes in non-obese diabetic mice by treatment with a class II major histocompatibility-blocking peptide. *J. Exp. Med.* **177:** 1499–1504.

Irwing BA, Weiss A. (1991) The cytoplasmic domain of the T cell receptor ζ chain is sufficient to couple to receptor-associated signal transduction pathways. *Cell* **64:** 891–901.

Ishioka GY, Lamont AG, Thomson D, Bulbow N, Gaeta FCA, Sette A, Grey HM. (1992) MHC interaction and T cell recognition of carbohydrates and glycopeptides. *J. Immunol.* **148:** 2446–2451.

Janeway CA. (1995) Ligands for the T-cell receptor: hard times for avidity models. *Immunol. Today* **16:** 223–225.

Janeway CA, Bottomly K. (1994) Signals and signs for lymphocyte responses. *Cell* **76:** 275–285.

Janeway CA, Medhzitov R, Pfeiffer C, Tao X, Bottomly K. (1995) Altered peptide ligands. Conformational changes in the TCR. *Immunologist* **3:** 41–44.

Jensen K, Sollid LM, Scott H, Paulsen G, Kett K, Thorsby E, Lundin KEA. (1995) Gliadin-specific T cell responses in peripheral blood of healthy individuals involve T cells restricted by the coeliac disease associated DQ2 heterodimer. *Scand. J. Immunol.* **42:** 166–170.

Jorgensen JL, Reay PA, Ehrich EW, Davis MM. (1992) Molecular components of T-cell recognition. *Annu. Rev. Immunol.* **10:** 835–873.

Kamikawaji N, Fujisawa K, Yoshizumi H, Fukunaga M, Yasunami M, Kimura A, Nishimura Y, Sasazuki T. (1991) HLA–DQ–restricted CD4+ T cells specific to streptococcal antigen present in low but not in high responders. *J. Immunol.* **146:** 2560–2567.

Karjalainen K. (1994) High sensitivity, low affinity – paradox of T-cell receptor recognition. *Curr. Opin. Immunol.* **6:** 9–12.

Katayama CD, Eidelman FJ, Duncan A, Hooshmand F, Hedrick SM. (1995) Predicted complementary determining regions of the T cell antigen receptor determine antigen specificity. *EMBO J.* **14:** 927–938.

Kearney ER, Pape KA, Loh DY, Jenkins MK. (1994) Visualization of peptide-specific T cell immunity and peripheral tolerance induction in vivo. *Immunity* **1:** 327–339.

Kelsoe A. (1995) Th1 and Th2 subsets: paradigms lost? *Immunol. Today* **16:** 374–379.

Khoury SJ, Hancock WW, Weiner HL. (1992) Oral tolerance to myelin basic protein and natural recovery from experimental autoimmune encephalomyelitis are associated with downregulation of inflammatory cytokines and differential upregulation of transforming growth factor β, interleukin 4, and prostaglandin E expression in the brain. *J. Exp. Med.* **176:** 1355–1364.

Kindred B, Shreffler DC. (1972) H-2 dependence of co-operation between T and B cells in vivo. *J. Immunol.* **109:** 940–943.

Kirberg J, Baron A, Jakob S, Rolink A, Karjalainen K, von Boehmer H. (1994) Thymic selection of CD8+ single positive cells with a class II major histocompatibility complex-restricted receptor. *J. Exp. Med.* **180:** 25–34.

Koelle DM, Corey L, Burke RL, Eisenberg RJ, Cohen GH, Pichyangkura R, Triezenberg SJ. (1994) Antigenic specificities of human CD4+ T-cell clones recovered from recurrent genital herpes simplex virus type 2 lesions. *J. Virol.* **68:** 2803–2810.

Kuchroo VK, Greer JM, Kaul D, Ishioka G, Franco A, Sette A, Sobel RA, Lees MB. (1994) A single TCR antagonist peptide inhibits experimental allergic encephalomyelitis mediated by a diverse T cell repertoire. *J. Immunol.* **153:** 3326.

Kumar V, Bhardwaj V, Soares L, Alexander J, Sette A, Sercarz E. (1995) Major histocompatibility complex binding affinity of an antigenic determinant is crucial for the differential secretion of interleukin 4/5 or interferon γ by T cells. *Proc. Natl Acad. Sci. USA* **92:** 9510–9514.

Lamb JR, Higgins JA, Hetzel C, Hayball JD, Lake RA, O'Hehir RE. (1995) The effects of changes at peptide residues contacting MHC class II T-cell receptor on antigen recognition and human Th0 cell effector function. *Immunology* **85:** 447–454.

Lanzavecchia A. (1989) Is suppression a function of class II-restricted cytotoxic T cells? *Immunol. Today* **10:** 157–159.

Lanzavecchia A. (1990) Receptor-mediated antigen uptake and its effect on antigen presentation to class II-restricted T lymphocytes. *Annu. Rev. Immunol.* **8:** 773–793.

Lanzavecchia A. (1993) Identifying strategies for immune intervention. *Science* **260:** 937–942.

Lanzavecchia A, Reid PA, Watts C. (1992) Irreversible association of peptides with class II MHC molecules in living cells. *Nature* **357:** 249–252.

Löhr H, Fleischer B, Gerken G, Yeaman SJ, Meyer zum Bueschenfelde KH, Manns M. (1993) Autoreactive liver-infiltrating T cells in primary biliary cirrhosis recognize inner mitochondrial epitopes and the pyruvate dehydrogenase complex. *J. Hepatol.* **18:** 322–327.

Lombardi G, Sidhu S, Batchelor R, Lechler R. (1994) Anergic T cells as suppressor cells in vitro. *Science* **264:** 1587–1589.

Lundin KEA, Sollid LM, Bosnes V, Gaudernack G, Thorsby E. (1990a) T-cell recognition of HLA class II molecules induced by γ-interferon on a colonic adenocarcinoma cell line (HT29). *Scand. J. Immunol.* **31:** 469–475.

Lundin KEA, Qvigstad E, Rønningen KS, Thorsby E. (1990b) Antigen-specific T cells restricted by HLA–DQw8: Importance of residue 57 of the DQβ chain. *Hum. Immunol.* **28:** 397– 405.

Lundin KEA, Scott H, Hansen T, Paulsen G, Halstensen TS, Fausa O, Thorsby E, Sollid LM. (1993) Gliadin-specific, HLA–DQ(α1*0501,β1*0201) restricted T cells isolated from the small intestinal mucosa of celiac disease patients. *J. Exp. Med.* **178:** 187–196.

Lundin KEA, Scott H, Fausa O, Thorsby O, Sollid LM. (1994) T cells from the small intestinal mucosa of a DR4,DQ7/DR4,DQ8 celiac disease patient preferentially recognize gliadin when presented by DQ8. *Hum. Immunol.* **41:** 285–291.

Mackay CR. (1993) Homing of naive, memory and effector lymphocytes. *Curr. Opin. Immunol.* **5:** 423–427.

Madden DR. (1995) The three-dimensional structure of peptide-MHC complexes. *Annu. Rev. Immunol.* **13:** 587–622.

Maggi E, Biswas P, Delprete G, *et al.* (1991) Accumulation of Th-2-like helper T cells in the conjunctiva of patients with vernal conjunctivitis. *J. Immunol.* **146:** 1169–1174.

Maggi E, Parronchi P, Manetti R, Simonelli C, Piccini MP, Rugiu FS, Carli MD, Ricci M, Romagnani S. (1992) Reciprocal regulatory effects of IFN-γ and IL-4 on the in vitro development of human Th1 and Th2 clones. *J. Immunol.* **148:** 2142–2147.

Mahon BP, Katrak K, Nomoto A, Macadam AJ, Minor PD, Mills KHG. (1995) Poliovirus-specific CD4+ Th1 clones with both cytotoxic and helper activity mediate protective humoral immunity against a lethal poliovirus infection in transgenic mice expressing the human poliovirus receptor. *J. Exp. Med.* **181:** 1285–1292.

Manca F, Habeshaw J, Dalgluish A. (1991) The naive repertoire of human T helper cells specific for gp120, the envelope glycoprotein of HIV. *J. Immunol.* **146:** 1964–1971.

Manickan E, Rouse RJD, Yu Z, Wire WS, Rouse BT. (1995) Genetic immunization against herpes simplex virus. Protection is mediated by CD4+ T lymphocytes. *J. Immunol.* **155:** 259–265.

Markovic-Plese S, Fukaura H, Zhang J, Al-Sabbagh A, Southwood S, Sette A, Kuchroo VK, Hafler DA. (1995) T cell recognition of immunodominant and cryptic proteolipid protein epitopes in humans. *J. Immunol.* **155:** 982–992.

Marrack P, Kappler J. (1988) The T-cell repertoire for antigen and MHC. *Immunol. Today* **9:** 308–315.

Matsuo H, Batocchi AP, Hawke S, Nicolle M, Jacobson L, Vincent A, Newson-Davis J, Willcox N. (1995) Peptide-selected T cell lines from myasthenia gravis patients and controls recognize epitopes that are not processed from whole acetylcholine receptor. *J. Immunol.* **155:** 3683–3692.

Meuer SC, Schlossman SF, Reinherz EL. (1982) Clonal analysis of human cytotoxic T lymphocytes: T4+ and T8+ effector T cells recognize products of different major histocompatibility complex regions. *Proc. Natl Acad. Sci. USA* **79:** 4395–4399.

Minutello MA, Pileri P, Unutmaz D, Censini S, Kuo G, Houghton M, Brunetto MR, Bonino F, Abrignani S. (1993) Compartmentalization of T lymphocytes to the site of the disease: Intrahepatic CD4+ T cells specific for the protein NS4 of hepatitis C virus in patients with chronic hepatitis C. *J. Exp. Med.* **178:** 17–25.

Möller E, Böhme J, Valugerdi MA, Ridderstad A, Olerup O. (1990) Speculations on mechanisms of HLA associations with autoimmune diseases and the specificity of 'autoreactive' T lymphocytes. *Immunol. Rev.* **118:** 5–19.

Möller G. (1988) Do suppressor T cells exist? *Scand. J. Immunol.* **27:** 247–250.

Mosmann TR, Coffman RL. (1989) Th1 and Th2 cells: Different patterns of lymphokine secretion lead to different functional properties. *Annu. Rev. Immunol.* **7:** 145–173.

Mouritsen S, Meldal M, Werdelin O, Hansen AS, Buus S. (1992) MHC molecules protect T cell epitopes against proteolytic destruction. *J. Immunol.* **149:** 1987–1993.

Müller I, Louis JA. (1989) Immunity to experimental infection with Leishmania major: generation of protective L3T4+ T cell clones recognizing antigen(s) associated with live parasites. *Eur. J. Immunol.* **19:** 865–871.

Mustafa AS, Lundin KEA, Oftung F. (1993) Human T cells recognize mycobacterial heat shock proteins in the context of multiple HLA-DR molecules: Studies with healthy subjects vaccinated with Mycobacterium bovis BCG and Mycobacterium leprae. *Infect. Immun.* **61:** 61–65.

Nelson CA, Roof RW, McCourt DW, Unanue ER. (1992) Identification of the naturally processed form of hen egg lysozyme bound to the murine major histocompatibility complex class II molecule I–Ak. *Proc. Natl Acad. Sci. USA* **89:** 7380–7383.

Nilsen EM, Lundin KEA, Krajci P, Scott H, Sollid LM, Brandtzaeg P. (1995) Gluten-specific, HLA-DQ-restricted T cells from coeliac mucosa produce cytokines with Th1 or Th0 profile dominated by interferon-gamma. *Gut* **37:** 766–776.

Obata F, Kashiwagi N. (1992) Sequence analysis of the T-cell receptor utilized for recognition of HLA-class II molecules. *In: HLA 1991* (eds K Tsuji, M Aizawa, T Saszuki). Oxford University Press, Oxford pp. 865–901.

Ono SJ, Bazil V, Sugawara M, Strominger JL. (1991) An isotype–specific trans-acting factor is defective in a mutant B cell line that expresses HLA–DQ, but not –DR or –DP. *J. Exp. Med.* **173:** 629–637.

Ottenhoff THM, Mutis T. (1990) Specific killing of cytotoxic T cells and antigen-presenting cells by CD4+ cytotoxic T cell clones. A novel potentially immunoregulatory T-T cell interaction in man. *J. Exp. Med.* **171:** 2011–2024.

Ottenhoff THM, Neuteboom S, Elferink DG, de Vries RRP. (1986) Molecular localization and polymorphism of HLA class II restriction determinants defined by Mycobacterium leprae-reactive helper T cell clones from leprosy patients. *J. Exp. Med.* **164:** 1923–1939.

Ottenhoff THM, Walford C, Nishimura Y, Reddy NBB, Sasazuki T. (1990) HLA-DQ molecules and the control of Mycobacterium leprae-specific T cell nonresponsiveness in lepromateous leprosy patients. *Eur. J. Immunol.* **20:** 2347–2350.

Parronchi P, Macchia D, Piccinni MP, Biswas P, Simonelli C, Maggi E, Ricci M, Ansari AA, Romagnani S. (1991) Allergen- and bacterial antigen-specific T-cell clones established from atopic donors show a different profile of cytokine production. *Proc. Natl Acad. Sci. USA* **88:** 4538–4542.

Paulsen G, Lundin KEA, Gjertsen HA, Hansen T, Sollid LM, Thorsby E. (1995) HLA-DQ2 restricted T cell recognition of gluten-derived peptides in celiac disease. Influence of amino acid substitutions in the membrane distal domain of DQβ1*0201. *Hum. Immunol.* **45:** 145–153.

Perkins DL, Berriz G, Kamradt T, Smith JA, Gefter ML. (1991) Immunodominance:

Intramolecular competition between T cell epitopes. *J. Immunol.* **146:** 2137–2144.
Pette M, Fujita K, Wilkinson D, Altmann DM, Trowsdale J, Giegerich G, Hinkkanen A, Epplen JT, Kappos L, Wekerle H. (1990) Myelin autoreactivity in multiple sclerosis: recognition of myelin basic protein in the context of HLA–DR2 products by T lymphocytes of multiple-sclerosis patients and healthy donors. *Proc. Natl Acad. Sci. USA* **87:** 7968–7972.
Przemioslo RT, Lundin KEA, Sollid LM, Nelufer J, Ciclitira PJ. (1995) Histological changes in small bowel mucosa induced by gliadin sensitive T-lymphocytes can be blocked by anti-interferon-γ antibody. *Gut* **36:** 874–879.
Quayle AJ, Wilson KB, Li SG, Kjeldsen-Kragh J, Oftung F, Shinnick T, Sioud M, Førre Ø, Capra JD, Natvig JB. (1992) Peptide recognition, T cell receptor usage and HLA restriction elements of human heat-shock protein (hsp) 65 and mycobacterial 65-kDa hsp-reactive T cell clones from rheumatoid synovial fluid. *Eur. J. Immunol.* **22:** 1315–1322.
Racioppi L, Ronchese F, Matis LA, Germain RN. (1993) Peptide-major histocompatibility complex class II complexes with mixed agonist/antagonist properties provide evidence for ligand-related differences in T cell receptor-dependent intracellular signaling. *J. Exp. Med.* **177:** 1047–1060.
Rammensee HG. (1995) Chemistry of peptides associated with MHC class I and class II molecules. *Curr. Opin. Immunol.* **7:** 85–96.
Robey E, Allison JP. (1995) T-cell activation: integration of signals from the antigen receptor and costimulatory molecules. *Immunol. Today* **16:** 306–309.
Rock KL, Benacerraf B. (1984) Inhibition of antigen-specific T lymphocyte activation by structurally related Ir gene-controlled polymers. II. Competitive inhibition of I-E- restricted, antigen-specific T cell responses. *J. Exp. Med.* **160:** 1864–1879.
Romagnani S. (1994) Lymphokine production by human T cells in disease states. *Annu. Rev. Immunol.* **12:** 227–257.
Rosenthal AS, Shevach EM. (1973) Function of macrophages in antigen recognition by guinea pig T lymphocytes. *J. Exp. Med.* **138:** 1194–1212.
Rudolphi A, Boll G, Poulsen SS, Claesson MH, Reimann J. (1994) Gut-homing CD4+ T cell receptor $\alpha\beta$+ T cells in the pathogenesis of murine inflammatory bowel disease. *Eur. J. Immunol.* **24:** 2803–2812.
Salazar M, Deulofeut H, Granja C, Deulofeut R, Yunis DE, Marcus–Bagley D, Awdeh Z, Alper CA, Yunis EJ. (1995) Normal HBsAg presentation and T-cell defect in the immune response of nonresponders. *Immunogenetics* **41:** 366–374.
Salemi S, Caporossi A, Boffa L, Longobardi MG, Barnaba V. (1995) HIVgp120 activates autoreactive CD4–specific T cell responses by unveiling of hidden CD4 peptides during processing. *J. Exp. Med.* **181:** 2253–2257.
Salgame P, Convit J, Bloom BR. (1991) Immunological suppression by human CD8+ T cells is receptor dependent and HLA–DQ restricted. *Proc. Natl Acad. Sci. USA* **88:** 2598–2602.
Sallusto F, Cella M, Danieli C, Lanzavecchia A. (1995) Dendritic cells use macropinocytosis and the mannose receptor to concentrate macromolecules in the major histocompatibility complex class II compartment: downregulation by cytokines and bacterial products. *J. Exp. Med.* **182:** 389–400.
Savage COS, Brooks CJ, Harcourt GC, Picard JK, King W, Sansom DM, Willcox N. (1995) Human vascular endothelial cells process and present autoantigen to human T cell lines. *Int. Immunol.* **7:** 471–479.
Schaeffer EB, Sette A, Johnson DL, Bekoff MC, Smith JA, Grey HM, Buus S. (1989) Relative contribution of 'determinant selection' and 'holes in the T-cell repertoire' to T-cell responses. *Proc. Natl Acad. Sci. USA* **86:** 4649–4653.
Schmid DS. (1988) The human MHC-restricted cellular response to herpes simplex virus type 1 is mediated by CD4+, CD8– T cells and is restricted to the DR region of the MHC complex. *J. Immunol.* **140:** 3610–3616.
Scott H, Sollid LM, Fausa O, Brandtzaeg P, Thorsby E. (1987) Expression of major histocompatibility complex class II subregion products by jejunal epithelium in patients with coeliac disease. *Scand. J. Immunol.* **26:** 563–571.
Sercarz EE, Lehmann PV, Ametani A, Benichou G, Miller A, Moudgil K. (1993) Dominance and crypticity of T cell antigenic determinants. *Annu. Rev. Immunol.* **11:** 729–766.
Serra HM, Crimi C, Sette A, Celis E. (1993) Fine restriction analysis and inhibition of antigen recognition in HLA-DQ-restricted T cells by major histocompatibility complex blockers and T cell receptor antagonists. *Eur. J. Immunol.* **23:** 2967–2971.
Shanafeldt MC, Hindersson P, Soderberg C, Mensi N, Turck CW, Webb D, Yssel H, Peltz G. (1991) T cell and antibody reactivity with the Borrelia burgdorferi 60-kDa heat shock protein in

Lyme arthritis. *J. Immunol.* **146:** 3985–3992.

Siliciano RF, Lawton T, Knall C, Karr RW, Berman P, Gregory T, Reinherz EL. (1988) Analysis of host-virus interactions in AIDS with anti-gp120 T cell clones: Effect of HIV sequence variation and a mechanism for CD4+ cell depletion. *Cell* **54:** 561–575.

Simitsek PD, Campbell DG, Lanzavecchia A, Fairweather N, Watts C. (1995) Modulation of antigen processing by bound antibodies can boost or suppress class II major histocompatibility complex presentation of different T cell determinants. *J. Exp. Med.* **181:** 1957–1963.

Sloan-Lancaster J, Allen PM. (1995) Significance of T-cell stimulation by altered peptide ligands in T cell biology. *Curr. Opin. Immunol.* **7:** 103–109.

Snoke K, Alexander J, Franco A, Smith L, Brawley JV, Concannon P, Grey HM, Sette A, Wentworth P. (1993) The inhibition of different T cell lines specific for the same antigen with TCR antagonist peptides. *J. Immunol.* **151:** 6815–6821.

Sollid LM, Thorsby E. (1993) HLA susceptibility genes in celiac disease: Genetic mapping and role in pathogenesis. *Gastroenterology* **105:** 910–922.

Sollid LM, Markussen G, Ek J, Gjerde H, Vartdal F, Thorsby E. (1989) Evidence for a primary association of celiac disease to a particular HLA–DQ $\alpha\beta$ heterodimer. *J. Exp. Med.* **169:** 345–350.

Spurkland A, Gedde–Dahl T, Hansen T, Vartdal F, Gaudernack G, Thorsby E. (1994) Heterogeneity of T cells specific for a particular peptide/HLA–DQ complex. *Hum. Immunol* **39:** 61–68.

Stalder T, Hahn S, Erb P. (1994) Fas antigen is the major target molecule for CD4+ T cell–mediated cytotoxicity. *J. Immunol* **152:** 1127–1133.

Steinman RM. (1991) The dendritic cell system and its role in immunogenicity. *Annu. Rev. Immunol.* **9:** 271–296.

Stern LJ, Brown JH, Jardetzky TS, Gorga JC, Urban RG, Strominger JL, Wiley DC. (1994) Crystal structure of the human class II MHC protein HLA-DR1 complexed with an influenza virus peptide. *Nature* **368:** 215–221.

Thompson CB. (1995) Distinct roles for the costimulatory ligands B7-1 and B7-2 in T helper cell differentiation. *Cell* **81:** 979–982.

Thorsby E. (1995) HLA-associated disease susceptibility - which genes are primarily involved? *Immunologist* **3:** 51–58.

Trentham DE, Dynesius-Trentham RA, Orav EJ, Combitchi D, Lorenzo C, Sewell KL, Hafler DA, Weiner HL. (1993) Effects of oral administration of type II collagen on rheumatoid arthritis. *Science* **261:** 1727–1729.

Trier JS. (1991) Celiac sprue. *N. Engl. J. Med.* **325:** 1709–1719.

van de Water J, Ansari A, Prindiville T, *et al.* (1995) Heterogeneity of autoreactive T cell clones specific for the E2 component of the pyruvate dehydrogenase complex in primary biliary cirrhosis. *J. Exp. Med.* **181:** 723–733.

Woolfrey AE, Nepom GT. (1995) Differential transcription elements direct expression of HLA–DQ genes. *Clin. Immunol. Immunopath.* **74:** 119–126.

Wraith DC, McDevitt HO, Steinman L, Acha-Orbea H. (1989) T cell recognition as the target for immune intervention in autoimmune disease. *Cell* **57:** 709–715.

Wucherpfennig KW, Strominger JL. (1995) Molecular mimicry in T-cell mediated autoimmunity: Viral peptides activate human T cell clones specific for myelin basic protein. *Cell* **80:** 695–705.

Yamamura M, Uyemura K, Deans RJ, Weinberg K, Rea TH, Bloom BR, Modlin RL. (1991) Defining protective responses to pathogens: Cytokine profiles in leprosy lesions. *Science* **254:** 277–279.

Yasukawa M, Zarling JM. (1984) Human cytotoxic T cell clones directed against herpes simplex virus-infected cells. I. Lysis restricted by HLA class II MB and DR antigens. *J. Immunol.* **133:** 422–427.

Zamvil SS, Steinman L. (1990) The T lymphocyte in experimental allergic encephalomyelitis. *Annu. Rev. Immunol.* **8:** 579–621.

Ziegler HK, Unanue ER. (1982) Decrease in macrophage antigen catabolism caused by ammonia and chloroquine is associated with inhibition of antigen presentation to T cells. *Proc. Natl Acad. Sci. USA* **79:** 175–178.

Zinkernagel RM, Doherty PC. (1974) Restriction of in vitro T-cell mediated cytotoxicity in lymphocytic choriomeningitis within a syngeneic or semiallogeneic system. *Nature* **248:** 701–702.

Zinkernagel RM, Doherty PC. (1987) MHC-restricted cytotoxic T cells: Studies in the biological role of polymorphic major transplantation antigens determining T-cell restriction – specificity, function, and responsiveness. *Adv. Immunol.* **27:** 51–177.

15

HLA and disease: from molecular function to disease association?

Frances C. Hall and Paul Bowness

15.1 Introduction

The association of the human leukocyte antigen (HLA) class I allele HLA-B27 with the rheumatic disease ankylosing spondylitis was first recognized over 25 years ago (Brewerton *et al.*, 1973). Many HLA associations with autoimmune diseases have subsequently been described, although the underlying pathogenic mechanisms remain unresolved. In this chapter, illustrative examples of disease associations will be used, but a more comprehensive listing can be found elsewhere (Tiwari and Terasaki, 1985). Historically, most HLA associations have been described with susceptibility to autoimmune disease. Recently, however, protection against severe malaria has been described for both class I and II alleles (Hill *et al.*, 1991a), and we would predict that controlled studies using large numbers will detect both protective and detrimental effects of different HLA alleles in many infectious diseases.

There is evidence that the pathogenesis of many autoimmune diseases involves an infectious trigger. Many hypotheses, not all mutually exclusive, have been put forward to explain the HLA associations of these diseases. In this chapter we will first describe the possible mechanisms by which HLA molecules might be involved in disease pathogenesis. We shall then describe the experimental strategies that may be used to elucidate the presence, pattern and possible pathogenic mechanisms underlying these disease associations. Finally a number of specific diseases will be used to illustrate current thinking on HLA disease associations.

15.2 Mechanisms of disease association

The mechanisms of HLA association with disease have not been adequately explained and, consequently, hypotheses abound (*Table 15.1*). Associations are usually described in terms of susceptibility, but negatively associated molecules are sometimes referred to as 'protective'.

HLA and MHC: genes, molecules and function, edited by M.J. Browning and A.J. McMichael.
© 1996 BIOS Scientific Publishers Ltd, Oxford.

Table 15.1. Hypotheses to explain the association of HLA with disease

Presentation of a pathogenic peptide by HLA molecules
Thymic selection of T cell receptor repertoire on HLA molecules with self-peptides
Peptide contribution by HLA molecules
Interaction of HLA (class II) molecules with superantigens
Chemical modification of HLA molecules, altering their peptide specificity
Specific binding and internalization of microorganisms
Linkage of HLA locus to a non-HLA disease-associated gene
Regulation of T-cell development
Cross-reactive antibody responses

15.2.1 Peptide presentation by HLA molecules

A number of hypotheses relate the role of HLA molecules in disease predisposition, including autoimmunity, directly to their physiological function. The natural function of HLA molecules is to bind and present peptides derived from both self and foreign proteins to T lymphocytes (as described in Chapters 9, 10 and 12). Thus they play a central role in the cell-mediated immune response to infections. It is now clear that HLA class I molecules bind short peptides of about 8–12 amino acids derived from intracellular proteins, and consequently are thought to be important in the immune response to viruses and other intracellular pathogens (see below). HLA class II molecules bind longer peptides of 10–34 amino acids in length, most of which are derived from extracellular and cell-surface proteins (Chicz *et al.*, 1993). HLA molecules are amongst the most polymorphic known; for example, the DRβ chain has over 100 allelic variants (Bodmer *et al.*, 1995). Much of this polymorphism involves the peptide-binding groove, and consequently different HLA class I and II molecules bind different sets of peptides (reviewed by Rammensee *et al.*, 1995 and by Davenport and Hill, Chapter 12 in this volume). It is likely that these differences in peptide binding result in differences in the immune response to infectious pathogens, and indeed there is evidence that MHC genes are under strong evolutionary pressure (see Chapter 1). It is attractive to hypothesize that the HLA associations with both infectious and autoimmune diseases stem directly from these differences in peptide selection. Possession of certain alleles may thus lead to either an absent or a particularly vigorous T-cell response to a given pathogen. A strong T-cell response might protect against disease or induce pathology, depending on the site and nature of antigen recognized, and on the effector function of the T cells involved. Disease might occur if a given HLA molecule presented a 'pathogenic peptide' to T cells, either as a function of the allele-specific peptide-binding motif or, perhaps, by access to peptides derived from unusual sites (as described for HLA-B27 below). Alternatively, disease might occur through failure to bind a protective epitope.

If a major histocompatibility complex (MHC) allele confers susceptibility by its gene product presenting a pathogenic peptide, then a dominant effect would be expected (although the products of different alleles might bind such a peptide with

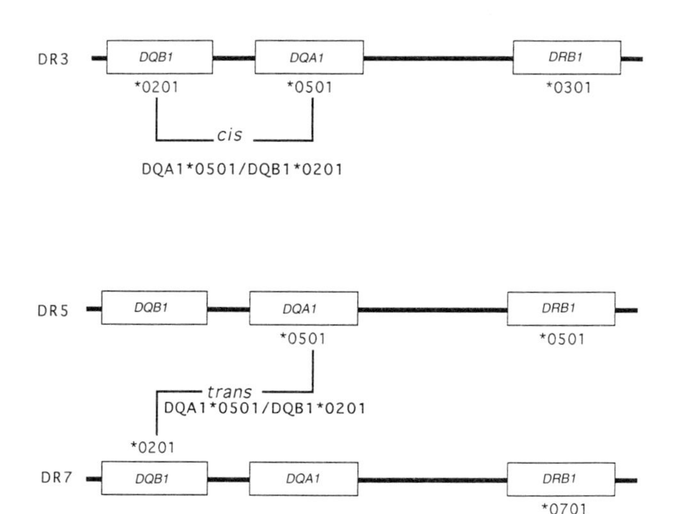

Figure 15.1. *Cis-* and *trans*-encoded DQ susceptibility in coeliac disease.

varying affinities, and therefore might differ in the strength of disease association). The penetrance of such a trait could be modified by competition for the pathogenic peptide by other MHC molecules, which present it benignly. This has been suggested in insulin-dependent diabetes mellitus (IDDM) where *DQB1*0302* confers dominant susceptibility, but penetrance is modified by the presence of various DR haplotypes (Nepom and Ehrlich, 1991). Alternatively, penetrance could be affected by competition for overlapping epitopes by other HLA alleles, as recently described for HLA-B8 and HLA-B27 (Tussey *et al.*, 1995). This picture might be further complicated by the presence of a huge excess of innocent or 'regulatory' peptides, possibly themselves MHC-derived (see below). In all cases, the outcome would depend principally on the affinity of the peptide–MHC interactions, but could also be affected by gene dosage and expression levels.

It is easy to envisage how all these factors affecting peptide selection by HLA molecules could account for the hierarchy of HLA associations seen in some diseases, ranging from strong predisposition, through neutral to protective effects. In IDDM, for example, DR3 and DR4 are strongly associated with susceptibility (the highest risk existing in those with genotype DR3/DR4); DR1 and DR8 are less strongly associated, DR5 mildly 'protective' and DR2 exhibits a marked negative association (Nepom and Ehrlich, 1991). A similar hierarchy exists for rheumatoid arthritis (Winchester, 1994) and early-onset pauciarticular juvenile chronic arthritis (Paul *et al.*, 1993).

Finally, in the case of the class II DQ isotype, both α and β chains are polymorphic and a hybrid susceptibility molecule with distinct peptide-binding properties may be encoded by different haplotypes. This mechanism has been proposed to explain the HLA associations in coeliac disease (*Figure 15.1*). Most coeliac patients are either DR3 positive or have the genotype DR5/DR7; 90% of patients carry *DQA1*0501* and *DQB1*0201*. These are both found together on the DR3 haplotype (*cis*), while, individually, the DR5 haplotype carries *DQA1*0501* and the DR7 haplotype *DQB1*0201*(*trans*). Interisotypic heterodimers have also been described; for example, DRα-DQβ, and such a mechanism could explain the interaction of *DP*0201* with disease-associated DRs in early-onset pauciarticular juvenile chronic arthritis (Paul *et al.*, 1993).

15.2.2 Thymic selection of T-cell repertoire

MHC molecules presenting self-peptides are thought to provide templates for T-cell receptor (TCR) repertoire formation in the thymus. Although the exact mechanisms involved and the relative contribution of peptide and MHC are not known, it is clear that one of three fates awaits the immature T cell entering the thymus (Ashton-Rickardt and Tonegawa, 1994). If the TCR fails to recognize self-MHC, the cell undergoes apoptosis. A high avidity interaction of the TCR with self-MHC and peptide also results in cell death ('negative selection'). Between these extremes, a range of avidities permit positive selection of the cell and egress into the periphery. Thymic selection could thus enable an MHC molecule either to increase or decrease disease susceptibility. Positive selection or deletion of the cross-reactive or auto-reactive cells specific to a particular disease will occur with varying efficiency according to HLA haplotype. Selection of cross-reactive T cells may even benefit the individual; for example, in DR4-positive individuals, the homology between the DR4 hypervariable region (HVR)3 and both Epstein–Barr virus (EBV) and *Escherichia coli* proteins might ensure a robust immune response to these common pathogens.

The level and distribution of HLA expression in the thymus will also contribute to the avidity of TCR interaction with MHC-peptide on presenting cells. MHC density is a product of gene dosage, allelic variation in transcription/translation efficiency and turnover rate. Homozygosity at a class II locus in mice approximately doubles the surface expression of the encoded molecule compared to the heterozygote. This difference is sufficient to alter efficiency of positive

Table 15.2. Genotype-specific relative risks for severe rheumatoid arthritis calculated relative to DRX/DRX (adapted from Wordsworth, 1995)

Genotypes	RR	*P* value
DR4/DR4	25	★★
Dw4/Dw4	15	★★
Dw4/Dw14	49	★★
Dw14/Dw14	14	
DR4/DR1	16	★★
Dw4/DR1	21	★★
Dw14/DR1	9	★★
DR4/DRX	5	★★
Dw4/DRX	6	★★
Dw14/DRX	4	★★
DR1/DR1	5	★
DR1/DRX	3	★
DRX/DRX	1	

*$P < 0.001$, **$P < 0.00001$.
DRX represents all non-DR1, non DR4 alleles.

selection, at least for certain receptor–ligand combinations (Berg *et al.*, 1990). Thus, in homozygotes, the loci could act co-dominantly to increase disease susceptibility or severity. In rheumatoid arthritis (RA), for example, expression of two DR4 alleles is associated with more severe disease (Salmon, 1992; Weyand *et al.*, 1992; Wordsworth *et al.*, 1992) and earlier disease onset (Nepom *et al.*, 1984). Interestingly, possession of two different DR4 alleles encoding the Dw4 and Dw14 cellular determinants ('compound heterozygotes') confers the highest relative risk (RR = 49, see *Table 15.2*). This finding is consistent with an effect of DR genotype on the development of a disease-susceptible repertoire in RA. Surface density of MHC allelic products also depends on the relative efficiencies of their promoters. Early-onset pauciarticular juvenile chronic arthritis has been associated with *DQA1* alleles present on both DR5 and DR8 haplotypes. The predisposing *DQA1* alleles each have a promoter mutation, which reduces transcription (Scholz and Albert, 1994). In this case we might envisage that low levels of DQα expression might permit survival of auto-reactive T cells, since negative selection of T cells requires a high avidity interaction.

15.2.3 Peptide contribution by HLA molecules

HLA class II antigens present peptides predominantly from proteins that intersect the endosomal/class II pathway (see Chapter 10). This includes not only extracellular organisms but also many self-proteins destined either for secretion, plasma membrane or endosomes themselves. Intriguingly, elution of naturally processed peptides from MHC molecules demonstrates that the majority of self-peptides are themselves derived from other MHC molecules, both class I and II (Chicz *et al.*, 1993; Ramensee *et al.*, 1995). HLA molecules therefore might regulate immune responses not only as restriction elements but also as a source of presented peptides. In this model, a 'protective' HLA molecule may provide a peptide that competes with stimulatory peptides. In contrast, an MHC-derived peptide may predispose to disease through structural similarity with a microbial-derived peptide. Infection with the microorganism may reverse tolerance to the MHC-derived peptide and initiate a true autoimmune response. Some intriguing homologies have been identified between HLA molecules and microbial proteins (discussed for B27 and DR4 later).

15.2.4 Superantigens

Superantigens are proteins that are able to stimulate large numbers of T cells by cross-linking their TCRs with the MHC class II molecules of presenting cells (reviewed by Webb and Gascoigne, 1994). Superantigens are produced by certain bacteria and viruses, and are also encoded by endogenous murine retroviruses. A given superantigen can bind to many class II molecules (usually DR) and to most, if not all, TCRs expressing particular Vβ families, ensuring that a large number of T cells are stimulated simultaneously. Superantigenic stimulation could, perhaps, break tolerance to a conventional self-antigen. However, superantigenic stimulation of T cells bearing the target Vβ family is often followed by clonal deletion. This effectively alters the TCR repertoire and may alter disease susceptibility.

15.2.5 Altered-self hypothesis

In general, mechanisms of central and peripheral tolerance are thought to maintain unresponsiveness to self-peptides presented on self-MHC molecules. However, if either MHC or self-peptide are modified at a later stage, tolerance may be broken. Modification of the MHC molecule has been implicated in a group of class I-associated diseases. It has been proposed that, in HLA-B27, the sulphydryl group of cysteine at position 67 may be oxidized in some tissues, altering the peptide-binding groove (Archer *et al.*, 1990). Chemical modification of cysteine could, theoretically, be important for some class II molecules as well; for example, the *DRB1*01* alleles have a cysteine at position 30 in the second hypervariable region, forming part of the floor of the peptide-binding groove. A similar mechanism has been advanced for berylliosis, a granulomatous lung disease resulting from inhalation of the metal beryllium. It is associated with *DPB1* alleles bearing glutamic acid at position 69, and it has been suggested that this negatively charged residue may bind beryllium and alter peptide-binding (Richeldi *et al.*, 1993).

Alteration of self-peptides could also occur, either by exposure of cryptic epitopes or by novel post-translational modification of proteins. Methimazole-induced insulin autoimmune syndrome may be an example of modification of a protein to reveal a cryptic epitope (Matsushita *et al.*, 1994). A peptide from the human insulin α-chain (^{8}TSICSLYQLE^{17}) binds well to disease-associated DRB1*0406 but not to the non-associated DRB2*0405. However, this peptide is usually unavailable for processing due to the disulphide bond between flanking cysteine residues at positions 6 and 11. Methimazole is a reducing compound which could cleave this bond, exposing the immunogenic peptide. It has recently been demonstrated that post-translational modification of proteins can affect immunogenicity. In a collagen type II-derived peptide (CII 256–270), removal of O-linked carbohydrate moieties abrogated recognition of the peptide by specific T-cell clones and increased the effectiveness of the peptide in inducing arthritis in susceptible mice (Michaelsson *et al.*, 1994). Interestingly, in some diseases, including RA, glycosylation of serum gamma immunoglobulin (IgG) is reduced (Parekh *et al.*, 1985), providing a possible mechanism for the induction of autoimmunity.

15.2.6 Receptor hypothesis

The receptor theory suggests that microorganisms could recognize specific MHC molecules and use them as a vehicle on which to enter the cell. A precedent for this exists in the blood-group system. The Duffy red blood cell antigen acts as a receptor for *Plasmodium vivax* and Nigerians lacking the Duffy antigens are resistant to this form of malaria (Miller *et al.*, 1975). More recently, another blood-group antigen, globoside, has been identified as the cellular receptor permitting infection with parvovirus B19 (Brown *et al.*, 1994).

15.2.7 Linkage to a disease-associated gene

In addition to class I and class II genes, the human MHC complex encodes over 70 other proteins, including complement components, tumour necrosis factor

(TNF) and peptide transporters involved in antigen processing (Campbell and Trowsdale, 1993). In some diseases, the associated HLA class I or II allele(s) may only be a marker for the true disease-associated gene lying in close proximity on the same haplotype; for example, association of HLA-*B47* with congenital adrenal hyperplasia has been explained by deletion of the gene encoding 21-hydroxylase (Manfras *et al.*, 1993) in linkage with *B47*. In systemic lupus erythematosus (SLE), the association with the *B8 C4AQ0 C4B1 DR3* haplotype may reflect the relative deficiency of complement component C4 due to the null allele *C4AQ0*, rather than a role for an MHC molecule (Arnett and Reveille, 1992). Lastly, totally unexpected associations have been described with diseases that have no apparent immunological basis. Perhaps the most perplexing is narcolepsy, where almost 100% of caucasian patients have DR2 compared to only 22% of controls and, as yet, no plausible candidate gene linked to DR2 has been identified.

15.2.8 Other mechanisms

There are many other possible mechanisms by which HLA might, theoretically, be involved in disease pathogenesis. Two mechanisms that are likely to have a significant effect are through variation in HLA expression and through regulation of functional T-cell differentiation. In rats transgenic for *HLA-B27*, development of a disease similar to the spondylarthropathies (see below) correlates with the level of expression of the *HLA-B27* transgene (Taurog *et al.*, 1994). Whilst HLA class I molecules are expressed on almost all nucleated cells, class II expression is tissue-specific. Constitutive expression occurs on B cells, but class II molecules can be induced by a variety of stimuli on other immune cells. Several cells of epithelial and mesenchymal origin, as well as tumours, can also be induced to express class II, and 'aberrant' expression in tissues such as pancreatic islet β cells (Campbell *et al.*, 1988) and synovium (Teyton *et al.*, 1987) may facilitate an autoimmune response. Class II expression is not a simple on/off phenomenon. Stimuli including cytokines, hormones, cross-linking of surface receptors and intracellular infection can up- or down-regulate class II density (Glimcher and Kara, 1992) and DR, DQ and DP are sometimes induced separately; for example, studies on epithelial tissues revealed that DR and DP but not DQ are responsive to interferon gamma (IFN-γ; Manyak *et al.*, 1988). In addition to such class II isotype-specific regulation, allelic differences in expression associated with promoter-region polymorphisms have been described (Louis *et al.*, 1994). If such differences resulted in qualitative or quantitative changes in expression, they could influence disease.

There is now evidence for functional differences in the T-cell response to different HLA class II isotypes (although not alleles). Thus whilst DR, DQ and DP are all known to present peptide in an MHC-restricted manner to T cells, superantigens bind preferentially to DR (Webb and Gascoigne, 1994). Distinct functional roles for the class II isotypes, DR, DQ and DP are suggested both by their specific patterns of expression and by important structural differences. The 'suppressor' phenomenon described in lepromatous leprosy may be an example of a DQ-restricted function (Bloom *et al.*, 1992).

In addition to their role in antigen presentation and thymic selection, HLA molecules have other functions that may be salient to some disease associations;

for example, class I molecules interact with CD8 and, in the case of some alleles, with natural killer (NK) cells (Yokoyama, 1995). Class II molecules bind to CD4 and exhibit allelic differences in this interaction (Fleury *et al.*, 1995). There is also evidence that class II antigens can themselves signal to the antigen-presenting cell (Scholl and Geha 1994).

15.3 Strategies for exploring class II disease association

15.3.1 Genetics of HLA association with disease

Most HLA-associated diseases almost certainly have a multifactorial aetiology, involving genetic, environmental and stochastic factors. The genetic components do not display simple Mendelian inheritance. HLA contributes only a part of the genetic susceptibility, and even within the HLA complex a number of separate loci may interact. Thus, careful genetic and epidemiological studies are essential to establish and define HLA associations with disease. The first problem relates to identifying precisely the location of the association within the HLA haplotype. In family studies, all alleles on the MHC haplotype will usually co-segregate and it is impossible to distinguish the effect of one allele from its neighbours. This phenomenon gives rise to linkage disequilibrium in population studies, where DR and DQ alleles are particularly tightly linked. Furthermore, since allelic HVRs have been 'shuffled' between alleles, it is important to dissociate the effect of a particular HVR from that of the whole allele. A second complexity is due to interactions between HLA molecules encoded on the same haplotype (*cis*) and between molecules on different haplotypes (*trans*). This could occur, for example, through competition for peptides, by presentation of peptides derived from the other HLA molecules and, in DQ, by association of different polymorphic α- and β-chains. Most early studies explored the role of HLA molecules simply as susceptibility determinants; for example, B27 in ankylosing spondylitis, DR4 in RA, and DR2 and DR3 in SLE. However, it is now clear that HLA antigens are also associated with both severity and distinct patterns of disease. Many HLA associations are described in clinically heterogeneous diseases, and the role of HLA molecules may differ in each associated disease. Consequently in each case a series of questions must be posed.

(i) What is the pattern of inheritance of the disease?
(ii) What is the relative contribution of HLA genes?
(iii) Does HLA influence disease susceptibility and/or severity?
(iv) Which HLA haplotypes are associated with disease?
(v) Which is the associated locus within the haplotype?
(vi) Which specific nucleotide substitutions are associated?
(vii) Do combinations of HLA alleles interact in disease?
(viii) What are the functional correlates of the HLA associations?

Family studies can indicate the number of genes involved, whether inheritance follows a classical Mendelian pattern (of dominant or recessive alleles) and the penetrance of the trait (Khoury *et al.*, 1993). The most important epidemiological parameter is the relative risk, λ_R, which specifies the recurrence risk of a relative (R) of the proband compared to the incidence of disease in the population as a whole. For siblings (S), therefore:

$$\lambda_S = \frac{\text{risk of recurrence in sibling of proband}}{\text{risk of disease in general population}}$$

Values of λ_S include 3.5 for type II diabetes, 53 for ankylosing spondylitis and between three and seven for RA, depending on the severity of disease in the proband (Lander and Schork, 1994). Twin studies comparing concordance in monozygotic twins (100% genetic identity assumed) and dizygotic twins (50% genetic identity assumed) can be used to construct models of inheritance, specifying number of genes involved and whether or not inheritance obeys Mendelian laws (reviewed in Winchester, 1994). The contribution of the HLA locus itself can be estimated by studying monozygotic twin concordance rates and HLA haplotype sharing among affected members of multicase families (Rotter and Landaw, 1984). However, twin concordance rates are unreliable since the possibility of an unaffected twin developing disease later in life cannot be excluded. A second method avoids this problem and, instead, estimates risk of disease recurrence in siblings sharing neither HLA haplotype with the proband compared with the population risk (Risch, 1987). In RA, for example, these strategies suggest that the inheritance of susceptibility does not obey simple Mendelian laws, but involves between two and four loci. While the total genetic component of susceptibility to RA amounts to between 12 and 50%, the HLA locus contributes between 25 and 50% of this (Winchester, 1994).

Four separate strategies are used in the genetic dissection of HLA association: linkage analysis, haplotype sharing, association studies and experimental crosses in animal models (Lander and Schork, 1994).

Linkage analysis. This is a statistical tool which tests models of disease inheritance. A model specifies location of the disease-associated locus, allele frequency and penetrance. In linkage analysis, the observed genotypes and phenotypes in a pedigree are used to test a variety of models ($M\phi$) against a null hypothesis (M0) which assumes no linkage to a susceptibility gene in the region of interest. The evidence for model M1 is expressed as either the likelihood ratio or the lod score:

$$\text{likelihood ratio} = \frac{\text{probability of data given model M1}}{\text{probability of data given model M0}}$$

$$\text{lod score (Z)} = \log_{10}(\text{likelihood ratio})$$

The model M1 is chosen from a family of models ($M\phi$) and the maximum likelihood (ML) estimate is that value of ϕ which makes the data most likely to have occurred. This technique has been widely applied to monogenic disorders but can also be extended to polygenic disease, an approach known as simultaneous search. In an analysis of Finnish multiple sclerosis pedigrees, linkage has been indicated both to the MHC complex on chromosome 6 and the myelin basic protein gene on chromosome 18 (Tienari *et al.*, 1994).

Haplotype sharing. These methods determine how often a particular copy of a chromosomal region (haplotype) is shared identical-by-descent, that is inherited from a common ancestor within a pedigree. This is compared with random expectations of haplotype-sharing. This method can be applied in the absence of any model of inheritance and will show excess haplotype-sharing even in the presence of incomplete penetrance, genetic heterogeneity and a high frequency of disease-associated haplotypes in the population. Sib-pair analysis is the simplest form of this approach, where the expected distribution of haplotype sharing is: zero, 25%; one, 50%; two, 25%. This technique was used to confirm the importance of the HLA complex in susceptibility to type I diabetes (e.g. Rich *et al.*, 1991). In some diseases, such as RA, sharing of two haplotypes is increased more than sharing of one haplotype and this has been advanced as evidence of a recessive mode of inheritance (Rigby *et al.*, 1991).

Association studies. These are case-control studies comparing unrelated affected and unaffected individuals drawn from the same population (Lander and Schork, 1994). The significance of the difference in HLA allele frequencies between patients and controls can be determined using Fisher's exact test or by applying the chi^2 test in a series of 2×2 contingency tables. The RR is also used to express the strength of the association (Haldane, 1955; Woolf, 1955). This strategy can be equally applied to study associations of HLA haplotypes, alleles or specific nucleotide sequences. Due to the proximity of the DR and DQ loci (or other loci in strong linkage disequilibrium), their relative effects are difficult to dissociate in a given population. However, this can be achieved by conducting association studies in various ethnic groups with different DR–DQ haplotypic combinations. Association studies can be applied to the problem of allelic interactions in class II-associated disease, for example the role of the DR genotype in RA (reviewed in Winchester, 1994) or the interplay of DR and DQ loci in IDDM (reviewed in Nepom and Ehrlich, 1991).

Experimental crosses. In mice or rats these provide a powerful tool for genetic analysis of animal models of MHC-associated disease; for example, non-obese diabetic (NOD) mice provide a model for IDDM (reviewed in Serreze and Leiter, 1994). In the NOD mouse, at least 50% of the genetic susceptibility is conferred by the unusual MHC haplotype of this strain, H-2g7, which encodes the rare *I-Abg7* allele and which fails to express any I-E molecules. IDDM is inhibited in NOD mice expressing transgenes encoding I-A or I-E alleles from diabetes-resistant strains. Interestingly, apart from the MHC complex, susceptibility to disease in this mouse involves at least 14 other loci, which each make a partial contribution to a threshold trait. After mapping susceptibility loci, their physiological effects can be studied by developing congenic mouse strains, differing only at the locus under study. Alternatively, a transgenic mouse can be made, which may express the gene in question in a tissue-specific manner.

Characterization of the HLA-encoded susceptibility to disease using this and the other strategies outlined above provides some insight into possible mechanisms of disease association. These mechanistic hypotheses can be pursued by three categories of study: analysis of peptide processing and presentation by HLA molecules, analysis of the effect of disease-associated HLA antigens on the TCR repertoire and lastly, functional T-cell assays.

15.3.2 Analysis of peptide processing and presentation by HLA molecules

Once genetic studies have identified disease-associated and non-associated HLA alleles, it is important to study whether these alleles differ in their peptide-binding properties. This does appear to be the case for the disease-associated HLA-B27 subtypes and the spondylarthropathies. For a number of HLA class II-associated autoimmune diseases, the strongest associations are with particular amino acids at certain polymorphic regions contributing to the peptide-binding groove, rather than with individual HLA alleles; for example, IDDM is associated with particular residues at DQβ position 57 (see below). RA is more strongly associated with the so-called 'shared epitope' in the third allelic hypervariable region of the *DRB1* locus than with any individual allele. The 'shared epitope' is encoded by disease-associated *DRB1* alleles *DRB1*0401, DRB1*0404, DRB1*0405* and *DRB1*0101* (*Figure 15.2*) and comprises part of the wall of the peptide-binding groove (Gregersen *et al.*, 1987). These associations further support the notion that peptide selection is important in disease pathogenesis.

Peptide selection by HLA molecules is discussed in Chapter 12. Briefly, it can be studied by the identification of naturally presented epitopes, by crystallography

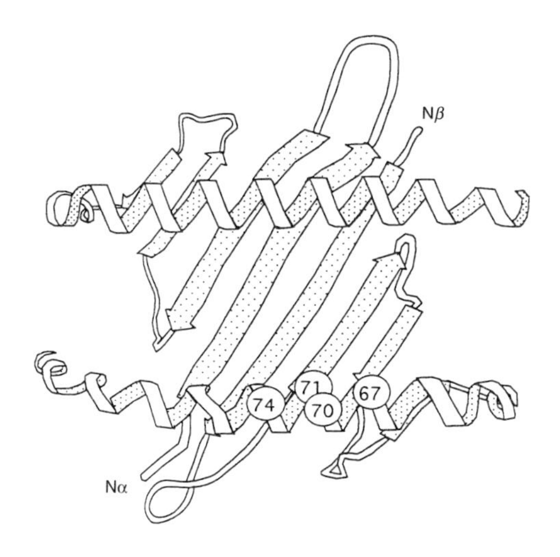

Serologic Specificity	T cell-defined determinant	Allele	Hypervariable Region 3				RA Associated
			67	70	71	74	
DR1	Dw1	*DRB1*0101*	L	Q	R	A	Y
DR4	Dw4	*DRB1*0401*	L	Q	K	A	Y
DR4	Dw14	*DRB1*0404*	L	Q	R	A	Y
DR4	Dw15	*DRB1*0405*	L	Q	R	A	Y
DR4	Dw14	*DRB1*0408*	L	Q	R	A	Y
DR10		*DRB1*1001*	L	R	R	A	Y
DR4	Dw10	*DRB1*0402*	I	D	E	A	N
DR4	Dw13	*DRB1*0403*	L	Q	R	E	N

Figure 15.2. The peptide-binding groove of an HLA class II molecule indicating the position of the third hypervariable region which forms the 'shared epitope' associated with RA.

of peptide–HLA complexes, by peptide elution and by studying peptide binding to HLA. Naturally processed peptides have been eluted from several class I and II molecules purified from cultured EBV-transformed B-cell lines (Chicz *et al.*, 1993; Ramensee *et al.*, 1995). This enables the definition of allele-specific peptide 'motifs', which in some cases have been validated using binding assays (Hammer *et al.*, 1995). Binding motifs have also been tested by site-directed mutagenesis of HLA genes (e.g. Fu *et al.*, 1995). It is hoped that the identification of 'motifs' for disease-associated HLA alleles will facilitate the search for the proposed pathological peptide(s) that trigger disease (discussed below for HLA-B27 and ankylosing spondylitis). Candidate peptides bearing the appropriate motifs have also been identified for class II molecules associated with both RA and pemphigus vulgaris (Hammer *et al.*, 1995; Wucherpfennig and Strominger, 1995a). This approach may be complemented by elution of peptides from diseased tissues, rather than from cultured cells (e.g. Gordon *et al.*, 1995) but peptides presented at low density remain difficult to detect.

15.3.3 T cell receptor repertoire analysis

There are now several examples of immune responses in which given HLA and peptide combinations are recognized by T cells bearing closely related TCRs (e.g. Bowness *et al.*, 1994). Nevertheless a clear correlation between MHC polymorphism and peripheral blood TCR V gene segment usage has proved surprisingly difficult to demonstrate. Recently, subtle differences have been described in the TCR repertoire of healthy individuals with DR1/DR4 genotypes as compared to DR3/DR7 individuals (Walser-Kuntz *et al.*, 1995). Perturbation of the TCR repertoire in disease states is studied, first, to identify additional disease susceptibility determinants and, second, in the hope that the triggering agent will have left its signature in the resultant pattern of clonality or Vβ family expansion. A superantigen would be expected to expand (or delete) a polyclonal cohort of T cells bearing related Vβ. A conventional antigen, conversely, should produce oligoclonality (reviewed in: Bowness and Bell, 1992; Gold, 1994). Unfortunately such studies have produced conflicting results.

15.3.4 Functional T cell studies

In certain murine models of autoimmune disease, the role of MHC can be clearly demonstrated by adoptive transfer of peptide-specific MHC-restricted T cells. Evidence that disease-associated HLA molecules present pathogenic peptides has been difficult to demonstrate in human diseases, even when candidate autoantigens are known, as in multiple sclerosis or coeliac disease. In multiple sclerosis T-cell clones have been grown from lesions, cerebrospinal fluid and blood (Lundin *et al.*, 1993; Tuohy *et al.*, 1994), and changes in T-cell reactivity to neural autoantigens have been seen as the disease progresses (Tuohy *et al.*, 1994). However, the fundamental difference between these T cells and those autoreactive T cells that can be isolated from healthy controls remains to be explained (see also Chapter 14).

Functional T-cell studies provide a tool for exploring the potential of microorganisms as environmental triggers of HLA-associated disease. In the

B27-associated reactive arthritides, certain microbial triggers have been identified and T-cell clones specific for bacterial peptides have been isolated from synovial fluid (discussed below). However, at present, their significance is unclear. The molecular mimicry hypothesis has been developed further in RA where the *E. coli* dnaJ heat-shock protein shares the amino acid sequence QKRAA of the 'shared epitope'. Patients exhibit enhanced proliferative responses of synovial T cells to the dnaJ protein, which is not evident in normal controls or patients with other autoimmune diseases (Albani *et al.*, 1995). Similarly, for multiple sclerosis, some cross-reactive clones have been identified which respond both to the putative autoantigen MBP(85-99) and to either bacterial or viral peptides (Wucherpfennig and Strominger, 1995b).

The interplay of MHC genotypes and haplotypes in disease susceptibility can be studied in animal models of, for example, collagen-induced arthritis or IDDM. Crosses between H-2 congenic strains and mice transgenic for various MHC antigens provide powerful tools for exploring mechanisms of T-cell regulation, including thymic selection, tissue-specific MHC expression and, possibly, distinct functional roles for HLA isotypes (Miller and Flavell, 1994).

15.4 Examples of HLA class I associations with autoimmune disease

15.4.1 HLA-B27 and the spondylarthropathies

Ankylosing spondylitis (AS) is a relatively common inflammatory rheumatic disease, affecting up to 0.5% of the population. Although the joints of the spine and axial skeleton are most commonly involved, AS is a multisystem disease and recognized features include asymmetrical peripheral arthritis, uveitis, aortic valve involvement and upper lobe lung fibrosis. The association of AS with possession of HLA-B27 is amongst the strongest described for an HLA locus. A recent study found that 94% of AS patients are HLA-B27 positive, compared to 9.4% of controls, giving an odds ratio of 161, with 95% confidence interval of 113–230 (Brown *et al.*, submitted). HLA-B27 is also less strongly associated with a clinically heterogeneous group of diseases, shown in *Table 15.3*. These conditions are all defined on clinical features alone. However, besides their HLA-B27 association, they share common features including arthritis of the spine and large joints, and

Table 15.3. Diseases associated with HLA-B27

Ankylosing spondylitis (AS)
Reactive arthritis (ReA)
sexually acquired
enteropathogenic
Uveitis
Arthritis associated with inflammatory bowel disease
Arthritis associated with psoriasis
Undifferentiated oligoarthritis

involvement of the skin, eye, genital mucosa and heart. Reactive arthritis (ReA) is less common than AS and less strongly HLA-B27-associated (Brewerton *et al.*, 1974; Laitinen *et al.*, 1977). ReA is, however, an important model for autoimmune disease in general because it follows some weeks after certain genital or gastrointestinal infections. Some, but not all, patients present with the classical triad of arthritis, urethritis and conjunctivitis described by Bauer and Engelman (1942). Only some infections predispose to ReA, with species of *Chlamydia* (usually genitally acquired), *Salmonella*, *Yersinia*, *Shigella* and *Campylobacter* most commonly implicated. All are Gram-negative bacteria that can live intracellularly. Spinal and peripheral joint involvement occurring together with psoriasis and inflammatory bowel disease are also HLA-B27-associated, as is isolated uveitis. A related but poorly understood clinical entity, 'undifferentiated spondylarthritis', is also recognized. We consider it likely that all these overlapping conditions, the spondylarthropathies, have a related aetiology.

15.4.2 Theories explaining the association of HLA-B27 with the spondylarthropathies

A number of different mechanisms have been put forward to explain the association of HLA-B27 with the spondylarthropathies. Many of these theories, summarized in *Table 15.1*, are applicable to other HLA-associated autoimmune diseases. In the context of the spondylarthropathies, all theories should attempt to explain not only the relationship with specific triggering organisms (found most clearly for ReA), but also the unique tissue distribution of the disease. It is also important to recognize, not only that disease can occur in the absence of HLA-B27, but also that other HLA alleles may predispose to these conditions, either independently or together with HLA-B27 (e.g. HLA-B60 together with HLA-B27; Brown *et al.*, submitted). Furthermore, family and twin studies have demonstrated that other, as yet unidentified, non-HLA genes make an even greater contribution to disease susceptibility than HLA-B27 (Calin *et al.*, 1983).

The finding that the natural role of HLA molecules is peptide binding and presentation to T cells led Benjamin and Parham (1990) to suggest that the spondylarthropathies result from HLA-B27's ability to bind a unique set of peptides. Their 'arthritogenic peptide' hypothesis proposes that disease resulted from a class-I-restricted cytotoxic T-cell (CTL) response to a peptide found only in joint tissues (and presumably also the other connective tissues involved in these diseases). This peptide or peptides could be bound and presented by all disease-associated HLA-B27 subtypes (see below), but not by other class I molecules. Priming of pathogenic T cells by previously cryptic epitopes could occur in the joint. Alternatively cross-reactive T cells might be primed at other sites such as the genital or gut mucosa by a peptide or peptides derived from one of the triggering pathogens.

What is known of the peptide-binding specificity of HLA-B27? Four complementary lines of evidence suggest that HLA-B27 does indeed bind a unique set of peptides. Natural epitopes presented by HLA-B27 during viral infection were found to share common structural features (Huet *et al.*, 1990). Solution of the structure of HLA-B27, crystallized with a variety of self-peptides (Madden *et al.*, 1991) showed a common arginine at the second position of bound

Table 15.4. Association of different HLA-B27 subtypes with ankylosing spondylitis (AS)

Subtype	Association with AS	Reference
★2701	Yes	1
★2702	*Yes*	1
★2703	Possibly not	2
★2704	Yes	3
★2705	*Yes*	1
★2706	Possibly not	3
★2707	Probably	3
★2708	Not known	
★2709	Possibly not	4

Common caucasian subtypes in bold.
HLA-B★2704 is common in Asians. *HLA-B★2703* appears confined to people of west African origin.
References: 1, Breur-Vriesendorp *et al.*, 1987; 2, Hill *et al.*, 1991b; 3, Lopez-Larrea *et al.*, 1995; D'Amato *et al.*, 1995.

peptide. The long side-chain of this arginine was accommodated in the 'B' or '45' pocket, comprising in HLA-B27 a unique combination of residues: 45E, 67C, 34V, 26G and 24T. Amino acid analysis of self-peptides eluted from HLA-B27 has confirmed the presence of an anchor arginine residue at the second position (Jardetzky *et al.*, 1991; Rötzschke *et al.*, 1994). There also appear to be preferences for particular amino acids at other positions of bound peptides, which differ between HLA-B27 subtypes. Thus HLA-B★2705 appears to bind peptides with C terminal amino acids that are either aromatic, hydrophobic or positively charged, whereas HLA-B★2702 can probably only accommodate aromatic or hydrophobic residues at this position (Rötzschke *et al.*, 1994). Finally, measurement of the ability of different peptides to bind to HLA-B27 has confirmed the importance of the P2 arginine (Bowness *et al.*, 1994), and also confirmed that different subtypes probably bind different but overlapping subsets of peptides (Tanigaki *et al.*, 1994). In particular HLA-B★2703 appears to bind only a subset of those peptides bound by HLA-B★2705 with a preference for a positively charged N-terminal amino acid (Colbert *et al.*, 1994; Tanigaki *et al.*, 1994). Given these differences in peptide binding we would predict that any arthritogenic peptide or peptides would bind with different affinity to different B27 subtypes, and there would consequently be differences in the strength of disease association (see below).

This leads to the question, are all HLA-B27 subtypes (shown in *Table 15.4*) associated with disease? Unfortunately the answer to this critical question has not yet been established, as the molecular typing methods (described in Chapter 6) have only recently become available, permitting large numbers of patients to be accurately subtyped. Such studies confirm the disease association of *HLA-B★2702, 04,* and *05* first described by Breur-Vriesendorp *et al.*, 1987). It has recently been suggested that *HLA-B★2703* and *HLA-B★2706* may not be associated with disease (Hill *et al.*, 1991b; Lopez-Larrea *et al.*, 1995). Larger confirmatory studies would provide strong support for the role of peptide binding in the aetiology of AS, as differences between subtypes are primarily confined to the peptide-binding groove. Combining this preliminary epidemiological evidence with knowledge of peptide-binding differences, we might envisage a candidate arthritogenic peptide containing an arginine at P2 and a hydrophobic or aromatic C terminal residue, but lacking a positive charged N terminal amino acid. Notably confirmation that certain subtypes are not disease-associated would also be compatible with aetiological theories in which HLA-B27 provides peptides (see below).

If HLA-B27's disease-association is indeed a consequence of its physiological role in peptide selection, HLA-B27-restricted CTL, specific for self or bacterial epitopes, should be demonstrable in the arthritic joints of patients with spondylarthropathies. Such responses have proved remarkably difficult to detect, in contrast to CD4 positive HLA class II-restricted cells (Hassell *et al.*, 1992; Viner *et al.*, 1991). However, Hermann and colleagues (1993) have recently isolated *Yersinia* and *Salmonella*-specific clones from two patients with ReA. We have identified a *Chlamydia* heat-shock protein peptide-specific T-cell line from another ReA patient (P. Bowness, R.L. Allen and A.J. McMichael, unpublished observations). An autoreactive HLA-B27-restricted type II collagen-specific T-cell line has also been isolated from the blood of a patient with ReA (Gao *et al.*, 1994). Whilst nothing is known of the function of these T cells, a pathogenic role would be suggested if they could be shown selectively to accumulate in affected joints and disappear during disease remission. More direct evidence could be obtained either by direct immunological intervention in patients, for example with monoclonal antibodies directed against the TCR, or by the use of animal models.

Both rats and mice transgenic for HLA-B27 and human β2-microglobulin have been studied and provide the best evidence that HLA-B27, and not a linked gene, is directly involved in disease pathogenesis. An increased frequency of the naturally occurring inflammatory joint disease ankylosing euthesopathy in transgenic mice (ANKENT) has recently been reported in mice transgenic for *HLA-B*2702*, although no obvious relationship with enteric infection could be identified (Weinreich *et al.*, 1995). Rats carrying a high copy number of *HLA-B*2705* transgenes develop an illness characterized by peripheral and axial arthritis, gut inflammation and both genital and skin lesions (Hammer *et al.*, 1990). Interestingly, rats kept in germ-free conditions do not develop the inflammatory intestinal or peripheral joint disease (Taurog *et al.*, 1994). It appears that this disease can be transferred by fetal liver cells alone, suggesting that antigen presentation by HLA-B27 in peripheral tissues such as joints is not essential for development of disease (Breban *et al.*, 1993). These data would favour theories involving T-cell education or presentation of HLA-B27-derived peptides by other HLA molecules.

The observation that HLA-B27 carries an unpaired cysteine residue at position 67 has led to the suggestion that this residue might be modified or disulphide-bonded to thiol-containing reagents. This might result in new immunological reactivity or 'altered self', and is discussed by Benjamin and Parham (1990). From what we now know of the structure of HLA-B27 it is likely that such a change would have a substantial effect on peptide binding. It should be pointed out that cysteine at position 67 is not unique to HLA-B27, also being present in, for example, HLA-B14. However, no other alleles also have the same amino acid residues at surrounding residues, in particular at positions 70 and 71.

Another possibility is that some feature of the biochemistry or cell biology of HLA-B27 predisposes to disease development; for example, Benjamin and colleagues (Benjamin *et al.*, 1991) observed that 'empty' cell-surface HLA-B27 molecules could present peptides to T cells, and suggested that this might lead to presentation of extracellular peptides not normally accessible to the class I processing pathway (or class II pathway).

It has recently been observed that amongst the most abundant naturally processed peptides eluted from HLA class II molecules are those derived from (other) HLA molecules (Chicz *et al.*, 1993). It is thus likely that HLA-B27-derived peptides are widely presented on the cell surface by class II molecules (Davenport, 1995), and it has also been suggested that HLA-B27 itself may be able to present B27-derived peptides (Scofield *et al.*, 1995). Whilst tolerance to such epitopes should normally be present, it is possible that certain HLA class II-restricted T lymphocytes, stimulated by bacterial infection, might have cross-reactive specificity for an HLA-B27-derived peptide presented by host cells. In this case, a strong HLA class II association with disease might be expected. However, this may not be evident as it is now clear that many peptides can promiscuously bind a variety of HLA class II molecules (Davenport, 1995). Although it is not clear how such a mechanism could explain the tissue specificity of disease, it has the advantage of being easily testable.

The 'molecular mimicry' hypothesis originally suggested that disease resulted from an antibody response cross-reactive between a unique portion of HLA-B27 and certain bacterial epitopes. Molecular mimicry between the *HLA-B*2705* amino acid positions 72–77 (QTDRED) and both the *Klebsiella* nitrogenase reductase enzyme and a *Shigella flexneri* plasmid gene product have been described, and cross-reactive antibodies can be found in patients with AS (Ebringer *et al.*, 1976; Tsuchiya *et al.*, 1990). However, this exact amino acid sequence is not shared by either *HLA-B*2702* or *HLA-B*2704*, both of which are also disease-associated. Furthermore, whilst antibodies reacting with candidate regions undoubtedly exist, many specificities are found when appropriate controls are included (Lahesma *et al.*, 1992).

15.4.3 Other HLA class I disease associations

Although the association of HLA-B27 with the spondylarthropathies is the most widely recognized, a number of other disease associations of HLA class I molecules with disease have been described. Their presence and nature argue in favour of an 'immunogenic peptide' mechanism for disease causation. Behcet's disease is an inflammatory multisystem disease of presumed autoimmune aetiology, characterized by arthritis, recurrent mouth and genital ulceration, arterial and venous thrombosis. Behcet's disease is common in Turkey and Asia and rare in the USA. A strong association with possession of the class I allele HLA-B51 has been described for a number of ethnic groups (Mineshita *et al.*, 1992; Mizuki *et al.*, 1992). Recently HLA-B51 has been subdivided into three alleles, *HLA-B*5101–3*, differing only at single amino acid substitutions at amino acids 167 and 171. A recent Japanese study has suggested that only *HLA-B*5101* is associated with Behcet's disease (Mizuki *et al.*, 1993). Since these amino acids border the peptide-binding groove, they are likely to affect peptide selection.

Takayasu's arteritis is an arteritis affecting large vessels, also relatively common in Japan, that is usually responsive to immunosuppressive treatment. In a study of 64 patients and 156 healthy controls using polymerase chain reaction (PCR) and sequence-specific oligonucleotide probes, two disease-associated alleles,

HLA-B52 and *HLA-B*3902* (EC39.2) have recently been described (Yoshida *et al.*, 1993). The authors note that the gene products of these two alleles share certain common amino residues (63E and 67S). If the peptide-binding specificities of these alleles are found to be similar, this would provide further support for a role in peptide selection in disease aetiology.

Other class I associations described include that of birdshot retinitis with *HLA-A*2902* (Lehoang *et al.*, 1992). Here the only difference between HLA-A*2901 and HLA-A*2902 is at amino acid position 12, with the apparently disease-resistant HLA-A*2901 bearing a histidine and HLA-A*2902 a glutamic acid. The strength of this association is comparable to that described for AS, with 54 of 58 subjects (93.1%) in this study having the allele (relative risk 157.3). In contrast to the examples described previously, this position is not obviously situated in the peptide-binding groove.

Finally some HLA class I associations are almost certainly markers for linked genes; for example, the association of HLA-B8 with the storage disease haemochromatosis is almost certainly due to linkage with an as yet unidentified gene located on chromosome 6 at 6p21.3 (Yaounq *et al.*, 1994).

Thus whilst the role of HLA class I molecules in disease pathogenesis remains unclear, we believe that for most of the associations described, available evidence is consistent with a key role of peptide selection, either in binding arthritogenic peptides or in determination of the T-cell repertoire. In the future, careful epidemiological and genetic studies are needed unequivocally to delineate subtype associations. Together with functional studies and use of transgenic models, rapid progress is to be hoped for.

15.5 HLA class II associations with autoimmune disease

HLA class II associations with disease are even more diverse than those described for class I. Given the pivotal role of class II molecules in immune regulation, associations with autoimmune diseases are not surprising. We will concentrate on three common diseases: RA, IDDM and systemic lupus erythematosus (SLE), although class II associations have been described for many other largely organ-specific autoimmune diseases, including forms of glomerulonephritis, multiple sclerosis, pemphigus vulgaris and pernicious anaemia.

15.5.1 Rheumatoid arthritis (RA)

RA is one of the most extensively studied class II-associated diseases. It is clinically heterogeneous, ranging from non-erosive synovitis with few systemic symptoms, to severe destructive joint disease and extra-articular involvement. Life-threatening complications include vasculitis, rheumatoid lung disease and cervical instability, predisposing to high cervical cord lesions. The first hint of HLA association was documented less than 25 years ago when lymphocytes from two-thirds of RA patients were found to be mutually non-reactive or poorly reactive when incubated together in mixed lymphocyte culture (MLC; Astorga and Williams, 1969). The major shared MLC type was later identified as Dw4 (Stastny, 1976) and the description of class II HLA molecules was a direct result of experiments designed to explain the altered MLC reactivity in RA and SLE patients.

Population studies using the serological specificities for class II molecules showed a striking association of RA with DR4 (reviewed in: Winchester, 1994; Wordsworth and Bell, 1992). This has been confirmed in the majority of populations studied and, in north-west European caucasians, confers a relative risk of between three and six. In contrast, some populations, including Israeli Jews, Greeks, Yakima Amerindians and Asian Indians in the UK, failed to exhibit a DR4 association. Population studies using the cellular determinant to type class II provided the explanation for this discrepancy. Of the five main subtypes of DR4, only Dw4, Dw14 and Dw15 were found in ethnic groups where DR4 was associated with RA. In populations where the predominant cellular subtype of DR4 was Dw10 or Dw13, no DR4 association was evident. Lesser associations with the specificities DR1 and DR10 were also demonstrated. When sequences from polymorphic regions of the *DRB1* alleles associated with RA were compared, a similarity was noted in the third hypervariable region (see *Figure 15.2*). The importance of this 'shared epitope' was particularly evident in the alleles encoding the DR4 specificity; for example, DRB1*0403 differs from DRB1*0401 by an alanine to glutamate substitution at position 74. This single non-conservative substitution was sufficient to abrogate association with RA (reviewed in Wordsworth and Bell, 1992).

The DR associations with RA could indicate either that DR is itself the susceptibility determinant or that it is linked to the risk locus. One candidate is DQ, which is in tight linkage disequilibrium with DR. Ethnic differences in the structure of class II haplotypes enabled a segregation-type analysis to map the precise location of the susceptibility determinant (Winchester, 1994). This identified *DRB1* as the susceptibility locus and, furthermore, confirmed the third hypervariable region encoding the 'shared epitope' as the most strongly associated nucleotide sequence.

The HLA-DR associations have proved more complex than would be expected in a dominant susceptibility model (reviewed in: Winchester, 1994; Wordsworth and Bell, 1992). First, a hierarchy of DR associations has been demonstrated. The most strongly associated alleles are *DRB1*0401* and *DRB1*0404*. Less strongly associated are DR1 and DR10 specificities. *DRB1*0402, DRB1*0403*, and DR2 and DR5 are negatively-associated (i.e. protective). Second, the strength of association with DR4 is greatest in severe disease, and weak or absent in mild seronegative RA. Finally, combinations of disease-associated *DRB1* alleles confer greater susceptibility than either alone and appear to be associated with more severe disease (Jawaheer *et al.*, 1994; Weyand *et al.*, 1992; Wordsworth *et al.*, 1992). The *DRB1* genotype *DRB1*0401/*0404* confers the surprisingly high relative risk of 49, compared to only about 5 for either *DRB1*0401* or *DRB1*0404* alone (see *Table 15.2*).

Family studies in RA support linkage to the DR4 haplotype. The recurrence rate in monozygotic twins is approximately 30% and 'homozygosity' for the shared epitope is the most important factor in determining concordance (Jawaheer *et al.*, 1994). The relatively sharp fall in recurrence risk from monozygotic twins to other first-degree relatives suggests polygenic susceptibility. However, the excess risk in siblings (λ_S) is still relatively high, at between three and 10 (Wordsworth and Bell, 1992). This suggests that the number of involved loci is small and it is estimated to be three or four. One or two of these may be contributed by the HLA complex.

The significance of the DR association with RA can be explored with studies of peptide binding, TCR repertoire formation and with assays of T-cell function. The peptide-binding motif has been identified for each of the RA-associated DR molecules (Rammensee, 1995). The 'shared epitope' determines the characteristics of the P4 pocket in the peptide-binding groove. Comparison of the DR4 alleles suggests that the fundamental difference between RA-associated and non-associated subtypes is the charge of this pocket. Disease-associated alleles selectively bind negatively charged residues. This information has been used to identify candidate peptides from putative autoantigens (Hammer *et al.*, 1995) but direct evidence of their involvement in pathogenesis is still awaited. This approach may be complemented by elution of peptides from diseased tissues (e.g. Gordon *et al.*, 1995). However, peptides presented at low density remain difficult to detect.

A model of molecular mimicry in RA has been supported by the similarity between the third hypervariable region of DRB1*0401 and sequences in both the 110 kDa late major capsid protein of Epstein-Barr virus and the *E. coli* dnaJ 40 kDa heat-shock protein (Albani *et al.*, 1992; Roudier *et al.*, 1989). Furthermore, synovial T cells from early RA patients mount strong proliferative responses to the *E. coli* dnaJ heat-shock protein, whereas T cells from normal subjects or controls with other autoimmune diseases do not (Albani *et al.*, 1995).

Since HLA-derived peptides can be presented on other HLA molecules, a peptide representing the third HVR of DRB1*0401, has also been investigated for stimulatory properties (Salvat *et al.*, 1994). In this study, most individuals mounted a DQ-restricted proliferative response to peptides representing the third HVR of their respective DRβ chains. However, subjects expressing DRB1*0401 were apparently tolerant to the DRB1*0401 peptide. This unresponsiveness could potentially extend to the similar EBV- or *E. coli*-derived peptides resulting in a defective immune response and, perhaps, persistent infection. Clearly this model is at variance with the model above in which RA T cells do respond to this epitope. This requires clarification.

The TCR repertoire in RA synovial fluid and peripheral blood has been extensively studied (reviewed in Bowness and Bell, 1992). Some studies have reported oligoclonal expansions and a few have suggested superantigenic involvement, but the results are inconclusive. Continuing refinement of fluorescence-activated cell sorting (FACS) analysis and cytokine assays offers the prospect of defining phenotypic and functional subgroups to which 'pathological' T cells belong. Isolation of such subgroups and investigation of TCRs and their interaction with HLA-peptide may prove more revealing than studies of unselected CD4 or CD8 T cells.

15.5.2 Insulin-dependent diabetes (IDDM)

In IDDM, T-cell infiltration of the pancreas is associated with destruction of the insulin-producing β-islets. Other autoimmune phenomena include the presence of anti-islet cell antibodies and an association with other organ-specific autoimmune diseases, such as thyroiditis. Family studies (Nepom and Ehrlich, 1991) demonstrate segregation of HLA haplotypes with disease and preferential sharing of certain HLA haplotypes between affected siblings. Monozygotic twins exhibit 50% concordance for disease but this drops to 20% in HLA-identical siblings. As in RA, the MHC complex encodes the major, but not the only genetic susceptibility factor. In population studies, a hierarchy of DR associations emerged. DR3 and DR4 are strongly associated with the disease and the highest risk is conferred by the DR3/DR4 genotype. DR1 and DR8 exhibit lesser positive associations, while DR2 and, to a lesser degree, DR5 are negatively associated.

Population studies have been used to localize the susceptibility determinant within the associated haplotypes (reviewed in Nepom and Ehrlich, 1991); for example, only DR4 haplotypes bearing *DQB1*0302* were strongly associated with IDDM. This *DQB1* allele is present on approximately 95% of IDDM DR4$^+$ haplotypes and is expressed by approximately 70% of all IDDM patients. However, 30% of patients do not express this *DQ* allele and, even in *DQB1*0302*-positive individuals, relative risk is modulated by other genes on the haplotype, including *DRB1* and *DQA1*. DR3 haplotypes are particularly interesting in this respect as they are 'extended haplotypes', showing unusual linkage disequilibrium between DR and both DP and class I loci.

Extending population studies beyond allelic susceptibility enables mapping of the susceptibility nucleotide residues or sequences themselves. In IDDM, this implicates position 57 of DQβ chain (at which aspartic acid confers protection) and arginine at position 52 of the DQα chain (which is associated with susceptibility; Khalil *et al.*, 1990; Todd, 1990). Homozygotes of an α52Arg$^+$/β57Asp$^-$ haplotype are extremely disease-prone, with a relative risk in excess of 40. However, the story is still incomplete and even the β57Asp-association is not absolute. The association between HLA and IDDM is dealt with in detail in the following chapter.

The complex interaction of class II loci in conferring susceptibility to IDDM has spawned several hypotheses. One model proposes that a diabetogenic peptide binds to the 'susceptibility' DQ but that other class II molecules can compete for this with varying efficiency, giving rise to the hierarchy of DR association (Nepom and Ehrlich, 1991). The synergy of the DR3 and DR4 haplotypes could be explained by a *trans*-complementing DQ heterodimer. Other hypotheses invoke effects of various alleles on thymic selection or, alternatively, dysregulation of class II expression, resulting in 'aberrant' expression on pancreatic islets.

Two excellent animal models of IDDM exist, the NOD mouse (discussed above) and the Biobreeding Lab (BB) rat. These permit an exploration of the intriguing class II locus interactions as well as identification of non-MHC genetic determinants of susceptibility (reviewed in Serreze and Leiter, 1994).

15.5.3 Systemic lupus erythematosus (SLE)

SLE is a multisystem autoimmune disease, characterized by the presence of a panoply of autoantibodies directed against a variety of intracellular and cell-surface antigens. The autoantibody profile varies between patients but tends to remain constant in an individual and is associated with particular clinical manifestations in this highly heterogeneous disease. Multiple loci within the HLA complex have been implicated in susceptibility. These include not only class II alleles but also complement and TNF loci.

The complement loci were the first within the HLA complex to be associated. Individuals with inherited complement deficiencies (particularly of C2 and C4 components) develop a syndrome resembling SLE (Morgan and Walport, 1991). Null alleles at either C2 or C4 loci were commonly found on HLA haplotypes associated with SLE. Heterozygous C2 deficiency (one null allele) is found in 6% of SLE patients, compared with 1–2% of the normal population (Glass *et al.*, 1976). The genetics of C4 deficiency is more complex and is reviewed by Arnett and Reveille (1992). Two isoforms of C4 exist, C4A and C4B, encoded by separate polymorphic loci (see *Figure 3.2*). Heterozygous deficiencies of either C4A or C4B are common in the normal population, occurring in 18% and 15%, respectively. Heterozygous C4A deficiency is associated with a relative risk of three for SLE, while homozygosity for C4A null alleles confers a relative risk of 17.

HLA DR2 and DR3 are both associated with SLE. Although the B8, DR3 haplotype also encodes a C4A null allele, it is probable that disease susceptibility is not simply due to linkage with C4. Population studies in ethnic groups where the two putative susceptibility alleles are on separate haplotypes have indeed demonstrated synergism between the loci. The relative risk of either DR2 or C4A null allele alone is approximately three, whereas inheritance of both DR2 and C4A null confers a relative risk of 25 (Arnett and Reveille, 1992).

SLE may be regarded as a composite disease of multiple autoantibody subsets, each of which is associated with certain clinical features. In fact, class II associations with presence of individual autoantibodies proved much stronger than association to SLE as a whole. The Ro and La autoantibodies, which occur in SLE and Sjögren's syndrome, are associated with DQw1 and DQw2 (Arnett and Reveille, 1992). These DQ alleles are in linkage disequilibrium with DR2 and DR3 respectively. Nucleotide sequence analysis has shown that all DQ molecules associated with Ro antibodies are encoded by a *DQA1* allele with a glutamine at position 34 and a *DQB1* allele with a leucine at position 26. Both of these residues lie on the floor of the peptide-binding groove. DQ allelic associations have also been described for autoantibodies directed against phospholipids, Sm antigen and dsDNA.

TNFα is implicated in animal models of SLE. The susceptible (NZB $\times$ NZW)F1 mice express low levels of TNFα and the disease improves following administration of TNF (Jacob *et al.*, 1990). In humans, the lowest TNF inducibility is associated with the DR2-DQw1 haplotype, which is associated with disease.

15.6 Other HLA disease associations

15.6.1 Infectious diseases

There have been surprisingly few reports of HLA associations with infectious diseases, but the distinction between infectious disease and autoimmune disease is rather tenuous. One possibility is that deleterious MHC alleles have been negatively selected through generations of infectious epidemics. The emergence of a new pathogen, such as the human immunodeficiency virus (HIV), may provide a good opportunity to study HLA-encoded susceptibility; for example, DR5 has been reported to be significantly increased in an HIV-infected African American population (Cruse *et al.*, 1991). Another possibility is that studies have been designed to detect only susceptibility determinants, rather than severity. Recently, possession of both class I and II alleles (*HLA-B53* and *HLA-DRB1*1302*) has been shown to confer protection against severe forms of malaria in Gambian children (Hill *et al.*, 1991a). HLA associations have also been described for forms of certain infectious diseases in which immune-mediated tissue damage is thought to play a key role. Thus certain DR haplotypes are associated with progression to tuberculoid leprosy (van Eden *et al.*, 1985), and progression to chronic Lyme disease is significantly associated with the serologic determinants DR4 and DR2 (Dwyer and Winchester, 1993).

15.6.2 Malignancies

Hodgkin's disease, a lymphoma, is associated with the A1–B8 haplotype and was one of the earliest diseases shown to be class I-associated (Hors and Dausset, 1983). Class II molecules may also influence susceptibility to tumours; for example, DQw3 is associated with a high risk of squamous cell carcinoma of the cervix following infection with human papilloma virus (Bavinck *et al.*, 1993). While associations of HLA molecules are unlikely to trigger development of a malignancy, they might determine the ease with which a tumour escapes immunological surveillance.

15.7 Conclusion

HLA class I and II associations have been described in a remarkable diversity of diseases, most of which are immunologically mediated. These include a myriad of organ-specific and non-specific autoimmune disorders as well as certain infectious diseases and malignancies. The genetics of autoimmune diseases are complex, involving multiple genetic loci. Family studies provide an estimation of the relative contribution of the HLA complex to susceptibility and also indicate the number of non-HLA loci involved. Sophisticated genetic mapping techniques identify specific alleles and even nucleotide sequences which account for the HLA association. In many diseases, the most strongly associated residues are known to be located in the wall or floor of the peptide-binding groove of HLA molecules. This supports the notion that HLA associations with disease reflect the distinct subsets of peptides presented by different HLA alleles.

The identification of allele-specific binding motifs has been used to screen self-proteins, such as MBP or collagen, for candidate peptides which would selectively bind to disease-associated HLA molecules. Clearly, the ability of an HLA molecule to bind a self-peptide is not sufficient to explain the development of autoimmune disease. It has been suggested that similarities between self-derived peptides and peptides from bacterial or viral proteins may trigger autoimmune disease. Alternatively, certain HLA molecules, such as B27, may be prone to modification by reducing agents or may have access to certain peptide populations not available to other alleles.

It is intriguing that combinations of HLA alleles may interact to modify disease susceptibility. This is particularly evident in class II-associated diseases.

Interactions may occur between alleles on the same haplotype (*cis*) or between the two HLA haplotypes (*trans*). This may reflect competition for peptides, an effect of thymic selection or, in the case of DQ, different combinations of polymorphic α- and β-chains. It is now also recognized that the majority of self-peptides eluted from HLA molecules are themselves MHC-derived. This adds a further dimension to the interaction between HLA molecules.

The most daunting task in the study of HLA disease associations is to characterize the significance of particular HLA combinations for T-cell behaviour and, therefore, for the development of the immune response. It is now possible to sequence and quantify TCRs, to label and to separate phenotypic subsets of T cells and to study clones of autoreactive cells. The major impediment to progress, however, is the inability to define or recognize a pathological T cell. This problem may be addressed using animal models, where certain T-cell clones are sufficient to transmit disease. In humans, HLA alleles or combinations of alleles may differ in their ability both to present peptides and also to regulate the class of the resulting immune response. Definition of a variant or 'pathological' immune response to self- or microbial peptides, in individuals with the disease-associated allele(s), may serve as a prelude to studying individual T cells. Perhaps the outmoded terminology of 'immune response' genes has been vindicated, as it is now clear that HLA molecules may be associated not only with disease susceptibility but also with severity and progression.

References

Albani S, Tuckwell JE, Esparza L, Carson DA, Roudier J. (1992) The susceptibility sequence to RA is a cross-reactive B cell epitope shared by the *Escherichia coli* heat shock protein dnaJ and the histocompatibility leukocyte antigen DRB1*0401 molecule. *J. Clin. Invest.* **89:** 327–331.

Albani S, Keystone EC, Nelson JL, Ollier WE, LaCava A, Montemayor AC, Weber DA, Montecucco C, Martini A, Carson DA. (1995) Positive selection in autoimmunity: abnormal immune responses to a bacterial dnaJ antigenic determinant in patients with early RA. *Nature Medicine* **1:** 448–452.

Archer JR, Whelan MA, Badakere SS, McLean IL, Archer IV, Winrow VR. (1990) Effect of a free sulphydryl group on expression of HLA-B27 specificity. *Scand. J. Rheumatol.* **87:** 44–50.

Arnett FC, Reveille JD. (1992) Genetics of systemic lupus erythematosus. *Rheum. Dis. Clin. N. Am.* **18**: 865–892.

Ashton-Rickardt PG, Tonegawa S. (1994) A differential avidity model for T-cell selection. *Immunol. Today* **15:** 362–366.

Astorga GP, Williams RC. (1969) Altered reactivity in mixed lymphocyte culture of lymphocytes from patients with RA. *Arthritis Rheum.* **12:** 547–554.

Bauer W, Engelman EP. (1942) Syndrome of unknown aetiology characterized by urethritis, conjunctivitis and arthritis (so-called Reiter's disease). *Trans. Ass. Am. Phys.* **57:** 7–313.

Bavinck JNB, Gissman L, Claas FHJ, *et al.* (1993) Relation between skin cancer, humoral responses to human papillomaviruses and HLA class II molecules in renal transplant recipients. *J. Immunol.* **151:** 1579-1586.

Benjamin R, Parham P. (1990) Guilt by association: HLA-B27 and ankylosing spondylitis. *Immunol. Today* **11:** 137–142.

Benjamin RJ, Madrigal JA, Parham P. (1991) Peptide binding to empty HLA-B27 molecules of viable human cells. *Nature* **351:** 74–77.

Berg LJ, Frank GD, Davis MM. (1990) The effects of MHC gene dosage and allelic variation on TCR selection. *Cell* **60:** 1043–1053.

Bloom BR, Modlin RL, Salgame P. (1992) Stigma variations: observations on suppressor T cells and leprosy. *Annu. Rev. Immunol.* **10:** 453–488.

Bodmer JG, Marsh SGE, Albert ED, Boomer WF, Bontrop RE, Charron D, Dupont B. (1995)

Nomenclature for factors of the HLA system, 1995. *Tissue Antigens* **46:** 1–18.

Bowness P, Bell J. (1992) T-cell receptors and rheumatic disease: approaches to repertoire analysis. *Br. J. Rheumatol.* **31:** 3–8.

Bowness P, Allen RL, McMichael AJ. (1994) Identification of TCR recognition residues for a viral peptide presented by HLA-B27. *Eur. J. Immunol.* **24:** 2357–2363.

Breban M, Hammer RE, Ricardson JA, Taurog JD. (1993) Transfer of the inflammatory disease of HLA-B27 transgenic rats by bone marrow engraftment. *J. Exp. Med.* **178:** 1607–1616.

Breur-Vriesendorp S, Dekker-Says A, Ivanyi P. (1987) Distribution of HLA-B27 subtypes in patients with ankylosing spondylitis: the disease is associated with a common determinant of the various B27 molecules. *Ann. Rheum. Dis.* **46:** 353–356.

Brewerton DA, Caffrey M, Hart FD, James DCO, Nichols A, Sturrock RD. (1973) Ankylosing spondylitis and HL-A27. *Lancet* **i:** 904–907.

Brewerton DA, Caffrey M, Hart FD, James DCO, Nichols A, Sturrock RD. (1974) Reiters disease and HL-A27. *Lancet* **ii:** 996–998.

Brown KE, Hibbs JR, Gallinella G, Anderson SM, Lehman ED, McArthy P, Young NS. (1994) Resistance to parvovirus B19 infection due to lack of virus receptor (erythrocyte P antigen). *N. Engl. J. Med.* **330:** 1192–1196.

Calin A, Marder A, Becks E. (1983). Genetic differences between B27 positive patients with ankylosing spondylitis and B27 positive healthy controls. *Arthritis Rheum.* **26:** 140.

Campbell IL, Oxbrow L, Koulmanda M, Harrison L. (1988) IFN- induces islet cell MHC antigens and enhances autoimmune streptozotocin-induced diabetes in the mouse. *J. Immunol.* **140:** 1111–1116.

Campbell RD, Trowsdale J. (1993) Map of the human MHC. *Immunol. Today* **14:** 349–352.

Chicz RM, Urban RG, Gorga JC, Vignali DAA, Lane WS, Strominger JL. (1993) Specificity and promiscuity among naturally processed peptides bound to HLA-DR alleles. *J. Exp. Med.* **178:** 27–47.

Colbert RA, Rowland-Jones SL, McMichael AJ, Frelinger JA. (1994) Differences in peptide presentation between B27 subtypes: the importance of the P1 side chain in maintaining high affinity peptide binding to B*2703. *Immunity* **1:** 121–130.

Cruse JM, Brackin MN, Lewis RE, Meeks W, Nolan R, Brackin B. (1991) HLA disease association and protection in HIV infection among African Americans and Caucasians. *Pathobiol.* **59:** 324–328.

D'Amato M, Fiorillo MT, Carcassi C, Mathieu A, Zuccarelli A, Bitti PP, Tosi R, Sorrentino R. (1995) Relevance of residue 116 of HLA-B27 in determining susceptibility to ankylosing spondylitis. *Eur. J. Immunol.* **25:** 3199–3201.

Davenport M. (1995) The promiscuous B27 hypothesis (letter). *Lancet* **346:** 500–501.

Dwyer E, Winchester R. (1993) Genetic basis of chronic lyme disease. In: *Immunogenetics of Lyme Disease* (ed. P. Coyle). Mosby Yearbook, St Louis, MO.

Ebringer A. (1991) Ankylosing spondylitis and Klebsiella – the debate continues. *J. Rheumatol.* **18:** 312–313.

Fleury S, Thibodeau J, Croteau G, Labrecque N, Aronson HE, Caubin C, Long EO, Sekaly RP. (1995) HLA-DR polymorphism affects the interaction with CD4. *J. Exp. Med.* **182:** 733–741.

Fu XT, Bono CP, Woulfe SL, Swearingen C, Summers NL, Sinigaglia F, Sette A, Schwartz BD, Karr RW. (1995) Pocket 4 of the HLA-DRB1*0401 molecule is a major determinant of T cell recognition of peptide. *J. Exp. Med.* **181:** 915–926.

Gao XM, Wordsworth P, McMichael A. (1994) Collagen-specific cytotoxic T lymphocyte responses in patients with ankylosing spondylitis and reactive arthritis. *Eur. J. Immunol.* **24:** 1665–1670.

Glass D, Raum D, Gibson D, Stillman JS, Schur PH. (1976) Inherited deficiency of the second component of complement. *J. Clin. Invest.* **58:** 853–861.

Glimcher LH, Kara CJ. (1992) Sequences and factors: a guide to MHC class II transcription. *Annu. Rev. Immunol.* **10:** 13–49.

Gold DP. (1994) TCR V gene usage in autoimmunity. *Curr. Opin. Immunol.* **6:** 907–912.

Gordon RD, Young JA, Rayner S, *et al.* (1995) Purification and characterization of endogenous peptides extracted from HLA-DR isolated from the spleen of a patient with RA. *Eur. J. Immunol.* **25:** 1473–1476.

Gregersen PK, Silver J, Winchester RJ. (1987) The shared epitope hypothesis - An approach to understanding the molecular genetics of RA susceptibility. *Arthritis Rheum.* **30:** 1205–1213.

Haldane JBS. (1955) The estimation and significance of the logarithm of a ratio of frequencies. *Ann. Hum. Genet.* **20:** 309–311.

Hammer J, Gallazzi F, Bono E, Karr RW, Guenot J, Valsasini P, Nagy ZA, Sinigaglia F. (1995)

Peptide binding specificity of HLA-DR4 molecules: correlation with RA association. *J. Exp. Med.* **181:** 1847–1855.

Hammer RE, Maika SD, Richardson JA, Tang J-P, Taurog JD. (1990) Spontaneous inflammatory disease in transgenic rats expressing HLA-B27 and human β2m: an animal model of HLA-B27-associated human disorders. *Cell* **63:** 1099–1112.

Hassell AB, Pilling D, Reynolds D, Life PF, Bacon PA, Gaston JSH. (1992) MHC restriction of synovial fluid lymphocyte responses to the triggering organism in reactive arthritis: Absence of a class I-restricted response. *Clin. Exp. Immunol.* **88:** 442–447.

Hermann E, Yu DT, Meyer zum Buscheufelde K, Fleischer B. (1993) HLA-B27-restricted CD8 T cells derived from synovial fluids of patients with reactive arthritis and ankylosing spondylitis [see comments]. *Lancet* **342:** 646–650.

Hill AVS, Allsopp CEM, Kwiatkowski D, Anstey NM, Twumasi P, Rowe PA, Bennett S, Brewster D, McMichael AJ, Greenwood BM. (1991a) Common West African HLA antigens are associated with protection from severe malaria. *Nature* **352:** 595–600.

Hill AVS, Allsopp CEM, Kwiatowski D, Anstey NM, Greenwood BM, McMichael AJ. (1991b) HLA class I typing by PCR: HLA-B27 and an African subtype. *Lancet* **337:** 640–642.

Hors J, Dausset J. (1983) HLA susceptibility to Hodgkin's disease. *Immunol. Rev.* **70:** 167–192.

Huet S, Nixon DF, Rothbard J, Townsend ARM, Ellis SA, McMichael AJ. (1990) Structural homologies between two HLA-B27 restricted peptides suggest residues important for interaction with HLA-B27. *Int. Immunol.* **2:** 311–316.

Jacob CO, Fronek Z, Lewis GD, Koo M, Hansen JA, McDevitt HO. (1990) Heritable major histocompatibility complex class II associated differences in production of tumor necrosis factor. *Proc. Natl Acad. Sci. USA* **87:** 1233–1237.

Jardetzky TS, Lane WS, Robinson RA, Madden DR, Wiley DC. (1991) Identification of self peptides bound to purified HLA-B27. *Nature* **353:** 326–329.

Jawaheer D, Thomson W, MacGregor AJ, Carthy D, Davidgon J, Dyer PA, Silman AJ, Ollier WER. (1994) 'Homozygosity' for the HLA-DR shared epitope contributes the highest risk for RA concordance in identical twins. *Arthritis Rheum.* **37:** 681–686.

Khalil I, d'Auriol L, Gobet M, Morin L, Lepage V, Deschamps I, Park MS, Degos L, Galiberl F, Hors J. (1990) A combination of HLA-DQ Asp57-negative and HLA-DQ Arg52 confers susceptibility to insulin-dependent diabetes mellitus. *J. Clin. Invest.* **85:** 1315–1319.

Khoury MJ, Beaty TH, Cohen BH. (1993) *Fundamentals of Genetic Epidemiology*. Oxford University Press, New York.

Lahesma R, Skurnik M, Granfors K, Mottonen T, Saario R, Toivanen A, Toivanen P. (1992) Molecular mimicry in the pathogenesis of spondyloarthropathies. A critical appraisal of cross-reactivity between microbial antigens and HLA-B27. *Br. J. Rheumatol.* **31:** 221–229.

Laitinen O, Leirisalo M, Skylv G. (1977) Relation between HLA-B27 and clinical features in patients with yersinia arthritis. *Arthritis Rheum.* **20:** 1121–1124.

Lander ES, Schork NJ. (1994) Genetic dissection of complex traits. *Science* **265:** 2037–2048.

Lehoang P, Ozdemir N, Benhamou A, Tabary T, Edelson C, Betuel H, Semiglia R, Cohen JHM. (1992) HLA-A29.2 subtype associated with birdshot retinochoroidopathy. *Am. J. Ophthalmol.* **113:** 33–35.

Lopez-Larrea C, Sujirachato K, Mehr N, *et al.* (1995) HLA-B27 subtypes in Asian patients with ankylosing spondylitis. Evidence for new associations. *Tissue Antigens* **45:** 169–176.

Louis P, Vincent R, Cavadore P, Clot J, Eliasu JF. (1994) Differential transcriptional activities of HLA-DR genes in the various haplotypes. *J. Immunol.* **153:** 5059–5067.

Lundin KEA, Scott H, Hansen T, Paulsen G, Halstensen TS, Fausa O, Thorsby E, Sollid LM. (1993) Gliadin-specific HLA-DQ(1*0501,1*0201) restricted T cells isolated from the small intestinal mucosa of celiac disease patients. *J. Exp. Med.* **178:** 187–196.

Madden DR, Gorga JC, Strominger JL, Wiley DC. (1991) The structure of HLA-B27 reveals nonamer self-peptides bound in an extended conformation. *Nature* **353:** 321–325.

Manfras BJ, Swinyard M, Rudert WA, Ball EJ, Lee PA, Kuhnl P, Trucco M, Bohm BO. (1993) Altered CYP21 genes in HLA-haplotypes associated with congenital adrenal hyperplasia (CAH) in a family study. *Hum. Genet.* **92:** 33–39.

Manyak CL, Tse H, Fischer P, Coker L, Sigal NH, Koo GC. (1988) Regulation of class II MHC molecules on human endothelial cells. Effects of IFN and dexamethasone. *J. Immunol.* **140:** 3817–3821.

Matsushita S, Takahashi K, Motoki M, Komoriya K, Ikagawa S, Nishimura Y. (1994) Allele specificity of structural requirement for peptides bound to HLA-DRB1*0405 and DRB1*0406 complexes: implication for the HLA-associated susceptibility to methimazole-induced insulin autoimmune syndrome. *J. Exp. Med.* **180:** 873–883.

Michaelsson E, Malmstrom V, Reis S, Engström A, Rurkhardt H, Holmdahl R. (1994) T cell recognition of carbohydrates on type II collagen. *J. Exp. Med.* **180:** 745–749.

Miller JFAP, Flavell RA. (1994) T-cell tolerance and autoimmunity in transgenic models of central and peripheral tolerance. *Curr. Opin. Immunol.* **6:** 892–899.

Miller LH, Mason SJ, Dvorak JA, McGinnis MH, Rothman IK.(1975) Erythrocyte receptors for (*Plasmodium knowlesi*) malaria: Duffy blood group determinants. *Science* **189:** 561–563.

Mineshita S, Tian D, Wang LM, et al. (1992) Histocompatibility antigens associated with Behcet's disease in northern Han Chinese. *Jpn J. Med.* **31:** 1073–1075.

Mizuki N, Ohno S, Tanaka H, Sugimura K, Seki T, Mizuki N, Kera J, Inaba G, Tsuji K, Inoko H. (1992) Association of HLA-B51 and lack of associated class II alleles with Behcet's disease. *Tissue Antigens* **40:** 22–30.

Mizuki N, Inoko H, Ando H, Nakamura S, Kashiwase K, Akaza T, Fujino Y, Masuda K, Takiguchi M, Ohno S. (1993) Behcet's disease associated with one of the HLA-B51 subantigens, HLA-B 5101. *Am. J. Ophthalmol.* **116:** 406–409.

Morgan BP, Walport MJ. (1991) Complement deficiency and disease. *Immunol. Today* **12:** 301–306.

Nepom B, Nepom GT, Schaller J, et al. (1984) Characterization of specific HLA-DR4-associated histocompatibility molecules in patients with juvenile rheumatoid arthritis. *J. Clin. Invest.* **74:** 287–291.

Nepom GT, Ehrlich H. (1991) MHC class II molecules and autoimmunity. *Annu. Rev. Immunol.* **9:** 493–525.

Parekh RB, Dwek RA, Sutton BJ, et al. (1985) Association of rheumatoid arthritis and primary osteoarthritis with changes in glycosylation pattern of total serum IgG. *Nature* **316:** 452–457.

Paul C, Schoenwald U, Truckenbrodt H, Bettinotti MP, Brunner G, Keller E, Nevinny-Stickel C, Yao Z, Albert ED. (1993) HLA-DP/DR interaction in early onset pauciarticular juvenile chronic arthritis. *Immunogenetics* **37:** 442–448.

Panayi GS, Wooley PH. (1977) B lymphocyte alloantigens in the study of the genetic basis of rheumatoid arthritis. *Ann. Rheum. Dis.* **36:** 365–368.

Rammensee HG. (1995) Chemistry of peptides associated with MHC class I and class II molecules. *Curr. Opin. Immunol.* **7:** 85–96.

Rammensee HG, Friede T, Stevanovic S. (1995) MHC ligands and peptide motifs: first listing. *Immunogenetics* **41:** 178–228.

Rich SS, Panter SS, Goetz FC, Hedlund B, Barbosa J. (1991) Shared genetic susceptibility of type 1 (insulin-dependent) and type 2 (non-insulin dependent) diabetes mellitus: contributions of HLA and haptoglobin. *Diabetologia* **34:** 350–355

Richeldi L, Sorrentino R, Saltani C. (1993) HLA-DPB1 Glutamate 69: a genetic marker of beryllium disease. *Science* **262:** 242–244.

Rigby AS, Silman AJ, Voelm L, Gregory JC, Ollier WER, Khan MA, Nepom GT, Thomson G. (1991) Investigating the HLA component in rheumatoid arthritis: an additive (dominant) mode of inheritance is rejected, a recessive mode is preferred. *Genetic Epidemiol.* **8:** 153–175.

Risch N. (1987) Assessing the role of HLA-linked and unlinked determinants of disease. *Am. J. Hum. Genet.* **40:** 1–14.

Rotter JI, Landaw EM. (1984) Measuring the genetic contribution of a single locus to a multilocus disease. *Clin. Genet.* **26:** 529–542.

Rötzschke O, Falk F, Stevanovic S, Gnau V, Jung G, Rammensee H-G. (1994) Dominant aromatic/aliphatic C-terminal anchor in HLA-B*2702 and B*2705 peptide motifs. *Immunogenetics* **39:** 74–77.

Roudier J, Petersen J, Rhodes GH, Luka J, Carson DA. (1989) Susceptibility to rheumatoid arthritis maps to a T cell epitope shared by the HLA-Dw4 DR -1 chain and the Epstein Barr virus glycoprotein gp110. *Proc. Natl Acad. Sci. USA* **86:** 5104–5108.

Salmon M. (1992) The immunogenetic component of susceptibility to rheumatoid arthritis. *Curr. Opin. Rheumatol.* **4:** 342–347.

Salvat S, Auger I, Rochell L, Begorich A, Geburher L, Sette A, Roudier J. (1994) Tolerance to a self-peptide from the third hypervariable region of HLA DRB1*0401 in rheumatoid arthritis patients and normal subjects. *J. Immunol.* **153:** 5321–5329.

Scholl PR, Geha RS. (1994) MHC class II signalling in B-cell activation. *Immunol. Today* **15:** 418–422.

Scholz S, Albert ED. (1994) Immunogenetic aspects of juvenile chronic arthritis. *Rheumatol. Europe*

23: 92–94.

Scofield R, Kuien B, Gross T, Warren W, Harley J. (1995). HLA-B27 binding of peptide from its own sequence and similar peptides from bacteria: implications for spondylarthropathies. *Lancet* 1542–1544.

Serreze DV, Leiter EH. (1994) Genetic and pathogenic basis of autoimmune diabetes in NOD mice. *Curr. Opin. Immunol.* **6:** 900–906.

Stastny P. (1976) Mixed lymphocyte cultures in rheumatoid arthritis. *J. Clin. Invest.* **57:** 1148–1157.

Steere AC, Dwyer E, Winchester R. (1991) Association of chronic Lyme arthritis with HLA-DR4 and HLA-DR2 alleles. *N. Engl. J. Med.* **323:** 219–223.

Tanigaki N, Fruci D, Vigneti E, Starace G, Rovero P, Londei M, Butler RH, Tosi R. (1994) The peptide binding specificity of HLA-B27 subtypes. *Immunogenetics* **40:** 192–198.

Taurog JD, Richardson JA, Croft JT, Simmons WA, Zhou M, Fernandez SJL, Balish E, Hammer RE. (1994) The germfree state prevents development of gut and joint inflammatory disease in HLA-B27 transgenic rats. *J. Exp. Med.* **180:** 2359–2364.

Teyton L, Lotteau V, Turmel P, Arenzana-Seisdos F, Virelizier JL, Pujol JP, Loyau G, Piatier-Tonneau D, Auffray C, Charron DJ. (1987) HLA DR, DQ and DP antigen expression in rheumatoid synovial cells: a biochemical and quantitative study. *J. Immunol.* **138:** 1730–1738.

Tienari PJ, Terwilliger JD, Ott J, Palo J, Peltonen L. (1994) Two-locus linkage analysis in multiple sclerosis (MS). *Genomics* **19:** 320–325

Tiwari JL, Terasaki PI. (eds) (1985) *HLA and Disease Association*. Springer, New York.

Todd JA. (1990) Genetic control of autoimmunity in type I diabetes. *Immunol. Today* **2:** 122–127.

Todd JA, Bell JI, McDevitt HO. (1987) HLA-DQ gene contributes to susceptibility and resistance to insulin-dependent diabetes mellitus. *Nature* **329:** 599–604.

Tsuchiya N, Husby G, Williams RJ, Stieglitz H, Lipsky PE, Inman RD. (1990) Autoantibodies to the HLA-B27 sequence cross-react with the hypothetical peptide from the arthritis-associated Shigella plasmid. *J. Clin. Invest.* **86:** 1193–1203.

Tuohy VK, Fritz RB, Ben-Nun A. (1994) Self-determinants in autoimmune demyelinating disease: changes in T-cell response specificity. *Curr. Opin. Immunol.* **6:** 887–891.

Tussey LG, Rowland-Jones S, Zheng TS, Androlewicz MJ, Cresswell P, Frelinger JA, McMichael AJ. (1995) Different MHC class I alleles compete for prevention of overlapping uralepitopes. *Immunity* **3:** 65–77.

van Eden W, Gonzalez NM, de Vries RRP, Conva J, van Rood JJ. (1985) HLA-linked control of predisposition to lepromatous leprosy. *J. Infect. Dis.* **151:** 9–14.

Viner NJ, Bailey LC, Life PF, Bacon PA, Gaston JSH. (1991) Isolation of Yersinia specific T cell clones from the synovial membrane and synovial fluid of a patient with reactive arthritis. *Arthritis Rheum.* **34:** 1151–1157.

Walser-Kuntz DR, Weyand CM, Weaver AJ, O'Fallon WH, Goronzk J. (1995) Mechanisms underlying the formation of the TCR repertoire in rheumatoid arthritis. *Immunity* **2:** 597–605.

Webb SR, Gascoigne NRJ. (1994) T-cell activation by superantigens. *Curr. Opin. Immunol.* **6:** 467–475.

Weinreich S, Euldrink F, Capkova J, *et al.* (1995) HLA-B27 as a relative risk factor in ankylosing enthesopathy in transgenic mice. *Hum. Immunol.* **42:** 103–115.

Weyand CM, Goronzy JJ. (1994) Functional domains on HLA-DR molecules: implications for the linkage of HLA-DR genes to different autoimmune diseases. *Clin. Immunol. Immunopathol.* **70:** 91–98.

Weyand CM, Xie C, Goronzy JJ. (1992) Homozygosity for the HLA-DRB1 allele selects for extraarticular manifestations in rheumatoid arthritis. *J. Clin. Invest.* **89:** 2033–2039.

Winchester R. (1994) The molecular basis of susceptibility to rheumatoid arthritis. *Adv. Immunol.* **56:** 389–466.

Woolf B. (1955) On estimating the relation between blood group and disease. *Ann. Hum. Genet.* **19:** 251–253.

Wordsworth BP, Bell JI. (1992) The immunogenetics of rheumatoid arthritis. *Springer Semin. Immunopathol.* **14:** 59–78.

Wordsworth BP, Pile KD, Buckey JD, Lanchrury JSS, Ollier B, Lathrop M, Bell JI. (1992) HLA heterozygosity contributes to susceptibility to rheumatoid arthritis. *Am. J. Hum. Genet.* **51:** 585–591.

Wordsworth P. (1995) Genes and arthritis. *Brit. Med. Bull.* **51:** 249–266.

Wucherpfennig KW, Strominger JL. (1995a) Selective binding of self peptides to disease-associated major histocompatibility complex (MHC) molecules: a mechanism for MHC-linked susceptibility to human autoimmune diseases. *J. Exp. Med.* **181:** 1597–1601.

Wucherpfennig KW, Strominger JL. (1995b) Molecular mimicry in T cell-mediated autoimmunity: viral peptides activate human T cell clones specific for myelin basic protein. *Cell* **80:** 695–705.

Yaounq J, Perichon M, Chorney M, et al. (1994) Anonymous marker loci within 400 kb of HLA-A generate haplotypes in linkage disequilibrium with the hemochromatosis gene (HFE). *Am. J. Hum. Genet.* **54:** 252–263.

Yokoyama WM. (1995) Natural killer cell receptors. *Curr. Opin. Immunol.* **7:** 110–120.

Yoshida M, Kimura A, Katsuragi K, Numano F, Sasazuki T. (1993) DNA typing of HLA-B gene in Takayasu's arteritis. *Tissue Antigens* **42:** 87–90.

16

HLA susceptibility to type 1 diabetes: methods and mechanisms

Francesco Cucca and John A. Todd

16.1 Introduction

The major histocompatibility complex (MHC) human leukocyte antigen (HLA) region of human chromosome 6p21.31 contains a densely packed array of at least 150 genes in 3500 kb of DNA. The products of these genes have a pivotal role in the regulation of the immune responses and influence susceptibility to a number of diseases. Among these, insulin-dependent diabetes mellitus (IDDM or type 1 diabetes) is one of the most thoroughly analysed. Recent work indicates that the MHC region contains the major genetic component of the disease (*IDDM1*), with the class II *HLA-DRB1*, *-DQA1* and *-DQB1* loci encoding the primary associated alleles, the protein products of which directly control a key checkpoint in the autoimmune aetiology of the disease. However, other MHC genes outside the class II region must be also involved, and, as shown first by genetic analysis of type 1 diabetes in mice, at least ten genes outside the MHC are also required to explain the inheritance of the disease.

The identification of the major mutations predisposing to disease and characterization of their biological effects in type 1 diabetes serves as a paradigm for the analysis of other multifactorial diseases.

16.2 An overview

Proteins encoded with the HLA region determine the way in which antigens are processed, translocated and presented to T lymphocytes. During development of the T-cell repertoire HLA class I and class II molecules control positive and negative selection of the T lymphocytes in the thymus. Not surprisingly, therefore, genes in the HLA region influence susceptibility to a large number of disorders, particularly autoimmune diseases.

The contribution of the HLA region to IDDM (*IDDM1*) was initially detected by association studies using a candidate gene approach and a case-control experimental design (Cudworth and Woodrow, 1974; Nerup *et al.*, 1974; Singal and Blajchman, 1973). This strategy utilizes knowledge of the pathophysiology of

HLA and MHC: genes, molecules and function, edited by M.J. Browning and A.J. McMichael.
© 1996 BIOS Scientific Publishers Ltd, Oxford.

the disorder, and tests whether allele frequencies of a gene, which has a known function and which might be directly involved in the aetiology of the disease, differ between patients and matched healthy controls. Subsequently, linkage analysis in families with multiple cases, in particular tests of allele-sharing by affected sib-pairs, proved that the case-control-based association studies were indeed indicative of a mutation or mutations in the HLA region predisposing to type 1 diabetes (Cudworth and Woodrow, 1975). However, linkage analysis, although robust, does not have a very high resolution owing to infrequent recombination between the major candidate HLA loci. Furthermore, allelic and locus heterogeneity and the low penetrance of *IDDM1* further complicated fine mapping using linkage analysis (Morton *et al.*, 1983). The identification of *IDDM1*, therefore, required linkage disequilibrium or allelic association mapping approaches, which use the recombination events within the HLA complex that have occurred during human history. This series of experiments, which began in 1973, has provided complicated results because it is evident that *IDDM1* is not one, but several mutations, within the HLA region, showing very strong disequilibrium between each other. In contrast, recent data from the insulin gene region on chromosome 11p15, using linkage disequilibrium mapping by plotting the degree of association of all the common polymorphisms across the region, defined the minimal region to 4.1 kb (Lucassen *et al.*, 1993). Subsequently, analysis of the transmission of different haplotypes in families, and population-based studies in other ethnic groups provided unequivocal assignment of the *IDDM2* mutation to the minisatellite repeat sequence 596 base pairs (bp) upstream of the insulin gene ATG codon (Bennett *et al.*, 1995; Julier *et al.*, 1994; Undlien *et al.*, 1995c).

The first associations of type 1 diabetes with HLA antigens involved the HLA class I alleles B8 and B15. Further analyses revealed that associations of the DR3 and DR4 alleles of the *HLA-DRB1* locus were significantly stronger than those at the class I loci (Cudworth and Woodrow 1974; Nerup *et al.*, 1974; Singal and Blajchman, 1973). In addition, HLA-DR2 and -DR5 alleles were decreased in frequency in patients compared to controls, implying that these alleles are associated with alleles of the disease loci that encode protection from, or resistance to, type 1 diabetes (Thomson *et al.*, 1988; Wolf *et al.*, 1983).

With the cloning of class II loci, restriction fragment length polymorphism (RFLP) analysis revealed that the *HLA-DQB1* locus was even more strongly associated with type 1 diabetes than *DRB1* (Festenstein *et al.*, 1986; Michelsen and Lernmark, 1987; Owerbach *et al.*, 1983). Sequence analyses revealed that alleles encoding aspartic acid at position 57 of the DQβ chain were associated with resistance, while alleles encoding a neutral residue at this position (valine, alanine or serine) conferred susceptibility (Todd *et al.*, 1987).

Subsequent papers reported that specific *HLA-DQA1* alleles and residues confer resistance, while others confer susceptibility (Khalil *et al.*, 1990; Todd *et al.*, 1989), implying the whole DQ molecule was important in susceptibility/resistance. Sheehy was the first to realize that there was differential susceptibility between individuals identical for the *DQA1* and *DQB1* loci, indicating that a gene, nearby in the class II region, also influenced susceptibility, namely *HLA-DRB1* (Sheehy *et al.*, 1989). This is hardly surprising, since both HLA-DQ and -DR are antigen presentation molecules. Finally, data based on haplotype and genotype analyses from different populations suggest that the *HLA-DRB1* and *-DQB1* loci encode primary aetiological components and influence type 1 diabetes risk in the same way (Cucca *et al.*, 1995; Erlich *et al.*, 1993; Sheehy *et al.*, 1989; Todd, 1994), namely by peptide-binding and T-cell recognition (Ellerman and Like, 1995; Kwok *et al.*, 1995; Nepom, 1990; Roustsias and Papadopoulos, 1995; Sanjeevi *et al.*, 1995; Sheehy, 1992; Todd *et al.*, 1987; Wucherpfennig and Strominger, 1995). However, as suggested by the presence of extended haplotypes associated with the disease, and by data from the non-obese diabetic (NOD) mouse model of type 1 diabetes (Ikegami *et al.*, 1995; Yamoto *et al.*, 1995), it is evident that other genes in the human MHC also influence the disease risk. The HLA class I region has been associated with the age-of-onset of the disease (Demaine *et al.*, 1995; Fujisawa *et al.*, 1995). An association, independent of the HLA class II genes, has been reported with the *LMP2* and *LMP7* genes (Deng *et al.*, 1995), which encode two subunits of the proteasome complex involved in the degradation of cytosolic proteins and generation of antigenic peptides, but this has not been replicated in others studies. Proof of a primary association of any HLA locus must take into account disequilibrium patterns and, therefore, requires very large clinical resources. Furthermore, associations must be replicated and should ideally be carried out in families. The putative independent associations of the transporter genes associated with antigen processing *TAP1* and *TAP2* (Caillat-Zucman *et al.*, 1993), 175 kb centromeric of the *HLA-DQB1* locus, are probably spurious (Caillat-Zucman *et al.*, 1995; Cucca *et al.*, 1994a; Ronningen *et al.*, 1993) and demonstrate the problems of delineating primary, aetiological associations from those owing to linkage disequilibrium with alleles at the *HLA-DQB1*, *-DQA1* and *-DRB1* loci.

16.3 Genetic analysis

16.3.1 Epidemiology versus whole-genome screening

Most cases of type 1 diabetes result from T-lymphocyte-dependent selective destruction (insulitis) of the insulin-producing pancreatic β cells and subsequent irreversible insulin deficiency. The incidence of type 1 diabetes varies widely in different ethnic groups, and between races. The highest incidence has been reported in Finns and in Sardinians (> 30 per 100 000 individuals per year in the age range 0–15 years), the lowest in the Japanese (1.8 per 100 000 in the same age range; Green *et al.*, 1992; Karvonen *et al.*, 1993). The 'global' disease frequency in European-derived Caucasian populations with any age-of-onset is 0.4%.

The average disease concordance, in siblings (6%), HLA identical siblings (14–25%) and monozygotic (MZ) twins (36–70%), suggests that both genetic and environmental factors are required to determine the disease (Kyvik *et al.*, 1995; Olmos *et al.*, 1988). These epidemiological observations made it possible to calculate a key parameter, the λ_S (S = sibling) ratio of familial clustering, defined as the recurrence risk for a sibling of an affected person divided by the risk in the general population. For type 1 diabetes $\lambda_S = 15$ (lifetime sib risk, 6%/population prevalence, 0.4%; Risch, 1987). Risch popularized the use of λ_S as a way of assessing the familial clustering of any disease, and showed that the available data for HLA and at the insulin locus could not explain the clustering of type 1 diabetes in families (Risch, 1987). The λ_S value is calculated for a specific disease locus from the ratio of the expected (0.25) and the observed proportion of sib-pairs sharing zero alleles identical by descent (IBD; Risch, 1987). The λ_S ratio therefore provides a simple but powerful mathematical framework for a whole-genome screen for genes predisposing to a multifactorial disease.

The relative contributions of nature (genetics) and nurture (environment) to familial clustering can be deciphered using epidemiological studies by analysing the frequency of disease in families with adopted or natural children. This sort of study has been carried out recently for multiple sclerosis, and the data indicate that the familial clustering of this autoimmune disease is genetic (Ebers *et al.*, 1995). Alternatively, one can search the entire genome by linkage analysis of large numbers of families with multiple affected cases to see if any other genes outside HLA and the insulin gene region are segregating, and, if they show significant evidence of linkage (i.e. an increase of sharing of alleles at the marker locus IBD in affected sib-pairs), the locus-specific λ_S ratios can be calculated and the total contribution of all loci to familial clustering estimated. Initially, this approach was carried out in mice, where genome screening by linkage analysis is greatly facilitated by the availability of inbred mice and large pedigrees. The results showed that there are many genes, at least ten, outside the MHC region that contribute to the frequency of disease in crosses of the diabetes-sensitive strain, NOD, with diabetes-resistant strains (Risch *et al.*, 1993; Todd *et al.*, 1991; Wicker *et al.*, 1995). The MHC genes were essential but so were the non-MHC genes. Nevertheless, neither the MHC nor the non-MHC genes alone were sufficient to cause the disease (Wicker *et al.*, 1995).

As further shown by Risch, the magnitude of the λ_S is the critical parameter in determining power to detect linkage by using affected sib-pairs, and thus it also defined the feasibility of whole-genome screening: the smaller the λ_S the more difficult it is to locate and clone the responsible gene(s) (Risch, 1990; Todd, 1995). The lack of concordance between the presence of the mutation in an individual and the diagnosis of the disease in that individual is the major complication in genetic analysis of complex diseases, and is responsible for the difficulty in mapping of susceptibility loci and proving that the candidate polymorphism is the aetiological mutation. Type 1 diabetes is more accessible to genetic analysis than other common diseases owing to its unequivocal diagnosis (absolute insulin dependence since diagnosis), strong clustering in families and childhood onset, which makes possible the collection of parental DNA.

Using locus-specific λ_S values and assuming a multiplicative model of epistasis (which fits with the rapid fall-off in recurrence risk from first- to second- to third-degree relatives, and is consistent with genetic analyses of NOD mice; Risch *et al.*, 1993) it is apparent that the familial clustering of human type 1 diabetes can be explained by the sharing of alleles at least at ten loci, with the HLA *IDDM1* locus contributing the major portion, about 32% (Cordell and Todd, 1995; Davies *et al.*, 1994; Todd, 1995). Interestingly, the relative contribution of the HLA region to familial clustering may differ between different countries: in families from the USA it appears that HLA contributes over 60% of the familial clustering of type 1 diabetes, almost twice that found in families from the UK. Obviously the greater the contribution from HLA the smaller the λ_S values will be for other loci, and the more difficult it will be to map and identify them. These interpopulation differences strongly suggest the existence of genetic or locus heterogeneity (contributions of different genes resulting in an identical phenotype), and can probably be explained, in the future, by the population-specific frequencies of predisposing and protective alleles at the various loci.

We can conclude that environmental factors are not responsible in a major way for familial clustering in type 1 diabetes (Todd, 1995). Nevertheless, identical twin disease discordance, evidence for seasonal variation in the incidence of type 1 diabetes and the increasing incidence of the disease in many countries over the last 30 years indicate the overall importance, albeit ubiquitous in nature, of environmental factors, and their role in influencing penetrance of the susceptibility alleles (Todd, 1991).

To investigate the genetic basis of a multifactorial disease such as type 1 diabetes, it is necessary to use methods that simultaneously take into account the joint effects of multiple loci. Using one of these methods (the two-locus version of the maximum lod score method of Risch) it has recently been demonstrated that the effects of *IDDM1/HLA* and *IDDM2/INS VNTR* together are best explained by a multiplicative model of epistasis in which the gene products from these loci interact. By contrast, the effects of *IDDM1* and *IDDM4* (linked to *FGF3* on chromosome 11q13) follow a heterogeneity model (lack of epistasis), although this depends on the population studied (Cordell *et al.*, 1995). It appears likely that the inheritance of type 1 diabetes involves more than ten loci which interact in very complex ways, a mixture of epistasis and heterogeneity, and which, in certain combinations, cause sufficient insulitis to exceed the threshold of β-cell destruction and precipitate disease. As found in NOD mice (McAleer *et al.*, 1995), it is unlikely that susceptibility alleles at all ten or so loci are required to be present at the same time. Instead, combinations of a smaller number of susceptibility alleles would be expected to be responsible for disease occurrence.

16.3.2 Twin concordance rates

Twin studies are particularly important for human geneticists because the risk in MZ twins provides a direct estimate of penetrance (the effect of the genotype on the phenotype) for a specific complement of susceptibility alleles at multiple loci. Penetrance of type 1 diabetes is clearly incomplete because in different studies the empirical risk for an MZ-twin of an affected patient is less than 100%. Even the penetrance of type 1 diabetes in NOD mice is less than 100%, unless the mice are reared in germ-free conditions (Todd, 1991). Deliberate (or accidental) infection of NOD mice with virus significantly reduces the penetrance of the disease. Note that the penetrance of insulitis is 100% in NOD mice bred under ordinary conditions, a finding that is probably related to the much higher penetrance of β-cell autoimmunity in human twin pairs (Verge *et al.*, 1995).

It is important to note that the current twin concordance rates are not based on any information concerning genotype status of the twins. A previous report has suggested that twin pairs with a DR3/4 genotype are concordant more often than the group of MZ twins as a whole (Johnston *et al.*, 1983). An obvious extension of this is that twins also carrying high-risk genotypes at MHC and non-MHC type 1 loci will be even more concordant. Finally, we can predict that individuals in the general population who carry susceptibility alleles at most of the major loci will be at high risk of developing the disease, although they will be rare (probably less than 0.1% of the general population; DR3/4-positive individuals are present at a frequency of approximately 2% in the UK).

Explanations for disease discordance in genetically identical twins include environmental factors (such as breast feeding or virus infection) and epigenetic mechanisms (such as differences in the T-cell receptors and immunoglobulin gene rearrangements, somatic mutations and, in female MZ twins, distinct patterns in X chromosome inactivation); for example, a study on multiple sclerosis, another HLA-associated multifactorial disease, indicates that the expressed T-cell receptor (TCR) repertoire differs within twin pairs discordant for the disease but not within concordant pairs (Utz *et al.*, 1993). However, in this kind of study it is extremely difficult to establish whether the T-cell repertoire is different before the disease manifests, and is thus a causal factor, or whether the disease process itself alters it.

Overall, therefore, environmental factors could be extremely heterogeneous and ubiquitous and this could make their identification extremely difficult (Dahlquist, 1995). Nevertheless, identification of the major susceptibility mutations should permit large prospective studies of genetically susceptible individuals and ultimately identification of key environmental factors.

16.3.3 Linkage disequilibrium

Recombination distances within the MHC are in good agreement with the physical map, where 1 cM (centimorgan) is roughly equivalent to 1 Mbp. In general, linkage disequilibrium decays with increasing recombination fractions across the 3 Mb of the MHC (Klitz *et al.*, 1995), and yet certain haplotypes exist in populations at a much higher frequency than expected. This unexpected disequilibrium in the case of the MHC is probably due to selection. The physical distance between *DQB1* and *DQA1* is 12 kb, and between *DQB1* and *DRB1* 175 kb, and these genes behave as a single 'evolutionary unit', within which extremely high linkage disequilibrium exists (Klitz *et al.*, 1995). This disequilibrium is higher than one would expect from the study of tightly linked markers in other chromosome regions (Jorde *et al.*, 1994).

In the numerous families that have been analysed to date no recombination events have been observed between *DQB1*, *DQA1* or *DRB1*. However, recombination events have occurred during human history and have been fixed by genetic drift in certain populations (and perhaps by selection); these population-specific haplotypes can be exploited to pinpoint susceptibility genes (Todd *et al.*, 1989).

The high degree of linkage disequilibrium, that is the absence of random association (equilibrium) between alleles, is at the same time one of the most important features of the HLA region and one of its most confounding features in trying to pinpoint primary aetiological mutations. Linkage disequilibrium keeps specific alleles of the HLA class I, II and III loci together, making it extremely difficult to attribute a primary role to an individual locus rather than to an HLA extended haplotype as a unit.

It is important to point out that patterns of disequilibrium in the MHC and in other regions of the genome have been shaped by the population history with a random fixation of limited sets of haplotypes during human evolution because of a small founding population size, bottlenecks, rapid population expansion and relatively recent origin of modern humans (Ayala, 1995). The history of human populations can account for ethnic-group-specific haplotypes, and the unexpected levels of disequilibrium that are being observed between polymorphic microsatellite marker loci, and between the latter and disease (Copeman *et al.*, 1995). Another (non-exclusive) explanation is that conserved haplotypes are held together owing to selective pressures: for example, a better coordination in the expression of the various HLA genes; functional advantages during the thymic selection of the T-cell repertoire; or preferential association of α- and β-chains from certain alleles (Gyllensten and Erlich, 1993). Selection pressure undoubtedly comes from infectious disease and there is now evidence for association of *DRB1*, *DQB1* and *HLA-B* alleles with resistance to malaria. Certain combinations of alleles at multiple loci may encode functional complementary molecules within and between pathways of antigen presentation and T-cell recognition. It has been argued, quite reasonably and with convincing experimental support (Oldstone *et al.*, 1991; Wucherpfennig and Strominger, 1995a), that the ability to respond to pathogens may lead to increased cross-reactivity with self-antigens and, in a genetic background of non-MHC-encoded defects in immune regulation, this could lead to autoimmune disease. Hence, certain haplotypes, such as the A1, B8, DR3 haplotype may predispose to autoimmunity with alleles at multiple loci contributing.

Linkage disequilibrium within the HLA region must therefore be controlled for before a primary association with HLA gene(s) can be declared. Note that the chance of success of such analyses critically depends on the number of individuals studied and on the genetic structure of the populations tested. Differences in the distribution of alleles and haplotypes make some populations more informative than others.

16.4 A primary aetiological role for *HLA-DQB1*, *-DQA1* and *DRB1* loci

In general an allele is said to be positively associated or predisposing when it occurs at a significantly higher frequency among affected than in control individuals. An allele is negatively associated or protective when it is found at a significantly higher frequency among controls compared with patients. Positive associations of genetic polymorphisms with disease can arise for three different reasons:

(i) the allelic marker is the variant directly responsible for disease;
(ii) the association is due to linkage disequilibrium with the primary associated mutation;
(iii) the association is an artefact due to population admixture because in mixed populations any trait that presents more frequently in a given ethnic group will be apparently associated with any allele that is more common in that group.

The presence of artefacts due to population admixture can be subtle and extremely confusing (Ewens and Spielman, 1995; Gough *et al.*, 1995). The problem of population stratification can be reduced to some extent by using genetically isolated populations for case-control studies, for instance the Finnish or the Sardinian populations. In mixed populations it is advisable to use family-association tests like the haplotype relative risk, the family-based control, or the

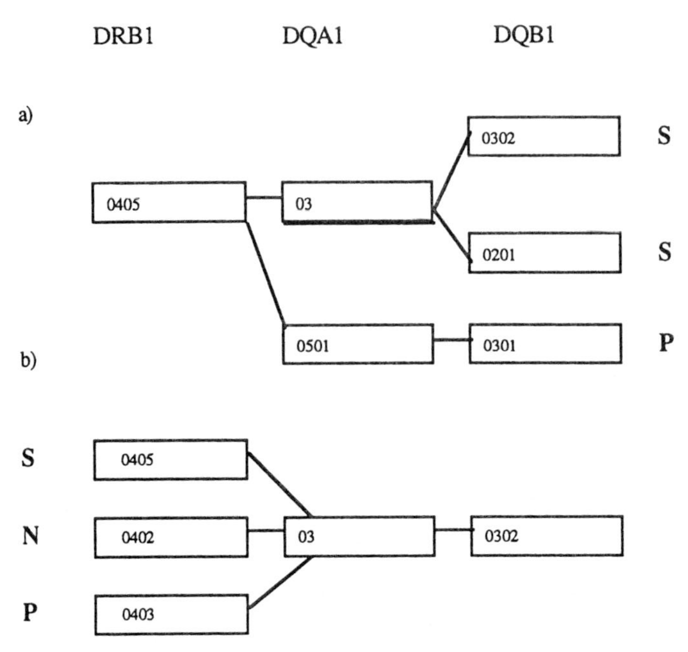

Figure 16.1. (a) Different splits of *DRB1*0405* and (b) of *DQA1*03, DQB1*0302* halpotypes in Sardinians. S, susceptibility, N, neutral and P, protective haplotypes.

transmission disequilibrium test (TDT) methods (Falk and Rubinstein, 1987; Spielman *et al.*, 1993; Thomson, 1988). Controls are assembled from the chromosomes not transmitted from the parents to the affected offspring. Nevertheless, it is apparent that some of these methods can be affected by recent admixture and the history of the population and, in general, the TDT is recommended (Ewens and Spielman, 1995). An additional advantage of family studies is that the analysis of parental DNAs allows unambiguous determination of the haplotypes (Bennett *et al.*, 1995). Moreover, in families the degree of association for each polymorphism can be assessed by calculating the percentage transmission, which is not sensitive, as is the relative risk, to the frequency of the allele under analysis.

In multifactorial diseases a positive association signifies susceptibility because the vast majority of the individuals who inherit predisposing alleles do not manifest the disease. Similarly, a negative association implies protection that is not absolute because it is possible to find affected individuals carrying protective alleles.

The Sardinians could be considered an ideal population to study the genetic association of type 1 diabetes with HLA, not only because the population has a low degree of genetic admixture, but also because several peculiarities in the nature and distribution of HLA haplotypes provide a valuable opportunity to solve the linkage disequilibrium in the MHC region and to clarify the role of individual loci (Cucca *et al.*, 1993, 1995); for instance, the presence in this population of DR4 haplotypes with several splits at the *DQB1* and *DRB1* loci offers a valuable opportunity for an unambiguous analysis of the effect of *DRB1* and *DQB1* loci in conferring type 1 diabetes risk (*Figure 16.1*). In particular, the effect of the *DRB1* locus was delineated by analysing the three *DRB1*04, DQA1*03, DQB1*0302* haplotypes common in this population and differing only at the *DRB1* locus (haplotype-matched analysis; *Figure 16. 1*). The different behaviour of these *DR4-DQB1*0302* haplotypes in conferring type 1 diabetes risk provided clearcut evidence that the *DRB1* locus was strongly associated with type 1 diabetes. In fact the *DRB1*0405, DQA1*03, DQB1*0302* haplotype was significantly predisposing, the *DRB1*0402, DQA1*03, DQB1*0302* haplotype was found to be neutral, and the *DRB1*0403, DQA1*03, DQB1*0302* haplotype showed a protective effect.

On the other hand the distribution of the *DRB1*0405* haplotypes in patients and controls was valuable in understanding the role of the *DQB1* locus and confirmed that this locus also influences disease risk because type 1 diabetes predisposition was restricted to *DRB1*0405* alleles in association with the susceptibility *DQB1*0302* or *DQB1*0201* alleles but not with the protective *DQB1*0301* allele. These results have been replicated in a second independent data set of 130 Sardinian families using the transmission disequilibrium test (F. Cucca, unpublished data).

Trans-racial analysis is a valuable approach not only to understanding the role of individual loci but also to providing independent replication of the data (Todd *et al.*, 1989); for example, observations from Black, Tunisian, Caucasian French, Mexican-American, African-American, Algerian, Greek and Chinese type 1 diabetic patients indicates the *DRB1*0405, DQA1*03, DQB1*0302* as a high-risk haplotype (Djoulah *et al.*, 1992; Erlich *et al.*, 1993; Fernandez *et al.*, 1993; Huang *et al.*, 1995; Ju *et al.*, 1991; Ronningen *et al.*, 1992). Different studies show that the *DRB1*0403, DQA1*03, DQB1*0302* haplotype is a protective haplotype: it has been found not associated with type 1 diabetes in Caucasian, Japanese and Chinese patients (Awata *et al.*, 1992; Huang *et al.*, 1995; Ronningen *et al.*, 1992; Van der Auwera *et al.*, 1995). Again note that, consistent with the multifactorial nature of type 1 diabetes, the protection was not absolute.

Haplotype analysis suggests that whether a particular *DRB1* allele confers diabetes risk depends on the nature of the allele at the *DQB1* locus and vice versa. In fact, among the various *DRB1–DQB1* allelic combinations, a single dose of a protective *DRB1* or *DQB1* allele (e.g. *DRB1*0403* or *DQB1*0301*) is sufficient to provide protection from type 1 diabetes, while susceptibility requires a combination of *DRB1* and *DQB1* susceptibility alleles (e.g. the *DR1*0405* and the *DQB1*0302* or *DQB1*0201* alleles).

A general guideline can be obtained from these results: a susceptibility genotype requires that all *DRB1* and *DQB1* alleles are predisposing (Huang *et al.*, 1995). The presence of only one protective allele at either locus results in a genotype that is not associated or protective overall. This does not entirely reflect the complexity of association between HLA and type 1 diabetes, particularly the presence of a hierarchy of protective and susceptibility alleles and haplotypes, but can help us to understand apparently contradictory data such as those from the Japanese population.

In the Japanese the *DQB1*0302* allele is not associated with type 1 diabetes (Todd *et al.*, 1990) only because the vast majority of the *DR4–DQB1*0302* haplotypes of the background population carry the *DRB1*0403* or the *DRB1*0406* protective alleles; these two alleles are structurally and functionally related, they differ only at codon 37 of the *DRB1* second exon (Awata *et al.*, 1992). However, the *DQB1*0302* variant is strongly and significantly associated with type 1 diabetes in the Japanese population when on a DR8 haplotype. The individual predisposing effect of *DQB1*0302* is not apparent because the associated *DR8–DQB1*0302* haplotype is very rare (1%) while *DRB1*0403/06–DQB1*0302* protective haplotypes are quite frequent (18%) in the general Japanese population (Awata *et al.*, 1992). A primary role for the *DQB1* locus in the Japanese is confirmed by the observation that the *DQB1*0602* allele that is strongly protective in Caucasians and in Blacks is also significantly protective in this ethnic group. The role of the *DRB1* locus is suggested not only by the protective behaviour of *DRB1*0403/06–DQB1*0302* haplotypes, but also by the observation that the predisposing *DRB1*0405* allele on *DQB1*0401* haplotypes, is also significantly associated with type 1 diabetes in the Japanese (Awata *et al.*, 1992).

The above serves as a paradigm of the complexity of disease association and of the difficulties created in genetic analysis of multifactorial disease by interactions between different loci and modes of transmission of disease susceptibility.

In trans-ethnic analysis patients should be matched for age of onset. First, some genotypes could be more frequent in patients with a specific age of onset. Second, and most confusing, clinical, immunological and genetic evidence suggest that patient groups with an older age of onset could contain a small but significant admixture of individuals with other forms of diabetes than type 1.

Another key concept in trans-ethnic analysis is that the predisposing effect of a haplotype is relative to the frequencies of haplotypes in a given population. In the absence of an associated haplotype a neutral one could emerge and confer an apparently higher relative risk than observed in populations that have high frequencies of highly predisposing haplotypes. As Sheehy suggested: "if one cuts the tip off the iceberg (the major susceptibility haplotypes), other haplotypes will become the new tip" (Sheehy, 1992). In the same way, in the absence of a strongly protective haplotype, a haplotype with a weaker protective effect could appear as the most protective. In Finns the *DRB1*0405* and *DRB1*0403* alleles are rare or virtually absent and the role of the *DRB1* locus in type 1 diabetes is suggested by the split of other *DR4-DQB1*0302* haplotypes, more frequent in that population. The *DRB1*0401-DQB1*0302* haplotype is positively associated with disease, while the *DRB1*0404–DQB1*0302* haplotype has a protective effect, although less substantial than that of *DRB1*0403–DQB1*0302* (Rejionen *et al.*, 1995). In *DQB1*0302*-positive Norwegian patients *DRB1*0401* and *DRB1*0405* were increased in frequency compared to subtypes *DRB1*0404* and *DRB1*0403* versus HLA-matched controls (Undlien *et al.*, 1995). By combining the various sets of available data, we suggest the following hierarchy of DR4 subtypes from most to least associated with type 1 diabetes:

*DRB1*0405* > *DRB1*0401* > *DRB1*0402* > *DRB1*0404* > *DRB1*0403* > *DRB1*0406*.

It is extremely difficult to evaluate the role of the *DQA1* locus in type 1 diabetes predisposition. The polymorphic products of *DQA1* and *DQB1* form a heterodimer and *in vivo* combinations of *DQB1* and *DQA1* alleles are subjected to functional constraints (preferential pairing of specific families of *DQB1* and *DQA1* alleles; Kwok *et al.*, 1993). Furthermore this locus is in strong linkage with the *DQB1* locus and recombinant haplotypes are rare.

A primary role for alleles at the *DQA1* locus, however, is strongly supported by a trans-racial gene mapping analysis (Todd *et al.*, 1989). The *DRB1*07–DQB1*0201* haplotype is positively associated with disease in Afro-Caribbean patients but not in Caucasians. The main difference between the two HLA class II haplotypes is the *DQA1* allele (which is *DQA1*0201* in Caucasians and *DQA1*0301* in Blacks). This is consistent with a recent observation indicating that the peptide derived from thyroid peroxidase (autoantibodies against this peptide are present in autoimmune thyroiditis and in about 20% of type 1 diabetic patients) can only bind to the *DQA1*0501–DQB1*0201* (which is present in the predisposing DR3 haplotype) and not to the DR7-related *DQA1*0201–DQB1*0201* dimer (Kwok *et al.*, 1995). *DQA1*0501* and *DQA1*0201* alleles differ at multiple positions and it is possible that some of these residues could affect disease predisposition and protection. Functionally, peptide-binding pockets of the class II binding site are made up of polymorphic residues from both chains (Roustsias and Papadopoulos, 1995) and both chains contribute to the electrostatic properties of the molecule (Sanjeevi *et al.*, 1995) and their pairing (Kwok *et al.*, 1993).

It is still a possibility that the *HLA-DPB1* locus (or a nearby locus) also contributes to susceptibility, but results, so far, are conflicting and not completely definitive (Buyse *et al.*, 1994; Easteal *et al.*, 1990a and b; Hu *et al.*, 1993; Magzoub *et al.*, 1992; Tait *et al.*, 1995). A primary effect of the products of the *DRB1* and *DQB1* loci is suggested by several lines of evidence:

(i) functional considerations related to the biological role of these molecules in (self)peptide presentation (Kwok *et al.*, 1995; Reich *et al.*, 1994; Wucherpfennig and Strominger 1995b);
(ii) the effect of *DRB1* and *DQB1* protective and susceptibility alleles is often haplotype independent; for example, the predisposing effect of *DRB1*0405* is present within *-DQB1*0302, -DQB1*0201,* and *-DQB1*0401* haplotypes, while the protective effect of *DRB1*0403* is detectable with *-DQB1*0302, -DQB1*0304* and *-DQB1*0305* haplotypes which have a different evolutionary history (Cucca *et al.*, 1994). In the same way, the protective effect of *DQB1*0301* is present in *-DRB1*11, -DRB1*12, DRB1*1402* and *-DRB1*04* haplotypes;
(iii) data from animal models (including transgenic animals) of type 1 diabetes strongly support a primary role for the products of *IA* (DQ homologue) and *IE* (DRB homologue; Wicker *et al.*, 1995);
(iv) certain amino acids in the DQ and DR β-chains correlate with disease susceptibility/resistance, and these amino acids are now known to be critical for the function of the class II molecule in peptide binding and T-cell recognition.

16.5 Amino acids and mechanisms

16.5.1 DQβ-chain position 57

Aspartic acid at position 57 of the DQβ-chain is encoded by *DQB* protective alleles, while an alanine at the same position is present in predisposing alleles (Todd *et al.*, 1987). If the protective alleles at *DRB1* are taken into account, then the presence/absence of DQβ chain Asp 57 explains the associations of *DQB1* alleles in a complete and satisfactory way, including data from different ethnic groups, in particular, results using Japanese subjects. Biochemical and structural studies prove the key functional role of DQβ-chain residue 57. Five peptide-binding pockets have been proposed for the DQ molecule (Roustsias and Papadopoulos, 1995). β57 is part of the fifth pocket, and in DR it has been shown that a β57Asp → Ser substitution (*DRB1*0401* → 0405) leads to a decrease in peptide affinity by three orders of magnitude (Marshall *et al.*, 1994). In DQ, β57Ala → Asp substitution prevented binding of a certain peptide (Kwok *et al.*, 1995). Note that βAla57 was not sufficient to permit binding of this peptide to other DQ molecules (Kwok *et al.*, 1995). β57Asp forms a salt bridge with DQα79Arg, and these residues form hydrogen bonds with the proximal amide nitrogen and carbonyl oxygen of the bound antigenic peptide (*Figure 16.2* see p. 258; Reich *et al.*, 1994; Wucherpfennig and Strominger, 1995b). β57 also plays an important role in the electrostatic properties of the DQ molecule (Sanjeevi *et al.*, 1995) as well in α/β-chain pairing (Kwok *et al.*, 1993). This residue is also known to be the only difference between certain alleles of *DQB1*, indicating that it is under selection. Its role in antigen- and allo-reactive T-cell recognition is well established. Other DQβ-chain residues are obviously important in peptide binding and in type 1 diabetes susceptibility, including those that make up the first pocket (positions 85, 86, 89 and 90) and the second pocket (position 13). β45 is important for electrostatic potential (Sanjeevi *et al.*, 1995) and chain pairing (Kwok *et al.*, 1993). Polymorphism at positions 45, 86 and 87 helps to explain the behaviour of some alleles such as *DQB1*0303* or the *DQB1*0401/02*. These alleles are Asp57 positive and are neutral or at least do not seem to confer the same protection as *DQB1*0301* and *DQB1*0602*. The *DQB1*0401/02* alleles also differ from more predisposing alleles at β45-56 dimerization patch, which, if hydrophobic, as in the case of the most predisposing alleles *DQB1*0201* and *DQB1*0302*, facilitates homodimerization and T-cell activation (Roustsias and Papadopoulos, 1995). Diabetes-resistant molecules, most notably, *DQB1*0602*, have hydrophilic and positively charged β49–56 dimerization patches, making them less prone to homodimerization and, accordingly, less able to activate cognate T-cell clones in the periphery (Roustsias and Papadopoulos, 1995).

16.5.2 DQα-chain

Polymorphic DQα-chain residues that form peptide-binding pockets include positions 9, 10, 27, 34, 35, 46. *DQA1*0501* (DR3, prediposing and DR5, non-predisposing) and *DQA1*0201* (DR7, non-predisposing) alleles differ at multiple residues (25, 34, 40, 47, 50, 51, 52, 53, 54, 75) and it is possible that some of these residues could affect disease predisposition and protection. Looking more carefully and including in the analysis *DQA1*0301* (DR4, DR7 and DR9 of Blacks) the only three residues that split *DQA1*0201* from *DQA1*03* and *DQA1*05* are positions 25, 52 and 54.

At position 25 the residues Tyr (*DQA1*05, -*03*) and Phe (*DQA1*0201*) are biochemically similar (relative hydrophobicity +2.3 and +2.5 respectively). Also at position 54 the residue Phe (DQA1*05, -*03) and Leu (DQA1*0201) are similar (+2.5 and +1.8, respectively). Substitutions at position 52 are, however, more striking: Ser is present in *DQA1*0101* (DR1), *DQA1*0102* (DR2) and

*DQA1*0103*; His is present in *DQA1*0201* (DR7); and an Arg is present in *DQA1*0301*, *DQA1*0501* and *DQA1*0401* (predisposing alleles). Relative hydrophobicities are: Ser −0.3, His +0.5, Arg −11.2. Also the fact that Arg52 is encoded by two different codons (CGC and AGA) provides indirect evidence of the functional relevance of that residue. On balance, therefore, αArg52, which has been used effectively in risk assessment but which may be associated with disease simply through disequilibrium with certain *DQB1* alleles (Sheehy, 1992; Todd, 1995), may well play a direct role in the mechanism of DQ-mediated type 1 diabetes susceptibility.

16.5.3 DRβ-chain

To identify residues that might be involved in diabetes susceptibility in the DRβ-chain the most informative approach is to compare the *DRB1*04* subtype alleles. *Table 16.1* compares the key residues that differ between the *DRB1*04* subtypes 05, 01, 02, 04, 03 and 06 (at positions 37, 57, 71, 74 and 86), their relative hydrophobicities and their associations with type 1 diabetes. Only position 86 distinguishes the most predisposing alleles *DRB1*0405* and *DRB1*0401* from the rest. Position 86, which is an important structural and functional feature of peptide-binding pocket 1, is specifically associated with rheumatoid arthritis and in T-cell recognition (Hammer *et al.*, 1995; Ong *et al.*, 1991; Wucherpfennig and Strominger, 1995b). Note that residue 71 does not correlate, at least in a simple way, with susceptibility and resistance, despite its utility for risk assessment (Zamani Ghabanbasani *et al.*, 1994): the most predisposing allele *DRB1*0405* and the most protective allele *DRB1*0406* both have Arg at this position (*Table 16.1*). Residue 74 may help explain the protective associations of *DRB1*0403* and *DRB1*0406*. Clearly no single residue explains the associations of all these alleles; for example, the combined presence of Ser, Ala and Gly at positions 57, 74 and 86, respectively, differentiate the highest risk allele *DRB1*0405* from the protective allele *DRB1*0403* (Cucca *et al.*, 1995). The *DRB1*0406*, which seems even more protective than *DRB1*0403*, differs from the *DRB1*0406* only at position 37 (Awata *et al.*, 1992; Erlich *et al.*, 1993; Huang *et al.*, 1995).

Table 16.1. Critical polymorphic residues of the *HLA-DRB1*04* β-chain that are associated with susceptibility and resistance to type 1 diabetes

*DRB1*04* alleles	Distinguishing polymorphic residues					Disease predisposition
	37	57	71	74	86	
0405	Tyr (2.3)	Ser (−0.3)	Arg (−11.2)	Ala (0.5)	Gly (0.0)	+++
0401	Tyr (2.3)	Asp (−7.4)	Lys (−4.2)	Ala (0.5)	Gly (0.0)	++
0402	Tyr (2.3)	Asp (−7.4)	Glu (−9.9)	Ala (0.5)	Val (1.5)	+/−
0404	Tyr (2.3)	Asp (−7.4)	Arg (−11.2)	Ala (0.5)	Val (1.5)	−
0403	Tyr (2.3)	Asp (−7.4)	Arg (−11.2)	Glu (−9.9)	Val (1.5)	−−
0406	Ser (−0.3)	Asp (−7.4)	Arg (−11.2)	Glu (−9.9)	Val (1.5)	−−−

Values in parentheses are the relative hydrophobicities (kcal/mol) for each amino acid. Negative values indicate preference for water and positive values a preference for non-polar solvent.

Recent crystallographic and functional studies also indicate that polymorphisms at positions 57, 74 and 86 of the DRβ-chain have a profound effect on peptide binding and can influence the stability of the DR α, β heterodimer (Matsushita *et al.*, 1994; Olson *et al.*, 1992, 1994; Ong *et al.*, 1991; Stern *et al.*, 1994; Verreck *et al.*, 1993). The importance of these DQβ and DRβ residues in peptide presentation is further suggested by the strong balancing selection at these positions in the general population.

Other residues are also implicated. DRβ13 is in the second pocket, which is the most prominent or peptide-anchoring pocket. This position is conserved in all DR4 alleles and could explain the general predisposing effect of DR4, an effect which could be enhanced or reversed by other residues. This position is highly polymorphic and the relative hydrophobicity goes from -11.2 of Arg present in DR2 (DR15 and DR16) to $+0.3$ and $+0.5$ of Ser and His present in DR3 and DR4 respectively.

16.6 A mechanism for HLA class II-associated susceptibility/ resistance

We will not attempt to formulate a 'unified model' for the association of HLA with type 1 diabetes because our understanding of the function and structure of class II molecules is still incomplete and we are still ignorant of what other susceptibility genes lie outside the class II region. Nevertheless, it is very evident that the *DQB1*, *DQA1* and *DRB1* loci are primary and aetiological, with specific residues playing key roles, particularly in peptide binding. Nepom proposed a model in which protective alleles bind the diabetogenic peptide with higher affinity than susceptible alleles of HLA class II loci, and, in peptide-limiting conditions, the susceptible alleles would have to compete for peptide (Nepom, 1990). These ideas are underpinned by work from Sercarz (Gammon *et al.*, 1991). Sheehy extended this model and focused instead on the function of class II molecules in tolerance (in the thymus or in the periphery and, used here in the broadest possible sense, including the induction of an immune response that switches off or suppresses another, e.g. polarization of the Th1 and Th2 phenotypes; Bohme *et al.*, 1995), where protective DQ and DR molecules (e.g. *DQB1*0602* and *DRB1*0406*) would bind diabetogenic peptides with high affinity thereby facilitating tolerance or non-responsiveness. Susceptible molecules would bind diabetogenic peptide with lower affinity than protective molecules, thereby allowing T cells with anti-β-cell reactivity to escape tolerance and to be present in the individual. Susceptible molecules carry out the antigen presentation during β-cell destruction in both models (Ellerman and Like, 1995). This model is consistent with Wraith's model of low affinity autoantigenic peptides and the presence in healthy individuals of self-reactive T cells (Fairchild and Wraith, 1992). DR3/4 individuals have about a 5% absolute risk of developing type 1 diabetes (Todd and Bain, 1992), and may well have low, but detectable, levels of anti-β-cell T cells in the circulation but remain healthy because they carry a sufficient complement of protective alleles at the non-MHC loci that prevent autoimmunity gathering any momentum or approaching the critical threshold to cause insulin-deficiency and diabetes. The mechanisms responsible for mouse IE or non-IA^{g7} (see below) or human *DQB1*0602*-mediated protection from type 1 diabetes are still incompletely understood [although many mechanisms of tolerance have been established in other systems (Bohme *et al.*, 1995)].

One of the most striking features of the HLA association with type 1 diabetes, and with coeliac disease (anti-gluten immune response) is the significantly increased risk associated with heterozygosity for certain HLA haplotypes. In type 1 diabetes DR3/4 is increased in frequency in most patient populations over the frequency expected by the Hardy-Weinberg equilibrium. In some populations, such as the Sardinians, the effect may not be as pronounced but the trend is still there (F. Cucca, unpublished information). It is noted that previously Rubinstein (1991) suggested that the heterozygous effect in type 1 diabetes may be due to preferential transmission of DR4 from fathers to type 1 diabetic offspring. This is not an adequate explanation because the observation that DR4 is transmitted to type 1 diabetic children more often from fathers than from mothers is not reported consistently and, in fact, a recent combined analysis shows that the effect, if it exists at all, is very small (Bain *et al.*, 1994; Undlien *et al.*, 1995b).

In coeliac disease, *DR3-DQB*0201–DQA1*0501* is the most associated haplotype. Most DR3-negative patients are DR5/7 heterozygotes in which the *DQA1*0501–DQB1*0201* heterodimer is present *in trans*. Thorsby and colleagues proposed that this heterodimer might be involved aetiologically (Sollid *et al.*, 1989). A secondary association is with the *DQA1*0301–DQB1*0302* heterodimer (Thorsby, 1995). Importantly, most gluten-specific $CD4^+$ T cells in the intestinal mucosa of coeliac disease patients preferentially recognize gluten antigens when presented by one of these two DQ molecules (Lundin *et al.*, 1993), providing a functional link between the genetic association of DQ and antigen presentation (see Chapter 14). In type 1 diabetes the heterodimers *DQA1*0301–DQB1*0201* and *DQA1*0501–DQB1*0302* have been implicated in several ethnic groups, including Caucasians (Nepom, 1990), Afro-Caribbeans (Todd *et al.*, 1989) and Chinese (Huang *et al.*, 1995, Penny *et al.*, 1992), with increased risk. This is presumably due to peptide-binding properties of these particular DQ molecules and perhaps other features, such as their ability to form homodimers and activate T cells (Roustsias and Papadopoulos, 1995). Taken together with the coeliac disease results, it seems likely that the heterozygous effect is the result of certain DQ heterodimers and their direct functional activity in the development of the disease. Additional explanations for heterozygous effects are plausible including a role for non-class HLA genes, and a dosage effect where the DQ molecules on the DR3 and DR4 haplotypes would bind different diabetogenic peptides thereby accelerating β-cell destruction.

16.6.1 Lessons from animal models about MHC class II molecules and type 1 diabetes

The MHC is linked to type 1 diabetes across species, with evidence for linkage in human, mouse and rat (Wicker *et al.*, 1995). In the NOD mouse the major determinant of the MHC-encoded component has been mapped to the class II region (*Idd1*). The IA^{g7} molecule, which is the homologue of DQ, is unique among common laboratory inbred strains in that it is the only one not to have Asp at position 57 in the β-chain. These data indicate that the βAsp/non-Asp57 correlation is conserved across species, which would be expected since the basic biochemical properties of class II molecules are unlikely to vary significantly between species. It is noted that results from certain transgenic NOD mice expressing non-IA^{g7} molecules are not interpretable, because McDevitt and colleagues have shown that one line of NOD mice transgenic for *IA*^{g7} itself (constructed as a negative control) is also protected from diabetes. Overexpression of IA in transgenic mice is associated with B-cell depletion and eosinophilia (H.O. McDevitt, personal communication). Another line of *IA*^{g7} NOD transgenic mice shows normal B cell number and a normal frequency of diabetes. *In vivo* analysis of the role of specific IA residues will require site-directed mutagenesis and replacement using embryonic stem cell technology.

Analysis of congenic strains indicates that the NOD allele at *Idd1* is an allele that is required to cause frequent and extensive insulitis and the progression from insulitis to diabetes. Interestingly, when the NOD MHC was replaced with an MHC containing the $H2^{h4}$ allele, which encodes the IA^{k} class II molecule and is permissive for experimentally induced thyroiditis (but protective for diabetes), a spontaneous frequency of thyroiditis was observed that was higher than that in NOD mice (Wicker *et al.*, 1995). These results indicate that *Idd1* controls the efficiency of presentation of tissue-specific autoantigens: β-cell peptides are presented if the IA^{g7} molecule is present, while IA^{k} is the class II molecule most efficient at binding and presenting a thyroglobulin-derived peptide (Wicker *et al.*, 1995).

NOD mice do not express IE, the homologue of DR, and transgenic and breeding experiments have shown that IE expression is protective for insulitis and type 1 diabetes, although it is not sufficient (Podolin *et al.*, 1993). Importantly, IE^{+} MHC heterozygous mice are no more protected from diabetes than are IE^{-} MHC heterozygotes.

Construction of chimeric mice and transgenic mice indicates that the protective determinants of a non-NOD MHC, which might include an expressed IE, requires expression in bone-marrow-derived cells; expression in the thymus alone is not sufficient (Bohme *et al.*, 1990; Podolin *et al.*, 1993). These results underpin the importance of peripheral mechanisms of non-responsiveness in resistance to type 1 diabetes (Bohme *et al.*, 1995).

16.6.2 Diabetogenic peptides

What kind of peptides might be involved? Certain antigens, such as glutamic acid decarboxylase (GAD), insulin (Eisenbarth, 1994), and, more recently, tyrosine kinase (37K antigen; Rabin *et al.*, 1994) have now been strongly implicated in disease development. There is evidence in support of cross-reactivity between GAD and viral proteins, supporting a molecular mimicry model for induction of autoimmune disease (Soliemna and de Camilli, 1995). Wucherpfennig and Strominger proposed that the P9 position of a diabetogenic peptide might be Asp or Glu since DQβ57 is Ala or Val in susceptibility molecules (Wucherpfennig and Strominger, 1995b), but this does fit easily the tolerance/low affinity diabetogenic peptide model described above. Alternatively, the protective alleles of DQ or IA could bind peptides and tip the T-cell balance to the Th2 phenotype (Bohme *et al.*, 1995; Tisch and McDevitt, 1996).

16.7 Other HLA susceptibility loci

Extended susceptibility haplotypes in human type 1 diabetes indicate that different haplotypes with apparently identical class II loci but different class I alleles predispose to disease to significantly different degrees. Loci in the class I region may affect the rate of pancreatic β-cell destruction and the age of onset of disease (Demaine *et al.*, 1995; Fujisawa *et al.*, 1995; Nakanishi *et al.*, 1993).

The best evidence so far comes from the NOD mouse, proving that an important determinant does indeed lie outside the class II region. Using a congenic NOD mouse strain that possesses a recombinant MHC from a diabetes-resistant sister strain, it has been demonstrated that the mouse *Idd1* consists of at least two components, one in and one outside the MHC class II region (Ikegami *et al.*, 1995). Candidate genes in the non-class II region include the class I MHC genes (Ikegami *et al.*, 1995).

Also, other non-DR, DQ genes in the class II region could be related to type 1 diabetes; for instance, a study has indicated a possible additive role of the *LMP2* and *LMP7* genes, which encode two proteasome subunits, in the determination of the HLA-encoded susceptibility to type 1 diabetes (Deng *et al.*, 1995; see Chapter 9 for a description of the role of *LMP2* and *7* in antigen processing). The independent effect of these genes was supported by significant differences in the comparison of *LMP* gene frequencies in HLA class II haplotype-matched patients and controls (Deng *et al.*, 1995). Nevertheless, these results will have to be replicated in other independent data sets and also DR4 subtypes must be taken into account.

The recently identified transporter associated with antigen-processing genes *TAP1* and *TAP2*, because of their role in antigen processing and their location within the HLA class II region, have been considered possible candidate genes. Although two early studies demonstrated an increased frequency of *TAP2A* among patients, with a corresponding decrease of the *TAP2B* allele, controversy has arisen as to whether these observations represented primary associations or a hitch-hiking effect due to linkage disequilibrium. Caillat-Zucman and colleagues interpreted the associations as an independent effect of this locus while Ronningen and co-workers suggested that they simply reflected linkage disequilibrium between *TAP2* and *HLA-DR* alleles (Caillat-Zucman *et al.*, 1993; Ronningen *et al.*, 1993). Another study that compared *HLA-DR* and *-DQ* genotype matched patients and controls from Sardinia supported the conclusion that there was no primary association between *TAP2* alleles and type 1 diabetes (Cucca *et al.*, 1994a). More recently Caillat-Zucman and colleagues, after increasing the number of individuals analysed, reviewed their data and confirmed the lack of an independent effect of *TAP2* genes (Caillat-Zucman *et al.*, 1995). Further negative results have been reported from Italian families (Esposito *et al.*, 1995). Similarly, evidence against an independent role for the human *TNFA* and the *TNFB* loci in type 1 diabetes has been reported (Cox *et al.*, 1994; Jenkins *et al.*, 1991).

Candidate genes must always be subjected to a rigorous evaluation before they are declared to be primarily associated. Studies for HLA candidate genes should take into account disequilibrium patterns and therefore require very large clinical resources. The following steps can represent a valuable tool to solve the effects due to linkage disequilibrium.

(i) Replication in different populations of any positive or negative association.
(ii) HLA haplotype-matched analysis: comparison of gene frequencies in extended haplotypes derived from patients and controls and differing only at one candidate locus. Allelic variation at the candidate locus should be able to modify the disease-predisposing effect of the haplotype, if an allele of the marker locus is in disequlibrium with an allele of the aetiological mutation, or is the mutation itself.
(iii) HLA genotype-matched case-control studies. These are based on statistical differences in the distribution of genotypes from large numbers of patients and ethnically matched controls homozygous for an associated combination of variants but heterozygous for a putative additional variant.
(iv) Transmission from parents, homozygous for a predisposing combination of variants but heterozygous for a putative additional variant, to affected children. If the transmission does not significantly exceed 50%, the additional disease-predisposing locus is not detected. An example would be the analysis of the transmission of different extended DR3-positive haplotypes in families that have parents who are identical at the DQ and DR loci but who differ in the class I region. Given the very large collections of multiplex (Bain *et al.*, 1990; Lernmark *et al.*, 1990) and, more recently, simplex families (Copeman *et al.*, 1995) that are available, these kind of studies should be feasible in the near future, and will yield definitive data as to presence and location of non-class II HLA type 1 diabetes susceptibility loci.
(v) Sequence analyses and functional studies.

16.8 Conclusion

The MHC region contains the major genetic component of type 1 diabetes (*IDDM1*). Taking all available genetic, structural and functional data into account, there is now little doubt that the HLA class II *DRB1*, *DQA1* and *DQB1* loci encode the primary aetiological mutations. HLA class II-encoded predisposition to type 1 diabetes is inherited in a recessive-like fashion, while HLA-encoded protection from disease is dominant-acting. A single dose of a strongly protective *DQB1* or *DRB1* allele bestows a protective status on the entire genotype. The functional dominance of protection is confirmed by the observation that the *DQB1*0602* allele confers protection even in islet cell antibody-positive relatives of type 1 diabetic patients (Pugliese *et al.*, 1995). An understanding of the mechanism of this lifelong protection is an important research goal. A combined *DQB1*, *DQA1* and *DRB1* model can explain results from different ethnic groups and the frequencies of protective and susceptible alleles at these loci correlates with the frequency of disease in different populations. The aetiological mechanism of the class II association involves peptide binding and T-cell recognition, and these properties of class II molecules are obviously constant among different ethnic groups. Collectively, the non-MHC genes, including the insulin gene locus *IDDM2*, have a combined effect equivalent to or greater than *IDDM1*. Identification of these mutations and an understanding of their functional consequences in immunity and in β-cell function and glucose homeostasis will, ultimately, lead to new strategies for the prevention of type 1 diabetes.

Acknowledgements

We thank our colleagues for their help and advice, and the European Biomed programme, the Wellcome Trust, the British Diabetic Association, the Juvenile Diabetes Foundation and the Medical Research Council for financial support. John Todd is a Wellcome Trust Principal Research Fellow.

References

Awata T, Kuzuya T, Matsuda A, Iwamoto Y, Kanazawa Y. (1992) Genetic analysis of HLA class II alleles and susceptibility to type 1 (insulin-dependent) diabetes mellitus in Japanese subjects [published erratum appears in *Diabetologia* **35(9):** 906]. *Diabetologia* **35:** 419–424.

Ayala FJ. (1995) The myth of Eve: molecular biology and human origins. *Science* **270:** 1930–1936.

Bain SC, Todd JA, Barnett AH. (1990) The British Diabetic Association - Warren Repository. *Autoimmunity* 7: 83–85.

Bain SC, Rowe BR, Barnett AH, Todd JA. (1994) Parental origin of diabetes-associated HLA types in sibling pairs with type 1 diabetes mellitus. *Diabetes* **43:** 1462–1468.

Bennett ST, Lucassen AM, Gough SC, *et al.* (1995) Susceptibility to human type 1 diabetes at IDDM2 is determined by tandem repeat variation at the insulin gene minisatellite locus. *Nature Genetics* **9:** 284–292.

Bohme J, Schuhbaur B, Kanagawa O, Benoist C, Mathis D. (1990) MHC-Linked Protection from Diabetes Dissociated from Clonal Deletion of T Cells. *Science* **240:** 293–295.

Bohme J, Brenden N, Rietz C, Pilstrom B. (1995) Dominant-disease protection in MHC-transgenic NOD mice. More than one mechanism? *The Immunologist* **3:** 45–50.

Buyse I, Sandkuyl LA, Ghabanbasani MZ, *et al.* (1994) Association of particular HLA class II alleles, haplotypes and genotypes with susceptibility to IDDM in the Belgian population. *Diabetologia* **37:** 808–817.

Caillat-Zucman S, Bertin E, Timsit J, Boitard C, Assan R, Bach JF. (1993) Protection from insulin-dependent diabetes mellitus is linked to a peptide transporter gene. *Eur. J. Immunol.* **23:** 1784–1788.

Caillat-Zucman S, Daniel SID-S, Timsit J, Garchon H, Boitard C, Bach J. (1995) Family study of linkage disequilibrium between TAP2 transporter and HLA class II genes: absence of TAP2 contribution to association with insulin-dependent diabetes mellitus. *Hum. Immunol.* **44:** 80–87.

Copeman JB, Cucca F, Hearne CM, *et al.* (1995) Linkage disequilibrium mapping of a type 1 diabetes susceptibility gene (IDDM7) to chromosome 2q31-q33. *Nature Genetics* **9:** 80–85.

Cordell HJ, Todd JA. (1995) Multifactorial inheritance of type 1 diabetes. *Trends Genet.* **11:** 499–504.

Cordell HJ, Todd JA, Bennett ST, Kawaguchi Y, Farral M. (1995) Two locus maximum lod score analysis of a multifactorial trait: joint consideration of IDDM2 and IDDM4 with IDDM1 in type I diabetes. *Am. J. Hum. Genet.* **57:** 920–934.

Cox A, Gonzalez AM, Wilson AG, Wilson RM, Ward JD, Artlett CM, Welsh K, Duff GW. (1994) Comparative analysis of the genetic associations of HLA-DR3 and tumour necrosis factor alpha with human IDDM. *Diabetologia* **37:** 500–503.

Cucca F, Muntoni F, Lampis R, Frau F, Argiolas L, Silvetti M, Angius E, Cao A, De VS, Congia M. (1993) Combinations of specific DRB1, DQA1, DQB1 haplotypes are associated with insulin-dependent diabetes mellitus in Sardinia. *Hum. Immunol.* **37:** 85–94.

Cucca F, Congia M, Trowsdale J, Powis SH. (1994a) Insulin-dependent diabetes mellitus and the major histocompatibility complex peptide transporters TAP1 and TAP2: no association in a population with a high disease incidence. *Tissue Antigens* **44:** 234–240.

Cucca F, Frau F, Lampis R, Floris M, Argiolas L, Macis D, Cao A, De Virgiliis S, Congia M. (1994b) HLA-DQB1*0305 and -DQB1*0304 alleles among Sardinians. Evolutionary and practical implications for oligotyping. *Hum. Immunol.* **40:** 143–149.

Cucca F, Lampis R, Frau F, *et al.* (1995) The distribution of DR4 haplotypes in Sardinia suggests a primary association of insulin dependent diabetes mellitus with DRB1 and DQB1 loci. *Hum. Immunol.* **43:** 301–308.

Cudworth A, Woodrow J. (1974) HL-A system and diabetes mellitus. *Diabetes* **24:** 345–349.

Cudworth AG, Woodrow JC. (1975) Evidence for HL-A-linked genes in 'juvenile' diabetes mellitus. *Br. Med. J.* **3:** 133–135.

Dahlquist G. (1995) Environmental risk factors in human type 1 diabetes – an epidemiological perspective. *Diabetes/Metabolism Reviews* **11:** 37–46.

Davies JL, Kawaguchi Y, Bennett ST, et al. (1994) A genome-wide search for human type 1 diabetes susceptibility genes. *Nature* **371:** 130–136.

Demaine AG, Hibberd ML, Mangles D, Millward BA. (1995) A new marker in the HLA class I region is associated with age at onset of IDDM. *Diabetologia* **38:** 623–628.

Deng GY, Muir A, Maclaren NK, She JX. (1995) Association of LMP2 and LMP7 genes within the major histocompatibility complex with insulin-dependent diabetes mellitus: population and family studies. *Am. J. Hum. Genet.* **56:** 528–534.

Djoulah S, Khalil I, Beressi JP, Benhamamouch S, Bessaoud K, Deschamps I, Degos L, Hors J. (1992) The HLA-DRB1*0405 haplotype is most strongly associated with IDDM in Algerians. *Eur. J. Immunogen.* **19:** 381–389.

Easteal S, Kohonen-Corish MR, Zimmet P, Serjeantson SW. (1990a) HLA-DP variation as additional risk factor in IDDM. *Diabetes* **39:** 855–857.

Easteal S, Viswanathan M, Serjeantson SW. (1990b) HLA-DP, -DQ and -DR RFLP types in south Indian insulin-dependent diabetes mellitus patients. *Tissue Antigens* **35:** 71–74.

Ebers GC, Sadovnick AD, Risch NJ, Group CCS. (1995) A genetic basis for familial aggregation in multiple sclerosis. *Nature* **377:** 150–151.

Eisenbarth GS. (1994) Mouse or man. Is GAD the cause of type I diabetes? *Diabetes Care* **17:** 605–607.

Ellerman KE, Like AA. (1995) A major histocompatibility complex class II restriction for BioBreeding/Worcester diabetes-inducing T cells. *J. Exp. Med.* **182:** 923–930.

Erlich HA, Zeidler A, Chang J, et al. (1993) HLA class II alleles and susceptibility and resistance to insulin dependent diabetes mellitus in Mexican-American families. *Nature Genetics* **3:** 358–364.

Esposito L, Lampasona V, Bosi E, Poli F, Ferrari M, Bonifacio E. (1995) HLA DQA1-DQB1-TAP2 haplotypes in IDDM families: no evidence for an additional contribution to disease risk by the TAP2 locus. *Diabetologia* **38:** 968–974.

Ewens WJ, Spielman RS. (1995) The transmission/disequilibrium test: history, subdivision and admixture. *Am. J. Hum. Genet.* **57:** 455–464.

Fairchild PJ, Wraith DC. (1992) Peptide-MHC interaction in autoimmunity. *Curr. Opin. Immunol.* **4:** 748–753.

Falk CT, Rubinstein P. (1987) Haplotype relative risks: an easy reliable way to construct a proper control sample for risk calculations. *Ann. Hum. Genet.* **51:** 227–233.

Fernandez VM, Ramirez LC, Raskin P, Stastny P. (1993) Genes for insulin-dependent diabetes mellitus (IDDM) in the major histocompatibility complex (MHC) of African-Americans. *Tissue Antigens* **41:** 57–64.

Festenstein H, Awad J, Hitman GA, Cutbush S, Groves AV, Cassell P, Ollier W, Sachs JA. (1986) New HLA DNA polymorphisms associated with autoimmune diseases. *Nature* **322:** 64–67.

Fujisawa T, Igekami H, Kawaguchi Y, Yamato E, Takekawa K, Nakagawa Y, Hamada Y, Ueda H, Shima K, Ogihara T. (1995) Class I associated with age-at-onset of IDDM, while Class II confer susceptibility to IDDM. *Diabetologia* **38:** 1493–1495.

Gammon G, Sercarz EE, Benichou G. (1991) The dominant self and cryptic self: shaping the autoreactive T-cell repertoire. *Immunol. Today* **12:** 193–195.

Gough SCL, Saker PJ, Pritchard LE, et al. (1995) Mutation of the glucagon receptor gene and diabetes mellitus in the UK: association or founder effect? *Hum. Mol. Genet.* **4:** 1609–1612.

Green A, Gale EA, Patterson CC. (1992) Incidence of childhood-onset insulin-dependent diabetes mellitus: the EURODIAB ACE Study. *Lancet* **339:** 905-909.

Gyllensten UB, Erlich HA. (1993) MHC class II haplotypes and linkage disequilibrium in primates. *Hum. Immunol.* **36:** 1–10.

Hammer J, Gallazzi F, Bono E, Karr RW, Guenot J, Valsasnini P, Nagy ZA, Sinigaglia F. (1995) Peptide binding to HLA-DR4 molecules: correlation with rheumatoid arthritis association. *J. Exp. Med.* **181:** 1847–1855.

Hu C-Y, Allen M, Chuang L-M, Lin BJ, Gyllensten U. (1993) Association of insulin-dependent diabetes mellitus in Taiwan with HLA class II DQB1 and DRB1 alleles. *Hum. Immunol.* **38:** 105–114.

Huang H-S, Peng J-T, She JY, Zhang L-P, Chao CCK, Liu K-H, She J-X. (1995) HLA-encoded susceptibility to insulin-dependent diabetes mellitus is determined by DR and DQ genes as well as their linkage disequilibria in a Chinese population. *Hum. Immunol.* **44:** 210–219.

Ikegami H, Makino S, Yamato E, Kawaguchi Y, Ueda H, Sakamoto T, Takekawa K, Ogihara T. (1995) Identification of a new susceptibility locus for insulin-dependent diabetes mellitus by ancestral haplotype congenic mapping. *J. Clin. Invest.* **96:** 1936–1941.

Jenkins D, Penny MA, Mijovic CH, Jacobs KH, Fletcher J, Barnett AH. (1991) Tumour necrosis factor-beta polymorphism is unlikely to determine susceptibility to type 1 (insulin-dependent) diabetes mellitus. *Diabetologia* **34:** 576–578.

Johnston C, Pyke DA, Cudworth AG, Wolf E. (1983) HLA-DR typing in identical twins with insulin-dependent diabetes: difference between concordant and discordant pairs. *Br. Med. J. (Clin. Res. edn)* **286:** 253-255.

Jorde LB, Watkins WS, Carlson M, Albertsen GH, Thliveris A, Leppert M. (1994) Linkage disequilibrium predicts physical distance in the adenomatous polyposis coli region. *Am. J. Hum. Genet.* **54:** 884–898.

Ju LY, Gu XF, Bardie R, Krishnamoorthy R, Charron D. (1991) A simple nonradioactive method of DNA typing for subsets of HLA-DR4: prevalence data on HLA-DR4 subsets in three diabetic population groups. *Hum. Immunol.* **31:** 251–258.

Julier C, Lucassen A, Villedieu P, et al. (1994) Multiple DNA variant association analysis: application to the insulin gene region in type 1 diabetes. *Am. J. Hum. Genet.* **55:** 1247–1254.

Karvonen M, Tuomilehto J, Libman I, La PR. (1993) A review of the recent epidemiological data on the worldwide incidence of type 1 (insulin-dependent) diabetes mellitus. World Health Organization DIAMOND Project Group. *Diabetologia* **36:** 883–892.

Khalil I, d'Auriol L, Gobet M, Morin L, Lepage V, Deschamps I, Park MS, Degos L, Galibert F, Hors J. (1990) A combination of HLA-DQ beta Asp57-negative and HLA DQ alpha Arg52 confers susceptibility to insulin-dependent diabetes mellitus. *J. Clin. Invest.* **85:** 1315–1319.

Klitz W, Stephens JC, Grote M, Carrington M. (1995) Discordant patterns of linkage disequilibrium of the peptide transporter loci within the HLA class region. *Am. J. Hum. Genet.* **57:** 1436–1444.

Kwok WW, Kovats S, Thurtle P, Nepom GT. (1993) HLA-DQ allelic polymorphisms constrain patterns of class II heterodimer formation. *J. Immunol.* **150:** 2263–2272.

Kwok WW, Nepom GT, Raymond FC. (1995) HLA-DQ polymorphisms are highly selective for peptide binding interactions. *J. Immunol.* **155:** 2468–2476.

Kyvik KO, Green A, Beck-Nielsen H. (1995) Concordance rates of insulin dependent diabetes mellitus: a population based study of young Danish twins. *Br. Med. J.* **311:** 913–917.

Lernmark A, Ducat L, Eisenbarth G, Ott J, Permutt MA, Rubenstein P, Spielman R. (1990) Family cell lines available for research. *Am. J. Hum. Genet.* **47:** 1028–1030.

Lucassen A, Julier C, Beressi J-P, Boitard C, Froguel P, Lathrop G, Bell J. (1993) Susceptibility to insulin dependent diabetes mellitus maps to a 4.1 kb segment of DNA spanning the insulin gene and associated VNTR. *Nature Genetics* **4:** 305–310.

Lundin KEA, Scott H, Hansen T, Paulsen G, Halstensen TS, Fausa O, Thorsby E, Sollid LM. (1993) Gliadin-specific, HLA-DQ (α1*0501, β1*0201) restricted T cells from the small intestinal mucosa of celiac disease patients. *J. Exp. Med.* **178:** 187–196.

Magzoub MM, Stephens HA, Sachs JA, Biro PA, Cutbush S, Wu Z, Bottazzo GF. (1992) HLA-DP polymorphism in Sudanese controls and patients with insulin-dependent diabetes mellitus. *Tissue Antigens* **40:** 64–68.

Marshall KW, Liu AF, Canales JC. (1994) Role of polymorphic residues in HLA-DR molecules in allele-specific binding of peptide antigens. *J. Immunol.* **152:** 4946–4957.

Matsushita S, Takahashi K, Motoki M, Komoriya K, Ikagawa S, Nishimura Y. (1994) Allele specificity of structural requirement for peptides bound to HLA-DRB1*0405 and -DRB1*0406 complexes: implication for the HLA-associated susceptibility to methimazole-induced insulin autoimmune syndrome. *J. Exp. Med.* **180:** 873–883.

McAleer MA, Reifsnyder P, Palmer SM, et al. (1995) Crosses of NOD mice with the related NON strain: a polygenic model for type 1 diabetes. *Diabetes* **44:** 1186–1195.

Michelsen B, Lernmark A. (1987) Molecular cloning of a polymorphic DNA endonuclease fragment that associates insulin-dependent diabetes mellitus with HLA-DQ. *J. Clin. Invest.* **79:** 1144–1152.

Morton N, Green A, Dunsworth T, Svejgaard A, Barbosa J, Rich S, Iselius L, Platz P, Ryder L. (1983) Heterozygous expression of insulin-dependent diabetes mellitus (IDDM) determinants in the HLA system. *Am. J. Hum. Genet.* **35:** 201–213.

Nakanishi K, Kobayashi T, Murase T, Nakatsuji T, Inoko H, Tsuji K, Kosaka K. (1993) Association of HLA-A24 with complete beta-cell destruction in IDDM. *Diabetes* **42:** 1086–1093.

Nepom GT. (1990) A unified hypothesis for the complex genetics of HLA associations with IDDM. *Diabetes* **39:** 1153–1157.

Nerup J, Platz P, Andersen OO, Christy M, Lyngsoe J, Poulsen JE, Ryder LP, Nielsen LS, Thomsen M, Svejgaard A. (1974) HLA antigens and diabetes mellitus. *Lancet* **2:** 864–866.

Oldstone MBA, Nerenberg M, Southern P, Price J, Lewicki H. (1991) Virus infection triggers insulin-dependent diabetes mellitus in a transgenic model: role of anti-self (virus) immune response. *Cell* **65:** 319–331.

Olmos P, A'Hern R, Heaton DA, Millward BA, Risley D, Pyke DA, Leslie RD. (1988) The significance of the concordance rate for type 1 (insulin-dependent) diabetes in identical twins. *Diabetologia* **31:** 747–750.

Olson RR, De MM, Di TA, Karr RW. (1992) Mutations in the third, but not the first or second, hypervariable regions of DR(beta 1*0101) eliminate DR1-restricted recognition of a pertussis toxin peptide. *J. Immunol.* **148:** 2703–2708.

Olson RR, Reuter JJ, McNicholl J, Alber C, Klohe E, Callahan K, Siliciano RF, Karr RW. (1994) Acidic residues in the DR beta chain third hypervariable region are required for stimulation of a DR(alpha, beta 1*0402)-restricted T-cell clone. *Hum. Immunol.* **41:** 193–200.

Ong B, Willcox N, Wordsworth P, Beeson D, Vincent A, Altmann D, Lanchbury JS, Harcourt GC, Bell JI, Newsom-Davis J. (1991) Critical role for the Val/Gly86 HLA-DR beta dimorphism in autoantigen presentation to human T cells. *Proc. Natl Acad. Sci. USA* **88:** 7343–7347.

Owerbach D, Lernmark A, Platz P, Ryder LP, Rask L, Peterson PA, Ludvigsson J. (1983) HLA-D region β-chain DNA endonuclease fragments differ between HLA-DR identical healthy and insulin-dependent diabetic individuals. *Nature* **303:** 815–817.

Penny MA, Jenkins D, Mijovic CH, Jacobs KH, Cavan DA, Yeung VTF, Cockram C, Hawkins BR, Fletcher JA, Barnett AH. (1992) Susceptibility to IDDM in a Chinese population. *Diabetes* **41:** 914–919.

Podolin P, Pressey A, DeLarato N, Fischer P, Peterson L, Wicker L. (1993) I-E^{+} nonobese diabetic mice develop insulitis and diabetes. *J. Exp. Med.* **178:** 793–803.

Pugliese A, Gianani R, Moromisato R, Awdeh ZL, Alper CA, Erlich HA, Jackson RA, Eisenbarth GS. (1995) HLA-DQB1*0602 is associated with dominant protection from diabetes even among islet cell antibody-positive first-degree relatives of patients with IDDM. *Diabetes* **44:** 608–613.

Rabin DU, Pleasic SM, Shapiro JA, Yoo-Warren H, Oles J, Hicks JM, Goldstein DE, Rae PM. (1994) Islet cell antigen 512 is a diabetes-specific islet autoantigen related to protein tyrosine phosphatases. *J. Immunol.* **152:** 3183–3188.

Reich E-P, von Grafenstein H, Barlow A, Swenson KE, Williams K, Janeway CA. (1994) Self peptides isolated from MHC glycoproteins on non-obese diabetic mice. *J. Immunol.* **152:** 2279–2288.

Rejionen H, Tuokko J, Ilonen J, Akerblom H. (1995) The effect of HLA-DR4 subtype on the risk of IDDM carried by DQB1**0302*. The 9th International Congress of Immunology, San Francisco, California.

Risch N. (1987) Assessing the role of HLA-linked and unlinked determinants of disease. *Am. J. Hum. Genet.* **40:** 1–14.

Risch N. (1990) Linkage strategies for genetically complex traits. II. The power of affected relative pairs. *Am. J. Hum. Genet.* **46:** 229–241.

Risch N, Ghosh S, Todd JA. (1993) Statistical evaluation of multiple locus linkage data in experimental species and relevance to human studies: application to murine and human IDDM. *Am. J. Hum. Genet.* **53:** 702–714.

Ronningen K, Spurkland A, Tait B, *et al.* (1992) HLA class II association in insulin-dependent diabetes mellitus among Blacks, Caucasoids and Japanese. *HLA 1991* (eds K. Tsuji, M. Aitzawa and T. Sasazuki). Oxford University Press, Oxford, pp. 713–722.

Ronningen KS, Undlien DE, Ploski R, *et al.* (1993) Linkage disequilibrium between TAP2 variants and HLA class II alleles; no primary association between TAP2 variants and insulin-dependent diabetes mellitus. *Eur. J. Immunol.* **23:** 1050–1056.

Roustsias J, Papadopoulos GK. (1995) Polymorphic structural features of modelled HLA-DQ molecules segregate according to susceptibility or resistance to IDDM. *Diabetologia* **38:** 1251–1261.

Rubinstein P. (1991) HLA and IDDM: facts and speculations on the disease gene and its mode of inheritance. *Hum. Immunol.* **30:** 270–277.

Sanjeevi CB, Lybrand TP, DeWeese C, Landin-Olsson M, Kockum I, Dahlquist G, Sundkvist G, Stenger D, Lernmark A, Study SCD. (1995) Polymorphic amino acid variations in HLA-DQ are associated with systematic physical property changes and occurence of IDDM. *Diabetes* **44:**

125–131.

Sheehy MJ. (1992) HLA and insulin-dependent diabetes. A protective perspective. *Diabetes* **41:** 123–129.

Sheehy MJ, Scharf SJ, Rowe JR, Neme dGM, Meske LM, Erlich HA, Nepom BS. (1989) A diabetes-susceptible HLA haplotype is best defined by a combination of HLA-DR and -DQ alleles. *J. Clin. Invest.* **83:** 830–835.

Singal DP, Blajchman MA. (1973) Histocompatability antigens, lymphocytotoxic antibodies and tissue antibodies in patients with diabetes mellitus. *Diabetes* **22:** 429–432.

Soliemna M, De Camilli P. (1995) Coxsackieviruses and diabetes. *Nature Medicine* **1:** 25–26.

Sollid LM, Markussen G, Ek J, Gjerde H, Vartdal F, Thorsby E. (1989) Evidence for a primary association of celiac disease to a particular HLA-DQ α/β heterodimer. *J. Exp. Med.* **169:** 345–350.

Spielman RS, McGinnis RE, Ewens WJ. (1993) Transmission test for linkage disequilibrium: the insulin gene region and insulin-dependent diabetes mellitus (IDDM). *Am. J. Hum. Genet.* **52:** 506–516.

Stern LJ, Brown JH, Jardetzky TS, Gorga JC, Urban RG, Strominger JL, Wiley DC. (1994) Crystal structure of the human class II MHC protein HLA-DR1 complexed with an influenza virus peptide. *Nature* **368:** 215–221.

Tait BD, Harrison LC, Drummond BP, Stewart V, Varney MD, Honeyman MC. (1995) HLA antigens and age at diagnosis of insulin-dependent diabetes mellitus. *Hum. Immunol.* **42:** 116–122.

Thomson G. (1988) HLA disease associations: models for insulin dependent diabetes mellitus and the study of complex human genetic disorders. *Annu. Rev. Genet.* **22:** 31–50.

Thomson G, Robinson WP, Kuhner MK, *et al.* (1988) Genetic heterogeneity, modes of inheritance, and risk estimates for a joint study of caucasians with insulin-dependent diabetes mellitus. *Am. J. Hum. Genet.* **43:** 799–816.

Thorsby E. (1995) HLA-associated disease susceptibility - which genes are primarily involved? *The Immunologist* **3:** 51–58.

Tisch R, McDevitt HO. (1996) Insulin dependent diabetes mellitus. *Cell* (in press).

Todd JA. (1991) A protective role of the environment in the development of type 1 diabetes? *Diabetic Med.* **8:** 906–910.

Todd JA. (1994) The Emperor's new genes: 1993 RD Lawrence lecture. *Diabetic Med.* **11:** 6–16.

Todd JA. (1995) Genetic analysis of type 1 diabetes using whole genome approaches. *Proc. Natl Acad. Sci. USA* **92:** 8560–8565.

Todd JA, Bain SC. (1992) A practical approach to the identification of susceptibility genes for type 1 diabetes. *Diabetes* **41:** 1029–1034.

Todd JA, Bell JI, McDevitt HO. (1987) HLA-DQ beta gene contributes to susceptibility and resistance to insulin-dependent diabetes mellitus. *Nature* **329:** 599–604.

Todd JA, Mijovic C, Fletcher J, Jenkins D, Bradwell AR, Barnett AH. (1989) Identification of susceptibility loci for insulin-dependent diabetes mellitus by trans-racial gene mapping. *Nature* **338:** 587–589.

Todd JA, Fukui Y, Kitagawa T, Sasazuki T. (1990) The A3 allele of the HLA-DQA1 locus is associated with susceptibility to type 1 diabetes in Japanese. *Proc. Natl Acad. Sci. USA* **87:** 1094–1098.

Todd JA, Aitman TJ, Cornall RJ, *et al.* (1991) Genetic analysis of autoimmune type 1 diabetes mellitus in mice. *Nature* **351:** 542–547.

Undlien DE, Akelsen HE, Knutsen I, Joner G, Dahl-Jorgensen K, Aagenaes O, Sovik O, Thorsby E, Ronningen KS. (1995a) DRB1*04 subtypes in IDDM. Implications for genetic risk assessment. *Autoimmunity* **21:** 17–18.

Undlien DE, Akselsen HE, Joner G, Dahl-Jorgensen K, Aagenaes O, Sovik O, Thorsby, E, Ronningen KS. (1995b) No difference in the parental origin of susceptibility HLA class II haplotypes among Norwegian patients with insulin-dependent diabetes mellitus. *Am. J. Hum. Genet.* **57:** 1511–1514.

Undlien DE, Bennett ST, Todd JA, Akselsen HE, Ikaheimo I, Reilonen H, Knip M, Thorsby E, Ronningen KS. (1995c) Insulin gene region-encoded susceptibility to IDDM maps upstream of the insulin gene. *Diabetes* **44:** 620–625.

Utz U, Biddison WE, McFarland HF, McFarlin DE, Flerlage M, Martin R. (1993) Skewed T-cell receptor repertoire in genetically identical twins correlates with multiple sclerosis. *Nature* **364:** 243–247.

Van der Auwera B, Van Waeyenberge C, Schuit F, Heimberg H, Vandewalle C, Gorus F, Flament J. (1995) DRB1*0403 protects against IDDM in Caucasians with the high-risk heterozygous DQA1*0301-DQB1*0302/DQA1*0501-DQB1*0201 genotype. *Diabetes* **44:** 527–530.

Verge C, Gianami R, Yu L, Pietropaolo M, Smith T, Jackson RA, Soelder JS, Eisembarth GS. (1995) Late progression of diabetes and evidence for chronic beta-cell autoimmunity in identical twins of patients with type I diabetes. *Diabetes* **44:** 1176–1179.

Verreck FA, Termijtelen A, Koning F. (1993) HLA-DR beta chain residue 86 controls DR alpha beta dimer stability. *Eur. J. Immunol.* **23:** 1346–1350.

Wicker LS, Todd JA, Peterson LB. (1995) Genetic control of autoimmune diabetes in the NOD mouse. *Annu. Rev. Immunol.* **13:** 179–200.

Wolf E, Spencer KM, Cudworth AG. (1983) The genetic susceptibility to type 1 (insulin-dependent) diabetes: analysis of the HLA-DR association. *Diabetologia* **24:** 224–230.

Wucherpfennig KW, Strominger JL. (1995a) Molecular mimicry in T cell mediated autoimmunity: viral peptides activate human T cell clones specific for myelin basic protein. *Cell* **80:** 695–705.

Wucherpfennig KW, Strominger JL. (1995b) Selective binding of self peptides to disease-associated major histocompatibility complex (MHC) molecules: a mechanism for MHC-linked susceptibility to human autoimmune diseases. *J. Exp. Med.* **181:** 1597–1601.

Yamoto E, Itoh N, Matsumoto E, Lund T, Yoshino M, Moriwaki K, Shiroishi T, Hattori M. (1995) Second MHC-linked diabetes gene located in the proximal side to LMP2 in the MHC class I-K region in NOD mice. *Diabetes* **44:** 18A.

Zamani Ghabanbasani M, Spaepen M, Buyse I, Marynen P, Bex M, Bouillon R, Cassiman JJ. (1994) Improved risk assessment for insulin-dependent diabetes mellitus by analysis of amino acids in HLA-DQ and DRB1 loci. *Eur. J. Hum. Genet.* **2:** 177–184.

17

MHC molecules in transplantation and tolerance

Rachel Hilton, Giovanna Lombardi and Robert Lechler

17.1 Introduction

Major histocompatibility complex (MHC)-incompatible tissues induce vigorous primary immune responses, manifest *in vitro* as the mixed lymphocyte reaction (MLR), and *in vivo* as allograft rejection. Indeed, it was the identification of the targets of graft rejection that led to the discovery of MHC molecules, initially called transplantation antigens.

The exceptional vigour of an alloresponse can, to some extent, be explained by the high precursor frequency of alloreactive T cells within the normal repertoire. Whereas the precursor frequency of T cells capable of responding to conventional antigens is low, of the order of 1:150 000–1:350 000 (Ford and Burger, 1983), as many as 1:1000 T cells may respond to alloantigen (Lindahl and Wilson, 1977). The basis for this extraordinarily high frequency of allorecognition has only recently become clear as knowledge of MHC structure and function, and the nature of the T-cell receptor (TCR) ligand, has developed. The molecular and cellular events that contribute to the generation of an alloresponse, and the means whereby these may be overcome to induce tolerance, form the basis of this chapter.

17.2 Molecular basis of allorecognition

There are two pathways whereby foreign MHC molecules can be recognized by T cells (*Figure 17.1*). In 'direct' allorecognition, T cells engage with complexes of intact allogeneic MHC molecules and bound peptide on the surface of cells of donor origin. Alternatively, alloantigen may be shed from the donor cell surface, processed and presented by recipient antigen-presenting cells (APCs) in association with self-MHC molecules, much like a conventional antigen. This is known as 'indirect' allorecognition (Lechler and Batchelor, 1982). Although indirect allorecognition may be of particular significance in chronic graft rejection (Cramer *et al.*, 1989), it is greatly overshadowed during acute allograft rejection by the direct pathway, due to the far larger numbers of T cells that can be activated by this route.

HLA and MHC: genes, molecules and function, edited by M.J. Browning and A.J. McMichael.
© 1996 BIOS Scientific Publishers Ltd, Oxford.

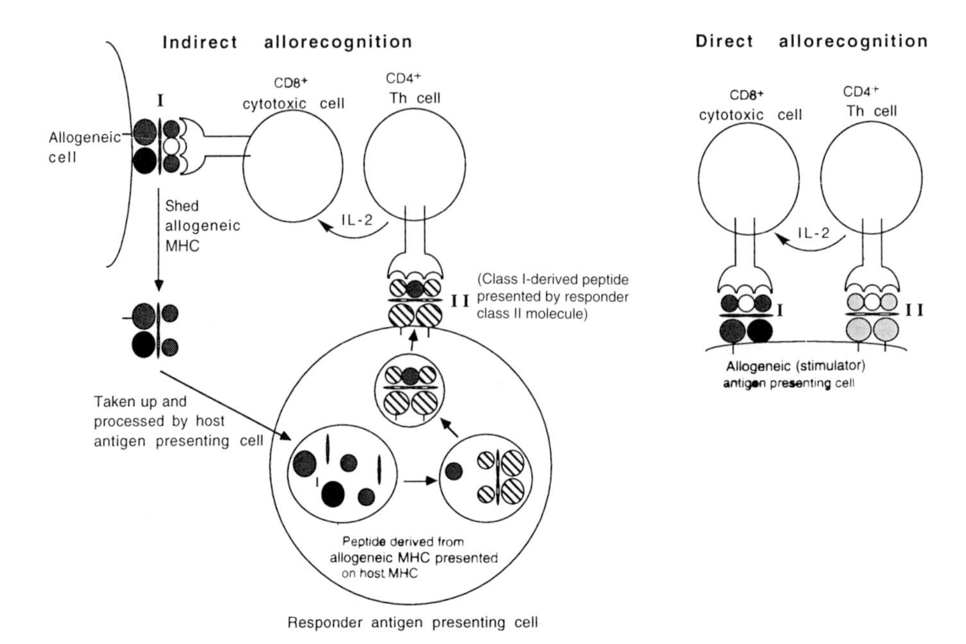

Figure 17.1. Indirect versus direct allorecognition. *Direct allorecognition.* Responder T cells recognize MHC directly on the surface of allogeneic antigen-presenting cells (APCs). Once activated, $CD8^+$ alloreactive cytotoxic cells will kill MHC class I-expressing allogeneic cells directly. *Indirect allorecognition.* Antigens shed from the allogeneic cell surface (primarily allogeneic MHC molecules) are processed and presented by professional APC of recipient origin to $CD4^+$ Th cells in association with host MHC class II molecules. This is, therefore, similar to self-restricted presentation and recognition of conventional foreign antigens. Th cells with indirect allospecificity may provide help for $CD8^+$ cytotoxic cells with direct anti-donor specificity and/or promote a delayed-type hypersensitivity response effected by activated macrophages.

Like the repertoire specific for conventional antigens, the TCR repertoire specific for a given alloantigen is diverse. Studies of the fine specificity of K^b-specific cytotoxic T lymphocyte (CTL) clones (Melief *et al.*, 1980; Sherman, 1980, 1981), and of TCR gene usage within an alloresponse (Bill *et al.*, 1989; Garman *et al.*, 1986), confirm extensive receptor diversity with recognition of many different epitopes on a single alloantigen. Furthermore, antigen-specific, self-restricted T-cell clones have been identified that also display alloreactivity (Braciale *et al.*, 1981; Hunig and Bevan, 1981; Lombardi *et al.*, 1989a; Sredni and Schwartz, 1980; Umetsu *et al.*, 1985), suggesting that the antigen-specific and allospecific repertoires are overlapping.

The reason such enormous diversity of antigenic epitopes can be displayed by a single MHC molecule, even when only a few amino acids differ between responder and stimulator MHC, can be deduced from knowledge of the structure of MHC molecules. MHC polymorphism focuses on residues that line the peptide-binding groove, as discussed elsewhere in this book. As this determines the range of peptide sequences that can bind to each MHC molecule, even a single amino acid change within the peptide-binding groove could potentially alter the range of peptides able to bind that molecule. Therefore, if peptide is involved as part of the epitope recognized by the T-cell receptor, this could explain the wide diversity of alloantigenic epitopes presented by a single MHC molecule.

17.2.1 Evidence for co-recognition of MHC-peptide complexes by alloreactive T cells

It is well established that self-restricted, antigen-specific T cells respond to peptide fragments of processed antigen in association with self-MHC. As the majority of cell-surface MHC molecules are occupied by self peptides, and as the antigen-specific and alloreactive T cell repertoires appear to overlap, it would seem reasonable to predict that alloreactive T cells also recognize peptide-MHC complexes (Sherman and Chattopadhyay, 1993). A number of observations support this prediction.

Incubation of MHC class II-expressing cells with an exogenous peptide capable of binding class II inhibits allorecognition by allospecific T cells (Eckels *et al.*, 1988; Rock and Benacerraf, 1984). Presumably the exogenous peptide competes with a peptide of 'self' origin for presentation to the allospecific T cell. In some instances this self-peptide can be identified. Alloreactive human T-cell clones have been reported, with specificity for MHC class I (Schendel, 1990; Sherman, 1981) or class II (Essaket *et al.*, 1990; Sherman, 1980) molecules, that require the expression of a second MHC molecule for allorecognition. The most likely explanation for these observations is that the T cells recognize the allo-MHC molecule together with peptides derived from the processing of the second MHC molecule.

There have been several descriptions of T cells that discriminate between MHC molecules differing only at residues predicted to lie within the peptide-binding groove (Dellabona *et al.*, 1991; Hunt *et al.*, 1990; Lombardi *et al.*, 1989a; Mattson *et al.*, 1989), and therefore inaccessible to the TCR. Unless alteration at these sites causes conformational changes elsewhere in the molecule, these observations are likely to be due to changes in the peptides bound to the MHC molecules.

The type of cell on which the MHC molecules are displayed can also influence allorecognition. Alloreactive T cells have been described that distinguish between the same MHC molecules displayed on murine or on human cells (Bernhard *et al.*, 1987; Heath and Sherman, 1991; Koller *et al.*, 1987; Lombardi *et al.*, 1989b; Mentzer *et al.*, 1986; Molina and Huber, 1990), even though these same transfectants can efficiently present peptide to antigen-specific T cells. One explanation could be the lack of display of a species-specific self-peptide, which has been demonstrated by restoration of allorecognition of K^b on a human cell by the addition of murine-derived peptides (Heath *et al.*, 1989). Cell-type-specific, rather than species-specific, allorecognition has also been observed (Heath and Sherman, 1991; Molina and Huber, 1990) whereby alloreactive T cells distinguish between the same MHC molecule expressed on cells of different lineage, again presumably because recognition depends on a peptide derived from a particular lineage of cell.

A number of mutant cell lines have been identified with defects in the pathways for antigen processing and presentation (Cerundolo *et al.*, 1990; Cotner *et al.*, 1990; Hosken and Bevan, 1990; Ljunggren and Karre, 1985; Mellins *et al.*, 1990; Schumacher *et al.*, 1990; Townsend *et al.*, 1989). These cells have abnormal MHC expression and are poorly recognized by allospecific T cells; for example, the antigen-processing mutant T2 transfected with the K^b gene was shown to express predominantly empty K^b molecules on its surface and was not recognized by K^b-specific CTL clones (Heath *et al.*, 1991). Allorecognition was restored by exposing the transfectants to cell-derived peptides, each clone recognizing a different peptide. Similarly, I-A^b transfected T2 cells were used in limiting dilution assays to determine the frequencies of allogeneic $CD4^+$ T cells capable of responding to these cells or to similarly transfected wild-type cells (Weber *et al.*, 1995). T2-H-$2A^b$ transfectants expressed only sodium dodecyl sulphate (SDS)-unstable (i.e. empty or invariant chain peptide-associated) class II MHC molecules and were poor stimulators of primary or secondary alloreactive $CD4^+$ T-cell responses. Incubation of these transfectants with an I-A^b-binding peptide derived from the I-Eα chain generated SDS-stable cell-surface I-A^b, and enabled these cells to stimulate alloreactive $CD4^+$ T cells that had been primed by the same peptide-MHC complex.

If the ligand for alloreactive T cells is allo-MHC complexed with a specific peptide, this could account for the observed high frequency of alloreactive cells (*Figure 17.2*). MHC molecules on the foreign cell surface are likely to be occupied by a wide variety of cell-derived peptides, current estimates ranging from 2000–10 000 (Engelhard, 1994). Since even minor differences in the MHC molecule allow binding of a different constellation of peptides, this would potentially generate a wide range of allo-MHC–peptide complexes, each of which might be recognized by a different alloreactive T cell. In this way, a single foreign MHC molecule would be able to stimulate a large number of different T cells, each with the same apparent allospecificity.

17.2.2 Evidence for direct recognition of allogeneic MHC

An alternative possibility is that alloreactive T cells directly recognize polymorphic residues on the allogeneic MHC molecule. From studies of cell lines expressing mutated MHC genes (Ajitkumar *et al.*, 1988; Clayberger *et al.*, 1990; Hogan *et al.*, 1989), it is clear that residues pointing up from the α-helix, which influence TCR contact but not peptide binding, are important in many alloresponses. In one such study, the peptides recognized in the alloresponse were associated with both wild-type (self) and mutant (allo) K^b molecules (Grandea and Bevan, 1992), showing that a self-peptide can be immunogenic when presented by allo-MHC.

A second line of evidence comes from the demonstration of peptide-mediated inhibition of allorecognition. If the allospecific T cell focuses on the α-helix of MHC, one might predict that peptides derived from this region would block TCR–MHC contact and thus inhibit allorecognition. This has been demonstrated both in the human (Parham *et al.*, 1987) and murine (Schneck *et al.*, 1989) systems for allorecognition of class I and class II MHC molecules (Lombardi *et al.*, 1991).

Finally, there is evidence that at least some allospecific T cells may not require endogenous peptide for allorecognition. This has been demonstrated in a cell-free system using class I molecules reconstituted from light and heavy chains in the absence of peptide (Elliott and Eisen, 1990), which induced proliferation in allospecific T cells. Similarly, some CTL clones recognize class I on the surface of antigen-processing mutants such as T2 (Heath *et al.*, 1991) or RMA-S (Townsend

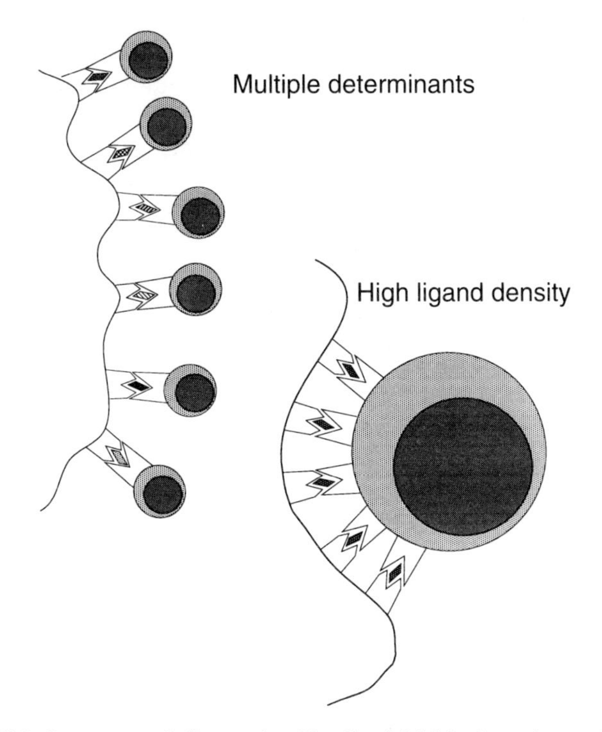

Figure 17.2. High frequency of alloreactive T cells. *Multiple determinants.* If the ligand for the alloreactive T cell is an allogeneic MHC molecule complexed with a specific peptide, and if the allogeneic MHC molecule can display a different constellation of peptides from self-MHC, then a wide range of allo-MHC–peptide complexes will be displayed by allogeneic cells, each of which could be recognized by a different alloreactive T cell. *High ligand density.* If the specificity of the alloreactive T cell is for exposed polymorphisms on the allogeneic MHC molecule, regardless of which peptides are bound, then the ligand for such T cells will be present at a far higher density than would be achieved for MHC complexed with a specific peptide. This would enable the recruitment of a wide range of alloreactive T cells into an alloresponse, including those with a lower specific affinity for their ligand than would normally contribute to a conventional antigen-specific response.

et al., 1989), although in one study the same clones showed more efficient lysis when peptide was present (Heath *et al.*, 1991).

Thus, in some circumstances, alloreactive T cells may interact directly with exposed regions of allo-MHC, bound peptide being of low importance. In such situations, the TCR ligand, which is the allo-MHC molecule itself, is present at a far higher density than would normally be achieved for MHC complexed with a specific peptide. Consequently, T cells with a lower specific affinity for their ligand than would be required for an antigen-specific response can be recruited for an alloresponse. This would provide an alternative explanation for the high precursor frequency of alloreactive T cells (*Figure 17.2*).

In order to fully understand the phenomenon of allorecognition, one needs to see how this can be accommodated within the framework of a T-cell repertoire positively selected to recognize peptide in the context of self-MHC. It is likely that the molecular basis of allorecognition is heterogeneous, and is probably

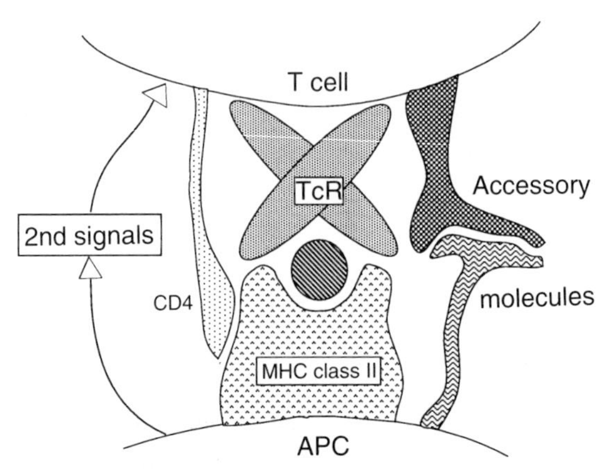

Figure 17.3. Two-signal model of T-cell activation. Signal one is supplied by the cognate interaction of the TCR–CD3 complex with its MHC–peptide ligand. In addition there are a variety of accessory molecular interactions. For activation of IL-2-secreting T cells there is a requirement for receipt by the T cell of 'second' or 'co-stimulatory' signals. The interaction between the B7 family of molecules on the APC and CD28/CTLA4 on the T cell plays an important role in the generation of signal two.

determined by the structural relationship between responder and stimulator MHC (Lechler and Lombardi, 1991). Where responder and stimulator MHC molecules are similar, sharing conserved sequences in the exposed TCR-contacting surface of the molecule, differences in the peptide-binding groove allow binding and display of different sets of peptides to which the responder T-cell repertoire is not tolerant. In this setting therefore, allorecognition mimics self MHC-restricted recognition of a novel series of peptides. Alternatively, where responder and stimulator MHC molecules are very different, allorecognition may result from a chance cross-reactivity between a TCR specific for a peptide in the context of self-MHC, and an allogeneic MHC molecule for which the TCR happens to have a better 'fit'. Support for this theory comes from the finding that alloreactive T cells in a responder-stimulator combination differing extensively at the TCR-contacting surfaces showed biased V_β gene usage (George *et al.*, 1994), as might be predicted from the fact that the CDR1 and CDR2 regions of the TCR, which contact the third variable regions of the MHC α- and β-chains, are encoded by germline V gene sequences. Conversely, alloreactive T cells from a responder-stimulator combination with very similar TCR-contacting surfaces, for which it would be predicted that bound peptide contributes significantly to allorecognition, showed no such bias. For many responder–stimulator combinations which fall between these two extremes, allorecognition will depend upon a combination of specificities for MHC polymorphisms and for differentially bound peptides. It is, therefore, possible to see how allorecognition arises within a self-restricted TCR repertoire.

MHC expression alone is not sufficient to elicit an alloresponse, however, as demonstrated by the observation that MHC-expressing liposomes or cell membranes failed to generate an alloresponse in rats (Batchelor *et al.*, 1978). This is consistent with a two-signal model of lymphocyte activation (*Figure 17.3*), whereby signal one is supplied by engagement of the TCR–CD3 complex with its ligand, but a second signal is required for full T-cell activation (Jenkins *et al.*, 1987). The consequence of receiving signal one alone is that the T cell is rendered non-responsive. This plays a significant role in the maintenance of self-tolerance, as well as providing a potential means whereby the vigorous response to alloantigen might be avoided.

17.3 Self-tolerance

17.3.1 Intrathymic selection

The mechanisms of induction and maintenance of self-tolerance provide a framework for understanding how tolerance to allografts may be achieved. The principal mechanism is generally accepted to be intra-thymic negative selection of differentiating self-reactive T cells (Burnet, 1959; Kappler *et al.*, 1987). There is some evidence that both clonal deletion and anergy are involved in negative selection (Antonia *et al.*, 1995; Ramsdell and Fowlkes, 1990), although deletion appears to be predominant. This can be demonstrated using TCR V_β-specific monoclonal antibodies; for example, T cells specific for the murine class II molecule I-E predominantly use the TCR $V_\beta17$ chain (Kappler *et al.*, 1987). Whereas I-E-negative mice have a significant proportion of $V_\beta17$-expressing peripheral T cells, I-E-positive mice delete nearly all such cells. Similar observations have been made using mice transgenic for TCRs specific for self-antigens present in the thymus, such as the male H-Y antigen (Kisielow *et al.*, 1988).

As both positive and negative intrathymic selection involve recognition of peptides associated with self-MHC, it is important to determine how signals transduced through the same TCR can lead to such different outcomes. One possibility is that the lineage of cell on which the peptide-MHC complex is displayed may be important. In addition to lymphocytes and epithelial cells, the thymus also contains bone-marrow-derived macrophages and dendritic cells, which are potent mediators of deletion, as demonstrated using bone-marrow chimaeras (Lo and Sprent, 1986; Marrack *et al.*, 1988). Early studies using I-E-expressing transgenic mouse lines suggested that thymic epithelial cells are poor mediators of deletion (Marrack *et al.*, 1988), but may induce anergy in $V_\beta17$-expressing cells (Ramsdell *et al.*, 1989). More recent reports suggest that all subsets of thymic epithelial cells are able to delete T cells expressing the appropriate TCR (Burkly *et al.*, 1993; Hoffmann *et al.*, 1992; Pircher *et al.*, 1993; Tanaka *et al.*, 1993), although dendritic cells remain the most efficient mediators of intrathymic deletion.

The processes of selection appear to be highly peptide dependent, as shown by studies using fetal thymic organ cultures derived from TCR-transgenic mice deficient in MHC class I expression (β2-microglobulin (β2-m) or transporter associated with antigen processing (TAP)-1 knockout mice; Ashton-Rickardt *et al.*, 1994; Hogquist *et al.*, 1994; Sebzda *et al.*, 1994). Addition of defined exogenous peptides to these cultures stabilized class I surface expression allowing the effects upon selection to be determined. Peptides mediating positive selection can be either TCR antagonists, or analogues of the wild-type peptide with low

TCR affinity. Furthermore, in two of these studies (Ashton-Rickardt *et al.*, 1994; Sebzda *et al.*, 1994), the same peptide that at low concentrations induced positive selection, at high concentrations led to negative selection. This suggests that the outcome of intrathymic antigen presentation to T cells depends on the avidity of TCR–ligand engagement, which is a function of both the affinity of the TCR for its ligand and the ligand concentration. This has been demonstrated *in vivo* in a TCR transgenic mouse model expressing an endogenous TCR antagonist ligand (Hsu *et al.*, 1995). In this model, transgenic TCR β-chains paired with endogenous α-chains to create a TCR with serendipitous reactivity to an altered peptide ligand of the epitope to which the parent clone was raised. Indeed, the wild-type peptide was a specific TCR antagonist for these T cells. Endogenous presentation of the antagonist ligand in the thymus eliminated the high-avidity T cells while sparing low-avidity cells. Further support for the avidity model of thymic selection comes from the finding that in TCR transgenic mice, thymocytes expressing a TCR at high density were negatively selected, whereas those with low density expression escaped (Allison *et al.*, 1994; Heath and Miller, 1993; Homer *et al.*, 1993). There is some recent evidence that, whereas high-affinity self-reactive T cells are deleted, lower-affinity T cells are rendered unresponsive but can subsequently respond to higher affinity cross-reactive foreign antigens (Kawai and Ohashi, 1995).

17.3.2 Peripheral tolerance mechanisms

Although many self-peptides are so abundant that they are easily presented within the thymus, it seems unlikely that thymic cells are capable of presenting every self-antigen. In particular, antigens with peripheral tissue-specific expression, peripheral antigens present in only limited amounts, or antigens that appear during sexual maturation are unlikely to mediate intrathymic negative selection. Peripheral mechanisms of tolerance are therefore necessary to avoid autoimmunity.

Self-ignorance. In some circumstances the T cell may simply ignore the antigen, due to failure to receive appropriate signals and/or help. This was demonstrated using double-transgenic mice expressing lymphocytic choriomeningitis virus glycoprotein (LCMV-GP) on pancreatic islet β cells, and a TCR specific for LCMV-GP (Ohashi *et al.*, 1991). Although present in sufficient numbers and able to respond to LCMV-GP *in vitro*, LCMV-GP-specific T cells did not cause autoimmune diabetes until the mice were infected with the appropriate virus. In this model, loss of self-ignorance was influenced by the MHC background of the mice and the number of self-reactive T cells present, as well as by the virus strain (Ohashi *et al.*, 1993). Similarly, transgenic mice expressing a TCR specific for the immunodominant peptide of myelin basic protein (MBP) failed to develop spontaneous experimental autoimmune encephalomyelitis (EAE) unless removed from a pathogen-free environment or immunized with MBP or pertussis toxin (Goverman *et al.*, 1993).

In other cases the T cell may be truly ignorant of its ligand because it is specific for a 'cryptic' epitope. Each self-protein contains a number of dominant determinants that will be preferentially processed and presented by APCs. These dominant epitopes are the ones to which self-tolerance is readily induced.

However, a majority of epitopes that are poorly processed and presented, and therefore 'cryptic', will be unavailable for tolerance induction, hence the normal T-cell repertoire will contain cells specific for such self-peptides (Mamula and Craft, 1994). These cells have not been made tolerant because they have never encountered their ligand. Such cells respond *in vitro* to the corresponding self-peptide, but not to the native self-protein. Under certain circumstances, these T cells can be activated *in vivo*, for instance, after priming by the appropriate cryptic self-peptide, or by a foreign antigen which mimics it, or by some modification of antigen processing that allows the normally cryptic epitope to be displayed (Lanzavecchia, 1995).

Anergy. Even when T cells recognize their ligand, such recognition does not necessarily lead to activation. Indeed, under some circumstances, antigen recognition can lead to a profound state of unresponsiveness, or anergy, in which the interleukin-2 (IL-2) gene is silenced and the T cell becomes refractory to further stimulation. This is the predominant extra-thymic mechanism of inducing self-tolerance. The conditions that lead to anergy are the receipt of signal one via the TCR–CD3 complex in the absence of a second signal. Conditions that lead to the delivery of signal one alone include antigen presentation by chemically fixed APC (Jenkins and Schwartz, 1987), by MHC class II molecules in artificial planar membranes (Quill and Schwartz, 1987), or by dimers of soluble class I MHC–peptide complexes (Abastado *et al.*, 1995). The same phenomenon has been observed using a variety of cells that would not normally function as APC, but which have been induced to express MHC class II molecules (Lo *et al.*, 1990; Warrens *et al.*, 1994), including keratinocytes (Bal *et al.*, 1990; Gaspari *et al.*, 1988), pancreatic β cells (Lo *et al.*, 1989) and activated human T cells (Sidhu *et al.*, 1992).

One of the key molecular interactions required for second signal generation involves the B7 family of molecules on the APC and their two ligands, CD28 and CTLA-4, on the T cell. B7-1 and B7-2 are expressed on cells with specialized antigen-presenting function, namely dendritic cells, activated B cells and macrophages (Robey and Allison, 1995). Indeed, an impotent APC can be made competent by transfection with B7 (Norton *et al.*, 1992), and co-stimulation by antibodies directed against CD28 can prevent tolerance induction (Harding *et al.*, 1992). Recent evidence suggests that CTLA-4 and CD28 are not equivalent, however, but may have opposing effects. Although anti-CTLA-4 antibodies augment T-cell proliferation in allogeneic MLRs, cross-linking these antibodies via their Fc regions inhibits proliferation (Krummel and Allison, 1995; Walunas *et al.*, 1994), implying that the enhanced proliferation results from blockade of an inhibitory effect rather than from stimulation. The implications of these observations are twofold: first, engagement of a TCR by its ligand on the surface of a non-professional APC without co-stimulatory function leads to T-cell non-responsiveness rather than activation, and second, the T cell must integrate the opposing signals delivered via CD28 and CTLA-4 in determining the outcome of stimulation.

The significance of the cell type on which antigen is presented may partly account for the well recognized phenomenon that the route of antigen administration determines whether responsiveness or tolerance is induced. Whereas local (subcutaneous) administration of antigen in adjuvant induces T-cell proliferation and subsequent immunity, systemic (intraperitoneal or intravenous) delivery without adjuvant favours specific T-cell unresponsiveness (Aichele *et al.*, 1995; Kearney *et al.*, 1994). It is likely that after subcutaneous administration, antigen is picked up locally by dendritic cells, which constitutively express B7 molecules and are the most potent stimulatory APC for T cells. Furthermore, the local inflammatory response to injection and to adjuvant is likely to up-regulate co-stimulatory molecules on other APCs. After systemic administration, however, antigen is rapidly distributed throughout the body where it will predominantly encounter less efficient APCs, including monocytes, macrophages and resting B cells, which do not constitutively express B7 and do not stimulate but rather make tolerant naive T cells (Ashwell *et al.*, 1985; Croft *et al.*, 1994; Krieger *et al.*, 1985).

Recently it has become clear that the 'two-signal' model is only part of the complex series of events leading to T-cell activation; for example, human T-cell clones express the full complement of co-stimulatory molecules, including high levels of B7, and yet are unable to activate responder T cells (Lombardi *et al.*, 1994b). Recent studies suggest that, rather than serving as a simple on-off switch, the TCR complex is sensitive to subtle changes in its ligand that may lead to different functional outcomes. Single-residue substitutions in antigenic peptides, which do not affect MHC binding, alter the T-cell response to stimulation in a variety of ways (Allen and Zinkernagel, 1994), including the positive selection of thymocytes (Hogquist *et al.*, 1994), cytokine production without proliferation (Evavold and Allen, 1991), anergy induction (Sloan-Lancaster *et al.*, 1993), and TCR antagonism (Evavold *et al.*, 1993). Thus, even in the presence of co-stimulation, presentation of a less than optimal ligand induces T-cell unresponsiveness, via altered signalling pathways (Madrenas *et al.*, 1995; Sloan-Lancaster *et al.*, 1994).

When the TCR engages an optimal ligand on the surface of a professional APC, the outcome, at least for IL-2-secreting T helper cells, is that the IL-2 gene is activated and IL-2 is secreted, with subsequent clonal proliferation and acquisition of effector function. IL-2-dependent proliferation thus seems to be a final common pathway for T-cell activation and is itself a critical event. Prevention of IL-2-driven proliferation by anti-IL-2 and anti-IL-2 receptor antibodies induces anergy in otherwise optimally stimulated $CD4^+$ T cells, which can be reversed by inducing proliferation with exogenous IL-2 (DeSilva *et al.*, 1991). T cells that have received an anergizing stimulus can also be rescued by antibody-mediated cross-linking of the common γ chain (γ_c) shared by the receptors for IL-2, IL-4 and IL-7 (Boussiotis *et al.*, 1994), suggesting that this signalling pathway may be involved in determining whether the outcome of TCR stimulation is proliferation or anergy.

TCR down-regulation. Although anergy is the predominant means of inducing peripheral self-tolerance, other mechanisms also play a significant role. One such mechanism is TCR down-regulation, which has been demonstrated in double-transgenic mice expressing a K^b-specific TCR together with K^b on various different tissues (Hammerling *et al.*, 1991). Although all mice could accept K^b-positive skin grafts, the manifestations of tolerance differed depending on the tissue expression of K^b, ranging from *in vivo* tolerance without TCR down-regulation that could be overcome *in vitro*, to permanent TCR and CD8 down-regulation that could not be reversed nor the T cells activated *in vitro*. From these observations it was proposed that TCR down-regulation, as a means of peripheral tolerance induction, is not a one-step process, but that tolerant T cells remain receptive to successive encounters with tolerogen, leading to a progressive deepening of the state of tolerance (Arnold *et al.*, 1993). This was confirmed in the same model using an inducible promoter (Ferber *et al.*, 1994). Minute amounts of K^b on hepatocytes rendered mice tolerant to K^b-positive skin grafts in association with partial down-regulation of the transgenic TCR, but after induction of the transgene and up-regulation of the level of K^b, TCR down-regulation was complete and there was some deletion of K^b-specific T cells.

Deletion/exhaustion. Peripheral deletion of mature T cells, or the related phenomenon of clonal exhaustion, has been demonstrated in a variety of other transgenic mouse models. Persistent high antigen dose seems to be a key feature in these models; for example, when LCMV-specific T cells from TCR-transgenic mice were transferred into LCMV-infected mice, they first proliferated then disappeared, failing to clear the virus (Moskophidis *et al.*, 1993). In another model, TCR transgenic T cells specific for the male-specific antigen H-Y were injected into bone marrow chimeras expressing different amounts of male antigen (Rocha *et al.*, 1995). At very high antigen concentrations T cells down-regulated surface CD8 and became anergic, but persisted *in vivo*, whereas at moderate antigen concentrations the population of male-specific T cells initially expanded then disappeared, even in the continued presence of male antigen. Similarly, in another transgenic mouse model, expression of K^b on hepatocytes caused the activation and subsequent deletion of K^b-specific TCR-transgenic T cells, those cells remaining showing an activated phenotype and partial TCR down-regulation (Bertolino *et al.*, 1995).

A different way in which clonal exhaustion may occur is by TCR interaction with superantigen. Exotoxins produced by certain bacterial strains, particularly *Staphylococcus aureus*, or endogenous murine retroviral superantigens are capable of activating all T cells expressing particular TCR Vβ regions, irrespective of their ligand specificity (Kappler *et al.*, 1989). This implies that superantigens interact with the TCR at a site shared by particular Vβ regions that is remote from the site involved in MHC-peptide binding. Furthermore, many MHC class II isoforms are able to present a particular superantigen to T cells, suggesting that superantigens bind to common sites on MHC class II which are remote from the peptide-binding groove. Superantigens can influence the TCR repertoire by intrathymic clonal deletion of entire Vβ families (White *et al.*, 1989), but similarly may interact with T cells in the periphery, initially to cause activation and expansion, but subsequently leading to anergy (Hewitt *et al.*, 1992; MacDonald *et al.*, 1991; Rammensee *et al.*, 1989).

Suppression. Another means of maintaining self-tolerance may be through suppression of potentially self-reactive cells. This was demonstrated by adoptive transfer studies in which T cells from tolerant animals were able to transfer tolerance to naive recipients when injected together with specific antigen (*Figure 17.4*; Batchelor *et al.*, 1984; Kilshaw *et al.*, 1975; Tilney *et al.*, 1978). This suppression may be mediated by T-cell-derived cytokines, such as IL-4 or IL-10, or, alternatively, anergic T cells may be directly suppressive by competing for ligand or for local cytokines. This was recently demonstrated *in vitro* using anergic

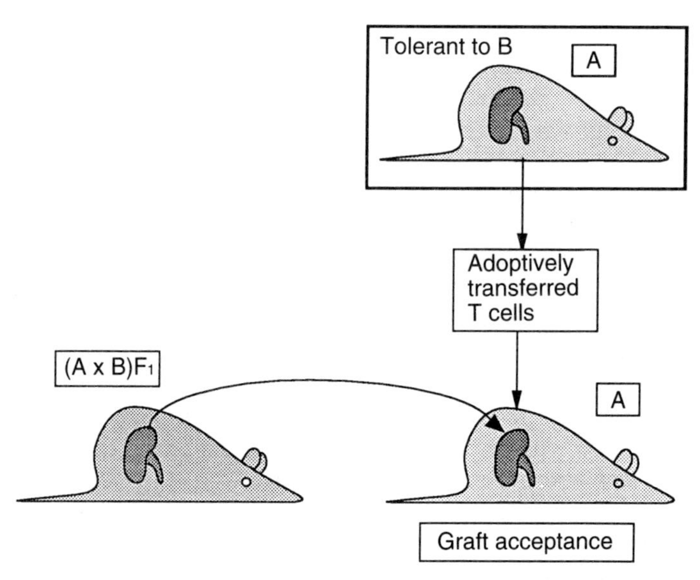

Figure 17.4. Adoptive transfer of tolerance. T cells from a tolerant animal are able to transfer specific tolerance to a naive animal when injected together with alloantigen.

human T-cell clones that inhibited both antigen-specific and allospecific T-cell proliferation by competing for the APC surface and for locally produced IL-2 (Lombardi *et al.*, 1994a). Other *in vivo* studies suggest that potentially pathogenic self-reactive T cells that have escaped deletion in the thymus may persist peripherally where they are suppressed by regulatory T-cell subsets (Tung, 1994). In these studies, $CD4^+$ thymocytes or neonatal splenocytes induced autoimmune disease on transfer to athymic syngeneic recipients, although only a small subset of adult splenocytes could do the same. Indeed, co-transfer of unseparated adult splenocytes, together with neonatal T cells, protected against autoimmunity, implying the presence of regulatory cells in the adult peripheral T-cell repertoire. In similar studies in rats, the pathogenic self-reactive $CD4^+$ T cells were identified as being of the Th1 phenotype, a subset of T-helper cells that secrete IL-2 and interferon-γ (IFN-γ) and provide help for cell-mediated immune functions such as cytotoxic T-cell activation (Fowell and Mason, 1993; Fowell *et al.*, 1991). The regulatory T cells were of the Th2 phenotype, which secrete IL-4, IL-5 and IL-10, and which provide help for humoral immune functions such as B cell differentiation. These T-helper cell subsets are known to be cross-regulatory.

It is more difficult to induce tolerance in Th2 than Th1 cells (Romball and Weigle, 1993), and, indeed, induction of anergy in human Th0 clones, which secrete cytokines characteristic of both Th1 and Th2 subsets and which are thought to represent pluripotential precursor cells, causes loss of Th1 characteristics and transition to a Th2 phenotype (Gajewski *et al.*, 1994). This phenomenon was recently demonstrated *in vivo* in a double-transgenic mouse model (Antonia *et al.*, 1995), where low-level expression of a transgenic protein on thymic epithelium caused anergy of thymocytes expressing the transgenic TCR, with selective loss of Th1 activity. This was mirrored peripherally and resulted in loss of CTL activity *in vivo*, although the CTL were fully functional *in vitro*. The conclusion was that this was due to the loss of Th1 activity, which normally would promote CTL maturation.

These observations suggest that the balance between self-tolerance and autoimmunity may depend upon the interplay between self-reactive and regulatory T cells. The bias is in favour of self-tolerance because potentially self-reactive Th1 cells are relatively sensitive to anergy induction, whereupon the anergic T cells may themselves play a suppressive role.

The idea that tolerant T cells may have a functional role is supported by the finding that these cells may have a long lifespan *in vivo*. This was shown in a double-transgenic mouse model using the K^b antigen in which the lifespan of tolerant T cells expressing the transgenic K^b-specific TCR was found to be at least eight weeks, comparable with that of non-tolerant T cells (Alferink *et al.*, 1995). In this model, the maintenance of tolerance was dependent upon the continued presence of antigen, as the T cells regained K^b-responsiveness within 2 weeks of transfer to K^b-negative athymic mice. Similar results were obtained in a different model using mice transgenic for a soluble form of a viral glycoprotein (Bachmann *et al.*, 1994). Tolerant T-helper cells regained normal responsiveness to the antigen within a week of transfer to athymic non-transgenic mice, and again seemed to be long-lived.

To summarize Section 3, deletion of self-reactive T cells in the thymus is the predominant mechanism for the establishment of self-tolerance, and in addition there are a variety of non-deletional mechanisms that operate both intra-thymically and peripherally, of which anergy is probably the most important. Knowledge of how self-tolerance is induced in mature, peripheral T cells is essential for understanding how alloreactivity can be overcome and tolerance to allografts can be achieved.

17.4 Allograft tolerance

Since the introduction of selective immunosuppressive drugs, considerable improvements in allograft survival have been made. However, the rate of graft rejection remains unacceptably high, with the loss of 10–20% of kidney grafts during the first year (Lechler *et al.*, 1991), and 3–4% per year thereafter (George *et al.*, 1995). Furthermore, immunosuppressive therapy is relatively non-specific and is associated with a number of complications, including increased susceptibility to infection and to malignancy. If specific allograft tolerance could be achieved, many of these problems could be eliminated.

Self-tolerance provides a useful model for understanding allotolerance, although there are several important differences between the two. Unlike a self-antigen which is present throughout ontogeny, an allograft presents an abrupt immunological challenge that is likely to require quite different tolerogenic mechanisms. Furthermore, the number of potentially alloreactive T cells in the normal repertoire greatly exceeds the number of cells specific for any conventional self or foreign antigen.

17.4.1 Central mechanisms

Central mechanisms of tolerance induction, which are predominant in the generation of self-tolerance, are less significant in allotolerance. Nevertheless, there are situations in which transplantation tolerance may result from intrathymic clonal deletion. One such method is the creation of allogeneic bone marrow chimeras, by reconstituting lethally irradiated rodents with allogeneic bone marrow. Such chimeras are tolerant of both donor and recipient antigens *in vitro* and *in vivo* (Rayfield and Brent, 1983). Recipient resistance to engraftment, which is largely T-cell-mediated, is overcome by total body irradiation, and graft-versus-host disease may be eliminated by marrow pretreatment with T-cell-depleting antibodies (Charlton *et al.*, 1994). There are two mechanisms of tolerance induction in this model: donor-origin bone marrow cells repopulate the recipient thymus and induce donor-specific tolerance by intrathymic deletion, just as occurs in self-tolerance; tolerance to recipient antigens, however, must be mediated by thymic epithelial cells and other non-bone-marrow-derived cells, and this may involve the induction of anergy (Schonrich *et al.*, 1992). A potential problem in this system is that T cells that are positively selected in the thymus for their ability to interact with foreign antigen in association with MHC expressed on thymic (i.e. recipient) epithelium, once out in the periphery must interact with donor-derived APCs. In the rarely achieved situation of 100% chimerism, peripheral T cell–APC interactions are poor due to this MHC disparity, and therefore allogeneic bone marrow chimaeras are immunologically compromised (Zinkernagel *et al.*, 1980). A more recent variation is to reconstitute the recipient with a mixture of both donor and syngeneic bone marrow to create a mixed bone marrow chimera (Ildstead and Sachs, 1984). This allows recipient APCs to remain in the periphery and so avoids the problems of T cell APC incompatibility. Donor-specific marrow infusion has been used in a clinical renal transplantation trial, with improved graft survival rates (Barber *et al.*, 1991). A more recent study by the same group correlates rejection-free graft survival with the presence of allogeneic chimerism in the recipients (McDaniel *et al.*, 1994).

A more direct route to tolerance via intrathymic deletion is to inject donor cells into the thymus, under cover of transient immunosuppression. This has been demonstrated with various tissues, including pancreatic islet cells (Posselt *et al.*, 1990) and renal glomeruli (Remuzzi *et al.*, 1991), with subsequent prolonged survival of the same tissue when transplanted extrathymically. Deletion of differentiating T cells was demonstrated by a reduction in precursor frequency of allospecific CTLs in the islet transplant recipients. However, tolerance was incomplete, as the islet recipients could reject a donor-MHC-bearing skin graft, which then precipitated rejection of the islets themselves (Posselt *et al.*, 1991). Subsequent studies have shown that the effect is not tissue-specific, as intrathymic spleen cells could prolong survival of skin allografts (Ohzato and Monaco, 1992), and intrathymic splenic T cells, but not B cells, macrophages or dendritic cells, could prolong cardiac allograft survival (Oluwole *et al.*, 1994). Similar results have been achieved using intrathymic autologous myoblasts transfected with donor MHC class I (Knechtle *et al.*, 1994), and also using soluble MHC-derived peptides (Oluwole *et al.*, 1993; Sayegh *et al.*, 1994), implying that the intrathymic allorecognition was indirect.

17.4.2 Peripheral mechanisms

Although intrathymic presentation of alloantigen may be an important feature of bone marrow transplantation, it is likely to be of less significance in solid organ transplantation, although some shed antigens might travel to the thymus in the circulation or in association with recirculating recipient APCs. In addition, it has been proposed by Starzl and colleagues that graft acceptance depends on a mutual host-graft leukocyte migration leading to a state of mixed chimerism in the recipient and in the graft (Starzl *et al.*, 1993), as will be discussed later. Nevertheless, it is likely that the most important mechanisms in the induction of allotolerance are mediated peripherally.

When considering allotolerance it is important to distinguish between graft antigenicity, denoting the display of antigens capable of being recognized by the immune system, and immunogenicity, denoting that such antigens are capable of eliciting a strong immune response. As already discussed, most mechanisms of tolerance require cognate interaction between the TCR and its MHC–peptide ligand, so for the acquisition of donor-specific tolerance antigenicity is required. However, graft MHC antigens are not themselves necessarily immunogenic, as discussed within the context of the two-signal model of T-cell activation. One approach, therefore, to favour peripheral tolerance is to modify the immunogenicity of the graft in some way so that a state of T-cell ignorance or even anergy results.

As already discussed, T-cell activation depends on presentation of peptide–MHC complexes together with appropriate co-stimulatory signals, and this is most efficiently achieved on the surface of a professional APC. Just as antigen presentation by non-professional APCs anergizes rather than activates antigen-specific T cells, so MHC antigens expressed on graft parenchymal cells may be ignored by or may anergize alloreactive T cells (Braun *et al.*, 1993). Thus, one strategy to reduce graft immunogenicity would be to remove professional APCs, and this can be achieved by a period of *in vitro* or *in vivo* culture prior to transplantation; for example, murine tissue grafts such as thyroid (Lafferty *et al.*, 1975, 1976) or islet cells (Lacy *et al.*, 1979) after a period of *in vitro* culture, are not rejected. The culture technique depletes professional APCs (passenger leukocytes) from the graft, with the result that recipient T cells encounter and ignore alloantigens expressed on graft parenchymal cells. In some of these models, tolerance can be overcome and rejection induced by administration of donor lymphoid cells in the early post-graft period (Batchelor *et al.*, 1979; Lafferty *et al.*, 1976; Talmage *et al.*, 1976), but this ability disappears with time, suggesting that other mechanisms of tolerance, such as anergy, emerge. Similarly, in a rat model, renal allografts transplanted under cover of a short course of immunosuppression are not rejected after immunosuppression is withdrawn. These grafts, if re-transplanted into a naive syngeneic recipient, are accepted without further immunosuppression in some strain combinations (*Figure 17.5*; Batchelor *et al.*, 1979; Welsh *et al.*, 1979). The mechanism is thought to be that while the graft is 'parked' in the first recipient, passenger leukocytes of donor origin are replaced by those of recipient origin (Braun *et al.*, 1993). If, however, the original recipient is lethally irradiated and reconstituted with donor-type bone marrow just prior to graft transfer, the retransplanted kidney is rejected (Lechler and Batchelor, 1982), presumably because donor-type passenger cells have repopulated the 'parked' graft. Unfortunately, the results of clinical trials of antibody-mediated depletion of professional APC (dendritic cells) from renal allografts prior to transplantation have so far proved disappointing (Brewer *et al.*, 1989).

The fate of the donor cells departing the graft has recently been addressed by Starzl, who found donor cell chimerism in patients up to 22 years after receiving a liver transplant, at diverse sites including skin, lymph nodes, spleen, bone marrow and thymus (Starzl *et al.*, 1993). It is proposed that this state of microchimerism, which is most readily observed after liver transplantation, leads to donor-specific tolerance, and that this accounts for the relative resistance of liver allografts to chronic rejection (Demetris *et al.*, 1995).

The ultimate goal of transplant physicians is long-lasting specific allograft tolerance, which can most safely be achieved by the induction of T-cell anergy or allospecific suppression. There are various ways in which donor antigens can be presented in a non-stimulatory fashion so that anergy results. Depleting the graft of dendritic cells so that alloreactive T cells meet their ligand on the surface of non-professional APCs has already been discussed in the context of T-cell ignorance, as tolerance induced by this means can be overcome by early challenge with donor lymphoid cells (Batchelor *et al.*, 1979; Lafferty *et al.*, 1976). However, after prolonged residence in the recipient, allografts become resistant to such rejection and the recipient develops donor-specific tolerance (Donohoe *et al.*, 1983). A

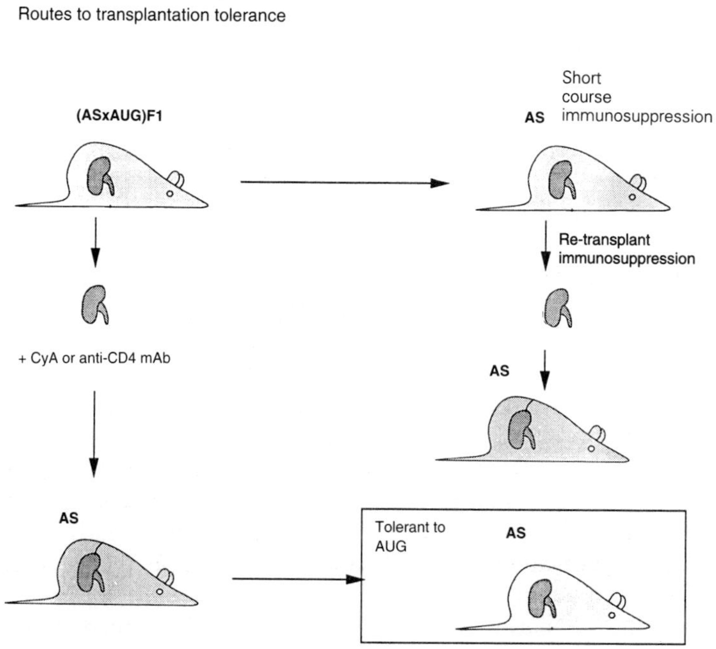

Figure 17.5. Routes to transplantation tolerance. *Graft 'parking'*. Renal allografts transplanted under cover of a short course of immunosuppression are not rejected after immunosuppression is withdrawn. After retransplantation into a naive syngeneic recipient, in the absence of further immunosuppression, the allograft is not rejected, because passenger leukocytes within the graft have been replaced by those from the first recipient. *Blockade of co-receptor molecules.* Exposure to donor antigen together with antibodies directed against adhesion molecules or T-cell surface molecules such as CD4 induces allospecific anergy.

different approach is to expose recipient T cells to alloantigen on non-professional APCs that are not part of the graft. This is the probable mechanism by which donor-specific blood transfusion promotes long-term survival of rat allografts (Armstrong *et al.*, 1987; Dallman *et al.*, 1987). Recipient T cells encounter a predominance of allogeneic B cells and monocytes rather than dendritic cells in the transfused blood, and are thus rendered anergic (Dallman *et al.*, 1991). This may also have reference to the microchimerism hypothesis, in that the majority of passenger leukocytes that populate an organ such as the liver are not dendritic cells but T, B, natural killer (NK) cells, or monocytes (Schlitt *et al.*, 1993), which migrate from the graft into host lymphoid tissues and cause anergy of alloreactive T cells. The other donor-derived population that has been detected in liver graft recipients is dendritic cell precursors and/or immature dendritic cells (Thomson *et al.*, 1995). These cells are non-immunogenic, thereby favouring the maintenance of tolerance, and may be longer lived than other bone-marrow-derived cell lineages. Thus there are two mechanisms by which an allograft may induce tolerance to itself: first through the non-stimulatory display of alloantigen on graft parenchymal cells, and second by seeding largely non-professional APC into distant host sites.

Another route to anergy is to block the co-stimulatory interactions between recipient T cells and donor APCs. In a number of studies, exposure to donor antigen together with antibodies directed against adhesion molecules or against other T-cell surface molecules induces allograft-specific anergy. The various antibodies used include those directed against ICAM-1, LFA-1 (Isobe *et al.*, 1992), and CD4 (*Figure 17.5*; Alters *et al.*, 1991; Pearson *et al.*, 1991). However, as the significance of the B7/CD28 interaction has become clearer, blocking this interaction using the soluble recombinant fusion protein CTLA4-Ig has been tried, and has been shown to be an efficient means of inducing allospecific tolerance *in vivo* in a variety of models (Baliga *et al.*, 1994; George *et al.*, 1995; Lenschow *et al.*, 1992; Lin *et al.*, 1993).

Suppressor cells have long been implicated in the maintenance of allotolerance, on the basis that T cells from animals with long-surviving allografts can, on adoptive transfer into a syngeneic host, prolong the survival of a fresh graft from the original donor (Batchelor *et al.*, 1984; Kilshaw *et al.*, 1975; Mottram *et al.*, 1990; Tilney *et al.*, 1978). The ability of anergic alloreactive T cells to inhibit allospecific T-cell proliferation has been demonstrated *in vitro* (Lombardi *et al.*, 1994a), as already discussed, which may explain the adoptive transfer of tolerance, as anergic T cells may act as inert competitors at sites of antigen presentation. Evidence for local suppression in allograft tolerance comes from the observation that, in tolerant animals, donor-specific tolerance *in vitro* could only be demonstrated in graft-infiltrating cells and not in peripheral blood leukocytes (Rosengard *et al.*, 1990).

17.5 Conclusion

In summary, a number of different mechanisms contribute to the maintenance of tolerance. In self-tolerance, the principal mechanism is intrathymic deletion of potentially self-reactive T cells, but a variety of peripheral mechanisms also play a significant role, the particular mechanism depending on a number of variables, as discussed. Although the immunological challenge presented by an allograft is very different from that presented by a self-antigen, an understanding of how self-tolerance is maintained forms the basis for understanding how the immune system may be manipulated so that the vigorous response elicited by an allograft may be avoided and the goal of long-term specific allograft tolerance may be attained. Recent insights into the key molecular interactions that determine the outcome of T-cell recognition provide exciting opportunities for more precisely targeted immunotherapy that may favour donor-specific tolerance.

References

Abastado J–P, Lone Y–C, Casrouge A, Boulot G, Kourilsky P. (1995) Dimerization of soluble major histocompatibility complex–peptide complexes is sufficient for activation of T cell hybridoma and induction of unresponsiveness. *J. Exp. Med.* **182:** 439–447.

Aichele P, Brduscha-Riem K, Zinkernagel RM, Hengartner H, Pircher H. (1995) T cell priming versus T cell tolerance induced by synthetic peptides. *J. Exp. Med.* **182:** 261–266.

Ajitkumar P, Geier SS, Kesari KV, Borriello F, Nakagawa M, Bluestone JA, Saper MA, Wiley DC, Nathenson SG. (1988) Evidence that multiple residues on both the α–helices of the class I MHC molecule are simultaneously recognized by the T cell receptor. *Cell* **54:** 47–56.

Alferink J, Schittek B, Schonrich G, Hammerling GJ, Arnold B. (1995) Long life span of tolerant T cells and the role of antigen in maintenance of peripheral tolerance. *Int. Immunol.* **7:** 331–336.

Allen PM, Zinkernagel RM. (1994) Promethean viruses? *Nature* **369:** 355–356.

Allison J, Georgiou A, Heath WR, Morahan G, Miller JFAP. (1994) Failure of allelic exclusion by rearranged T cell receptor chain transgenes. *Transgene* **1:** 135–146.

Alters SE, Shizuru JA, Ackerman J, Grossman D, Seydel KB, Fathman CG. (1991) Anti–CD4 mediates clonal anergy during transplantation tolerance induction. *J. Exp. Med.* **173:** 491–494.

Antonia SJ, Geiger T, Miller J, Flavell RA. (1995) Mechanisms of immune tolerance induction through the thymic expression of a peripheral tissue–specific protein. *Int. Immunol.* **7:** 715–725.

Armstrong HE, Bolton EM, McMillan I, Spencer SC, Bradley JA. (1987) Prolonged survival of acutely enhanced rat renal allografts despite accelerated cellular infiltration and rapid induction of both class I and class II MHC antigens. *J. Exp. Med.* **164:** 891–907.

Arnold B, Schonrich G, Hammerling GJ. (1993) Multiple levels of peripheral tolerance. *Immunol. Today* **14:** 12–14.

Ashton-Rickardt PG, Bandeira A, Delaney JR, van Kaer L, Pircher H-P, Zinkernagel RM, Tonegawa S. (1994) Evidence for a differential avidity model of T cell selection in the thymus. *Cell* **76:** 651–663.

Ashwell JD, DeFranco AL, Paul WE, Schwartz RH. (1985) Can resting B cells present antigen to T cells? *Fed. Proc.* **44:** 2475–2479.

Bachmann MF, Rohrer UH, Steinhoff U, Burki K, Skuntz S, Arnheiter H, Hengartner H, Zinkernagel RM. (1994) T helper cell unresponsiveness: rapid induction in antigen-transgenic and reversion in non-transgenic mice. *Eur. J. Immunol.* **24:** 2966–2973.

Bal V, McIndoe A, Denton G, Hudson D, Lombardi G, Lamb J, Lechler R. (1990) Antigen presentation by keratinocytes induces tolerance in human T cells. *Eur. J. Immunol.* **20:** 1893–1897.

Baliga P, Chavin KD, Qin L, Woodward J, Lin J, Linsley PS, Bromberg JS. (1994) CTLA4Ig prolongs allograft survival while suppressing cell–mediated immunity. *Transplantation* **58:** 1082–1090.

Barber WH, Mankin JA, Laskow DA, Deierhoi MH, Julian BA, Curtis JJ, Diethelm AG. (1991) Long term results of a controlled prospective study with transfusion of donor-specific marrow in 57 cadaveric renal allograft recipients. *Transplantation* **51:** 70–75.

Batchelor JR, Welsh KI, Burgos H. (1978) Transplantation antigens per se are poor immunogens within a species. *Nature* **273:** 54–56.

Batchelor JR, Welsh KI, Maynard A, Burgos H. (1979) Failure of long-surviving, passively enhanced kidney allografts to provoke T-dependent alloimmunity. I. Retransplantation of (AS $\times$ Aug)F_1 kidneys into secondary AS recipients. *J. Exp. Med.* **150:** 455–464.

Batchelor JR, Phillips BE, Grennan D. (1984) Suppressor cells and their role in the survival of immunologically enhanced rat kidney allografts. *Transplantation* **37:** 43–46.

Bernhard EJ, Le AX, Yanelli JR, Holterman MJ, Hogan KT, Parham P, Engelhard VH. (1987) The ability of cytotoxic T lymphocytes to recognize HLA-A2.1 or HLA-B7 antigens expressed on murine cells correlates with their epitope specificity. *J. Immunol.* **139:** 3614–3621.

Bertolino P, Heath WR, Hardy CL, Morahan G, Miller JFAP. (1995) Peripheral deletion of autoreactive $CD8^+$ T cells in transgenic mice expressing $H\text{-}2K^b$ in the liver. *Eur. J. Immunol.* **25:** 1932–1942.

Bill J, Yague J, Appel VB, White J, Horne G, Erlich HA, Palmer E. (1989) Molecular genetic analysis of 178 $I\text{–}A^{bm12}$–reactive T cells. *J. Exp. Med.* **169:** 115–133.

Boussiotis VA, Barber DL, Nakari T, Freeman GJ, Gribben JG, Bernstein GM, D'Andrea AD, Ritz J, Nadler LM. (1994) Prevention of T cell anergy by signalling through the γ_c chain of the IL-2 receptor. *Science* **266:** 1039–1042.

Braciale TJ, Andrew ME, Braciale VL. (1981) Simultaneous expression of H-2-restricted and alloreactive recognition by a cloned line of influenza virus-specific cytotoxic T lymphocytes. *J. Exp. Med.* **153:** 1371–1376.

Braun MY, McCormack A, Webb G, Batchelor JR. (1993) Evidence for clonal anergy as a mechanism responsible for the maintenance of transplantation tolerance. *Eur. J. Immunol.* **23:** 1462–1468.

Brewer Y, Palmer A, Taube D, *et al.* (1989) Effect of graft perfusion with two CD45 monoclonal antibodies on incidence of kidney allograft rejection. *Lancet* **2:** 935–937.

Burkly LC, Degermann S, Longley J, Hagman J, Brinster RL, Lo D, Flavell RA. (1993) Clonal deletion of $V_\beta 5^+$ T cells by transgenic I-E restricted to thymic medullary epithelium. *J. Immunol.* **151:** 3954–3960.

Burnet FM. (1959) *The Clonal Selection Theory of Acquired Immunity.* Cambridge University Press, Cambridge.

Cerundolo V, Alexander J, Anderson K, Lamb C, Cresswell P, McMichael A, Gotch F, Townsend A. (1990) Presentation of viral antigen controlled by a gene in the major histocompatibility complex. *Nature* **345:** 449–452.

Charlton B, Auchincloss HJ, Fathman CG. (1994) Mechanisms of transplantation tolerance. *Ann. Rev. Immunol.* **12:** 707–734.

Clayberger C, Rosen M, Parham P, Krensky AM. (1990) Recognition of an HLA public determinant. (Bw4) by human allogeneic cytotoxic T lymphocytes. *J. Immunol.* **144:** 4172–4176.

Cotner T, Mellins E, Johnson AH, Pious D. (1990) Mutations affecting antigen processing impair class II-restricted allorecognition. *J. Immunol.* **146:** 414–417.

Cramer DV, Qian S, Harnaha J, Chapman FA, Estes LW, Starzl TE, Makowka L. (1989) Cardiac transplantation in the rat. 1. The effect of histocompatibility differences on graft arteriosclerosis. *Transplantation* **47:** 414–419.

Croft M, Bradley LM, Swain SL. (1994) Naive versus memory CD4 T cell response to antigen. Memory cells are less dependent on accessory cell costimulation and can respond to many antigen-presenting cell types including resting B cells. *J. Immunol.* **152:** 2675–2685.

Dallman MJ, Wood KJ, Morris PJ. (1987) Specific cytotoxic T cells are found in the nonrejected kidneys of blood-transfused rats. *J. Exp. Med.* **165:** 566–571.

Dallman MJ, Shiho O, Page TH, Wood KJ, Morris PJ. (1991) Peripheral tolerance to alloantigen results from altered regulation of the interleukin 2 pathway. *J. Exp. Med.* **173:** 79–87.

Dellabona P, Wei BY, Gervois N, Benoist C, Mathis D. (1991) A single amino acid substitution in the A^k molecule fortuitously provokes an alloresponse. *Eur. J. Immunol.* **21:** 209–213.

Demetris AJ, Murase N, Delaney CP, Woan M, Fung JJ, Starzl TE. (1995) The liver allograft, chronic (ductopenic) rejection, and microchimaerism: what can they teach us? *Transplant. Proc.* **27:** 67–70.

DeSilva DR, Urdahl KB, Jenkins MK. (1991) Clonal anergy is induced in vitro by T cell receptor occupancy in the absence of proliferation. *J. Immunol.* **147:** 3261–3267.

Donohoe JA, Andrus L, Bowen KM, Simeonovic C, Prowse SJ, Lafferty KJ. (1983) Cultured thyroid allografts induce a state of partial tolerance in adult recipient mice. *Transplantation* **35:** 62–67.

Eckels DD, Gorski J, Rothbard J, Lamb JR. (1988) Peptide-mediated modulation of T cell allorecognition. *Proc. Natl Acad. Sci. USA* **85:** 8191–8195.

Elliott TJ, Eisen HN. (1990) Cytotoxic T lymphocytes recognize a reconstituted class I histocompatibility antigen (HLA-A2) as an allogeneic target molecule. *Proc. Natl Acad. Sci. USA* **87:** 5213–5217.

Engelhard VH. (1994) Structure of peptides associated with class I and class II molecules. *Annu. Rev. Immunol.* **12:** 181–207.

Essaket S, Fabron J, de Preval C, Thomsen M. (1990) Corecognition of HLA-A1 and HLA–DPw3 by a human CD4^{+} alloreactive T lymphocyte clone. *J. Exp. Med.* **172:** 387–390.

Evavold BD, Allen PM. (1991) Separation of IL-4 production from Th proliferation by an altered T cell receptor ligand. *Science* **252:** 1308–1310.

Evavold BD, Sloan-Lancaster J, Allen PM. (1993) Antagonism of superantigen–stimulated helper T-cell clones and hybridomas by altered peptide ligand. *Proc. Natl Acad. Sci. USA* **91:** 2300–2304.

Ferber I, Schonrich G, Schenkel J, Mellor AL, Hammerling GJ, Arnold B. (1994) Levels of peripheral T cell tolerance induced by different doses of tolerogen. *Science* **263:** 674–676.

Ford D, Burger D. (1983) Precursor frequency of antigen-specific T cells: effects of sensitization *in vivo* and *in vitro*. *Cell. Immunol.* **79:** 334–344.

Fowell D, Mason D. (1993) Evidence that the T cell repertoire of normal rats contains cells with the potential to cause diabetes. Characterisation of the CD4^{+} T cell subset that inhibits this autoimmune potential. *J. Exp. Med.* **177:** 627–636.

Fowell D, McKnight AJ, Powrie F, Dyke R, Mason D. (1991) Subsets of CD4 T cells and their roles in the induction and prevention of autoimmunity. *Immunol. Rev.* **123:** 37–64.

Gajewski TF, Lancki DW, Stack R, Fitch FW. (1994) "Anergy" of T_H0 helper T lymphocytes induces down–regulation of T_H1 characteristics and transition to a T_H2–like phenotype. *J. Exp. Med.* **179:** 481–491.

Garman RD, Ko JL, Vulpe CD, Raulet DH. (1986) T cell receptor variable gene usage in T cell populations. *Proc. Natl Acad. Sci. USA* **83:** 3987–3991.

Gaspari AA, Jenkins MK, Katz SI. (1988) Class II MHC–bearing keratinocytes induce antigen–specific unresponsiveness in hapten–specific Th1 clones. *J. Immunol.* **141:** 2216–2220.

George A, Dazzi F, Lynch J, Sidhu S, Marelli F, Batchelor RJ, Lombardi G, Lechler RI. (1994) Biased TCR gene usage in alloreactive T cells specific for a structurally dissimilar MHC alloantigen. *Int. Immunol.* **6:** 1785–1790.

George AJT, Ritter MA, Lechler RI. (1995) Disease susceptibility, transplantation and the MHC. *Immunol. Today* **16:** 209–211.

Goverman J, Woods A, Larson L, Weiner LP, Hood L, Zaller DM. (1993) Transgenic mice that express a myelin basic protein-specific T cell receptor develop spontaneous autoimmunity. *Cell* **72:** 551–560.

Grandea AGI, Bevan MJ. (1992) Single-residue changes in class I major histocompatibility complex molecules stimulate responses to self peptides. *Proc. Natl Acad. Sci. USA* **89:** 2794–2798.

Hammerling GJ, Schonrich G, Momburg F, Auphan N, Malissen M, Malissen B, Schmitt-Verhulst AM, Arnold B. (1991) Non-deletional mechanisms of peripheral and central tolerance: studies with transgenic mice with tissue-specific expression of a foreign MHC class I antigen. *Immunol. Rev.* **122:** 47–67.

Harding FA, McArthur JG, Gross JA, Raulet DH, Allison JP. (1992) CD28–mediated signalling co-stimulates murine T cells and prevents induction of anergy in T-cell clones. *Nature* **356:** 607–609.

Heath WR, Miller JFAP. (1993) Expression of two α chains on the surface of T cells in TCR transgenic mice. *J. Exp. Med.* **178:** 1807–1811.

Heath WR, Sherman LA. (1991) Cell–type–specific recognition of allogeneic cells by alloreactive cytotoxic T cells: A consequence of peptide–dependent allorecognition. *Eur. J. Immunol.* **21:** 153–159.

Heath WR, Hurd ME, Carbone FR, Sherman LA. (1989) Peptide-dependent recognition of H–$2K^b$ by alloreactive cytotoxic T lymphocytes. *Nature* **341:** 749–752.

Heath WR, Kane KP, Mescher MF, Sherman LA. (1991) Alloreactive T cells discriminate among a diverse set of endogenous peptides. *Proc. Natl Acad. Sci. USA* **88:** 5101–5105.

Hewitt CR, Lamb JR, Hayball J, Hill M, Owen MJ, O'Hehir RE. (1992) Major histocompatibility complex independent clonal T cell anergy by direct interaction of *Staphylococcus aureus* enterotoxin B with the T cell antigen receptor. *J. Exp. Med.* **175:** 1493–1499.

Hoffmann MW, Allison J, Miller JFAP. (1992) Tolerance induction by thymic medullary epithelium. *Proc. Natl Acad. Sci. USA.* **89:** 2526–2530.

Hogan KT, Clayberger C, Bernhard EJ, Walk SF, Ridge JP, Parham P, Krensky AM, Engelhard VH. (1989) A panel of unique HLA-A2 mutant molecules define epitopes recognized by HLA-A2-specific antibodies and cytotoxic T lymphocytes. *J. Immunol.* **142:** 2097–2104.

Hogquist KA, Jameson SC, Heath WR, Howard JL, Bevan MJ, Carbone FR. (1994) T cell receptor antagonist peptides induce positive selection. *Cell* **76:** 17–27.

Homer RJ, Mamalaki C, Kioussis D, Flavell RA. (1993) T cell unresponsiveness correlates with quantitative TCR levels in a transgenic model. *Int. Immunol.* **5:** 1495–1500.

Hosken NA, Bevan MJ. (1990) Defective presentation of endogenous antigen by a cell line expressing class I molecules. *Science* **248:** 367–370.

Hsu BL, Evavold BD, Allen PM. (1995) Modulation of T cell development by an endogenous altered peptide ligand. *J. Exp. Med.* **181:** 805–810.

Hunig T, Bevan MJ. (1981) Specificity of T cell clones illustrates altered self hypothesis. *Nature* **294:** 460–462.

Hunt HD, Pullen JK, Dick RF, Bluestone JA, Pease LR. (1990) Structural basis of Kbm8 alloreactivity. Amino acid substitutions on the beta-pleated floor of the antigen recognition site. *J. Immunol.* **145:** 1456–1462.

Ildstad ST, Sachs DH. (1984) Reconstitution with syngeneic plus allogeneic or xenogeneic bone marrow leads to to specific acceptance of allografts or xenografts. *Nature* **307:** 168–170.

Isobe M, Yagita H, Okumura K, Ihara A. (1992) Specific acceptance of cardiac allograft after treatment with antibodies to ICAM–1 and LFA–1. *Science* **255:** 1125–1127.

Jenkins MK, Schwartz RH. (1987) Antigen presentation by chemically modified splenocytes induces antigen-specific T cell unresponsiveness in vitro and in vivo. *J. Exp. Med.* **165:** 302–319.

Jenkins MK, Pardoll DM, Mizuguchi J, Quill H, Schwartz RH. (1987) T-cell unresponsiveness in vivo and in vitro; fine specificity of induction and molecular characterisation of the unresponsive state. *Immunol. Rev.* **95:** 113–135.

Kappler J, Kotzin B, Herron L, Gelfand EW, Bigler RD, Boylston A, Carrel S, Posnett DN, Choi Y, Marrack P. (1989) V-beta specific stimulation of human T cells by Staphylococcal toxins. *Science* **244:** 811–813.

Kappler JW, Roehm N, Marrack P. (1987) T cell tolerance by clonal elimination in the thymus. *Cell* **49:** 273–280.

Kawai K, Ohashi PS. (1995) Immunological function of a defined T-cell population tolerized to low-affinity self antigens. *Nature* **374:** 68–69.

Kearney ER, Pape KA, Loh DY, Jenkins MK. (1994) Visualisation of peptide-specific T cell immunity and peripheral tolerance induction in vivo. *Immunity* **1:** 327–339.

Kilshaw PJ, Brent L, Pinto M. (1975) Suppressor T cells in mice made unresponsive to skin allografts. *Nature* **255:** 489–491.

Kisielow P, Bluthmann H, Staerz UD, Steinmetz M, von Boehmer H. (1988) Tolerance in T cell receptor transgenic mice involves deletion of non-mature $CD4^{+}8^{+}$ thymocytes. *Nature* **333:** 742–746.

Knechtle SJ, Wang J, Jiao S, Geissler EK, Sumimoto R, Wolff J. (1994) Induction of specific tolerance by intrathymic injection of recipient muscle cells transfected with donor class I major histocompatibility complex. *Transplantation* **57:** 990–996.

Koller TD, Clayberger C, Maryanski JL, Krensky AM. (1987) Human allospecific cytolytic T lymphocyte lysis of a murine cell transfected with HLA-A2. *J. Immunol.* **138:** 2044–2049.

Krieger JI, Grammer SF, Grey HM, Chesnut RW. (1985) Antigen presentation by splenic B cells: resting B cells are ineffective whereas activated B cells are effective accessory cells for T cell responses. *J. Immunol.* **135:** 2937–2945.

Krummel MF, Allison JP. (1995) CD28 and CTLA-4 have opposing effects on the response of T cells to stimulation. *J. Exp. Med.* **182:** 459–465.

Lacy PE, Davie JM, Finke EH. (1979) Prolongation of allograft survival following *in vitro* culture (24 degrees C) and a single injection of ALS. *Science* **204:** 312–313.

Lafferty KJ, Cooley MA, Woolnough J, Walker KZ. (1975) Thyroid allograft immunogenicity is reduced after a period in organ culture. *Science* **188:** 259–261.

Lafferty KJ, Bootes A, Dart G, Talmage DW. (1976) Effect of organ culture on the survival of thyroid allografts in mice. *Transplantation* **22:** 138–149.

Lanzavecchia A. (1995) How can cryptic epitopes trigger autoimmunity? *J. Exp. Med.* **181:** 1945–1948.

Lechler RI, Batchelor JR. (1982) Immunogenicity of retransplanted rat kidney allografts. *J. Exp. Med.* **156:** 1835–1841.

Lechler RI, Batchelor JR. (1982) Restoration of immunogenicity to passenger cell-depleted kidney allografts by the addition of donor strain dendritic cells. *J. Exp. Med.* **155:** 31–41.

Lechler R, Lombardi G. (1991) Structural aspects of allorecognition. *Curr. Opin. Immunol.* **3:** 715–721.

Lechler RI, Gallagher RB, Auchincloss H. (1991) Hard graft? Future challenges in transplantation. *Immunol. Today* **12:** 214–216.

Lenschow DJ, Zeng Y, Thistlethwaite JR, Montag A, Brady W, Gibson MG, Linsley PS,

Bluestone JA. (1992) Long term survival of xenogeneic pancreatic islet grafts induced by CTLA4-Ig. *Science* **257:** 789–792.

Lin H, Bolling SF, Linsley PS, Wei RQ, Gordon D, Thompson CB, Turka LA. (1993) Long-term acceptance of major histocompatibility complex mismatched cardiac allografts induced by CTLA4-Ig plus donor-specific transfusion. *J. Exp. Med.* **178:** 1801–1806.

Lindahl KF, Wilson DB. (1977) Histocompatibility antigen-activated cytotoxic T lymphocytes. II. Estimates of the frequency and specificity of precursors. *J. Exp. Med.* **145:** 508–522.

Ljunggren HG, Karre K. (1985) Host resistance directed selectively against H-2 deficient lymphoma variants. Analysis of the mechanism. *J. Exp. Med.* **162:** 1745–1759.

Lo D, Sprent J. (1986) Identity of cells that imprint H-2-restricted T-cell specificity in the thymus. *Nature* **319:** 672–675.

Lo D, Burkly LC, Flavell RA, Palmiter RD, Brinster RL. (1989) Tolerance in transgenic mice expressing class II major histocompatibility complex on pancreatic acinar cells. *J. Exp. Med.* **170:** 87–104.

Lo D, Burkly LC, Flavell RA, Palmiter RD, Brinster RL. (1990) Antigen presentation in MHC class II transgenic mice: stimulation versus tolerization. *Immunol. Rev.* **117:** 121–134.

Lombardi G, Sidhu S, Batchelor JR, Lechler RI. (1989a) Allorecognition of DR1 by T cells from a DR4-DRw13 responder mimics self-restricted recognition of endogenous peptides. *Proc. Natl Acad. Sci. USA* **86:** 4190–4194.

Lombardi G, Sidhu S, Lamb JR, Batchelor JR, Lechler RI. (1989b) Co-recognition of endogenous antigens with HLA-DR1 by alloreactive human T cell clones. *J. Immunol.* **142:** 753–759.

Lombardi G, Barber L, Sidhu S, Batchelor JR, Lechler RI. (1991) The specificity of alloreactive T cells is determined by MHC polymorphisms which contact the T cell receptor and which influence peptide binding. *Int. Immunol.* **3:** 769–775.

Lombardi G, Sidhu S, Batchelor R, Lechler R. (1994a) Anergic T cells as suppressor cells *in vitro*. *Science* **264:** 1587–1589.

Lombardi G, Sidhu S, Dodi T, Batchelor R, Lechler R. (1994b) Failure of correlation between B7 expression and activation of interleukin-2-secreting T cells. *Eur. J. Immunol.* **24:** 523–530.

MacDonald HR, Baschieri S, Lees RK. (1991) Clonal expansion precedes anergy and death of $V\beta8^+$ peripheral T cells responding to staphylococcal enterotoxin B in vivo. *Eur. J. Immunol.* **21:** 1963–1966.

Madrenas J, Wange RL, Wang JL, Isakov N, Samelson LE, Germain RN. (1995) ζ phosphorylation without ZAP–70 activation induced by TCR antagonists or partial agonists. *Science* **267:** 515–518.

Mamula MJ, Craft J. (1994) The expression of self-antigenic determinants: implications for tolerance and autoimmunity. *Curr. Opin. Immunol.* **6:** 882–886.

Marrack P, Lo D, Brinster R, Palmiter R, Burkly L, Flavell RH, Kappler J. (1988) The effect of thymic environment on T cell development and tolerance. *Cell* **53:** 627–634.

Mattson DH, Shimojo N, Cowan EP, Baskin JJ, Turner RV, Shvetsky BD, Coligan JE, Maloy WL, Biddison WE. (1989) Differential effects of amino acid substitutions in the β-sheet floor and α-2 helix of HLA-A2 on recognition by alloreactive viral peptide-specific cytotoxic T lymphocytes. *J. Immunol.* **143:** 1101–1107.

McDaniel DO, Naftilan J, Hulvey K, Shaneyfelt S, Lemose JA, Lagoo–Deenadayalan S, Hudson S, Diethelm AG, Barber WH. (1994) Peripheral blood chimaerism in renal allograft recipients transfused with donor bone marrow. *Transplantation* **57:** 852–856.

Melief CJM, de Waal LP, van der Meulen MY, Melvold RW, Kohn HI. (1980) Fine specificity of alloimmune cytotoxic T lymphocytes directed against H-2k. A study with K^b mutants. *J. Exp. Med.* **151:** 993–1013.

Mellins E, Smith LL, Arp BA, Cotner T, Celis E, Pious D. (1990) Defective processing and presentation of exogenous antigens in mutants with normal HLA class II genes. *Nature* **343:** 71–74.

Mentzer SJ, Barbosa JA, Strominger JL, Biro PA, Burakoff SJ. (1986) Species-restricted recognition of transfected HLA-A2 and HLA-B7 by human CTL clones. *J. Immunol.* **137:** 408–413.

Molina IJ, Huber BT. (1990) The expression of a tissue-specific self-peptide is required for allorecognition. *J. Immunol.* **144:** 2082–2088.

Moskophidis D, Lechner F, Pircher H, Zinkernagel RM. (1993) Virus persistence in acutely infected immunocompetent mice by exhaustion of antiviral cytotoxic effector cells. *Nature* **362:** 758–761.

Mottram PL, Mirisklavos A, Dumble LJ, Clunie GJA. (1990) Suppressor cells induced by transfusion and cyclosporine. Studies in the murine cardiac allograft model. *Transplantation* **50:** 1033–1037.

Norton SD, Zuckerman L, Urdahl KB, Shefner R, Miller J, Jenkins MK. (1992) The CD28 ligand, B7, enhances IL-2 production by providing a costimulatory signal to T cells. *J. Immunol.* **149:** 1556–1561.

Ohashi PS, Oehen S, Buerki K, Pircher HP, Ohashi CT, Odermatt B, Malissen B, Zinkernagel R, Hengartner H. (1991) Ablation of "tolerance" and induction of diabetes by virus infection by viral antigen transgenic mice. *Cell* **65:** 305–317.

Ohashi PS, Oehen S, Aichele P, Pircher H, Odermatt B, Herrera P, Higuchi Y, Buerki K, Hengartner H, Zinkernagel RM. (1993) Induction of diabetes is influenced by the infectious virus and local expression of MHC class I and tumour necrosis factor–alpha. *J. Immunol.* **150:** 5185–5194.

Ohzato H, Monaco AP. (1992) Induction of specific unresponsiveness (tolerance) to skin allografts by intrathymic donor–specific splenocyte injection in antilymphocyte serum–treated mice. *Transplantation* **54:** 1090–1095.

Oluwole SF, Chowdhury NC, Jin MX, Hardy MA. (1993) Induction of transplantation tolerance in rat cardiac allografts by intrathymic inoculation of allogeneic soluble peptides. *Transplantation* **56:** 1523–1527.

Oluwole SF, Chowdhury NC, Jin MX. (1994) The relative contribution of intrathymic inoculation of donor leukocyte subpopulations in the induction of specific tolerance. *Cell. Immunol.* **153:** 163–170.

Parham P, Clayberger C, Zorn SL, Ludwig DS, Schoolnik GK, Krensky AM. (1987) Inhibition of alloreactive cytotoxic T lymphocytes by peptides from the $\alpha2$ domain of HLA-A2. *Nature* **325:** 625–628.

Pearson TC, Madsen JC, Wood K. (1991) Effect of anti-CD4 monoclonal antibody dosage when combined with donor antigen for the induction of transplantation tolerance. *Transplant. Proc.* **23:** 565–566.

Pircher H, Brduscha K, Steinhoff U, Kasai M, Mizuochi T, Zinkernagel RM, Hengartner H, Kyewski B, Muller KP. (1993) Tolerance induction by clonal deletion of $CD4^+8^+$ thymocytes in vitro does not require dedicated antigen-presenting cells. *Eur. J. Immunol.* **23:** 669–674.

Posselt AM, Barker CF, Tomaszewski JE, Markmann JF, Choti MA, Naji A. (1990) Induction of donor-specific unresponsiveness by intrathymic islet transplantation. *Science* **249:** 1293–1295.

Posselt AM, Naji A, Roark JH, Markman JF, Barker CF. (1991) Intrathymic islet cell transplantation in the spontaneously diabetic BB rat. *Ann. Surg.* **214:** 363–371.

Quill H, Schwartz RH. (1987) Stimulation of normal inducer T cell clones with antigen presented by purified Ia molecules in planar lipid membranes: specific induction of a long-lived state of proliferative non-responsiveness. *J. Immunol.* **138:** 3704–3712.

Rammensee HG, Kroschewski R, Frangoulis B. (1989) Clonal anergy induced in mature V beta 6^+ T lymphocytes on immunizing Mls-1b mice with Mls-1a expressing cells. *Nature* **339:** 541–544.

Ramsdell F, Fowlkes BJ. (1990) Clonal deletion versus clonal anergy: the role of the thymus in inducing self-tolerance. *Science* **248:** 1342–1348.

Ramsdell F, Lantz T, Fowlkes BJ. (1989) A non-deletional mechanism of thymic self-tolerance. *Science* **246:** 1038–1041.

Rayfield LS, Brent L. (1983) Tolerance, immunocompetence, and secondary disease in fully allogeneic radiation chimaeras. *Transplantation* **36:** 183–189.

Remuzzi G, Rossini M, Imberti O, Perico N. (1991) Kidney graft survival in rats without immunosuppressants after intrathymic glomerular transplantation. *Lancet* **337:** 750–752.

Robey E, Allison JP. (1995) T-cell activation: integration of signals from the antigen receptor and costimulatory molecules. *Immunol. Today* **16:** 306–309.

Rocha B, Grandien A, Freitas AA. (1995) Anergy and exhaustion are independent mechanisms of peripheral T cell tolerance. *J. Exp. Med.* **181:** 993–1003.

Rock KL, Benacerraf B. (1984) Selective modulation of a private I–A allostimulating determinant(s) upon association of antigen with an antigen-presenting cell. *J. Exp. Med.* **159:** 1238–1252.

Romball CG, Weigle WO. (1993) In vivo induction of tolerance in murine $CD4^+$ cell subsets. *J. Exp. Med.* **178:** 1637–1644.

Rosengard BR, Kortz EO, Guzetta PC, Sundt TMI, Ojikutu CA, Alexander RB, Sachs DH. (1990) Transplantation in miniature swine: analysis of graft infiltrating lymphocytes provides evidence for local suppression. *Hum. Immunol.* **28:** 153–158.

Sayegh MH, Perico N, Gallon L, Imberti O, Hancock WW, Remuzzi G, Carpenter CB. (1994) Mechanisms of acquired thymic unresponsiveness to renal allografts. Thymic recognition of immunodominant allo-MHC peptides induces peripheral T cell anergy. *Transplantation* **58:** 125–132.

Schendel DJ. (1990) On the peptide model of allorecognition: cytotoxic T lymphocytes recognize an alloantigen encoded by two HLA-linked genes. *Hum. Immunol.* **27:** 229–239.

Schlitt HJ, Raddatz G, Steinhoff G, Wonigeit K, Pichlmayr R. (1993) Passenger lymphocytes in human liver allografts and their potential role after transplantation. *Transplantation* **56:** 951–955.

Schneck J, Maloy WL, Coligan JE, Margulies DH. (1989) Inhibition of an allospecific T cell hybridoma by soluble class I proteins and peptides: estimation of the affinity of a T cell receptor for MHC. *Cell* **56:** 47–55.

Schonrich G, Momburg F, Hammerling GJ, Arnold B. (1992) Anergy induced by thymic medullary epithelium. *Eur. J. Immunol.* **22:** 1687–1691.

Schumacher TNM, Heemels MT, Neefjes JJ, Kast WM, Melief CJ, Ploegh HL. (1990) Direct binding of peptide to empty MHC class I molecules on intact cells and in vitro. *Cell* **62:** 563–567.

Sebzda E, Wallace VA, Mayer J, Yeung RSM, Mak TW, Ohashi PS. (1994) Positive and negative thymocyte selection induced by different concentrations of a single peptide. *Science* **263:** 1615–1618.

Sherman LA. (1980) Dissection of the B10.D2 anti-H-2-K^b cytolytic T lymphocyte receptor repertoire. *J. Exp. Med.* **151:** 1386–1397.

Sherman LA. (1981) Mutationally derived H-2 antigenic differences as defined by cytotoxic T lymphocyte clones. *J. Immunol.* **127:** 1259–1260.

Sherman LA, Chattopadhyay S. (1993) The molecular basis of allorecognition. *Annu. Rev. Immunol.* **11:** 385–402.

Sidhu S, Deacock S, Bal V, Batchelor JR, Lombardi G, Lechler RI. (1992) Human T cells cannot act as autonomous antigen-presenting cells, but induce tolerance in antigen-specific and alloreactive responder cells. *J. Exp. Med.* **176:** 875–880.

Sloan-Lancaster J, Evavold BD, Allen PM. (1993) Induction of T-cell anergy by altered T-cell receptor ligand on live antigen-presenting cells. *Nature* **363:** 156–159.

Sloan-Lancaster J, Shaw AS, Rothbard JB, Allen PM. (1994) Partial T cell signalling: altered phospho-ζ and lack of Zap70 recruitment in APL-induced T cell anergy. *Cell* **79:** 913–922.

Sredni B, Schwartz RH. (1980) Alloreactivity of an antigen-specific T cell clone. *Nature* **287:** 855–857.

Starzl TE, Demetris AJ, Murase N, Thomson AW, Trucco M, Ricordi C. (1993) Donor cell chimaerism permitted by immunosuppressive drugs: a new view of organ transplantation. *Immunol. Today* **14:** 326–332.

Talmage DW, Dart G, Radovich J, Lafferty KJ. (1976) Activation of transplant immunity: effect of donor leukocytes on thyroid allograft rejection. *Science* **191:** 385–388.

Tanaka Y, Mamalaki C, Stockinger B, Kioussis D. (1993) In vitro negative selection of $\alpha\beta$ T cell receptor transgenic thymocytes by conditionally immortalized thymic cortical epithelial cell lines and dendritic cells. *Eur. J. Immunol.* **23:** 2614–2621.

Thomson AW, Lu L, Subbotin VM, Li Y, Qian S, Rao AS, Fung JJ, Starzl TE. (1995) In vitro propagation and homing of liver-derived dendritic cell progenitors to lymphoid tissues of allogeneic recipients. *Transplantation* **59:** 544–551.

Tilney NL, Graves MJ, Strom TB. (1978) Prolongation of organ allograft survival by syngeneic lymphoid cells. *J. Immunol.* **121:** 1480–1482.

Townsend A, Ohlen C, Bastin J, Ljunggren HG, Foster L, Karre K. (1989) Association of class I major histocompatibility heavy and light chains induced by viral peptides. *Nature* **340:** 443–448.

Tung KS. (1994) Mechanism of self-tolerance and events leading to autoimmune disease and autoantibody response. *Clin. Immunol. Immunopathol.* **73:** 275–282.

Umetsu DT, Yunis EJ, Matsui Y, Jabara HH, Geha RS. (1985) HLA-DR4-associated alloreactivity of an HLA-DR3-restricted human tetanus toxoid-specific T cell clone: Inhibition of both reactivities by an alloantiserum. *Eur. J. Immunol.* **15:** 356–361.

Walunas TL, Lenschow DJ, Bakker CY, Linsley PS, Freeman GJ, Green JM, Thompson CB, Bluestone JA. (1994) CTLA-4 can function as a negative regulator of T cell activation. *Immunity* **1:** 405–413.

Warrens AN, Zhang JY, Sidhu S, Watt DJ, Lombardi G, Sewry CA, Lechler RI. (1994) Myoblasts fail to stimulate T cells but induce tolerance. *Int. Immunol.* **6:** 847–853.

Weber DA, Terrell NK, Zhang Y, Strinberg G, Martin J, Rudensky A, Braunstein NS. (1995) Requirement for peptide in alloreactive $CD4^+$T cell recognition of class II molecules. *J. Immunol.* **154:** 5153–5164.

Welsh KI, Batchelor JR, Maynard A, Burgos H. (1979) Failure of long-surviving, passively enhanced kidney allografts to provoke T-dependent alloimmunity. II. Retransplantation of

(ASxAug)F_1 kidneys from AS primary recipients into (ASxWF)F_1 secondary hosts. *J. Exp. Med.* **150:** 465–470.

White J, Herman A, Pullen AM, Kubo R, Kappler JW, Marrack P. (1989) The V-beta specific superantigen Staphylococcal eneterotoxin B: stimulation of mature T cells and clonal deletion in neonatal mice. *Cell* **56:** 27–35.

Zinkernagel RM, Althage A, Callahan G, Welsh RMJ. (1980) On the immunocompetence of H-2 incompatible irradiation bone marrow chimaeras. *J. Immunol.* **124:** 2356–2365.

Index